What the Experts Say About this Book

(continued from the book's back cover)

"Keeping the logic of organic chemistry, Professor Green leads the reader through the most important topics of this field of science in an unusual fashion. Reading the manuscript allows the knowledge to be absorbed without an awareness that one is learning. The book is therefore not only very useful, but even very entertaining. Important parts of the history of chemistry are embedded in an excellent manner into the appropriate places of the text allowing the subject to be presented in a broad sensible context. I recommend this book to all students and teachers dealing with organic chemistry."

—*Peter Huszthy, Budapest University of Technology and Economics, Hungary*

"This unusual textbook boldly questions our current approach to teaching organic chemistry and provides an alternative that is both unique and sensible. All too often, textbooks of organic chemistry present context-less elementary principles that rely on rote memorization, and only later do the "cool" and breathtaking applications of those principles come to be discussed. By drawing on riveting examples, this book reverses that approach by discovering the elementary principles in the wonderful applications of organic chemistry in our lives and uses this context to spur student learning. Such an approach, which more closely aligns with the natural learning process, could well be the answer to teaching this fascinating subject in a fun and effective way."

—*Dasan M. Thamattoor, Colby College*

"I looked at this book out of pure curiosity. I opened the book at random and started to read. After a while I became so interested that I read on and on and missed a prior appointment. The book describes organic chemistry, the way it came about in the last 200 years. It is an irresistible read."

—*Arnost Reiser, Polytechnic Institute of New York University*

"The idea of your book is new and revolutionary. It may take time for many people to accept it, but I consider your book highly valuable. I would encourage you to publish it and believe that eventually many people would like it."

—*Lin Pu, University of Virginia, Charlottesville*

"This is an organic chemistry textbook that deviates from the traditional bottom-up approach, which begins with atoms and ends with biomolecules. In stark contrast, this book takes us first to the real molecular world through an active dialog that illustrates the importance of organic chemistry to our lives—what organic chemistry deals with. Perhaps, many students will then grasp the basic concepts for the first time. The book should be a useful reference and a gem for years to come"

—*Pedro Cintas, Facultad de Ciencias-UEX, Badajoz, Spain*

D0841775

"You have confronted, in the specific case of organic chemistry, the two big problems in the teaching of experimental sciences in the University at the twenty first century.

1) How is it possible to learn the permanently increasing amount of knowledge necessary to achieve expertise in a discipline of science, which is additionally including information from other scientific fields?

2) How is it possible for this learning to occur by real understanding, which is the only path to true expertise, and not by simply overcoming evaluations and examinations?

Organic Chemistry Principles in Context, in starting from a complex relevant topic, which is the final objective of learning, dissects the elements and basic scientific knowledge necessary to explain the topic. Taking a story telling historical approach attracts the student's attention, which together with starting with an attractive topic is very probably the only way to explain complementary scientific disciplines in superior education."

> —*Ribo, JM, Department of Organic Chemistry and Institute of Cosmos Science, University of Barcelona, Catalonia, Spain*

"This book is anything but traditional. It opens with carbohydrate chemistry, a subject often relegated to the end of a beginning organic course because it is 'so complicated'. Mark Green makes in a few beginning pages this "complicated" subject simplicity itself and moves effortlessly on into stereochemistry, organic reaction mechanisms and pretty much everything else that belongs in an organic chemistry course. The difference is that he tells organic chemistry as an adventure story. Everything is there. It's fun. It's interesting. It's about chemistry and people and how it all came about and what it means. Surely this is why students (should) go to the university — to learn about ideas rather than only facts. The good student will learn organic chemistry the way it should be learned from this book. Curriculum committees are likely to find this book a square peg in a round hole. Maybe we need a bit more of that for good teaching?"

> —*Richard M. Kellogg, University of Groningen (retired), Syncom Corporation, The Netherlands.*

"Starting with the pictures of the scientists that significantly contributed to our knowledge as a human factor, organic chemistry is brought to us as an adventure, an exciting story. Almost all important issues dealt with in organic chemistry appear in this book, however, not in the conventional order. With complex, real life examples, all fundamentals of organic chemistry are explained. The way the references to the scientists are made makes the book a report of a human endeavor coherent in time and place and not simply a collection of facts. The book is an entertaining, context-based treatise of organic chemistry that is very rich for students and teachers with at least the basic knowledge presented in general chemistry. The book is decorated with more than 250 figures and includes more than 640 problems. The textbook is written by a well-documented and extremely knowledgeable organic chemist."

> —*J. A. J. M. Vekemans, Eindhoven University of Technology, The Netherlands*

"This book should be read by every organic chemist, academic or industrial."

Harold Wittcoff, Process Evaluation and Research Planning, Nexant, Inc. (ret.)

"For beginning students, it is not necessary to study all the details and all the reactions, old and new, in organic chemistry. The important thing is to study the fundamental principles, which brings the student to understand how the science is the product of human works and thoughts, the art and culture of organic chemistry. Your textbook just fits to this objective, I believe.

The book starts with: "Both cellulose and starch are polymers". At first students might ask why the book starts with this sentence. As they are reading Chapter 1, they see that an organic molecule is an artistic composition in three dimensions and come to understand the beauty of this three dimensional character, which is well represented by the difference between cellulose and starch. Finally their study will lead them to understand and even create new molecules using the art and culture of organic chemistry.

This book is not an accumulation or a compilation of organic reactions but shows an interesting series of historical stories or victories and how organic chemistry has progressed. Nylons, elastomers and polyolefins are important stories of macromolecular chemistry from both a scientific and industrial point of view, with attention to scientists who played important roles. Your narrative description and writing style makes it easy for the students to understand the principle and importance in our life of the area which they are studying. The developments of these macromolecules are good examples of the fusion of science and engineering. I can turn over every page excitingly imagining what is written on the next page. The book is helpful and useful for every student to find the ways of the futures which they should follow."

—Koichi Hatada, Professor Emeritus of Osaka University

"Any serious students or practitioners of Organic Chemistry will realize significant benefits and deepen their understanding of this beautiful science by reading this book."

—James A. Moore, Rensselaer Polytechnic Institute

"The book's one-of-a-kind approach to teaching organic chemistry gets rid of the fears that usually come with a college organic chemistry textbook. The historical accounts, along with important organic chemistry principles, are narrated in such a unique way that makes the whole subject fun to learn! Prof. Green's book prepares students interested in pursuing science by teaching the fundamental ideas in chemistry and the end-of-the-chapter questions guide students through thinking like an organic chemist. This is so unlike all of the other textbooks that teach the subject only through pages and pages of reactions to be memorized! "

—Jinhui Zhao, Biomolecular Science B.S., Class of 2012, Polytechnic Institute of NYU

" *Organic Chemistry Principles in Context* **is a wonderful textbook for any student of organic chemistry. This textbook harmoniously combines fundamental chemistry principles with the historical context of their development, allowing the student to understand not only the chemical mechanisms, but also the social and scientific context of the development of organic chemistry.** But most importantly, this textbook manages to avoid all of the clutter seen in conventional organic chemistry textbooks — given by the huge lists of chemical reactions that students have to memorize, along with their catalytic conditions — and focuses the students' attention on the basic mechanisms that underlie this wonderful scientific field. Personally, I think that by doing this, Professor Mark Green has managed to remove the fear of memorizing organic chemistry from the hearts of the students and replace that fear with a desire to understand organic chemistry. I have used this textbook during my two semesters of Organic Chemistry with Professor Green and it has helped me understand organic chemistry at a level which allowed me to pursue a Masters degree in Chemistry and also obtain a high score on the MCAT exam. **"**

—*Radu Iliescu, Biomolecular Science B.S./Chemistry M.S., Class of 2013, Polytechnic Institute of NYU*

ORGANIC CHEMISTRY

Principles in Context

A Story-Telling Historical Approach

MARK M. GREEN

ORGANIC CHEMISTRY Principles in Context

ISBN 978-0-615-70271-1

Published By:
ScienceFromAway Publishing
New York, NY 10014
w12thstreet@gmail.com

Book Designer, Robert L. Lascaro
www.lascarodesign.com
Typeset in Minion Pro
Display type: Helvetica Neue
Printer: CreateSpace, a divison of Amazon.com Inc.

Library of Congress Cataloging-in-Publication Data

Printed in the United States of America
on acid-free paper

"Those ignorant of the historical development of science are not likely ever to understand fully the nature of science and scientific research."

Sir Hans Adolf Krebs, 1970.

WITH GRATITUDE AND LOVE TO MY PARENTS, who opened the door to accomplishment for their children by making so much more out of life than they were given, and to Ruth Schulman for demonstrating the value of strength in adversity and her love and support, and always to my many students over the years who showed me the treasures accessible to a teacher's life.

To my wife, children, sons-in-law and grandchildren—thank you for family life and all its wonders, which continue to supply the foundation.

Finally, to my teachers for showing me the way, Kurt Mislow, Carl Djerassi, Herbert Morawetz, Arnost Reiser and Harold Wittcoff.

About The Author:

MARK M. GREEN is a 1958 graduate of the City College of New York. He received his Ph.D. from Princeton University working with Kurt Mislow followed by a National Institutes of Health postdoctoral fellowship with Carl Djerassi at Stanford University. He served as professor of chemistry at several universities with long experience in teaching organic chemistry to students of widely varying abilities. He has been at his current position at the Polytechnic Institute of New York University since 1980. Professor Green's over 40 year career of academic research has been widely recognized. He was awarded a National Science Foundation "Special Creativity Award" in 1995, elected chair wof the Polymer Chemistry Gordon Conference for the year 2000, elected a "Fellow of the Japan Society for the Promotion of Science" in 2003 and was named a winner of the Society of Polymer Science of Japan award for "Outstanding Achievement in Polymer Science and Technology" in 2005. He has been elected as a "Fellow of the American Association for the Advancement of Science" for "pioneering work in important new areas of polymer science." He serves on the editorial board of "Topics in Stereochemistry," and has served on the editorial board of the American Chemical Society journal "Macromolecules." Professor Green received a Jacobs' Excellence in Teaching Award by the Polytechnic Institute of NYU in 2006. His interest in communicating science to general audiences has led to several years of writing columns for two newspapers, which are published in a blog, sciencefromaway.com.

In recent years Professor Green has turned his attention to further developing his long interest in teaching organic chemistry in context by using a story-telling historical approach. His first book, *Organic Chemistry Principles and Industrial Practice* (2003 Wiley-VCH) written with Harold A. Wittcoff, has been widely praised as a resource for chemistry teachers seeking material to enhance their classes and has been used as a text for both chemical engineering students studying beginning organic chemistry as well as for graduate courses in the chemical sciences.

Organic Chemistry Principles in Context, **designed for the motivated student and to motivate students, has been used successfully in manuscript form as a primary text for beginning organic chemistry classes at the Polytechnic Institute of New York University.**

Rather than accepting offers for traditional publication the author has maintained control of the copyright to set an affordable price as a primary text — or to also allow *Principles in Context* **to be used as an adjunct text along with more conventional textbooks.**

Organic Chemistry Principles in Context **has been written with the intent to increase the author's own appreciation and love for the subject. As Mark Van Doren of Columbia University pointed out in 1964: "A teacher can fool his colleagues; he may even fool his president; but he never fools his students. They know when he loves his subject and when he does not."**

Books Co-Authored and Co-Edited:

Organic Chemistry Principles and Industrial Practice,
Mark M. Green and Harold A. Wittcoff, Wiley-VCH, 2003.

Materials-Chirality, edited by Mark M. Green, Roeland Nolte and Bert Meijer,
Volume 24 in the series, *Topics in Stereochemistry,* Wiley-Interscience, 2003.

Popular Science Articles:

Sciencefromaway.com

Advice to students using
Organic Chemistry Principles in Context

READ EACH CHAPTER'S SECTIONS WITH A PENCIL IN HAND to redraw the molecular structures, putting in all the atoms and electrons, including lone pairs until you feel these drawings are second nature to you and you can use just the line drawings. Organic chemistry is a combination of the image with the idea and facility with drawing organic chemical structures is key to understanding the concepts of the science.

LOOK FOR FUNCTIONAL GROUPS AND REACTIONS on pages xiii-xv that correlate with what you've just read and learn to draw the structures so you can easily recognize a functional group. As you read the sections, imagine new molecules that can demonstrate the principles discussed and draw their structures. In general, a pencil and paper should be in hand whenever you are studying organic chemistry.

TRY THE "STUDY GUIDE QUESTIONS" FOLLOWING EACH SECTION. We have attempted to use the "Study Guide Questions" to guide you as to what is expected from each section. The term study guide is also consistent with the nature of some of the questions, which often contain information that amplify the text or ask you to reason about subject matter that is about to be discussed in a subsequent section. At the same time some of these questions are designed to help you to dig deeper into the subject, to take the material further along. This latter aspect is supported by a downloadable answer book, which can be seen in part, as an extension of the material presented in the text. *(see source information on the right)*

FOR ANSWERS
to all of the problems in this textbook go to:
OrganicChemistryPrinciplesInContext.com

READ THE "CHAPTER SUMMARY OF THE ESSENTIAL MATERIAL" at the end of each chapter and make certain you can reproduce, using that pencil, the images and ideas noted in this summary. When it is not clear, then go back and reread the section of the chapter about that area and get it down until you are certain of it – using that pencil. The purpose of the summary is to point to the material that should be known when the work on the chapter is over.

ENJOY THE HISTORICAL MATERIAL AND THE STORIES AND THE PICTURES of the scientists as you go, realizing that you are not responsible for reproducing that information, although I hope that the flow and context – the stories - will help you to remember why you are learning this subject and will help you to remember it. Read the **Introduction** on pages 2-4, which although intended more for the teacher in the course will nevertheless give you an idea of what the book is trying to do.

AND ONE FINAL NOTE: use those curved arrows to follow the electrons. So much can be figured out about the reactions and mechanisms in organic chemistry by making certain your drawings show all the electrons involved, bonding and nonbonding, and where they are going in the transformation you are following.

Functional Groups
& Chemical Reactions
DISCUSSED IN THE TEXT

FUNCTIONAL GROUPS

Acetal	$R-\underset{H}{\overset{OR'}{\underset{|}{\overset{|}{C}}}}-OR'$	**Chloride**	$R-Cl$	**Nitrile**	$R-C\equiv N$
Aldehyde	$R-\overset{O}{\overset{\|}{C}}-H$	**Enols and enediols**		**Nitro**	$R-NO_2$
Alkene		**Eneamine**		**Phenyl (arene)**	
Alkyne		**Epoxide**		**Phosphate**	
Amide	$R-\overset{O}{\overset{\|}{C}}-NH_2$; $R-\overset{O}{\overset{\|}{C}}-NHR'$; $R-\overset{O}{\overset{\|}{C}}-NR'_2$	**Ester**	$R-\overset{O}{\overset{\|}{C}}-O-R'$	**Phosphine**	R_3P
Amine	$R-NH_2$; $R-NHR'$; $R-NR'_2$	**Ether**	$R-O-R'$	**Sulfonamide**	$R-\overset{O}{\underset{O}{\overset{\|}{\underset{\|}{S}}}}-NHR'$
Azo	$N=N$	**Hemiacetal**	$R-\underset{H}{\overset{OH}{\underset{\|}{\overset{\|}{C}}}}-O-R'$	**Sulfonyl chloride**	$R-\overset{O}{\underset{O}{\overset{\|}{\underset{\|}{S}}}}-Cl$
Brosylate	$Br-\overset{O}{\underset{O}{\overset{\|}{\underset{\|}{S}}}}-O-$	**Hemiketal**	$R-\underset{R}{\overset{OH}{\underset{\|}{\overset{\|}{C}}}}-O-R'$	**Sulfoxide**	$R-\overset{O}{\overset{\|}{S}}-R'$
Carbamate	$R-\underset{H}{\overset{\|}{N}}-\overset{O}{\overset{\|}{C}}-O-R'$	**Hydroxyl**	$R-O-H$	**Thioester**	$R-\overset{O}{\overset{\|}{C}}-S-R'$
Carbonate	$RO-\overset{O}{\overset{\|}{C}}-OR'$	**Isocyanate**	$R-N=C=O$	**Thiol**	$R-SH$
Carboxylic acid	$R-\overset{O}{\overset{\|}{C}}-O-H$	**Ketal**	$R-\underset{R}{\overset{O-R'}{\underset{\|}{\overset{\|}{C}}}}-O-R'$	**Tosylate**	CH_3-
Carboxylic acid anhydrate	$R-\overset{O}{\overset{\|}{C}}-O-\overset{O}{\overset{\|}{C}}-R$	**Ketone**	$R-\overset{O}{\overset{\|}{C}}-R'$	**Urea**	
Carboxylic acid chloride	$R-\overset{O}{\overset{\|}{C}}-Cl$	**Lactone**			

CHEMICAL REACTIONS

Outline of examples of some of the types of chemical reactions discussed in the text
(not all the reactants and products are shown)

Acetal formation

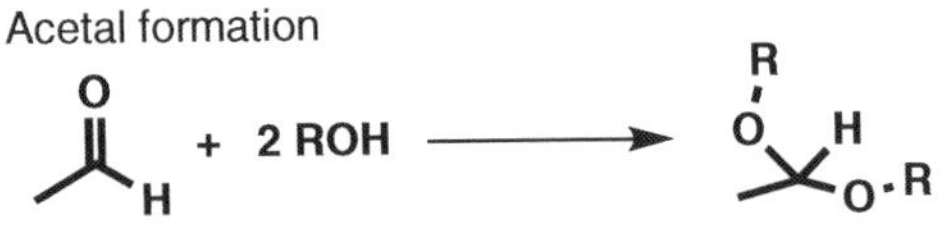

Aldol condensation

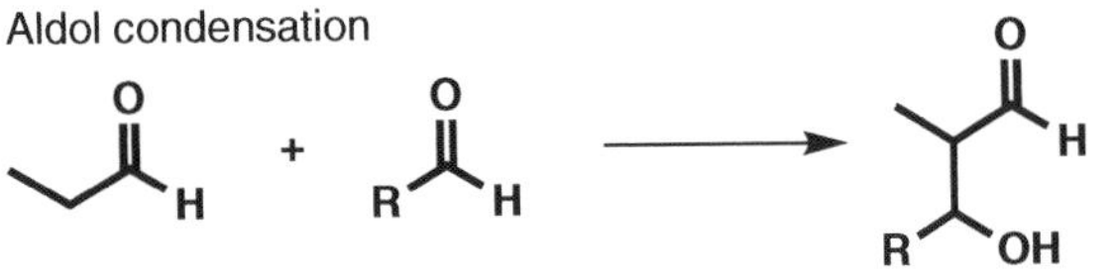

Amide formation

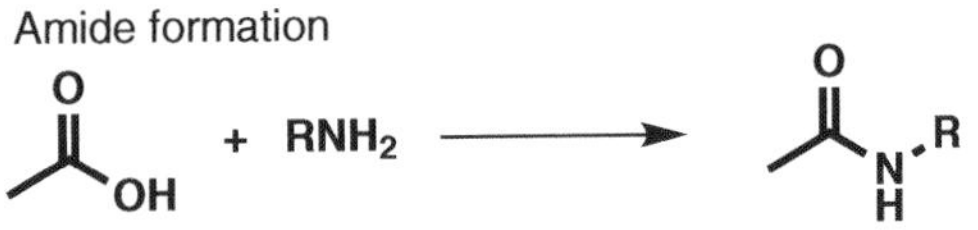

Baeyer-Villiger reaction

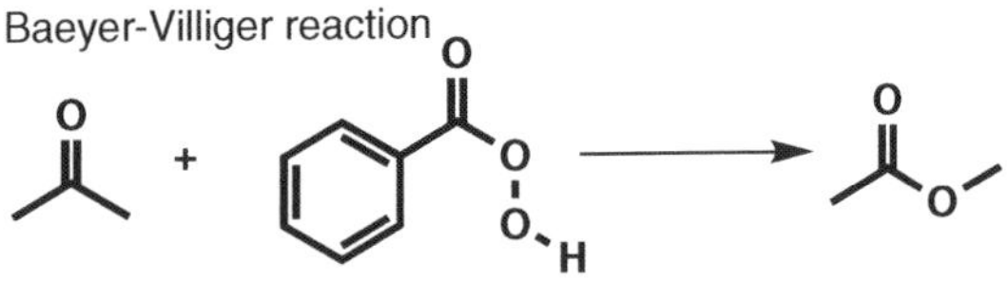

Carbocation 1,2 shift

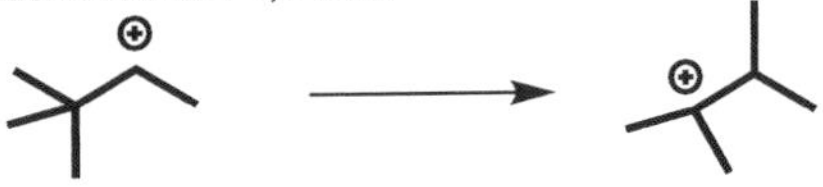

Claisen reaction

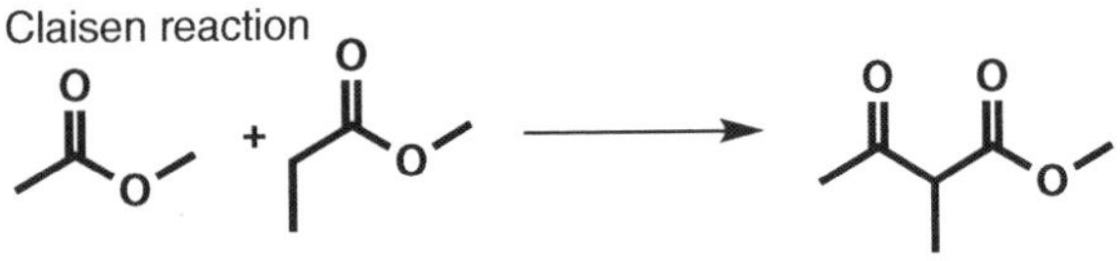

Conjugate addition

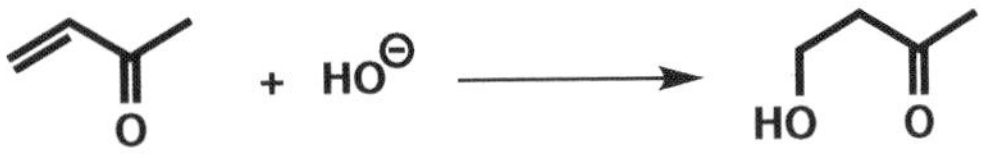

Diels-Alder reaction

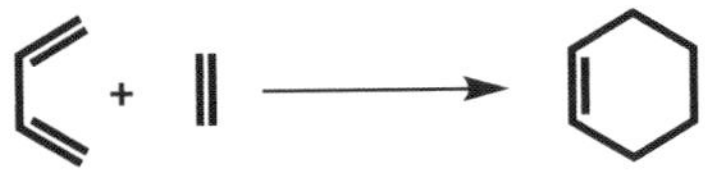

Electrophilic 1,2 and 1,4 addition of halogen to conjugated double bonds

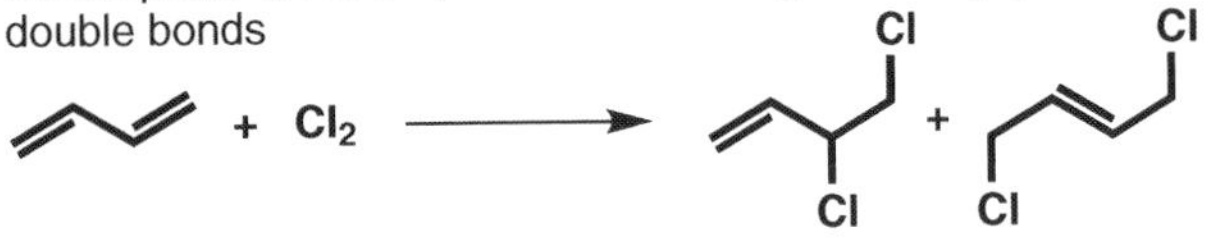

Electrophilic addition to double bonds

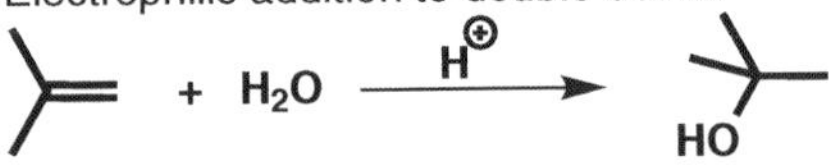

Electrophilic aromatic substitution

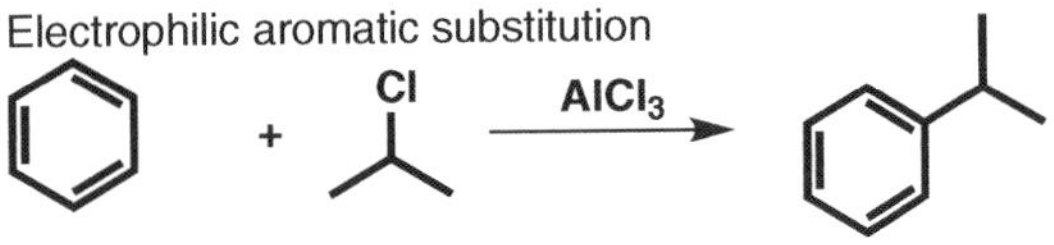

Elimination - acid catalyzed alkene formation (E-1 reactions)

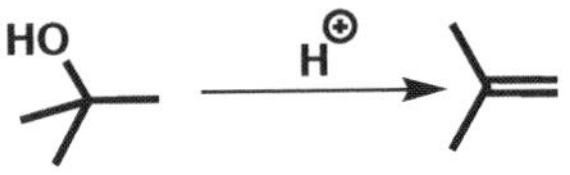

Elimination - base catalyzed alkene formation (E-2 reactions)

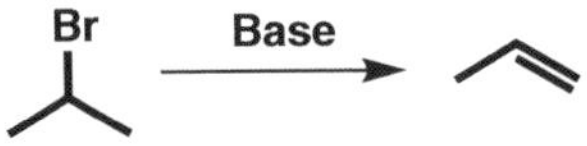

Enolate formation

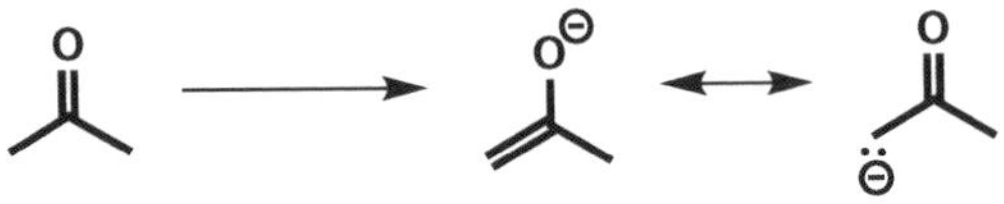

Ester hydrolysis (saponification)

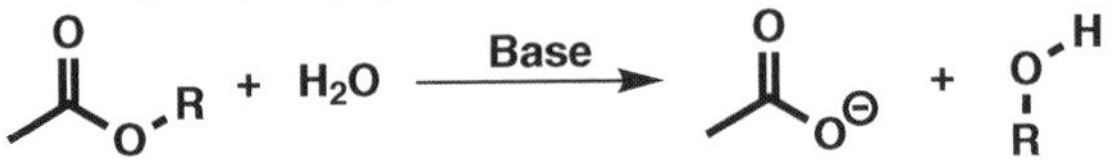

Esterification

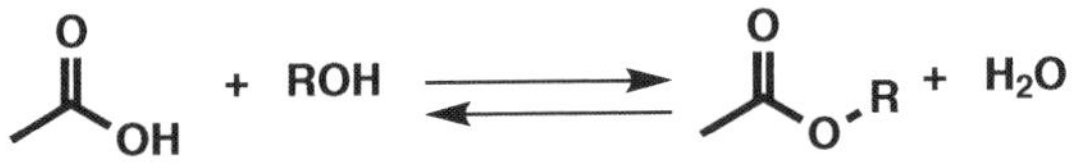

Free radical addition to double bonds

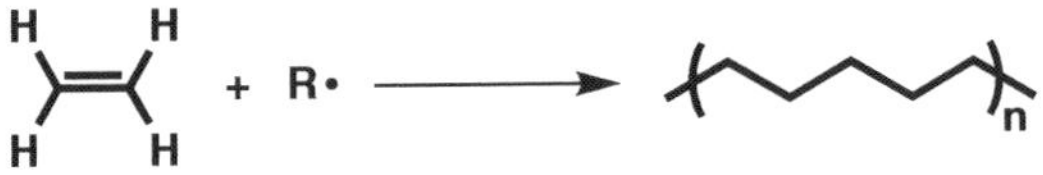

Free radical carbon-carbon bond formation

Free radical hydrogen abstaction

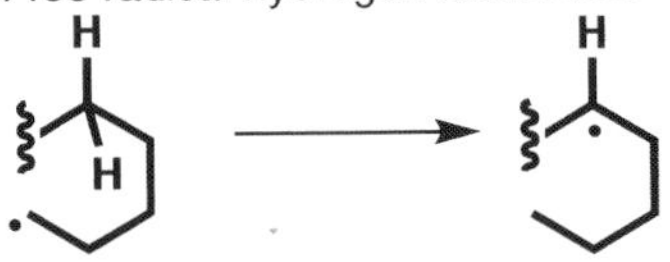

Free radical β-cleavage

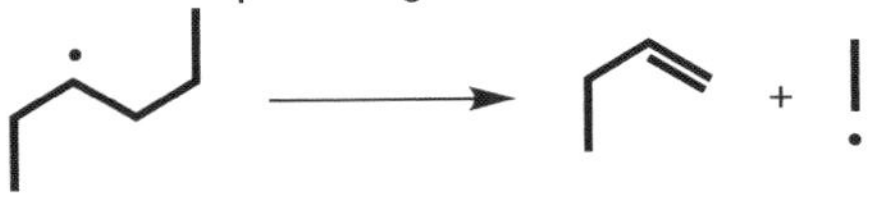

Grignard reaction

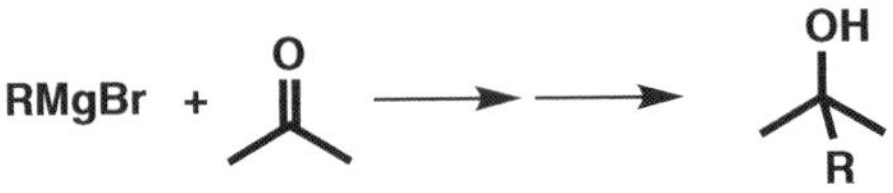

Hydroboration

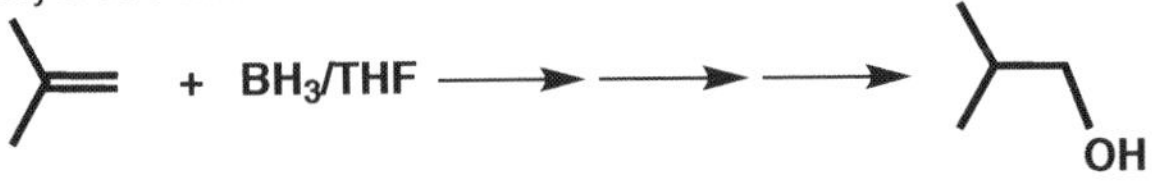

Hydrogenation of benzene rings

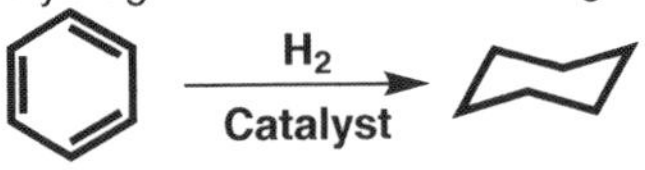

Hydrogenation of double bonds

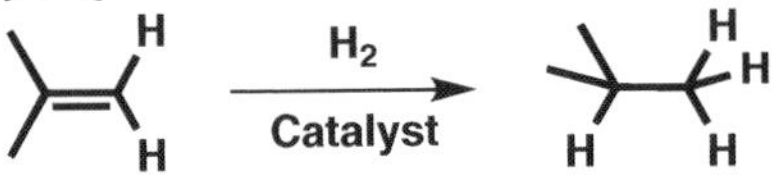

Ketal formation

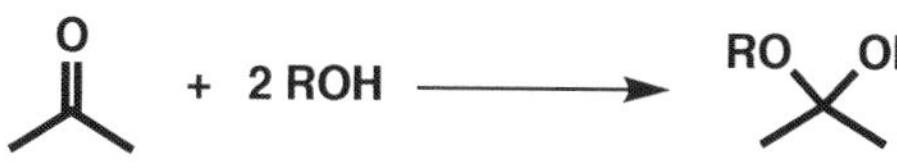

$$\text{O} + 2\ ROH \longrightarrow RO\quad OR$$

Lewis acid base reaction

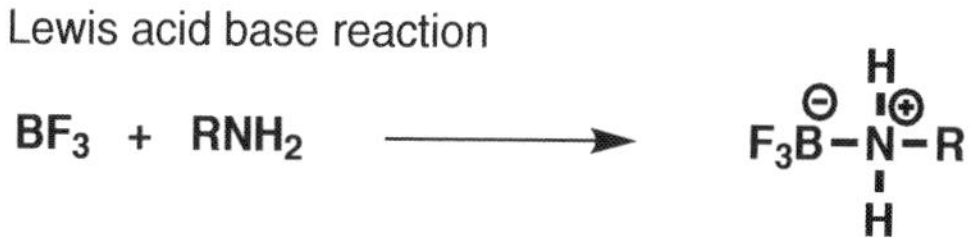

$$BF_3 + RNH_2 \longrightarrow F_3B-N-R$$

Loss of carbon dioxide

$$\longrightarrow + CO_2$$

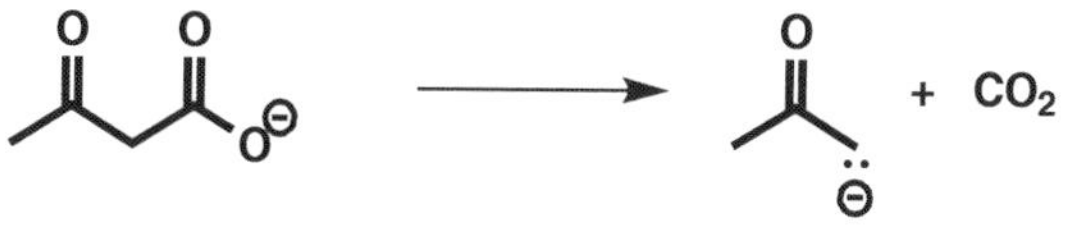

Lowry-Brønsted acid base reaction

$$HCl + RNH_2 \longrightarrow RNH_3^{\oplus} + Cl^{\ominus}$$

Michael addition

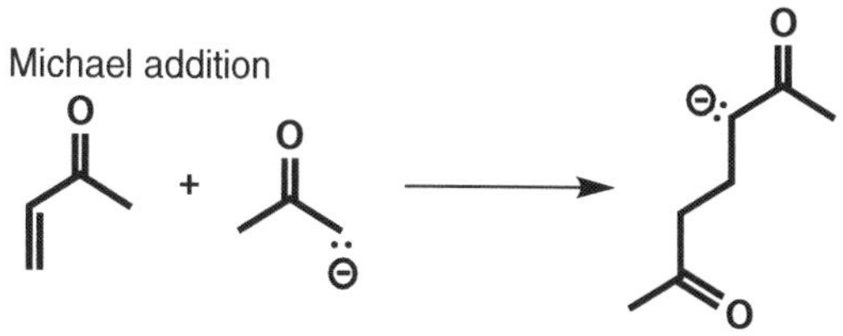

Nucleophilic addition to carbonyl by acetylide anion

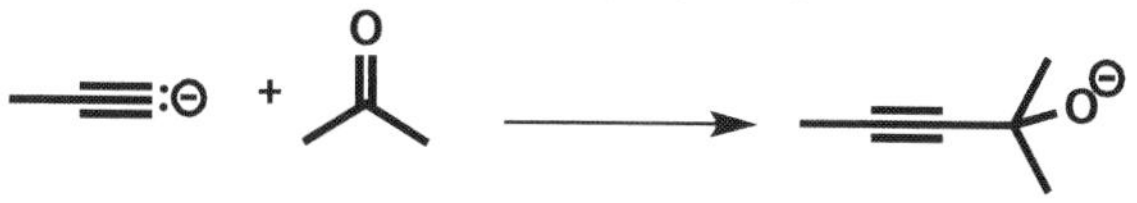

Nucleophilic addition to isocyanate

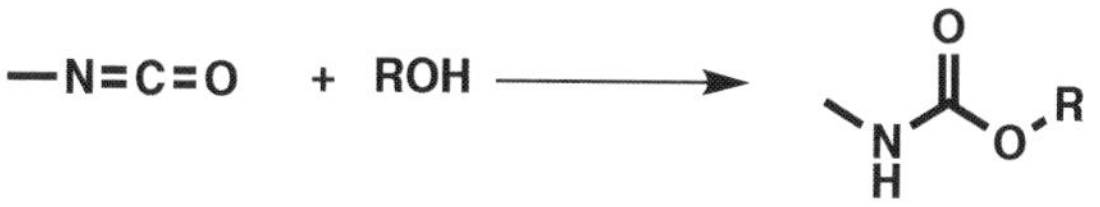

$$-N=C=O + ROH \longrightarrow$$

Nucleophilic addition at carbonyl carbon

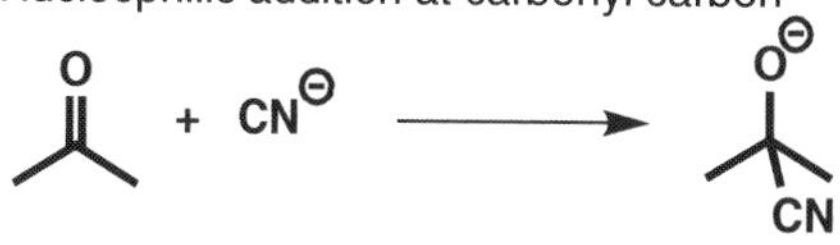

$$+ CN^{\ominus} \longrightarrow$$

Nucleophilic substitution at acyl carbon

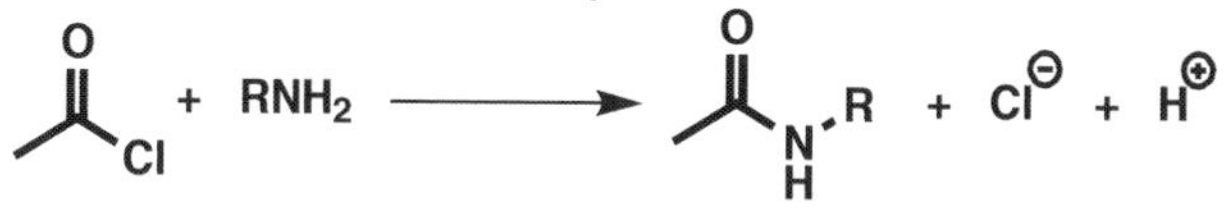

$$+ RNH_2 \longrightarrow + Cl^{\ominus} + H^{\oplus}$$

Nucleophilic substitution at saturated carbon (S_N2)

$$Br + CN^{\ominus} \longrightarrow CN$$

Oxidation

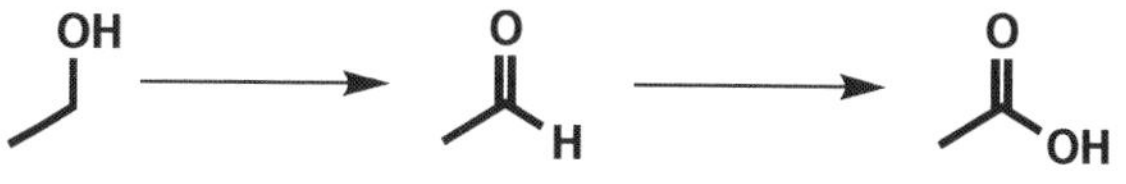

Oxymercuration

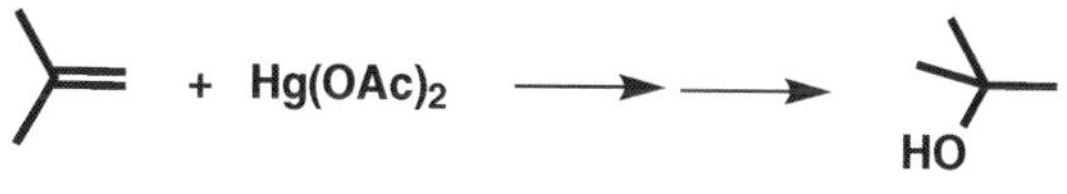

$$+ Hg(OAc)_2 \longrightarrow$$

Protecting group formation

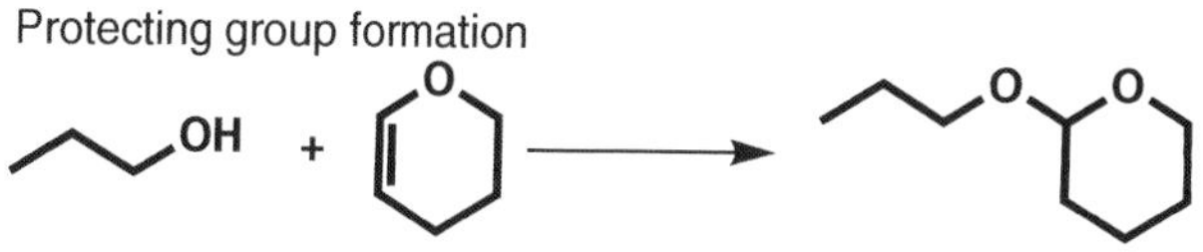

Reduction

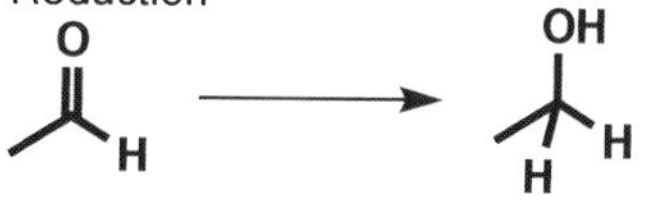

Retro-Claisen reaction

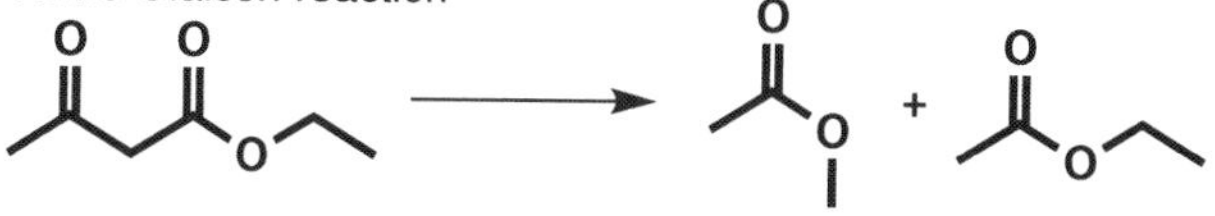

Ring closure reaction

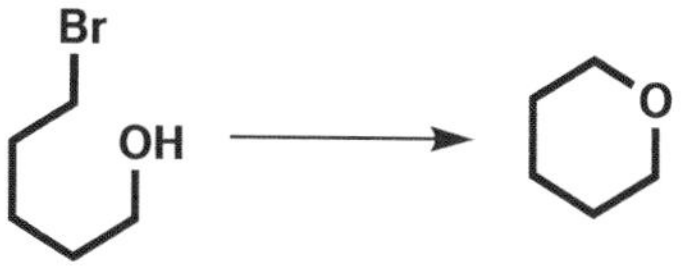

Robinson ring annulation

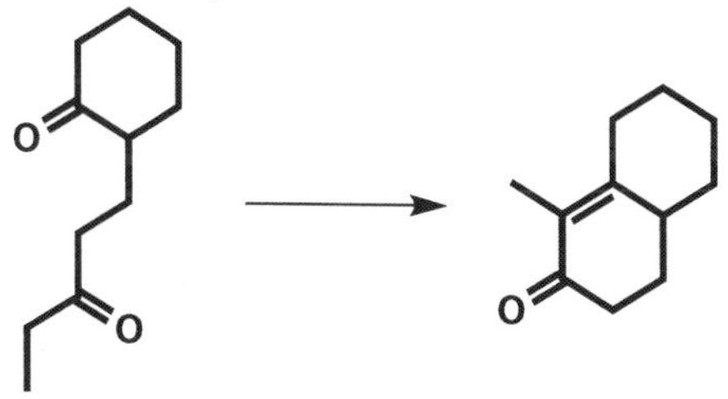

S_N1 reaction

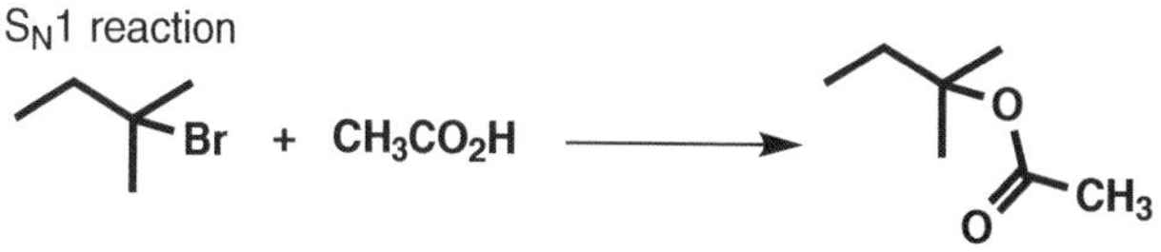

$$Br + CH_3CO_2H \longrightarrow$$

S_N2 reaction

$$Br + RNH_2 \longrightarrow$$

Tautomerism: enols and enediols

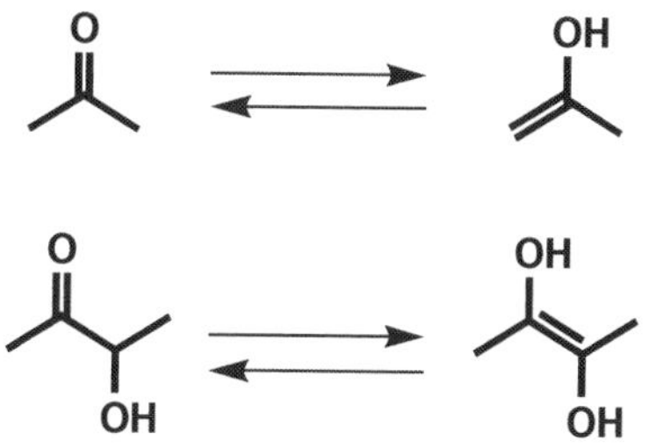

Wittig reaction

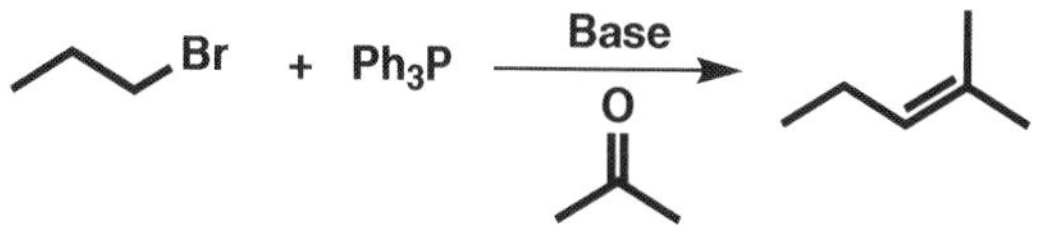

$$Br + Ph_3P \xrightarrow[\text{O}]{\text{Base}}$$

CONTENTS

- **Advice to Students using this book** ...*page xi*
- **Functional Groups & Chemical Reactions** —discussed In the text ...*page xii–xv*

INTRODUCTION 2-4

CHAPTER 1:
From Cellulose and Starch to the Principles of Structure and Stereochemistry 6–43

1.1: Starch and cellulose are polymers............. 6–7

1.2: Organic chemical structures are presented in ways where all the atoms in the formula may or may not be shown....................... 7–8

1.3: How can starch and cellulose have such similar chemical structures and yet have such different properties?....................... 8–9

1.4: Why do molecules have three dimensional structures?....................... 9–12

1.5: There is more to understand: electrons, structure, formal charge and the octet rule........... 12–16

1.6: The mirror images of glucose are different; they differ as we differ from our mirror image. What is the consequence of this fact at the molecular level?...................... 17–19

1.7: Stereoisomers are pairs of molecules, which although having the same formula and identical bonding, nevertheless differ from each other............................ 19–21

1.8: To understand diastereomers we have to understand isomers that are not stereoisomers, isomers that we call constitutional or structural isomers........................ 21–25

1.9: Chirality and handedness and how two molecules that are mirror image related can be distinguished from each other....................... 25–28

1.10: The experiments of Biot and Pasteur in the nineteenth century led to the first realization that molecules can exist in mirror image forms and that molecular mirror images could be studied with light, that is, optical activity could be measured from such molecules. 28–33

1.11: Eventually, as the three dimensional structure of molecules came to be understood, it became clear which structural features of a molecule could lead to mirror image isomerism, to enantiomeric pairs of molecules........................ 33–34

1.12: As experiments arose that could portray the three dimensional structures of mirror image molecules, it becpme necessary to develop a nomenclature that could distinguish left from right......................... 34–38

1.13: A molecule can rapidly change its shape by motions about the bonds that hold the atoms together; and the differing shapes of a single molecule are, by definition, stereoisomerically related to each other........................ 38–42

Chapter Summary of the Essential Material........................ 43

CHAPTER 2:

A Survey of the Experiments Usually Performed by Chemists to Understand the Structures of Organic Molecules: Mass Spectrometers, Infrared Spectrometers and Nuclear Magnetic Resonance Spectrometers

.... 44–69

2.1: Mass Spectra 44–49

2.2: Infrared Spectra 49–52

2.3: Nuclear Magnetic Resonance Spectrometry (NMR) 52–56

2.4: NMR Chemical Shift 56–58

2.5: Spin-spin Coupling in Proton NMR 59–68

Chapter Summary of the Essential Material 68–69

CHAPTER 3:

From Galactosemia to the Properties of Six-membered Rings:
An Introduction to the Mechanisms of Chemical Reactions

.... 70–103

3.1: What is the childhood malady called galactosemia? 70–71

3.2: To understand the molecular basis of galactosemia we have to
understand the nature of six-membered rings 71–74

**3.3: It took many years for chemical science to accept early ideas that rings did
not have to be flat** and that acceptance of this idea could explain many aspects of the
chemical behavior of cyclic molecules. An important advance, as is often the situation in
science, was the use of a new kind of instrument applied to the problem. 74–76

3.4: The Conformational Properties of Cyclohexane 77–79

3.5: The conformational properties of n-butane permit judging the relative
energies of the equatorial versus axial methyl cyclohexane: torsional and steric strain 79–82

3.6: Why should the difference between an equatorial and an axial bond on a six-member
ring sugar molecule be the difference between life and death for a stricken infant? 83–84

3.7: A background in the sugars, including their history, will help to set the stage for
understanding the fundamental difference between glucose and galactose and
ltherefore galactosemia. 84–85

3.8: Solving the wide variety of problems glucose presented, in order to come to a full
understanding of its structure, was a central theme in the development of chemistry 85–86

**3.9: We need a slight diversion from our story to understand the concept
of functional groups.** 87

3.10: There were two kinds of problems with the first structure proposed for glucose.
One of these problems could not be solved until it was realized that glucose was a cyclic
molecule. The second problem could not be solved until a chemist with extraordinary
experimental skills took up the task of figuring out the stereochemistry 87–89

3.11: The Second Problem in Determining the Structure of Glucose 89–92

3.12: How does glucose differ from the other seven diastereomers shown
in Figure 3.12? The answer can be found in the cyclic structure formed.
Glucose is the fittest molecule in the Darwinian sense 92–95

3.13: The Aldehyde Functional Group: π-Bonds and the Consequences of Electronegativity 95–97

**3.14: Reactive Characteristics of Aldehydes and other Carbonyl containing
Functional Groups:** Mechanism, Curved Arrows, Nucleophiles and Electrophiles 97–100

3.15: Galactosemia is caused by the reactivity of an aldehyde functional group.
A healthy infant supplies an enzyme to convert a derivative of galactose to a derivative
of glucose to avoid the reactivity of an exposed aldehyde functional group. 100–101

3.16: What can we now understand about the difference between cellulose and starch? 101–102

Chapter Summary of the Essential Material 103

CHAPTER 4:

Understanding Carbocations: From the Production of High Octane
Gasoline to the Nature of Acids and Bases

.... 104–131

4.1: What did Eugene Houdry do that revolutionized the petroleum industry
and had an important effect on the outcome of World War II? 104–105

4.2: What's happening in these catalysts? 105–107

**4.3: It took a great deal of time before chemists allowed the possibility that the carbon
skeleton of a molecule could change,** and then even longer to realize that the agent of
change was a chemical intermediate with positively charged carbon, a carbocation 107–110

**4.4: What are carbocations and what is the basis of their ability to
rearrange molecular structure?** It's all about that empty p-orbital. 110–113

**4.5: We are shortly going to find it convenient to name the hydrocarbons
involved in gasoline production.** Let's therefore take a moment to
step into the nomenclature of these molecules. 113–114

4.6: How do carbocations produced in catalytic cracking
increase the octane number of gasoline? 114–116

4.7: Why do carbocation rearrangements lead to branched structures?
The answer has to do with how the stability of carbocations
varies with molecular structure. 116–118

**4.8: Getting the lead out of gasoline made the problem of producing better
fuels even more critical** and therefore it became essential to understand what
structural features were necessary to produce higher octane number hydrocarbons. 116–120

**4.9: Industrial chemists invented an efficient reaction path to high octane
gasoline** using chemicals obtained in large quantities from the catalytic cracking
of petroleum. To understand how this was accomplished requires some
understanding of the behavior of acids and bases 120–122

4.10: Chain Mechanisms and the Rule of Vladimir Vassilyevich Markovnikov 123–126

4.11: The Brønsted-Lowry concept of acidity and basicity is too narrow and needs to be broadened to understand the industrial process that produces high octane gasoline. One of the great chemists of the twentieth century, G. N. Lewis, took the idea further. 127–130

Chapter Summary of the Essential Material 131

CHAPTER 5:
Carbocations in Living Processes 132–163

5.1: We've seen the chemical properties of carbocations to be essential for the industrial production of high octane gasoline. Now we'll discover that these identical chemical properties are of no less use for nature's purposes - terpenes to steroids. 132–134

5.2: **Terpenes and the Terpene Rule:** The treasures of our existence, color, odor and taste, are greatly dependent on a class of molecules, the terpenes, which derive from a single five carbon molecule, isopentenyl diphosphate, and if this were not enough this molecule is also the building block of the steroids that control our sex, our nature and our behavior. 005–138

5.3: Carbocations may arise by the breaking of a chemical bond with the two electrons in that bond leaving with one of the participants of the bond. The participant that gets the bonding electrons is appropriately called the leaving group. Leaving groups act as an important driving-force in biological pathways. 139–140

5.4: Resonance is the word used when a single molecular representation, a structural drawing for example, is inadequate to describe the distribution of electron density in a molecule. We compensate for this inadequacy by drawing multiple representations in which the atoms do not move but we draw the electrons as distributed differently. When multiple representations are necessary, when resonance is necessary, the actual molecule is more stable than that of any single representation resonance stabilization. 141–144

5.5: Carbocations are the key to the synthesis of terpenes and steroids, but not without enzyme catalysis. Markovnikov's rule is demonstrated in vivo. 144–146

5.6: Just as two molecules, which are constitutionally identical, can have a stereoisomeric relationship, **two parts of a single molecule, which are constitutionally identical, can also have a stereoisomeric relationship.** 147–149

5.7: Why is a five carbon entity with the carbon skeleton of isoprene so well suited to produce such a wide variety of biologically important chemicals, the terpenes? 150–153

5.8: Nature chooses the terpene route to gain entry to the family of steroids. 154–156

5.9: The conversion of the open chain 30 carbon molecule to a molecule with many fused rings requires the open chain to fold into a state bringing many atoms in close proximity and as well requires the presence of a small strained molecule, which springs open to start the process. 157–160

5.10: Given the proper conformation of oxidosqualene, the derived carbocation simply has to add to double bonds and carry out 1,2 shifts to produce lanosterol. 160–162

Chapter Summary of the Essential Material 163

CHAPTER 6:

Aromatic–A Word that Came to Mean Something Other
than Odor in the Chemical Sciences 164–204

6.1: **The Discovery of Benzene** 164–165

6.2: **A Short Diversion about the Ratio of Hydrogen to Carbon in Various
Organic Molecules** 165–167

6.3: **When Faraday discovered benzene, the formula for a molecule was
a key piece of information**–really the most important, if not the only piece
of information available. 167–168

6.4: **The stage was now set to propose a structure for benzene
that would explain its properties.** 168–172

6.5: **A Brief Stop for Benzene Nomenclature** 173

6.6: **Objections to Kekulé's hexagonal ring structure for benzene** required an
explanation that was equivalent to the concept of resonance. 173–175

6.7: **Hydrogenation of benzene yields a quantitative measure of the
aromatic stability of benzene** 176–179

6.8: **Understanding Benzene: Erich Hückel's Theory** 179–185

6.9: **Applications of Hückel's Theory to Biologically Important Molecules** 186–187

6.10: **Cumene, the common name for isopropyl benzene, is produced by
the world chemical industry at the level of billions of pounds.** The industrial
process introduces us to electrophilic aromatic substitution and the
Friedel-Crafts reaction and a confrontation between industry's goals and
organic chemistry principles. 187–193

6.11: **Energy of Activation, Reaction Rate Constants, and Reaction
Coordinate Diagrams** 193–197

6.12: **Resonance Resurrected** 197–200

6.13: **Application of the Ideas of Resonance Stabilization of Wheland
Intermediates in Electrophilic Aromatic Substitution** 200–203

Chapter Summary of the Essential Material 203–204

CHAPTER 7:

Fatty Acid Catabolism and the Chemistry of the Carbonyl Group 206–243

7.1: **The fatty acids in living organisms are saturated and unsaturated.** 206–207

7.2: **Fatty Acids.** 207–210

7.3: **Saponification** 210–213

7.4: **Similarities and Differences between Ketones and Aldehydes
and Derivatives of Carboxylic Acids: Mechanism of Saponification** 213–216

7.5: Hydrolysis of the Triglyceride Ester Bonds: Nature's Path. 216–220

7.6: Biochemical Conversion of Fatty Acids to their Thioesters
with Coenzyme A: The Key Role of Leaving Groups 221–226

7.7: Breaking a fatty acid down into two carbon piece first requires
introducing a double bond using an oxidizing coenzyme. 226–231

7.8: The Next Step in the Catabolism of the Fatty Acid:
Conjugate Addition to a Double Bond. 231–233

**7.9: Oxidation of β-Hydroxyl Fatty Acyl Coenzyme A Using an
Enzyme and an Oxidizing Coenzyme** 233–237

**7.10: Cleaving a Two Carbon Fragment from the Fatty Acid
Chain:** The Retro-Claisen Reaction. 237–241

Chapter Summary of the Essential Material 242–243

CHAPTER 8:

Carbanions and Carbonyl Chemistry: Sugar Catabolism,
Isopentenyl Diphosphate Synthesis and the Citric Acid Cycle

.... **244–274**

8.1: Nature's Problem with the Catabolism of Glucose and its Solution 244–249

**8.2: Tautomerism: Enediols are a special case of the dynamic
interconversion between enol and keto tautomers.** 249–253

**8.3: We've seen how the reverse of the Claisen condensation
in the catabolism of both fats and sugars causes breaking of
carbon-carbon bonds.** Let's see how nature uses the Claisen
reaction in the other direction, to make carbon-carbon bonds 253–256

8.4: The Aldol Condensation 256–257

8.5: Continuing on the Path to Isopentenyl Diphosphate 257–260

8.6: The Citric Acid Cycle: what is it about? 261–264

8.7: The Organic Chemistry of the Krebs Cycle 264–267

8.8: Stereochemistry: Why Krebs' proposal was thought to be impossible. 267–269

8.9: Why is adenosine triphosphate, ATP, life's way of storing energy?
In organic chemical terms we find an answer in the concept of leaving groups. 269–273

Chapter Summary of the Essential Material 273–274

CHAPTER 9:

Investigating the Properties of Addition and Condensation Polymers:
Understanding more about Free Radicals, Esters and Amides

.... **276–305**

9.1: "If we knew what we were doing, it wouldn't be called research, would it?" 276

9.2: Polyethylene: The Background Story. 276–278

9.3: The Mechanistic Path to LDPE–Free Radicals 279–284

9.4: An important reaction of free radicals is responsible for the production of ethylene and other alkenes in large volumes from the steam cracking of petroleum fractions. 284–286

9.5: Contrasting Thermodynamic Factors Control Polymerization of Ethylene and Steam Cracking of the Naphtha Fraction of Petroleum 287–288

9.6: Resonance works against the chemical industry again. 288–290

9.7: A Short Story about a Nobel Prize 290–292

9.8: We've followed the polyethylene thread that led from ICI's foray into basic research. **Now let's follow the nylon fiber that unwound out of DuPont's move in the same direction: Polyesters first.** 293–296

9.9: Nylon. But first let's take a look at proteins on which the nylons are modeled. 296–298

9.10: Nylon 6,6 298–301

9.11: Hexamethylene diamine and adipic acid react together in the industrial process to produce nylon 6,6. 301–303

9.12: Why is nylon such an excellent fiber forming substance? Because it mimics a property of silk – interchain hydrogen bonds. 303–305

Chapter Summary of the Essential Material 305

CHAPTER 10:

The Industrial Road Toward Increasing Efficiency in the Synthesis of Hexamethylene Diamine with Stopovers at Kinetic Versus Thermodynamic Control of Chemical Reactions, Nucleophilic Substitution, and with a Side Trip to Laboratory Reducing Agents

.... 306–331

10.1: Benzene to Adipic Acid 306–308

10.2: Nylon 6,6: Hexamethylene Diamine–The Classic Route From Adipic Acid 308–309

10.3: A Side Trip to Laboratory Reducing Agents 309–312

10.4: Hexamethylene Diamine – An Attempt at a Better Route. 312–315

10.5: How industry overcomes a supposedly insurmountable problem arising from thermodynamic versus kinetic control in addition of halogen to double bonds, to invent an elegant and commercially viable route to two commercial polymers, only **to finally fail because of an unforeseen environmental consequence of their path**. 315–319

10.6: The reactions of the isomeric dichlorobutenes with cyanide ion leads us to investigate one of the most studied reactions in organic chemistry, nucleophilic substitution at saturated carbon, which can take place at the extremes via **the S_N1 or S_N2 mechanism.** 319–328

10.7: Stereochemical Probes of Nucleophilic Displacement 328–331

Chapter Summary of the Essential Material 331

CHAPTER 11:

Much can be learned about organic chemistry from
the study of natural rubber and other elastomers

Much can be learned about organic chemistry from
the study of natural rubber and other elastomers 332–368

11.1: Two Different Trees 332–333

11.2: Cis and Trans Alkenes 334–336

11.3: Why should the difference between a cis and trans double bond
make the difference between an inelastic and an elastic material? 336–338

11.4: Why does rubber get hotter when stretched and why does rubber
get stiffer at higher temperatures? The answer increases our knowledge of
thermodynamics. 338–339

11.5: Crosslinking of rubber is necessary. 339–340

11.6: How do crosslinks form when rubber is heated with sulphur? 341–344

11.7: Synthetic elastomers: Hypalon–crosslinking without double bonds
requires introducing a functional group to a polyethylene chain. 345–348

11.8: Crosslinking of Hypalon: The Parallel Reactive Character of
Carboxylic Acid Chlorides and Sulfonyl Chlorides 348–349

11.9: A Review of Nucleophilic Attack at Carbonyl and Sulfonyl and the
Role of Leaving Groups 349–354

11.10: Sulfonamides: Crosslinking of Hypalon and Sulfa Drugs 354–355

11.11: Industrial tradition rejects a perfectly good elastomer: more about
free radicals. 356–359

11.12: Elastomers without Covalent Crosslinks–The Glassy State 359–363

11.13: A thermoplastic elastomer that is not based on a glassy state: Spandex. ... 363–367

Chapter Summary of the Essential Material 367–368

CHAPTER 12:

Synthesis Part One

Synthesis Part One 370–403

12.1: Synthesis is important. 370–371

12.2: R. B. Woodward 371–374

12.3: Cholesterol: The First Step 374–378

12.4: Cholesterol: Adding the Third Fused Ring 378–382

12.5: Cholesterol: Setting the Stage for Adding the Fourth Fused Ring 382–389

12.6: Woodward uses a Grignard reagent to form the fourth fused ring. 389–392

12.7: A diversion from the synthesis of cholesterol to understand
how Woodward used a ketal to protect a double bond. 393–395

12.8: The End Game 395–399

12.9: Addition, Substitution and Elimination Reactions–Paying More
Attention to the Latter 400–403

Synthesis Part Two 403–443

12.10: Elias J. Corey 403

12.11: Prostaglandin, the Beginning Steps 404–406

**12.12: Two remarkable rearrangements: The Baeyer-Villiger reaction forms a
lactone, which is then rearranged to another lactone.** 406–408

12.13: A Diversion into Ring Closing Chemistry 409–411

12.14: Boron and Phosphorus: Useful Elements in Synthetic Chemistry 412

12.15: The Wittig Reaction 412–416

12.16: Hydroboration and Oxymercuration 416–419

**12.17: The Importance of the Wittig Reaction to Corey's Synthesis
of the Prostaglandins** 419–422

12.18: Protecting groups are necessary. 422–425

**12.19: The End Game–The Wittig Reaction One More Time and a
Protecting Group in Disguise** 425–428

12.20: Retrosynthesis 428–435

12.21: The Mechanism of No-Mechanism Reactions–Frontier Molecular
Orbitals Applied to the Diels-Alder Reaction, and Other Pericyclic Reactions 435–441

Chapter Summary of the Essential Material 441–443

Index: 444–452

Introduction

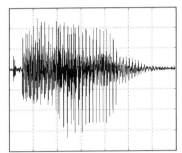

"We don't learn the alphabet before we hear people speaking."

IN THE ACADEMIC STUDY OF THE ARTS, the principles necessary to create a work of art such as a painting or a poem or a musical composition are discovered by studying the completed work. In this way the student encounters the beauty arising from the use of these principles at the very beginning, with the pleasure of this encounter stimulating the desire to understand what stands behind such an accomplishment. The method of learning of the arts is close to how we learn outside of the academic world, how a child learns from the start. We don't learn the alphabet before we hear people speaking. We don't learn the colors or the shapes of common objects before seeing the world around us. The wonders of sound and shape and color intrigue us and stimulate our desire to figure out what is going on and what it all means.

The study of science rarely takes this context-based path found in the study of the arts, insisting instead that the student learn the principles, and only later see how these principles lead to the complexity of, for example, the production of an industrial product such as nylon, or the in vivo catabolism of a fatty acid. Although we may point to the complex result of the use of the principles we teach as we go along, we don't use this result as a template for introducing these principles.

The intention in writing this book is to demonstrate that the approach taken in the arts can be successfully used in organic chemistry and perhaps in other disciplines of science as well and can act to enhance the learning experience of students of the subject. For this approach to be successful, both the teacher and student must be willing to allow that material will be presented that it is not possible to fully explain. Normally, the material in a chapter in an organic chemistry textbook can be bounded, that is, the author creates a logical framework and chooses material for each chapter so that it can be explained in full detail at the level of the book. Here, a chapter may contain varied subjects that would not usually be treated within a single unit but are held together by a narrative.

The material is often presented as a complex application of the science, in a story telling historical context with particular attention paid to scientists who have played important roles. Subjects are treated that could have been the focus of a large number of lectures, if not a large part of a course. However, such a subject will be presented if it offers a source for a principle of organic chemistry that the student is ready for. The criterion is that the principle appears in an understandable manner at the level of beginning students, even if only a general understanding is offered of the larger picture. That general understanding then supplies the context, which we feel is valuable to the learning process.

We have taken this new approach at the Polytechnic Institute of New York University in teaching organic chemistry to sophomores majoring in chemical engineering and chemistry by utilizing complex processes from industry and biochemistry and academic laboratories. In this manner we are allowed a wide range for choosing what works best for the principles we present. Our experience shows that students will accept and even treasure this approach as long as they know what is expected. The student sees the big picture, understands its importance and even beauty, and hopefully is stimulated to work hard at learning some of the principles of our science that contribute to this picture. It is the acceptance and enthusiasm, and even gratitude of our students, that encouraged writing this text.

The approach taken here allows the opportunity to present the same principles multiple times in different contexts, therefore reinforcing and demonstrating the wide ranging importance of these principles. Consider stereochemistry for one example: its principles can be found throughout the complexity of organic chemistry. But rather than waiting to present these complex phenomena until the principles of stereochemistry have been demonstrated in simpler molecular terms we attempt to find the principles of stereochemistry, for just a few examples, in the structure of glucose and its polymers, in the formation of isotactic polypropylene, in the prochiral specificity of enzyme catalyzed reactions and in the difference between *gutta percha* and *Hevea* rubber.

Aromaticity is another among the many examples. The struggle of the bulk chemical industry to avoid multiple alkylation of benzene on the route to starting materials for important plastics offers a perfect template to understand the power of aromatic stabilization and resonance. On the other hand, biology's co-enzymes that sit on a knife edge of aromatic stabilization to allow their reversible use, such as $NAD^+/NADH$ and $FAD/FADH_2$, are no less valuable as a means to appreciate the nature of aromaticity. And then there is the special stability of the nucleotide bases, which not only yield lessons in aromaticity but can as well be used to reinforce the ideas of hybridization of atomic orbitals, to understand which electrons contribute to the aromatic character.

An advantage of presenting fundamental principles in differing contexts is that the text is written in a manner to allow choosing among many of these context-based discussions to reduce the course content, if that is desirable, and/or to reserve some material for independent study or special topics. Our intention is to demonstrate that the principles of the science, rather then being presented, can instead be discovered in what is important in our lives.

In other words, one does not need to know any principle or nomenclature until it is necessary. In place of a section early in the study of the subject devoted to memorizing nomenclature, the student gradually becomes familiar with nomenclature as the subject moves. Following the same approach, why learn about enols and enediols until one learns about glycolysis? Why learn about carbocations until one comes across the catalytic cracking of petroleum fractions or the synthesis of terpenes, or electrophilic aromatic substitution?

We have attempted to use the "Study Guide Questions" which follow each of the sections within each chapter to guide the students to what is expected of them, and to answer the inevitable question in one form or another: what am I responsible for? The term study guide is also consistent with the nature of many of the questions, which often contain information that amplify the text and ask the student to reason about subject matter that is about to be discussed in a subsequent section. At the same time some of these questions are designed to look into areas not covered in the text and also to help students to dig deeper into the subject, to take the material further along. This latter aspect is focused on in the tutorial for the book, which can be seen, in part, as an extension of the material presented in the text.

In addition, each chapter ends with a summary section outlining the essential ideas taken from the study of the material in that chapter (Chapter Summary of the

Essential Material). This important narrative at chapters' ends is an opportunity for the students to test themselves. The narrative is written in a general manner. If the chapter has been well understood by the student, the narrative will make sense and the student will be able to fill in the details left out. Otherwise, the narrative will point the student to the areas where another look at the material is necessary.

The basic idea is that learning the fundamental language gives the student the power to comprehend any aspect of organic chemistry, while the context-based story telling historical approach points to the importance and intrinsic interest of the subject. This is accomplished while covering essentially the same material that is commonly found in organic chemistry textbooks, while allowing far fewer pages in this book. Moreover, much important material is covered that is not usually treated in other texts designed for beginning students.

The approach of this book is certainly a radical departure from what we all have done for many years but perhaps a quote from one of the most famous American inventors of the 20th century may pertain: *"The world hates change, but it is the only thing that has brought progress."* Charles Franklin Kettering.

Much is owed to the students in the organic chemistry classes for which the manuscript in its various stages was used and whose suggestions and complaints were so important to improving the work. Some of these students appear on the cover as follows from left to right: Benjamin Osei-Bonsu; Stephany Paulette Torres; Tina Xiong; Joseph Asad; Jerome Fineman; Radu Gabriel Iliescu; Jinhui (Liz) Zhao. A special thank you goes to Radu Iliescu who spent a great deal of time gathering and organizing the photographs of the scientists shown in the book and even more time helping me with computer issues.

Critically important to the work was the precise editing work of James Moore of the Rensselaer Polytechnic Institute and Harold Wittcoff of Nexant Consultants. The book owes much to the helpful criticism and encouragement of Ian Fleming of Cambridge University and to the critical reading of J.A.J.M. Vekemans of Eindhoven Technical University. Their extensive comments were of great importance. I am grateful to Jerome Berson of Yale who supplied important historical information and made necessary corrections when I was mistaken. And thank you to Dr. Andrea Kover, graduate of Universitat de Barcelona and Dr. Filbert Totsingan, graduate of Università degli Studi di *Parma for working along with me and for* their skill with ChemDraw.

Mark M. Green

Polytechnic Institute of
New York University
September 2012, New York City

"The basic idea is that learning the fundamental language gives the student the power to comprehend any aspect of organic chemistry,"

Chapter 1

From Cellulose and Starch to the Principles of Structure and Stereochemistry

1.1

Starch and cellulose are polymers.

BOTH *CELLULOSE* AND *STARCH* ARE *POLYMERS*. Poly means many and because polymers are made of a very large number of small molecules connected together, polymers are usually very large molecules, macromolecules. Polymers are critical to life. DNA and proteins as well as cellulose and starch belong to the polymer class and are made from large numbers of small molecules connected together. Polymers are also critical to the chemical industry and to our life style. They are too numerous to mention here, but consider just polyethylene, polypropylene and nylon, which are made in the billions of pounds each year around the world.

Each kind of polymer, if it be essential to life, such as DNA, or to how we live our life, such as Spandex or nylon, is made of its own unique smaller molecule components. In both starch and cellulose, the small molecule components are based on the structure of *glucose*. In other words, if one were to take starch and cellulose apart by adding water, that is, to *hydrolyze* these polymers, both would yield only glucose. Starch and cellulose are made of the same small molecule, glucose, but put together in a different manner. That's pretty unusual in the world of polymers. Usually different polymers are made of different small molecules, called monomers. A copolymer may be made of different small units. But nature has found a way to make two very different materials, one that is a food, starch, and the other a construction material, cellulose, from the same building block. Interesting!

Figure **1.1** shows a molecular picture of a portion of cellulose and also a particular kind of starch known as *amylose*. Also included in Figure 1.1 are the molecular

FIGURE 1.1 ▶

Chemical Structures of Glucose and the Polymers Derived from Glucose

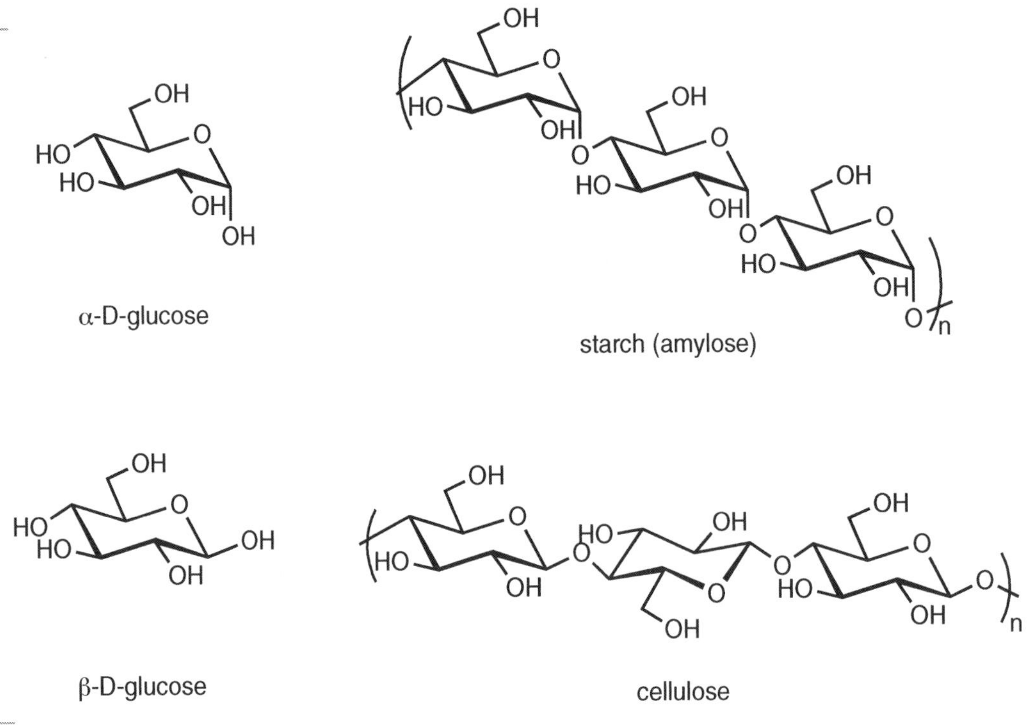

α-D-glucose

starch (amylose)

β-D-glucose

cellulose

structures of two forms of glucose, which as mentioned above is the molecule from which both cellulose and starch are made. In organic chemistry when we present the structure of a polymer we may show only a portion of the molecular structure. This is okay for starch and cellulose because of the repetitive nature of these polymers. In a small molecule we have to show the entire structure to get the whole picture. This way of presenting structures is followed in Figure 1.1 for the two polymers, cellulose and starch and for the small molecules, the two forms of glucose.

▶ **PROBLEM 1.1**

Redraw the chemical structures in Figure 1.1 showing the carbon and hydrogen atoms with the realization of a carbon atom where the lines meet and that each carbon atom has to have four bonds to it and that bonds not shown are always to hydrogen atoms.

▶ **PROBLEM 1.2**

What is meant by the statement that every carbon and oxygen atom in the structures in Figure 1.1 obeys the octet rule while every hydrogen atom obeys the equivalent of the octet rule for a first row element? How do the lines between the atoms, which represent the bonds, contribute to the answer to the question about the octet rule?

▶ **PROBLEM 1.3**

Explain the fact that there are electrons associated with every oxygen atom in the structures in Figure 1.1 that are not shown, whereas all the electrons associated with the carbon and hydrogen atoms are shown.

▶ **PROBLEM 1.4**

If you have a set of models (which is highly recommended as noted in the text) construct the two glucose structures shown in Figure 1.1 and point to the difference in three dimensions. Now do the same for three units of the cellulose and three units of the amylose structures and look for their three dimensional differences.

▶ **PROBLEM 1.5**

Do you see any relationship between the two glucose structures and the structures of cellulose and starch?

GLUCOSE IS MADE ENTIRELY OF carbon, hydrogen and oxygen and has the formula, $C_6H_{12}O_6$. This formula can be expressed in a different way, as $C_6(H_2O)_6$, which accounts for the fact that glucose belongs to a class of molecules know as carbohydrates. In the situation of glucose, six carbon atoms and six water molecules, or in other words, hydrated carbon. The molecular structures for the two forms of glucose in Figure 1.1 don't seem to fit the formula just given. Yes, the oxygen atoms are there, as O, and there are the six of them as required by the formula. And there seem to be the necessary hydrogen atoms, H, in each of the two glucose molecules shown, but only 5 H, not the 12 H in the formula. Moreover, the symbol for carbon, C, does not appear anywhere in either of the two glucose molecules shown or in the structures for cellulose or amylose as well. Not showing the symbol for carbon comes from a tradition of how organic chemical structures are sometimes presented in which the straight lines in the structure represent covalent bonds, to be discussed below in section 1.4, and where these lines meet is the site of a carbon atom.

Now we can see from the number of sites in the structures in Figure 1.1 where lines meet that each of the glucose structures and each of the glucose derived units in both cellulose and amylose contain six carbon atoms as required by the formula. However inspection of the two polymer structures in Figure 1.1 show that not only are the carbon

1.2

Organic chemical structures are presented in ways where all the atoms in the formula may or may not be shown.

atoms, C, not shown. but most of the hydrogen atoms, H, are not shown as well. We know where the carbon atoms are now, at the angled junction between the covalent bonds, but the placement and number of the hydrogen atoms is more of a problem. For now you'll get the right number and placement of hydrogen atoms on the carbon atoms in each polymer structure by simply making certain that each angled junction is surrounded by four covalent bonds, four lines. If not, then simply add as many lines as necessary with hydrogen atoms terminating these lines, C-H bonds. We'll say more about the reason for this in section 1.4. That is, we'll see why carbon is tetravalent.

▶ **PROBLEM 1.6**
Determine the molecular weights for the formula you determined for α and β-D-glucose.

▶ **PROBLEM 1.7**
How do the molecular weights of cellulose and starch depend on the n in Figure 1.1?

1.3

How can starch and cellulose have such similar chemical structures and yet have such different properties?

THIS WOULD BE AN EXCELLENT TIME to buy a set of molecular stick models, which are inexpensive, and use them to build parts of the molecular structures shown in Figure 1.1 and then use your set of stick models to build and therefore help to understand many molecular structural ideas that we are going to come across.

The clue to the answer to the question posed in the title to this section comes from simply looking at the molecular structures in Figure 1.1. These molecules, cellulose, starch and glucose, exist in three dimensions with the darkened edge of the ring in each structure coming out toward the viewer. And the difference between α and β glucose is simply a change in shape as is the difference between cellulose and amylose. In fact, looking closely at the molecular structures in Figure 1.1, you could observe that the difference between α and β glucose resembles the difference between amylose and cellulose. The difference is found in only one oxygen atom, which is pointing downward in α-D-glucose and in amylose. In β-D-glucose and in cellulose this same oxygen atom is pointing upward.

We have a name for two molecules that only differ in their shape - stereoisomers. Stereochemistry focuses on stereoisomers. The word stereochemistry makes sense because stereo points to three dimensions, three-dimensional-chemistry. Stereochemistry is the study of how molecules differ in their three-dimensional shapes.

Cellulose and starch are stereoisomers, as are α and β glucose. Later, we'll discover that the details of the stereochemical differences observed in Figure 1.1 make perfect sense with regard to nature making food out of one of these polymers and wood out of the other. It may seem remarkable that what looks like the small differences in molecular structure make such a big difference in properties. But it is a fact. However, for now let's pay more attention to certain details of the molecular structures in Figure 1.1 and discover why molecules have defined shapes, that is, why there is such a thing as stereochemistry.

▶ **PROBLEM 1.8**
Must stereoisomers have the same molecular weight?

PROBLEM 1.9
Must stereoisomers have the same connection between the atoms in their structures?

PROBLEM 1.10
Must stereoisomers have different shapes?

PROBLEM 1.11

Using the rule that each carbon atom must have four bonds to it, make up as many structures as you can imagine for the formula C_8H_{18}. Pick pairs of two of the structures you have created and ask the question if they are stereoisomers. What is the relationship between a pair of structures that are different but are not stereoisomers? Is this a different kind of isomer and if so what would be the rule for the existence of this different kind of isomer?

N O ONE KNOWS HOW IT ALL BEGAN, but in about 5 billion years our sun is predicted to enter a red giant phase in which there will not be enough hydrogen to sustain the fusion to helium that yields the sun's energy now. Our sun, or any sun, is predicted to eventually collapse and become far hotter in its core, hot enough that fusion of helium to carbon and oxygen will take place. Presumably, this is the manner in which the elements that make up our life came to be.

Any carbon atom produced by fusion will eventually cool enough to take on the necessary six electrons to neutralize the six protons in the nucleus of the atom, the reason we assign atomic number six to carbon. These six electrons will distribute themselves in the orbitals that quantum mechanics teaches us about. This is the state of an isolated carbon atom and, as we'll see, this is the beginning of our story of why molecules are precisely geometrically defined. However, in the situation of an isolated carbon atom the six electrons are distributed in a manner that is not suitable for the bonding characteristics known from our experimental investigations of molecules containing carbon.

Carbon is tetravalent, which means that in all stable molecules the carbon atoms are involved in four electron pair bonds. This arrangement applies to all the structures in Figure 1.1. But as seen in **Figure 1.2,** only two of the four electrons in its outer orbital are free for covalent bonding. Why is that?

A covalent bond, that is the lines between the atoms in Figure 1.1, contains two electrons. One electron in the covalent bond comes from each of the atoms that are bonded together by the covalent bond. This means that for each carbon atom in the structures shown in Figure 1.1, the carbon atom must have available for bonding four unpaired electrons. However, in the isolated carbon atom in Figure 1.2, two of the four electrons in the outer orbital are paired with each other and therefore not available for forming covalent bonds with another atom. The two unpaired electrons in the isolated carbon atom in Figure 1.2 can be assigned to any two of the p orbitals, p_x, p_y, or p_z. Therefore if two covalent bonds were formed with these two unpaired electrons, the molecule formed could only exist in two dimensions, x and y (for the situation shown in Figure 1.2), x and z, or y and z. The problem therefore with the electron configuration of the isolated carbon atom is that carbon would be predicted to form divalent molecules in two dimensions rather than forming tetravalent three dimensional molecules as is known from experiment and as we've seen for every carbon atom in the structures for cellulose, starch and glucose in Figure 1.1.

The solution to the problem of how to reconcile the three dimensional structures experimentally observed for many molecules containing carbon atoms was invented long ago by **Linus Pauling,** a chemist who won two Nobel Prizes, in 1954 for chemistry and in 1962 for peace, neither of which was for his solution to this problem. Pauling was a young man when he first began thinking about this problem of the four-coordinate three dimensional bonding of carbon, and published, before he was thirty years old, the critical paper outlining his concept of hybridization of carbon.

This was just the beginning of his contributions to the foundations of chemistry. Just a few years later Pauling introduced another fundamental idea on which chemistry is based, the concept of electronegativity, a characteristic of atoms that accounts for

1.4

Why do molecules have three dimensional structures?

Linus Pauling

Isolated carbon atom

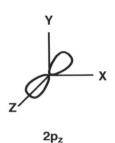
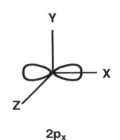
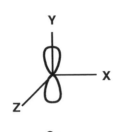

FIGURE 1.2 ▶

Atomic Orbitals of Carbon Hybridized for Four Coordinate Bonding

Hybridized carbon - four coordinate bonding

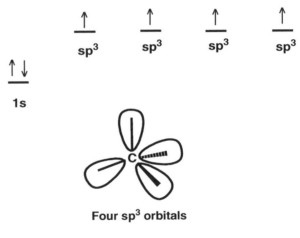

Four sp³ orbitals

β-D-glucose   or   methane

many properties of molecules and something we will study when we talk first about a technique used to explore molecular structure, nuclear magnetic resonance (section 2.3), and later about chemical reactions (section 3.10). And we still are not mentioning Pauling's contribution that won him his first Nobel Prize, for proposing a structural element of proteins, the α-helix.

The idea of hybridization of atomic orbitals seems a simple one in retrospect. Mix, that is, hybridize the orbitals to make new ones that are suitable for the bonding observed around carbon, or for that matter around any atom where hybridization is required to create suitable bonding arrangements. This mixing of orbitals necessary for the bonding arrangements of the carbon atoms in the structures shown in Figure 1.1, that is, three dimensional tetravalency, are shown in Figure 1.2. The hybridized orbitals now suitably describe how the electrons in the valence shell of carbon, the second quantum level, are consistent with the array of four atoms around carbon seen in the experimental observations responsible for the structural drawings in Figure 1.1.

Pauling did not consider that an isolated carbon atom spontaneously hybridized the orbitals but rather that in the process of forming the chemical bonds such a reorganization of the electron occupation of the orbitals took place. In that way, the extra energy necessary to move the orbital arrangement out of the state shown for the isolated atom in Figure 1.2 to the hybridized orbital arrangement was more than compensated for by the energy gained by the formation of the four bonds.

Many experimental techniques have demonstrated that when carbon is bonded to four atoms, whatever those atoms may be, such as four hydrogen atoms in methane, CH_4, or various carbon, oxygen and hydrogen atoms as is the situation for each carbon in glucose, cellulose or starch, the four surrounding atoms are arrayed as in the approximate geometry of a tetrahedron. If all the atoms surrounding the carbon atom are identical, then the average geometry will be a perfect tetrahedron with an angle of 109.47° as shown in Figure 1.2 for methane. If the atoms surrounding the central carbon differ from each other, as is the situation for every carbon atom in glucose, starch and cellulose (Figure 1.1) then a somewhat distorted tetrahedral array of these atoms will be formed with differing angles between any two of these surrounding atoms. But the overall geometry will still be approximately tetrahedral.

The structure for methane in Figure 1.2 uses lines to show bonds in the plane of the paper and a slashed line and a darkened wedge to represent those bonds going behind the page and in front of the page respectively. These representations of the three-dimensional directions of chemical bonds are widely used and will be used throughout this book.

The picture of a tetrahedral or approximately tetrahedral array of atoms around a central carbon atom, which we obtain from our experimental studies, fits the orbital picture arising from Pauling's hybridization idea shown in Figure 1.2. The hybridization, that is, the mixing of the four atomic orbitals of the second quantum level leads to four equivalent orbitals, each with one part s and three parts p, therefore designated sp^3. Four orbitals, each of which contain 1/4 s adds up to the s orbital and four orbitals each of which contain 3/4 p adds up to the 3 p orbitals that we started with.

Ronald Gillespie

Because the three p orbitals that make up part of each of the four hybridized orbitals, sp^3, are arrayed along the x, y and z directions as shown in Figure 1.2, the hybridized orbitals must exist in all three of these dimensions as also shown in Figure 1.2. Moreover, because each of the four bonds surrounding the central carbon atom contains two electrons, the bonds, with their negative charges, repel each other. Geometry teaches that surrounding a central point with four points in a tetrahedral array keeps the surrounding points as far apart as possible. Hybridization of orbitals in combination with relief of the electron repulsions among the four bonds surrounding the central carbon atom are the sources of the three-dimensional character of all the molecular structures shown in Figure 1.1.

Another view, which led to the same conclusions about the geometry of molecules, was developed at University College London by **Ronald Gillespie,** an Englishman who is now emeritus professor at McMaster University in Canada and **Sir Ronald Nyholm,** a distinguished Australian born inorganic chemist, who was a leader in the development of the profession of chemistry in England (for which he was knighted) but unfortunately died too young in a car accident. Their focus is on electron pair repulsion, the VSEPR theory (valence shell electron pair repulsion) without necessarily considering hybridization of orbitals. Using VSEPR theory one simply counts the number of electron pair domains around the carbon atom, independent of the electrons constituting or not constituting a covalent bond–a lone pair of electrons (section 1.5) counts the same as a covalent bond. For a carbon atom with two domains the geometry will be linear, for example a carbon atom in acetylene, three domains planar trigonal, for example the carbon atom in formaldehyde, four

Sir Ronald Nyholm

domains tetrahedral, for example the carbon atom in methane, and so on for more complex geometries associated with other central atoms, which takes VESPR theory into the realm of inorganic chemistry.

Both men were associated with Christopher Ingold early in their careers where they met after WW II. Ingold was a pioneer in the overlap of physical and organic chemistry, which led to the field we now call physical organic chemistry. Gillespie's work with Ingold involved early detailed investigations of reactions occurring in superacid media, media that were used so effectively by Geoge Olah in Nobel prize winning work involving NMR investigations of the properties of carbocations (Chapter 4, section 4.4). Professor Gillespie's views on education, a long standing focus of his interests, have also seen the importance of context, in line with the approach taken in this book.

We'll hear much about the contributions of Ingold in Chapter 10 (section 10.6) concerning one of the most studied and important reactions in organic chemistry, nucleophilic substitution, and even later in this chapter in Ingold's contributions to the nomenclature of chiral molecules.

Whatever may be our explanation of the fact, nature creates an ability of carbon to bond with other elements using all four electrons in the valence shell of carbon. A food and a building material are created from the same starting material (Figure 1.1). And there is so much more. The entire structure of organic chemistry becomes possible and all that is described and predicted by this science–life itself.

PROBLEM 1.12
Why do the arrangements of electrons in the valence shell orbitals of the isolated atoms not allow carbon to be tetravalent but do allow oxygen to be divalent and hydrogen to be monovalent?

PROBLEM 1.13
(a) Justify the statement that each hybridized orbital in tetracoordinate carbon is identical with 25% s and 75% p.
(b) Why does the incorporation of all three p orbitals in each hybrid orbital make it necessary for the hybridized orbitals to occupy three dimensional space?
(c) Why is a tetrahedron the best geometry for tetravalent carbon?

PROBLEM 1.14
How does hybridization of atomic orbitals offer an explanation for the fact that the bond angle between the oxygen and the two hydrogen atoms in water is approximately 105°?

PROBLEM 1.15
Use a set of models to construct a glucose molecule from Figure 1.1 knowing that each carbon atom should have an approximate tetrahedral geometry. Can the six atoms of the ring be in a plane?

1.5

There is more to understand: electrons, structure, formal charge and the octet rule.

IN ALL MOLECULES CONTAINING CARBON, the carbon atom will have no formal charge only if there are equal numbers of negative and positive charges, that is, six electrons "associated" with the carbon atom to counter the six protons in the nucleus of that atom. Carbon has an atomic number of six. Whatever is the bonding state of carbon, or in other words, independent of the hybridization of the orbitals, the two electrons in the first quantum level, $1s^2$ (Figure 1.2), act to counter two of the protons in the nucleus of the atom.

In carbon, the four remaining electrons in the second quantum level of each carbon atom, the electrons in the valence shell, act to counter the positive charge of the remaining four protons in the nucleus of the carbon atom. In an isolated carbon atom (Figure 1.2) the balance of positive charge from the nucleus of the atom and negative charge from the surrounding electrons makes for no net charge. However, this situation becomes more complicated when bonding with another atom is involved. The two inner electrons in the 1s orbital are always uninvolved in bonding and therefore countering the positive charge of two of the nuclear protons, so that we must only consider the remaining four protons in consideration of a bonded carbon atom.

Half the numbers of electrons in each of these bonds count toward neutralizing the remaining four nuclear proton charges. For example, in methane (**Figure 1.3**) or for that matter in all carbon atoms in the structures in Figure 1.1, there are four bonds to each carbon atom. The line (—) representing each bond designates two electrons in that bond. One half of those two electrons in the bond may be considered as "belonging to" or "associated with" each atom at the end of that bond. Four bonds

H—C—H (with H above and below)

Accounting: 1 electron from each bond = 4
2 electrons in the 1s orbital = 2
Total = 6

Atomic number of carbon is 6, therefore 6 protons in the nucleus of the atom. No formal charge.

---

oxygen atom in glucose   **formaldehyde**

Accounting: 1 electron from each bond = 2
4 non-bonding electrons = 4
2 electrons in the 1s orbital = 2
Total = 8

Atomic number of oxygen is 8, therefore 8 protons in the nucleus of the atom. No formal charge.

H—N—H (with H below)
ammonia

Accounting: 1 electron from each bond = 3
2 non-bonding electrons = 2
2 electrons in the 1s orbital = 2
Total = 7

Atomic number of nitrogen is 7, therefore 7 protons in the nucleus of the atom. No formal charge.

H—N⁺—H (with H above and below)
ammonium

Accounting: 1 electron from each bond = 4
2 electrons in the 1s orbital = 2
Total = 6

Atomic number of nitrogen is 7, therefore 7 protons in the nucleus of the atom. One positive (+) formal charge on nitrogen.

◄ **FIGURE 1.3**

Counting Electrons to Judge Formal Charge and the Octet Rule for Second Row Elements

supply therefore four electrons to each carbon atom connected by these bonds to other atoms. When added to the two electrons in the first quantum level noted above, we add up to the six electrons necessary for the carbon atom in question to have no charge, what is termed no formal charge.

Regarding the accounting considerations of the last paragraph, it makes no difference if the bond is a single bond as we've come across so far, or a double bond or a triple bond, which we'll come across later. For example in an aldehyde, such as formaldehyde, a bonding arrangement we'll take up in detail in Chapter 3 (Figure 3.14) the carbon atom is linked by two single bonds (—) (—) and one double bond (=). There are eight electrons in these four lines and, as above, half "belong" to the carbon atom. These four electrons then are added to the two electrons in the first quantum level ($1s^2$) leading to no formal charge for that carbon atom.

The considerations outlined above apply to all bonded atoms. Consider any of the oxygen atoms in the cyclic structure of glucose (Figure 1.1) or the oxygen atom in formaldehyde (Figure 1.3). Oxygen has an atomic number of eight, which means each oxygen atoms brings eight electrons to the structure it is part of, in contrast to carbon, which brings six electrons to each structure it is part of. Oxygen has eight protons in the nucleus of each atom, while carbon has six protons in the nucleus of each atom.

Again there are two electrons in the inner shell, the first quantum level. In every oxygen atom in glucose, there are two single bonds (—) (—) to the oxygen atom. Each single bond (—) contains two electrons half of which belong to each of the atoms bonded by that single bond. Therefore for each oxygen atom we can account for one electron each from these single bonds for a total of four electrons, two from the inner shell and one from each bond (2x(—)) . The remaining four electrons out of the eight electrons that oxygen originally brought to the table, so to speak, are not taking part in the bonding. These four electrons are therefore reasonably designated nonbonding electrons.

We'll talk more about these nonbonding electrons later, but for now let's note that nonbonding electrons appear in pairs and may or may not be, shown in chemical structures as a line next to the atom or as two dots (..). In the structures in Figure 1.1 these nonbonding electron designations are not shown. If the nonbonding electrons are shown in the structure, or not shown, makes no difference to their accounting. For now let's pay attention to this accounting to judge the presence or absence of formal charge.

We are now ready to evaluate if any of the oxygen atoms in the structures in Figures 1.1 or 1.3 have a formal charge. We have two electrons from the first quantum level plus two electrons, one each from the two single bonds surrounding each oxygen atom. Since the nonbonding electrons are not participating in bonding, all four "belong" to the oxygen atom for a total of eight electrons to counter the eight protons in the nucleus of each oxygen atom.

If you apply the above accounting to an oxygen atom in formaldehyde you will again come up with four nonbonding electrons and the absence of a formal charge on oxygen. As for each oxygen atom in the structures in Figure 1.1, the oxygen atom in formaldehyde has two nonbonding electron pairs – four electrons that belong to the oxygen atom, electrons not involved in bonding (Figure 1.3).

As you come across organic chemical structures with atoms bearing nonbonding electrons, such electrons are only sometimes shown as the lines or dots mentioned above. Nevertheless, nonbonding electrons are important and often participate in chemical reactivity. This makes sense doesn't it? The nonbonding electrons are in the outer quantum level of the atom but are not involved in a bond and therefore are easily available to form a bond. We'll first see more about this characteristic of nonbonding electrons in detail in Chapter 3 (section 3.10 and Figure 3.15) and then all throughout the book.

Let's now look at a situation where there is a formal charge. Consider the reaction between ammonia and hydrogen chloride. Did you ever have the experience of just cracking open the cap on a bottle of ammonia and a bottle of concentrated hydrogen chloride in the same laboratory, even quite a distance apart, and see a trail of smoke connecting the two bottles? The smoke trail is ammonium chloride formed as the vapors meet each other. The accounting shown in Figure 1.3 leads to a conclusion of no formal charge on ammonia but a formal charge on the nitrogen atom of the ammonium ion, a product of the reaction with HCl. Aside from our main focus here on formal charge, in the reaction of ammonia and hydrochloric acid we see the role of a lone pair of electrons in ammonia.

Why the word formal before the word charge? The answer is that what we designate as a formal charge is not an actual picture of the charge relationships in the molecule. That is far more complicated than simply accounting for electrons and protons.

Certainly, there is a positive charge on the ammonium ion, but how that charge is distributed among the atoms that make up the ammonium ion, the four hydrogen atoms and the nitrogen atom, can not be discerned easily, although it is important in understanding the most detailed nature of the molecule, which is somewhat advanced for us at this point.

With all the electron counting in this section it is reasonable to look into the electron counting for evaluation of formal charge, as we have done, and the electron counting for the octet rule. A stable arrangement for an atom of the second row of the periodic table is one in which eight electrons in the valence shell (the second quantum level) surround the atom. This is the criterion for obeying the octet rule. However, the electrons that contribute to this octet do not necessarily "belong" to the atom. In methane for example, CH_4, the four bonds between the central carbon atom and the four hydrogen atoms contain eight electrons. Even if only half of these electrons contribute to the absence of a formal charge, all of the eight contribute to evaluation of obeying the octet rule.

Every atom in every structure in Figure 1.3 obeys the octet rule, a rule that arose from experimental observations of bonding starting in the late 1800s until its complete formulation by about 1920. G. N. Lewis, whom we will hear much more about and his ideas of bonding was central to the development of what we know now as the octet rule.

PROBLEM 1.16
(a) Draw the structure for α-D-glucose, showing not only all the atoms but as well any nonbonding valence electrons on each atom.
(b) Use your answer to (a) to determine if there is a formal charge on any of the atoms in the structure of α-D-glucose.

PROBLEM 1.17
What does the number of protons in the nucleus of an atom have to do with formal charge of that atom in a bonded state?

PROBLEM 1.18
How does a tetracoordinate carbon atom avoid formal charge by matching the six protons in the nucleus of that atom?

PROBLEM 1.19
How does a divalent oxygen atom avoid formal charge by matching the eight protons in the nucleus of the atom?

PROBLEM 1.20
Ammonium ion, NH_4^+, is a tetravalent state of nitrogen with a positive charge. Is this charge a formal charge and how do the number of protons in the nucleus of nitrogen contribute to that charge?

PROBLEM 1.21
The formula for α-D-glucose in Figure 1.1 is $C_6H_{12}O_6$. Imagine other ways to organize these 24 atoms into a molecule without considering how reasonable the structure seems. Determine the formal charges and violations of the octet rule, if any, for each of the atoms in the structures you have imagined.

PROBLEM 1.22
Among the atoms in the second row of the periodic table, nitrogen and boron bond to three other atoms, for example ammonia, NH_3, and boron trifluoride, BF_3. Using Pauling's hybridization idea and taking account of the atom numbers of these elements show how the electronic configuration of nitrogen and boron atoms could lead to this bonding. Try to predict the geometry of ammonia and boron trifluoride based on the p orbitals utilized in the hybridization and the repulsion of the electrons in the chemical bonds in these molecules. You are using VSEPR theory to answer this question.

PROBLEM 1.23
In general, an atom with four "entities" around it, such as CH_4, is sp^3 hybridized, that with three "entities" around it, such as for BF_3, or for each carbon atoms in ethylene, $H_2C=CH_2$, is sp^2 hybridized, and that with two "entities" around it, such as each of the carbon atoms in acetylene H-C≡C-H, is sp hybridized. What geometry around each of the atoms in the above examples would you predict based on the differing hybridizations?

PROBLEM 1.24
Counting the lone pair of electrons in ammonia, NH_3, the nitrogen atom has four "entities" surrounding it, three bonds to hydrogen atoms and a lone pair of electrons. The hybridization of ammonia is generally not assigned to exactly sp^3. Experiment shows that the angle between the three N-H bonds to be about 107 degrees, less than the tetrahedral angle. Offer an explanation for this experimental fact.

PROBLEM 1.25
Considering your answer to problem 1.24 and counting the two lone pairs of electrons on oxygen in water, H_2O, the oxygen atom has four "entities" around it. Offer an explanation for the fact that the angle between the two O-H bonds is about 105 degrees, less than that of ammonia and less than the tetrahedral angle.

PROBLEM 1.26
Each of the oxygen atoms in the structures in Figure 1.1 has four entities around it, two bonded atoms and two lone pairs of electrons, just as in water. Yet the angle between the atoms bonded to the oxygen atom can be predicted to be more than 105 degrees but still less than the tetrahedral angle. Offer a possible explanation of this prediction.

PROBLEM 1.27
What is the correlation between the number of atomic orbitals that hybridize and the number of "entities" that surround the hybridized atom?

PROBLEM 1.28
In problem 1.11 you generated differing structures that fit the formula C_8H_{18}. Do any of the carbon or hydrogen atoms in these structures have formal charges and if so which ones and why? If none of your structures have formal charges on any atom then create structures that do show formal charges for one or more atoms.

PROBLEM 1.29
Which atoms in the differing structures you imagined for the formula C_8H_{18} disobey the octet rule? If none do, then imagine structures in which the octet rule is disobeyed for some of the atoms. Why does the octet rule not apply to atoms that are in other than the second row of the periodic table?

WE COULD HAVE DRAWN THE STRUCTURES of the α-glucose and β-glucose molecules in a different way from that in Figure 1.1. This has been done in **Figure 1.4**. A mirror could be drawn down the middle of Figure 1.4. The structures on the right in Figure 1.4 are mirror images of the structures on the left of this figure, which are the two glucose stereoisomers first shown in Figure 1.1.

To ease discussion of these structures on the right and left side of Figure 1.4, let's designate them as α-L-glucose and β-L-glucose on the right and α-D-glucose and β-D-glucose on the left. What's the difference between these L and D structures? Are they the same and if not do they both exist? Does it matter if cellulose and starch are made of one or the other, L or D types of glucose, or mixtures of the two?

Consider a tea cup unadorned with any markings. If you inspect its mirror image and imagine removing the mirror image from the mirror and placing it next to the original object, you'll discover that the object and its mirror image are identical. What is meant by identical is that the object and its mirror image are superimposable. In other words one can not see any difference between the two. You could replace one with the other. You can set one down and put the other on top of it. They are the same – period.

Now try this experiment on yourself, on your own image, and discover that you and your mirror image can not be superimposed. Imagine your mirror image stepping out of the mirror and standing next to you. You could point to many differences between the two of you. Everything on your left side would be on your mirror image's right side and vice versa. If the right and left sides of your body were identical, which they are not, then you and your mirror image could be superimposed.

Now let's try this game with the molecular pairs in Figure 1.4, α-D-glucose / α-L-glucose and β-D-glucose / β-L-glucose. These D and L structures, these mirror images, are different. The D and L structures are not superimposable just as we are not superimposable with our mirror images. Prove this to yourself with the new set of molecular stick models you purchased. Or use your imagination to pick each structure up out of the page. Keep in mind not to change anything other than rotating the whole structure in space, as if you picked up any object and simply looked at it from all angles. Try to superimpose one on the other. You won't be able to do it.

As we are going to discover as we get further into this world of mirrors in organic

1.6

The mirror images of glucose are different; they differ as we differ from our mirror image. What is the consequence of this fact at the molecular level?

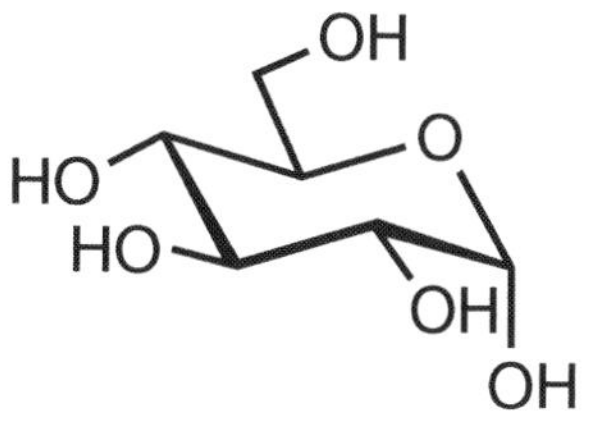

α-D-glucose

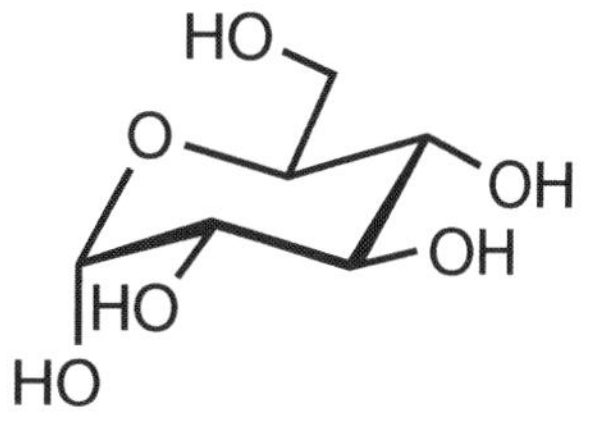

α-L-glucose

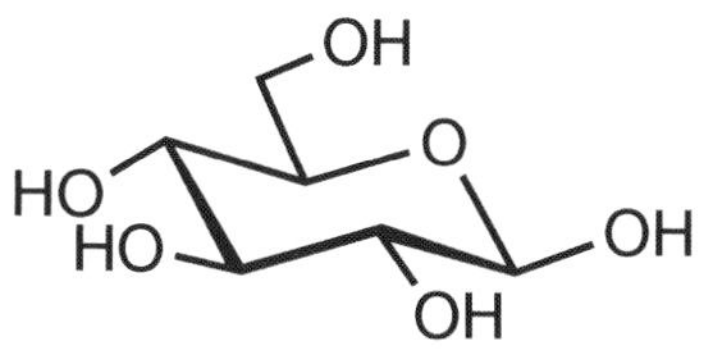

β-D-glucose

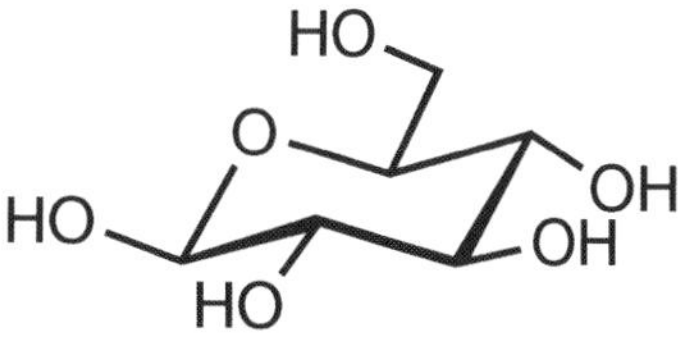

β-L-glucose

◀ FIGURE 1.4

Stereoisomers Possible for α and β-Glucose

chemistry, life is replete with molecules for which the mirror images are not identical, that is, can not be superimposed on each other. The way our entire body is related to its mirror image turns out to teach us something about the way that molecules in our body relate to their mirror images, that is, as non-identical molecules. The α and β forms of glucose in Figure 1.4 (which are not mirror reflections of each other) each has its own non-identical mirror image. The α D and L and β D and L sugars are just examples of numerous biologically important molecules that are not identical to their mirror images.

Glucose is certainly biologically important. It is the sole source of energy for the function of the brain, in its being able to cross into the brain from the rest of the body, to cross the "blood-brain barrier." In fact, glucose is the most important source of energy to all functions of life from bacteria to human beings. Through glycolysis and the citric acid cycle (Chapter 8), glucose is converted to carbon dioxide and water yielding the energy that sustains our life.

Consider the beautiful symmetry of how life is sustained. These very molecules produced when glucose yields its energy to sustain life on earth, carbon dioxide and water, are the same molecules used by the photosynthetic activity of green plants to produce glucose using the sun's energy. This mystical cycle of carbon dioxide, water, green plants and sun producing glucose and then glucose and life on earth producing carbon dioxide and water is enabled by organic chemistry. The principles of the science of organic chemistry are found in every nook and cranny of the cycle connecting the sun's energy to our life.

For now we will focus on a single key fact - a fact that seems remarkable on first hearing of it. All of this biology relates only to the D-glucose. L-glucose is not produced by photosynthesis, the process described above, and is not recognized by life for any of the functions for which D-glucose is critical, in spite of how similar these mirror image molecules appear to be (Figure 1.4).

PROBLEM 1.30
Inspect common objects around you and determine if they are, or are not, identical with their mirror images. What does the ability to bisect an object by a mirror plane, so that one half of the object is reflected perfectly by the other half of the object, have to do with an object being identical with its mirror image? If you slice the tea cup example used in this section by a two sided mirror, are the two images identical? How about if you try this on either α or β-glucose, L or D?

PROBLEM 1.31
Build molecules with your model set, or draw three dimensional structures, with the following formulas to discover what kind of tetrahedral structure will be non-identical with its mirror image, that is, not superimposable: $HCCl_3$; $HCCl_2Br$; $HCClBr_2$; $HCClBrI$. How would your answer differ if the arrangement around the central carbon atom was in a single plane rather than in a tetrahedral array? Try the mirror slicing technique from problem 1.30 and see how the result correlates with your answer.

PROBLEM 1.32
Consider this: just as you reach out with your right hand to shake the right hand of a friend – your left hand would not work with your friend's right hand, your body can only deal with one handedness, D, of glucose. Consider what this means about the mechanisms that sustain your life?

PROBLEM 1.33
A symmetric object (or molecule) has one of possible three elements of symmetry: a plane of symmetry; a center of symmetry; or an n fold alternating axis of symmetry where n is an even number. Why are all molecules that are devoid of symmetry, that is, asymmetric, necessarily capable of having non-identical mirror image forms? Try out these symmetry tests on the molecules in problem 1.31 and on the glucose structures in Figure 1.4. Using a set of ball and stick molecular models is the best way to look at these symmetry questions.

PROBLEM 1.34

Why is it obvious that chemists have long used a word related to the Greek word for hand, (cheir), chiral, to describe a molecule that is capable of having mirror images that are not identical?

PROBLEM 1.35

Does the word chiral describe the properties of a single molecule and if so why? Consider that both a molecule and its non-identical mirror image are present in precisely equal amounts in a sample. Can the word chiral be used to describe more than a single entity, a single molecule? Can a mixture of non-identical mirror image molecules be described with the word chiral? Would your answer to the last question depend on the ratio of the non-identical mirror image molecules?

FOLLOWING THE DEFINITION OF STEREOISOMERS as two molecules of identical formula and identical bonding but that differ from each other (section 1.3) leads to the conclusion that comparison of any two structures in Figure 1.4 is a comparison between stereoisomers.

There are four molecular structures in Figure 1.4, all with the identical formula, $C_6H_{12}O_6$ and all with identical bonding, that is, identical in the connections between the atoms. Let's see how differences can arise in spite of the identical features. Two pairs are mirror images, α-D-glucose / α-L-glucose and β-D-glucose / β-L-glucose, and the two members of each pair, D/L, can not be superimposed. As designated in **Figure 1.5**, these pair wise relationships are enantiomeric, a word derived from the Greek word enantios, which means opposites, such as opposite sides of the mirror.

However, all other pair wise relationships in Figure 1.5 are not mirror images. Pairs of stereoisomers that are not mirror images differ fundamentally from those which are mirror images. These four pair wise stereoisomeric relationships that are not mirror images, which are therefore not enantiomers, are: α-D-glucose / β-D-glucose; α-L-glucose / β-L-glucose; α-D-glucose / β-L-glucose; α-L-glucose / β-D-glucose. Each of these pairs is designated as a pair of diastereomers. For example, α-D-glucose is a diastereomer of β-D-glucose or in other words, α-D-glucose and β-D-glucose have a diastereomeric relationship to each other.

The word chosen to describe such stereoisomers, incorporates the suffix, dia, which comes from the Greek word diakritikos, that which separates or distinguishes, which clearly describes two molecules that are stereoisomers but which are not mirror image

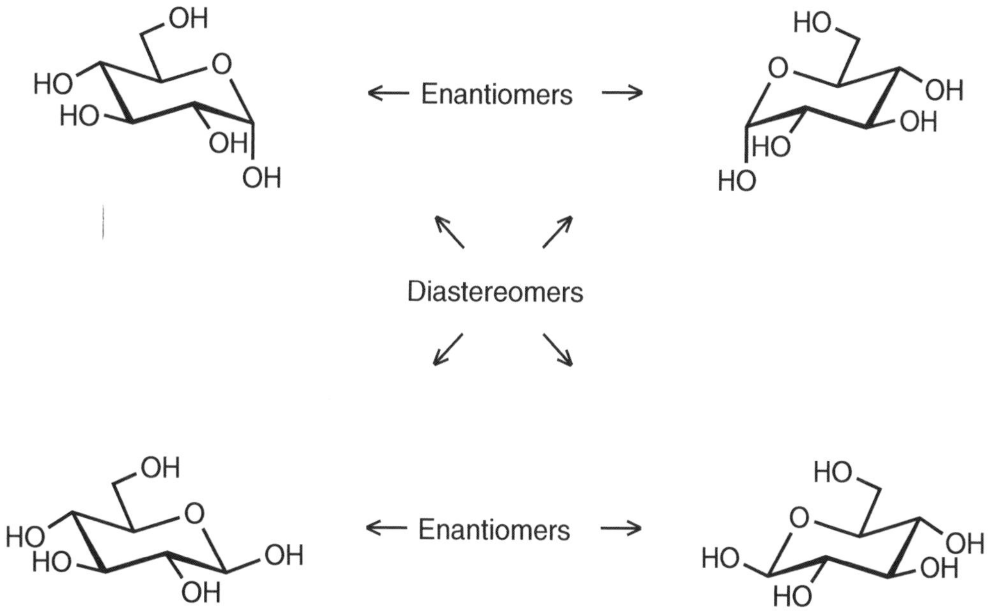

1.7

Stereoisomers are pairs of molecules, which although having the same formula and identical bonding, nevertheless differ from each other.

← Enantiomers →

Diastereomers

← Enantiomers →

◄ FIGURE 1.5

Identification of the Enantiomers and Diastereomers of Glucose

stereoisomers. Any two molecules that are stereoisomers but are not mirror images of each other are diastereomers.

In other words, there are two different kinds of stereoisomeric relationships possible between two molecules. The two molecules of the pair may be related as enantiomers or diastereomers and these relationships are demonstrated using the basic bonding structure of glucose in Figure 1.5.

There is a fundamental difference in the relationship between two molecules that are enantiomers, such as α-D-glucose / α-L-glucose, and two molecules that are diastereomers, such as α-D-glucose / β-D-glucose.

Two molecules that are enantiomers are identical in all respects and can not be told apart except by a probe which is itself capable of existing in enantiomeric forms, that is, capable of mirror image isomerism. Because you and I, and all living entities, are capable of such isomerism, we can distinguish α-D-glucose and α-L-glucose or for that matter any two molecules that are mirror image related, that is, which are enantiomers. It is our absence of symmetry, the fact that each of us is chiral (a word we'll come upon shortly), that allows you to look at the enantiomeric structures in Figure 1.5 and see that they are different. In fact, there are even certain volatile molecules in which the enantiomers smell differently to us. Carvone is an example. One enantiomer smells like spearmint and the other enantiomer like caraway.

Let's take this idea a bit further. Just as you exist only on one side of the mirror, your mirror image is simply a reflection, the enzymes that catalyze the biochemical processes within you also exist on only one side of the mirror. Following the same principle that allows you and me to distinguish enantiomers, those enzymes involved with glucose distinguish one enantiomer from the other and moreover with such specificity that the enzymes only catalyze reactions involving one of the glucose enantiomers, the D-glucose to the absolute exclusion of L-glucose. This is the molecular basis of the insistence in all biological processes on the use of the D enantiomer to the exclusion of the L enantiomer of glucose, as pointed out in the last paragraph of section 1.6.

On the contrary, if one studied two enantiomers by a property that was not capable of mirror image isomerism, such as melting point or boiling point, or dissolving the molecules of the pair in a common solvent such as water, the enantiomers would behave identically. More will be said about this apparently peculiar characteristic of enantiomers in section 1.8 to follow. But on reflection it is not so peculiar. After all, every time you try to put a left handed glove on your right hand or vice versa you are coming across the same principle that is at work in biochemical processes that distinguish enantiomers. On the other hand, you can identically pick up a piece of paper, or to use the analogy above, an unadorned tea cup, with a right or left handed glove.

PROBLEM 1.36
You can not shake your friend's right hand equally well with your left hand or your right hand. How might this common experience relate to the fact that life can use only one of the enantiomers of glucose? However, does this fact about shaking hands predict that D-glucose but not L-glucose is used rather than L-glucose but not D-glucose?

PROBLEM 1.37
Parts of the structure of most enzymes are helical. Would you expect experiments to show that these helical regions of the structure are found to be both right and left handed?

PROBLEM 1.38
Why do all aspects of stereoisomerism relate to comparisons between two molecules and not to relationships between more than two molecules?

PROBLEM 1.39
Two molecules that are enantiomers must have identical solubilities in water but two molecules that are diastereomers may have differing solubilities in water. Why?

PROBLEM 1.40

Consider the following experiment: α-D-glucose is bonded in some way (it does not matter how) to α-L-glucose. Now α-L-glucose is bonded in the same way to another α-L-glucose. Will the two bonded pairs have the same or different properties and why?

PROBLEM 1.41

Looking ahead to the next section, are there any pairs of constitutional isomers in Figure 1.5?

PROBLEM 1.42

How are the stereochemical assignments in Figure 1.5 related to the experimental data in Figure 1.7?

1.8

To understand diastereomers we have to understand isomers that are not stereoisomers, isomers that we call constitutional or structural isomers.

To understand the difference between two molecules that are diastereomers, we have to understand how any two molecules may differ, molecules that are not related as stereoisomers. Naturally, if you have two molecules with different formulas you can expect the two molecules will have different properties. In fact, the chemical industry would be an impossible endeavor if molecules with different formulas did not differ in their properties. How would the petroleum companies fractionate crude oil into the fractions necessary to produce the wide variety of molecules that are then used to produce everything from plastics to pharmaceuticals to gasoline?

This fractionation is accomplished because crude oil contains hundreds of molecules with differing formulas, that is, made up of differing numbers and even kinds of atoms. But crude oil also contains hundreds of pairs of molecules with the same formula, which are not stereoisomers. These pairs of molecules with the same formula, which are not stereoisomers, are still related as isomers. The two molecules of the pair differ from being stereoisomers in that the atoms in each isomer are connected together in different ways. Such compared molecules within a pair of molecules are called structural or constitutional isomers.

These molecules with either their different formulas or identical formulas but different structures have different properties including vapor pressures and this is the basis of those huge fractionating towers found in all petroleum refining facilities. Separation of these molecular components of crude oil is what these fractionating towers are all about. For example, light naphtha, which is used to make gasoline as we'll learn about in Chapter 4, is composed of molecules with far fewer carbon atoms than the molecules that are used to make Diesel fuel. But in a barrel of crude petroleum, all these molecules are mixed together, those that could be converted to gasoline and those that can be converted to Diesel fuel. These structurally different molecules have to be separated.

Separation of molecules is big business. Consider the example of separation of water, H_2O, from ethanol, CH_3CH_2OH, hardly an academic exercise considering that billions of dollars is now spent to produce ethanol from corn and other crops. Ethanol is a biofuel. But the fermentation process that produces ethanol produces it in a mixture with water and this mixture is impossible to use as an additive with gasoline. The ethanol must be separated from the water. Ethanol and water have different formulas and therefore have different properties. Their boiling points are widely different allowing ethanol and water to be separated by distillation to their azeotropic mixture, 96:4 respectively.

Even molecules that have the same formula but different structures, different ways in which the atoms are connected to each other, can have widely varying properties allowing them to be easily separated. Examples are shown in **Figure 1.6** for two sets of

Boiling points of constitutional isomers

FIGURE 1.6 ▶

Molecules with the same formulae but different bonding arrangements have different properties.

69 °C   50 °C   60 °C   (C_6H_{14})

or

158 °C   132 °C   91 °C   ($C_6H_{14}O$)

■ **Joseph Louis Gay-Lussac**

three molecules each of this type and their differing boiling points. Such molecules, as noted above, are called by organic chemists, structural isomers, or constitutional isomers.

Internal combustion engines and therefore automobiles bring up the best examples of the differences between structural isomers that I am aware of, that is, the difference in octane numbers of isomeric molecules. Look ahead in Chapter 4 to Figure 4.10 and sections 4.6 and 4.8. Notice in Figure 4.10 that nine molecular structures in this figure, each with the same formula, C_7H_{16}, vary from fuels that will destroy your automobile engine with extreme knocking, to fuels that will help the engine power your car smoothly up the highest grades.

The idea of what we now call isomeric molecules was greatly resisted in the early part of the 19th century when analysis began to reveal that many molecules with apparently the same ratio and kinds of atoms nevertheless had different properties. This was a time when there were no certain ways that could connect this analysis of the atomic composition of a molecule to its structure. How could two molecules made of the same number and kinds of atoms have different properties? This was the question of the day.

At first, prominent chemists were certain that the identical formulas were artifacts of bad analyses. If the analyses were carried out more accurately, they claimed, there would appear differences in the atomic makeup of these compared molecules. But the opposite happened. The more careful the analysis, the closer became the atomic composition of these molecules with different properties. This finally reached a head when a famous French chemist of that time, **Gay-Lussac,** in 1814, pointed out that the atomic compositions of acetic acid and glucose, the sugar from which starch and cellulose is constructured, could both be expressed by the formula $C_2(H_2O)_2$. Of course having studied the contents of Figure 1.1 we are wise to the problem. Each molecule of acetic acid, CH_3CO_2H, which is the small molecule constituent of vinegar, is composed of two carbon atoms, two oxygen atoms and four hydrogen atoms, the elements in its formula and can not be compared to glucose. Glucose is composed of larger numbers of carbon, hydrogen and oxygen atoms even if their ratio is also $C_2(H_2O)_2$, which is simply the formula of glucose, $C_6(H_2O)_6$, divided by three.

In those early days although the ratio of the atoms could be determined, the absolute numbers of each atom could not be easily known. The molecular weight could not be reliably determined, although the chemists around the time of Gay-Lussac's work certainly must have noticed that acetic acid is a liquid while cellulose is a solid.

But even small molecules were being discovered that had the same formula, in other words, molecules that were actually entirely made of the identical elements in identical numbers. The most famous of these pairs of isomers were ammonium cyanate and urea, both with the formula CH_4N_2O. These isomers were in the spotlight arising from

Friedrich Wöhler's breakthrough experiment in 1828 in converting ammonium cyanate to urea. In this single experiment this young man, then only 28 years old, demonstrated that different molecules could exist with identical formulas. But even more important he rang in the death knell of the necessity of belief in a "vital force" for the production of molecules found in nature. He did this by synthesizing urea, a molecule arising from life, from a molecule not derived by a biological process, ammonium cyanate. Wöhler's famous experiment came to change the minds of many scientists in those days and in the years to follow. Not bad work for someone not yet 30 years old.

No one could define what this "vital force" was but it was thought to be connected to why things are alive - some kind of indefinable characteristic. Don't snub your nose at these nineteenth century musings. Although we have gained a great deal of insight into the molecular workings of life and discovered that the fundamental principles of organic chemistry are equally at work in living systems as they are in the laboratory or the chemical plant, a fact that will be demonstrated beyond question in this book, we continue to be mystified as to how life began and a significant number or scientists are at work in a field that is generally termed "Origin of Life."

Nearly two hundred years have passed since science focused on the question of isomerism. And Figure 1.6 is a small demonstration of how far we have come in realizing the nature of structural isomerism. In Figure 1.6 we find a very few examples of representative molecules of the same formula and see how different their properties can be. Inspecting Figure 1.6 might help you to reinforce the discussion in section 1.2 about how organic chemistry has a variety of ways of presenting molecular structure. Take the trouble of redrawing the structures in Figure 1.6 showing all the atoms, which is a perfectly acceptable way these molecules could have been presented.

■ **Friedrich Wöhler**

Stereochemical ideas were very far from being considered in these early days of the 19th century. But now we realize that even molecules that have the same formula and the same numbers and kinds of chemical bonds, which the molecules in Figure 1.6 do not, can still have widely differing properties. Such compared pairs of molecules as long as they are not enantiomers are called diastereomers. These diastereomeric relationships were seen in Figure 1.5.

The relationships among the varying glucose molecules in Figure 1.5 are based on the rule that stereoisomers that are mirror images of each other are enantiomers, while stereoisomers that are <u>not</u> mirror images of each other are diastereomers. We offered no experimental evidence for the prediction that enantiomers have identical properties in a symmetrical environment (an environment that offers no nonidentical mirror image possibility) (section 1.6) while diastereomers, as is the situation for structural isomers, will have different properties in all environments. Let's now look at some experiments to test these predictions. The prediction is clear. The crystals of the designated enantiomeric pairs in Figure 1.5 should melt identically. The crystals of the designated diastereomeric pairs in Figure 1.5 should melt differently. The experimental results for the melting of these crystals are exhibited in **Figure 1.7** *(see page 24)* inaddition to the results of another kind of experiment, optical rotation, which we'll discuss in the next section.

The crystalline form of α-D-glucose monohydrate, which means that each molecule of glucose has a single water molecule associated with it, is 83° C. α-L-glucose is identically hydrated in the crystal and melts at an identical temperature. According to Figure 1.5, these are enantiomers. β-D-glucose forms a different crystal than the crystal formed by α-D-glucose, which melts at a different temperature, not 83° C but rather at near to 150° C. According to Figure 1.5, these are diastereomers. β-D-glucose forms an identical crystal to that formed by β-L-glucose both melting at the same temperature, in the range of 150° C. According to Figure 1.5, these are enantiomers.

The prediction was clear and the experimental results confirm this prediction. The α-glucose enantiomers melt identically and at a different temperature than the β-glucose enantiomers, which have identical melting points. The examples given in Figure 1.7 for these sugars demonstrating the identity of enantiomers to a symmetrical probe such as heat and the difference in diastereomers to this symmetrical probe are exemplary.

FIGURE 1.7 ▶

Melting Points and Optical Rotations of the Enantiomers and Diastereomers of Glucose

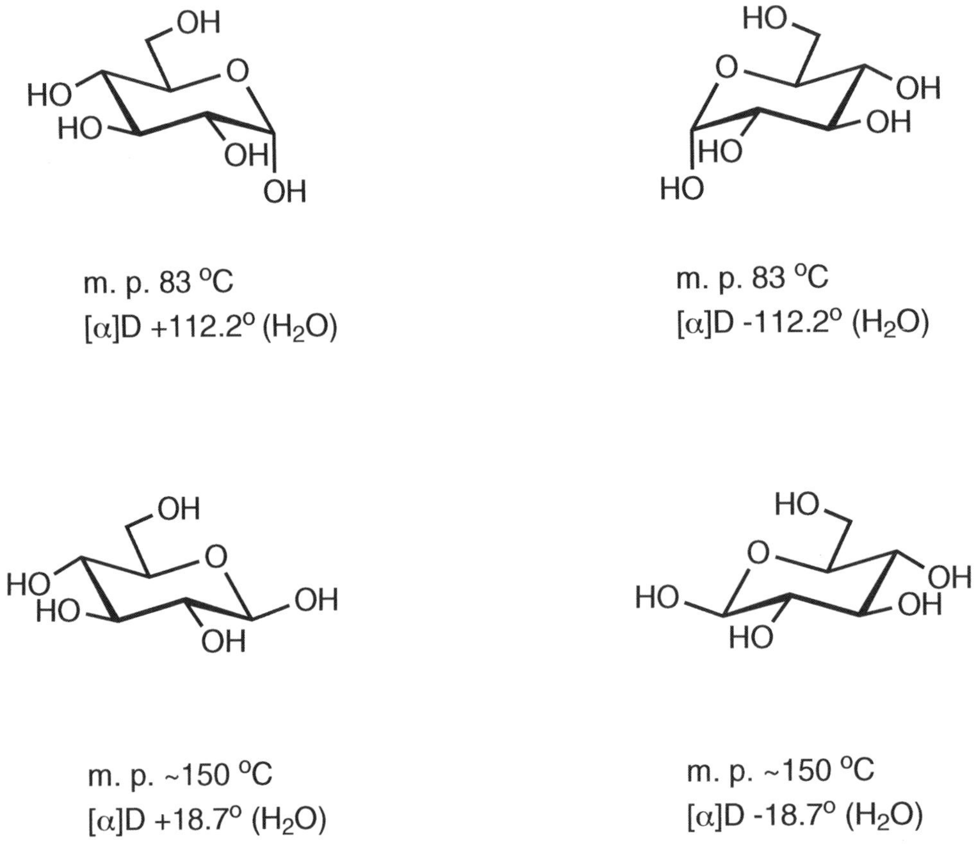

m. p. 83 °C
[α]D +112.2° (H₂O)

m. p. 83 °C
[α]D -112.2° (H₂O)

m. p. ~150 °C
[α]D +18.7° (H₂O)

m. p. ~150 °C
[α]D -18.7° (H₂O)

The results are a model for all enantiomeric pairs and diastereomeric pairs. Melting temperature is a molecular probe without the ability to distinguish enantiomers but as for any molecular probe, diastereomers are always different, will be distinguished.

Application of heat and the melting temperature of molecules are certainly interesting, but very important to chemistry is the symmetry of a probe molecule. By a probe molecule, what is meant is a molecule that interacts in some manner, either by a physical interaction or by a chemical reaction, with each of the two enantiomers or with each of the two diastereomers. Such a physical interaction can be simply a contact in some manner, for example a collision, after which the probe departs, or alternatively, the interaction can lead to a chemical change. If such a probe molecule is capable of mirror image isomerism, then enantiomeric pairs will behave differently and as for any probe, so will also the diastereomeric pairs.

If such a probe molecule is incapable of mirror image isomerism then only diastereomers will behave differently from each other. For example, water will react differently with cellulose and amylose because these polymers are diastereomers. But water will react identically with any pair of molecules designated as enantiomers in Figure 1.5 or for that matter with any enantiomeric pair of molecules. Let's discover more about this in the next section. It is a subject with quite some interest considering that distinguishing enantiomers and diastereomers sits at the foundation of the chemistry of life.

PROBLEM 1.43

Use model building or your ability to draw molecules and visualize molecular structures in three dimensions to test the following statement: A series of molecules with increasing numbers of carbon atoms in which each atom in every structure has no formal charge with all hydrogen atoms monovalent and every carbon atom tetravalent will have the following numbers of isomers (n): CH_4 (0); C_2H_6 (0); C_3H_8 (0); C_4H_{10} (2); C_5H_{12} (3); C_6H_{14} (5); C_7H_{16} (9).

PROBLEM 1.44

As we have been discussing, two structures with identical formulas can be compared as identical or isomers and isomers can be either constitutional or stereoisomeric. Stereoisomers can be diastereomeric or enantiomeric. Use these terms to describe the isomers you discovered in answering Problem 1.43. For which isomeric pairs of molecular structures could physical properties such as boiling or melting point be different? For which isomeric pairs of molecular structures must these physical properties be identical?

Pairs of isomers with different physical properties can be constitutional isomers or stereoisomers (diastereomers). Assign these designations to the pairs of isomers picked that you predicted to have different physical properties.

PROBLEM 1.45

Imagine a series of molecular structures all with the formula C_6H_{14} and then another series with the formula C_6H_{12}. For each formula draw as many structures as you can think of with the restriction that each carbon atom is bonded to four other atoms and that no formal charges or deviations from the octet rule are allowed. Designate all pairs of molecules that are constitutional isomers and all pairs of molecules that are stereoisomers.

PROBLEM 1.46

The names of the constitutional isomers on the top line of Figure 1.6 in order are: n-hexane; 2,2-dimethylbutane; 2-methylpentane. The names of the molecules in the second line are named in order as: 1-hydroxyhexane; 3-hydroxyhexane; dipropylether. Can you use these names in combination with the information given below to develop the beginning of a set of nomenclature rules for organic molecules even if your ability to memorize such a set of rules will not be a focus of this book?

Now, try to create a series of hydrocarbon molecular structures of all varieties including different numbers of carbon atoms. Give names to your structures. For this purpose, carbon chains of increasing numbers of atoms are called in order of 1-10: methane, ethane, propane, butane, pentane, hexane, heptane, octane, nonane and decane. If a chain of four carbon atoms is attached to a longer chain one uses the term butyl for the four carbon piece, for one carbon methyl, for two carbons ethyl and so on. You can apply the use of the yl replacing ane for any chain length.

1.9

Chirality and handedness and how two molecules that are mirror image related can be distinguished from each other.

FORMALDEHYDE, CH_2O, A MOLECULE that is identical with its mirror image, is therefore an example of a molecule that can <u>not</u> interact differently with two molecules that are mirror images of each other. Just as the melting points of each of an enantiomeric pair of molecules are identical, as we've seen in Figure 1.6, an interaction of any kind of an enantiomeric pair of molecules with a molecule like formaldehyde is identical as shown for α-D-glucose and α-L-glucose in **Figure 1.8**.

Let's expand a bit on the use of the word interaction introduced in the preceding section. An interaction is meant to designate any contact between the probe molecule and each of the enantiomers. This could mean simply mixing formaldehyde with each enantiomer in turn of a pair of enantiomers and measuring the change in melting behavior or the mutual miscibility, or in a different kind of test, allowing some chemical reaction to take place between formaldehyde and each enantiomer of the enantiomeric pair of molecules and measuring the equilibrium constant or the rate of the reaction. The interaction can therefore vary from a chemical reaction between the probe molecule and each of the two enantiomers to a fleeting interaction in which the probe molecules simply collide with each of the two enantiomers. The results, in all regards, would be identical for both enantiomers. The characteristic one measured for the interaction of formaldehyde with each of the enantiomers would be identical.

Formaldehyde, by-the-way, as you can see from the formula, is the simplest carbohydrate, $C(H_2O)$, and it has been found in space causing some to think that

FIGURE 1.8 ▶

The Interaction of the Enantiomers of α-Glucose with Two Achiral and One Chiral Probe Molecules

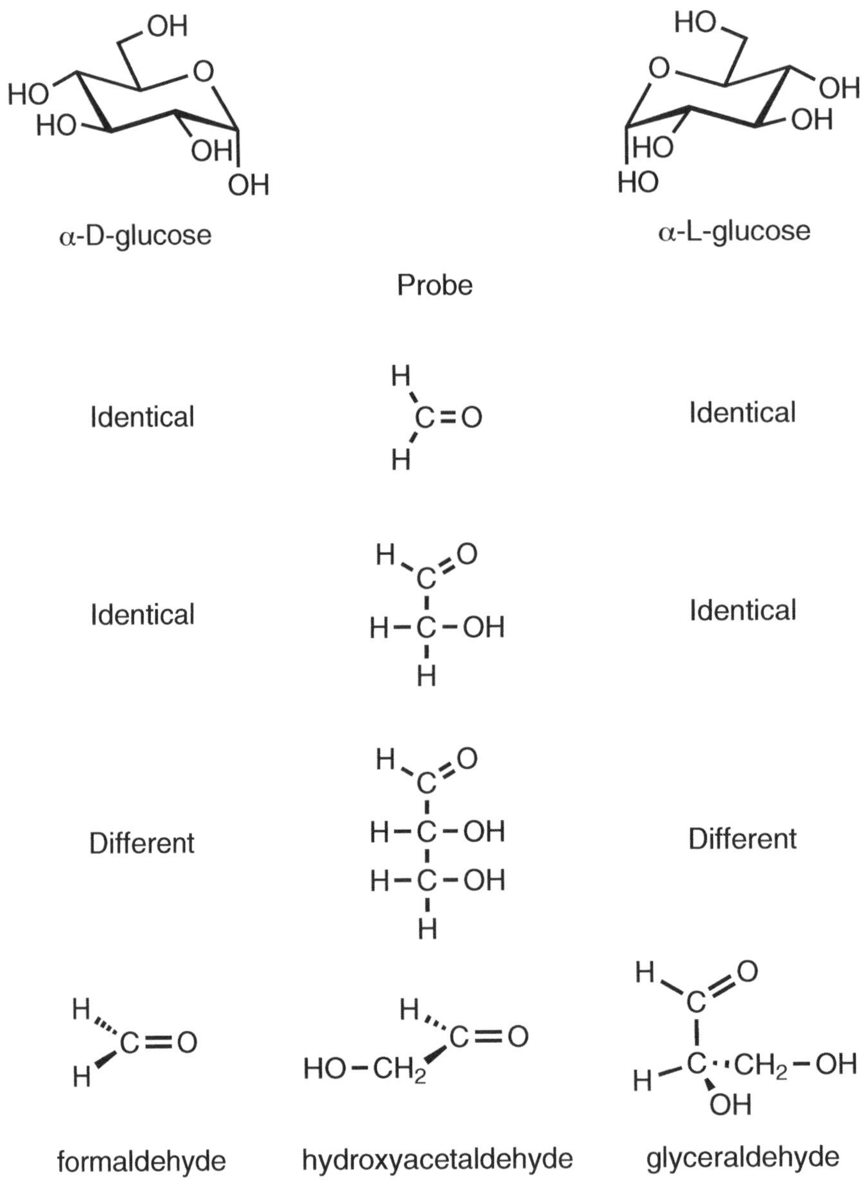

formaldehyde is one of the building blocks of organic matter and of life. Well, that may or may not be true, but what is certainly true is that formaldehyde and its mirror image are identical. Following on the lesson we have just learned, we could conclude that if formaldehyde underwent a chemical reaction with α-D-glucose, it would undergo an identical chemical reaction with α-L-glucose.

The second probe molecule in Figure 1.8, hydroxyacetaldehyde, is an important biological intermediate and is used in certain food preparations and, just as for formaldehyde, is not capable of mirror image isomerism, that is, is not chiral. Any interaction, as defined above, of hydroxyacetaldehyde with the enantiomers of glucose would therefore be identical.

So how can mirror image related molecules be distinguished from each other?

We've seen in section 1.8 how the difference between one's right and left hands is useful for understanding the nature of mirror image isomerism and how enantiomers behave compared to each other. Simply reaching out with your right hand to shake hands with someone demonstrates that you are not able to grip the other person's right and left hands equivalently. This familiar concept of shaking hands involves the identical principle demonstrated in Figure 1.8 and discussed above. It's hardly surprising that chemists apply the term chiral to any molecule that is not superimposable with its mirror image, any molecule as we have seen that can exist in enantiomeric forms, any molecule to come to the point, which is handed. Chiral is derived from the Greek word for hand, χειρ (cheir). The probe molecules in Figure 1.8 and in fact every molecule in this figure is a carbohydrate, $C_n(H_2O)_n$, but only the last probe, $C_3(H_2O)_3$, glyceraldehyde, could cause some interaction to be different for α-D-glucose and α-L-glucose. Of the three probe molecules shown, only glyceraldehyde has a structure that makes it non-identical from its mirror image. The glyceraldehyde shown in Figure 1.8 is D-glyceraldehyde. Its mirror image is L-glyceraldehyde - just like the D/L nomenclature that distinguishes the glucose enantiomers (Figures 1.3 and 1.4). Glyceraldehyde is chiral.

The principle that only a chiral probe can distinguish enantiomeric molecules is the basis of an important chemical industry. Because the biochemical mechanisms that sustain life respond differently to enantiomeric molecules, the pharmaceutical industry has to test enantiomers separately as proposed drugs.

A chiral separation industry has evolved to address this separation problem with chromatography playing a major role. Here one arranges that the enantiomers to be separated come into a very large number of fleeting interactions with a single enantiomer of a chiral molecule or macromolecule. If the fleeting interactions differ to a large enough extent, and the interactions are numerous enough, the two initially not separated enantiomers, dissolved in a solvent, will pass along a tube filled with the chiral entity at different rates, and therefore appear at the end of the tube at different times. The most important of these chiral entities, which fill the tube at this time are polymers closely related to cellulose (Figure 1.1). The cellulose derivative is immobile and fills the tube, and the enantiomers to be separated, dissolved in a solvent, pass through the tube coming into uncountable numbers of fleeting interactions, collisions, with the cellulose derivative filling the tube.

Wouldn't it be interesting if molecules were found in space that could exist in mirror image forms? And if such molecules existed in space with an excess of one enantiomer over the other that would be even more interesting because molecules exist in life on earth with an excess of one enantiomer over the other or as we've seen with the enantiomers of glucose, to the complete exclusion of the other enantiomer (section 1.5). Well, some scientists have become quite excited on discovering that amino acids, another critically important class of biological molecules, which except for one amino acid, glycine, exist in mirror image forms, have been found in meteorites with a small excess of the same enantiomer (enantiomeric excess) as that found in life on earth.

The term enantiomeric excess (e.e.) is widely used by organic chemists although it is gradually being replaced by a related term e.r., the enantiomeric ratio. Enatiomeric excess (e.e.) is defined as the fractional difference between the two enantiomers times one hundred. For example if a sample of a chiral molecule has 0.7 of one enantiomer and 0.3 of the other, then the e.e. is 40% of the excess enantiomer. Express as enantiomeric ratio (e.r.) this sample would be designated as 3/7. An e.e. differing from zero or an e.r. differing form unity (5/5) is so closely associated with life on earth that finding evidence of unequal concentrations of mirror image molecules in outer space stimulates the imagination to consider the possibility of life beyond earth's boundaries, or even that space has been a source of life on earth, the concept of Panspermia (a fascinating idea to look up on Google).

While we are focused on space let's note that like formaldehyde, hydroxyacetaldehyde is also found in space including on planets and in the Milky Way. It has been called the "sugar in space" with a quote in June, 2000 from the American National Aeronautics and Space Administration, NASA, *"The prospects for life in the Universe just got sweeter."* Well, hydroxyacetaldehyde is a carbohydrate but to call it a sugar may be stretching it. Nevertheless, it certainly is interesting that a molecule we associate with life and find in our own bodies is found in the clouds of the Milky Way from which stars are formed.

PROBLEM 1.47

Draw structures of the three probe molecules shown in Figure 1.8 putting in all non- bonded electrons and specifying the hybridization at every carbon and every oxygen atom in each structure.

PROBLEM 1.48

Using molecular models and/or structural drawings in three dimensions test each of the probe molecules for chirality by trying the superimposition test and also the symmetry test as from the information in problem 1.33.

PROBLEM 1.49

Explain the following statements: when an achiral probe molecule interacts with each of two enantiomers, the enantiomeric relationship is maintained; when a chiral probe interacts with each of two enantiomers, the enantiomeric relationship is changed to a diastereomeric relationship.

PROBLEM 1.50

How does your understanding of the statements in problem 1.49 explain the fundamental basis of all processes that separate enantiomers?

PROBLEM 1.51

Convince yourself of the truth of the statements in problem 1.49 using appropriate structural drawings.

1.10

The experiments of Biot and Pasteur in the nineteenth century led to the first realization that molecules can exist in mirror image forms and that molecular mirror images could be studied with light, that is, optical activity could be measured from such molecules.

A COROLLARY OF THE STEREOCHEMICAL LESSON we've just learned from the information in Figure 1.8 is that discerning the difference between enantiomers, let alone separation of two molecules that are enantiomers of each other, is an impossible job without using another chiral entity, which, as we'll see from the story below, could actually be a human hand.

Louis Pasteur, the great French chemist, whose name most of us first became familiar with as children on learning about pasteurized milk, was the first scientist to learn this lesson. Certain salts of an acid derived from "cream of tartar," a related salt obtained from the wine casks used to make wine from grapes in France, were highly crystalline. Pasteur, when a young man of 26, took up the study of these salts and noticed that the crystals took forms that appeared to be mirror images of each other, left handed and right handed. Such crystals are called hemihedral. The Latin word for grapes is racemus and so the substance that made these crystals had been called racemic and coming from tartar the acid was named tartaric acid.

As discussed in section 1.7, many of the molecules that make up living systems are chiral, and these chiral molecules exist in life in a single enantiomeric form. In fact, all individuals of all living species are chiral and, parallel to many of the molecules within us, each of us is not superimposable with our mirror image. We exist but our mirror images do not. This fact gives each of us the ability to look at the structures of enantiomeric pairs in the figures in this text and to see the difference. If this were not so, Pasteur would not have been able to see the difference between the left and right handed crystals of this salt of tartaric acid. He would have been like the molecules in Figure 1.8 that could not distinguish the enantiomers of glucose, formaldehyde and hydroxyacetaldehyde. Such molecules are achiral, that is, not chiral.

Pasteur decided to separate the left handed crystals from the right handed crystals, which he accomplished by simply inspecting each crystal under a microscope and using a pair of tweezers to pick up each crystal and place it in the appropriate pile. When Pasteur carried out this experiment, he was carrying out the first example of a resolution of a mixture of enantiomers, which is now, more than 150 years later, the basis of a multibillion dollar component of the pharmaceutical industry.

Crystals are precise arrangements of many usually identical molecules. Dissolving a crystal in a solvent such as water, as Pasteur did, breaks up the arrangement among the many molecules in the crystal and the molecules separate from each other. Dissolving a crystal therefore shifts our observation from the property of the crystal to the property of the individual molecules, although surrounded by the solvent water molecules in Pasteur's experiment.

Pasteur therefore had two aqueous solutions, one from the left handed crystals and one from the right handed crystals. In these early years in the development of chemistry, it was not possible to know anything about the structure of the molecules that were released into the water solution by dissolution of the crystal. But Pasteur was aware of the work of a famous physicist of the time, a much older man, **Jean-Baptiste Biot**, who was working with polarized light, a special kind of light obtained by passing light through certain crystals. Biot was observing effects on polarized light that were ascribed to what he called molecular asymmetry.

Louis Pasteur

There's a story about Biot that took place in 1817 that demonstrates the intensity of his interest in science and polarized light and also sets the stage for the story we are hearing about Pasteur, which took place in 1848. Biot had already observed that liquid turpentine had an effect on his polarized light and wanted very much to know if this was an effect of the liquid or the molecules in the liquid. Would vapor of turpentine also have an effect? But to carry out this experiment he needed a very long tube filled with turpentine to get the light to pass through enough of the vapor to hope to see anything. Biot was quite a distinguished scientist with considerable influence, enough influence to convince the peers of an ancient church to allow him to use their cloister. He needed a very large space to construct his apparatus and of course a boiler to heat the turpentine to get it to vaporize. You can imagine the result. Turpentine is combustable and this led to an explosion, and yes he burned down the church. But before Biot fled, he observed the effect he was looking for—the vapor of turpentine did have an effect on the polarized light. What exactly was Biot doing with polarized light? What is polarized light and what does it have to do with chirality?

In plane polarized light, the electric vector traces out a sine curve in two dimensions, in a plane. But chirality is a phenomenon that requires three dimensions for its observation. How can a light effect existing in two dimensions have any possibility of measuring something to do with chirality?

The answer is that plane polarized light can be considered to be the vector sum of a mixture of right and left handed circularly polarized light. Imagine the electric vectors of propagating light rotating clockwise and counterclockwise. The electric field of the propagating light would trace out left and right handed helices, helices that can not be superimposed on each other. Adding these vectors together as they move in precise concert with each other yields a sum in a single plane. In other words, and strongly connected to precisely what we have been discussing, the propagating light, although not a material entity such as a chiral molecule or Pasteur's hands, is nevertheless a mixture of mirror image related rotating electric field vectors, a symmetry characteristic that is chiral.

Jean-Baptiste Biot

Take two identical lengths of wire and form right and left handed helical arrays of each by turning the wires clockwise and counterclockwise, respectively, around a cylindrical object like a pencil or a flashlight. You will not be able to superimpose the two helical forms you have produced. Each one is a representation of the left and right handed circularly polarized light whose vector sum constitutes plane polarized light. As long as the left and right handed circularly polarized light pass through the medium identically, the experimental observation, the vector sum of these circular polarizations, is plane polarized light in the identical plane as the incoming light (before it passes through the medium).

However, if one of the circularly polarized beams of light differs in its index of refraction from the other, so that the two circular polarizations no longer precisely cancel each other, the summation of their vectors, although remaining plane polarized, will vibrate in a plane altered from the original plane. The plane of polarization will be *rotated* by an angle to the right or the left of the original plane of polarization.

This is the source of the word rotation in the term optical rotation, which are the data, $[\alpha]_D$, in Figure 1.7. The numerical values of $[\alpha]_D$ in Figure 1.7 are the number of degrees that the plane of polarized light is rotated away from the original plane of polarization for a certain number of molecules. The specification of the number of molecules through which the light passes is determined by the concentration of these molecules in the solution under observation and the path the light takes in passing through the solution.

In addition, the effect depends on the wavelength of the light, which therefore must be specified. The greater the concentration of the solution, and the longer the path through which the light passes, the larger the effect. In the early years of spectroscopy, in the days of Biot and Pasteur, the yellow light of a candle was a convenient reliable color of light, which was designated the D-line or in modern terms the sodium D-line. This is the $_D$ in the equation below, where α is the observed rotation of the plane of polarized light for the solution used and c is the concentration of that solution in grams per 100 cubic centimeters of solution and l is the path through which the light travels expressed in decimeters. The observation also depends on temperature, which is often shown (not below) as a superscript to $[\alpha]$.

$$[\alpha]_D = 100\alpha/cl$$

Although the numerical value of the rotation, α, depends on the concentration, the wavelength and the path, the plus and minus signs in front of the numbers in Figure 1.7 are not affected by the concentration of the solution or the length of the path through which the light passes. These signs, + or - , inform us of the direction of the change in the plane of polarization from its original position. The negative sign stands for rotation to the left, *levo*, while the positive sign stands for rotation to the right, *dextro*. These signs are often designated therefore by "l" and "d," and as seen in Figure 1.7 correspond to the effect of enantiomers on the plane of polarized light. If one enantiomer of a pair is l, then the other enantiomer must be d. One enantiomer will speed one of the circularly polarized vectors faster than the other, and the other enantiomer will do precisely the opposite.

In section 1.5 we discovered that D and L designated the enantiomeric glucose molecules. But while d and l are experimental characteristics, a consequence of the differing indexes of refraction of left and right circularly polarized light at some specified wavelength of light, D and L correspond to information in the drawings we use to represent molecular structure. As we'll see in section 1.12, a molecular structure designated as D may exhibit dextrorotatory, d, or levorotatory, l, optical activity properties.

With this background we have about polarized light and, in modern terms, the chirality that Biot was looking for in the turpentine vapor in that long tube in the church cloister, let's follow what Louis Pasteur did in 1848 after separating the left and right handed crystals into two piles. He passed the plane polarized light, which we now understand consists of left and right handed circularly polarized light, through an aqueous solution made from one of the handed crystals. Then he carried out the identical experiment using the aqueous solution made from the opposite handed crystals, followed by a third experiment that involved shining the mixture of left and right circularly polarized light through a solution made from the crystals of tartaric acid that had not been separated. Pasteur carried out all three of these experiments and observed that the solution made from the crystals that had not been separated caused no change in the plane of the polarized light. Could you have predicted this result and if so what would you predict for the two solutions made from the separated mirror image crystals?

Let's get the answer to this question from the man who did the experiment. Here are Pasteur's words in translation from the French. "*I carefully separated the crystals which were hemihedral to the right from those hemihedral to the left, and examined their solutions separately in the polarizing apparatus. I then saw with no less surprise than pleasure that the crystals hemihedral to the right deviated the plane of polarization to the right, and that those hemihedral to the left deviated to the left.*" And even more revealing in Pasteur's own words is what happened next: "*I remember hurrying from the laboratory and grabbing one of my chemistry assistants and excitedly telling him that 'I have made a great discovery...I am so happy that I am shaking all over and am unable to set my eyes against the polarimeter.' At this time, I was twenty-five years old and had only been doing research for two years. "*

In 1848 it was impossible to understand the molecular basis of what was then a startling experimental result. But Pasteur had the right idea when he offered the following possibilities to explain what he had observed. "*Are the atoms of the dextro-acid* (levo-acid) *grouped on the spirals of a dextrogyrate* (levogyrate) *helix, or placed at the summits of an irregular tetrahedron, or disposed according to some particular dissymmetric grouping or other?*"

We can answer Pasteur's question, emphatically yes, now that we understand the structures of molecules and therefore the structure of the tartaric acid salts Pasteur had studied. We see this dissymmetry, this irregular tetrahedron, by the inability to superimpose the mirror images of the structures exhibited in **Figure 1.9** where we have shed the sodium and ammonium counterions of the salt structures and look at the structures of the tartaric acid stereoisomers directly in the various ways that chemists present such structures. In this figure the solid wedges represent the bond coming out of the paper toward the viewer with the other direction designating the bond going behind the paper. The last presentation in Figure 1.9 follows on a method developed by Emil Fischer, the great German chemist who worked out the stereochemistry of glucose. We will look into the details of these "Fischer Projections" in Chapter 3 (section 3.11).

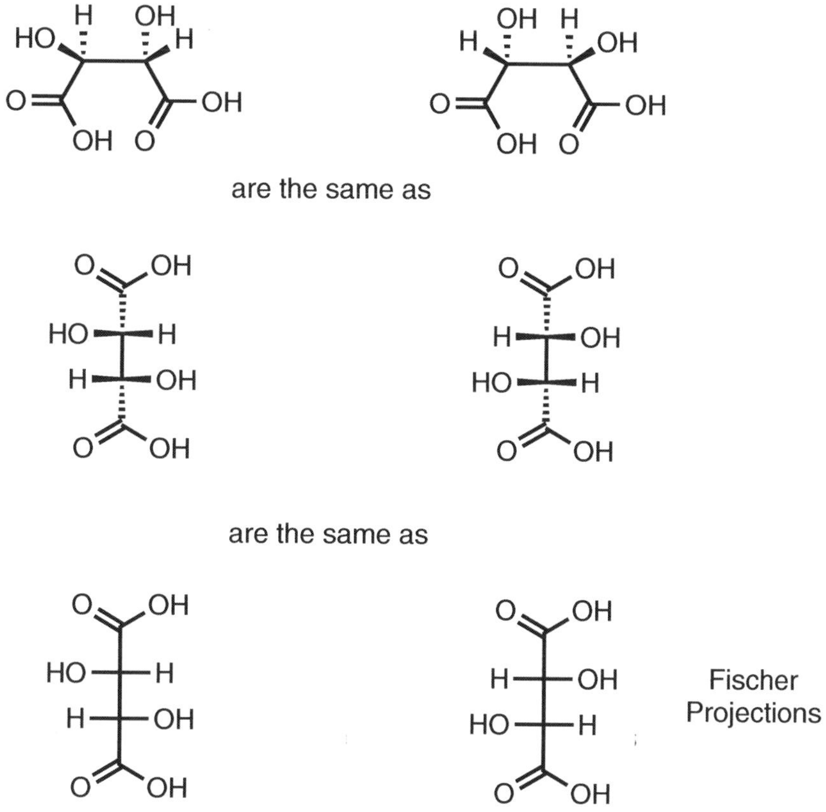

are the same as

are the same as

Experimental facts:
$[\alpha]_D$ - 12.0° (H_2O)
m. p. 168 - 170 °C

$[\alpha]_D$ + 12.0° (H_2O)
m. p. 168 - 170 °C

Fischer Projections

Nomenclature:
    D-tartaric acid
    l-tartaric acid
    (*S*)(*S*)-tartaric acid

    L-tartaric acid
    d-tartaric acid
    (*R*)(*R*)-tartaric acid

◄ FIGURE 1.9

Different Structural Representations and Names for the Enantiomers of Tartaric Acid

In Pasteur's observation, the two solutions, one made from each kind of handed (hemihedral) crystal, contain the salts of the tartaric acid enantiomers shown in Figure 1.9. Please forgive me for putting this in such simplistic terms but each solution Figure 1.9 may be thought of as if it were either a large number of right or left handed gloves. And the incoming plane polarized light, finds analogy if it were a mixture of two hands, one right and one left. Clearly the interaction of the right and left hands with the gloves would differ.

The consequence of this chiral interaction is that the index of refraction for left and right circularly polarized light will differ and differ in precisely opposite ways for the solutions of the enantiomeric molecules. The result is rotation of the plane of polarization away from its original position with optical rotation resulting, as Pasteur excitedly observed. One solution will rotate light in the positive direction, designated, d, while the other will rotate light in the negative direction, designated, l.

On the other hand, the third solution, which contained the mixture of enantiomers of the tartrate salt, the racemic solution made from the unseparated crystals, would not cause the right and left handed circularly polarized beams of light to differ from each other. Whatever effect one of the enantiomers had on the light, the other enantiomer would have the opposite effect and these effects would cancel each other. This will be the experimental observation for a racemic mixture of any chiral molecule.

We can go further. Consider making a solution of formaldehyde in water or hydroxyacetaldehyde in water, the two achiral molecules in Figure 1.8 or for that matter any achiral molecule in water or for that matter how about just water. You get the idea - nothing chiral. We would get the same result as the result for the racemic mixture of a chiral molecule – the left and right handed circularly polarized light would pass unchanged through the solution so that their vector sum, the plane of the polarized light, would be unchanged from the original angle. These solutions would not show optical activity. The specific rotation, [α], for any wavelength and for any concentration and path would be zero.

■ **Friedrich August Kekule**

■ **Archibald Scott Couper**

PROBLEM 1.52

(a) Build-D-glucose and -L-glucose from the structures in Figure 1.5 with your set of molecular models, which, <u>hopefully</u>, you have purchased. Pick each model up with your right hand, one at a time, and determine if it is possible for each model to sit in your hand identically. What does this experiment have to do with Pasteur's experiment discussed in section 1.10?

(b) Now take each of the enantiomers in turn you built in (a) and allow each of the molecular models you built in Problem 1.43 to approach from the same side in the same way. For example, allow the C-H bond of the approaching molecule to come to the middle of the six member ring of one of the glucose models. How does this experiment relate to the results presented in Figure 1.8 and the discussion in section 1.9?

PROBLEM 1.53

Take a cylinder shaped object, such as a flashlight or a candle and wind a stiff piece of wire around it to make a helix, either right or left handed. Slip the wire off the cylinder. Repeat the kind of experiment you conducted in Problem 1.52 (b) by allowing the helical wire to approach the six member ring of each enantiomer in turn in the same way. How does this experiment relate to the phenomenon of optical activity discussed in section 1.10?

PROBLEM 1.54

Why is plane polarized light not affected by a solution of a racemic mixture of chiral molecules even though the individual molecules in the mixture are affecting the left and right handed circularly polarized light differently? Does the reason for there being no effect on plane polarized light from a solution of a racemic mixture of a chiral molecule differ from the reason behind the observation with a solution of an achiral substance?

PROBLEM 1.55

Would Pasteur have drawn a different conclusion from his famous experiment with the salts of tartaric acid if the solution of the crystal hemihedral to the left had been dextrorotatory and the hemihedral crystal to the right had been levototatory?

PROBLEM 1.56

Use the equation for specific rotation to calculate the observed rotation for a solution of a molecule that is dextrorotatory with a specific rotation of 100° with a path of 1 decimeter and a concentration of one gram per 100 ml. Carry out the same calculation for the enantiomer.

PROBLEM 1.57

Does the fact that the molecule in problem 1.56 is dextrorotatory mean that it is the D enantiomer?

PROBLEM 1.58

Might it be possible to design a device for separating enantiomers in which a glass tube is irradiated with left handed circularly polarized light as the racemic mixture passes through the tube?

TO DISCOVER IF A MOLECULE IS CAPABLE of mirror image isomerism, you could always draw the structure of the molecule and its mirror image, both in three dimensions or construct the mirror image molecules from a set of molecular models, and then try to superimpose the two structures to determine if the molecules are enantiomers. If they are enantiomers, then the molecule is chiral. This works well and we used the three dimensional drawings for α and β-glucose in Figure 1.4 and determined these isomers to be chiral, each one existing as one of a mirror image pair.

If you tried this method of comparing structural models on the three probe molecules by looking at the three dimensional structures drawn on the bottom of Figure 1.8 you would discover that formaldehyde and hydroxyacetone superimpose on their mirror images, while glyceraldehyde does not. Good!

But you would save the trouble of trying to superimpose molecular structures and get the same answer by inspecting the three structures in Figure 1.8 or, for that matter, any organic chemical structure and looking for at least one carbon atom in the molecule with four different groups bonded to it. Only one of the three, glyceraldeyde, fits this requirement of having four different groups around any of the carbon atoms in the structure. Although a tetrahedral carbon with four different groups is not the only structural characteristic consistent with chirality it is the most common direction to gain chirality and is the basis of chiral molecular structures in biology.

The seeds for the realization that molecular mirror images could be related to tetrahedral carbon bonded to four different groups, was planted when Pasteur, as quoted in section 1.10, suggested that his experimental results led to the idea that tartaric acid could have a structure described by an "irregular tetrahedron". This suggestion by Pasteur implied that carbon was four coordinate, a characteristic of the bonding of carbon we know very well now but was hardly clear at the time of Pasteur's work.

Although accurate formulae for organic molecules were increasingly available from experiment throughout the first half of the nineteenth century, there was no clear understanding of how the elements were combined with each other. The insight that carbon was four coordinate was first published by **Friedrich August Kekulé** in 1857, when he was 28 years old. The clarity this brought to organic chemical studies brought Kekulé fame but began a life of tragedy for a young Scottish chemist, **Archibald Scott Couper**, who had written what has been called an amazing paper with the same insights about carbon but presented in brilliant and clearer terms causing some to call it a work of genius. Couper had prepared his paper to be published in the same year, 1857, but the head of his laboratory, **Charles Adolph Wurtz**, who was a powerful person in the world of chemistry and the head of an important laboratory in Paris, delayed the publication. We'll never know if this was the reason, but it is reported that Wurtz and Kekulé developed a friendship when they both worked in Paris in the early 1850s. Couper's paper did not

1.11

Eventually, as the three dimensional structure of molecules came to be understood, it became clear which structural features of a molecule could lead to mirror image isomerism and therefore to enantiomeric pairs of molecules.

■ **Charles Adolph Wurtz**

■ **Alexander Michailovich Butlerow**

■ **Joseph Achille Le Bel**

appear until 1858 by which time Kekulé had received all the credit for the idea of the tetravalence of carbon. It is written that Couper lost his temper with his powerful boss and was expelled from the laboratory forcing his return to Edinburgh where he had a nervous breakdown. Couper never published another paper spending the last 30 years of his life at the home of his mother in Glasgow. In science we talk of "being scooped," which perfectly describes what happened to A. S. Couper. He was scooped by Kekulé.

The tetravalent nature of carbon was advanced even further in 1862 by **Alexander Michailovich Butlerow(v)** who was a distinguished professor of botany and chemistry in Kazan in Russia. Butlerow(v) had spent time in France and Germany where he met and had discussions with both Kekulé and Couper at the time they were both preparing their theory of tetravalency of carbon. But Butlerow(v) advanced these ideas further when he proposed for the first time that the tetravalency of carbon took the form of a tetrahedron.

Butlerow(v) was known as a spiritualist, someone who believed in supernatural phenomena, connections with the dead, seances, premonitions, the irrational and the esoteric, things that were believed to be beyond science. There was an intense interest in the spiritual at this time in Russia that extended into considerable influence in the royal court, an influence that was thought to have contributed to the end of the Russian empire in the early 20th century. Perhaps Butlerow(v)'s capacity to accept what could not be proved helped in his proposal that four coordinate carbon was tetrahedral, an assumption, for which there could be no proof at that time.

In this assumption of a tetrahedral geometry for four-coordinate carbon, Butlerow(v) put down the foundation for the effort of two young men, one French and one Dutch, who were children when Pasteur offered his stimulating idea of an irregular tetrahedron. The year was 1874 when **Joseph Achille Le Bel,** and **Henricus Jacobus van't Hoff**, both in their twenties, publishing in French and in Dutch respectively, took the idea of tetrahedral carbon a step further and showed how tetrahedral carbon with four different groups would give rise to mirror image isomerism as the structural basis of the phenomenon of optical activity, or as we understand it in structural terms, the basis of chirality.

PROBLEM 1.59
Why was the suggestion that carbon bonds to four different entities not sufficient to account for the phenomenon of mirror image isomerism? Could four different groups bond to carbon be possible without mirror image isomerism?

1.12

As experiments arose that could portray the three dimensional structures of mirror image molecules, it became necessary to develop a nomenclature that could distinguish left from right.

WHILE THERE IS ONLY ONE CARBON atom in glyceraldehyde with four different groups (Figure 1.8), there are two such carbon atoms in tartaric acid (Figure 1.9) and five such carbon atoms each in α and β-glucose in Figure 1.1. There is a way that organic chemists use to describe how the four different groups are arrayed around the central carbon atom, no matter how many such carbon atoms there may be in a single molecule. Using this method you can assign a name to each enantiomer of a pair instead of having to draw the three dimensional structure to figure out which enantiomer one is talking about. This is very useful especially for biologically important organic molecules, which are very often chiral and where only one of the two possible enantiomers is found in living systems.

An early version of this system of nomenclature was published in 1950 by two Englishmen, **R. S. Cahn**, the editor of the most important chemical journal in England, The Journal of the Chemical Society, and by **Christopher Ingold** of University College, London, who was a leading chemist of the 20th century whose work we will cover later in the book and finally with the contribution in 1955 of **Vladimir Prelog**, a Croatian who spent most of his scientific career in Zurich, Switzerland. Prelog was a great organic chemist who won the Nobel Prize in 1975 for "his research into the stereochemistry of organic molecules and reactions."

A photograph is shown of the three scientists when they attended a conference on stereochemistry in Switzerland in 1966 about a decade after their nomenclature was fully developed and accepted. This conference, which takes place once a year in Bürgenstock, is an important forum for stereochemical results.

To appreciate the value of the nomenclature, known as the CIP convention, it is helpful to look back at Figure 1.9. Before Cahn, Ingold and Prelog developed their nomenclature the mirror image forms of tartaric acid, shown in Figure 1.9, were known as D and L tartaric acid. These capital letters, which are derived from a convention for naming stereoisomers of glucose, a subject we'll look into in more detail in Chapter 3, are not clearly related to the structures shown in Figure 1.9. Nor are D and L related to the experimental result for the direction of rotation of the plane of polarized light determined from the optical activities of the enantiomers of tartaric acid (section 1.9). Whereas rotation to the right is designated dextrorotatory, d, and rotation to the left is designated levorotatory, l, D-tartaric acid rotates the plane of polarization to the left, while L-tartaric acid rotates the plane of polarization to the right. The experimental results for optical activity account for the D-tartaric acid being l-tartaric acid, while L-tartaric acid is d-tartaric acid.

A large number of compounds were found to be optically active in the development of chemistry up until the middle of the twentieth century without any experiment existing to know the precise three dimensional structures corresponding to the sign of the rotation of each compound. This situation changed abruptly in 1950-51 when a Dutch physical chemist, **Johannes Martin Bijvoet** (pronounced bifoot) discovered a way to use the diffraction of X-rays to distinguish enantiomers and determine the precise mirror form of an enantiomer. This was not an easy task considering that X-rays are not chiral and therefore in the usual experiment where X-rays are used to determine the structure of a chiral molecule, that is, how the molecule is put together including all the bond angles and bond distances, the two enantiomers give identical results. Bijvoet found a way to alter the X-ray experiment to allow distinguishing the enantiomeric crystals and to determine how the atoms were arrayed in three dimensions.

Appropriately, Bijvoet worked at an institute in Utrecht, Holland named after van't Hoff, one of the two chemists (section 1.11) who connected the tetrahedral array around carbon to optical activity, that is, to chirality. In Bijvoet's publication of his new kind of X-ray experiment, he noted the necessity for a nomenclature to distinguish and show the precise chemical structure of an enantiomeric molecule. Bijvoet realized that the old method of D and L or d, and l would no longer be adequate and in one of only three references in his breakthrough paper he referred to a new idea for nomenclature that would satisfy the need, a paper published in 1950 by Cahn and Ingold.

In one of those coincidences in science that seem almost to arise from something in the air that tells scientists that something new is afoot, Cahn and Ingold had been thinking of the need for a nomenclature that would connect a name for a molecule with its enantiomeric structure. For example, none of the designations, D, L, d, or l can allow one to draw the enantiomeric structure of tartaric acid shown in Figure 1.9. In their 1950 paper Cahn and Ingold pointed out that an experiment yielding structural information where mirror images might be distinguished, might arise in the future. They were in fact presciently pointing to a paper that would appear the following year, the paper in Nature by Bijvoet.

The fundamental idea behind the CIP convention is to observe the tetrahedral array, as seen in **Figure 1.10**, and give a priority, 1, 2, 3, and 4 to each group bonded to the central carbon. The priorities would be judged by the atomic numbers of each of the bonded groups, with the smallest atomic number given priority 4. One then is instructed to observe the tetrahedral array with the priority 4 group furthest away and determine if groups with priority 1, 2 and 3 trace out a clockwise, or counterclockwise arrow.

If you compare the four groups around the central carbon atom in the structure of D-glyceraldehyde (adapted from the Cahn and Ingold 1950 paper (Figure 1.10) according to the atomic number of the atom connected to the central carbon, then H is clearly the lowest and therefore priority 4, while O is clearly the highest and therefore

■ **Henricus Jacobus van't Hoff**

■ *(l to r)* **R.S. Cahn, Christopher Ingold, Vladimir Prelog**

■ **Vladimir Prelog**

■ **Johannes Martin Bijvoet**

FIGURE 1.10 ▶

Cahn-Ingold-Prelog (R) and (S) Nomenclature for the Chiral Carbon Atoms of Various Chiral Molecules

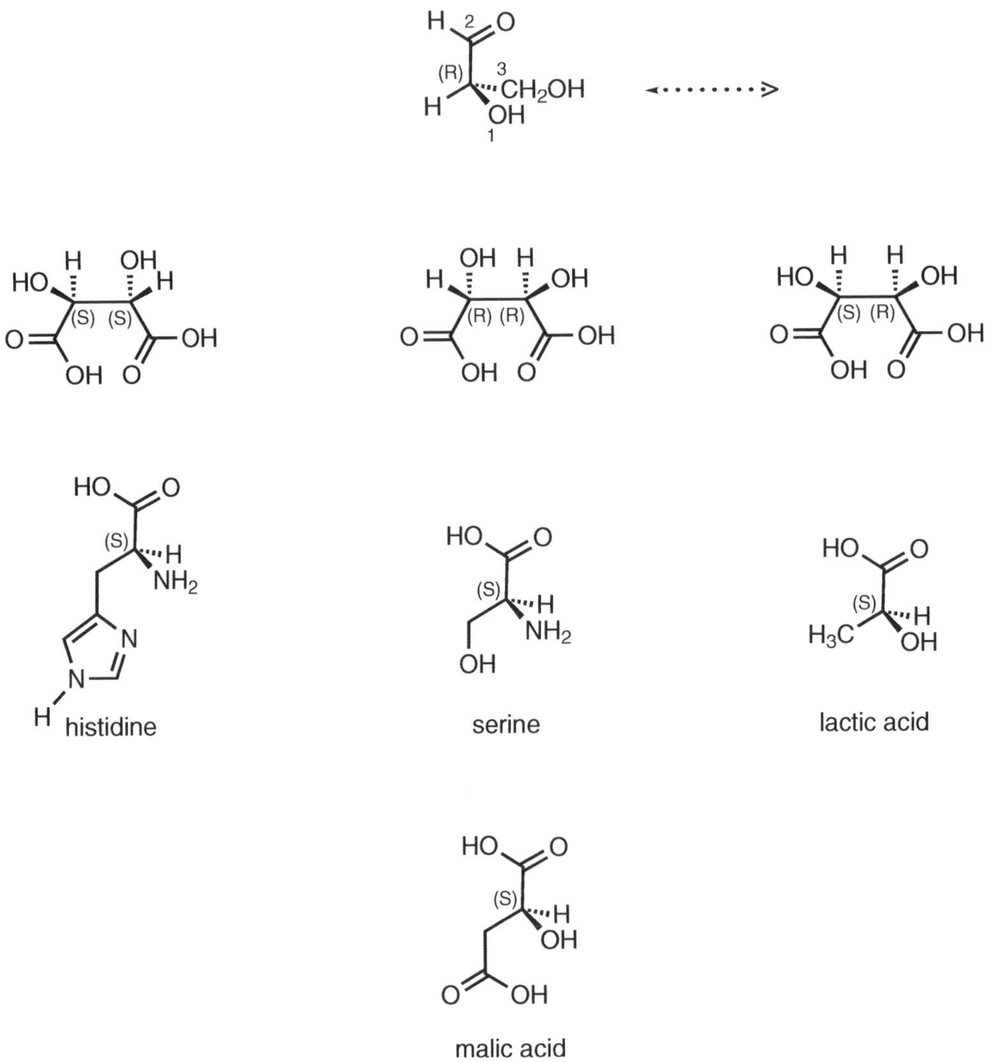

of priority 1. But what about the other two groups, which both have C attached to the central carbon. Cahn and Ingold made the reasonable suggestion to look to the groups bonded to this atom, which for one of these is H and a doubly bonded O and for the other is H, H and a singly bonded O (Figure 1.10). They counted a double bond to represent two of the atoms at the end of the double bond so that now we are comparing O, O and H, to H, H and O. Clearly the first has a larger total atomic number and so we give priority 2 to CHO and priority 3 to CH_2OH. As the "eye" shows in this figure, taken from the Cahn-Ingold publication, going from priority 1 to 2 to 3 with priority 4 behind traces out a clockwise direction.

By the time Cahn and Ingold became involved with Prelog a next step was taken. If 1, 2 and 3 traced clockwise, the new carbon configuration was R and if counterclockwise, S. The timing was perfect because Bijvoet's X-ray crystallography work was yielding the precise arrangement of the four different groups around tetrahedral carbon allowing therefore use of the CIP R, S nomenclature to name these arrangements.

Although there is no evidence of any influence, it is interesting that the initials of one of the authors, R. S. Cahn, became the designated way to name the path of the traced arrow. These initials proved handy considering that the words *rectus* and *sinister* are derived from the Latin for right and left. However, I have been told corrrectly that rectus in Latin refers to behavior (right or proper) rather than direction.

Figure 1.10 shows the use of the CIP nomenclature for the tartaric acid enantiomers in Figure 1.9 and another tartaric acid stereoisomer you have not seen before and also several chiral molecules, picked out among many possibilities as examples that are important in biochemistry. These are: the amino acids histidine and serine, important in enzymes that hydrolyze certain kinds of bonds (section 7.5); lactic acid, which is produced in muscle with inadequate oxygen; malic acid, an intermediate in the citric acid cycle, a series of reactions that produce carbon dioxide and reduced coenzymes with the latter acting as critical inputs into the biochemical production of energy. We'll be studying this cycle, also known as the Kreb's cycle in Chapter 8 (sections 8.6-8.8).

Two of the tartaric acid stereoisomers shown in Figure 1.10 are the same as those seen in Figure 1.9, the two whose salts produced the left and right handed crystals that Pasteur separated. But the new stereoisomer of tartaric acid is distinguished from the others by being superimposable on its mirror image and therefore is not chiral. Try this, and check on the claim above that this tartaric acid is achiral, by drawing the mirror image of this molecule, which is the tartaric acid designated (S, R). Did it bother you that this achiral form of tartaric acid still has carbon atoms that can be assigned R or S designations?

Applying the CIP convention to this achiral stereoisomer of tartaric acid, (S, R), shows that the two carbon atoms with four different groups have opposite assignments. One is R and the other S, and because the four differing groups on each of the two carbon atoms are identical there is therefore a mirror plane bisecting the molecule. If you imagine a mirror slicing through the middle bond of the molecule, the two images would be identical. Such a mirror plane in a molecule demonstrates that the molecule and its mirror image would also be identical. Such molecules, in which there are more than one carbon with the same four different groups, and with planes of symmetry, are called *meso*, and this is meso tartaric acid. Such achiral molecules have sometimes been called internally compensated.

There is a lesson here related to the nature of isomers that are diastereomers. We learned in sections 1.7 and 1.8 that while enantiomers have identical properties in the absence of a chiral entity, diastereomers have inherently different properties. Meso tartaric acid is certainly a stereoisomer of d or of l tartaric acid or as we are able to call these molecules now, R,R-tartaric acid and S,S-tartaric acid (Figures 1.9 and 1.10). But the stereoisomeric relationship between R,R-tartaric acid and R,S-tartaric acid and as well between S,S-tartaric and R,S-tartaric acid is not mirror related. Therefore these stereoisomeric relationships are diastereomeric. Returning to the data in Figure 1.9, the melting points of R,R and S,S-tartaric acids are identical, 168-170° C. The melting point of meso tartaric acid is about 140° C. And every other physical property will also distinguish meso tartaric acid from R,R or S,S-tartaric acid.

PROBLEM 1.60

(a) Using Google under the heading thalidomide will bring up a tragic situation, which is the best lesson one can have about the differing human response to mirror image molecules. Study the structure of thalidomide and identify the bonding sources of its chirality. Draw both mirror image forms and identify the enantiomer that caused the tragic response. Assign (R) or (S) to appropriate carbon atoms in each enantiomer.

(b) The structure of carvone can be found on Google. One enantiomer smells like spearmint gum and the other enantiomer like caraway. Draw both mirror images assigning (R) and (S) nomenclature to appropriate atoms. Is there any relationship, in principle, between the differing odors of the carvones and the differing physiological responses of the thalidomide enantiomers?

PROBLEM 1.61

Imagine that equal amounts of the enantiomers of thalidomide were dissolved in a solvent, which was then caused to flow down a tube containing cellulose (Figure 1.1). What experiment could you design to test what is occurring in the tube?

PROBLEM 1.62

Answer true or false (if false then write the true statement):

1-all chiral molecules designated D must also be d and (R);

2-the determination that a chiral molecule is D or L depends on the index of refraction with circularly polarized light;

3-only chiral molecules can be optically active;

4-optical activity arises from the difference between the index of refraction of left and right handed circularly polarized light;

5-d and l molecules always have a higher index of refraction for the opposite handedness of circularly polarized light;

6-while d and l connect a chiral molecule to a physical property, D, L, (R) and (S) are simply nomenclature.

PROBLEM 1.63

Redraw the structures in Figure 1.10 showing all atoms and all lone pairs of electrons, evaluating the structures for the formal charge and accordance with the octet rule. Use the structures you redrew and assign (R) or (S) configuration when possible. Check your answers against the assignments in the figure.

PROBLEM 1.64

Pick a web site on Google under the heading amino acids where you will find a table of the structures of the twenty natural amino acids. Draw a three dimensional structure for each, which corresponds to the (S) configuration, the enantiomer nature uses for the *in vivo* synthesis of all natural proteins. Are there any amino acids for which this assignment is not possible?

PROBLEM 1.65

(a) In Figure 1.10 there is a molecule that is not chiral although assignments of (R) and (S) could be made. Explain how this can happen and the meaning of the assignment "meso." Can you imagine the structures of other molecules that would fit into the meso designation? (b) Explain the response of left and right handed circularly polarized light to a meso molecule versus to another kind of achiral molecule in which it is not possible to assign (R) or (S) configuration to any carbon atom in the structure.

PROBLEM 1.66

Assign (R) or (S) configuration where ever possible to the carbon atoms in the structures in Figure 1.1.

1.13

A molecule can rapidly change its shape by motions about the bonds that hold the atoms together; and the differing shapes of a single molecule are, by definition, stereoisomerically related to each other.

IN CHAPTER 3 WE LOOK INTO THE SHAPES of six membered rings. We'll also see that molecules are not static either in their position in space or in their individual shapes. Not being static in space is quite reasonable. After all a molecule in the gas phase or in a liquid may be expected to be in motion and there is an energy associated with this motion, aptly named kinetic energy. And the speed with which molecules move about increases as the temperature increases, which increases their kinetic energy.

Motion of the entire molecule will have no affect on anything said in section 1.9. After all, if (S,R) tartaric acid, that is, meso tartaric acid, as presented in Figure 1.10, moves about, the identity with its mirror image will be maintained. However, molecular motion does not only involve constant movement of the entire molecule from one place to another, molecular motion also involves motion about the bonds that connect the atoms to one another within each molecule. This internal motion allows the atoms within a molecule to move with respect to each other and also increases in frequency as the temperature increases.

Without changing the fundamental shape of the molecule, the atoms are in constant vibration. The length of the bond between two atoms, such as one of the carbon atoms and the oxygen of the OH group in meso tartaric acid constantly changes around an equilibrium length. This vibration takes place at exceptionally high frequencies corresponding to the infrared region of the electromagnetic spectrum (section 2.2). Similarly, the angles between the atoms change, as for example the angle that is made between that same oxygen of the OH group with the hydrogen atom connected to the same carbon atom, that is, the angle made at the carbon atom in the array, H-O-C-H.

What this means is that the lengths of bonds between atoms, and the angles made between atoms, are constantly changing and the values reported, as for example the 109.47° tetrahedral angle for methane, CH_4, is the equilibrium value of the constantly changing angle, rather than some fixed angle.

While motion of the entire molecule from place to place and vibrational and angular motions make no difference to the stereochemical conclusions for the structure of the meso tartaric shown in Figure 1.10, such as the judgment that meso tartaric acid is achiral, there is another motion that does make a difference, which is called torsional motion involving rotation of one part of a molecule with respect to another part of the same molecule. It took quite awhile after the connection between tetrahedral carbon and stereoisomerism and optical activity was introduced by van't Hoff and Le Bel in 1874 (section 1.10) to understand that certain chemical reactions could not be understood without considering the movements of different parts of a molecule with respect to other parts of the same molecule—a dynamic picture.

Johannes Wislicenus

An early champion of this dynamic view was **Johannes Wislicenus**, a German of Polish descent, whose interesting early life involving his family having to leave Europe for the United States because of his father's radical religious and political views is worth looking into on the web. Johannes Wislicenus overcame many obstacles to attain one of the highest positions in German science, Professor at the University of Leipzig. Most of the obstacles arose from his own character, including his outspoken ways and politically incorrect views, which brought the displeasure of powerful chemists – I guess like father like son.

Earlier in his career, Wislicenus immediately accepted van't Hoff's theory of tetrahedral carbon and even organized and wrote the introduction to the German translation of van't Hoff's book in 1877 when the theory was quite controversial. This activity caused him to be rebuked by the great German chemist, **Adolf Kolbe**, an enmity that could have caused harm to Wislicenus' career. After all, Kolbe had just famously criticized van't Hoff's treatise, the very volume that Wislicenus had translated:

"A Dr. H. van 't Hoff of the Veterinary School at Utrecht has no liking, apparently, for exact chemical investigation. He has considered it more comfortable to mount Pegasus (apparently borrowed from the Veterinary School) and to proclaim in his 'La chimie dans l'espace' how the atoms appear to him to be arranged in space, when he is on the chemical Mt. Parnassus which he has reached by bold flight."

Adolph Kolbe

Kolbe even flung one at Wislicenus, accusing him of having lost his scientific senses.

Ironically, when Kolbe died in 1884, Wislicenus accepted his professorial chair and headed up Kolbe's former laboratory in Liepzig.

The concept that single bonds allowed what was called "free rotation" gained increasing acceptance in the waning years of the nineteenth century as it was realized that Wislicenus' thesis was correct. The current view that single bonds allow rotation of the connected groupings is expressed, in what is very close to modern terms, from quantum mechanical considerations by Linus Pauling, whom we've seen in section 1.4 as responsible for the idea of hybridization of orbitals. Here are Pauling's words in 1931: *"Each of these tetrahedral bond eigenfunctions is cylindrically symmetrical about its bond direction. Hence the bond energy is independent of orientation about this direction, so that there will be free rotation about a single bond, except in so far as rotation is hindered by steric effects, arising from interactions of the substituent atoms or groups."* What is the stereochemical consequence of rotation around the bond connecting the two tetrahedral carbon atoms in the isomers of tartaric acid shown in Figure 1.9? We can use the kinds of structural drawings in this figure and other derived drawings as shown in **Figure 1.11** to answer this question.

First in Figure 1.11 we reproduce the structural drawing of meso tartaric acid (S, R) from Figure 1.10 and designate this structure as **1**. Now we make two changes. We turn the molecule over without changing anything else, and also replace the solid and dashed connections to solid lines of different lengths, but still designating the same information. This produces **2**. Now we turn the molecule again without changing anything else and get **3**. The changes from **1** to **2** to **3** simply involve moving the entire unchanged molecule to different frames of reference - as if we were looking at the same object from different directions.

Now we are ready to carry out the kinds of changes that Wislicenus proposed as possible, and which Pauling justified with quantum mechanical considerations, that is, rotations of one part of the molecule with respect to another, rotations around single bonds. For this internal motion we imagine two arrows circling the rotating bonds but pointing in opposite directions, designating a change that is subject to the laws of equilibrium and rate. Rotations of the kinds we are describing here are exceptionally rapid in their rate, taking place in minute fractions of a second, even microseconds. The structures produced by these rotational motions about the designated bonds, **3**, **5**, and **7**, are certainly stereoisomers. They are identical in formula and bonding but differ in their shape. Because they are not mirror image related they are diastereomers. But they have to be given a special designation because of the ease with which they can change into each other.

The rapidly interchanging stereoisomers **3**, **5** and **7** in Figure 1.11, and many others that are possible, are called conformational isomers to distinguish them from the (R, R) and (S, S) and (R, S) isomers of tartaric acid, which are called configurational isomers. Bonds must be broken to interconvert (R, R) to (S, S) or to (R, S) tartaric acid, which means that the interconversion must be inherently slow and often requires special

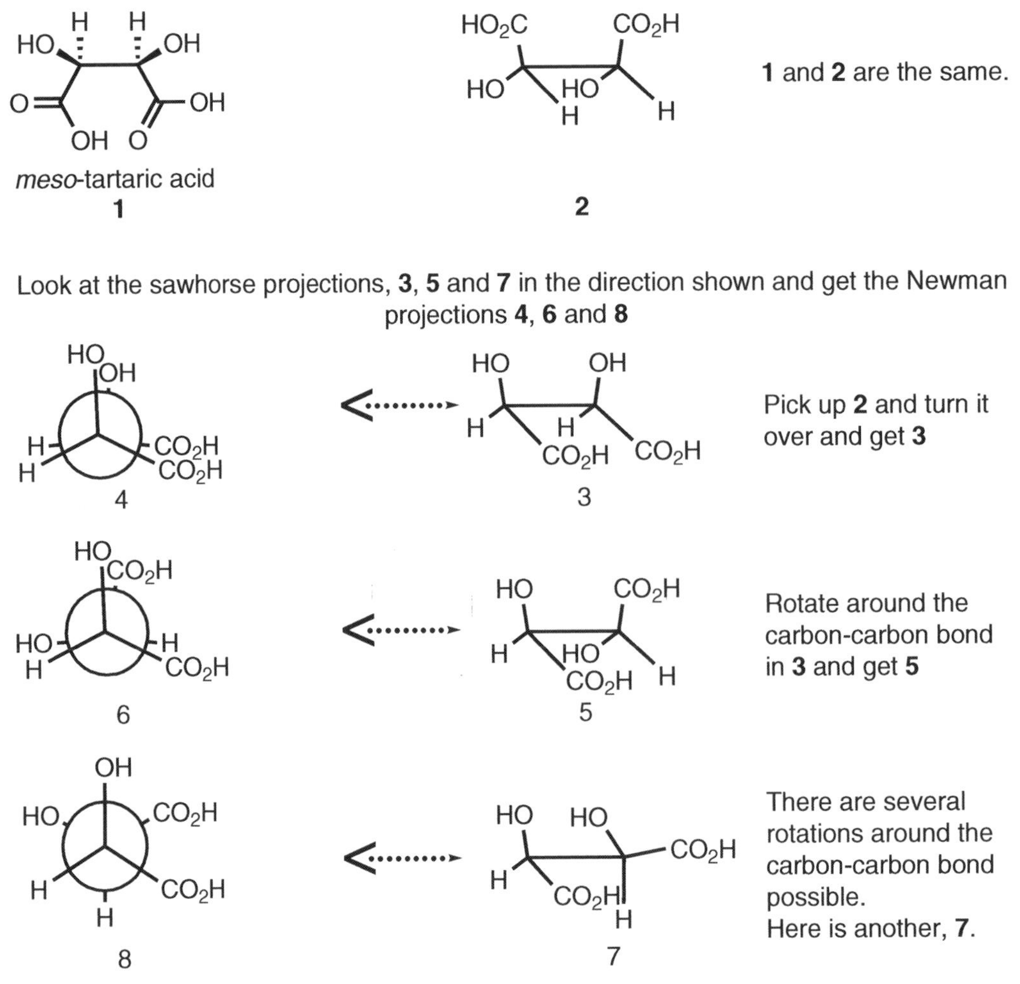

FIGURE 1.11 ▶

Sawhorse and Newman projections wconformational possibilities for meso-tartaric acid.

meso-tartaric acid
1

2

1 and 2 are the same.

Look at the sawhorse projections, **3**, **5** and **7** in the direction shown and get the Newman projections **4**, **6** and **8**

4

3

Pick up **2** and turn it over and get **3**

6

5

Rotate around the carbon-carbon bond in **3** and get **5**

8

7

There are several rotations around the carbon-carbon bond possible.
Here is another, **7**.

chemical procedures. However, only rotational motion about bonds is necessary to interconvert **3**, **5**, and **7** and other conformational isomers that can be formed from the basic structure of meso tartaric acid.

This difference in rate of interconversion means that configurational isomers, as is also the situation for constitutional isomers, can be separated from each other and kept for long periods of time. Conformational isomers, on the other hand, can not be isolated under normal conditions and any liquid or gaseous sample of a molecule subject to this kind of isomerism will exist as a mixture of rapidly interconverting conformational isomers.

Visualization has always been a key to advancing the science of chemistry as can be seen from the very origin of the science of chemistry, as far back as alchemy. Interesting things can be found by consulting Google under the heading alchemy symbols. In 1952, one of the distinguished chemists of the twentieth century, **Melvin S. Newman** of Ohio State University, helped advance understanding of conformational isomerism by creating a method of projection of these isomers, which has come to be called Newman projections.

Professor Newman was a greatly loved teacher and researcher of organic chemistry known for his wit and enthusiasm. He was a great fan of jazz and having lived in New Orleans he had made a friend of Louis Armstrong. One time Armstrong had a jazz concert in Cleveland around the time that the great synthetic chemist and Nobel Prize winner **R. B. Woodward** from Harvard University was giving a lecture. Newman brought Woodward to meet Satchmo after the concert and introduced the great chemist to the great jazz musician by saying that Professor Woodward is to chemistry what you are to jazz. Armstrong is reported to have responded: "Gee! Mr. Newman - this cat must really be something!" I don't know if Louis Armstrong, who played the blues knew that Woodward's favorite color was blue, a color the famous chemist was well known to indulge at every opportunity (section 12.2).

▪ **Melvin S. Newman**

The conformational pictures discussed so far in Figure 1.11 are called sawhorse projections in which the structure is viewed from the side. Newman suggested viewing the structure along the carbon-carbon bond of interest, which in our situation of tartaric acid is the central carbon-carbon bond in the structure. In the Newman projections, **4**, **6**, and **8** in Figure 1.11 the front four-coordinate-carbon-atom is centered at the junction of the Y and precisely in front of the rear carbon. The three ends of the Y at the front and back circles connect the three groups bonded to the front and back carbons respectively. The bond one is looking along, that is, between the front and back carbons, is not shown.

The Newman projection allows a helpful view of conformation as can be seen in the comparison between the sawhorse models, **3**, **5**, and **7** and their Newman equivalents, **4**, **6**, and **8**. In Newman projections the complete overlap of the front and back Ys is called the eclipsed conformation as seen in **4** and **6** while in the conformation that is described as anti or gauche, as seen in **8**, the bonds on the connected carbon atoms do not eclipse each other. In all anti and gauche conformations, the front and back Ys overlap as in **8** with anti ascribed to the situation in which the largest groups are opposite to each other and gauche when these groups are adjacent (60° apart in the Newman projection). **8** is a gauche conformation.

▪ **Robert Burns Woodward**

An entire specialization of chemistry is of great importance to biochemical phenomena, and using powerful computer-based methods, evaluates the energies of conformational isomers based on attractive and repulsive characteristics of the groups that approach each other in the conformation under consideration. This is a complicated business that can be applied to molecules as complex as proteins, which find their shape, which is critical to their function, by torsional motions around the multitude of single bonds in these polymers. More will be said later (first in Chapter 5) about the critical shapes that proteins take.

Discussion of the energies of conformational isomers, which depend on relative torsional and steric strain (section 3.5) relates to equilibrium. At any instant the relative proportions of conformational isomers, no matter how rapidly they are interconverting with each other, will be determined by their energies according to the normal thermodynamic analysis for determination of equilibrium constants. This is the same analysis applied to any chemical reaction, $\Delta G = -RT\ln K$. Even if conformational change is not a chemical reaction, because no bonds are made or broken in conformational change, and even if the conformational isomers are interconverting on the time scale of microseconds, the identical rules of thermodynamics apply allowing connection between energy difference, temperature and equilibrium. If we take a snapshot allowing an instantaneous view, which is possible with certain kinds of spectroscopy, we find that the numbers of each kind of conformational isomer are determined by their relative energies.

Let's stop now with this introduction to conformational isomerism and Newman projections to pick up the subject again in Chapter 3, sections 3.3 to 3.5, where we'll discover that these ideas are absolutely necessary to understand the nature of six membered rings and therefore to understanding why galactosemia is such a deadly disease to infants.

PROBLEM 1.67

In Problem 1.43 you discovered that the molecule with the formula C_4H_{10} has two constitutional isomers. In one of these isomers, n-butane, H_3C-CH_2-CH_2-CH_3 (a), the four carbon atoms are connected in a row, while in the other isomer, 2-methylpropane (b), three of the carbon atoms are each connected to the fourth carbon atom $(CH_3)_3C$-H. Using the concept of conformational isomerism discussed in this section and exhibited in Figure 1.11, and using Newman projections, look along any of the three carbon-carbon bonds in 2-methylpropane to convince yourself that there are only two conformational isomers possible.

There are also three carbon-carbon bonds in n-butane. But one, the central carbon-carbon bond, differs from the other two. Draw a Newman projection along the central carbon-carbon bond and convince yourself that there are numerous conformational isomers. Are there any enantiomeric pairs among these conformational isomers?

PROBLEM 1.68

Are there any conformational isomers of meso tartaric acid that have an identical mirror image? Are there conformational isomers that do not have an identical mirror image?

PROBLEM 1.69

Determine if the (R) and (S) designations for the chiral carbon atoms for the meso, the d and the l tartaric acids (Figures 1.9 and 1.10) change for the various conformational isomers you considered in Problem 1.68. Did you expect them to change?

CHAPTER ONE SUMMARY of the Essential Material

THE CHAPTER IS MEANT TO INTRODUCE THE FUNDAMENTAL IDEAS of structure and stereochemistry with the expectation that finishing this chapter will yield an understanding of bonding in organic molecules, including ideas of hybridization of orbitals and how the three dimensional character of organic molecules arises. A subset of this understanding is to be able to evaluate reasonable structures based on ideas of formal charge and octet rule. There is an expectation that structures can be understood in the various ways they can be presented, often without the atom symbol or the lone pairs of electrons shown. Although the focus is on sugars and their polymers, the purpose is to be able to apply this understanding to organic molecules in general. It is important that a structure presented in one way can be translated to another representation.

Given the understanding of molecular structure noted above, the expectation is that such understanding will form a foundation for the ideas of isomerism. The differences between constitutional and stereoisomers are important as are the differences between the kinds of stereoisomers–diastereomers and enantiomers. It is important to realize that judgments of isomers of any kind always relate to pairs of molecules and to the ways that two isomeric molecules may differ from each other–diastereomers and constitutional isomers having different properties while enantiomers have different properties only in a chiral environment. The ability to judge if a molecule is chiral or achiral is essential as is understanding what is necessary to allow separation of enantiomers or to understand when and how enantiomers may differ.

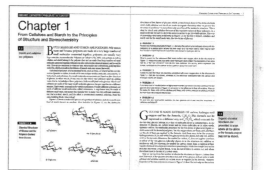

One important subject regarding chirality is the nomenclature used for such molecules including the different kinds of information given by the terms, d versus l, and the terms D and L, and then the nature of (R) and (S) and how these terms arise from structure or from experiments carried out on chiral molecules, such as optical activity. It's expected that although the symbols (R) and (S) usually are associated with chiral molecules they may not as in, for example, meso tartaric acid.

Although in this chapter the issue of nomenclature in general is not a focus, It is expected that the rules of nomenclature, addressed in one of the study guide problems, will become increasingly familiar as study of the subject of organic chemistry goes on.

Finally, the chapter brings up the subject of molecular motion both of the entire molecule and of portions of the molecule with respect to each other, or in other words, the concept of conformational isomerism. Understanding Newman Projections is important and how they relate to other kinds of representation such as Sawhorse Projections. The concept of conformation will play an important role in Chapter 3 where these ideas will be extended.

Chapter 2

A Survey of the Experiments Usually Performed by Chemists to Understand the Structures of Organic Molecules: Mass Spectrometers, Infrared Spectrometers and Nuclear Magnetic Resonance Spectrometers

2.1
Mass Spectra

THE ELECTRON IMPACT MASS SPECTRUM of one of the molecules, n-hexane, used as one example of three constitutional isomers with the formula C_6H_{14} in Figure 1.6, is reproduced in **Figure 2.1**. In the spectrum we see a series of lines rising from the abscissa, which is labeled m/z. The letter m stands for mass and that of z for charge. Every line in the spectrum corresponds to an entity with a single positive charge. The height of the lines corresponds to the number of each of these entities. What's going on in this instrument?

Normal hexane, n-hexane is volatile and as is the situation for all routine electron impact mass spectra, the sample is passed into a high vacuum chamber and subjected to bombardment with electrons with energies far in excess of that necessary to cause ejection of an electron from the molecule to be analyzed. A molecular cation radical, M^+, is produced with a great deal of energy, enough to cause the molecule to break into pieces, into fragments. Some of these fragments are charged and some are not. The charged fragments appear as lines in the mass spectrum because the instrument sends the various charged entities into an analyzer that can take several different forms according to the type of instrument, but all with the common feature of distinguishing the charged entities

FIGURE 2.1 ▼
Electron Impact Mass Spectrum of n-Hexane

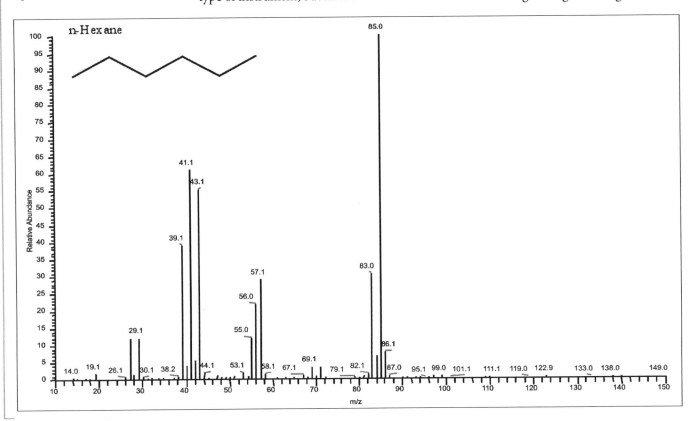

according to their momentum. By accounting for the time it takes for a fragment to reach a detector or based on the changing value of an electric or magnetic field necessary to bring a fragment to a slit leading to the detector, a value of m/z is assigned and from this information there arises the abscissa in the spectrum shown in Figure 2.1.

For the mass spectrum in Figure 2.1 the signal with the largest mass to charge ratio appears at m/z 86, which would correspond to the molecular weight of the analyzed molecule, C_6H_{14} except for the fact that most of this line intensity corresponds to $C_6H_{13}^+$ by loss of a hydrogen atom from the molecular ion, $C_6H_{14}^{+}$ but with one ^{12}C replaced by a ^{13}C, an event with a probability of 1.1% of the intensity of the signal at m/z 85. This means that hardly any unfragmented molecules of the charged species produced from n-hexane are stable enough to reach the detector of the mass spectrometer. Other fragments form and these are detected with lower values of m/z. For example, the large signal at 57 corresponds to the loss of 29 from 86, which means that M^+ has found a way to eject an ethyl group, C_2H_5.

The mass spectrum of one of the other two constitutional isomers of n-hexane, 2,2-dimethylbutane, is reproduced in **Figure 2.2**. In this spectrum we don't observe any signal at m/z 86. The large array of signals in the range from 39 to 43 means that charged fragments with three carbon atoms are readily formed with varying numbers of hydrogen atoms attached to each.

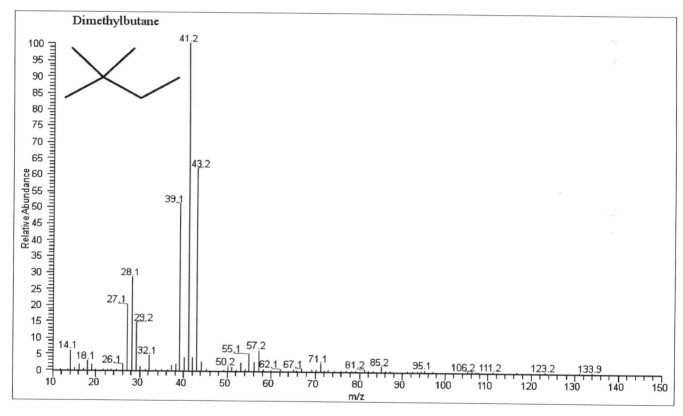

The electron impact mass spectrum of another of the constitutional isomers from Figure 1.6 with the formula C_6H_{14}, 2-methylpentane, is shown in **Figure 2.3**. In this spectrum we again see almost no signal at m/z 86 while maintaining the strong signal at 41. In all three mass spectra (Figures 2.1-2.3) there are many fragments formed and although there are possibilities of connecting the pattern of fragmentation to the structure of these constitutional isomers, it would be difficult to make a firm assignment from first principles of which structure is which just given the three mass spectra or even to be certain of the molecular weight considering that the m/z value for M^+ may not appear with certainty, depending on the structure. This is often the situation for the mass spectra obtained from small volatile molecules subjected to impact with high energy electrons. The spectrum is often correlated with the structure of the molecule not by a precise prediction but rather by a previously known fingerprint pattern of peaks associated with that particular structure.

◄ **FIGURE 2.2**

Electron Impact Mass Spectrum of 2,2-Dimethylbutane

However, because there are well researched rules for how certain kinds of molecules fragment in electron impact mass spectrometers, this information can sometimes be used to assign a structure to the analyzed molecule. As well there are certain structural features, such as aromatic rings (discussed in Chapter 6) for one example, that are likely to yield easily discernable molecular ions in the mass spectrum so that the molecular weight and even the atomic composition can be determined.

In recent years, mass spectra have been taken in a different manner in the attempt to make the technique of greater use to large polar molecules and especially to biological molecules. The problem with the kind of spectra shown in Figure 2.1-2.3 is twofold: the molecule is volatile, which is not the situation for biologically interesting organic molecules; the energy necessary to impart the necessary charge required is so high that the molecule fragments to an extent that the critically important molecular weight information is lost.

These problems have been solved in two different ways leading to a shared Nobel Prize to **John Fenn** and **Koichi Tanaka** for these techniques. For our purposes let's focus on the former technique, which is called electrospray mass spectrometry. The spectra for D-glucose and L-tartaric acid taken by this method are shown in **Figure 2.4**

Electron impact mass spectra for the molecules whose electrospray mass spectra are shown in Figure 2.4, and other polar molecules of relatively high melting point, are made difficult by the low volatility of these molecules. The high temperatures necessary to vaporize the sample leads to thermal decomposition and although special techniques can help, in general polar molecules are troublesome. Moreover, the difficulty increases with increasing molecular weight because volatility decreases, blocking wide use of mass spectrometry for biologically interesting molecules, such as proteins, polynucleotides and polysaccharides among many other large molecules encountered in biological work.

The two mass spectra in Figure 2.4 are strikingly different in their fundamental nature compared to the mass spectra in Figures 2.1-2.3 even if the instrumental method for determining the weight of the charged entities is related. The electrospray method gives rise to a single intense signal appearing at m/z 148.6 for L-tartaric acid and at 202.6 for D-glucose.

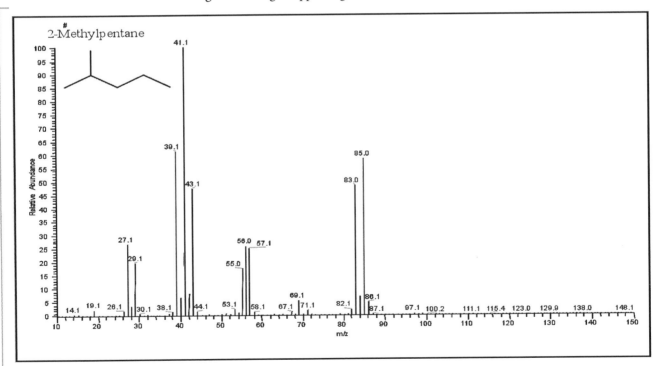

FIGURE 2.3 ▲

Electron Impact Mass Spectrum of 2-Methylpentane

As for electron impact mass spectra and other mass spectral methods, the charge in the electrospray method can be positive or negative. The negative charge in the mass spectrum of L-tartaric acid arises by loss of a proton (H^+) from the molecule while the positive charge in the spectrum of D-glucose arises from addition of a sodium ion (Na^+) to the molecule. Rounding off the experimental value for L-tartaric acid comes within experimental precision for the theoretical molecular weight of this molecule, minus-one-hydrogen, 149.1. Similarly, for D-glucose, rounding off the experimental value comes within experimental precision for the theoretical molecular weight of this molecule adding a sodium ion, 203.1. The charged entities in each situation find their way into the gas phase and somehow make their way down a long tube to the detector without breaking into pieces. What's going on with this method?

John Fenn's Nobel lecture on December 8, 2002 was titled: "Electrospray Wings for

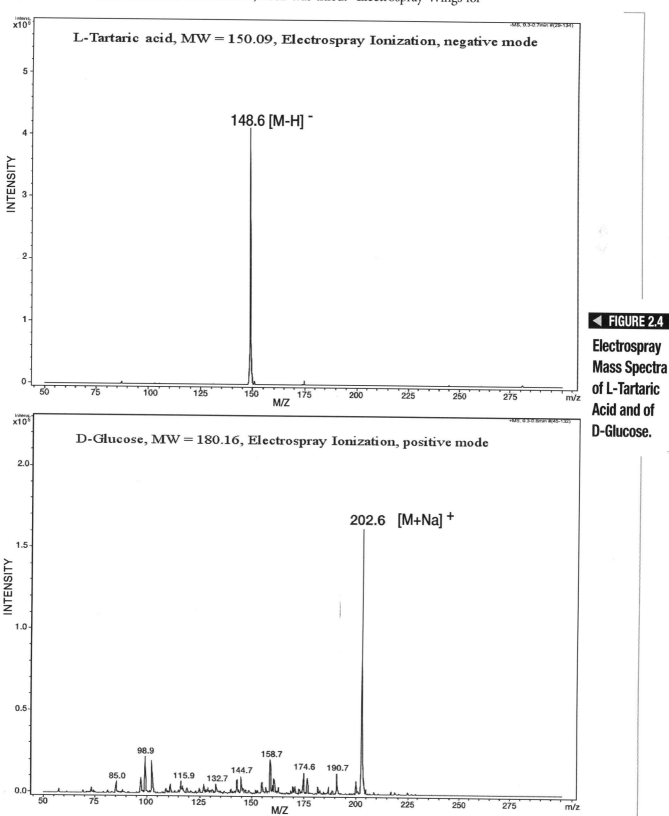

◀ FIGURE 2.4

Electrospray Mass Spectra of L-Tartaric Acid and of D-Glucose.

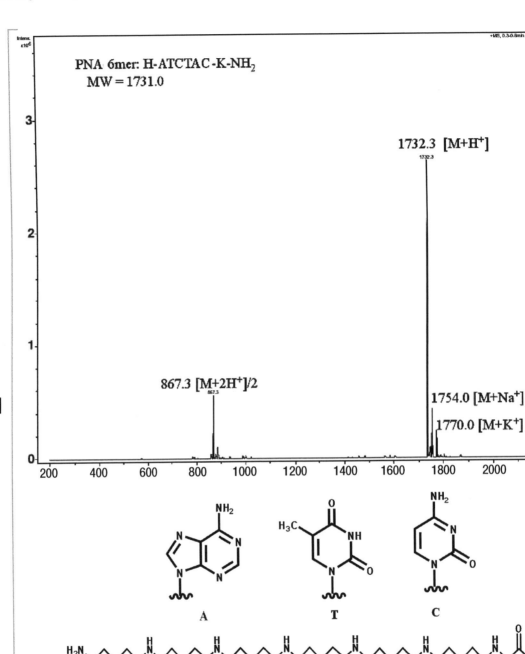

FIGURE 2.5 ▶

Electrospray Mass Spectrum of a Peptide Nucleic Acid

Molecular Elephants." The title fits because although we gave examples of the method above with two molecules that are hardly very large, the method works as well for molecules as large as proteins with molecular weights in the tens and even hundreds of thousands – truly elephantine molecules, yielding molecular weight information and more. Such large molecular weights exceed the range of the detecting apparati of the mass spectrometers, a problem that is overcome by the fact that multiple charges are produced in such large molecules, causing the instrument to "see" a molecular weight that is a fraction (m/z, where $z = n$ = integer larger than 1) of the true molecular weight. Computational methods are then used to determine the true molecular weight, that is, m for m/z, $z = 1$.

If **Malcolm Dole** of Northwestern University had lived it is possible that he would have stood on that stage in Stockholm with John Fenn. It was Dole's originating insight, published in 1968, that led to the concept of vaporizing molecules that are too large to gain the vapor state by ordinary methods. Dole devised a system that is at the heart of the electrospray method by realizing that as drops of a liquid, with a charged molecule solute, grew smaller and smaller by evaporation of the liquid molecules, water for example, a size would be reached in which the drop would have to explode into a large number of smaller drops. The inevitable change to smaller and smaller drops is a variation on a prediction made by **Lord Raleigh** in the late 1800s and finally leads to the unsolvated ion, which Fenn adopted for the electrospray method by inventing a process where the solvent molecules did not return.

Koichi Tanaka, the other scientist awarded the Nobel Prize with Fenn for work on mass spectrometry had independently put down the foundation for a technique widely known as MALDI, matrix assisted laser desorption ionization, which also gives intact molecular ions for large biologically interesting molecules.

An example of an electrospray mass spectrum for a somewhat large molecule is a polymer, which was of interest in our laboratory, a peptide nucleic acid (PNA) as shown in **Figure 2.5**. These chimeras composed of peptide like backbones with pendant nucleotide bases are of interest for medical science via what is called antisense technology in which the peptide nucleic acid with the proper nucleotide base sequence can complex with a DNA strand to shut down an undesirable in vivo process. For example the PNA shown in Figure 2.5 has a base sequence complementary to a DNA strand with the sequence, TAGATG. More will be said about the aromatic character and structure of nucleotide bases in Chapter 6.

As a final note, methods to fragment the gaseous charged proteins, produced by the electrospray method, under controlled conditions but before initiating their journey to the detector in the mass spectrometer has become a powerful method for determining the sequence of amino acids in a protein and even in complex biological systems such as ribosomes where the molecular weights are astonishingly high for a mass spectrometer. The varieties of amino acid based fragments that come apart from the whole protein reveal the sequence by which they are linked in the original structure. More will be said about proteins in their role as enzymes and the dependence of this function on the sequence of amino acids in Chapter 7 and 8.

■ **Malcolm Dole**

■ **Lord Raleigh**

T HE THIRD PARAGRAPH IN SECTION 1.13, as an introduction to the concept of conformational isomerization, noted that the atoms in a molecule are in constant vibration. The frequencies of these motions for organic molecules correspond to the infrared region of the electromagnetic spectrum in the wavelength range from about 3×10^{-4} to 3×10^{-3} centimeters (3 to 30 μ). The energies corresponding to this region of the electromagnetic spectrum range from about 1 to 10 kcal/mole, which correspond to the differences in vibrational energy states for common motions of the atoms of organic molecules. If the frequency of a bond vibration or other internal motion in an organic molecule falls in the infrared range, 3333- 333 cm^{-1}(wavenumbers), energy of that frequency, υ (E = hυ), will be absorbed by the molecule causing promotion to a higher vibrational state of that motion, a state that differs in energy from the ground state by the energy absorbed. An infrared spectrometer is designed to irradiate a sample with infrared radiation from some heat source and then determine the wavelengths (or frequencies or both) that have been diminished on reaching a detector. For this purpose a prism or grating is necessary to separate the wavelengths of the infrared radiation.

Figure 2.6 shows infrared spectra of two constitutional isomers shown in Figure 1.6, di-n-propyl ether and 1-hexanol. Reduced transmission on the ordinate of these spectra measures loss of the radiant energy, while the abscissa yields the frequency of that absorbed energy.

2.2

Infrared Spectra

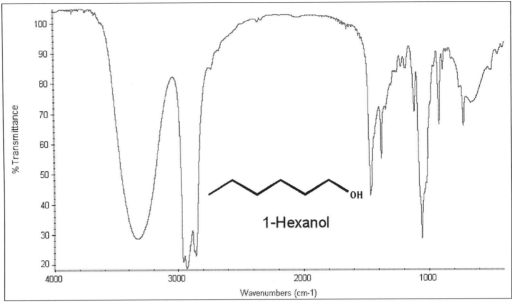

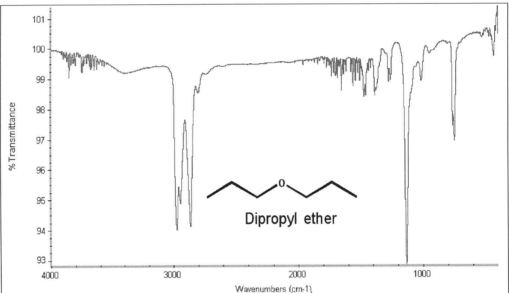

FIGURE 2.6 ▶

Comparison of the Infrared Spectra of the Constitutional Isomers 1-Hexanol and di-n-Propyl Ether

The spectra in Figure 2.6 demonstrate the large information content in the frequency dependence of the absorption of infrared light by organic molecules. In molecules as complex as those in Figure 2.6, although it is not reasonably possible to connect all the IR bands in the spectrum to the particular responsible motions, many of the bands can be identified with precise elements of the molecular structure.

At a first level, the spectra clearly show that molecules of similar structure, such as the constitutional isomers in this figure, are easily distinguished and that the spectra could be used as fingerprints to identify molecules given standard spectra. But much more is possible.

From experience over many decades, in combination with theoretical effort, assignment of particular IR frequencies to certain functional groups (section 3.9) and other molecular moieties has become possible. Reasonable expectations are met such as for just two examples: double bonds vibrate at higher frequencies than single bonds, if the double bonds exist between carbon and carbon, or between carbon and oxygen; heavier atoms vibrate at lower frequencies than lighter atoms.

There are scientists who specialize in vibrational spectroscopy and much is understood. But for our introductory purposes general correlations of molecular structure to absorption of infrared frequencies is adequate. Tables of these correlations

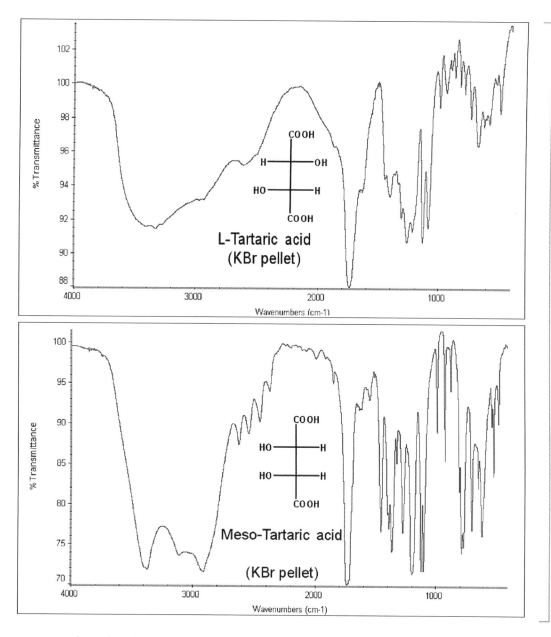

◀ **FIGURE 2.7**

Comparison of the Infrared Spectra of L-Tartaric Acid and Meso-Tartaric Acid

appear on the web and in many textbooks. One excellent source on the web is given here but much can be found by simply using Google under the heading infrared frequencies: http://www2.ups.edu/faculty/hanson/Spectroscopy/IR/IRfrequencies.html

In such correlations one can discover that OH groups exhibit a broad strong band between 3200 and 3500 cm^{-1}, precisely in line with the spectra shown in Figure 2.6. The grouping of C-O-C atoms undergoes a vibration leading to absorption of IR light between 1050 and 1150 cm^{-1}, as seen in the spectrum of dipropyl ether in Figure 2.6.

The spectra of the L and meso isomers of tartaric acid (**Figure 2.7**) are especially informative about the value of this spectroscopy and as well further reinforce the principle predicting that meso and either D or L tartaric acid are related as diastereomers and therefore the expectation that their IR spectra, as in fact all their physical and chemical properties, should differ, as seen in Figure 2.7.

The tartaric acids are solids and there are various ways to take the IR spectra of solids including the one used here in which the sample is ground up in KBr powder and then subjected to high enough pressure to form a pellet that does not scatter infrared light. KBr does not absorb IR light in the region of interest.

Checking a table of IR frequencies connected to organic structure such as the web site noted above informs us that the carbonyl group (C=O) of a carboxylic acid group,

FIGURE 2.8 ▶

**Infrared
Spectrum of
D-Glucose**

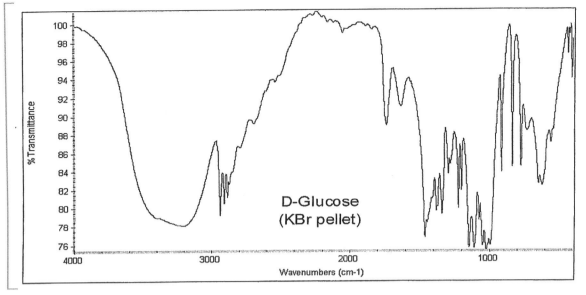

D-Glucose
(KBr pellet)

% Transmittance

Wavenumbers (cm-1)

a prominent feature of tartaric acid (OH-C=O) absorbs between 1700 and 1725 cm^{-1} with the OH group of the carboxylic acid frequency absorbing IR radiation with a strong broad band around 3000 cm^{-1}. The OH (hydroxyl) functional group found in 1-hexanol (Figure 2.6 also absorbs infrared radiation in this region of the spectrum as does do the hydroxyl groups of glucose as seen in **Figure 2.8**. However, in neither 1-hexanol nor glucose is there a strong band around 1700 cm^{-1} – no carbonyl groups (C=O) in these structures.

Infrared spectrometry is a powerful widely used method to identify certain arrangements of atoms, what are called functional groups. We've seen in the spectra discussed in this section three functional groups prominently displayed (section 3.9), ether, hydroxyl and carboxylic acid (carbonyl). We'll be coming across many functional groups (page ix) as we progress in the study of this science and a focus on the role of functional groups in organic chemistry.

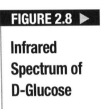

2.3

Nuclear Magnetic Resonance Spectrometry (NMR)

ONE OF THE IMPORTANT ADVANCES in atomic physics in the twentieth century was the understanding of the magnetic properties associated with the precessing of atoms, the change in the direction of the axis of the atom's rotation, as in a gyroscope. **Isidor Isaac Rabi,** who was born in Galicia in Austria-Hungary in what is now Poland but was brought to the United States as an infant, and lived most of his life in New York City where he died at the age of 89 in 1988, won the Nobel Prize in 1944 for his demonstration that the precession of the atoms in a beam of atoms could be detected with radio waves of a frequency matching the frequency of the precession.

Although not realized at the time, Rabi's experiment set the stage for one of the great advances in understanding the nature of organic matter for which another Nobel Prize was awarded eight years later to **Edward M. Purcell** (1912-1997) and **Felix Bloch** (1905-1983). Purcell was a physicist who was born in Illinois and worked at Harvard. Felix Bloch was driven from Europe by Hitler's policies and worked at Stanford University. Within weeks of each other in December 1945 and January 1946 and without knowing about the others work, both Purcell and Bloch and their collaborators demonstrated that the interaction of radio waves with the precession frequencies of atoms could be detected in samples of paraffin and water respectively. When the frequency of the precession and that of the radio wave were the same, at a specified magnetic field strength, energy could be transferred causing a shift in the angle of precession and causing a signal to be detected in the amplitude of the radio signal. Nuclear magnetic resonance (NMR) spectroscopy was born.

■ **Isidor Isaac Rabi**

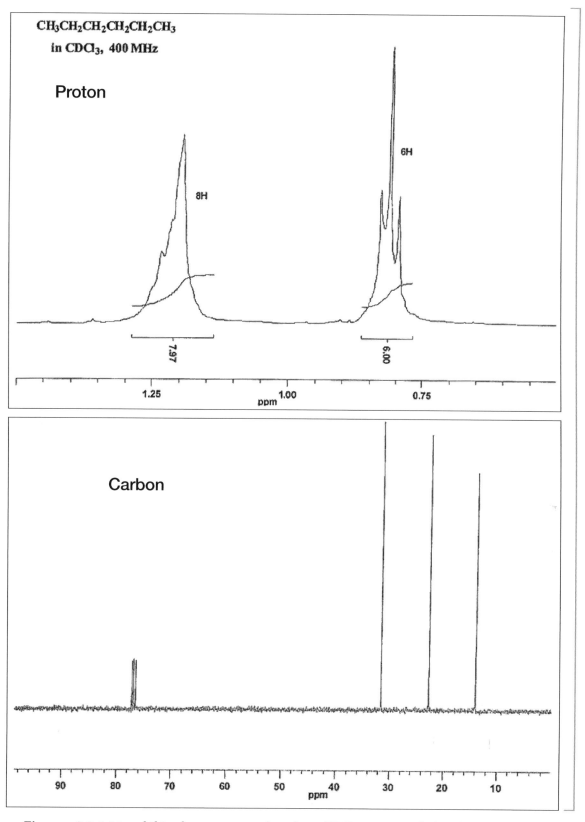

◀ **FIGURE 2.9**

Proton and Carbon NMR Spectra of n-Hexane

Figures 2.9-2.11 exhibit the proton and carbon NMR spectra of the three constitutional isomers, n-hexane, 2,2-dimethylbutane and 2-methylpentane shown in Figure 1.6 and whose mass spectra were exhibited in Figures 2.1-2.3.

In practice, NMR spectrometers may operate by varying either the magnetic field while keeping a constant radio frequency or the reverse, which is the mode of operation of modern Fourier transform (FT) instruments, but whichever way the instrument operates the x-axis is represented as if the magnetic field is varied. Let's

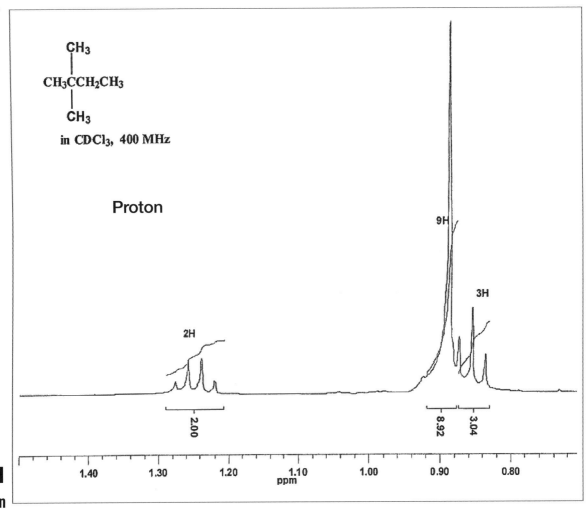

Proton

FIGURE 2.10 ▶

Proton and Carbon NMR Spectra of 2,2-Dimethylbutane

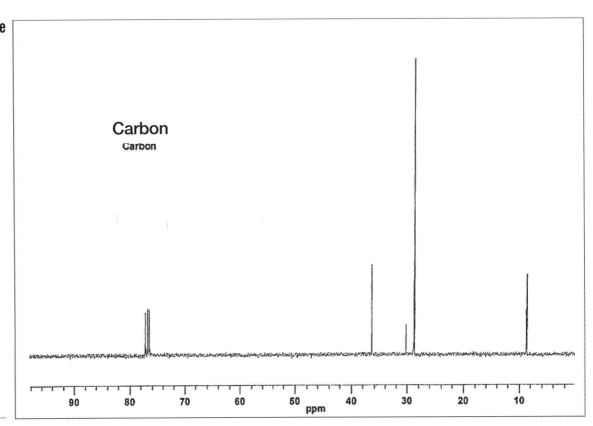

Carbon

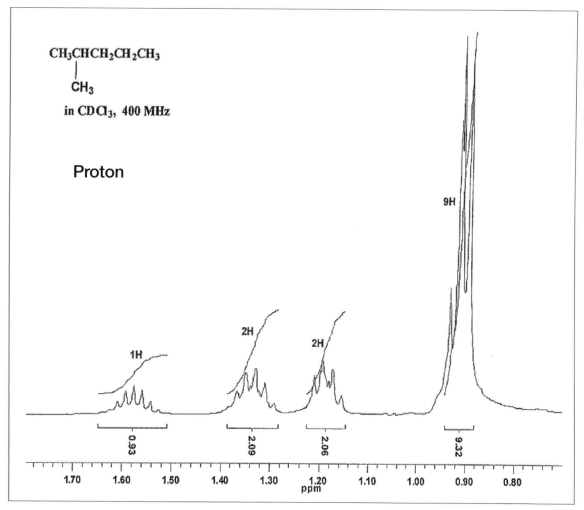

CH₃CHCH₂CH₂CH₃
|
CH₃

in CDCl₃, 400 MHz

Proton

9H

1H

2H

2H

0.93

2.09

2.06

9.32

1.70 1.60 1.50 1.40 1.30 1.20 1.10 1.00 0.90 0.80
ppm

◀ **FIGURE 2.11**

Proton And Carbon NMR Spectra Of 2-Methylpentane

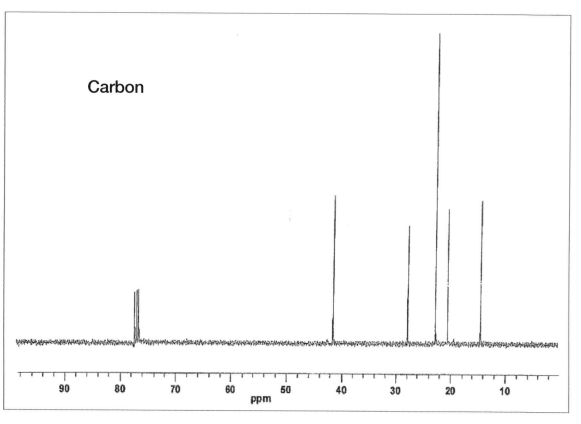

Carbon

90 80 70 60 50 40 30 20 10
ppm

follow that idea. Consider that a molecule is exposed to the instrument's constant radio frequency and that the magnetic field is adjusted to a field strength to bring the observed nucleus into the resonance condition.

The x-axes in Figures 2.9-2.11 inform us of the parts per million (ppm) of the change in the magnetic field from a standard value that is necessary to bring the various nuclei in the NMR spectra into resonance with a constant radio frequency. For each molecule the upper spectrum shows the various hydrogen atoms in the molecule, while in the lower spectrum, the various carbon atoms are brought to this resonance condition.

Happily, for analysis of organic molecules, hydrogen has a spin allowing observation by NMR as also does one of the isotopes of carbon, ^{13}C. Although ^{13}C is only present to the extent of about 1% compared to ^{12}C (section 2.1), the most abundant isotope of carbon, and the isotope of carbon, which is invisible for NMR measurements, the proportion of ^{13}C is large enough to be measured by modern instruments using special techniques. For this reason organic molecules can be studied by NMR observations of both carbon and hydrogen.

When the radio frequency is in resonance with the atom under observation in a strong magnetic field, this frequency corresponds to the energy difference between two allowed spin states of the atom, which can be thought of as with and opposed to the external magnetic field. The strength of the radio frequency signal will be slightly diminished at the point when the external magnetic field allows the resonance condition, which is detected by a signal in the NMR spectrum, such as the lines we see in Figures 2.9-2.11.

2.4

NMR Chemical Shift

NOW PLACE IN THIS RADIO FREQUENCY, a sample molecule such as those in Figures 2.9-2.11, and we discover that a single magnetic field strength is not correct to attain the resonance condition for the various atoms, hydrogen atoms for example, in the sample. There is some chemical effect causing the structurally different hydrogen atoms in the structure in the upper spectra, or the carbon atoms in the lower spectra, to come into resonance to absorb the radio frequency of the instrument. This is called the "chemical shift," which is measured along the x-axis of the NMR spectrum.

If the electron density around a particular atom (hydrogen or carbon) is increased because of this chemical effect, the atom will be shielded from the magnetic field of the instrument (the external magnetic field) and therefore a higher external field will be necessary to bring the atom into the resonance condition with the radio waves. These atoms are called "shielded." On the other hand, the atoms in the sample with the lower electron densities because of some chemical effect are what is called "deshielded" and the magnetic field necessary for the resonance condition will be lower. This is the source of the chemical shift.

The chemical shift, therefore, is an extremely sensitive measure of the intimate electron properties around the observed atom. With this in mind, chemists have chosen a standard molecule, tetramethyl silane, $(CH_3)_4Si$, (TMS), against which to measure all chemical shifts. All twelve hydrogen atoms in this molecule are identical to each other, as are the four carbon atoms to each other, which gives rise to one peak in either the proton NMR or carbon NMR spectra, a great value because a single reference point of high intensity exists. In addition, TMS is unreactive to the extreme, so it can be mixed with a wide variety of other molecules and solvents, and is also highly volatile so it can easily be removed and is also generally soluble – an ideal substance for the job.

But most important is that the four methyl groups in TMS are connected to an atom, silicon, with a very low electronegativity compared to carbon and to hydrogen, so that the electron density around the four methyl groups (CH_3) is high. This means that the magnetic field necessary to bring these atoms into resonance will be higher for the hydrogen atoms in that methyl group than for virtually all hydrogen atoms in any molecule under analysis and similarly the carbon atoms in TMS will require the highest possible field compared to any molecule under analysis for its carbon NMR spectrum. The atoms in TMS, the twelve hydrogen atoms and the four carbon atoms, are highly shielded.

We've just noted, with the word electronegativity, another of Pauling's contributions to modern chemistry, which was first mentioned in section 1.4. When Pauling was considering the nature of chemical bonding leading to the idea of hybridization of orbitals (section 1.4) he was faced with the fact that only identical atoms shared electrons equally in a balanced covalent relationship. But many different atoms bonded with varying degrees of equivalent sharing of the two electrons involved even to the extent of no sharing of electrons as in an ionic bond (Na^+ Cl^-). This led Pauling to develop the concept of electronegativity as a measure of the ability of an element to attract or hold onto electrons. In the example of sodium chloride the electronegativity difference is so extreme that the sodium atom gives up both otherwise shared electrons to the chlorine atom. This idea plays a critical role in chemical reactions, as will be introduced in section 3.13 and also in understanding the chemical shift phenomenon we are focused on here. The electron density around an atom, which depends strongly on the electronegativity of that atom compared to atoms it is bonded to, determines the shielding of that atom and therefore the magnetic field strength "felt" at that atom and therefore the chemical shift.

Based on the discussion above, this means that the peak for the atoms in TMS which are highly shielded will occur at the equivalent of the highest external magnetic field. By convention, this corresponds to the lowest value on the ppm scale, a value set arbitrarily to zero, which appears at the right end of the x-axis. Unfortunately when the spectra in Figures 2.9-2.11 were measured the computer did not output the x-axis for zero ppm and therefore the signal for TMS, is not shown at the extreme right.

This zero value will correspond to entirely different absolute values of the corresponding magnetic fields and radio frequencies for hydrogen and carbon. But for our purposes, these absolute values are not necessary to consider. We only need consider the values on the ppm scale measuring the difference between the atoms in our analyzed sample with those in TMS in either the proton or the carbon NRM spectra.

The zero value on the ppm scale corresponds to the most shielded atom with increasing values of ppm corresponding to decreased shielding. The scale is therefore called delta, δ, for deshielding. The lower the electron density about a hydrogen or carbon atom, the larger is the ppm value at which it will appear in the NMR spectrum.

The spectra in Figures 2.9-2.11 are measured in solution in $DCCl_3$, deuterochlorofom, so that there is no peak for the solvent in the proton NMR spectra but there is a peak in the carbon NMR spectra that appears near 78 ppm – the solvent molecules still contain their approximately 1% of ^{13}C.

In general the larger the nucleus under observation, the larger is the range of chemical shifts as a function of its chemical bonding and environment within the molecule, that is, those factors affecting the electron density around the nucleus and therefore the shielding, as discussed above. However, this correlation is not a simple matter and is based in quantum mechanical theory. Nevertheless, the expected large difference in the range of chemical shifts is seen in the x-axes for the proton NMR for the three molecules, compared to the carbon NMR spectra for these molecules (Figures 2.9-2.11). For hydrogen, the range is from about 0 to 10 ppm while for carbon the range is about 0 to 100 ppm even if the atoms in the hydrocarbons in Figures 2.9-2.11 do not have chemical shifts over the entire range.

As makes sense, in the NMR spectra of atoms bonded to electronegative atoms such as oxygen, the chemical shifts appear at larger ppm values than those seen in Figure 2.9-2.11 for hydrogen and carbon. Oxygen is a highly electronegative element compared to both carbon and hydrogen so that when these elements are bonded to oxygen their electron densities are diminished leading to deshielding and therefore higher chemical shifts – larger values of ppm along the x-axis – lines in the spectrum further to the left – or as is said in the parlance of NMR, lower field.

Here are some examples of the effect of electronegativity and the differing range of proton and carbon NMR. The carbon NMR signals for the same structures are in parentheses. The proton NMR signal for hydrogen on a carbon atom with no influencing electronegative atom will appear in the range of 0-2 (0-40) ppm, introducing into the structure electronegative atoms moves the ppm value to higher values as: $R-O-CHR_2$, 3-4 (50-90) ppm; $Br-CHR_2$, 2.5-4 (10-50) ppm; $Cl-CHR_2$, 3.5-4 (30-50) ppm; $F-CHR_2$, 4-5 (60-100) ppm.

Carbon NMR spectra, because of the low abundance of the necessary ^{13}C isotope, are measured in a manner leading generally to single lines for each carbon atom. As a consequence of the necessary instrumental method to compensate for the low abundance of ^{13}C, there is not a simple correlation of the intensity of the line to the number of carbon atoms contributing to that line. The intensity of the line therefore can not be used as a measure of the number of carbon atoms contributing to the signal. In proton NMR the intensity of the signal is, in contrast, a measure of the number of hydrogen atoms contributing to that signal, an exceptionally useful insight into chemical structure.

However, the single line for each type of carbon atom in the molecule is a great advantage in making a simple relationship of the carbon NMR spectrum to the molecular structure. This is beautifully seen in Figures 2.9-2.11 and you can advance your knowledge of chemical structure by counting the numbers of different carbon atoms in each structure and comparing your number to the number of lines in each carbon NMR spectrum.

Let's try this out on n-hexane, $CH_3CH_2CH_2CH_2CH_2CH_3$ where the structure shows three different carbon atoms and the carbon NMR spectrum in Figure 2.9 correspondingly shows three lines. Counting the number of different kinds of carbon atoms can be carried out for all molecules and the correspondence between the number expected from the structure and the number of lines in the carbon NMR spectrum is strong evidence (although not certain evidence) that the structure proposed is correct.

This game doesn't work with the proton NMR spectra in Figure 2.9-2.11 because these spectra are considerably more complicated than the carbon NMR spectra. There are two reasons for this: The far smaller chemical shift range in proton NMR cause overlap of signals. There is spin-spin coupling in proton NMR (see below).

Multitudes of web sites are brought up on Google by typing in proton and carbon NMR chemical shifts. In the data on these web sites we find that hydrogen and carbon atoms in a range of organic molecules are consistently found at predictable chemical shifts as discussed above. Detailed chemical shift data found in these sources inform us that the ppm range in going from CH_3, to CH_2 to CH precisely fits that proton NMR data seen in the three isomeric hydrocarbons in Figures 2.9-2.11.

Similarly in carbon NMR, the signals of such sp^3 hybridized carbon atoms, as in the three molecules in Figures 2.9-2.11, are seen in the higher field part of the spectrum, a region now expanded to the range of 50 ppm. Moreover, if we looked further into more detailed tables on the web, we would discover, as for the proton NMR, that the ppm values scale to larger values of ppm within this general range in going from CH_3, to CH_2 to CH. Clearly, NMR is a powerful molecular probe.

HE PROTON NMR SPECTRA IN FIGURES 2.9-2.11 show features not seen in the carbon NMR spectra. The various proton signals appear as multiple lines. In addition, as noted above, the number of hydrogen atoms contributing to each signal can be counted.

The reason this multiplicity of the signal is not seen in the carbon NMR spectra as noted in section 2.4, is that the instrumental procedures are organized in a manner to erase the information that leads to the multiplicity in order to maximize the signal. This is done to compensate for the low abundance of the ^{13}C isotope. Moreover, this instrumental procedure is also responsible for the absence of a connection between the strength of the signal and the number of carbon atoms contributing to the carbon NMR signal. We'll not go into the detail of how this is carried out but it is routine in carbon NMR where the chemical shift of the line for each carbon atom in the structure is the primarily sought information. The wide chemical shift range in carbon NMR normally allows separate chemical shifts for every structurally different carbon atom in the molecular structure.

The multiplicity phenomenon seen in the proton NMR is called spin-spin splitting and yields information about the relationships between structurally different hydrogen atoms that are separated by three bonds, such as H_a-C-C-H_b.

Every hydrogen atom in the structure of the molecule is spinning and therefore producing a small magnetic field, which is felt by those hydrogen atoms nearby in the structure. The magnetic field produced by each of these nearby hydrogen atoms has two states in the external magnetic field, and therefore can add to, or subtract from, the large magnetic field produced by the instrument's magnet. As a consequence, more than a single resonance condition to match the radio wave frequency is produced for the hydrogen atom that is affected by the nearby atom. Therefore multiple lines are seen in the proton NMR spectra in Figures 2.9-2.11.

For this effect to be observed in the NMR spectrum the nearby hydrogen atom that is affecting the magnetic field strength of a hydrogen atom near to it (and therefore the resonance condition of this hydrogen atom) must have a different chemical shift. It is therefore necessary, although it may not be sufficient, for the two hydrogen atoms to be structurally different. For example the three hydrogen atoms of a CH_3 group, which are structurally identical, do not interact in this way. In addition, the effect is transferred through chemical bonds so that, with some exceptions, only those hydrogen atoms connected to adjacent carbon atoms affect each others magnetic field.

The spin properties of the hydrogen atoms can take two states, which can be designated +1/2 and -1/2, or in another way in line with the external magnetic field (↑), or against this field (↓), with these spin states either adding to or subtracting from the magnetic field at an adjacent hydrogen atom. Therefore, if a hydrogen atom in the structure is in resonance at δ ppm in the absence of spin-spin coupling, a single hydrogen atom on an adjacent carbon atom would add to and subtract from the instrumental magnetic field determining that δ value. Instead of a single line at δ, one would find two lines of equal intensity, one slightly higher ppm and one slightly lower ppm than δ, that is, a doublet.

The distance in ppm between the two lines is called the coupling constant, J, which is related to the geometry and other factors between the nuclei which are coupled. In addition, the intensity of the doublet is identical to what the intensity of the single line at δ would be in the absence of the coupling. **Figure 2.12** represents the nature of the coupling with various numbers of hydrogen atoms where the simple arithmetic underlying the observed coupling can be understood: the number of lines in the splitting will be one more than the number of coupled hydrogen atoms with the relative intensities corresponding to Pascal's triangle as shown in Figure 2.12.

In addition, if the hydrogen atom under observation, that with a chemical shift δ, as noted above, is coupled with another hydrogen atom with chemical shift γ causing the δ signal to appear as a doublet, then the hydrogen atom with the γ chemical shift will also be split into a doublet by the δ chemical shift hydrogen. If these atoms are not coupled to any other hydrogen atoms in the structure, but just to each other, then the spectrum for these two hydrogen atoms will be what we would call a doublet of doublets – two doublets, one for the δ hydrogen atom and one for the γ hydrogen atom and both with identical coupling constants, J.

Although the hydrocarbon spectra in Figures 2.9-2.11 are not ideal molecules for demonstrating this phenomenon (see below), the general principles can be seen. Spin-spin coupling in proton NMR spectra can be quite complicated if the coupled hydrogen atoms have similar chemical shifts and especially if the hydrogen under observation is coupled to several different hydrogen atoms that differ from each other, as is the situation in the proton spectra in Figures 2.9-2.11. Nevertheless, we can see the consequences of spin-spin coupling in the two identical CH_3 groups in n-hexane (Figure 2.9). Both methyl groups have identical chemical shifts and would appear as a single line with a relative intensity of six hydrogen atoms near to 0.8 ppm in the absence of coupling. However, the two identical hydrogen atoms in the adjacent CH_2

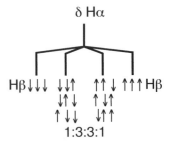

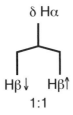

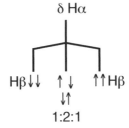

FIGURE 2.12 ▶

The Source of the Spin-Spin Splitting Patterns in Proton NMR Spectra

and so on following Pascal's triangle

```
        1
       1 : 1
      1 : 2 : 1
     1 : 3 : 3 : 1
    1 : 4 : 6 : 4 : 1
        etc.
```

(methylene) groups are located three bonds removed from each of the three hydrogen atoms in the methyl groups. The two prerequisites for spin-spin coupling are observed. The hydrogen atoms on the methyl group and the hydrogen atoms on the methylene group are structurally different and are located three bonds apart. Inspection of Figure 2.12 for the situation of splitting by two hydrogen atoms should pertain causing the methyl hydrogen atoms to appear as a triplet with an intensity ratio of 1:2:1, close to but not exactly what is seen.

The fact that the three lines for the methyl hydrogen atoms are not symmetrical arises from the fact that the chemical shifts of the coupled hydrogen atoms, the CH_3 and CH_2 groups, are not far enough apart. This problem reaches an extreme in the coupling of the two sets of structurally different CH_2 groups in this molecule (Figure 2.9). Here the chemical shifts are close enough that a clear coupling pattern is not observed.

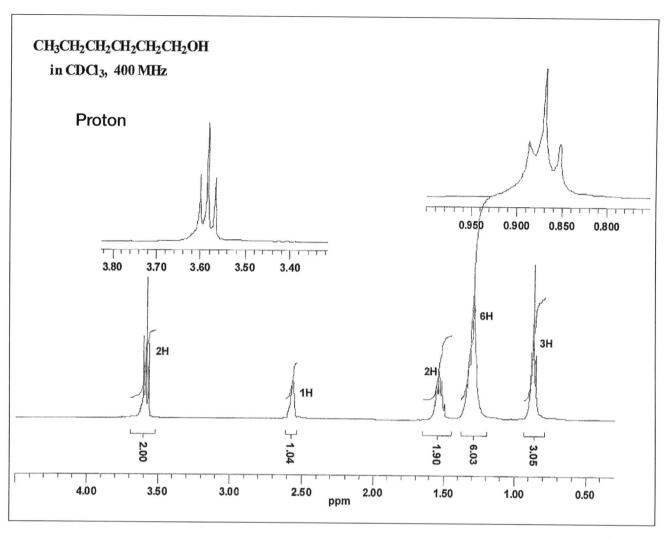

CH₃CH₂CH₂CH₂CH₂CH₂OH

in CDCl₃, 400 MHz

Proton

Proton NMR Spectrum of 1-Hexanol

In the proton NMR spectra of the three structural isomers shown in Figures 2.9-2.11, the most ideal spin-spin coupling is seen in 2,2-dimethyl butane (Figure 2.10) for the CH₂ group. The two identical hydrogen atoms on this carbon atom are coupled only to the adjacent methyl group, which from Figure 2.12 should give rise to four lines for the CH₂ hydrogen atoms in the ratio of 1:3:3:1, which is close to that observed centered at near to 1.25 ppm.

Figure 2.13 exhibits the proton NMR spectrum of CH₃CH₂CH₂CH₂CH₂CH₂-OH (1-hexanol) in which one observes both simpler and more complex spin-spin splitting patterns, which well demonstrate the chemical shift effects on this phenomenon. The distinctly different chemical shift centered at 3.59 ppm appears at low field arising from the high electronegativity of the oxygen atom and therefore deshielding of the adjacent CH₂ group. The large difference in chemical shift of the CH₂ group adjacent to the OH group with the adjacent CH₂ (1.51 ppm) causes the coupling to be close to ideal.

As predicted from the considerations in Figure 2.12, three lines should appear for this CH₂ group, a triplet, with a ratio of 1:2:1 as is seen. A less ideal triplet is seen for the terminal methyl (CH₃) group (near to 0.87 ppm), which is coupled to the adjacent methylene (CH₂) group. This chemical shift of this methylene group is buried in the multiplet of six hydrogen atom in the three CH₂ groups around1.3 ppm.

NMR yields amazing detailed information about chemical structures, doesn't it? Now let's try to use these ideas.

PROBLEM 2.1

The two infrared spectra shown below are of molecules that differ from the molecules whose infrared spectra are shown in the infrared spectra (Figures 2.6 and 2.7} discussion in the text. Draw chemical structures of two molecules whose spectra fit those shown below.

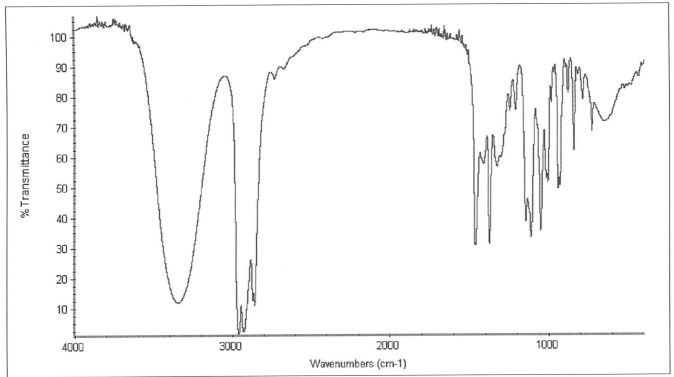

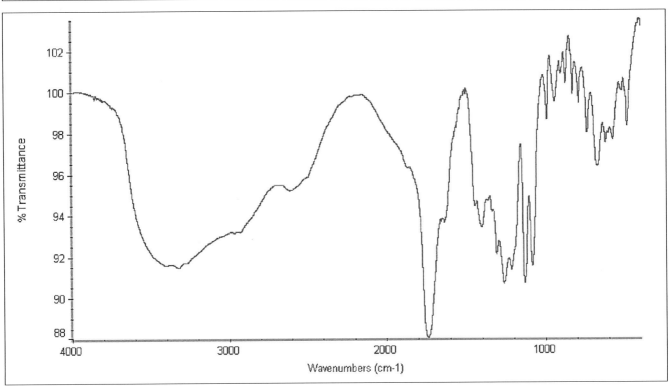

PROBLEM 2.2

Which different molecules would give identical electrospray mass spectra to the electrospray spectra shown in Figure 2.4?

PROBLEM 2.3

Draw the structure showing all atoms, all nonbonding electrons and designating any formal charges for A, T and C in Figure 2.5. For each, how many lines would appear in the Carbon NMR spectrum.

PROBLEM 2.4

Which molecule whose infrared spectrum was shown in this chapter would give the carbon NMR spectrum shown?

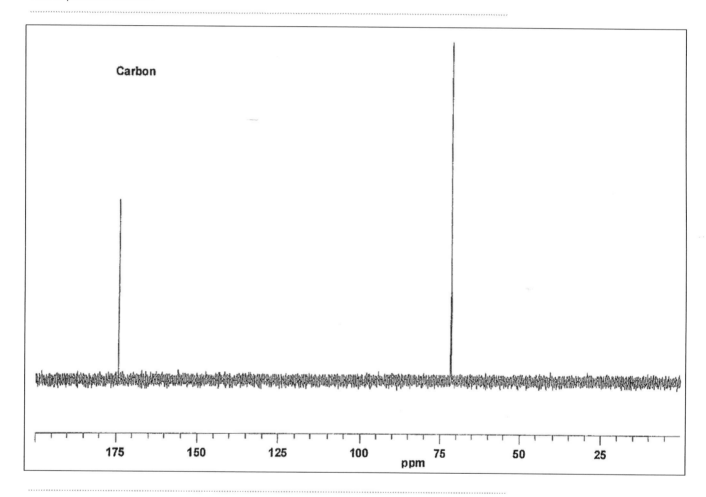

PROBLEM 2.5

Use the web site given in the text for infrared frequencies to identify as many of the bands as possible for the infrared spectrum of D-glucose shown in Figure 2.8.

PROBLEM 2.6

In problem 1.11 you were asked to create as many isomers of C_8H_{18} as possible. Predict how many lines would be obtained in the carbon NMR spectrum for each of these isomers. Estimate the relative chemical shifts of the differing carbon atoms.

PROBLEM 2.7

Assign the lines in the carbon NMR spectra for the hydrocarbons in Figures 2.9-2.11 to the carbon atoms in the structures.

PROBLEM 2.8
Explain why increasing values along the ppm scale in both proton and carbon NMR spectra correspond to lower field signals?

PROBLEM 2.9
In the proton NMR spectrum of 2-methylpentane in Figure 2.11 the lowest field signal appearing in the range of 1.55 to 1.60 ppm should have nine lines if the resolution where high enough. Why? Assuming the resolution of the proton NMR spectra were high enough to observe all the lines arising from spin-spin coupling, assign the number of lines that would be observed for all hydrogen atoms in the three hydrocarbons whose spectra are shown in Figures 2.9-2.11.

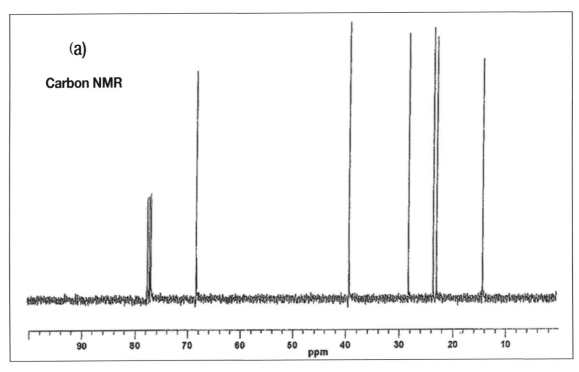

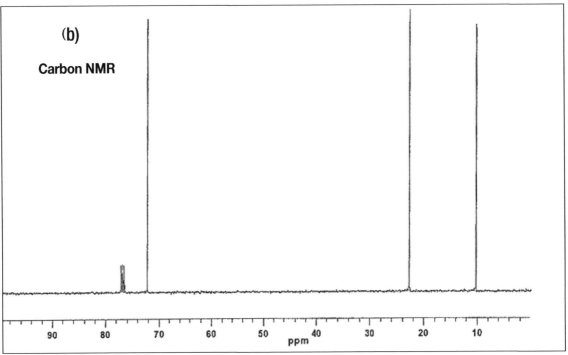

PROBLEM 2.10

Use the process outlined in Figure 2.12 to justify the relative heights of the lines arising from the spin-spin coupling leading to five lines as shown in Pascals' triangle.

PROBLEM 2.11

Connect each of the carbon NMR spectra (a, b, c, and d) shown below to one of the following molecular structures: racemic 2-hexanol (CH_3-CH(OH)-CH_2-CH_2-CH_2-CH_3); di-n-propyl ether; D-glucose; any isomer of tartaric acid.

Assign each line in the carbon NMR spectrum to a carbon atom in the chosen structure. The signal near to 78 ppm in (a) and (b) is from the carbon atom of the solvent, $CDCl_3$, deuterochloroform. The spectra for (c) and (d) were taken in water.

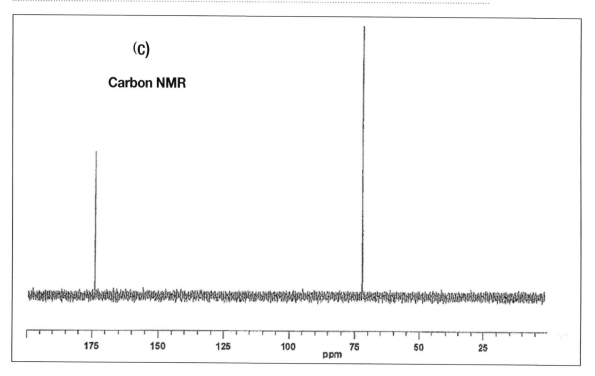

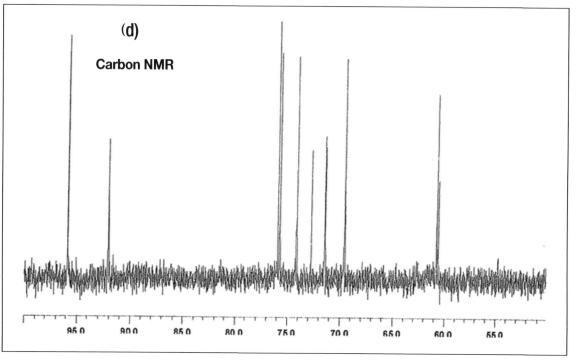

PROBLEM 2.12

Answer problem 2.11 for the proton NMR spectra shown below with the addition of addressing the spin-spin coupling assignments. For 1 and 2 the solvent is deuterochlorform, $CDCl_3$, while for 3, 4 and 5 the solvent is heavy water, D_2O, which leads to a single large signal between 4.5 and 5.0 ppm for HDO. How do you think HDO might be produced in the solutions of samples 3, 4 and 5?

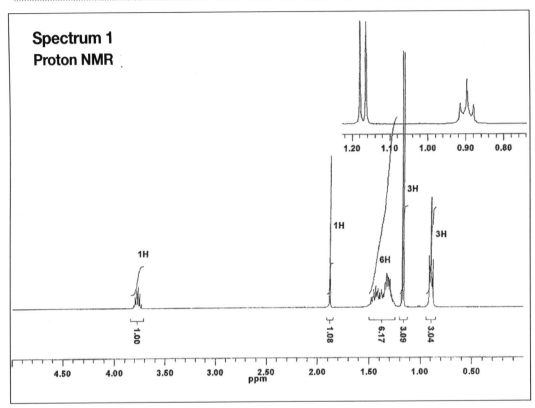

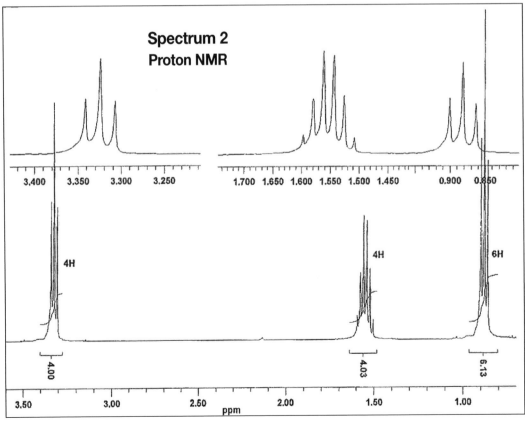

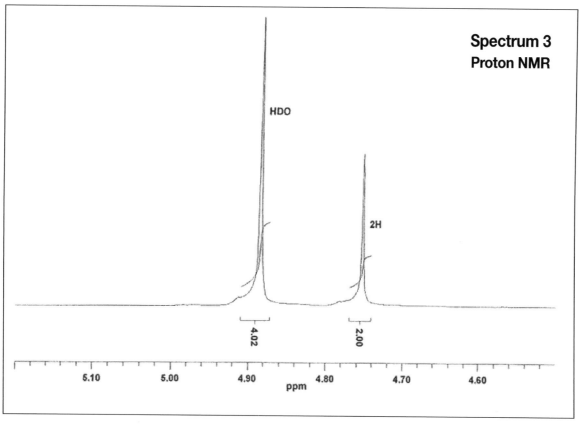

Spectrum 3
Proton NMR

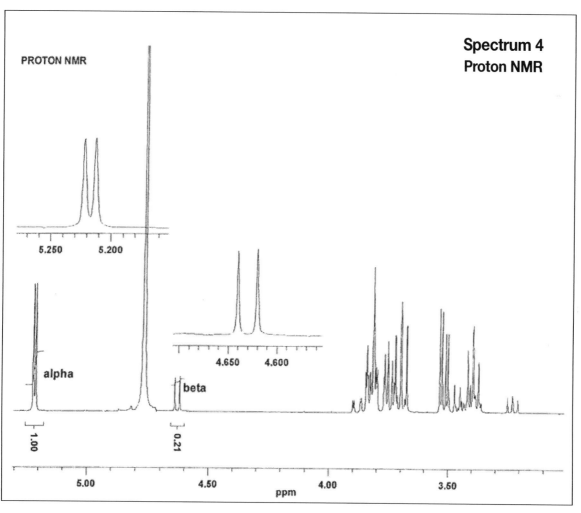

Spectrum 4
Proton NMR

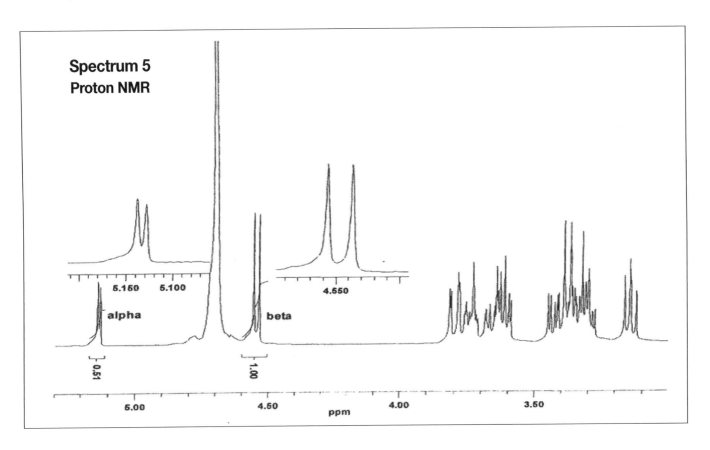

Spectrum 5
Proton NMR

5.150   5.100

alpha

0.51

4.550

beta

1.00

5.00   4.50   4.00   3.50

ppm

CHAPTER TWO SUMMARY of the Essential Material

THIS CHAPTER COVERS THE FUNDAMENTAL USE of the standard instrumental techniques used by chemists to probe the structures of organic molecules. Although not discussed, the first step in any question of molecular structure is that the sample be pure, that is, that there is only one kind of molecule present. Naturally, the molecule may be a mixture of conformational isomers, but what is meant by pure is that the no other molecules are present with different bonding arrangements or numbers and kinds of atoms and molecular weights, or in the situation of stereoisomers, no other molecules that are diastereomers. Purification of organic molecules can be accomplished, as it has been, even reaching into the nineteenth century by methods of distillation and crystallization but also by newer methods based on chromatography, which is not discussed.

The methods discussed in this chapter will give identical results for the enantiomers of chiral molecules although by using special techniques for NMR, enantiomers can be distinguished by the addition of chiral agents, a methodology also not discussed here.

Electron impact is the traditional method used in mass spectrometry and as can be seen by the examples in the chapter is typically used as a fingerprint of the molecule so that assignments of structure can be made based on comparisons to standard spectra. Sometimes one can obtain a molecular ion so that the molecular weight can be known. By use of mass spectrometers not discussed here, high resolution mass spectrometers, the m/z value can be known to enough significant figures to allow assignment of the numbers and kinds of atoms present. There have been many advances in mass spectrometry over the years and a very important one, electrospray mass spectrometry, which allows obtaining a signal for the unfragmented molecule, even of very large molecules and biological arrays, is discussed.

For most organic chemistry work infrared (IR) and nuclear magnetic resonance spectrometry (NMR) yield the most detailed information about molecular structure. The chapter goes over the basis of IR spectrometry and how it measures internal vibrational motions of the molecule, which occur with frequencies in the infrared region of the spectrum. The technique allows determination of specific arrangements of atoms and therefore can identify functional groups. IR can also be used as a fingerprint for the molecule, which is useful because in only special cases can every band in the IR spectrum be assigned to a specific part of the molecule. Nevertheless, using tables of typical IR frequencies, which appear in many places including on the web, one can often determine the presence or absence of particular structural features. Examples of this are discussed in the chapter.

The most powerful structural probe method is NMR, which is shown to involve the precession of atoms with certain spin properties. Fortunately for organic molecules, the hydrogen atom has this property and therefore is active in NMR spectra, which is therefore called proton NMR. Two important pieces of information come from proton NMR spectra, chemical shift and spin-spin coupling. The nature of chemical shift is understood to arise from the electron density around the hydrogen atom and this property is associated with connections between each of the hydrogen atoms in the structure with the atom they are bonded to. The electron density depends primarily on the concept of electronegativity, which plays an important role in determining chemical shift. In this way, different hydrogen atoms within a structure can show distinguishable chemical shifts in a predictable manner, a great aid to structural determination. For this it is important to understand the different kinds of hydrogen atoms in the structure of an organic molecule.

Spin-spin coupling in proton NMR can be understood from understanding the basis of the working of the instrument and how the magnetic field of the instrument, the external magnetic field, can be added to or subtracted from, by hydrogen atoms nearby to the hydrogen atom under observation. These nearby hydrogen atoms supply a small magnetic field associated with their spin, which is responsible for this effect. Specific rules allow understanding of how spin-spin coupling works and how it proves to be an important tool of proton NMR spectrometry.

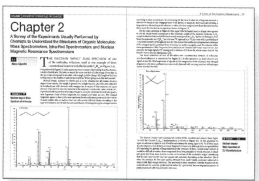

In addition, the signal strength for each hydrogen atom in a proton NMR spectrum gives the relative number of this hydrogen atom in the structure compared to other hydrogen atoms in this structure.

Although the common isotope of carbon, ^{12}C, is not active in the NMR, the ^{13}C isotope does have the property of giving an NMR signal. Although there is only about 1% of the active isotope for each ^{12}C, instruments have been designed that allow easy detection of the isotope but at the expense of giving up spin-spin coupling and also the correlation between the intensity of the signal and the numbers of each kind of carbon atom in the structure. A big advantage of carbon NMR is that the chemical shift range for the kinds of carbon atoms in organic molecules is at least ten times the range for the kind of hydrogen atoms in organic molecules. The chemical shift, as for proton NMR, is also greatly dependent on the electronegativity difference between the carbon atom observed and atoms it is connected to. This means that a carbon NMR spectrum normally gives a separate single signal for each carbon atom in the structure. When one determines the numbers of different carbon atoms in a chemical structure an important correlation can be made with the spectrum.

Using these three techniques in their various forms, MS, IR and NMR, the modern organic chemist has powerful techniques for determining the structures of organic molecules. The primary purposes as discussed in the chapter are for molecular weight (MS), functional group (IR) and molecular skeleton (NMR) although each technique overlaps with the others to some extent.

Not discussed is the use of the ultraviolet (UV) part of the electromagnetic spectrum, which is most useful for molecules with p-electrons, which can be promoted to higher energy states with the energy of UV light. UV spectrometry, which measures the absorbance of UV light, is well suited for the study of carbonyl containing molecules and molecules with double bonds and aromatic rings.

Chapter 3

From Galactosemia to the Properties of Six Membered Rings: An Introduction to the Mechanisms of Chemical Reactions

3.1

What is the childhood malady called galactosemia?

IN 1908, A MEDICAL DOCTOR REPORTED the case of a breast fed infant who was brought to him with new kinds of symptoms. The infant failed to thrive, and exhibited an enlarged liver and spleen. Moreover an unusual sugar, galactose, which is a stereoisomer of glucose, was found in the urine. The baby stopped excreting galactose in the urine when milk products were removed from the diet. This was the first report of galactosemia, a disease that can lead to mental retardation and death. The disease, which is devastatingly apparent shortly after the birth of a stricken infant, arises from an inability of the body's biochemistry to convert D-galactose to D-glucose.

This is a very serious matter considering that galactose is part of the lactose in mother's milk. Lactose is enzymatically broken down in the body to both glucose and galactose as shown in **Figure 3.1**. Normally, the galactose does not build up in the body because it is converted to glucose by another enzyme. But the absence of this enzyme blocks the conversion from galactose to glucose causing the disease.

It is apparent that the only change in going from D-galactose to D-glucose is at a single carbon atom. Using the CIP nomenclature from section 1.12 shows that the difference between life and death focuses on the inability to convert a single chiral carbon from (S) in D-galactose to (R) in D-glucose.

Why should a stereochemical difference of a single carbon atom in two structures, which are otherwise identical, wreak such havoc in the newborn's biochemistry? To answer this question we have to understand more about six membered rings, a molecular class that both glucose and galactose belong to.

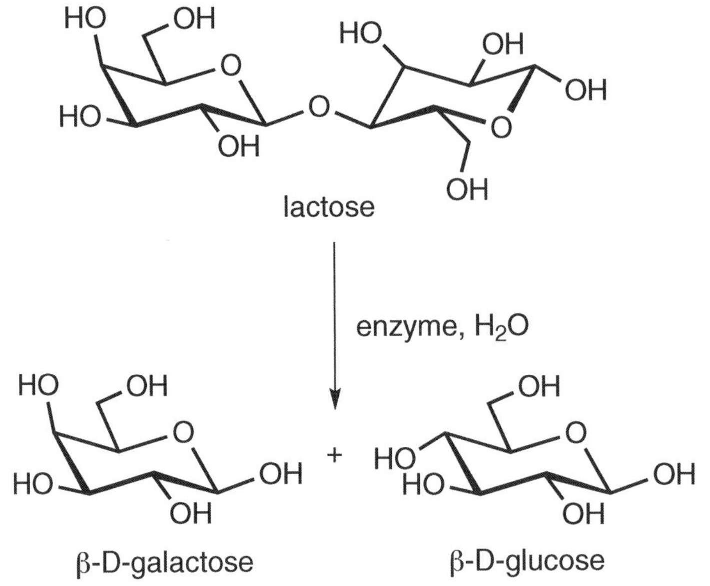

FIGURE 3.1 ▶

Lactose is enzymatically hydrolyzed to D-glucose and D-galactose.

lactose

enzyme, H_2O

β-D-galactose   +   β-D-glucose

PROBLEM 3.1

Redraw all the structures in Figure 3.1 showing all the atoms and lone pairs of electrons and testing for formal charge and the octet rule.

PROBLEM 3.2

D-Galactose and D-glucose, as shown in Figure 3.1, both have the same formula. Are these sugars constitutional isomers or stereoisomers? If the latter, what kind of stereoisomers are they?

PROBLEM 3.3

Explain the following statement: Even if these isomers were enantiomers there would still be expected to be a difference between them in vivo.

PROBLEM 3.4

Assign absolute configurations to the chiral carbon atoms that differ between D-glucose and D-galactose.

PROBLEM 3.5

Considering only β-isomers, how many stereoisomers would be possible for the structure of D-glucose.

U NDERSTANDING RINGS OF CARBON atoms brings us back to **Johann Friedrich Adolf von Baeyer**, born in 1835 in Berlin. He was the son of Major General Johann Jakob Baeyer, of the Prussian General staff, who was distinguished as the originator of the European system of geodetic measurements, a field which yields the distance between places on the earth and therefore certainly of great importance to military activities (consider aiming a cannon) among its other critical aspects.

Following on his father's interests, but shedding a couple of his given names, and moving from miles to **A**ngstroms, the son, Aldoph von Baeyer, as history records him, turned out to be one the great chemists of the nineteenth century, winning the Nobel Prize in 1905 for his contributions to the interaction between theory and practice. As the President of the Royal Swedish Academy put it on December 10, 1905: "Among the living research workers who have contributed directly or indirectly to the unique development of the tar-dyestuffs industry the place of honour goes to the Professor at Munich University, Aldoph von Baeyer, for his researches into the composition of indigo as well as into the triphenyl methane dyestuffs."

The industrial synthesis of dyes from components of coal, from tar, which Baeyer accomplished, was a major force behind the rapid growth of the chemical industry in the nineteenth century. Taking a vile, smelly, sticky, material like tar and finding a way to use it to make the colors that adorn our clothing was certainly a way to show the value of chemistry. That was a big enough accomplishment, but our interest in Baeyer focuses on another of his interests.

Baeyer was a student of Kekulé in Heidelberg working on his doctoral degree around the time Kekulé was formulating his ideas on tetracoordination of carbon (section 1.11). After Baeyer left Heidelberg in 1858 he took up his own independent research and although it did not involve any theory of

3.2

To understand the molecular basis of galactosemia we have to understand the nature of six-membered rings.

■ **Johann Friedrich Adolf von Baeyer**

bonding, these were the years when tetracoordinate carbon was proposed to take a tetrahedral geometry and when van't Hoff and LeBel proposed the structural basis of optical activity and isomerism, as discussed in section 1.11. This is certainly the foundation that Baeyer turned to in his attempt to judge the nature of rings of carbon, which as we'll see, will bring us back to understanding the dangerous nature of galactose compared to glucose.

It was becoming apparent after 1880 that molecules could be synthesized of rings of carbon atoms. Contributing to this synthetic effort on rings was a young man working in Baeyer's laboratory, **William Henry Perkin Jr.** Perkin's father had something in common with Baeyer. These great chemists set the stage in England and Germany, respectively, for a chemical industry stimulated by the ability to synthesize dyes from coal products. For Perkin Sr., this occurred in 1856 when he was only eighteen years old by his commercialization of the first synthetic dye, mauve. The result was an inexpensive purple dye so that the masses could wear a color that had before this been worn only by royalty. Perkin synthesized the dye when he was only seventeen!

■ **William Henry Perkin Jr.**

Although the son, Perkin junior, did not attain the stature of his father, he nevertheless was a distinguished chemist whose interest in rings of carbon stimulated Baeyer's interest in the subject. Perkin junior remembers Baeyer speaking to him in November, 1882 and pointing out that if substances from such rings are possible and have reasonable stability, how is it that they had never been met with in Nature?

Well, we now know that such rings are in fact found in nature as we've seen in glucose and for that matter in a myriad of other substances including the steroids and terpenes, molecules that we will be meeting later (Chapter 5). None of this was known when this question stimulated Baeyer's curiosity who saw all these ring structures as flat, so that the required tetrahedral angle would be impossible to attain. For example, if a six-membered ring of carbon were flat, a regular hexagon, the angle at each atom would be 120°, quite a deviation from the tetrahedral angle.

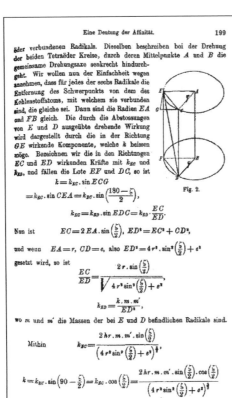

■ **Hermann Sachse attempted to use geometry and trigonometry to prove his theory.** [Z. phys. chem. 11, 185-219) There is no known image of Sachse.

Baeyer published a paper in 1885, three years after his conversation with Perkin junior. The paper (translated from the German) is entitled: *Theory of Ring Closure and the Double Bond.* Baeyer pointed out the following: *The four valencies of the carbon atom act in the directions that connect the center of a sphere with the corners of a tetrahedron and that form an angle of 109° 28' with each other. The direction of attraction can experience a deviation that will, however, cause an increase in strain correlating with the degree of this deviation."*

This statement is now the source of the Baeyer Strain Theory, one of his important legacies to ideas of chemical structure that finds application to the current day in the chemical sciences far beyond studies of cyclic molecules. We know very well now that molecules become increasingly unstable when the internal angles deviate from the tetrahedral angle for any tetracoordinate carbon. If the deviation from near to 109 degrees to 120 degrees bothered Baeyer, consider a ring of three carbon atoms where the angle would have to be 60 degrees or a ring of four carbon atoms where the angle would be 90 degrees. Yet rings of these sizes in which each carbon atom in the ring is tetracoordinate are well known. And the strain that Baeyer proposed so many years ago is well understood to be playing a role in the properties of these so-called strained rings. In Chapter 5 (section 5.9) we'll see that nature takes advantage of this strain in a three membered ring to accomplish a critical step on the path to the synthesis of the steroids and in Chapter 12 (section 12.13) we'll see how ideas of ring strain play a role in synthetic chemistry.

But another idea involving model building would arise to relieve six-membered carbon rings of the strain that Baeyer hypothesized. Use of models is central to the understanding of chemistry and we've suggested more than once that you buy a set of ball and stick models. Van't Hoff used models to help convince chemists that tetrahedral carbon could indeed lead to isomerism and optical activity (section 1.11) and his ideas led to explanations of many difficult to understand experimental results. In fact a young chemical assistant in Berlin, **Herman Sachse**, in 1890, is credited with being the first to realize by using ball and stick models that rings need not be flat at all, but could be puckered, or bent. Sachse realized there was a way that six membered rings could be formed without Baeyer's strain. Such puckered six-membered rings in particular allowed the angles about the six carbon atoms to come very close to the ideal tetrahedral angle.

Unfortunately, Sachse died quite young (at the age of 31) in 1893 and never came to see his seminal insights and predictions confirmed. He foresaw with his models that cyclohexane, C_6H_{12}, (hexane for six and cyclo for ring) would take various non-flat shapes and that it was likely that these shapes could change into each other without breaking any chemical bonds. As discussed in section 1.13 such differing shapes are conformational isomers and one of Sachse's conformational predictions is now known as the chair form of this six-membered ring of carbon atoms.

Hermann Sachse was prescient in understanding how the tetrahedral carbon angle could be incorporated in a stable structure for a six-membered ring. But he presented this insight using mathematical arguments and published this in a manner that organic chemists then, and now, find very difficult to follow. And so his insights were ignored for many years by Baeyer and others to a time well beyond Sachse's death. It was not until 1918 that Ernst Mohr, professor of chemistry at Heidelberg, published his findings on the X-ray crystallographic studies of diamond and found support for Sachse's picture of cyclohexane.

Diamond is made up of an infinite number of six-membered rings of carbon atoms fused together so that there are no hydrogen atoms. A portion of the diamond structure is shown in **Figure 3.2.**

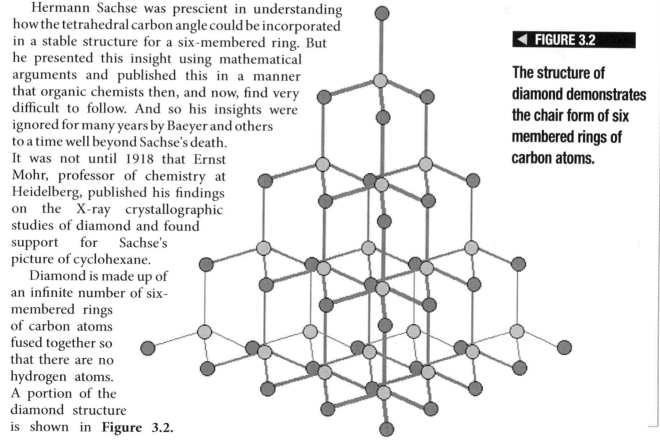

◀ **FIGURE 3.2**

The structure of diamond demonstrates the chair form of six membered rings of carbon atoms.

The shape taken by these rings is precisely that predicted by Sachse for what we now call the chair form of cyclohexane, a fact realized by Mohr. But Sachse, who would have been only 56 years old at this time and well able to take pleasure in his discovery, had already been dead for 25 years. James Moore of Rensselear Polytechnic Institute pointed out the following highly informative web site: https://webspace.yale.edu/chem125/125/history99/6Stereochemistry/Baeyer/Sachse.html

PROBLEM 3.6
Why would Baeyer have had far less of an objection to a ring of five carbon atoms than a ring of six carbon atoms in which every carbon atom in both rings was tetracoordinate?

PROBLEM 3.7 (Look ahead to section 3.5 for problems 3.7-3.9).
Although Baeyer introduced the idea of strain caused by deviation from the tetrahedral angle, what we now call angle strain, there are other kinds of strain. Considering that covalent bonds contain electrons causing nearby bonds to repel each other, can you understand a strain that would be found in all rings of carbon that are flat, a strain called torsional strain arising from eclipsing (Section 1.13, Figure 1.11) of covalent bonds? Draw a structural diagram of a flat cyclohexane ring that demonstrates how torsional strain might arise. Use a Newman projection, introduced in section 1.13 to demonstrate the torsional strain that would be present in hexagonal cyclohexane.

PROBLEM 3.8
How is torsional strain related to the stability of the gauche and anti conformations compared to eclipsed conformations seen in Newman projections and discussed in section 1.13?

PROBLEM 3.9
The simplest organic molecule in which torsional strain is found is ethane, $H_3C\text{-}CH_3$. The central bond connecting the two carbon atoms allows the two methyl groups (CH_3) to rotate easily, millions of times each second, as if pinwheels on a stick. Use Newman projections to show how this rotation causes the ethane molecule to exist momentarily in repeating conformations with and then without torsional strain.

▪ **Odd Hassel**

3.3

It took many years for chemical science to accept the idea that rings did not have to be flat and further that acceptance of this idea could explain many aspects of the chemical behavior of cyclic molecules. An important advance, as is often the situation in science, was the use of a new kind of instrument applied to the problem.

IN SPITE OF MOHR'S WORK (Figure 3.2) it was not until many years later that chemists fully realized that the isolated cyclohexane structure was Sachse's structure. This is characteristic of the chemical profession and in fact of human beings in general who often will not accept an idea until it proves useful in some way. This path to utility for Sachse's idea began with the efforts of a great Norwegian physical chemist working at the University of Oslo in the years just before World War II. **Odd Hassel** carried out convincing experiments using electron diffraction of vapors, which revealed that the structure Sachse predicted for cyclohexane was correct and moreover that Sachse's mathematical modeling studies had also correctly predicted the potential flexible nature of the structure of cyclohexane. Hassel's experiments were able to verify that the cyclohexane ring was changing extremely rapidly, millions of times each second, between two identical forms, which we now call chair conformations.

Although **Odd Hassel** was the son of a medical doctor and had studied the scientific prerequisites as an undergraduate that would have allowed him to follow his father's profession, he became intrigued by chemistry as early as high school and this became his major at the University of Oslo. Many students in those years when Hassel graduated from university, in 1920, went to Germany for further study and Hassel was no different, receiving a doctorate degree from the University of Berlin in 1924. It was in Germany, where he was introduced to x-ray crystallography, that he became interested in the molecular structure of molecules, an interest that became the focus of his career at the University of Oslo after he returned to Norway in 1925. This background and the ability to afford an electron diffraction instrument for his laboratory in 1938 led to his investigations on cyclohexane.

Normally, considering Hassel's background as a student in Germany, his important results on cyclohexane structure would be expected to be published in German, one of the three important languages for scientific exchange at that time. But Hassel reported his findings instead in Norwegian journals that were not widely read. This was the time of World War II and Germany's invasion of Norway, perhaps had something to do with Hassel's decision. In fact, the University of Oslo was shut down by the Germans during part of the war and Hassel spent a couple of years in a Nazi concentration camp.

Consistent with our discussion about acceptance of ideas and demonstration of utility, Hassel's confirmation of Sachse's picture of the structure of cyclohexane did not have an influence on organic chemistry until several years later.

In 1949 a young assistant lecturer at Imperial College in London was invited by a distinguished organic chemist interested in steroids to lecture for a year at Harvard University in the United States. **Derek Barton** took up the offer, which led to what is generally regarded as the origin of conformational analysis (section 1.13).

■ **Derek Barton**

Barton was invited to Harvard by **Louis Fieser**, an organic chemist who was working on steroids. Barton was aware, from his wide reading of the literature, of the work of Hassel and realized that many mysterious aspects of the chemistry of steroids could be understood by abandoning the two dimensional way organic chemists were still looking at these structures and instead substituting a three dimensional picture based on the geometric properties of cyclohexane rings.

Barton published his insight in 1950 in a short paper ("Experientia" Volume VI/8 page 316). **Figure 3.3** exhibits, with the permission of Springer Verlag, the first page of this paper, which is now regarded as a classic that led to an explosion of interest in this stereochemical approach to understanding the properties of organic molecules. Organic chemists were not much interested in the work of Sachse or of Mohr or of Hassel until they could see something practical come out of it. This Barton supplied when he showed how these ideas on the shape of six-membered rings could explain properties of molecules found in nature.

The Nobel committee in Sweden awarded their prize to Barton and Hassel in 1969 for a collaborative insight in which the two participants, a physical chemist and an organic chemist did not work directly together but came together through the literature. Barton felt that his success was based in his scholarly approach to the literature, to the time he spent in reading and Barton's students told stories about how Barton insisted that the progress of science required a thorough knowledge of the literature.

■ **Louis Fieser.**

Sachse was the first to realize from his studies that in a non-planar form of cyclohexane (now designated the chair form) that the two hydrogen atoms on each of the six carbon atoms were geometrically distinguished. Six hydrogen atoms are parallel and stick up and down from the structure, three up and three down, while the other six hydrogen atoms are splayed out from the structure, also three tending upward and three tending downward. The former six hydrogen atoms we now designate as axial, while the six latter hydrogen atoms we now designate equatorial. In Barton's paper (Figure 3.3) the term equatorial has its current meaning while polar was used for what we now term axial.

STUDIORUM PROGRESSUS

The Conformation[1] of the Steroid Nucleus

By D. H. R. BARTON[2], Cambridge, Mass.

In recent years it has become generally accepted that the chair conformation of cyclohexane is appreciably more stable than the boat. In the chair conformation it is possible[3,4] to distinguish two types of carbon-hydrogen bonds; those which lie as in (Ia) perpendicular to a plane containing essentially the six carbon atoms and which are called[3] *polar* (p), and those which lie as in (Ib) approximately in this plane. The latter have been designated[3] *equatorial* (e).

The notable researches of HASSEL and his collaborators[5,6] on the electron diffraction of cyclohexane derivatives have thrown considerable light on these more subtle aspects of stereochemistry. Thus it has been shown[5] that monosubstituted cyclohexanes adopt the equatorial conformation (IIa) rather than the polar one (IIb). This is an observation of importance for it indicates that the equatorial conformations are thermodynamically more stable than the polar ones. It should perhaps be pointed out here that although one conformation of a molecule is more stable than other

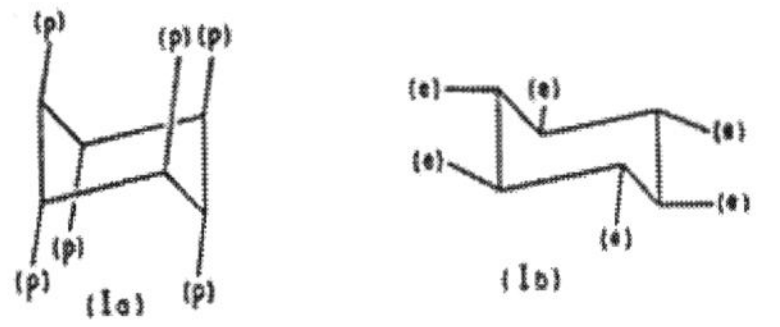

possible conformations, this does *not* mean that the molecule is *compelled* to react as if it were in this conformation or that it is rigidly fixed in any way. So long as the energy *barriers* between conformations are small, separate conformations cannot be distinguished by the classical methods of stereochemistry. On the other hand a small difference in free energy content (about one kilocal. at room temperature) between two possible conformations will ensure that the molecule appears by physical methods of examination and by thermodynamic considerations to be substantially in only *one* conformation.

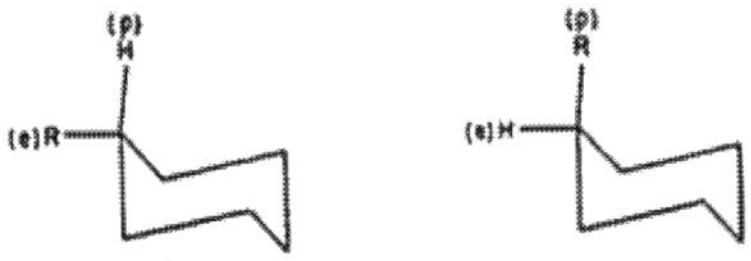

The equatorial conformations are also the more stable in both cis-1:3- and trans-1:4- disubstituted cyclohexanes[1]. Thus cis-1:3-dimethylcyclohexane adopts the diequatorial conformation (IIIa) rather than the dipolar one (IIIb), whilst trans-1:4-dimethylcyclohexane exists as (IVa) rather than (IVb).

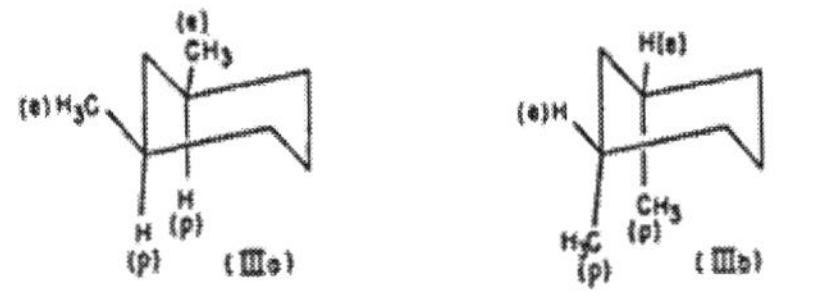

Thermodynamic calculations[1] show that trans-1:2-dimethylcyclohexane takes up the diequatorial conformation (V; R=CH$_3$) rather than the dipolar one (VI; R=CH$_3$). For cis-1:2-disubstituted cyclohexanes there are two possible conformations. In both of these one of the substituents forms an equatorial bond, the other a polar one. Since these differences in thermodynamic stability between equatorial and polar conformations are presumably of steric origin[1], it would appear logical to make the larger substituent form the equatorial bond.

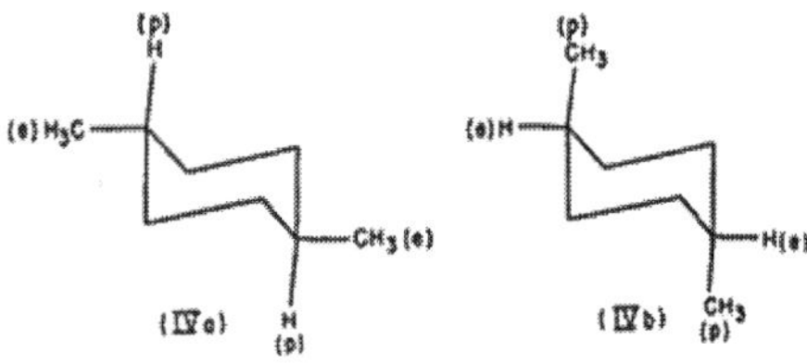

Considerations of the same type can be extended to 2-substituted cyclohexanols. Thus[2,3] the cis-alcohols (VII; R= alkyl), on equilibration by heating with sodium, furnish almost entirely the trans-isomers (VIII; R= alkyl). In the former one substituent is polar, one equatorial; in the latter both are equatorial. The same conclusion on relative stability is reached from a consideration of thermochemical data[4]. Similarly[5] the 2:6-disubstituted cyclohexanol (IX), with two equatorial and one polar substituents, is isomerized to (X) on equilibration. The situation is the same[3] with the bicyclic trans-α-decalol. Here the isomer (XI) is isomerized to (XII) on equilibration.

A consideration of the conformations[6] (XIII) and (XIV), assumed by the steroid nucleus when the A/B ring fusion is respectively trans- and cis-, provides a striking illustration of the usefulness of the concept of

---

[1] The word conformation is used to denote differing strainless arrangements in space of a set of bonded atoms. In accordance with the tenets of classical stereochemistry, these arrangements represent only one molecular species.

[2] Harvard University Visiting Lecturer, 1949–50, Harvard University, Cambridge 38, Mass.

[3] C. W. BECKETT, K. S. PITZER, and R. SPITZER, J. Amer. Chem. Soc. 69, 2488 (1947).

[4] O. HASSEL's nomenclature[5] is different, but the distinction remains the same.

[5] O. HASSEL and H. VIERVOLL, Acta Chem. Scand. 1, 149 (1947).

[6] See O. HASSEL and B. OTTAR, Acta chem. Scand. 1, 929 (1947) for a summarizing paper and references to earlier work.

[1] C. W. BECKETT, K. S. PITZER, and R. SPITZER, J. Amer. Chem. Soc. 69, 2488 (1947).

[2] G. VAVON, Bull. Soc. Chim. [4], 49, 937 (1931).

[3] W. HÜCKEL, Ann. Chem. 533, 1 (1937).

[4] A. SKITA and W. FAUST, Ber. Dtsch. Chem. Ges. 64, 2878 (1931).

[5] G. VAVON and P. ANZIANI, Bull. Soc. Chim. [5], 4, 1080 (1937).
In connection with the conformations of poly-substituted cyclohexanes it should be mentioned that O. BASTIANSEN, O. ELLERSEN, and O. HASSEL, (Acta chem. Scand. 3, 918 [1949]) have recently shown that the five stereoisomeric benzene hexachlorides assume, in agreement with our general argument, those conformations which have the maximum possible number of equatorial carbon-chlorine bonds.

[6] Conformations (XIII) and (XIV) are unambiguous representations of the steroid nucleus provided that rings A, B, and C are chairs. This is almost certainly true for a trans-A/B ring fusion (compare the X-ray evidence of C. H. CARLISLE and D. CROWFOOT (Proc. Roy. Soc. A 184, 64 [1945]) on the conformation of cholesteryl iodide) and a similar situation, at least in solution, probably holds for a cis-A/B fusion. The justification for the latter has been more

---

FIGURE 3.3 ▶

The First Page of Barton's Nobel Prize Winning Paper on the Conformation of Cyclohexane Rings

S ACHSE'S MATHEMATICAL AND MODELING formulations (section 3.2) also predicted an interconversion process, or flipping of chair forms, which although leading to identical cyclohexane conformations, would cause the axial and equatorial hydrogen atoms to change places.

3.4

The Conformational Properties of Cyclohexane

Figure 3.4 demonstrates, using two differing structural representations, the widely used chair form, and a combination of two Newman projections (equivalent representations of the conformation of a cyclohexane ring), that exchange of the six equatorial hydrogens with the six axial hydrogens occurs as the ring undergoes what is described as a flip. We now understand this flipping to describe the characteristics of cyclohexane and as well other six-membered rings such as the sugars discussed in this chapter.

The conformational process shown in Figure 3.4 takes place through a series of intermediate higher energy conformations, designated as half-chair, twist and boat. These conformational states will not be described here but you can get a chance to explore these conformational states by using your model set and answering problem 3.10.

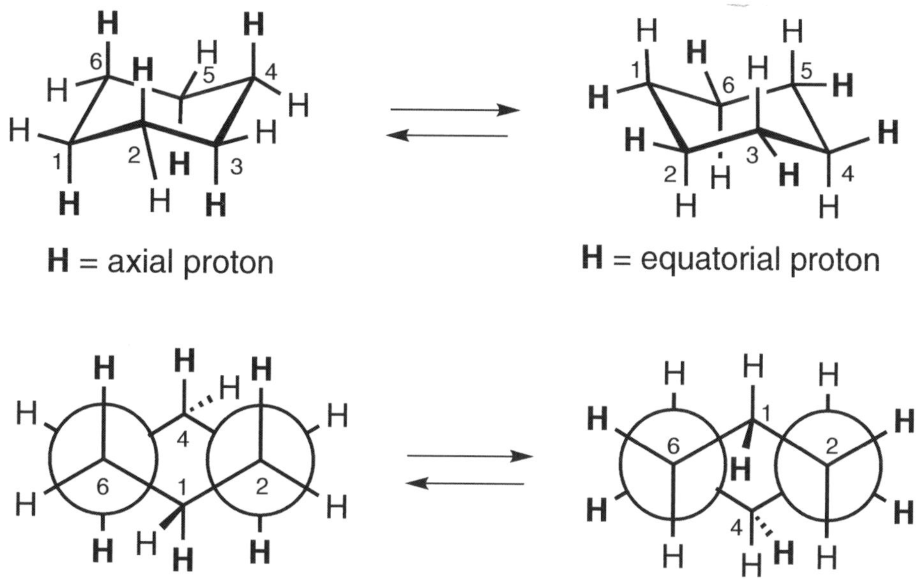

H = axial proton   **H** = equatorial proton

◀ **FIGURE 3.4**

The Interconversion of Axial and Equatorial Hydrogen Atoms as a Consequence of Chair Conformation Flipping in Cyclohexane

In cyclohexane, the flip of the ring shown in Figure 3.4 exchanges two conformations that are identical in energy, that are, in fact, simply identical - superimposable. But this is not the situation if one of the hydrogen atoms is replaced by a larger group, such as for one example of many, a methyl group, CH_3. **Figure 3.5** shows that the substituted cyclohexane, which we name methyl cyclohexane, in which one of the twelve hydrogen atoms is replaced by a methyl group, also undergoes the flip, but that the process produces conformations that are clearly different, which are not superimposable. The identical structural information is again presented in two different ways in Figure 3.5 – again, first in the conventional chair form structures widely used and second as a combination of two Newman projections (section 1.13).

The two structures shown in Figure 3.5 certainly fit the definition of stereoisomers (section 1.3). They have an identical formula, C_7H_{14}, and identical bonding, that is, all the atoms are connected to each other in the same way in both molecules. In addition, the fact that the two molecules can interconvert rapidly via ring flipping justifies our use of the designation conformational isomers.

These rapidly interconverting stereoisomers are not mirror image of each

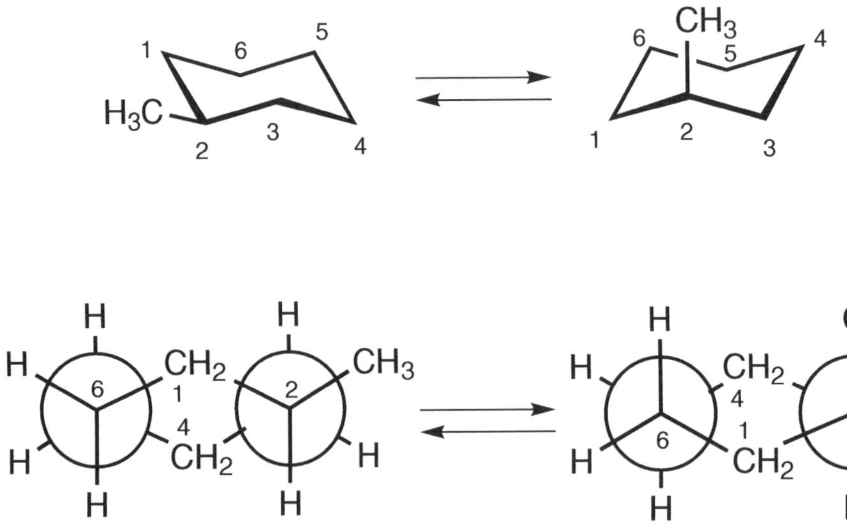

The Relative Stability of Axial And Equatorial Methyl Cyclohexane

other and therefore they must be diastereomers. We've learned that diastereomers (sections 1.7 and 1.8) differ from each other in all circumstances, which means that one will be more stable than the other. Let's discover the source of the difference between these conformational isomers, one with the CH_3 group equatorial and the other with the CH_3 group axial.

In the Newman projections (Figure 3.5) one is looking at the molecule with carbon atom 1 facing the viewer so that carbon atom 4 is furthest away. The bonds making up the front carbon atoms and the rear carbon atoms (which are not shown) of the two circles are C_5-C_6, and C_2-C_3, so that carbon atoms 3 and 5 are hidden behind carbon atoms 2 and 6 respectively. The flipping of the ring shown with the chair forms in Figure 3.5 is then reproduced in the changes shown in the two Newman projections. Let's now see how the Newman projections allow judgment of the relative energies of the equatorial versus the axial CH_3 (methyl) groups. To judge the relative energies, although it may seem surprising, we need to turn our attention to the conformational properties of n-butane ($CH_3CH_2CH_2CH_3$).

PROBLEM 3.10
Draw and/or make a model of the chair form of cyclohexane as in Figure 3.4. The conformational motion shown for cyclohexane, which allows the six axial hydrogen atoms to exchange positions with the six equatorial hydrogen atoms, also allows a conformation that is what is called a boat form. In this boat form carbon atoms 1 and 4 could be imagined as the bow and stern of a boat. Using Newman projections along the bonds between carbon atoms 2 and 3 and as well 5 and 6, judge if the boat conformation suffers from torsional strain. Use your models to try, also, to show the twist and half-boat conformations noted in this section. The names aptly apply. Skip ahead to problem 11.8 in Chapter 11 for further understanding of these structures.

PROBLEM 3.11
Might it be possible to accomplish a separation of conformational isomers using low temperatures: and if yes why? What kinds of conformational isomers could not be separated by such an experiment?

PROBLEM 3.12
Construct the chair form of cyclohexane from your set of models, and draw the structure as well, showing clearly the six equatorial and six axial hydrogen atoms attached to the ring carbons. Use the model to flip the ring and observe how the equatorial and axial hydrogen atoms switch positions. Are the flipped forms stereoisomers?

FIGURE 3.6 EXHIBITS THE STRUCTURE of n-butane as a sawhorse projection and as Newman projections around the central carbon-carbon bond, the bond between the two CH_2 groups in the structure. Three important conformations of this molecule are shown.

In section 1.13 the complex structure of tartaric acid made it difficult to predict the relative energies of the various conformations shown as Newman projections in Figure 1.10. However, in a molecule like n-butane, assignment of relative energies of conformational isomers becomes more straightforward.

In general it is reasonable that gauche and anti conformations, where the overlap of groups is reduced, will likely be of lower energy, that is, more favorable than conformations in which the groups are eclipsed. Why are eclipsed conformations of higher energy than gauche or anti conformations?

One answer is fundamentally the same as the answer to the question as to why tetracoordinate carbon takes a tetrahedral shape – electron pair repulsion as developed by Gillespie and Nyholm in the VSEPR theory (section 1.4). Keeping the electrons in the four bonds around a carbon atom as far apart as possible is attained by a tetrahedral array. Keeping the electrons in the bonds on adjacent carbon atoms as far apart as possible is attained by avoiding eclipsing of these bonds. The avoiding of eclipsing of electron pairs in bonds is a fundamental reason why eclipsed conformations are of higher energy than gauche and anti conformations. Eclipsed bonds cause **torsional strain** (Problems 3.7-3.9).However, another reason for the higher energy of eclipsed conformations depends on the size of the groups that are attached to the eclipsed bonds. This interaction is called **steric or van der Waals strain**.

Although, in a molecule such as n-butane, the gauche and anti conformations (Figure 3.6), both avoid eclipsing bonds, the two conformations greatly differ in the relationship of the two methyl (CH_3) groups. The CH_3 groups take up space— they have what are called van der Waals radii. When atomic groups or atoms come closer than their van der Waals radii they repel each other, and this repulsive force increases greatly as the distance decreases. When the distance between the methyl groups is greater than the summation of the van der Waals radii, the interaction is not repulsive and can even be attractive, depending on the space between the

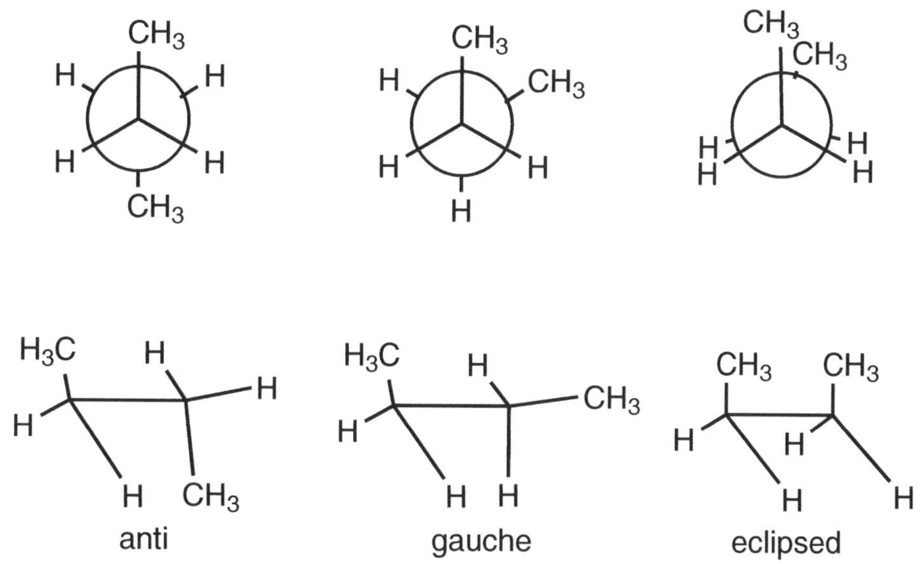

3.5

The conformational properties of n-butane permit judging the relative energies of the equatorial versus axial methyl cyclohexane: torsional and steric strain.

◄ **FIGURE 3.6**

Newman Projections of n-Butane for the Eclipsed, Gauche and Anti Conformations

interacting methyl groups. Although the distance between the groups is least in the eclipsed conformation the distance remains close enough in the gauche conformations to cause steric strain depending on the size of the interfering groups.

Understanding these attractive and repulsive forces that depend on distance between atoms or groups of atoms contributed to an early understanding of the relationships between gases and liquids and an early contributor to this understanding was the great Dutch physicist, **Johannes Diderik van der Waals,** who won a Nobel Prize in 1910 for his contributions. van der Waals' name can be added to van't Hoff (section 1.11) as another nineteenth century scientist from the Netherlands who laid the foundation of the chemical sciences.

The van der Waals radius for methyl is about two Ångstroms causing the two methyl groups in both the eclipsed and gauche conformation of n-butane to repel each other, a repulsion that is not present in the anti conformation where the two methyl groups are further apart than the sum of their van der Waals' radii. The repulsion is worse in the eclipsed conformation, which also suffers from torsional strain. But this repulsion is still present in the gauche conformation, which does not suffer from torsional strain.

Let's now return to the Newman projections of the methyl substituted cyclohexane ring in Figure 3.5 and relate the equatorial and axial CH_3 groups to what we've learned about the relative conformational energies of n-butane.

Looking along the C_2-C_3 bond in the Newman projection of the axial conformation in Figure 3.5 shows a gauche interaction between the CH_3 group on C_2 and the C_4 carbon of the ring with its two attached hydrogen atoms, which is called a methylene group (CH_2). This gauche interaction is relieved in the equatorial conformation where the Newman projection shows these groups, the pendant CH_3 on carbon 2 of the ring and the CH_2 (at position 4) group, to enjoy an anti relationship, therefore removing all the van der Waals' strain of the axial conformation.

We have just discovered the fundamental source for the preference for equatorial conformations in six membered rings, an analysis that applies equally well to OH groups on six membered rings and, therefore, to the properties of glucose and galactose.

The first page of Barton's paper (Figure 3.3), is the prediction of what can be expected about the conformational isomers shown in Figure 3.5, an expectation that Barton informs us comes from Odd Hassel. We quote the critical sentence in Barton's paper:

Thus it has been shown[5] that monosubstituted cyclohexanes adopt the equatorial conformation (IIa) rather than the polar (axial) one (IIb).

The two isomers in Figure 3.5 are stereoisomers that can be rapidly changed into one another by motions around bonds and therefore are conformations (footnote 1 in Barton's paper (Figure 3.3)). The word rapid hardly describes the time scale involved in the process. A fraction of a millionth of a second is more than enough for multiple exchanges back and forth between the two conformations.

What Hassel published in 1947 and what Barton is reminding us about in the sentence quoted above is that the conformational diastereomers in Figure 3.5 differ greatly in their stability. The equatorial conformation on the left in both presentations in Figure 3.5 is greatly preferred, that is, has a lower energy. Although the two conformations are changing back and forth into each other rapidly, if one could stop the clock and look at the ensemble of molecules, at the array of molecules in the sample, you would find that overwhelming numbers are in the form of the equatorial conformational isomer.

In **Figure 3.7** the structures of β-D-glucose and α-D-glucose are reproduced on the left and their conformational isomers on the right. From Hassel's finding and all discussed above, the structures on the right would be of higher energy, that is, unstable compared to those on the left - too many axial hydroxyl groups

■ **Johannes Diderik van der Waals**

in the conformational isomers on the right. This means that in a sample of a large number of these molecules, only a minute fraction of these molecules would exist as the conformations shown on the right. Given the choice, the ring will flip so that a group that is pendant to a six-membered ring will move into the equatorial conformation. We now can begin to understand the difference between glucose and galactose.

In the light of this discussion the difference between β-glucose and β-galactose is seen as the difference between an equatorial and an axial hydroxyl group on one of the carbon atoms of these otherwise identical sugar molecules (Figure 3.1). As Barton discussed (Figure 3.3), Hassel's experiments had shown that an equatorial group pendant to the cyclohexane ring is more stable than an axial pendant group. Barton and Hassel would immediately have predicted that the more stable of these two diastereomers is glucose. Now the question arises: so what if the six membered ring structure of glucose is more stable than the comparable ring of galactose? Why should that have anything to do with the health of a newborn?

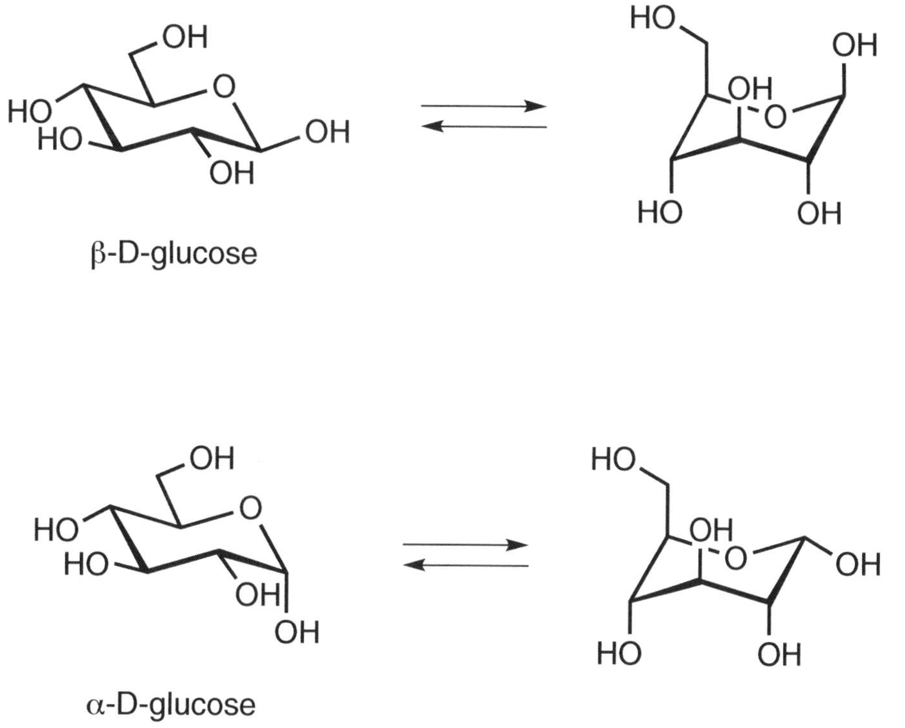

β-D-glucose

α-D-glucose

◄ FIGURE 3.7

Ring flipping of the chair forms of β and α-D-glucose.

PROBLEM 3.13
In comparing the conformational structures in Figures 3.5 and 3.6, redraw the structures to discover how the atoms of n-butane find parallel in some of the atoms in the methyl cyclohexane structure.

PROBLEM 3.14
Describe the basis of the strain that causes equatorial methyl cyclohexane to be favored over the axial isomer.

PROBLEM 3.15
In Figure 3.6 three conformational isomers of n-butane are shown as Newman projections. Are there other conformational isomers not shown here and if so present these isomers as Newman projections, assign each one to the designation, eclipsed,

anti and gauche, and predict their relative energies, showing which isomers suffer from torsional and/or van der Waals' strain.

PROBLEM 3.16
Why are there two gauche butane interactions for axial methyl cyclohexane although the Newman projection in Figure 3.5 shows only one?

PROBLEM 3.17
Explain why only one of the eclipsed forms of n-butane is chiral.

PROBLEM 3.18
The gauche conformation of n-butane shown in Figure 3.6 is chiral. However, the two enantiomers form a rapidly exchanging racemic mixture. Explain this statement. Does this rapid exchange between mirror image conformations apply also to one of the eclipsed forms of n-butane?

PROBLEM 3.19
(a) Explain the fact that although α and β-D-glucose differ by the configuration of a single OH group, which is equatorial in β-D-glucose and axial in α-D-glucose, a ring flip can not interconvert these diastereomers.

(b) Explain the fact that although D-glucose and D-galactose differ by the configuration of a single OH group, which is equatorial in D-glucose and axial in D-galactose, a ring flip can not interconvert these diastereomers.

(c) Contrast your answers to (a) and (b) with the fact that a ring flip can interconvert the two diastereomers that differ by an equatorial and an axial CH_3 group in Figure 3.5.

PROBLEM 3.20
Are glucose and galactose configurational or conformational isomers and why? Answer this question for α and β-D-glucose. Are the conformational isomers shown in Figure 3.5 for methyl cyclohexane, diastereomers or constitutional isomers?

PROBLEM 3.21
Give several examples of two molecular structures that are configurational diastereomers and several examples of two molecular structures that are conformational diastereomers. For each isomeric set, show the change necessary for their interconversion.

ALTHOUGH GLUCOSE IS A SIX-MEMBERED RING of five carbon atoms and an oxygen atom the ideas discussed in sections 3.2 to 3.5 apply just as well. Glucose takes a chair conformation with equatorial placement of all the pendant groups in the β-isomer.

We learned that axial and equatorial pendant groups can exchange positions by a flipping of the ring (Figure 3.5). Would it be possible to convert galactose into glucose by a conformational motion, to convert the axial hydroxyl group in galactose to the equatorial position, as this group exists in glucose (Figure 3.1)? The answer is clearly no. Although flipping the ring would convert the axial hydroxyl group, which is causing the difference from glucose, into an equatorial position, such a flip would cause all the other equatorial pendant groups in galactose to become axial making the situation even worse in forming a molecular structure of even higher energy (**Figure 3.8**).

Converting galactose into glucose requires a chemical change in which bonds are broken and remade. Such a change is not a conformational change but rather a configurational change, a change that can not be accomplished simply by twisting about bonds but instead must involve breaking of bonds.

The kinds of twisting motions involved in conformational motions, such as in Figure 3.5, involve far smaller energy changes than the breaking of chemical bonds and therefore, bond breaking is far slower, more difficult. Changing galactose into glucose is not easy and in fact this is why an enzyme is necessary to catalyze the process, to lower the energy of activation for the process, to lower the energy that is otherwise necessary to accomplish the change. We'll learn more about energy of activation in Chapter 6 (section 6.11) but when the energy of activation is too high, when the required enzyme is not available, or is unable to do its job because of a biochemical abnormality, galactosemia follows.

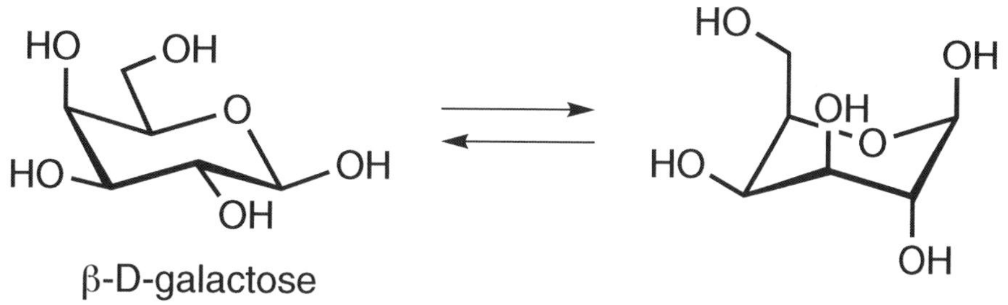

3.6

Why should the difference between an equatorial and an axial bond on a six-membered ring sugar molecule be the difference between life and death for a stricken infant?

β-D-galactose

◀ **FIGURE 3.8**

Ring Flipping between Chair Forms of β-D- Galactose

We've seen diastereomers of glucose, over and over again in Figures 1.1, 1.3, 1.4 and 1.6, which are configurational rather than conformational. α and β-Glucose are configurational diastereomers. You can not change one into the other by flipping the six-membered ring. Try it and convince yourself. To interconvert α and β-glucose a chemical bond has to be broken and then remade. In other words, the interconversion of α and β-glucose requires a chemical change, the making and breaking of chemical bonds.

In summary, a chemical change is necessary to interconvert the diastereomers α and β-glucose and a chemical change is necessary to interconvert the diastereomers glucose and galactose. However, while α and β-glucose are necessary for life's functions, galactose is deadly to life. When the interconversion from galactose to glucose is blocked a deadly disease ensues. What makes galactose so deadly! To answer this question, it would be helpful if we learn more about sugars in general and glucose in particular.

Andreas Marggraf

PROBLEM 3.22
Use a set of ball and stick models to explore the strain necessary in the conformational changes necessary for flipping a cyclohexane ring.

PROBLEM 3.23
Flipping methyl cyclohexane is an easy conformational motion that can take place millions of times a second while flipping of galactose or glucose is so difficult as to not occur. Why?

PROBLEM 3.24
Equatorial groups, such as three methyl groups, in alternate positions along a cyclohexane ring, such as in positions 1, 3 and 5 all point in the same direction. By pointing in the same direction is meant pointing either above or below the approximate plane of the ring. While three methyl groups in positions 1, 2 and 4 can not all point in the same direction unless one is axial. Draw structures to support this statement.

PROBLEM 3.25
Why is a cyclohexane ring with methyl groups all pointing in the same direction and located at positions 1, 2 and 4 easier to flip back and forth than a cyclohexane ring with methyl groups pointing in the same direction at positions 1, 3 and 5?

3.7

A background in the sugars, including their history, will help to set the stage for understanding the fundamental difference between glucose and galactose and therefore, galactosemia.

SUGAR IS LONG KNOWN TO HUMAN BEINGS. It is reported that when the Persians invaded India 500 years before the birth of Christ they discovered *the reed which gives honey without bees*. Apparently the Persians greatly appreciated the taste because about 1200 years later when the Arabs invaded Persia sugar cane was growing in what is now Iran and a technology existed for extracting the sweet principle of the plant, what we now know to be sucrose (**Figure 3.9**). You can note in Figure 3.9 that sucrose consists of two different carbohydrate molecules bonded together and that the one on the left looks like a unit in starch (Figure 1.1), a unit based on D-α-glucose. The carbohydrate on the right is D-fructose, which is not a six but rather a five-membered ring. We're going to see in Chapter 8 that D-fructose plays a critical role in the catabolism of glucose.

War follows war and after another approximately 500 years, European Christian crusaders returning from their battles talked of a new spice they'd come across, which led to a large increase of sugar consumption in Europe. Production of sugar increased over the intervening centuries while the price of sugar fell. In the fourteenth century sugar is reported to have sold in current prices in the range of $50 a pound - quite a bit more than we pay now. In fact it was so expensive that only nobility could afford it and, in an ironic twist, it seems that rotten teeth came along with great wealth, quite a change from today. Having rotten teeth, shown by a mouth full of black teeth, was a sign that you were wealthy and in fact people were said to be proud of their black teeth, which makes me wonder about tooth aches in those days. That must have been the time when sugar's effect on dental health first manifested itself by excessive consumption.

Getting used to sugar caused great difficulties for the French during the Napoleonic Wars of the 1800s. Their source of sugar was the West Indies, and British naval power shut down this connection. Moreover, the climate in Europe was not suited to grow sugar cane. Lucky for the French sweet tooth, a German chemist, **Andreas Marggraf,** who was trained in chemistry and medicine to replace his father as chief apothecary of the court in Berlin, and who made a distinguished career for himself in the 1700s, had discovered in 1747 that sugar

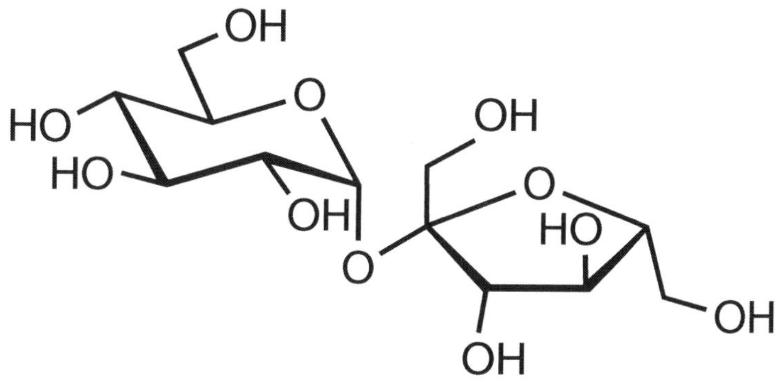

sucrose

could be obtained from beets. This discovery led to Napoleon's development, many years later, of an industry based on obtaining sugar from beets, a hardy plant suitable to the European climate. Marggraf is also credited with discovering glucose in raisins and therefore setting in motion a chemical interest in this most fundamental of the sugars. However, many years passed until the chemistry of glucose was finally fully understood.

PROBLEM 3.26
Redraw the structure in Figure 3.9 including all atoms and all non bonding electrons. Assign (R) or (S) configuration to each chiral atom. Test the structure for formal charges and obeying the octet rule.

THE PROBLEM CHEMISTS encountered all during the nineteenth century with understanding sugar, and glucose in particular, arises from the complexity of the structures (Figures 1.1 and 3.9). Although the formula for glucose was early determined to be $C_6H_{12}O_6$, chemists were faced with the question of how the carbon atoms were arranged. The structures in Figure 1.6, when we were first introduced to the idea of constitutional isomerism, show how the same number of carbons can be arranged in different ways in molecules of identical formula. Chemists asked how the six carbon atoms in glucose were arranged.

The answer to this question arose by using chemical reactions that transformed glucose into related molecules in which the six carbon atoms were discovered to be arranged in a row, just as in the isomer of hexane that boils at 69° C shown in Figure 1.6. Chemists then assumed that the six carbons in glucose were also in a row, that is, in a linear arrangement.

This was the classic comparative approach taken to understanding the structures of new chemical compounds: use a chemical reaction to convert a molecule that was understood, or thought to be understood, into the new unknown molecule or a molecule derived from the unknown molecule and then assume that the two molecules shared a common structural feature; convert the molecule of unknown structure into a molecule of known structure, which also indicated a shared common feature - at least until proven otherwise. And proven otherwise was not uncommon in those early days when structural uncertainty was common.

Structural problems, especially of complex molecules, which are now solved

3.8

Solving the wide variety of problems glucose presented, to come to a full understanding of its structure, was a central theme in the development of chemistry.

almost instantly by the methods discussed in Chapter 2, could take decades to come to a firm structural conclusion using comparative methods. We have the classic examples of Hassel's electron diffraction experiment on cyclohexane (section 3.3) and Bijvoet's x-ray scattering in the determination of enantiomeric structure (section 1.12) of how technological advances continue to play a critical role in advancing the chemical sciences.

The six carbon atoms making up glucose are not bonded to fourteen hydrogen atoms as in normal hexane (n-hexane Figure 1.6) but rather to twelve hydrogen atoms. In addition, glucose has six oxygen atoms bonded to these six carbon atoms. How in the world was all this arranged? Eventually it was determined that five of these oxygen atoms made up five hydroxyl groups, OH, and one each of these hydroxyl groups was on each of five carbon atoms. The sixth oxygen atom differed. The carbon atom to which it is attached is in an oxidized state. What's that?

We've seen how oxygen can be in an oxidized state in Figure 1.8 where we find the grouping of atoms H-C=O and in Figure 1.9 where we find a grouping of atoms HO-C=O. The former is a grouping designated an aldehyde group, while the latter is the arrangement of atoms corresponding to a carboxylic acid. Eventually chemists determined that the sixth oxygen atom, the one not part of a hydroxyl group, was part of an aldehyde group, H-C=O.

When oxygen is bonded to carbon, the fewer hydrogen atoms bonded to that same carbon atom the higher the oxidation state. We, therefore, order the following arrangements from lower oxidation state to higher oxidation state: (refer to the table of functional groups in the inside cover of the book) hydroxyl; aldehyde; carboxylic acid. Any chemical reaction that causes a change to a higher oxidation state is an oxidation while change to a lower oxidation state is a reduction. We'll see many examples of oxidation and reduction reactions in the course of studying organic chemistry. Much more will be said about oxidation and reduction as we go further into the book. However, now something has to be said about the concept of functional groups in organic chemistry, which is the subject of the next section.

PROBLEM 3.27

Making certain to introduce no formal charges and obeying the octet rule, propose as many structural isomers as you are able to for the formula of glucose, $C_6H_{12}O_6$, but rejecting all possibilities that are not chiral, a characteristic of glucose demonstrated by the experimental observation of optical activity.

CHEMISTS IN THE NINETEENTH CENTURY had realized that certain groupings of atoms within a molecule, such as shown on pages xiii to xv of the book, were what are called functional groups: hydroxyl, aldehyde, carboxylic acids among many other arrangements of atoms we will be coming across are just three of numerous functional groups found in organic molecules.

The value of recognizing certain arrangements of atoms as a functional group is that chemists have learned by long experience that a functional group has similar properties no matter what structure it is part of. For example, the hydroxyl groups (OH) in glucose or in cellulose or starch or for that matter ethanol, CH_3CH_2OH, or cholesterol or estradiol (Figure 5.1 and 5.2) behave in a similar manner, have similar chemical properties and therefore allow predictions of physical-molecular characteristics and chemical reactions — very useful indeed. We'll introduce more functional groups as you need to know about them.

3.9

We need a slight diversion from our story to understand the concept of functional groups.

PROBLEM 3.28
Go back through the text up to this point and list the names and structures of every functional group shown. Compare your findings to the page of functional groups In the inside cover of the book.

THE FIRST PROBLEM IN SOLVING THE STRUCTURE of glucose was to understand its cyclic structure.

Eventually, chemists in the nineteenth century determined that the arrangement of atoms in **Figure 3.10** corresponded to the overall structure of glucose, which you'll notice does not resemble the structure you first encountered for glucose in Figure 1.1. In fact, it soon became apparent to those nineteenth century researchers that the structure in Figure 3.10 can not be the whole story, can not completely describe glucose. Even more difficult problems stood in the way of a complete understanding of this molecule, a rather important molecule considering it is nature's choice as the receptacle of the sun's energy responsible for life on earth.

There were two general problems that arose from the structure of glucose determined by the decades of effort made in the nineteenth century (Figure 3.10).

The first is an experimental problem. In a surprising finding, a chemist working at the end of the nineteenth century, C. J. Tanret, a Finnish chemist interested in natural materials, was able to isolate two highly purified crystalline forms of natural glucose with different optical rotations (section 1.10). Usually when the optical activity of a pure chemical dissolved in a suitable solvent is measured the value does not change with time.

But in the situation of dissolving one or the other crystalline forms of glucose in water the optical activity ($[\alpha]_D$) slowly changed from the initial value. The initial value for one of the isomers, which we now know to be α-D-glucose (Figure 1.7), is +112.2° and changes slowly over a period of hours to +52.6° and the initial value for the other isomer, which

3.10

There were two kinds of problems with the first structure proposed for glucose. One of these problems could not be solved until it was realized that glucose was a cyclic molecule, a structural feature related to galactosemia.

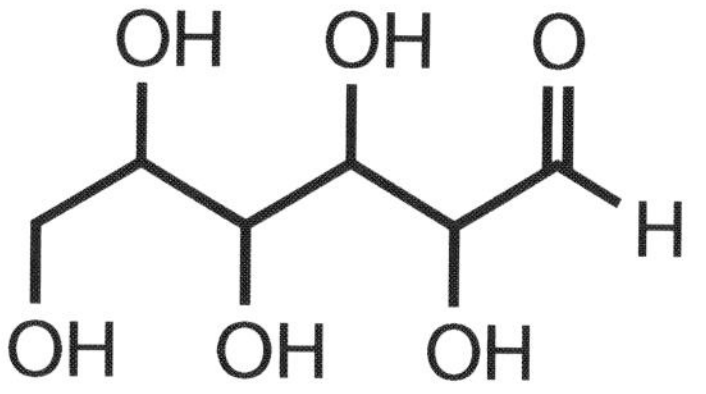

◄ **FIGURE 3.10**

Basic Connections between the Atoms In Glucose In Two Dimensions

we now know to be β-D-glucose (Figure 1.6), changes slowly from +18.7° also to +52.6°. This long known slow change of the optical activity of glucose solutions in water is called mutarotation.

The two forms of glucose came to be called α and β, although no one knew how these isomers differed. Over the early years of the twentieth century, chemists gradually came to realize that these were six-membered ring isomeric structures, which slowly interconverted in aqueous solution. The formula for both isomers was identical to that of the molecule that glucose was thought to be, $C_6H_{12}O_6$, which is the open form shown on the left in **Figure 3.11** (although the stereochemical information shown in this figure was not known at that time).

It is now known that glucose exists almost entirely in the six-membered ring structure, which when dissolved in water undergoes opening and closing, a process that allows a molecule of α-glucose to change into a molecule of β-glucose and vice versa (Figure 3.11). This bond breaking and making allows the configurational change to interconvert α and β-glucose (configurational diastereomers). At any instant, each of the uncountable numbers of molecules in a solution in water are in either of the closed forms, α or β-glucose, with each isomer occasionally changing into the other through an open form. This is the source of the slow change in optical rotation, which is known as mutarotation. Finally an equilibrium state is reached where although the two forms are still

FIGURE 3.11 ▶

Ring Opening and Closing Mechanism for Mutarotation in D-Glucose

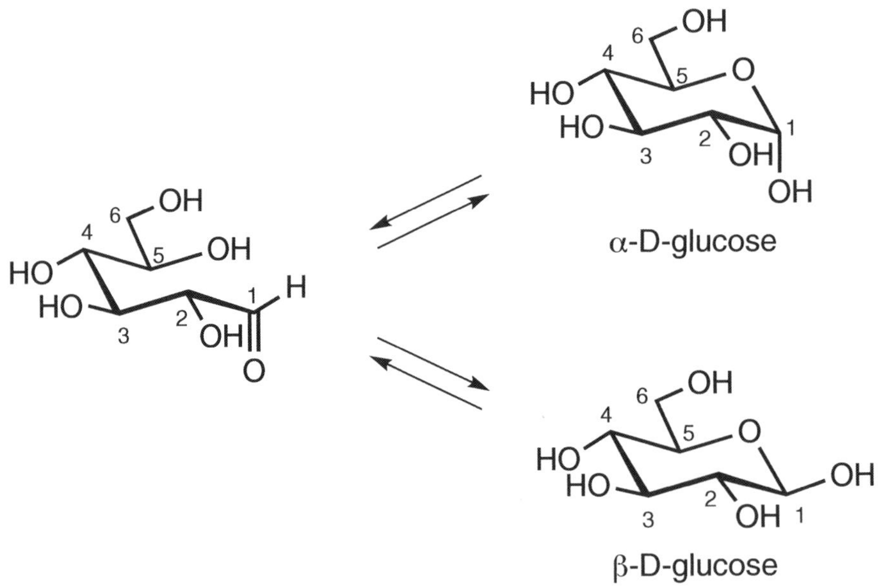

interconverting, the ratio of the two remains constant. This ratio determines the final optical activity, 52.6°.

The open form only exists for a very short period of time. It is an intermediate state between the more stable closed forms. It is a transitory state between the closed forms of glucose.

PROBLEM 3.29

Pure α-D-glucose has a specific rotation of +112°, while β-D-glucose has a specific rotation of +18.7°. If either of these forms of glucose is dissolved in water the specific rotation changes to a value of +52.7°, which does not change further. Describe what is happening in quantitative detail.

PROBLEM 3.30

Explain how the diastereomeric change between axial and equatorial positions on a six membered ring that is pointed to in problem 3.29 takes place without a conformational motion.

IT WAS REALIZED THAT GLUCOSE, as known in the 1880s in the open form (Figure 3.11), was a molecule in which isomerism had to play a large role. Why?

Glucose has six carbon atoms, which are numbered in Figure 3.11. Only carbon atoms 2, 3, 4 and 5 in the open form with four different groups around each of these carbon atoms are capable of isomerism. As we've seen, the difference between D-glucose and D-galactose arises from isomerism of carbon 4 (Figure 3.1). The source of this isomerism is the positions of two of the four different groups on carbon-4, the H and the OH.

Glucose is a great deal more complicated than tartaric acid, a molecule that only has two carbon atoms capable of isomerism. Moreover, both of these carbon atoms in tartaric acid are identically substituted (Figures 1.8 and 1.9). This is not the situation in glucose. All the isomeric carbon atoms in the structures of the open forms of glucose (Figures 3.10 and 3.11) differ from each other. If you inspect the groupings around each of the four carbons in the structure you discover that each carbon is differently substituted. A carbon NMR spectrum of either α or β glucose will show six different chemical shifts, which will be different for each diastereomer (problem 2.12, section 2.4).

In a general rule for the situation in glucose or other molecules with multiple carbon atoms, which differ from each other each, with four different groups, the number of isomers can be shown by the methods of statistics to be 2^n. In glucose, $n = 4$ for the four carbon atoms that can contribute to isomerism in the open form. Theoretically there should therefore be 16 isomers.

Which one of the 16 is the glucose found in nature? The answer to this question and many other questions about glucose and about other sugars comes from the work of one chemist, **Emil Fischer**. It's not that other chemists were not involved in the important problem of figuring out the structural mysteries of the stereoisomers of glucose, it is simply that Fischer did the experiments that put the whole story together. In fact, Fischer, in a tour-de-force of chemical research leading to the award of the Nobel Prize in 1902, (the second Nobel Prize given (the first was to van't Hoff (section 1.11)) was able to bring order to the structures of the various carbohydrates and to sort out these stereoisomeric possibilities for glucose.

Starting from small molecules using chemical reactions he had discovered and perfected, Fischer synthesized all sixteen of the glucose isomers. Imagine that he had these isomers in sixteen separate flasks on his laboratory bench, each with their individual melting and optical activity properties. Your imagination is probably close to the truth.

The first step in Fischer's success in connecting the physical samples to stereochemical structures depended on his finding a way of representing the possible structures in an orderly manner **(Figure 3.12)**.

Fischer created these two dimensional representations of the three dimensional reality of the molecular structures to be able to easily see the complex stereoisomeric issues he was dealing with. He could manipulate these projections in his mind as long as he followed certain rules. These rules can be understood from inspection of the Fischer projections of the tartaric acid stereoisomers in Figure 1.9. By comparison of the three dimensional structural models in Figure 1.9 with the Fischer projections one can determine that the backbone carbon chain must be vertical and that the horizontal groupings pendant to the backbone are coming out of the page toward the viewer. This means that the projections must not be turned over. The only movement possible is that of rotating the projection in the plane of the paper by 180 degrees.

When Fischer created the representations of the stereoisomeric possibilities for glucose shown in Figure 3.12 the cyclic structures of the sugars were not clear, and certainly not understood. Nevertheless, representing the various sugars as

3.11

The second problem in determining the structure of glucose brings us closer to understanding galastosemia

■ Emil Fischer.

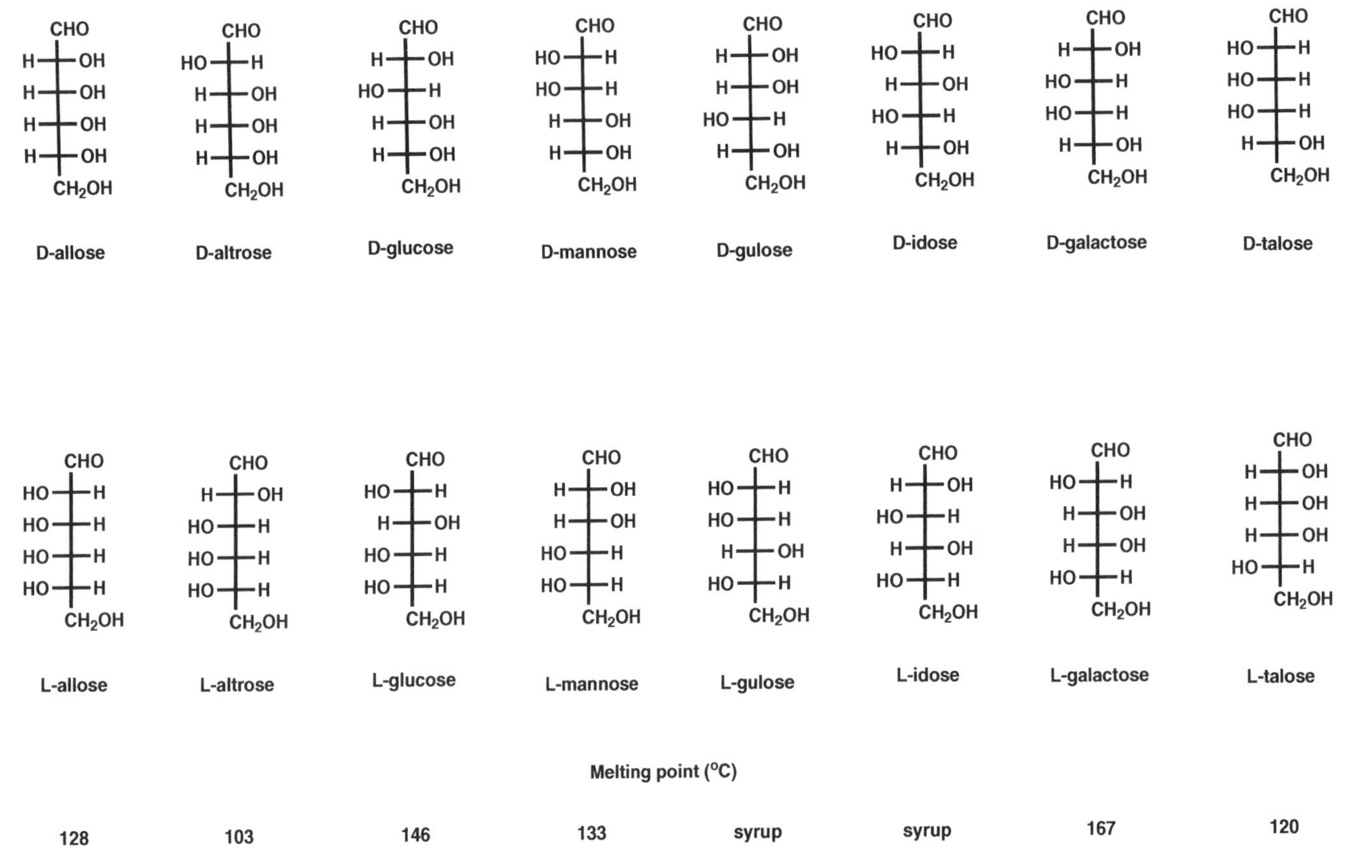

D-allose — CHO; H—OH; H—OH; H—OH; H—OH; CH₂OH

D-altrose — CHO; HO—H; H—OH; H—OH; H—OH; CH₂OH

D-glucose — CHO; H—OH; HO—H; H—OH; H—OH; CH₂OH

D-mannose — CHO; HO—H; HO—H; H—OH; H—OH; CH₂OH

D-gulose — CHO; H—OH; H—OH; HO—H; H—OH; CH₂OH

D-idose — CHO; HO—H; H—OH; HO—H; H—OH; CH₂OH

D-galactose — CHO; H—OH; HO—H; HO—H; H—OH; CH₂OH

D-talose — CHO; HO—H; HO—H; HO—H; H—OH; CH₂OH

L-allose — CHO; HO—H; HO—H; HO—H; HO—H; CH₂OH

L-altrose — CHO; H—OH; HO—H; HO—H; HO—H; CH₂OH

L-glucose — CHO; HO—H; H—OH; HO—H; HO—H; CH₂OH

L-mannose — CHO; H—OH; H—OH; HO—H; HO—H; CH₂OH

L-gulose — CHO; HO—H; HO—H; H—OH; HO—H; CH₂OH

L-idose — CHO; H—OH; HO—H; H—OH; HO—H; CH₂OH

L-galactose — CHO; HO—H; H—OH; H—OH; HO—H; CH₂OH

L-talose — CHO; H—OH; H—OH; H—OH; HO—H; CH₂OH

Melting point (°C)

| 128 | 103 | 146 | 133 | syrup | syrup | 167 | 120 |

FIGURE 3.12 ▲

Fischer Projections for the Sixteen Stereoisomers, Eight Diastereomers each with an Enantiomer, One of Which is D-Glucose

open chains in his projections led to no disadvantage. As Fischer gradually was able to assign the particular chemical samples on his laboratory bench, each with their melting and optical activity properties, to the various stereoisomeric representations in Figure 3.12, the ring structures would later be defined as well because there is a unique relationship between each of the stereoisomers in Figure 3.12 and its cyclic form. Once the stereochemistry of the open form is known the structure of its closed form is known as we've seen already in Figure 3.11.

Fischer's projections are a perfect example of the use of models, but hardly the first. For example when van't Hoff and Le Bel came to understand the role of tetrahedral carbon in isomerism, van't Hoff made paper models and sent them to leading chemists of the day greatly helping the acceptance of these ideas and greatly advancing his own career by having chemists accept his ideas. We've seen the opposite of that happening in the story of Hermann Sachse (section 3.2) who presented his insights into the structure of cyclohexane in a mathematical rather than a visual format. Nevertheless, there was a problem facing Fischer, which could not be resolved even given his exceptional skills and insights. It was a problem that was impossible to solve at the time and would have to wait for the work of Bijvoet we discussed in section 1.12.

Fischer realized at the beginning of his investigations that although he could distinguish stereoisomers that were not mirror image related, that is, diastereomers, he could not distinguish mirror image stereoisomers. There was no experimental technique available at that time. Certainly enantiomers gave rise to opposite signs of optical activity, as we've noted with the d, l nomenclature (section 1.10 Figure 1.9) but there was no way to connect these signs with a chemical structure. It would be many years, more than a half century, before Bijvoet (section 1.12) discovered how scattering of X-rays could be modified in order to distinguish enantiomeric crystals.

So, the first task Fischer faced required that he make a decision. He had two sets of eight diastereoisomeric possibilities, the mirror image sets in Figure 3.12, and knowing he could not distinguish one set from the other by the experiments he was going to use, he arbitrarily chose one of the sets to consider, the one on the top of Figure 3.12. He realized that whatever physical properties he assigned to each of the structures in the set he chose, the mirror image molecule he was not considering would have the identical physical property. The only difference would be in the sign of the optical rotation.

Although Fischer's experiments could not distinguish mirror image isomers, he could distinguish the eight diastereomers from each other within one mirror image set. The set of eight diastereomers he chose were eventually designated D-sugars (Figure 3.12). In every one of the eight diastereomers within these D-sugars, the hydroxyl group on the next to the last carbon from the bottom of each stereoisomer in the set, as he wrote his projections, and as presented in Figure 3.12, was to the right. In the set he did not consider, the equivalent carbon had the hydroxyl group to the left and this set came to be designated L.

Fischer realized that if the single structure his experiments led to as natural glucose was, in fact, the mirror image of natural glucose, if his choice of the top set in Figure 3.12 instead of the bottom set, had been wrong, then chemists who followed him could simply replace their structural diagrams with mirror image diagrams. But, as we've seen, when Bijvoet carried out his experiments (section 1.11) it turned out that Fischer's guess was correct. The glucose found in nature was in fact one of the eight diastereomers on the top of Figure 3.12, therefore D-glucose.

PROBLEM 3.31
There are sixteen stereoisomers shown in Figure 3.12 for the non-cyclic structures of these constitutionally identical sugar molecules. If the cyclic forms were used, how many stereoisomers would be possible?

PROBLEM 3.32
Use the structures of the glucose stereoisomers to test the formula for calculating the number of isomers as 2^n where n designates the number of chiral carbon atoms. Does the formula work for tartaric acid and if not why?

PROBLEM 3.33
Why are the structures in figure 3.12 designated as pairs of isomers? How many pairs of isomers and what kinds of isomers are there in this figure?

PROBLEM 3.34
Assign (R) and (S) configuration to each stereocenter in the open forms of the sugars in Figure 3.12 and then use this information to assign axial and equatorial groupings to drawings of the chair conformations of the cyclic forms of these sixteen sugar molecules. Ignore the stereocenter at the carbon atom that was derived from the aldehyde group in each structure. Remember that each horizontal bond in these Fischer projections rises out of the page.

PROBLEM 3.35
Why do Fischer projections lose their meaning in translation to three-dimensional structures when rotated 90° but not 180°?

PROBLEM 3.36
Why are there only eight experimental parameters shown in Figure 3.12.

PROBLEM 3.37

Why is the unique use of glucose over all other diastereomers shown in Figure 3.12 an obvious choice by nature while unique use of the D over the L isomer rather than the L over the D isomer is a mystery?

PROBLEM 3.38

Imagine the figurative experimental problem faced by Emil Fischer. He looks at a laboratory bench containing eight vials, each with a substance, two of which are syrups and six of which are crystalline (Figure 3.12). Because of the synthetic source of these samples, which he knows because the work was done in his own laboratory by his students, he knows that each sample represents a pure substance of what he has designated the D series. The purity represents not only chemical purity, but also diastereomeric purity and enantiomeric purity. Now which vial corresponds to which Fischer projection? If at some point in your life you want to know how he did it, you can find the information in many textbooks and as well via Google on the web under the heading: Fischer's proof of the structure of glucose.

3.12

How does glucose differ from the other seven diastereomers shown in Figure 3.12? The answer can be found in the cyclic structure formed. Glucose is the fittest molecule in the Darwinian sense.

WHY OF ALL THESE eight diastereomeric possibilities for the basic structure is glucose chosen? Why does D-glucose alone of all the stereoisomers in Figure 3.12 play the central role as we've seen in Figure 1.1 in the formation of both starch and cellulose, and as the only sugar taking part in the biochemical mechanisms responsible for supplying energy and the building blocks for life processes? In fact D-glucose in all the forms in Figure 1.1, polymeric or not, is the most abundant organic molecule on earth, estimated to be more than 50% of the dry biomass on the earth.

Why did evolving life processes select only glucose of all the eight diastereomeric possibilities shown in Figure 3.12 to be created by the action of light acting to combine carbon dioxide and water, in what we call photosynthesis? There is a clue. Inspection of Figure 3.12 shows that there are only three stereoisomers of D-glucose that differ from it by change of R and S at a single carbon, D-galactose, D-mannose and D-allose. D-galactose and D-mannose are the only two stereoisomers of glucose that participate in our biochemical processes when we, respectively, drink milk or milk products and when we eat meat. However, in the absence of disease, enzymes erase the difference between these carbohydrates and D-glucose.

The other stereoisomer of D-glucose that differs from D-glucose by a single change between R and S is D-allose, which although found in nature, is quite rare, and which certain microorganisms are capable of converting to D-fructose. But allose is not a sugar found in our systems.

All the four remaining D-carbohydrates, D-altrose, D-gulose, D-idose and D-talose are not found naturally. These sugars could only be converted to D-glucose or D-fructose by interchanging R and S at more than a single carbon - a bigger job and one apparently Nature does not take on.

Throughout Chapters 1 and 3 we've seen that D-glucose exists in a cyclic form. **Figure 3.13** presents the six-membered structures that are formed in the ring closing of the D-carbohydrates in Figure 3.12. Let's focus for now, however, only on carbon atoms 2-5 in each of the sugars, the carbon atoms for which stereochemical change can not occur by ring opening and closing. The stereochemical configurations (R) or (S) at these carbon atoms are invariant within each diastereomer in contrast to the stereochemical configuration at carbon-1, the α or β configuration, which can

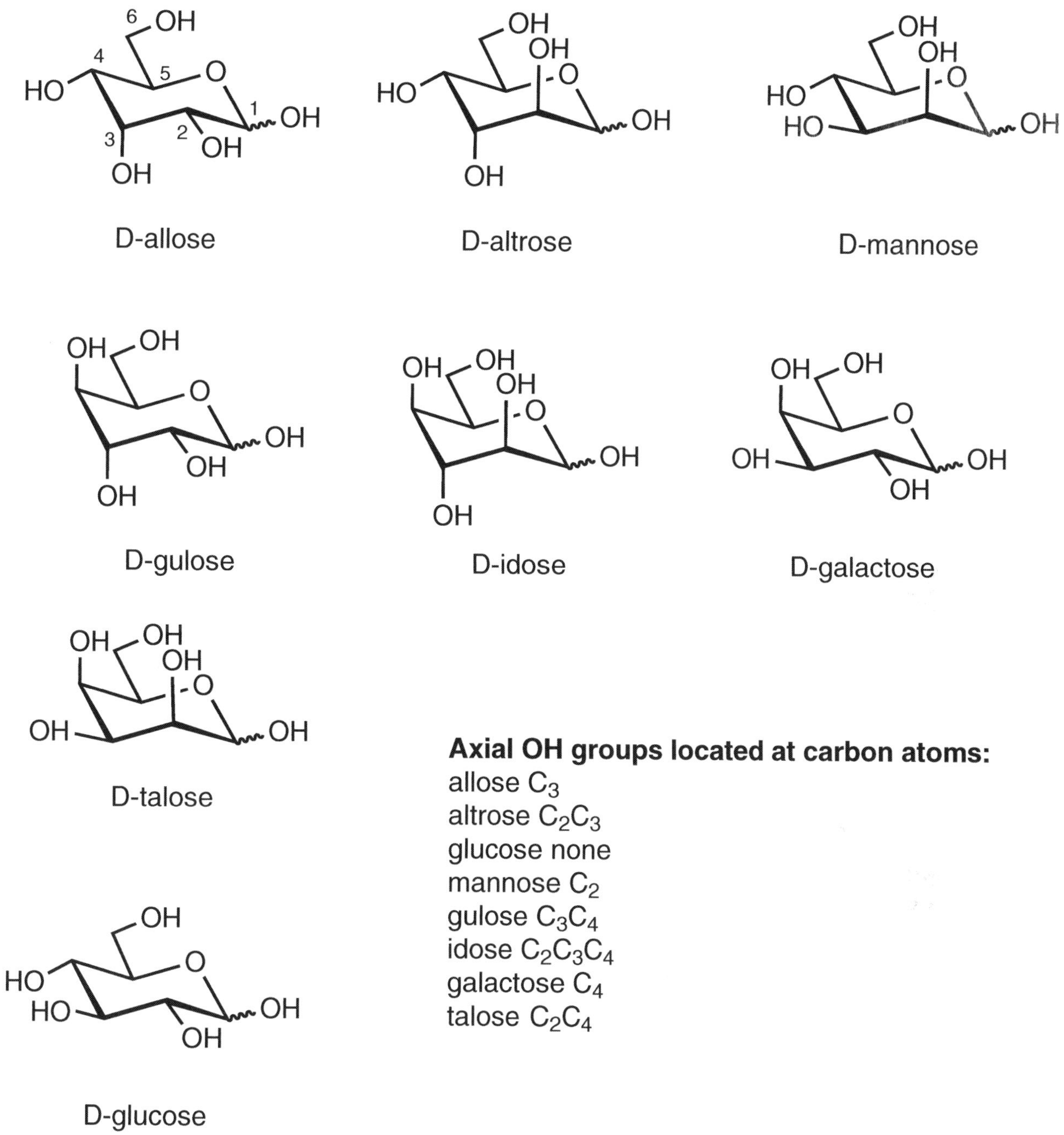

D-allose

D-altrose

D-mannose

D-gulose

D-idose

D-galactose

D-talose

Axial OH groups located at carbon atoms:
allose C_3
altrose C_2C_3
glucose none
mannose C_2
gulose C_3C_4
idose $C_2C_3C_4$
galactose C_4
talose C_2C_4

D-glucose

be changed by ring opening and closing, as in mutarotation (Figure 3.11).

In every sugar where the configurations at carbon atoms 2-5 differ from D-glucose as seen in comparing Figures 3.12 and 3.13, the pendant groups on these ring carbon atoms will take an axial instead of an equatorial position. And, the three to one ratio of equatorial to axial pendants in allose, mannose, and galactose in the conformations shown would disfavor flipping to the other chair form to escape the higher energy of the axial pendants. Moreover, in altrose, gulose, iodose and talose there are equal numbers of equatorial and axial pendant groups on C2 to C5 in both chair forms, making this conformational change more likely since it would simply switch equatorial to axial as a price for switching axial to equatorial.

Whatever conformational motions these eight diastereomeric sugars may undergo, axial pendant groups on six-membered rings cause the ring to be less stable, as we've discussed in detail in sections 3.3, 3.4 and 3.5, and glucose as shown in Figure 3.13 is the only one of the eight diastereomers with the possibility of all

▲ **FIGURE 3.13**

Chair Forms for Eight Diastereomers, One of which Is D-Glucose

equatorial pendants in a single conformation. Glucose therefore has the most stable closed form of all the diastereomers shown in Figures 3.12 and 3.13.

The structures in Figure 3.13 and the absence of an axial group on the fixed configuration carbons of D-glucose, C2-C5 (Figure 3.12), make it difficult to see glucose as anything but the fittest molecule - demonstrating Darwinian evolution at the molecular level. And this favoring of glucose over all the other seven diastereomers can not be understood without knowing that these sugars tend to form a six membered ring structure. In the absence of the ring, the configurations at carbons 2-5 in glucose would have no advantage over the other seven diastereomers.

Emil Fischer determined which diastereomer is glucose without using the information that the sugars form rings, let alone anything about the nature of those rings and the concept of axial and equatorial. Yet if given this problem today we could have predicted which diastereomer was glucose because it is the one that forms the most stable ring structure. Add this to Fischer's lucky (?) guess of which enantiomer was natural glucose, D in his nomenclature. Professor Fischer was prescient on all counts.

All this bring us back to galactosemia. Ring formation in the seven other diastereomers is less favorable than in glucose because of the presence of one or more axial pendant groups. In the array of molecules in aqueous solution, in blood for example, the proportion of molecules in the closed form would be largest for glucose. The open form of the sugar, which we've seen in Figure 3.11 as responsible for the mutarotation, would be more favorable in the other diastereomers where the closed form is destabilized by axial pendant groups.

We are almost ready to understand galactosemia but first must be understood the chemical nature of the aldehyde functional group, which is exposed in the open form of all the sugars in Figure 3.13 and therefore least accessible for undergoing chemical reactions in glucose. In glucose, the highly favored cyclic form ties up the aldehyde group as part of the ring structure. Now we have a clue as to the molecular source of galactosemia.

PROBLEM 3.39
Why does D-gulose exist with two axial pendant hydroxyl groups in either possible chair conformation? Ignore the OH group from carbon-1.

PROBLEM 3.40
Gulose will exist in aqueous solution as a mixture of configurational and conformational diastereomers as well as a pair of constitutional isomers. Show all these structures including equilibrium arrows designating the interconversions among them.

PROBLEM 3.41
Would ring flipping from the chair forms shown in Figure 3.13 increase the stability of any of the diastereomers and if so which would benefit most? Ignore the OH group from carbon-1.

PROBLEM 3.42
Would the idea that D-glucose is an example of Darwinian evolution hold up as well if natural glucose had been the L isomer: and if so, why?

PROBLEM 3.43
Would it not have been better if evolution had constructed mothers' milk entirely from glucose? Can you find any advantage in the use of galactose?

PROBLEM 3.44
How would the proportion of the open form of the sugar, with the aldehyde group exposed, be related to the axial versus equatorial placement of the pendants on the cyclic form and why?

PROBLEM 3.45

Use Newman projections, as used for cyclohexane and methyl cyclohexane in Figures 3.4 and 3.5, to redraw the structures of several of the sugars shown in Figure 3.13. Use the Newman presentations to evaluate the source of the instability in these sugars making the approximation that an OH group can be treated as if it were sterically similar to a methyl group.

LDEHYDES, SUCH AS THE SIMPLEST ALDEHYDE, formaldehyde, $H_2C=O$, and also the aldehyde functional group (CHO)found in all the diastereomers of $C_6H_{12}O_6$ (Figure 3.12), are two dimensional structures, which exist, therefore, in one of either the x,y, or the x,z, or the y,z plane, that is, in one plane. Hybridization of atomic orbitals, which accounted so well for the geometry of four-coordinate carbon (Figure 1.2) can account just as well for the three-coordinate central carbon atom in an aldehyde or for that matter any other of the functional groups (section 3.9) featuring a carbon-oxygen double bond (a carbonyl function) such as ketones and derivatives of carboxylic acids, which we'll come across later in the book.

The orbital picture for formaldehyde, which applies to any carbonyl compound, is shown in **Figure 3.14** where we observe that rather than mixing 2s, $2p_x$, $2p_y$ and $2p_z$ leading to sp^3 hybridization and tetrahedral bonding we instead leave one of the p orbitals out of the mix. For the representation of the structure in Figure 3.14, which finds all the atoms directly connected to the central carbon of the aldehyde in the xz plane, the p_y orbital is not hybridized. The mix of the second quantum level p_x, p_z and s orbitals forms three hybridized sp^2 orbitals.

The result of the sp^2 hybridization allows three bonds to form between the central carbon of the aldehyde group with its three sp^2 orbitals and the three connected atoms, the oxygen, and the two other groups, which in any of the sugars in Figure 3.12 would be a hydrogen atom and carbon-2 of the sugar.

In formaldehyde, as shown in Figure 3.14, the carbonyl carbon is bonded to two hydrogen atoms and the necessary oxygen atom. All three sp^2 bonds must exist in the xz plane as shown in Figure 3.14 as required by the use of the p_4 and p_z orbitals in forming the hybrid orbital sp^2 orbitals.

In the sp^3 hybridization of tetracoordinate carbon the four bonds formed a tetrahedral array with ideal bond angles, which allowed the greatest separation between the four bonds, a separation required by the fact that each bond contained two electrons, with these electron pairs repelling each other (section 1.4). Driven by the repulsion of the electron pairs in the three sp^2 bonds, the greatest separation between the three bonds in a plane would occur in a trigonal array with bond angles of 120 degrees. And these indeed are the bond angles found by experiment between the atoms connected to a carbonyl carbon.

The three bonds arising between the central carbon atom of the carbonyl group and the surrounding atoms are similar in nature to the bonds formed in tetracoordinate carbon, that is, sigma (σ) bonds. In this bonding situation there is maximum overlap between the orbitals forming the bond as shown in Figure 3.14. Such overlap allows the highest bond strength because the electron density in the bond penetrates maximally to the positively charged nuclei. The bond between the sp^2 hybridized carbon and the carbonyl oxygen atom is, however, a double bond. One bond of the two is a σ bond as just described. The second bond of the double bond is an entirely different kind of bond, a weaker bond, termed a pi (π) bond.

3.13

The Aldehyde Functional Group: π-Bonds and the Consequences of Electronegativity

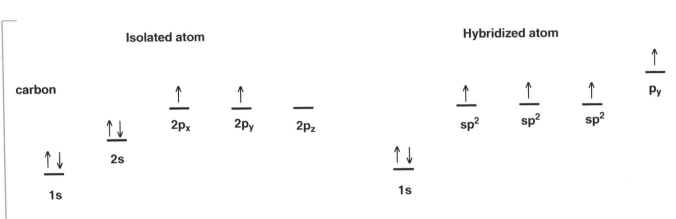

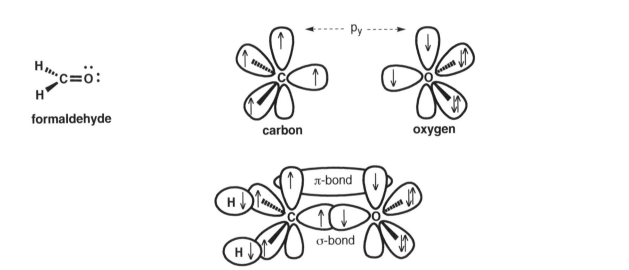

Atomic Orbitals and Hybridized Orbitals for the Carbon and Oxygen Atoms Forming a Carbonyl Group, Applied to Formaldehyde

If the oxygen atom is also hybridized sp^2, as is the carbon it is bound to, then the eight electrons will be distributed as shown in Figure 3.14: two electrons in the 1s orbital; one electron in the p_y orbital; and one electron in one of the three sp^2 orbitals. The two remaining sp^2 orbitals would contain two electrons each, two lone pairs of electrons (section 1.5). The single electron in the sp^2 orbital overlaps in a σ bond with the central carbon sp^2 bond.

The single electron in the p_y orbital on oxygen can not overlap in a σ bonding relationship with the single electron in the p_y orbital on the carbonyl carbon. The two p_y orbitals, one on oxygen and the other on carbon, do not lie along a line but rather are parallel to each other and separated at their regions of maximum electron density at the position of the two atoms forming the bond. This picture is shown in Figure 3.14. The poorer overlap arising from the π bond that forms gives the carbonyl group its special reactivity to be discussed shortly.

Another factor contributes to the properties of the carbon oxygen double bond in addition to the poor overlap of the π bond, electronegativity (section 1.4), which was shown to play a role in chemical shifts in both proton and carbon NMR (section 2.4). This difference in electronegativity between carbon and oxygen is the basis of a bond dipole between the carbon and oxygen atoms of the carbonyl group. The bond dipole represents an unequal distribution of the electrons in the bond. Electron density is attracted away from carbon and toward oxygen further weakening the bond between these atoms.

Tables of electronegativity can be widely found on the web by searching under electronegativity where it can be seen that the electronegativity of carbon is less, around 2.5, than the electronegativity of oxygen, around 3.4. This is consistent with the general trend of increasing electronegativity as one moves from left to right in the same row across the periodic table, a concept that is learned at the very beginning levels of the study of chemistry.

THE LOSS OF ELECTRON DENSITY at the carbon atom end of the carbonyl group arising from the electronegativity difference between carbon and oxygen discussed in the last section, acts to attract electron-rich moieties to interact with this carbon atom. Such electron-rich entities are given the name **nucleophiles** for seekers of the nucleus, seeker of the positive. The seeker of the electron-rich, on the contrary, is called an **electrophile**. Much more use will be made of these terms and their wide application in organic chemistry in later chapters.

A typical reactivity pattern always follows from the weak π bond and the bond dipole in carbon oxygen double bonds, which can be applied to a wide range of chemical reactions at carbonyl carbon. This pattern of reactivity is responsible for the formation of the closed form of glucose, and for a myriad of other chemical reactions, which we'll come across as study of the subject of organic chemistry moves ahead.

An electron rich group, the nucleophile, adds to the carbonyl carbon, the electrophile, with a breaking of the π-bond of the carbonyl group. For example, the ring closing of sugars occurs via the oxygen of the hydroxyl group at C-5, which is a nucleophile, with the four nonbonding electrons (electron rich), acting as the moiety attacking the electron poor carbonyl carbon, C-1, of the aldehyde group (Figure 3.15).

In **Figure 3.15** we come across, for the first time, a way that organic chemistry expresses a feature of chemical reactivity, the curved arrow. Such an arrow points to the way we hypothesize, that is, how we believe, a chemical reaction takes place, what we call a mechanism

To understand what the curved arrow represents, that is, the movement of pairs of electrons in the direction of the arrow head we have to look again at sections 1.4 and, especially, 1.5 from Chapter 1 where the ideas about bonding and

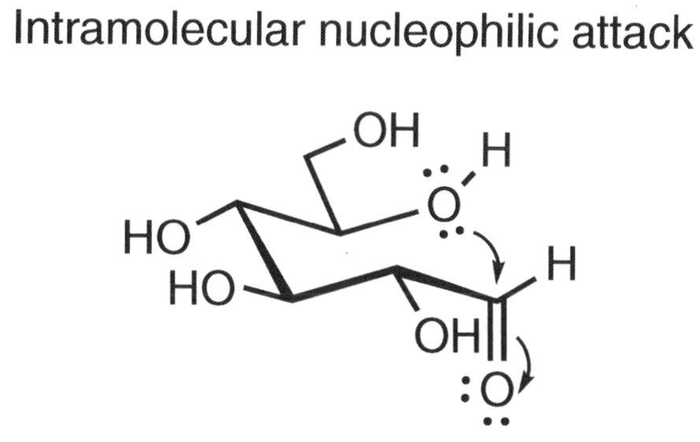

3.14
Reactive Characteristics of Aldehydes and other Carbonyl Containing Functional Groups: Mechanism, Curved Arrows, Nucleophiles and Electrophiles

Intramolecular nucleophilic attack

◀ **FIGURE 3.15**

Representation of the Movement of Electrons Involved in the Ring Closing of D-Glucose demonstrating Nucleophilic Attack at Carbonyl Carbon

▲ FIGURE 3.16

Mechanism Involving Water Molecules for the Ring Closing of D-Glucose

hybridization of carbon were introduced and where the concept of nonbonding electrons was introduced. Nonbonding electrons are most available for making new bonds. All chemical reactions, without exception, involve the movement of electrons, requiring a change in electron density, and organic chemistry reserves the curved arrow seen in Figures 3.15 to show this electron "movement."

In his classic text "Chemistry of Organic Compounds," written nearly half a century ago, **Carl R. Noller** wrote the following sentence: *The path that molecules are assumed to follow in order to account for the various factors on the course and rate of a reaction is called the **mechanism of the reaction**.*

The bold and italicized words are Noller's and the statement is the best I am aware of to describe what mechanism is. Chemical reactions of all varieties involve the movement of electrons and because electrons make up the bonds between atoms, chemical reactions also involve the movement of atoms. Therefore understanding the idea of mechanism, which is necessary to understand why galactose is a dangerous sugar in our bloodstreams, must involve understanding chemical bonding and mechanism.

The reaction shown in Figure 3.15 does not involve the isolated molecule but rather a water solution and a more fully described mechanism certainly involves water molecules as shown in **Figure 3.16**. Here we see that the nucleophilic character of the hydroxyl group and the electrophilic character of the carbonyl group are both enhanced by proton transfers with nearby water molecules. There are several variations on how this can take place. Although we don't know with absolute certainty the precise motions of the interaction between the water molecules and the open form of the glucose, we have chosen a most reasonable one to present in Figure 3.16. Other possibilities are that the proton transfers occur independently of the formation of the bond between the hydroxyl group and the carbonyl carbon or variations with mixtures of these two extremes. There can be a great deal of uncertainty in the extreme details of the mechanisms of organic chemical reactions.

▪ **Carl R. Noller**

PROBLEM 3.46

Consider boron trifluoride, which is a molecule in which the boron atom is bonded to three atoms, fluorine in this situation. Aldehydes are also bonded to three atoms, which in the open form of glucose are carbon, oxygen and hydrogen. What analogies exist between the orbitals' arrangement in boron trifluoride and in the aldehyde of glucose and what differences as well?

PROBLEM 3.47

Explain how the p-orbital not used in the hybridization of orbitals for bonding in boron trifluoride and forming the π-bond in aldehydes determines the plane of the bonding among the atoms.

PROBLEM 3.48

Show how the orbitals involved on both oxygen and carbon lead to the difference between the π and σ bonds of the double bond linking these two elements in the aldehyde functional group. Why is the π-bond only about 2/3 as strong as the σ-bond? Now try to describe the orbital arrangement for ethylene, $H_2C=CH_2$, which forms a similar bond, a double bond, between the carbon atoms, as the bond between the carbon and oxygen atoms in the aldehyde.

PROBLEM 3.49

What fundamental characteristics of carbon and oxygen and the double bond between them lead to the reactivity of the aldehyde functional group with electron rich groupings, such as the oxygen of -OH and the nitrogen of -NH2? Why would such reactivity characteristics differ for a carbon-carbon double bond?

PROBLEM 3.50

Why is there no molecular dipole moment for carbon tetrachloride, CCl_4, while for methyl chloride, H_3CCl, a large dipole moment can be experimentally determined although both molecules have large bond dipoles between the carbon and chlorine atoms?

PROBLEM 3.51

The reactivity of a carbonyl functional group manifests itself by addition of an electron rich group (a nucleophile) to the central carbon of the carbonyl function causing the bonding around this carbon atom to change from approximate sp^2 to sp^3 hybridization with the change in geometry from three to four coordinate. Given this information can you think of a reason why aldehydes are so much more reactive than ketones? In ketones the central carbon atom doubly bonded to oxygen is bonded to two other carbon groups instead of, as in the aldehyde, one carbon group and a hydrogen atom.

PROBLEM 3.52

Alkenes such as ethylene, $H_2C=CH_2$, a functional group we will discuss in great detail in later chapters, also exhibit a reactivity associated with the weakness of the π bond. Why might you expect that this reactivity characteristic would differ greatly from that of an aldehyde or a ketone?

PROBLEM 3.53

What does formal charge have to do with atomic number? Use this relationship to show that while aluminum bonded to four oxygen atoms is negatively charged, silicon bonded to four oxygen atoms is neutral.

PROBLEM 3.54

Propose an alternative mechanism for that shown in Figure 3.16 in which the proton transfers between water molecules and glucose occur prior to the ring closing step. An intermediate is therefore produced. Draw the structure of this intermediate including all formal charges. To aid in answering this question and for determining the formal charges it is best to show all nonbonding electrons on the atoms involved in the reactions.

PROBLEM 3.55

Some ascribe the explosive power of molecules that contain the nitro functional group, $-NO_2$, to a functional group with a formal charge. One example is trinitrotoluene (TNT). Find this formal charge in the nitro group and hypothesize how this idea connecting formal charge to explosive power could be correct.

PROBLEM 3.56
How might the frequency region around 1700 cm-1 in the infrared spectra of the diastereomers of glucose reveal fundamental properties of their structure? How might this frequency region differ for glucose versus galactose?

3.15

Galactosemia is caused by the reactivity of an aldehyde functional group. A healthy infant supplies an enzyme to convert a derivative of galactose to a derivative of glucose to avoid the reactivity of an exposed aldehyde functional group.

THE BIOCHEMICAL BASIS OF GALACTOSEMIA points to the source of the toxicity of galactose as arising from a higher proportion of the open form of this sugar than for glucose, which means an aldehyde group is available to react with intermolecular nucleophiles **(Figure 3.17)**. **Intermolecular** designates interactions between molecules rather than within the same molecule, for which the word **intramolecular** is used (Figure 3.15).

Aldehyde is a highly reactive functional group and aldehydes react with many functional groups with available electrons, nucleophiles, some of which are commonly found in vivo, such as, in particular, the amino groups, -NH$_2$, on proteins. Such uncontrolled reactivity of sugars is given the name glycation in contrast to the controlled biochemical reactions of proteins with sugars called glycosylation.

Glycation causes disease by altering the structure of proteins in ways not intended for their normal biochemical function. Glycosylation of proteins is enzyme-controlled, such as in posttranslational modification of proteins, a biochemical reaction that acts to expand the functions of proteins beyond the structural capacity of the basic structure of the protein.

However, the posttranslational modifications of a protein are narrowly prescribed. Posttranslational modifications of a protein, which are not intended by the information in DNA, can let loose in the biochemical machinery a substance that can carry out chemical and physical changes that can damage the organism. Glycation is such a damaging posttranslational modification and available aldehydes can take part in such damaging in vivo chemistry.

FIGURE 3.17 ▶

Intermolecular Nucleophilic Attack of the Side Chain Amino Group of Lysine at the Carbonyl Group of D-Galactose

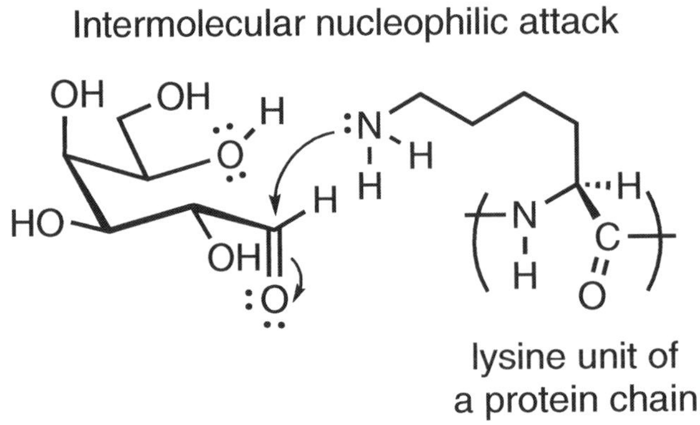

Intermolecular nucleophilic attack

lysine unit of
a protein chain

When ring closing of a sugar leads to a higher energy structure because of the axial pendant groups that are produced, then the aldehyde group seeks other sources of electron density, which can be found, among other sources, in the abundant amino functional groups, -NH$_2$, found in proteins (Figure 3.17).

The nucleophilic character of amino groups that makes them available partners for aldehydes arises from the lone pair of electrons on the nitrogen atom. The nitrogen atom must surround itself with seven electrons (the atomic number) to avoid a net charge (section 1.5). In Figure 3.17 an amino group, -NH$_2$ is shown as part of a protein structure. This amino group is bonded to a carbon atom and two hydrogen atoms and therefore is surrounded by three electrons, half of each of the bonds to the two hydrogen atoms and the carbon atom. Two more electrons from the valence shell must be added to the two electrons in the first quantum level to make up the seven electrons to avoid a net charge on the nitrogen and as in oxygen, these two electrons find themselves in a single orbital, a lone pair of electrons. The lone pair of electrons is shown on one of the nitrogen atoms in Figure 3.17.

If the oxygen atom on carbon number 5 of the sugar resists closing the ring because an axial group will be generated, then the lone pair of electrons on the amino group of the protein (Figure 3.17) may take the opportunity to react with the now available aldehyde of the sugar, a glycation event. In this manner an unwanted posttranslational modification of the protein takes place, therefore leading to results nature did not intend, disease in the form of galactosemia.

PROBLEM 3.57

Write an "essay" using only chemical structures outlining your understanding of the source of galactosemia.

A GREAT DEAL IN THIS CHAPTER has been written about axial versus equatorial bonds on six-membered rings and we've seen that much can be understood based on the structural insights of Hassel and Barton (Figure 3.3). The difference between cellulose and starch are equatorial linkages between the glucose units in cellulose and axial linkages in starch.

It makes sense, doesn't it? If nature needs to construct a macromolecule that will be most resistant to destruction, a material that is used for the construction of trees and of cell walls in plants, would it not make sense to choose the equatorial linkage between the glucose units, the more stable of the two possibilities? And if a material were to evolve from the identical units used in the construction of cellulose but a material that is designed to ease breaking the units apart so that they can be broken down in an energy yielding process, does it not make sense to link the units together with an inherently less stable bond, an axial bond?

There is much involved in the difference between cellulose and starch including important properties that have to do with how the macromolecular chains are put together, in the ability of the polymer chains to form crystals. Strong relationships among the chains in cellulose act to resist breaking down the individual chains because access by destructive groups, groups that act to break the linkages between the glucose units have restricted access. If starch is exposed to water, the granules swell while cellulose under the same conditions resists interacting with water.

Equatorial linkages act together with the equatorial pendant groups on the glucose units to make an overall more symmetrical structure for the cellulose and therefore better packing relationships among many chains, which leads to crystal stability. In general for all of chemistry, crystallization and higher melting points occur for more symmetrical molecules and macromolecules.

Nature uses this symmetry factor in the opposite direction for starch. The highest proportion of starch, amylopectin, is not fully linear (not shown in Figure 1.1) but has branches of linked glucose units going off in all directions. This increases the disorder of starch, retards crystallization, and increases access

3.16

What can we now understand about the difference between cellulose and starch.

of the enzymes designed to break down the starch to smaller units. In contrast, cellulose chains are unbranched and tend to be longer than the chains in starch, which means higher strength. In this manner cellulose is protected against breakdown both by the stability of the equatorial linkage within the chains of glucose units and by the crystal stability arising from the symmetry of the all equatorial relationships and the far larger number of units in each chain arranged in an orderly linear manner.

There it is.

CHAPTER 3 SUMMARY of the Essential Material

THE ESSENTIAL CONCEPT IN THIS CHAPTER is to understand the properties of six membered rings, first cyclohexane and then, by extrapolation, glucose and its diastereomers. To incorporate tetrahedral angles in such rings, the rings can not be flat but must be puckered and the favorable shape is termed a chair form. Such conformations of six membered rings cause pendant hydrogen atoms or other groups attached to the carbon atoms of the ring to take an axial or equatorial position, while the flexibility of these rings allows a motion called flipping in which the axial and equatorial groups can interconvert. These motions occur extremely rapidly.

The conformational properties of cyclohexane and by extension six membered ring-forming sugars such as glucose can be well represented by Newman projections, which allow comparison between the six member ring of carbon, cyclohexane with a pendant methyl group, and conformations of n-butane. Such comparisons reveal the source of the stability of equatorial versus axial pendant groups on the rings by comparing the conformational details to gauche and anti forms of n-butane. In this way we understand how concepts of steric strain affect the stability properties of six membered rings.

The material in the chapter then focuses on the properties of glucose and its diastereomers. We discover that there are eight diastereomers including glucose and how they differ and understand the general approach Emil Fischer took to understand the stereochemical properties of these diastereomers. An essential feature of this approach was the development of an extremely useful way of presenting the structure, which is now called a Fischer projection. Because the sugars form six membered rings the configurations at each of the carbon atoms determines if the group will be an axial or equatorial pendant to the six membered ring formed. It is only glucose out of the eight diastereomers with all pendant groups equatorial, which causes the ring form of glucose to be most stable of all other diastereomeric sugars.

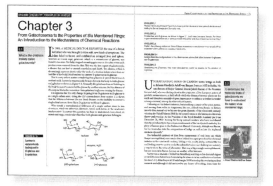

The chapter introduces the idea of functional groups and the prominent functional groups in glucose and its diastereomers are hydroxyl, and aldehyde. We discover how the bonding in the aldehyde group is responsible, in addition to the electronegativity difference between carbon and oxygen, for the reactive properties of this functional group. This focus on reactivity introduces the concepts of nucleophile and electrophile, or that is, the electron rich and the electron poor and how this difference accounts for the reactive characteristics often seen in organic chemistry. The idea of mechanism is introduced, which is the descriptive term for how chemical reactions take place. We discover the role of the curved arrow and see its use in the intramolecular reaction between electron rich oxygen in OH, with its lone pairs of electrons, and electron poor carbon in the carbonyl group of the aldehyde causing the ring closing of the sugars. We investigate this ring closing reaction and discover the source of a phenomenon called mutarotation in which the optical activity of glucose in water changes with time. In this way we reinforce the difference between configurational and conformational isomers.

Finally we see how these ideas, the combination of the conformational properties of six membered rings of sugars and the reactive properties of the aldehyde functional group lead to understanding a disease of infants, galactosemia, which arises from an intermolecular reaction between electrophilic carbon in an aldehyde group and nucleophilic carbon of an amino group, NH_2, on a protein.

Last, the structural ideas discovered in this chapter help understanding of the differing properties of cellulose and starch.

Chapter 4

Understanding Carbocations: from the Production of High Octane Gasoline to the Nature of Acids and Bases

4.1

What did Eugene Houdry do which revolutionized the petroleum industry and had an important effect on the outcome of World War II?

▪ **Eugene Houdry**

"*ON MARCH 31, 1937, at Sun Oil's Marcus Hook plant hard by the Delaware River, engineers charged Houdry Unit 11-4 with 15,000 barrels of sloppy residuum after Sun's thermal cracking refiners thermal cracking refiners had squeezed every drop of gasoline they could from the crude. Up went the heat to 900°. Pressure was applied. And as still men and panel men anxiously watched the gauges, the vaporized residuum was forced through the macaroni-shaped catalyst of silica and alumina. When 11-4 had done its work, yield sheets showed that the waste oil had given 7,200 barrels of gasoline. Furthermore, the gasoline had an octane (antiknock) rating of 81, compared with the octane-60 which average crudes yield under present processes.*"

Eugene Jules Houdry was one of the great inventors of the twentieth century, responsible not only for our focus here on the invention of catalytic cracking of petroleum but as well for finding a way to make a key molecular component of synthetic rubber. And if that was not enough he also invented the first catalytic converter for cars to help to reduce exhaust contribution to smog.

Houdry was born near the end of the nineteenth century in France, the son of Jules Houdry and his wife, the former Emilie Thais Julie Lemaire the owners of a steel manufacturing facility. The young Houdry wanted to enter the family business and therefore decided to study mechanical engineering. In those years the work carried out by a modern chemical engineer was the job of a team - a mechanical engineer and a chemist so that young Houdry's decision connected him to the world of chemistry. Houdry's life's work proved that he encompassed both professions within his own abilities, which were considerable. In a portent for his future accomplishments, when he graduated from the Ecole des Arts et Métiers in Paris in 1911, he won a gold medal for receiving the best grades in his class.

Houdry became interested in gasoline driven engines from his time in the tank corps of the French army in World War I, an interest that lead to a life-long love affair with race cars and speed, which also led him to try to find better methods to produce gasoline. All this eventually led Houdry to accept an offer from American oil companies to carry on his research in the United States where he discovered the catalytic process described in general terms in the *Time Magazine* article copied above.

Houdry had tried hundreds of substances to try to alter the structure of the molecules in petroleum to improve the rather poor performing gasoline available at that time. The idea was to increase the octane number so as to avoid knocking, that is, premature ignition and, therefore, yield more power in each explosion taking place in the cylinders of the internal combustion engine. Among these were clays, which in modern terms have evolved to be the most important catalysts in the petroleum industry, the zeolites.

Zeolites are networks of aluminum oxides and silicon oxides in which both metal atoms are dispersed throughout the network, each bound to four oxygen atoms. An example of this bonding is shown in **Figure 4.1**. The network itself is very interesting in consisting of channels of the dimensions of small molecules, that is, in the range of several to tens of Angstroms. On entering these channels the molecular structures of the petroleum molecules are transformed from low to high octane hydrocarbons.

How this happens will be one of our focuses in this chapter.

The greater yield and the higher octane of the gasoline obtained by the Houdry process is given credit for a considerable advantage of the Allies over the Axis powers in the second world war including in the critically important Battle of Britain when German air power tried to pound the British into submission. Although the German planes were at least as well engineered if not superior to the British planes, the gasoline from Houdry's catalytic cracking process is reported to have given the British planes far greater engine power for both take-off and climbing, for maximum speed and for carrying a load. When the British planes took off to meet the air attack these advantages are given some credit for the failure of Germany to win the Battle of Britain.

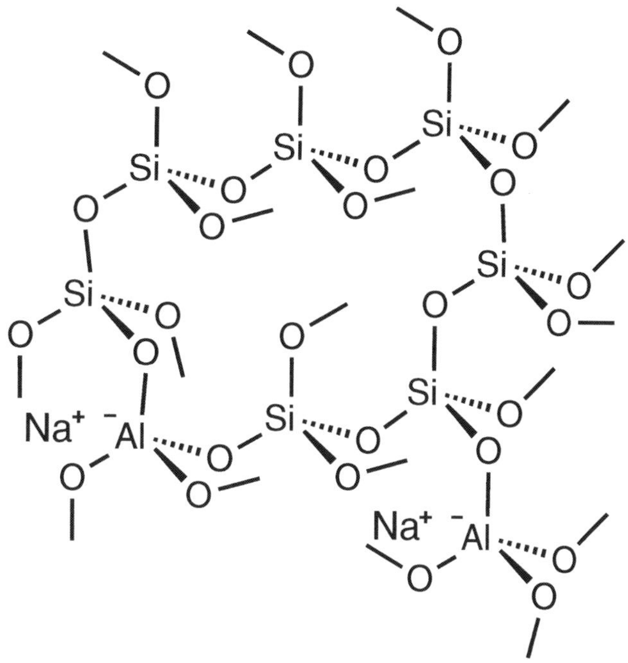

Zeolite Structure Showing the Charged Aluminum Atoms

PROBLEM 4.1

The ability of zeolites to act as catalysts depends on the formal charge on the occasional aluminum atoms in the network as shown in Figure 4.1. Account for the fact that silicon in the network has no formal charge while aluminum does.

PROBLEM 4.2

Replace one of the silicon atoms by a phosphorus atom and another of the silicon atoms by a boron atom in the structure shown in Figure 4.1 and calculate the formal charges, if any, on these replacement atoms. The incorporation of boron and phosphorus atoms in a silicon network is the basis of materials used for solar cells to generate electricity.

PROBLEM 4.3

The electronic structure of carbon monoxide, CO, and carbon dioxide, CO_2, two products of combustion of gasoline in the internal combustion engine can be arranged both with and without formal charge. Show the bonding in these structures and account for all electrons. Are you able to find a bonding arrangement that obeys the octet rule for both molecules?

4.2

What's happening in these catalysts?

THE KEY TO HOUDRY'S INVENTION is found in the words "catalyst of silica and alumina." In section 1.5 we learned something about the necessity for atoms in molecules to be surrounded with as many electrons as there are protons in the nucleus of the atom. Carbon needs six, oxygen eight, nitrogen seven and so on, where the number of electrons is determined by the atomic number of the atom in question. We saw that each single bond, each line of the structure drawn, contributes one electron, half of the two electrons in that bond, to the accounting. In tetrahedral carbon this means that half of the electrons in four single bonds to the four groups that are bonded to the carbon atom contribute to the accounting, that is, four electrons. These four electrons are then added to the two electrons in the uninvolved $1s^2$ orbital to make up the necessary six electrons to match the atomic number of

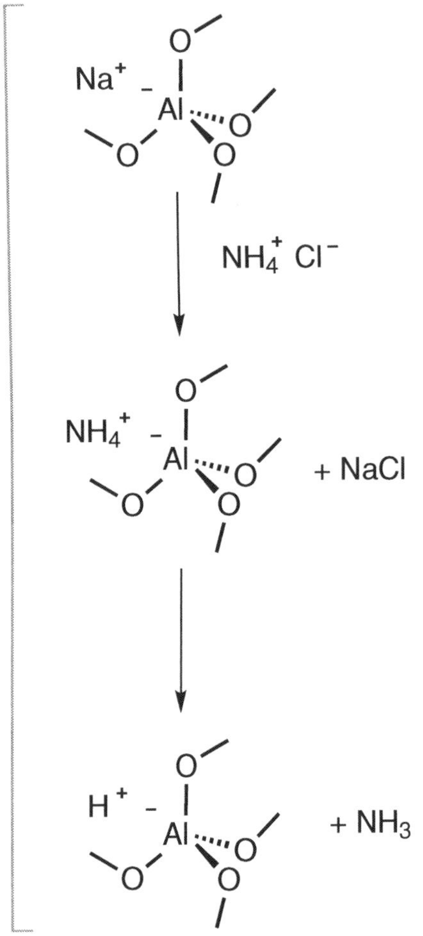

FIGURE 4.2 ▶

Replacement of the Sodium Counterions for the Charged Aluminum Atoms in a Zeolite with Protons

■ Auguste Laurent

carbon. As a result there is no formal charge on the bonded carbon atom.

A similar accounting works for carbon that is doubly bonded to oxygen, as in any carbonyl compound, such as an aldehyde (section 3.13). Again here there are four bonds to the carbon atom. Although three of these bonds are σ bonds and one is a π bond we still come up with four electrons "belonging" to the carbon atom to be added to the two electrons in the uninvolved $1s^2$ orbital.

Applying this accounting to the structural picture in Figure 4.1 shows that an aluminum atom bonded to four oxygen atoms, as is encountered in the structure of the zeolite, requires that the aluminum atom has a negative charge, which can not be avoided by any bonding scheme or arrangement of electrons. Not so for each of the far more abundant silicon atoms, which have no formal charges. A negative charge must have a positive countercharge to balance it and in a zeolite this positive charge is usually supplied by a nearby sodium ion, which is shown in Figure 4.1. In the petroleum industry today the sodium counterions in modern zeolites are replaced with ammonium ions, NH_4^+, and then heated to eject ammonia, NH_3, leaving a proton behind . This means that along the molecular size channels, the petroleum molecules find themselves in proximity to a charged environment filled with protons, in other words, an acidic environment. This transformation is shown in **Figure 4.2**.

In the highly polar environment of the zeolite cavities bristling with protons, which are acting as counterions to the negatively charged AlO_4^- moieties interspersed within the more numerous SiO_4 frameworks, the hydrocarbon fractions from

FIGURE 4.3 ▶

Positively Charged Carbon Atom in a Hydrocarbon Structure, a Carbocation

A carbocation is formed along the n-octane chain

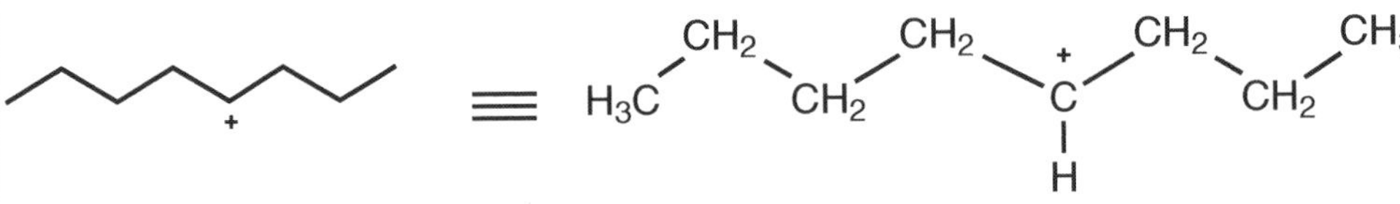

petroleum encounter an exceptionally reactive environment at high temperature. Even today petroleum chemists can not define precisely every reaction that takes place but one certainty is that carbocations are formed, that is, trivalent positively-charged carbon atoms (**Figure 4.3**), which are believed to be the species responsible for the changes that Houdry produced with his catalyst.

··

PROBLEM 4.4

In carbocations (also called carbenium ions) there are three groups bonded to the positively charged carbon atom. Count the electrons in a carbocation, as shown for example in Figure 4.3, to account for the positive charge. Is there any difference between this charge and what we have been calling a formal charge?

PROBLEM 4.5

In Chapter 3, we came across the aldehyde functional group in which carbon was bonded to three groupings. Compare the orbitals and the hybridization in a carbocation with that in an aldehyde and show how this information leads to the same geometry for both.

PROBLEM 4.6

How do the two p orbitals that participate in the hybridization for a carbocation determine the plane in which the charged carbon and the three atoms bonded to it exist, xy, xz, or yz?

PROBLEM 4.7

If a carbon with four different groups attached, a chiral carbon, lost one of these groups (with the two electrons that bound it to the carbon) to become positively charged, what would be the stereochemical consequence if the lost group eventually returned to reform the chiral carbon?

▪ **Egor Egorevich Vagner**

FOR ABOUT A HUNDRED YEARS, for the entire nineteenth century, chemists could not imagine the possibility that what was called the carbon skeleton of a molecule could change. The bonding arrangements of the carbon atoms within that skeleton were believed to be fixed. All chemical reactions were thought to take place in a way that left that part of the structure, the carbon skeleton, unchanged.

According to an article published by Ludmila Birladeanu in the Journal of Chemical Education in 2000, the great French chemist **Auguste Laurent**, a man whose career and in fact life were greatly limited by his difficulty in getting along with others, expressed this belief as the principle of "least structural change," a concept articulated earlier by Kekulé, stating that molecules tend to undergo the fewest possible changes in structure during a chemical reaction.

However, experimental results had accumulated throughout the nineteenth century, which, if understood, would have disproved this principle. These reactions, however, were not understood well enough to realize that structural changes were taking place, as was especially the situation with chemical changes of molecules obtained from natural sources, the terpenes. The principle of least structural change could not be eliminated from the thinking of chemists until someone came along with a new insight.

Figure 4.4 shows a transformation between two terpene molecules, which was observed as early as 1802, but without understanding the chemical structures, without understanding that the carbon atoms in the skeleton of isoborneol and camphene are connected differently. These experiments, if understood at the time, would have absolutely demonstrated that carbon skeletons can undergo change. But the tools for structural analysis now available, such as nuclear magnetic resonance, would show in an instant the changes in the carbon skeleton. But NMR would not be available to chemists for over 150 years.

It took nearly one hundred years until a Russian chemist came to fully understand the structural difference, the skeletal difference between what we now know to be isoborneol and camphene (Figure 4.4). This man, **Egor Egorevich Vagner**, who studied at the University of Kazan under another well know organic chemist (**Alexander Zaytsev**, who in turn had worked for Alexander Michailovich Butlerow(v) (section 1.11), was professor in Warsaw. He called himself Georg Wagner, I guess to better fit in with the German dominated field, and is well remembered as the Wagner in the Wagner-Meerwein rearrangement, a name chemists give to the kind of changes in structure seen in Figure 4.4.

Wagner's insight, which had been earlier developed as the concept of molecular rearrangement by Butlerow(v), his scientific grandfather, was first revealed in

4.3

It took a great deal of time before chemists accepted the possibility that the carbon skeleton of a molecule could change, and then, even longer to realize that the agent of change was a chemical intermediate with positively charged carbon, a carbocation.

▪ **Alexander Zaytsev**

publications in the 1890s. However, although Wagner correctly understood the structural differences shown between the two molecules in Figure 4.4, he could not explain how it occurred, that is, the mechanism (section 3.15). This feat was accomplished later by **Hans Meerwein**'s hypothesis in 1922 of the intervention of a carbocation, but too late for Wagner, who died in 1903.

The transformation of the carbon skeleton of organic molecules seen in Figure 4.4 is not an exception but, in fact, is commonly observed when organic molecules are subjected to certain reactive conditions, especially reactive conditions that form carbocations.

I once saw an undergraduate student in Professor **Paul von R. Schleyer**'s lab at Princeton University sitting in front of a large flask fitted with a long air condenser to trap whatever volatile substance might come off the black liquid that filled half the flask. The flask was heated to high temperature with a Bunsen burner. The student was repeating an experiment carried out in 1957 by Schleyer, an experiment that was based on the power of carbocations to undergo rearrangements of their carbon skeletons. From time to time I would drop in to see if anything was happening. On one of those visits there was considerable excitement in the lab, needle-like crystals had appeared, and in rather considerable quantity, crystals of a molecule that is found in only minute amounts in nature - adamantane, a molecule with a diamond-like structure.

The highly symmetrical structure of adamantane (**Figure 4.5**) had earlier attracted the attention of Vladimir Prelog, whom we have met in Chapter 1 (section 1.12), the chemist who participated in the creation of the basic (R) and (S) nomenclature necessary to describe chiral molecules. Prelog had synthesized adamantane in 1941 long before his interest in chiral nomenclature but it was a synthesis requiring a complex series of chemical reactions taking much time and effort, hardly an undergraduate student simply sitting quietly in front of a flask.

What magic was going on inside that student's flask and why did Professor Schleyer hypothesize that such an amazing transformation could take place? The answer is that carbocations were at work.

When you look at the transformation occurring in that student's flask (Figure 4.5) or in that nineteenth century transformation of two terpenes (Figure 4.4) we'll

FIGURE 4.4 ▶

Rearrangement Of The Carbon Skeleton In Going From Isoborneol To Camphene Caused By Carbocation Intermediates

isoborneol

- H$_2$O | H$^+$

camphene

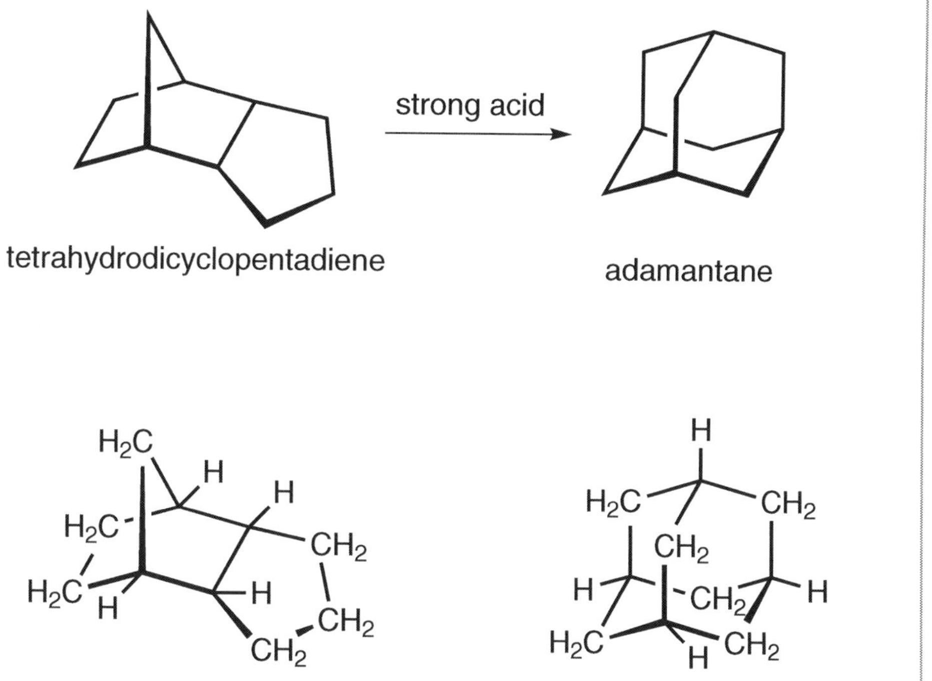

tetrahydrodicyclopentadiene

strong acid

adamantane

◀ **FIGURE   4.5**

Carbocation Rearrangement from Tetrahydrocyclopentadiene to Adamantane

come to see that rearranging linear to branched hydrocarbons to raise the octane number (look ahead to sections 4.5 and 4.6) of a fuel is a piece of cake once a carbocation is involved.

Branched hydrocarbons are defined as structures in which the skeleton of carbon atoms is not linear. In Figure 1.6 two of the isomers of C_6H_{14} are branched. Forming carbocations from the molecules in petroleum to cause branching of the linear hydrocarbons is precisely what zeolites can accomplish, which is what Houdry discovered.

PROBLEM 4.8

Are isoborneol and camphene (Figure 4.4) isomers?   Answer this question for tetrahydrodicyclopentadiene and adamantane in Figure 4.5.

PROBLEM 4.9

Redraw all the structures in Figures 4.4 and 4.5 in three dimensions showing all atoms. If possible, construct molecular models of these compounds. If you remove a molecule of water (H_2O) from the structure of isoborneol, which in fact is chemically possible inorder to produce a carbon-carbon double bond, does this change cause an isomeric relationship between the two structures?

PROBLEM 4.10

Six membered rings can be found within the structures in Figures 4.4 and 4.5. Identify the atoms making up these rings and determine if chair forms of these rings are present. Evaluate the strain present in other six membered arrangements that are not chair forms but exist within these structures. Is there torsional strain or angle strain?

PROBLEM 4.11

Is there any structure in Figures 4.4 and 4.5 which does not suffer from torsional strain (Problems 3.7-3.9)? Give a reason for your answer.

PROBLEM 4.12
Is there any opportunity for conformational isomerism with the structures in Figures 4.4 and 4.5 and if not, why not?

PROBLEM 4.13
What functional groups, if any, can be found in the structures in Figures 4.4 and 4.5?

4.4

What are carbocations and what is the basis of their ability to rearrange molecular structure? It's all about that empty p-orbital.

MANY CHEMICAL REACTIONS PASS THROUGH several states before arriving at the final product or products, that is, those chemical structures that we can isolate. If a chemical can be isolated it means that the chemical has to have a life time that is long on the time scale of our measurements, which is generally many hours if not days and longer. This requirement does not mean that we have to be unaware of shorter-lived entities in the chemical processes we study. Using modern methods based on lasers, for one example, fleeting species existing for fractions of a microsecond can be studied, but not isolated, not put in a bottle, so-to-speak.

We've already seen in section 1.13 that some molecules exist as mixtures of conformational isomers, which interconvert with each other so quickly there is no chance to isolate one from the other. Carbocations are another example of molecular species that almost always can not be isolated because they exist for too short a length of time. Nevertheless carbocations are important because they often intervene in the transformation of one molecule into another. Such changes brought about by the presence of carbocations confused nineteenth century researchers (Figure 4.4) but, as we've seen, fascinated twentieth century researchers in transforming a rather disorderly structured molecule into a molecular model for diamond (Figure 4.5 compare Figure 3.2).

Carbocations are called intermediates, a word chemists use to point to molecular species that are not isolated but nevertheless play important roles in chemical transformations.

Other short-lived chemical entities are conformational isomers which, although also not isolated because of their short lifetime are, however, not necessarily involved in chemical reactions. They are, therefore, not necessarily intermediates. Conformational isomers may or may not be especially reactive.

Intermediates are always highly reactive. Although the lifetime of an intermediate is very short, in the range for example of milliseconds, this is long enough to carry out the molecular changes they are responsible for. The fact that carbocations are exceptionally reactive makes sense. Look at the structure of a carbocation in Figure 4.3. A carbocation disobeys all of the rules for molecular stability. They have a charge arising from the fact that there are only five electrons belonging to the carbon, two from the usual place, the 1s orbital, and three from half the electrons in the three bonds to that charged carbon atom. Carbon needs six electrons "belonging" to it and hence comes the positive charge.

Moreover, stable molecules of the second row of the periodic table obey the octet rule, meaning that all the second quantum level electrons, the valence electrons surrounding the atoms, add up to eight, the number of second row electrons in neon. In a carbocation, the total number of valence electrons surrounding the carbon atom adds up only to six.

The carbocation is, therefore, both charged and disobeys the octet rule predicting an unstable species. And instability means the potential for change, which translates to chemical reactivity. The orbital picture of the carbocation in Figure 4.3 is shown in **Figure 4.6** where the three bonds surrounding the charged carbon are in the x,z plane, and, from this picture, we can understand the basis of carbocation reactivity.

We've seen that carbon bonded to four groups, as are all the carbon atoms in the

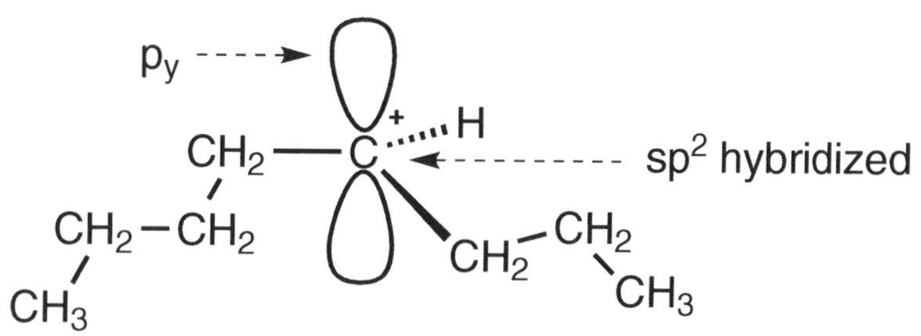

◀ **FIGURE   4.6**

Orbital Hybridization and Geometry of a Carbocation

cyclic form of glucose, tends to a tetrahedral geometry and can be understood by Pauling's theory as sp³ hybridized (Figure 1.2). We've also seen that carbon bonded to three groups, as in the aldehyde group in the open form of sugars, can be understood as sp² hybridized (Figure 3.13). The charged carbon in a carbocation (Figure 4.3) is bonded to three groups and as in an aldehyde can be well described as sp² hybridized with internal angles of 120°. This hybridized state is shown in Figure 4.6 where the orbitals hybridized to form the sp² orbital are 2s and $2p_x$ and $2p_z$ with the remaining empty p orbital assigned to $2p_y$. This choice of orbitals participating in the sp² hybridization is the reason that the plane of the three groups surrounding the carbon of the carbocation in Figure 4.6 can be defined (section 1.4) - the x, z plane.

In Figure 4.6 we discover the driving force behind Houdry's invention, which is the empty p orbital. This empty orbital space, one might say, is "hungry" for electron density, that is, the carbocation is electrophilic. Two electrons "pouring" into this empty p orbital to be shared with an incoming atom, would satisfy the octet rule for this carbon and as well remove the positive charge by satisfying all the charges of protons in the nucleus of the carbon atom.

Finding electron density for this empty p orbital (**Figure 4.7**) by causing another carbon atom to take on the destabilizing characteristics of a carbocation does not deter the

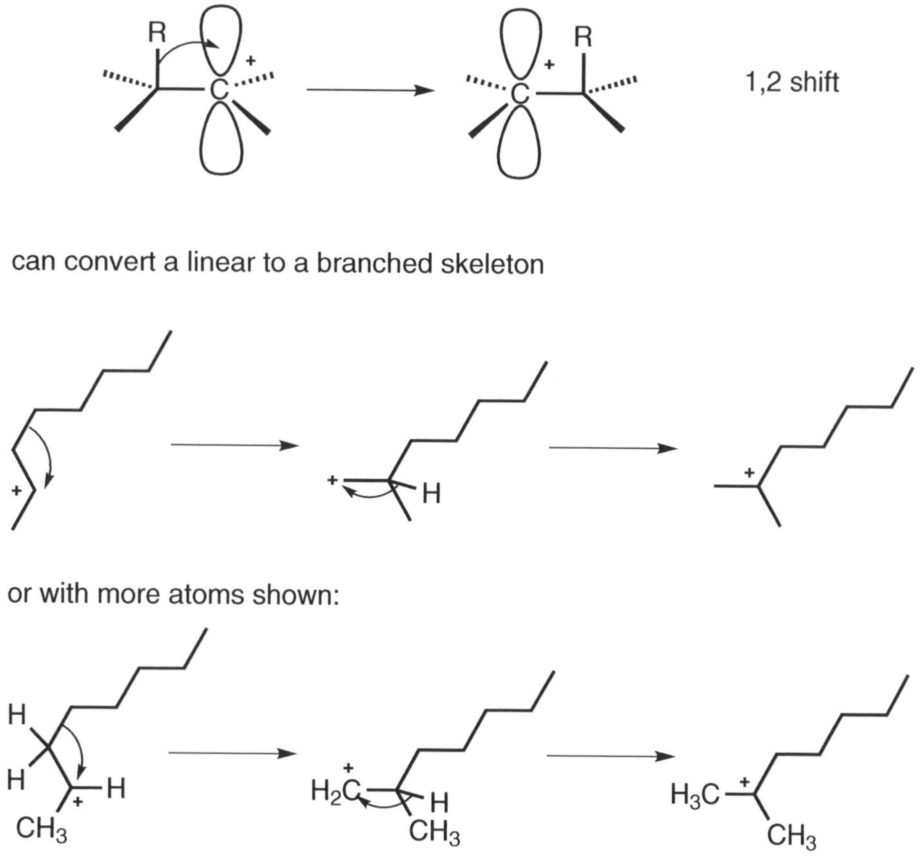

1,2 shift

can convert a linear to a branched skeleton

◀ **FIGURE   4.7**

Examples of How the 1,2 Shift of a Carbocation can Convert a Linear to a Branched Hydrocarbon

or with more atoms shown:

process. Carbocations accomplish this transfer of the charge to another atom in the same molecule by a reaction path that is characteristic for these intermediates, the 1,2 shift. As seen in the 1,2 shift (Figure 4.7), a reaction path of carbocations that takes place with great speed, passing the instability to another carbon atom is quite acceptable. One could see this as the acceptability of "passing the buck," a characteristic of chemical reactivity that is quite common, as we'll come across many times in the study of organic chemistry.

It is this 1,2 shift of carbocations, this "passing the buck," that is responsible for the transformations taking place in catalytic cracking of petroleum fractions and the other carbon skeleton rearrangements discussed above, and to be discussed in the following chapter when we turn our attention to biological phenomena.

In Figure 4.7 we have shown two 1,2 shifts that transform a hydrocarbon with a linear skeleton to a hydrocarbon with a branched skeleton.

For many years the intervention in chemical reactions of carbocations as intermediates was a hypothesis that was difficult to prove, even if chemists were certain that such species were responsible for the changes in structure that were seen, as in Figures 4.4, 4.5 and 4.7. The problem is that carbocations change and disappear as rapidly as they are formed. Finding an experiment to detect, let alone capture such a fleeting intermediate, was destined to change when a greatly talented young chemist in October 1956 saw that the tanks entering Hungry to quell the revolt against Soviet domination (a brutal action causing much loss of life) was reason to move to a western country. He did this with his family and much of his research group, first moving to England and then in the following year to Canada, where the Dow Chemical Company was establishing a new laboratory that took in the young researcher.

George Olah was already known for his work. Starting in the late 1940s, while still in Hungary, Olah began to investigate reactions in which carbocations were hypothesized to be intermediates. For example, he investigated the Friedel-Crafts reaction, which we are going to look at in detail (Chapter 6), a reaction that is a stalwart of the chemical industrial synthesis of important plastics and a reaction that also involves carbocations.

Olah's early papers and his interest in molecules that contained fluorine, an element that was to prove exceptionally useful in studies of carbocations, attracted the attention of Hans Meerwein, the German chemist who proposed carbocations as responsible for the rearrangements and other changes seen in the terpenes (Figure 4.4), the Meerwein of the Wagner-Meerwein rearrangement (section 4.3). This attention and that of other prominent chemists who had noticed Olah's work in Hungary certainly helped in his finding a way to continue his work and his finding suitable positions after escaping from Hungary.

Olah eventually established his research efforts in the United States where his work on carbocations led to the Nobel Prize in chemistry in 1994, awarded for finding ways to observe carbocations using spectroscopic techniques such as by nuclear magnetic resonance (NMR) (section 2.3). He accomplished this by extending the life of carbocations. This approach involved keeping them at exceptionally low temperature and by forming and maintaining carbocations in a "super acid" environment in the absence of reactive nucleophiles forcing the carbocation to remain in the charged state.

■ **George Olah**

PROBLEM 4.14
Atoms in the first row of the periodic table do not obey the octet rule but instead a rule that derives from the same basic idea. Describe this rule for hydrogen and see if it fits the hydrogen atoms in the various structures you've come across.

PROBLEM 4.15
If a positively charged carbon atom is hybridized sp^2 using the p$_x$ and p$_z$ orbitals, the remaining p orbital must exist along the y axis. Use this information to make three dimensional drawings using both the sawhorse and Newman projections showing how this remaining orbital can be anti, gauche or eclipsed with an atom on a carbon atom adjacent to the positively charged carbon atom.

PROBLEM 4.16
Can the concepts of conformational analysis (section 1.13) be applied to problem 4.15 and if so how?

PROBLEM 4.17
Draw a Newman projection (a sawhorse projection would work as well) to ask the question if rotation around the σ-bond between a carbocation carbon and an adjacent carbon has any effect on the 1,2 shift described in Figure 4.7. Answer the question using the terms, anti, gauche and eclipsed.

PROBLEM 4.18
Use a series of 1,2 shifts starting from a carbocation site anywhere along the chain of n-nonane, C_9H_{20}, to produce a branched isomeric carbocation. Can you imagine a more branched hydrocarbon with the formula C_9H_{20}.

H AVE YOU'VE TRIED YOUR HAND at problem 1.46? You've also been exposed to names of organic molecules as we've gone along up to now and, therefore, have some idea of how organic molecules are named and gained some knowledge in the area of nomenclature even without focusing on it. In this chapter we are about to come across a wide variety of hydrocarbons in the study of petroleum cracking and octane rating and related subjects concerning hydrocarbons. Some insight into the nomenclature that is used by organic chemists would help. Here are some essential points, some of which were noted in problem 1.46.

All linear hydrocarbons are named by the number of carbon atoms in the linear skeleton so that going from 1 to 10 we have: methane, ethane, propane, butane, pentane, hexane, heptane, octane, nonane and decane. Moreover, if the carbon skeleton is linear, then we call the hydrocarbon normal with the designation n before the name, as for example, n-pentane.

When presented with a structure to name we search for the longest string of carbon atoms and name the structure as a derivative of that number of carbon atoms. Let's see how this works. If the "ane" in each of these names is replaced by a "yl," such as methyl or ethyl and so on, this change designates a group with one carbon or two carbons and so on that is bonded to another atom. In this way, methyl chloride is H_3C-Cl. Or 2,2,4-trimethyl pentane would be a linear chain of five carbon atoms in which there are methyl groups, CH_3, attached at positions 2 and 4 of the chain, two methyl groups bonded to carbon 2 and one methyl group bonded to carbon 4. Later in this chapter we are going to look at terpenes, which are molecules based on a building block named isoprene. The carbon skeleton of isoprene has five carbon atoms, four connected in a chain and a methyl group on the second carbon of the chain. The name for such a skeleton is 2-methyl butane. That'll get you started and you'll progress with exposure, as for any language, as noted in the Introduction of the book.

4.5
We are shortly going to find it convenient to name the hydrocarbons involved in gasoline production. Let's therefore take a moment to step into the nomenclature of these molecules.

PROBLEM 4.19
Draw as many constitutional isomers of the formula C_9H_{20} as you can imagine and use your nomenclature skills to name them. Look ahead to section 4.6 to evaluate which of these isomers would have the highest and which the lowest octane numbers?

PROBLEM 4.20
Answer problem 4.19 but for structures with the formula $C_7H_{15}OH$ where the oxygen always is part of an OH group, a hydroxyl functional group.

4.6

How do carbocations produced in catalytic cracking increase the octane number of gasoline?

THE ANSWER TO THIS QUESTION, as noted above, is the 1,2 shift, by which a positively charged carbon atom shifts its positive charge burden to another carbon (Figure 4.7). But before we look into how these 1,2 shifts accomplish the increase in octane number observed in Houdry's cracking process, first let's find out some things about octane number.

As early as 1882 engineers were aware that internal combustion engines, which are defined by a spark-initiated fuel explosion in the cylinder, were subjected to premature ignition. Explosion of the fuel before ignition by the spark could be detected by knocking or pinging of the engine, which in some extreme cases could damage the engine and, in all cases robbed the engine of power. Because gasoline is a mixture of hydrocarbons it became interesting to determine which hydrocarbons were better or worse at causing this unwanted premature ignition.

In 1927 all parties interested in automobiles as well as fuel production formed what was called the Cooperative Fuel Research Committee. This lead to the concept of "octane number" based on the fact that the worst hydrocarbon for knocking was the linear hydrocarbon heptane while the best fuel, available at the time, which did not cause knocking, was 2,2,4-trimethyl pentane also known as isooctane. Octane number came to be defined as the percentage of 2,2,4-trimethyl pentane that had to be added to n-heptane to give the same knocking characteristics as the fuel being tested. For example, a fuel with an octane number of 84, typical regular gasoline we buy at the pump, would be a mixture of hydrocarbons with the same knocking characteristics as a mixture of 84% 2,2,4-trimethyl pentane and 16% n-heptane.

2,3,4-trimethyl pentane, with a very closely related structure to 2,2,4-trimethyl pentane, is also a great fuel for an internal combustion engine, with an octane rating of 97. n-Octane, a structural isomer of these branched hydrocarbons, in contrast, has an octane number near to zero, a horrible fuel to be kept away from your car's engine.

Now take a look at Figure 4.3 and 4.6. Here we see the structure of n-octane and the structure of the carbocation formed at carbon atom 4 in the chain. In this figure we saw how two sequential 1,2 shifts from the carbocation in this figure could produce a branched hydrocarbon, 2-methylheptane, with a positive charge at carbon 2 in the chain. Let's now see (**Figure 4.8**) how several more 1,2 shifts could convert the n-octane all the way to that great fuel, 2,3,4-trimethyl pentane, which has the same formula, C_8H_{18}, as n-octane. These two hydrocarbons, only one appropriate for the internal combustion engine are structural isomers (section 1.8).

Every transformation between chemical structures in Figure 4.8 is a 1,2 carbocationic shift with the curved arrows within each structure (section 3.15) showing the movement of the electrons responsible for the 1,2 shift. These were the chemical changes going on in the catalyst invented by Houdry and reported in 1939 (section 4.1). And these are the changes going on in modern catalytic crackers in the zeolite catalysts that have evolved from Houdry's invention, all driven by that empty p-orbital seeking electron density to rid itself of charge and gain the same number of electrons as in neon.

However, there are two reactions between the branched positively charged carbocations with the neutral molecule designated by R-H as shown in Figure 4.8, which are not 1,2 shifts - reactions that form neutral 2,3,4-trimethyl pentane. These are examples of the step that transfers the positive charge between molecules, an intermolecular change, rather than the 1,2 shift, which transfers the positive charge between carbon atoms within a molecule, an intramolecular change.

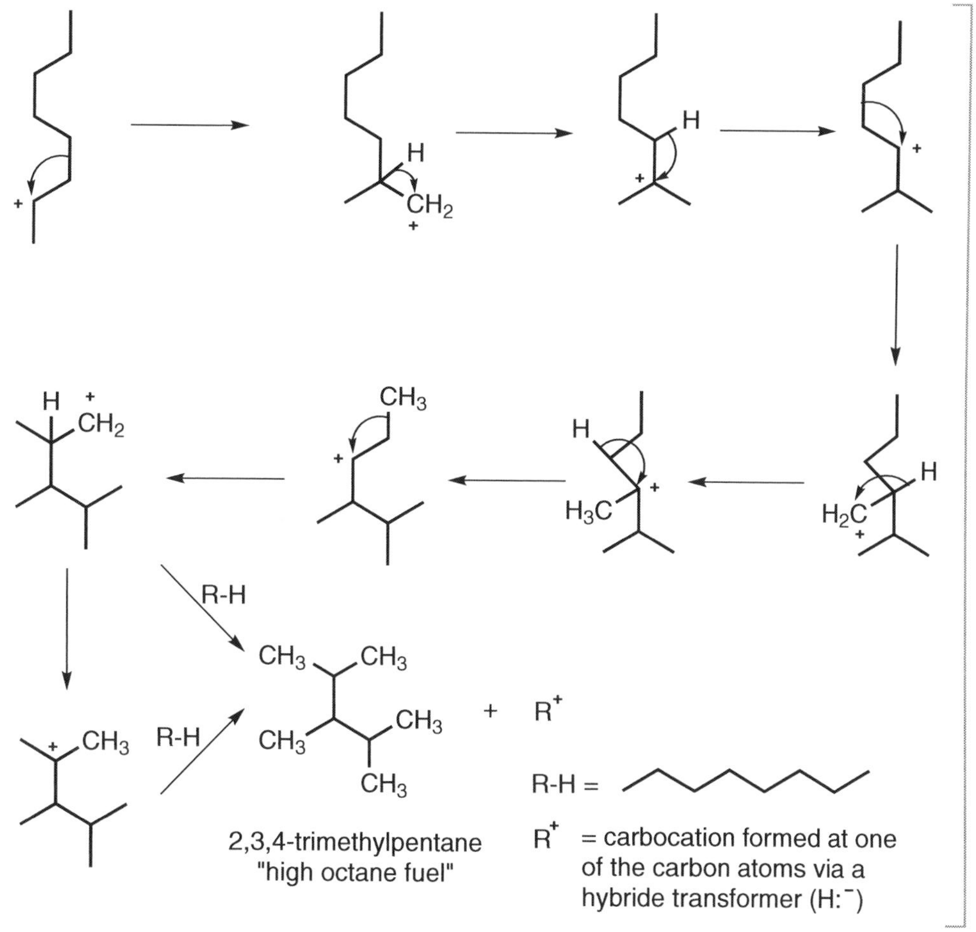

◄ FIGURE 4.8

Carbocation rearrangements convert n-octane to 2,3,4-trimethyl pentane. A poor fuel with a very low octane rating is converted to an excellent fuel.

2,3,4-trimethylpentane
"high octane fuel"

R⁺ = carbocation formed at one of the carbon atoms via a hybride transformer (H:⁻)

R-H =

The intermolecular reaction plays another role in converting a neutral n-octane molecule to the carbocation state initiating another series of 1,2 shifts that will transform the linear into the branched isomer. We've come across these commonly used terms in organic chemistry, inter and intramolecular, in section 3.15 (Figures 3.15 — 3.17) with regard to the chemistry of sugars.

The chemical reactions, the structural transformations, shown in Figure 4.8 define the essential chemical transformations in the catalytic cracking process. First of all, every reaction is reversible, a 1,2 shift can proceed in either direction. In addition, there are other 1,2 shifts that are possible and not shown here leading eventually to still other C_8H_{18} structural isomers. And as well, any neutral molecule in the mass of petroleum molecules from the fraction subjected to the cracking process can be a source of the intermolecular transfer of the hydride, the H⁻, from R-H, to a carbocation site in a charged molecule. The neutral molecule, which happened to collide with the charged molecule, is then itself transformed to a positively charged molecule and begins undergoing 1,2 shifts. But the key point is that all processes lead in the direction of forming branched structures. Why branched structures?

PROBLEM 4.21
Describe the electron occupation of the orbitals leading to the hydride ion structure, H⁻. Is there any relationship between the hydride ion and a noble gas?

PROBLEM 4.22
Consider five carbon atoms linked in a row, n-pentane. Now remove a hydrogen atom with the two bonding electrons from one of the terminal carbon atoms. A positive charge is therefore produced on this carbon atom. Propose a series of 1,2 shifts that would convert this carbocation to a carbocation with the structure $(CH_3)_2C^+(CH_2CH_3)$.

PROBLEM 4.23

Working with the carbon skeleton of n-octane, with a carbocation site at any carbon atom in the structure, explore all the 1,2 shifts you can imagine to generate a wide variety of branched carbocation structures. At any point in the path of the changing positively charged molecule, allow an intermolecular hydride shift from another n-octane molecule.

PROBLEM 4.24

Is there really any difference between an intramolecular 1,2 shift of a hydride and an intermolecular shift of a hydride? Are not both really the same process in principle? What do you think about this?

4.7

Why do carbocation rearrangements lead to branched structures? The answer has to do with how the stability of carbocations varies with molecular structure.

THE ANSWER TO THIS QUESTION is that carbon in an electron deficient state, that is, carbon with a positive charge and without an octet number of electrons, is most stable when most substituted. What does this mean, most substituted and why does substitution affect carbocation stability? And how is stability determined?

One approach, which gives an answer to the latter question, is to study a series of molecules in which a chlorine atom is bonded to a carbon atom that is differently substituted. Such a series is shown in **Figure 4.9.** In each one of these alkyl chlorides, a reaction can be made to occur in the gas phase in which the chlorine-carbon bond is broken so that the two electrons in the bond leave with the chlorine atom producing chloride anion and a carbocation site. Such a bond breaking is called heterolytic cleavage because the two atoms bonded end up in different charged states. Because the chloride anion is identical in each reaction compared, the endothermic enthalpy change of the reaction, ΔH, the energy necessary to break the bond, is a measure of the stability of the different carbocations produced. The more energy that is necessary to break the bond, the larger the ΔH, the less stable is the carbocation.

The information in Figure 4.9 makes sense. A carbocation is exceptionally electron deficient and relieving this electron deficiency is of primary importance to stability. It makes sense, therefore, that a carbocation site in a larger molecule, a molecule with more atoms and more electrons would be more stable than a carbocation site in a smaller molecule. After all, although we write the structure of the positively charged molecule as if the carbocation site were isolated on a single carbon atom, this idea can not be the whole story. If a molecule has a carbon atom within it that is positively charged, doesn't it make sense that this perturbation, this disruption, this "irritation," would be transmitted beyond the formal site of the positive charge? Or given multiple sites for a carbocation within a molecule does it not make sense that a carbocation site would be more stable on a carbon atom that is more highly substituted, one with more electrons around it?

Even if electrons are part of bonds not directly associated with the positively charged carbon atom in the molecule, the electron distribution within the molecule will be distorted by the positive charge. Every bond in the molecule will respond to the presence within the molecule of an atom that because it is positively charged is extremely electronegative (sections 1.4 and 2.4), more electronegative in fact than any atom without a charge. Every bond in the molecule and especially those bonds nearest to the carbocation will be polarized so that the electrons in that bond will be pulled in the direction of the carbocation. Such stabilization in quantum mechanical terms belongs to the realm of hyperconjugation or alternatively as no-bond resonance. But that is getting us into the idea of resonance, which is ahead of our story right now.

The larger the molecule and the more deeply buried the positive charged carbon is within the molecule, the greater are the number of electrons within that molecule to be perturbed by the positive charge, to reduce the "tension" of that positively charged site. The data in Figure 4.9, and so many more observations of carbocationic organic molecules, demonstrate the essential truth of this qualitative statement.

The qualitative prediction between carbocation stability and substitution outlined above is precisely in line with theoretical calculations based on quantum mechanics and most importantly is seen experimentally in the enthalpic data in Figure 4.9, which shows a rather large effect. The energy differences are huge. The most substituted, tertiary butyl (3º), is most stable, with isopropyl, the secondary carbocation (2º) next, followed by the primary carbocation ethyl (1º), and least stable, methyl.

So what does all this have to do with catalytic cracking induced rearrangements producing branched hydrocarbons? Once a carbocation is formed with the capacity for structural reorganization allowed by the 1,2 shifts (Figures 4.7 and 4.8), these rearranging shifts will continue to change the structure trying to find a tertiary (3º) site for the positively charged carbon. This effect is seen in Figure 4.8 and although the rearranging structure may sometimes go back and forth between 3º, 2º, and 1º sites the trend is always toward the tertiary site. This result is the basis by which catalytic cracking produces branched hydrocarbons.

The relationship between structure and stability in carbocations discussed here is the driving force in catalytic cracking production of branched structures, and contributes

$$R\text{-}Cl \longrightarrow R^+ + :\overset{..}{\underset{..}{Cl}}:^-$$

R	$+\Delta H$ (kJ/mole)	Carbocation class
H_3C	950	methyl
H_3C-CH_2	800	primary
$H_3C-\underset{\underset{CH_3}{\vert}}{CH}$	720	secondary
$H_3C-\overset{\overset{CH_3}{\vert}}{\underset{\underset{CH_3}{\vert}}{C}}$	650	tertiary

◄ **FIGURE   4.9**

Carbocation Stability Measured by the Endothermic ΔH Change on Loss of Chloride Anion from Various Alkyl Chlorides

to producing even higher octane number fuels using processes not involving catalytic cracking. 2,2,4-Trimethyl pentane with an octane number of 100 is produced by the petroleum industry using an approach that is also based on carbocation stability - more on this shortly.

PROBLEM 4.25
Bromomethyl cyclobutane has a -CH₂Br group bonded to one of the carbon atoms of a ring of four carbon atoms, cyclobutane. What reasonable chemical steps might rearrange the structure of this molecule following the loss of a bromide ion, Br⁻, from the original point of attachment to form a carbocation and then return of the Br⁻ to form a neutral rearranged structure?

PROBLEM 4.26
Might your answer to the question in problem 4.25 change if the –CH₂Br was bonded to one of the six carbon atoms of cyclohexane instead of cyclobutane? What's the difference?

PROBLEM 4.27

Consider the following molecule, $(CH_3)_3C-CH_2Br$. Now consider a related molecule, $(CH_3)_3C-C(CH_3)_2Br$. First draw the complete Louis structures, showing all bonds. Allow loss of the bromide ion, Br^-, to initially form the carbocation from each structure. Now allow return of the bromide ion to capture the carbocation by forming a carbon to bromine bond. What would you predict about the likely chemical changes?

PROBLEM 4.28

Consider a molecule with a basic structure of four carbon atoms in a row with a bromine atom attached to the first carbon and a methyl group attached to the next to the last carbon, the penultimate carbon atom: 1-Bromo-3-methyl butane, $(CH_3)_2CH-CH_2-CH_2-Br$. Show how loss of Br^- from its original point of attachment could, on return of the Br^- to the rearranged structure, lead to 2-bromo-2-methyl butane, $CH_3CH_2C(CH_3)_2Br$. What do the data in Figure 4.9 have to do with your answer? What roles do 1,2 shifts play in the transformation? Do any of the 1,2 shifts involve shifting of carbon atoms?

PROBLEM 4.29

Look back at the structural change in Figure 4.4. Take the starting structure, add a proton to the OH group and then lose water to form the carbocation. Are you able to carry out the transformation using 1,2 shifts?

PROBLEM 4.30

Consider the following alkene: 2-methyl-2-butene, $(CH_3)_2C=C(CH_3)H$,. Account for the fact that adding first H^+ and then Br^-, that is, the acid HBr, will produce the same product produced in Problem 4.28.

4.8

Getting the lead out of gasoline made the problem of producing better fuels even more critical and therefore it became essential to understand what structural features were necessary to produce higher octane number hydrocarbons.

I N THE LAST YEARS OF THE NINETEENTH CENTURY and into the twentieth century as automobile use increased, and engines became more powerful, the problem of finding a proper fuel for the internal combustion engine became a prime focus of automobile manufacturers. The often variable mixtures of linear hydrocarbons obtained from the distillation of petroleum, known as straight run gasoline or light naphtha were poor fuels as discussed in section 4.6. We learned in section 4.1 how Houdry's invention of catalytic cracking supplied large quantities of higher octane number fuel for internal combustion engines. But Houdry's process did not come along until the late 1930s, in time for the Second World War but late in the game for solving the problems automobiles faced.

Long before catalytic cracking, **Thomas Midgley, Jr.**, a mechanical engineer/chemist, who came from a family of inventors, and had graduated from Cornell University in 1911, was working for a subsidiary of General Motors, the Dayton Research Laboratories. In this position in 1921 Midgley discovered that an organic derivative of lead, tetraethyl lead, $(C_2H_5)_4Pb$, when added to the poor gasoline that existed at that time, was able to greatly reduce knocking. Midgley was working for **Charles Kettering**, the man who invented the first electric starter for automobiles, which was introduced in the 1910 Cadillac, and the man who founded the Dayton Engineering Laboratories Company, DELCO. In addition, the Memorial Sloan-Kettering Cancer Center in New York City is named after Kettering.

Midgley greatly helped the automobile industry in its fuel problem, but he doesn't come off too well in the historical record. On one web site I found, there is a quote from the Georgetown University historian, J.R. McNeill, with the claim that Midgley "had more impact on the atmosphere than any single organism in Earth's history." McNeill was not praising Midgley who had not only invented tetraethyl lead but also Freon, both of which we now know are serious environmental hazards, the former as a poison and the second for destroying the earth's ozone layer. But

Midgley can not be condemned for, at the time, unforeseeable consequences.

The discovery about the danger of tetraethyl lead followed on the work of **Clair C. Patterson,** a geochemist, who was studying lead to understand the geologic record of the age of the earth by using the known decay constant for the breakdown of uranium into lead. His estimate of close to 4.5 billion years still stands. In these studies Patterson discovered that a surprisingly large amount of lead contamination existed in his samples and was able to demonstrate that the tetraethyl lead added to gasoline was the source of this contamination, a contamination of the environment and all of us as well. Patterson took a great deal of abuse for this conclusion, especially from Ethyl Corporation, which was marketing tetraethyl lead but his view eventually prevailed with the passage of the Clean Air Act in 1970. While Ethyl Corporation had a great investment in tetraethyl lead it also was the company that employed Graham Edgar, the chemist who gave us octane rating. And also, Edgar was the chemist who first synthesized all nine structural isomers of n-heptane and, very importantly, the isomer of n-octane, 2,2,4-trimethyl pentane, the basis of determining octane number. It was this synthetic work and the subsequent testing of the physical properties and knocking characteristics of these isomers that quantitatively taught the petroleum and automobile industries of the importance of branching in hydrocarbons used for fuels.

▪ **Clair C. Patterson**

The isomers synthesized by Edgar are shown in **Figure 4.10** with their octane numbers and, also, other hydrocarbons for which octane numbers are readily available. You can expand your increasing familiarity with nomenclature (section 4.5) using this figure.

In the modern petroleum industry, catalytic cracking is not the only route to branched hydrocarbons with high octane numbers. Various catalysts that produce carbocation intermediates have been invented that are responsible for the conversion of isobutene and isobutane into Graham Edgar's high octane 2,2,4-trimethyl pentane (Section 4.10). When we look at the mechanism of the reactions that produce this 100 octane number fuel, we'll see another aspect of carbocation reactivity that makes sense based on what we've learned above.

But before we look into the mechanism of the combination of these two four carbon molecules, isobutane, $(CH_3)_4C$, and isobutene, $(CH_3)_2C=CH_2$, to the eight carbon molecule with the 100 octane number,

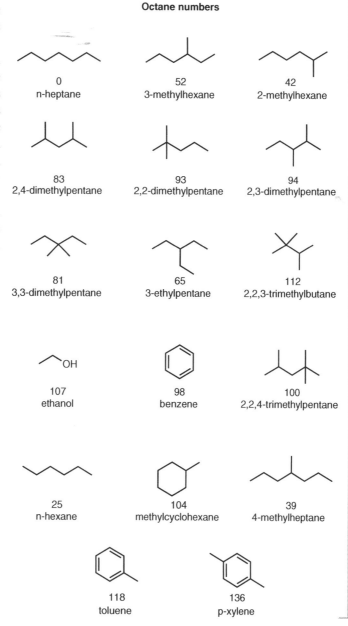

Octane numbers

0
n-heptane

52
3-methylhexane

42
2-methylhexane

83
2,4-dimethylpentane

93
2,2-dimethylpentane

94
2,3-dimethylpentane

81
3,3-dimethylpentane

65
3-ethylpentane

112
2,2,3-trimethylbutane

107
ethanol

98
benzene

100
2,2,4-trimethylpentane

25
n-hexane

104
methylcyclohexane

39
4-methylheptane

118
toluene

136
p-xylene

◀ **FIGURE   4.10**

Octane Numbers of Various Hydrocarbons

Johannes Nicolaus Brønsted

2,2,4-trimethyl pentane (Figure 4.10), we need to take a short diversion into what acids and bases are all about. Acid catalysis is the critical element in the reactions that produce high octane gasoline, which we are about to consider.

PROBLEM 4.31

Here's one to chew on. Use the following pieces of information to attempt to understand the basis of the difference between Diesel fuel and gasoline: (1) The structural variations of the C_7 and C_8 hydrocarbons in Figure 4.10 and their octane numbers; (2) In excellent fuels for Diesel engines, the fuel explodes in a cylinder from a compressive force rather than a spark, and fuels with low octane numbers are better Diesel fuels; (3) Fuels with low octane numbers in an internal combustion engine are characterized by premature ignition, that is, before the spark fires; (4) The C-H bonds in CH_3 are stronger than other kinds of C-H bonds and branched hydrocarbons have more CH_3 groups than linear hydrocarbons; (5) The premature ignition causing knocking in poor fuels in internal combustion engines is caused by explosive reactions initiated by free radicals; (6) Tetraethyl lead $(C_2H_5)_4Pb$ with weak bonds between carbon and lead, which produces free radicals that quench other free radicals, reduces knocking on addition to gasoline.

4.9

Industrial chemists and chemical engineers invented an efficient reaction path to high octane gasoline using chemicals obtained in large quantities from the catalytic cracking of petroleum. To understand how this was accomplished requires some understanding of the behavior of acids and bases.

Thomas Martin Lowry

WE'VE SEEN THE WORD ACID QUITE A BIT, although not spelling out what this term precisely means. There are different definitions of acids and in this chapter we have so far seen both the Brønsted-Lowry and Lewis varieties. Brønsted acids are readily recognized as molecules that give up a proton, H^+. But give up a proton to what? The answer is to a base, that is, to a molecule that can accept a proton. In some situations, the accepting partner for a proton, the base, is a molecule of the solvent such as the case when HCl, hydrochloric acid, or a carboxylic acid, such as acetic acid, H_3C-CO_2H, are dissolved in water.

In other situations the entity accepting the proton is a molecule that may not be part of the medium, such as the situation we'll discuss in the formation of the 100 octane fuel production (Section 4.10).

First, let's note that the development of the theory of acidity by **Brønsted** and **Lowry**, published in 1923, was built on an earlier proposal by Arthur Lapworth. Dr. Lapworth was Professor of Organic Chemistry at Manchester University in England. He proposed in 1908 that acids are donors of hydrogen ions and bases are acceptors of hydrogen ions, apparently first clearly stating the basis of our current understanding. However, another important contribution of Professor Lapworth is that **J. R. Partington**, who wrote the volumes, "A History of Chemistry," which I so much appreciate and use in writing this book, and which anyone interested in the history of chemistry must consult for complete and accurate information on the subject, was Professor Lapworth's first doctoral student in 1909.

The propensity of a molecule to donate a proton determines the acidic strength of this molecule and two or more molecules may be compared in this regard if the base is kept constant. The comparison can be expressed in quantitative fashion using the concept of pK_a, which is simply a version of the equilibrium shown in **Figure 4.11**, designed to place the comparison of acid strength on a logarithmic scale: $pK_a = -\log K_a$.

In Figure 4.11, the acid strengths of hydrochloric and acetic acids are compared, with water, the solvent, acting as the base. The true equilibrium constant, K, and K_a differ by the molar concentration of H_2O (about 56 Molar at 25° C), which hardly changes because it is the solvent, therefore being effectively a constant allowing it to be taken out of the equilibrium ratio and multiplied with K.

The information in Figure 4.11 shows that the stronger the acid, corresponding to larger values of K_a, the smaller is the value of pK_a, with the strongest acids having negative values for this parameter. Negative values for pK_a arise from the very large values of K_a.

The range of acid strengths among organic molecules is enormous. Methane, **H-CH₃**, for example could donate a proton to a water molecule only in principle. The K_a is estimated in the range of 10^{-48}, or therefore a pK_a as large as +48.

Ethanol, CH_3CH_2OH, which as in acetic acid, CH_3CO_2H, has a potential proton donor, **H⁺**, attached to oxygen, is not such an impossible proton donor to water, with a pK_a of about 16. The pK_a for ethanol, however, is still far larger than the pK_a of acetic acid, corresponding to a many orders of magnitude smaller equilibrium constant for loss of the proton to the surrounding water molecules. The pK_a of acetic acid is far larger, in turn, than the pK_a for HCl, corresponding to about 12 orders of magnitude difference in K_a. The structural basis of these differences resides in fundamental principles of organic chemistry, which will be dealt with shortly.

The other player in this acid base interplay is, naturally, the base, and bases have their own scale of ability to accept a proton from an acid. The weaker the base, the stronger must be the acid to be able to donate a proton to this base and vice versa. All bases, whatever their molecular structure, according to the Brønsted-Lowry definition, must have electrons available to form a bond with the donated proton. It makes sense therefore that a weak acid such as ethanol with a pK_a of nearly +16 may not be able to donate a proton to a base that would be able to accept a proton from hydrochloric acid with its pK_a of -7.

■ **James Riddick Partington**

$$HCl + H_2O \text{ (solvent)} \rightleftharpoons H_3O^+ + :\overset{..}{\underset{..}{Cl}}:^-$$

$$K[H_2O] = \frac{[H_3O^+][Cl^-]}{[HCl]} = K_a$$

◀ **FIGURE 4.11**

Definition of pK_a Applied to a Hydrochloric and Acetic Acids

$$K[H_2O] = \frac{[H_3O^+][CH_3\text{-}CO_2^-]}{[CH_3CO_2H]} = K_a$$

$$pK_a = -\log K_a = -7 \quad HCl$$

$$pK_a = -\log K_a = +4.8$$

The strength of a base, the ability of a molecule to accept a proton, can be related to the strength of an acid that could yield that base. What do we mean by that? Well, consider HCl with its pK_a of -7 in water. Loss of H⁺ from HCl yields Cl⁻, which is called the conjugate base of HCl. Cl⁻ is certainly a base with electrons available to accept a proton to reform HCl but it is a weak base considering that the stability of Cl⁻ is responsible for the acid strength of HCl. HCl is termed, in turn, the conjugate acid of Cl⁻. A far stronger acid than HCl, that is, an acid with a lower pK_a (larger negative number) than HCl, would be necessary to convert Cl⁻ to HCl.

Consider another example, at the opposite extreme. If one could somehow extract a proton from methane, CH_4, or form CH_3^- by some other means, which organic chemists know how to do, you would have in CH_3^- the conjugate base of an exceptionally weak acid. In fact CH_3^- is the conjugate base of methane, which could hardly be called an acid with a hardly measureable but estimated pK_a of near to +50. Exceptionally weak acids, to say the least, would be capable of converting CH_3^- to CH_4. CH_3^- is a powerfully strong base.

Completely at the opposite end of the spectrum, really in a different universe, is therefore, CH_3^- from Cl^-. The former is the strongest base one could imagine, which in a manner of speaking is wildly anxious to add a proton and return to CH_4, while the latter is one of the weakest bases known, perfectly stable and, in a manner of speaking, having no regret at having given up the proton by which it was formed from HCl. What this result means is that, while few molecules are strong enough acids to convert Cl^- to HCl, many molecules are capable of donating a proton to CH_3^- to form CH_4. Ethanol, let alone acetic acid, is a far more than a strong enough acid for CH_3^- but far too weak an acid to yield a proton to Cl^-.

Let's now try out some of these ideas in the problems below and also in helping to understand the chemistry in the next section by which 2,2,4-trimethyl octane is industrially synthesized.

PROBLEM 4.32
Hydrocyanic acid, $HC\equiv N$, has a pK_a of about 9. Write an equation for the chemical reaction that occurs when this acid is dissolved in water and assign the equilibrium constant for the reaction.

PROBLEM 4.33
The pK_a of the hydronium ion, H_3O^+, is nearly -2, well below zero. One could say that in an aqueous solution, a weak acid can be distinguished from a strong acid by having a pK_a above zero. Can you support this statement by observing the behavior of a weak versus a strong acid in water in quantitative terms using acetic acid versus hydrochloric acid?

PROBLEM 4.34
Would it be correct to say that an acid can be characterized as weak or strong depending on the base it reacts with and, if so, why? And if this conclusion is correct, does it mean that pK_a is solvent dependent, so that the values given in this section apply only to aqueous solutions?

PROBLEM 4.35
Go to the web or any text book of chemistry and you can easily find tables of pK_a values. Can you figure out which acids would donate a proton to the conjugate bases of other acids?

IN THE INDUSTRIAL SYNTHESIS of 2,2,4-trimethyl pentane (Graham Edgar's 100 octane gasoline, section 4.8), acid catalysis is absolutely essential. Strong acids have been traditionally used in this process, sulfuric acid, H_2SO_4 and hydrofluoric acid, HF. The conjugate bases of these acids, HSO_4^-, and F^-, are exceptionally weak bases, consistent with the discussion just above in section 4.9.

In the industrial synthesis, exceptionally strong acids are necessary because the base that must accept the protons from these acids is a rather weak base, although not as weak a base as Cl^- or either HSO_4^- or F^-. The molecule that must accept the proton, that is, act as the base, is isobutene (**Figure 4.13**). Isobutene is produced in petroleum processing in abundance making isobutene a reasonable starting material to produce a component of modern gasoline. Isobutene itself is not suitable as a gasoline or as a component of gasoline. It is too small of a molecule, making it too volatile and containing too little energy. But we'll discover how simple it is, in a clever industrial process, to convert two four carbon molecules, isobutene and a molecule easily obtained from it, isobutane, to the branched octane we need, 2,2,4-trimethyl pentane (Figure 4.13).

In isobutene we are introduced to a new functional group (section 3.9) to add to our growing list of functional groups. Isobutene is an alkene, a molecule that contains a carbon-carbon double bond. We've come across double bonds before this in the aldehyde functional group on carbon-1, the anomeric carbon of glucose and its diastereomers (section 3.13, Figure 3.14) and in carboxylic acids such as, for one example, acetic acid (Figure 4.11).

Aldehydes and carboxylic acids, among many other functional groups belong to the class of molecules containing a carbonyl group in which carbon is doubly bonded to oxygen. In Figure 3.14 we looked at the orbital picture of the carbonyl group in formaldehyde, the simplest aldehyde and discovered that the double bond is described as made up of two very different kinds of covalent arrangements, a sigma (σ) bond and a pi (π) bond. This combination is no less the situation for the double bond in an alkene as shown in **Figure 4.12**.

The linear overlap of orbitals in a σ-bond lends strength to this bond, that is, makes it more difficult to break, compared to the orbital overlap in a π-bond. It is this inherent weakness of the π-bond in the aldehyde carbon oxygen double bond in the open form of galactose that gives rise to the reactivity responsible for the tragedy of galactosemia (section 3.15). And similarly, this inherent weakness of the π-bond, even if the bond is between two carbon atoms rather than a carbon and an oxygen atom, allows isobutene to act as a base for the industrial acids used in the synthesis of 2,2,4-trimethyl pentane.

The acid base reaction, (1), that initiates the other reactions in Figure 4.13 produces a carbocation. Looking back at Figure 4.11, the acid base reaction shown between HCl

4.10

Chain Mechanisms and the Rule of Vladimir Vassilyevich Markovnikov

Figurative description of orbital accounting for ethylene applies to all alkenes

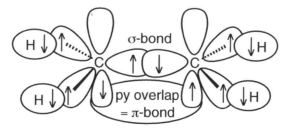

ethylene

sp^2 hybridization using 2s, 2px, 2pz

◀ FIGURE 4.12

Orbital Hybridization and Geometry of Ethylene

and H_2O also lead to a positive charge, but in that situation H_3O^+. In Figure 4.13 the base is also a source of electrons to make the bond with the proton. However, here we are not using a lone pair of electrons, as on the oxygen atom of water, but rather the two electrons that form the π-bond of the double bond of isobutene. There is another big difference in having the π-bond supply the electrons to form the bond to H^+. The proton can become covalently bound to either of the two carbon atoms of the π-bond with the other carbon atom becoming positively charged.

Why is the carbocation shown in Figure 4.13 produced, rather than the carbocation that could be produced by addition of the proton to the other end of the double bond? The answer can be found in the discussion in section 4.7 (Figure 4.9) where we discovered that the success of catalytic cracking of petroleum fractions arises from carbocations rearranging to the most substituted site in the molecule. If the proton

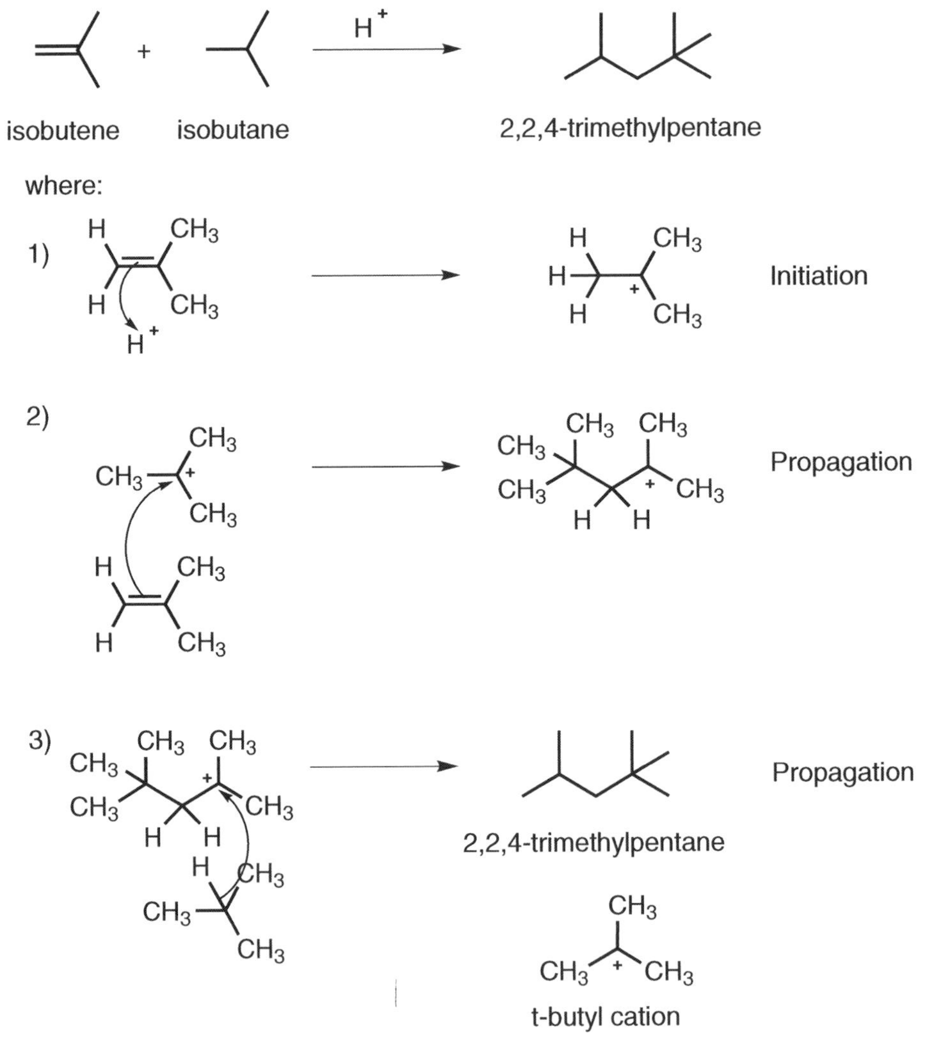

FIGURE 4.13 ▶

Carbocation Chain Mechanism for the Industrial Formation of 2,2,4-Trimethyl Pentane from Isobutene and Isobutane

addition to isobutene (Figure 4.13, (1)) had added to the most substituted carbon of the double bond, the carbocation site produced would have to reside at the primary carbon, at the $=CH_2$ end of the double bond, rather than at the tertiary carbon, the $=C(CH_3)_2$ end of the double bond. Formation of $-CH_2^+$ is the less stable situation. We'll say more about this shortly.

The carbocation produced by addition of the proton contributed by sulfuric or hydrofluoric acid to the $=CH_2$ end of the isobutene in reaction (1) (Figure 4.13) is called tertiary (t-) butyl cation. The proton addition is the initiation step of what is a chain mechanism in which this step leads to propagation step, (2) (Figure 4.13), which produces the product, 2,2,4-trimethyl pentane, and a new t-butyl cation.

The newly produced t-butyl cation from steps (2) and (3) then undergoes another reaction (2) with a new isobutene and begins the process again, hence the designation "chain reaction mechanism." This process will continue until one or both of the reactants, isobutene and isobutane are consumed or if some other step occurs that terminates the chain. If such a termination occurs, a new chain can begin by a new proton reacting with isobutene.

The t-butyl cation produced in step 3 occurs by a parallel reaction to one we've seen before in the mechanism of catalytic cracking shown in Figure 4.8 and discussed in section 4.6. In the catalytic cracking process, the 1,2 shifts that transformed the linear hydrocarbon to the branched hydrocarbons were intramolecular reactions, reactions that occur within a molecule, as defined in section 3.14. However, the branched carbocation produced by these intramolecular reactions is released from the rearrangment process by taking a hydride ion, H$^-$, from an uncharged molecule, R-H, in the petroleum fraction. This is a reaction between molecules and is called an intermolecular reaction. The production of the t-butyl cation in step 3 in Figure 4.13 is also the transfer of a hydride ion between molecules, another example of an intermolecular reaction transforming the branched carbocation to a neutral product molecule, 2,2,4-trimethyl pentane.

In all three steps in Figure 4.13 the carbocation produced is tertiary (Figure 4.9, section 4.7), which, as we've just discussed, is the most stable carbocation state we've discussed so far in the book. The tertiary carbocation is produced in reaction (1) by addition of the proton to the terminal carbon of isobutene, as noted above. Another tertiary carbocation is produced in reaction (2) by the addition of the t-butyl cation produced in reaction (1) to the terminal carbon atom of another isobutene molecule. In reaction (3) the new t-butyl cation is produced by transfer of a hydride ion, H$^-$, that is, transfer of a hydrogen atom carrying both electrons that had bound this hydrogen atom to the tertiary carbon in isobutane.

In this manner, 2,2,4-trimethyl pentane is produced and the tertiary carbocation that propagates the process, turns, so-to-speak, the wheel for another spin.

Remember the great Russian chemist Alexander Michailovich Butlerow(v), who contributed important advances to structural theory (section 1.11) and whose student Egor Egorevich Vagner we learned started the road to revealing the existence of carbocations, the Wagner in Wagner-Meerwein (section 4.3). Well, Butlerow(v) had another student whose name is known to almost all students of organic chemistry, **Vladmir Markovnikov.** Markovnikov was so highly thought of when he graduated with his doctoral degree in 1869 that he succeeded his mentor as professor at the University of Kazan. Although he lasted there for only two years before he left, he generated a rule that has become famous among students of organic chemistry"Markovnikov's Rule."

For reasons that are now seen as an "inspired guess," as put in an article about this rule by Peter Hughes from Westminster School in England, Markovnikov, in translation from the original German, wrote: "When an unsymmetrical alkene combines with a hydrohalic acid, the halogen adds on to the carbon atom containing the fewer hydrogen atoms, that is the carbon that is more under the influence of other carbons."

Precisely correct Professor Markovnikov, and we know the reason why, a reason the Russian professor could not have known in those early days of the science. In the addition of a hydrohalic acid to an alkene, the proton is added first, forming the most stable and therefore the most substituted carbocation, which then adds the negatively charged halide that the proton left behind.

For a full discussion of Markovnikov's rule look ahead to the discussion

■ **Vladimir Vassilyevich Markovnikov**

in Chapter 6 around **Figure 6.13**, In observing the reactions (1) and (2) of the chain mechanism in Figure 3.13, we could see the results as fitting into a slightly broadened application of Markovnikov's rule as he stated it so many years ago.

PROBLEM 4.36

An allene is a molecule in which two double bonds are connected to the same atom. Allene itself has the structure $H_2C=C=CH_2$. Propose hybridized orbitals consistent with this structure and predict the geometry of allene based on your proposed hybridization.

PROBLEM 4.37

Use hybridization of orbitals to describe acetylene, $HC\equiv CH$, and predict its geometry. Now use this hybridization to describe another molecule with a triple bond, hydrocyanic acid, $HC\equiv N$, making certain that you take account of all electrons associated with the nitrogen atom.

PROBLEM 4.38

Draw two dimensional structures for ethylene, $CH_2=CH_2$, changing the sp^2 hybridized orbitals from x and z, to x and y.

PROBLEM 4.39

We've seen that the reactivity of aldehydes arises from the weakness of a π bond in combination with the electronegativity difference between carbon and oxygen. In an alkene we have the π bond but not the electronegativity difference. How do the reactivity properties of alkenes still arise from the weakness of the π bond but differ from aldehydes because of the absence of the electronegativity difference?

PROBLEM 4.40

In addition reactions to double bonds the product of the reaction shows unequivocally if the mode of addition follows Markovnikov's rule. Draw the structures of alkenes in which addition of HCl could test this statement. Now draw the structures of alkenes where addition of HCl would not reveal if the addition followed Markovnikov's rule.

PROBLEM 4.41

In Figure 4.13 in step 1 the proton adds to the double bond of isobutene to produce the most substituted carbocation, obeying Markovnikov addition. In step 2 the carbocation produced in step 1 adds to the double bond of another isobutene to again produce the most stable carbocation. These two modes of addition, steps 1 and 2, produce the carbon skeleton of 2,2,4-trimethyl pentane. Now carry out steps 1 and 2 in conflict with Markovnikov's rule. Show the carbon skeleton this mode of addition leads to and name the neutral molecule that would arise from an intermolecular hydride transfer.

PROBLEM 4.42

The alkene, 3-methyl-1-butene, $CH_2=CH-CH(CH_3)_2$, in the presence of a Brønsted-Lowry acid is transformed to 2-methyl-2-butene, $CH_3-CH=C(CH_3)_2$. How does the rule of Markovnikov play a role in this transformation from a terminal double bond to an internal double bond? If the alkene, 4-methyl-1-pentene, $H_2C=CH-CH_2-CH(CH_3)_2$, is treated in the same way, it is transformed to 2-methyl-2-pentene, $H_3C-CH_2-CH=C(CH_3)_2$. What do these reactions reveal about the relationship between structure and stability of double bonds, a subject we will take up in later chapters?

PROBLEM 4.43

In work carried out by M. S. Kharash and F. R. Mayo at the University of Chicago in 1933, addition of HBr to certain alkenes seemed to go with the rule and against the rule of Markovnikov in a random way over many experimental trials. What would they have observed if HBr were added to isobutylene, $CH_2=C(CH_3)_2$? The result was finally attributed to the fact that instead of a polar addition, that is, first H^+, and then Br^-, the reaction path sometimes followed a free radical path with first addition of $Br\cdot$ and then addition of $H\cdot$. Write out a chain mechanism (as for carbocations in Fig.4.13) for this free radical path and account for the so-called anti-Markovnikov behavior. We'll learn about free radicals and their capricious behavior in Chapter 9.

REACTIONS (2) AND (3) IN FIGURE 4.13 are defined by the idea of acid base reactions just as much as the proton donation to water that occurs when hydrochloric or acetic acid is dissolved in water (Figure 4.11) or when a proton from a strong acid is added to isobutene (Figure 4.13, reaction (1)). But reactions (2) and (3) do not belong to the narrow definition of acids and bases within the Brønsted-Lowry concept. A proton is not added to a base in these reactions, a carbocation is added. A broader idea of acid base reactivity was proposed by **Gilbert Newton Lewis** in the same year, 1923, that the Brønsted-Lowry definition was published. It's about time you were introduced to G. N. Lewis, who took giant steps in the progress of chemistry.

Although here we will focus on Lewis' idea about what constitutes an acid and a base, G. N. Lewis's most well known contribution to chemistry was his understanding that the line drawn by organic chemists to designate the bonded relationship between two atoms in a molecule was, in fact, a shared pair of electrons. It's difficult to believe that what we now see as so simple an idea was not understood until Lewis' influential paper published in the Journal of the American Chemical Society in 1916. In this paper Lewis not only proposed the shared electron pair bond but also showed how this concept fit into a continuum of bonding under his heading of polar and nonpolar.

At the polar extreme, Lewis saw the electron pair as being transferred to one of the atoms, giving this atom a negative charge and therefore necessarily its bonding partner a positive charge - consider NaCl, that is, Na$^+$ and Cl$^-$. At the other extreme using Lewis' own example would be the carbon-carbon bond in a hydrocarbon, such as the ones we have studied in this chapter. He used n-hexane (section 4.5) for his particular example of a molecule in which the carbon-carbon bonds were formed by equal sharing of the electrons, one from each carbon atom.

And then there were all the bonds in between these extremes such as water, in which the electrons in the bond between oxygen and hydrogen are pulled in the direction of the oxygen atom and away from the hydrogen atom forming a polar bond. But in water we do not reach the extreme of a fully polar bond, what we now would call an ionic bond, as in NaCl.

It is a remarkable experience to read this 1916 publication to see what we now consider obvious to be presented for the first time. In fact Lewis' bonding concepts are considered to be the foundation on which the understanding of chemical reactivity and mechanism were built in the years to follow. Drawing a line between two atoms in a molecule was well enough for structural ideas at that time, which only required knowing where the atoms were, but knowing what that line meant in terms of electrons was necessary to understand the changes occurring in chemical reactions, which are almost always represented in organic chemistry by the curved arrow, that is, by the movement of electrons (section 3.14, Figures 3.15-3.17).

Moreover, the concept of the electron pair bond was also the foundation of structural chemistry as this insight was extended by quantum mechanical ideas into the realm of orbitals and then to the idea of Linus Pauling about hybridization of orbitals and the geometry of molecules (section 1.4). It is interesting to read letters, which can be found on the web, between Lewis and Pauling to see how much the older man at Berkeley appreciated the work of his younger colleague at the California Institute of Technology.

In section 4.7 the word resonance was introduced as a concept describing the distortion of electron density in a molecule with a positively charged carbon atom. Although we are not yet ready to delve deeply into this important concept of chemistry let's say just a bit more about the origin of the idea. Pauling contributed greatly to this area but again here it was Lewis who first introduced the idea that a single structural representation (drawing) may not in many situations properly describe the bonding characteristics of a molecule. Lewis suggested that more than one representation (drawing) may be necessary in which the atoms' positions are

4.11

The Brønsted-Lowry concept of acidity and basicity is too narrow and needs to be broadened to apply to the industrial process that produces high octane gasoline. One of the great chemists of the twentieth century, G. N. Lewis, took the idea further.

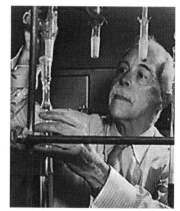

Gilbert Newton Lewis

not changed but the electrons are shown in a different arrangement.

Lewis did not use the word resonance, a word that evolved later, but the original concept was his. Let's say no more about this until it is necessary, a bit later (section 5.4). You don't need to know about the details of resonance just yet.

G. N. Lewis, a chemist of constant innovation and original thought, was also a great educator. His brilliance was quickly recognized and before graduating from the University of Nebraska, he was sent to Harvard University to finish his education. His teachers in the Midwestern part of the United States recognized that this young man needed the influence of the leading scientists of his era.

Lewis was successful in his studies at Harvard and in fact was asked to become an instructor at this famous school. But this did not last for reasons that are unknown now, although Lewis is reported to have "bragged" later in his life, that he had been fired from Harvard. At any rate he was sought out by the University of California at Berkeley which gave him the position and the resources to build the chemistry program on this west coast bastion of excellence. And this Lewis did, building this department to one of the best in the world of science, which position Berkeley still maintains to this day. It is generally agreed that Lewis was a historical figure who brought America to the foremost place in chemistry both in his research, his theories and his teaching.

Let's now hear Lewis' ideas about acids and bases as expressed by **Gerald E. K. Branch**, one of Lewis's first hires at Berkeley and regarded as the colleague he was closest to in the development of his ideas about chemistry.

"Among Lewis' own application of this theory of the electron pair bond to chemistry was his generalized concept of acids and bases. In this theory, the base has a pair of electrons to share with the acid which has room for such a pair. Lewis' definition of an acid was therefore based on phenomena as well as theory. Thus an acidic hydrogen compound was classed as an acid, not only for its ability to form an addition compound with a base by a hydrogen bond, but also because it gives the proton to a base in an almost instantaneous process."

Well, Professor Branch's view of Lewis' acid base concept is an interesting statement to mull over for anyone, and especially for a beginning student of the subject. Let's translate the concept into more accessible language, language that will show us that Lewis would consider all three reactions (1), (2) and (3) in Figure 4.13 to be acid base reactions.

Professor Lewis was not aware of the concept by Brønsted and Lowry when he published his idea in the same year. He was thinking of the concept of acidity and basicity independently and realized some broad definition was necessary. Lewis considered that any molecule or an ion that lacked a pair of electrons in its valence shell should be considered an acid. And that any entity that can supply the electron pair to fill that empty shell of what he considered an acid, would be defined as a base. Typical examples of what Lewis had in mind as acids, which would not be acids in the Brønsted-Lowry definition, are shown in **Figure 4.14** in their reactions with a Lewis base. None of these Lewis acids would be considered acids in the Brønsted-Lowry definition. In the latter definition of an acid, the proton is not the acid, but rather the molecule that supplies the proton, for example HCl. In the Lewis definition the proton or the t-butyl cation (Figure 4.13) are both acids.

With the information in Figure 4.14 in mind look again at reactions (1), (2) and (3) in Figure 4.13. Reaction (1) fits both definitions if we consider the source of H^+, for example H_2SO_4, the Brønsted-Lowry acid and isobutene the base. In the Lewis definition, H^+ is the acid. Reactions (2) and (3) fit only the Lewis definition with the acids being the carbocations on the left side of these reactions. For reaction (2) the base is isobutene again, as it is the base in reaction (1). In reaction (3) the base is isobutane, which contributes two electrons forming a hydride ion, to satisfy the acid, the carbocation at carbon-4 of 2,2,4-trimethyl pentane.

In many situations in the material to follow in the book, the acid-base concepts of both Brønsted-Lowry and Lewis will offer understanding of the many reactions of

⁕ **Gerald E. K. Branch**

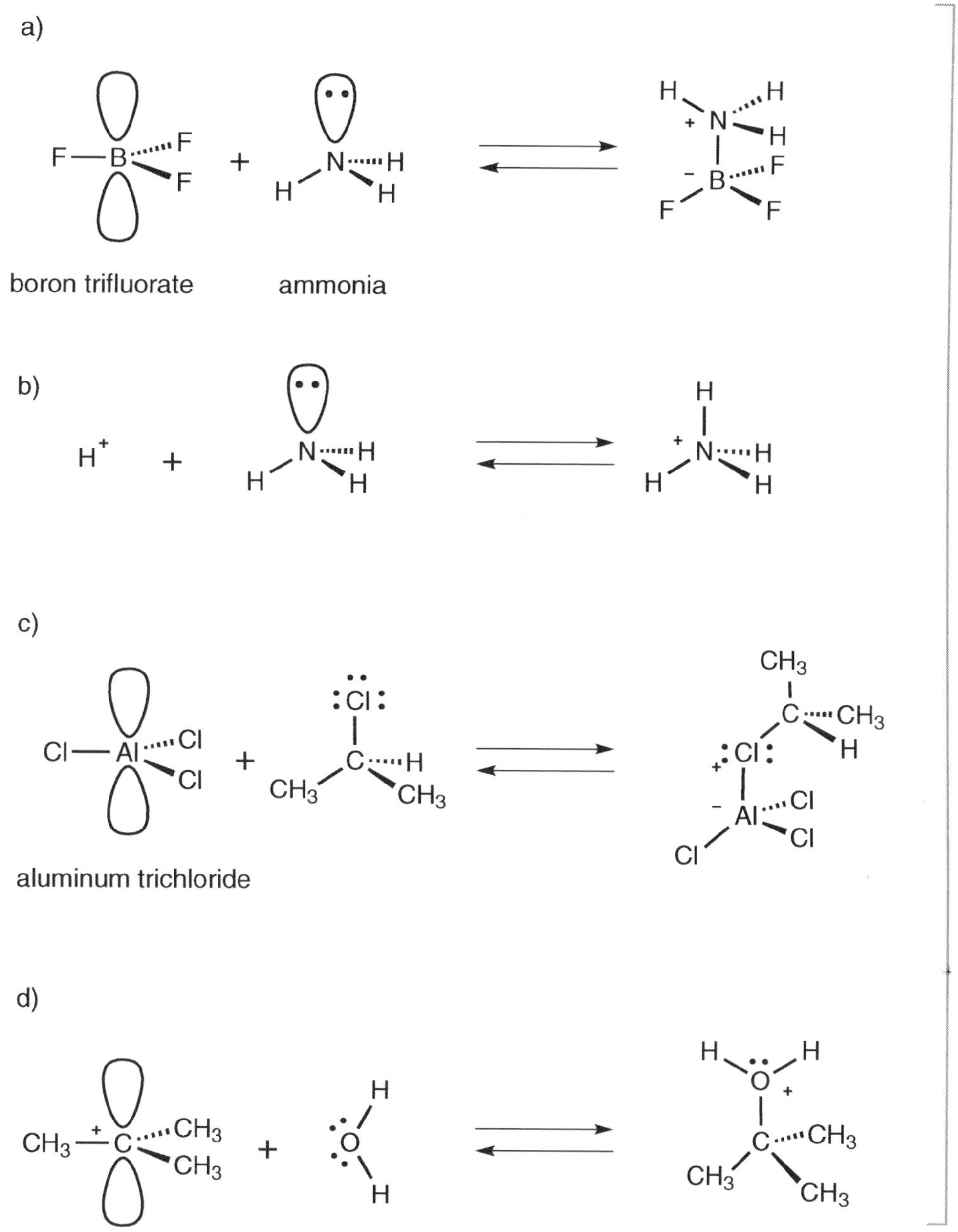

a)

boron trifluorate   ammonia

b)

◀ **FIGURE   4.14**

Examples of Lewis Acid/Base Reactions

c)

aluminum trichloride

d)

organic molecules from both biochemical and industrial examples. In fact, in some ways of thinking, as we'll discover, many reactions of organic molecules can be seen as belonging to one or the other category of acid base reactions.

PROBLEM 4.44

Answer true or false and explain your answer: (a) The conjugate base of a neutral acid will always be negatively charged; (b) HCl is not a Lewis acid; (c) H^+ is not a Lewis acid; (d) BF_3 is Lewis acid because of its empty p orbital; (e) HBr is a Brønsted-Lowry acid; (f) a molecule with a large pK_a is necessarily a strong base; (g) a conjugate base formed from an acid with a small pK_a will always be a weak base; (h) pK_a can be a negative or a positive number; (i) a double bond can act as a weak base; (j) the transformations of hydrocarbons in a zeolite cavity are not acid base reactions; (k) the conjugate bases of strong acids are unstable; (l) a negative pK_a corresponds to a small equilibrium constant for ionization of a Brønsted-Lowry acid; (m) the protonated conjugate base of a weaker acid will have a larger pK_a than the protonated conjugate base of a stronger acid.

PROBLEM 4.45
Assign acid or base designations to the reactants in Figures 4.13 and 4.14 according to the two definitions, Brønsted-Lowry or Lewis.

PROBLEM 4.46
Could the 1,2 shifts observed for carbocations, as in Figures 4.7 and 4.8, be considered to be acid base reactions within a single molecule, that is, intramolecular acid base reactions? If so, identify the acid and the base according to either the definition of Brønsted-Lowry or Lewis.

PROBLEM 4.47
Look ahead to Chapter 5. G. N. Lewis pointed out that a single structural representation (drawing) may not in many situations properly describe the bonding characteristics of a molecule and more than one representation (drawing) may be necessary in which the atoms' positions are not changed but the electrons are shown in a different arrangement. Are you able to imagine the structures of several molecules where multiple representations are necessary and several molecules where multiple representations are not necessary?

PROBLEM 4.48
Would it be correct to say that a single representation of the structure of the t-butyl cation in Figure 4.13 does not represent the complete characteristics of this ion? If you believe the answer to this question is yes, then how would you represent the structure of the t-butyl cation? Would resonance, that is, drawing multiple structures in which the electrons change but the atoms do not, help and if so how would you apply this concept of resonance to this situation? Read ahead to section 5.4 to help to answer this question.

PROBLEM 4.49
In the structure of benzene, six carbon atoms form a regular hexagonal ring, each bearing a single hydrogen atom. There is no formal charge and the octet rule is obeyed for every carbon atom in the structure. Is it possible to describe benzene by a single structural drawing (representation) considering that double and single bonds normally have different lengths? If more than one drawing is necessary, what do you suggest? Here we are looking ahead to Chapter 6.

CHAPTER 4 SUMMARY of the Essential Material

THE FOCUS OF THIS CHAPTER is the nature of carbocations, trivalent carbon with a formal positive charge. We learn how the empty p orbital in these sp² hybridized reactive intermediates is the basis of the powerful electrophilic properties of carbocations. The electrophilic nature of carbocations drives two kinds of chemical reactions, 1,2 shifts and addition to double bonds. The 1,2 shift is the reaction responsible for the rearrangements of chemical structure that occurs when carbocation intermediates are present. These kinds of rearrangements are responsible for the chemical changes occurring in many chemical reactions including in cracking of petroleum fractions, which leads to the change from the linear hydrocarbons found in petroleum to the branched hydrocarbons necessary to gain high octane ratings for gasoline. Addition of carbocations to double bonds is the basis of a reaction between two four carbon products of petroleum, isobutene and isobutane, which is used to synthesize a highly branched hydrocarbon that is a standard for defining octane number.

The foundation of the chemistry of hydrocarbons is the relationship between carbocation structure and stability and we spend time understanding why more substituted carbocations are most stable and how this manifests itself including the rule generated long ago by Markovnikov about addition to double bonds. We are introduced, in the industrial process for addition of carbocations to double bonds, to the chain mechanism, a favorite of industry, which we'll see several examples of later in the book.

Understanding these industrial processes requires learning about the differences between the Brønsted-Lowry definition and the Lewis definition of acids and bases and the fundamental nature of acid-base chemistry and how acidity and basicity is measured and defined.

We learn how increasing familiarity with aspects of nomenclature helps in discussion of the changes in structure focused on in this chapter.

Chapter 5
Carbocations In Living Processes.

We've seen the chemical properties of carbocations to be essential for the industrial production of high octane gasoline. Now we'll discover that these identical chemical properties are of no less use for nature's purposes-terpenes to steroids

IN POPULAR INTERPRETATIONS OF GENESIS, God is said to have created Eve from Adam's rib. In other words, man came to life before woman. Genesis was written quite a long time before the twentieth century when it came to be understood that the biological pathway to the sexual hormone that defines the female, estradiol, starts with the male hormone, testosterone, a kind of molecular confirmation of an ancient idea. Steroids are clearly very important molecular components of life and **Figure 5.1** shows the structures of these hormonal steroids.

As we're about to learn, all steroids are synthetically derived in vivo from a class of molecules known as terpenes, which arise by the linking together of molecules with five carbon atoms each based on the carbon skeleton of isoprene. The linking chemistry uses the same intermediate we've learned is responsible for the production of high octane gasoline, that is, the carbocation. Let's now discover how the behavior of carbocations in life's chemistry is driven by the identical forces at work in the laboratory and in the chemical industry.

Testosterone's source is another steroid, cholesterol, which brings us to the chemist who discovered cholesterol, **Michel Eugène Chevreul**, who was born in

FIGURE 5.1 ▶

Estradiol and Testosterone, the Female and Male Sex Hormomes

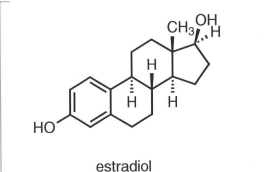

estradiol   testosterone

Angers, France in 1786, and when he died in 1889, yes, at nearly the age of 103, a public funeral was held in Paris demonstrating again how much this great chemist was known and appreciated. Three years earlier, medals with his image were issued on his one hundredth birthday. This was a greatly beloved man. Certainly important to this appreciation was not so much that Chevreul had lived through almost all the developments that took chemistry to the edge of modern structural theory but as much also for what he did that affected people's lives. For example, he transformed the manufacture of candles so that the average person could have reliable and pleasant light. And pleasant his candles were. Chevreul's research formed the basis of constructing candles without using wax derived directly from living sources, and therefore without unpleasant smells when they burned. His scientific work also set the stage for understanding the use of color in the arts and in the dye industry and not to be diminished, he possessed a jovial and generous outsized personality - a likeable person of great accomplishments, a winning combination.

Chevreul was born to a family of physicians but turned down a career in

■ **Michel Eugène Chevreul**

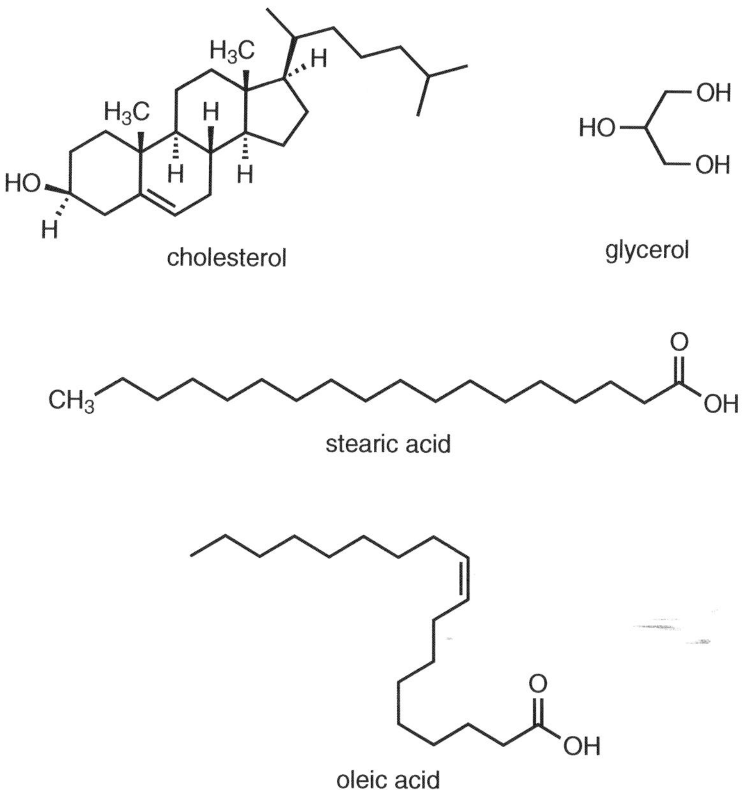

◀ **FIGURE   5.2**

Structures of Cholesterol, Glycerol, Stearic and Oleic Acids

medicine to enter the world of chemistry. He was a man of immense curiosity about the natural world, which probably accounted for his decision to enter the world of science. His investigations touched many areas and near the end of his life he even began to study the nature of aging, a topic which likely interested a man who had lived more than one hundred years and a topic of enduring interest.

Our interest here is about the work for which Chevreul is best remembered, his research on soaps and animal fats. These investigations among other advances led to his discovery of cholesterol, and several fatty acids, including most prominently, stearic and oleic acids and as well glycerol (**Figure 5.2**). Moreover, his efforts led to the understanding that fats are, in the words of that time, anhydrides of fatty acids and glycerol. Now we use the word ester to describe the functional group (section 3.9) formed by the combination of a carboxylic acid and an alcohol with elimination of water, that is, the functional group that defines the combination of a fatty acid and glycerol. More will be said about fatty esters later in Chapter 7 when we focus on how biological processes break them down with subsequent conversion to the energy that sustains us and, as well, builds them up to create our body's fat. But for now our focus is on the route that leads to the family of steroids.

The biological processes that lead to the molecules which Chevreul investigated, cholesterol and the fatty acids (Figure 5.2), begin with the same two carbon moiety, an acetyl group, in which a carbon with three hydrogen atoms, a methyl group, CH_3, is bonded to a carbon with a double bond to oxygen, a carbonyl group, C=O. The biological molecule that begins both the synthesis of fatty acids and of cholesterol is acetyl coenzyme A, abbreviated acetyl CoA, which is shown in **Figure 5.3**. Although acetyl CoA is the starting point on the route to the steroids and to the fatty acids, the first steps taken on the two paths quickly diverge. We are going to look carefully at biological syntheses in Chapters 7 and 8 where we will investigate the properties of negatively charged carbon, the carbanion. But for now let's focus on a key intermediate produced in the path to the steroids, isopentenyl diphosphate also shown in Figure 5.3.

FIGURE 5.3 ▶

Structures of Acetyl Coenzyme A and Isopentenyl Diphosphate

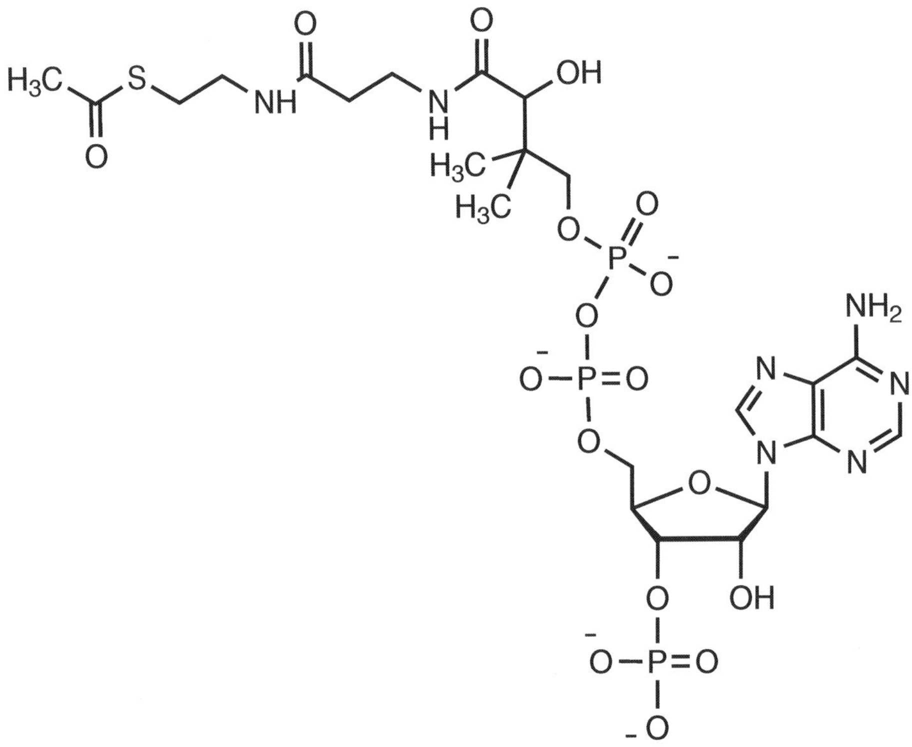

acetyl coenzyme A

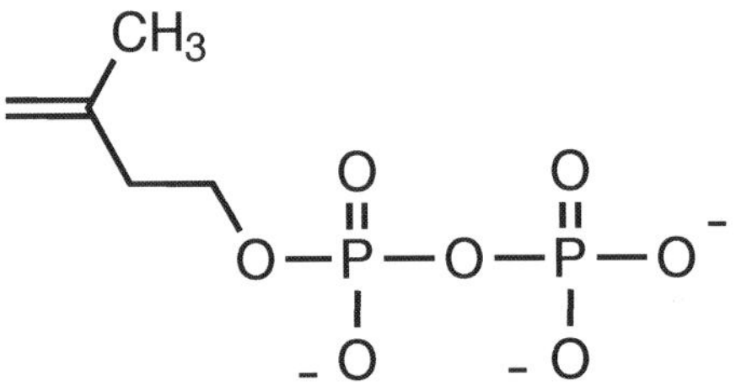

isopentenyl diphosphate

PROBLEM 5.1

In all structures in Figures 5.1-5.3 account for all lone pairs of electrons, including on phosphorus, and for the presence or absence of formal charge and attention to the octet rule where applicable. How many different functional groups can you find in these structures? Make a list of these functional groups and discover which ones you know the names for.

PROBLEM 5.2

If you have a set of molecular models and can share with others who also have a set, then you may have enough carbon and hydrogen atoms to construct cholestanol, which has the same structure as cholesterol (Figure 5.2) but with two hydrogen atoms added to the double bond. Construct the three six member rings in the structure in their chair forms so that the two sites of fusion of the three rings are made with all equatorial bonds within each of the three rings. Are the two methyl groups that sit at the two fusion sites in equatorial or axial positions? We will be studying the synthesis of cholesterol in Chapter 12.

MANY THOUSANDS OF MOLECULES have been isolated from natural sources in which the arrangement of the five carbon atoms in isopentenyl diphosphate can be recognized. **Figure 5.4** shows examples of these molecules, which are given the name terpenes and vary from so-called monoterpenes with ten carbon atoms, to sesquiterpenes with fifteen carbon atoms, diterpenes with twenty and so on up to rubber with uncountable numbers of carbon atoms.

According to many sources, the connection of human beings to terpenes derives from a tree common to the Mediterranean region, the terebinth tree. It's difficult to judge the truth of that claim because what we call terpenes are found in all plant matter, as well as in the animal and insect world. In a walk in any forest we are calmed by the odor of the multitude of terpenes exuding from the plants surrounding us. Whatever may be the history of our ancestors who took those first pleasant breaths and who eventually thought up the idea of creating perfumes from these sources, it is agreed that the terebinth tree is the source of the name terpene.

Near Istanbul in the Sea of Marmara there is a small island whose name in Turkish translates to Mother-of-Pearl Island and whose ancient Greek name, Terebinthos, translates to turpentine, a liquid, composed of various molecules of approximate formula $C_{10}H_{16}$, corresponding to that of monoterpenes. One doesn't have to travel to Istanbul and take a ferry to obtain turpentine. Copious amounts of terpenes are obtained on steam distillation of the exudate of all conifers and the oils of numerous plants including citrus fruits and eucalyptus trees. I remember my first visit to the

5.2

Terpenes and the Terpene Rule: the treasures of our existence, color, odor and taste, are greatly dependent on a class of molecules, the terpenes, which derive from a single five carbon molecule, isopentenyl diphosphate, and if this were not enough this molecule is also the building block of the steroids that control our sex, our nature and our behavior.

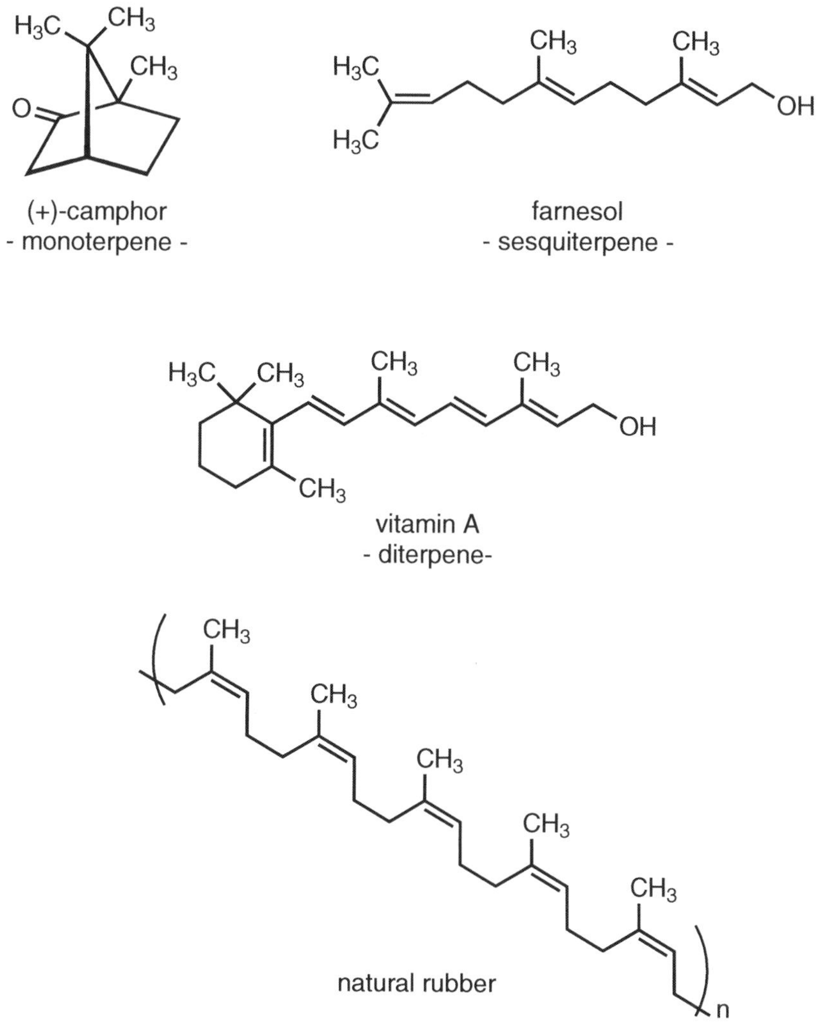

(+)-camphor
- monoterpene -

farnesol
- sesquiterpene -

vitamin A
- diterpene-

natural rubber

◄ **FIGURE 5.4**

Several Terpenes Varying from a Monoterpene to Rubber

University of California at Berkeley and the overwhelming wonderfully distinctive smell of the eucalyptus forest surrounding the campus.

At the other end of the molecular weight spectrum, but nevertheless also a terpene and where enough volatility to reach us is impossible, is natural rubber (Figure 5.4). Nature's elastomer is also obtained from a tree, *Hevea brasiliensis*, whose name derives from the place of its discovery by Europeans in the Amazon Rain Forest of Brazil. We'll be discovering much about rubber and other elastomers and the organic chemistry principles associated with this subject in Chapter 11.

FIGURE 5.5 ▶

Various Monoterpenes with Attractive Odors

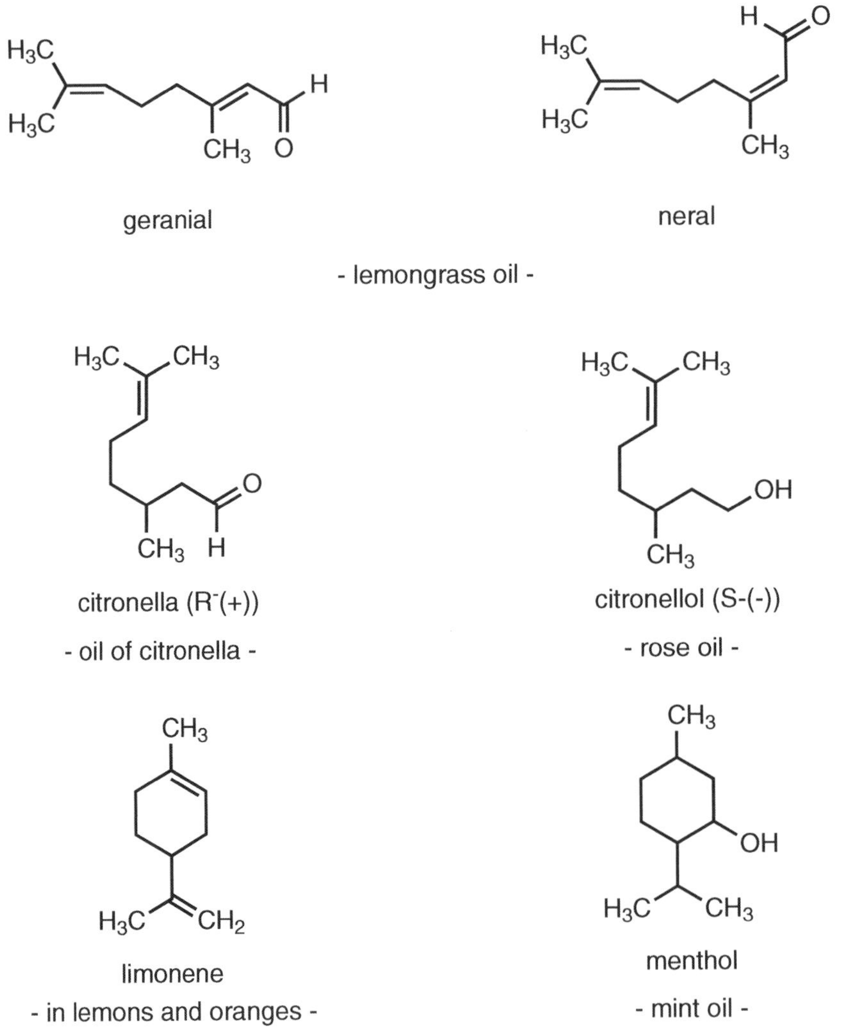

geranial

neral

- lemongrass oil -

citronella (R⁻(+))

- oil of citronella -

citronellol (S-(-))

- rose oil -

limonene

- in lemons and oranges -

menthol

- mint oil -

The historical roots of organic chemistry find their origin in investigations into our own lives and the plant and animal life around us and therefore it is hardly surprising that terpenes were a subject of investigation even before modern scientific methods were introduced, during the centuries when alchemy reigned,. In those ancient studies the word terpene didn't exist nor did any understanding of the molecular structure of these marvelous molecules but now we understand that many of the odors that intrigue us and form the basis of the perfume industry are terpenes. **Figure 5.5** shows several of these chemical structures, which are critical components of familiar scents: geranial and neral - lemongrass oil; α and β-ionones - odor of violets; geraniol - palmarosa oil of gingergrass; nerol - orange blossom; R-(+)-citronellal - oil of citronella; S-(-)-citronellol - rose oil; limonene - component of lemon and orange; (-)-menthol - mint oil. And these are only exemplary of the multitude of terpenes that contribute to the symphony of odors that envelope us from natural sources. When one speaks of the chemistry of terpenes there immediately comes to the fore the great chemist, **Lavoslav Stjepan Ruzicka**, who was born in 1887 in what is now Croatia. He was said to be the first in the history of his family of craftsmen and farmers to be educated beyond a few years of schooling. This turned out to be a good idea considering that Ruzicka won a Nobel Prize in 1939. The path to this prize began with his work with another Nobel Prize winner, Hermann Staudinger in Germany, with whom he investigated a natural material of use for insect control, Dalmation insect powder. Ruzicka took a life-long interest in what are called in organic chemistry, natural products or more particularly, essential oils.

Terpenes are the essence of essential oils in the plant world, which although investigated by chemists for over one hundred years before Ruzicka entered the field, were not understood to be built from a common starting material and with a recurring structural pattern. From his investigations over many years, mostly at the Eidgenossische Technische Hochschule in Zurich, Switzerland, Ruzicka saw, with a few understandable exceptions, that the multitudes of terpenes were all built of five carbon units based on isoprene, an insight he termed the "isoprene rule" (**Figure 5.6**). According to this rule, terpenes can be recognized by being able to break the carbon skeleton down to isoprene units. How this is done for several terpenes is shown in Figure 5.6 where a dashed line and numbered carbon atoms demonstrate that these molecules fit the isoprene rule.

■ **Lavoslav Stjepan Ruzicka**

◀ **FIGURE 5.6**

Demonstration of Ruzicka's Terpene Rule

In Ruzicka's Nobel lecture, the delivery of which was delayed until December, 1945, after the end of World War II, he pointed to the larger picture that would emerge from his studies, a picture that could reveal fundamental principles of how nature synthesizes the chemicals he had studied, and therefore would give insight into the workings of life. Let's use his words:

"*Attempts may be made to interpret the isoprene rule, not only as a working hypothesis in the laboratory, but also as a structural principle employed by nature. The structural similarities of the higher terpenes raise the question as to whether these compounds may have been formed according to a uniform principle in nature. At the present time, however, there is no point of reference which might lead to an interpretation of the mechanism of this biochemical process, which is so widely distributed in nature.*"

The mechanism of this "biochemical process," as Ruzicka put it, would later connect the terpenes, which so fascinated Ruzicka, to the world of steroids. But let us put this story aside for a moment to realize that the biochemical process Ruzicka was so interested in understanding brings us full circle to the creation of high octane gasoline. This connection is best expressed by an outstanding French chemist, **Guy Ourisson**, who summarized the biochemical synthesis of the terpenes in 1990 as follows (Ourisson used the term carbenium ion for what we have called carbocation and the term polyprenol for terpenes with multiples of five carbons):

"*The Biogenetic Isoprene Rule of the Zurich School has given us a universal interpretation of the structure of all known terpenoids: they derive from the simply acyclic polyprenols by "normal" chemistry, mediated by enzymes but involving "normal" reaction mechanisms. Of these mechanisms, the most important ones for the formation of the skeleton are the addition of a carbenium ion on a double bond, and the 1,2 migration on a carbenium ion: these insure not only the formation of the simpler acyclic precursors, the polyprenols, but also their cyclizations and the rearrangements of cyclized products.*"

Ourisson followed this statement with structural drawings showing the same 1,2 shift we have seen in Figure 4.7 and the addition of a carbocation to a double bond, which we have seen in reaction (2) in Figure 4.13. Apparently the chemical reactions we discover in our laboratories and our industries have been going on right under our noses, so to speak. And the carbocations are only one of numerous examples supporting this statement, as we shall see later in the book.

■ **Guy Ourisson**

PROBLEM 5.3
For the structures in Figure 5.6 identify all stereochemical possibilities and name the stereoisomeric pairwise relationships.

PROBLEM 5.4
Use numbering of carbon atoms to test the terpene rule for all the chemical structures in Figures 5.4 and 5.5.

PROBLEM 5.5
What conclusion can you draw about the biological source of stearic and oleic acids (Figure 5.2) by their structures? Are these fatty acids synthesized biologically by parallel mechanisms which create the terpenes?

PROBLEM 5.6
Does the structure of cholesterol obey the terpene rule? How can you correlate your answer with the fact that the source of cholesterol is isopentenyl diphosphate?

5.3

Carbocations may arise by the breaking of a chemical bond with the two electrons in that bond leaving with one of the participants of the bond. The participant that gets the bonding electrons is appropriately called the leaving group. Leaving groups act as an important driving-force in biological pathways.

THERE ARE FIVE CARBON ATOMS both in isopentenyl diphosphate (Figure 5.3) and in isoprene (Figure 5.6), and the two molecules have the same carbon skeleton. Isoprene however is not a contributor to the natural world of terpenes. Isopentenyl diphosphate stands in for isoprene, a fact that Ruzicka could not have known in those early days of the understanding of biological mechanisms. In those years the other important biological contributor to the terpene-steroid story, acetyl CoA (Figure 5.3), an essential building block of isopentenyl diphosphate, was also unknown.

We now understand that isopentenyl diphosphate is derived in vivo from three acetyl CoA molecules with the extra carbon atom being lost as carbon dioxide in the synthetic path, a process we are going to study in Chapter 8 . For now let's see how isopentenyl diphosphate can be the source of the terpenes shown in Figures 5.4-5.6 and further how the terpenes can lead to a molecule that is the precursor to all steroids, lanosterol. Carbocations, as we shall see, are the key.

In section 4.9, in our introduction to the properties of acids and bases, we introduced the concept of the conjugate base, the molecule produced when a Brønsted-Lowry acid loses a proton. A conjugate base produced by loss of a proton from a strong acid, an acid with a low positive or negative pK_a, must be a stable molecule. Otherwise, why would the acid easily give up the proton? And alternatively, a conjugate base arising by loss of a proton from a weak acid must be relatively unstable. The conjugate base instability retards the loss of the proton leading to a higher positive pK_a.

Consider Cl^- the conjugate base of HCl, or HSO_4^-, or $H_2PO_4^-$, the conjugate bases of sulfuric or phosphoric acids. The structural origin of the stability of each conjugate base may differ. For example Cl^- finds stability in attaining the electron configuration of argon while satisfying a highly electronegative character. The negative charges of sulphate and phosphate anions may be spread among three oxygen atoms in the former and two oxygen atoms in the latter, a favorable electronic arrangement, which falls under the heading of a subject we have not yet addressed, but which we will shortly discuss, resonance stabilization (section 5.4).

Whatever the source of a conjugate base, its stability resides within its structure, not its history. What this means, is that the molecule we are calling a conjugate base may not derive from an acid at all. The moiety we have been calling a conjugate base, for example Cl^- in Figure 4.9, may derive from the breaking of a bond in which a proton is not lost. The "conjugate base" may simply be a group that leaves with the two electrons that had connected it to another atom, whatever that atom may be – a carbon atom in the examples in Figure 4.9.

This possibility that the conjugate base structure may not arise by loss of a proton from an acid brings us to the idea of leaving group. For example, Figure 4.9 reveals the energy necessary, expressed as the endothermic change, for ionization of various alky chlorides. The ionizations produce different carbocations while always yielding the same anion, Cl^-. Now Cl^- is the conjugate base of HCl but here this anion is not produced from HCl but rather by heterolytic breaking of a bond between carbon and chlorine. where the two electrons in that bond go with the chlorine, heterolytic bond breaking.

Organic chemists use the term "leaving group" to describe Cl^- in this situation: the entity that has left the positively charged carbon atom behind. Cl^- is an excellent leaving group for the same reason that it is the conjugate base of a strong acid, HCl. In other words, if a molecular moiety that would form an excellent leaving group is bonded to carbon, then that bond is weakened and a carbocation is produced. If this leaving group is bonded to hydrogen, then that bond is also weakened and a proton is produced. When the bond breaks delivering the electron pair to the leaving group a stable anion is formed.

Biochemists have traditionally called weak chemical bonds high energy bonds and high energy bonds (section 8.9) are necessary to most biochemical transformations and therefore to life. High energy bonds require excellent leaving groups for the reasons just given and nature has focused on just a few excellent leaving groups to do the numerous biochemical jobs essential to life. The best among these are derivatives of phosphoric acid.

The structure of isopentenyl diphosphate in Figure 5.3 shows a molecule with an excellent leaving group, which would form $P_2O_7^{-4}$, on breaking the bond to the CH_2 group. This leaving group is essential to the role isopentenyl diphosphate plays in the biological production of the terpenes and the derived steroid family.

There are two players in the breaking of a chemical bond in which the electrons are unevenly distributed, what we have called a heterolytic cleavage. There is the moiety that takes the electron pair, the leaving group, and the moiety that is left behind, the positively charged carbon atom, as we've seen in Figure 4.9.

It makes sense that the more stable is the positively charged carbon left behind, the weaker is the bond to the leaving group. We see this in the ΔH values in Figure 4.9. The more stable is the carbocation produced when Cl^- leaves, the smaller the enthalpic change – the easier is the breaking of the bond. Applying this idea to isopentenyl diphosphate (Figure 5.3), does not point to a weakened bond in this biological molecule. If $P_2O_7^{-4}$ left, the carbocation site would be a $-CH_2$ group, a relatively unstable carbocation (Figure 4.9). To understand how nature solves this problem we have to learn about the concept of resonance, which we'll see has to do both with the stability of the leaving group, $P_2O_7^{-4}$, and as we shall see is the phenomenon which nature uses to change a $-CH_2$ group from a poor site to an accceptible site for positive charge.

PROBLEM 5.7

Bond energies are defined with negative H values. In other words the stronger the bond the larger is $-\Delta H$ or therefore the larger is the exotherm when the bond is formed from the elements that comprise the bond. Since smaller values of $-\Delta H$ may be considered higher energy, weaker bonds are called high energy bonds. When such weaker bonds are formed, a smaller exotherm is produced. Make a graph to demonstrate these ideas by looking up bond energy values on the web and notice that weaker bonds have smaller bond energies. An excellent site is http://www.cem.msu.edu/~reusch/OrgPage/bndenrgy.htm

PROBLEM 5.8

Must leaving groups always be associated with the formation of trivalent carbocations?

RESONANCE IS A SUBJECT alluded to in section 4.11 where we discovered that G. N. Lewis realized that molecular representations, the structures we now draw or make models of on computers or elsewhere, may not tell the whole structural story. He suggested that in the description of a single molecule it may be necessary to present the electrons in different ways while not varying the positions of the atoms to gain a complete, or that is to say, a true picture of the molecule. In other words, a single drawing although describing the positions of the atoms, may not adequately describe the bonding in that molecular structure. This idea was developed in its present form by Linus Pauling.

We've already heard about Pauling in Chapter 1 (section 1.4) on hybridization of orbitals to account for the bonding of carbon, but Linus Carl Pauling was so much more. He has been called the "Scientist for the Ages," and is recognized as one of the greatest scientists of the twentieth century. Pauling was born in 1901 and died in 1994 and his long life, his extraordinary intelligence and energy, and his social consciousness, touched almost all aspects of life in the century his life came close to spanning. Pauling felt keenly about the role of scientists in society and acted accordingly with the view that science has much to say beyond the boundaries of science, to the nature of how we conduct ourselves, our social behavior and politics, a view that led to a second Noble Prize, not, as for his first Noble Prize for his scientific work, but the Peace Prize in 1962.

Pauling's interest in the molecular basis of biology led him to propose, later in his life, the importance of nutrition in health and to create a field termed orthomolecular medicine. The most famous example of this view was in his proposal that large amounts of vitamin C could be important for human health. As with many of his ideas, including his opposition to testing of atom bombs in the atmosphere, which led the government to take away his passport during the early 1950s and therefore his ability to travel out of the United States, his ideas on orthomolecular medicine were attacked as quackery by the medical establishment.

I remember reading, when I was a young professor, a paper by Pauling in defense of using Vitamin C, which intrigued me for being so reasonable. Pauling proposed that, in contrast to other species, the behavior of human beings over the millennia has changed too quickly for our evolutionarily determined biochemistry to keep up with and particularly with the way we eat. For someone planning a life in science and wanting to know how the great man conducted himself I recommend reading about his life: http://lpi.oregonstate.edu/lpbio/lpbio2.html.

In a remarkable series of insights submitted for publication to the Journal of the American Chemical Society between February, 1931 and May, 1932 under the general heading of "The Nature of the Chemical Bond," Pauling at an age between 30 and 31, set the foundations for the thinking of all scientists dealing with chemical phenomena to this day. We've already noted two of these concepts, hybridization and electronegativity (sections 1.4, 2.4 and 3.13).

In the third of these contributions Pauling titled a subsection of the paper under the heading "The Transition from One Bond Type to Another." Here, he discussed how to deal with the possibility that a chemical bond may have both ionic and covalent character. Based on quantum mechanics, Pauling proposed that more than a single electronic structure may be necessary to describe the bonding, as G. N. Lewis had suggested earlier (section 4.11) and that if this were the situation, it would follow that the molecule would be more stable than if a single electronic bonding picture sufficed. In his words: "*The molecule could be described as fluctuating rapidly between two electronic formulas, and achieving stability greater than that of either formula through "resonance energy" of this fluctuation.*"

We understand now that the molecule is **not** fluctuating except in a theoretical sense and so the word resonance is certainly not the best word to use. In fact, there is a proposal for a better word, mesomerism, a word that however is not, unfortunately,

5.4

Resonance is the word used when a single molecular representation, a structural drawing for example, is inadequate to describe the distribution of electron density in a molecule. We compensate for this inadequacy by drawing multiple representations in which the atoms do not move but we draw the electrons as distributed differently. When multiple representations are necessary, when resonance is necessary, the actual molecule is more stable than that of any single representation— resonance stabilization.

used (section 10.6). The bonding situation in a single representation of a molecule may not, for certain molecules, be adequate to describe the true character of the bonding of the molecule. Without moving the atoms we attempt to draw multiple bonding arrangements and if they are all reasonable, as for example without excessive formal charge or not obeying the octet rule, then we can assume that the true picture of the bonding of the molecule is some hybrid of these structures, yielding an increased stability, a resonance stability for the molecule.

An excellent example of a molecule that is <u>not</u> subject to this stabilization is a saturated hydrocarbon such as isobutane (Figure 4.13) or for other examples, the hydrocarbons in this class in Figure 4.10; there is no better way to present these hydrocarbons than with all covalent bonds as shown. Other ways to present the distribution of electrons in isobutane are all unreasonable from the view of formal charge and octet rule.

The more equally the resonance structures contribute to the overall structure of the molecule the greater is the stabilization from this resonance. Benzene, with equally contributing resonance structures, is a molecule where representation by resonance is critical, a subject we'll take up in Chapter 6.

In Figure 4.11 we found that acetic acid is a moderate acid with a pK_a of 4.8. The conjugate base of acetic acid, acetate ion, is shown in **Figure 5.7** where we see that two equally contributing resonance structures are necessary for a true picture of the bonding–stabilization results, causing acetic acid to yield the proton more easily to a base. No resonance structures, for example, are possible for the conjugate base of ethanol $H_3C\text{-}CH_2\text{-}O^-=$. Ethanol has a pK_a of about 16, many orders of magnitude a weaker acid than acetic acid, even though the proton in both ethanol and acetic acid arises from an OH group.

Resonance stabilization plays an important role in the structures of HSO_4^- and $P_2O_7^{-4}$, which are the conjugate bases of strong acids and, therefore, act as excellent leaving groups (section 5.3). Figure 5.7 shows how resonance is used for a full description of these molecules.

Although we may realize the inadequacy of a single representation of the electron distribution of a molecular structure, we nevertheless often make do. Only when the representation is egregiously off in describing the properties of the molecule do we resort to the use of multiple descriptions (resonance structures) in which the electrons are placed differently while the atoms remain fixed. For example, in the structures in Figure 5.3 several resonance structures could have been drawn for both acetyl CoA and isopentenyl diphosphate. Many of the bonds in these molecules could have been represented with the electrons distributed differently.

In some situations instead of using multiple resonance structures, we add certain symbols, which are intended to add information about the electron distribution without drawing an entirely new resonance structure. An example is the symbol δ with a plus or minus sign, as seen for tertiary butyl chloride in Figure 5.7.

An example can be found in Chapter 3 (section 3.14) where the reactivity of the carbonyl group is discussed as arising from two factors, the weakness of the π-bond and the electronegativity difference between carbon and oxygen. These factors, important for the properties of carbonyl groups can be represented, as shown in Figure 5.7, by drawing resonance structures, or alternatively by using the δ nomenclature discussed above.

Traditionally, the double headed arrow is used to designate resonance structures, and to distinguish these structural representations from molecules that are in equilibrium with each other such as in a chemical reaction or with conformational isomers. In such equilibria, the positions of the atoms always change. Resonance is only a means to describe the distribution of the electron density within the molecule because our structural representations fail to adequately describe the electron distribution for some situations.

As you make your way through the maze of organic chemistry structures and the ways of chemists to present the behavior of these molecules, these methods will gradually become more and more familiar. Just as exposure to a language gradually increases your confidence in the use of that language, exposure to these kinds of structural representations will work in the same manner.

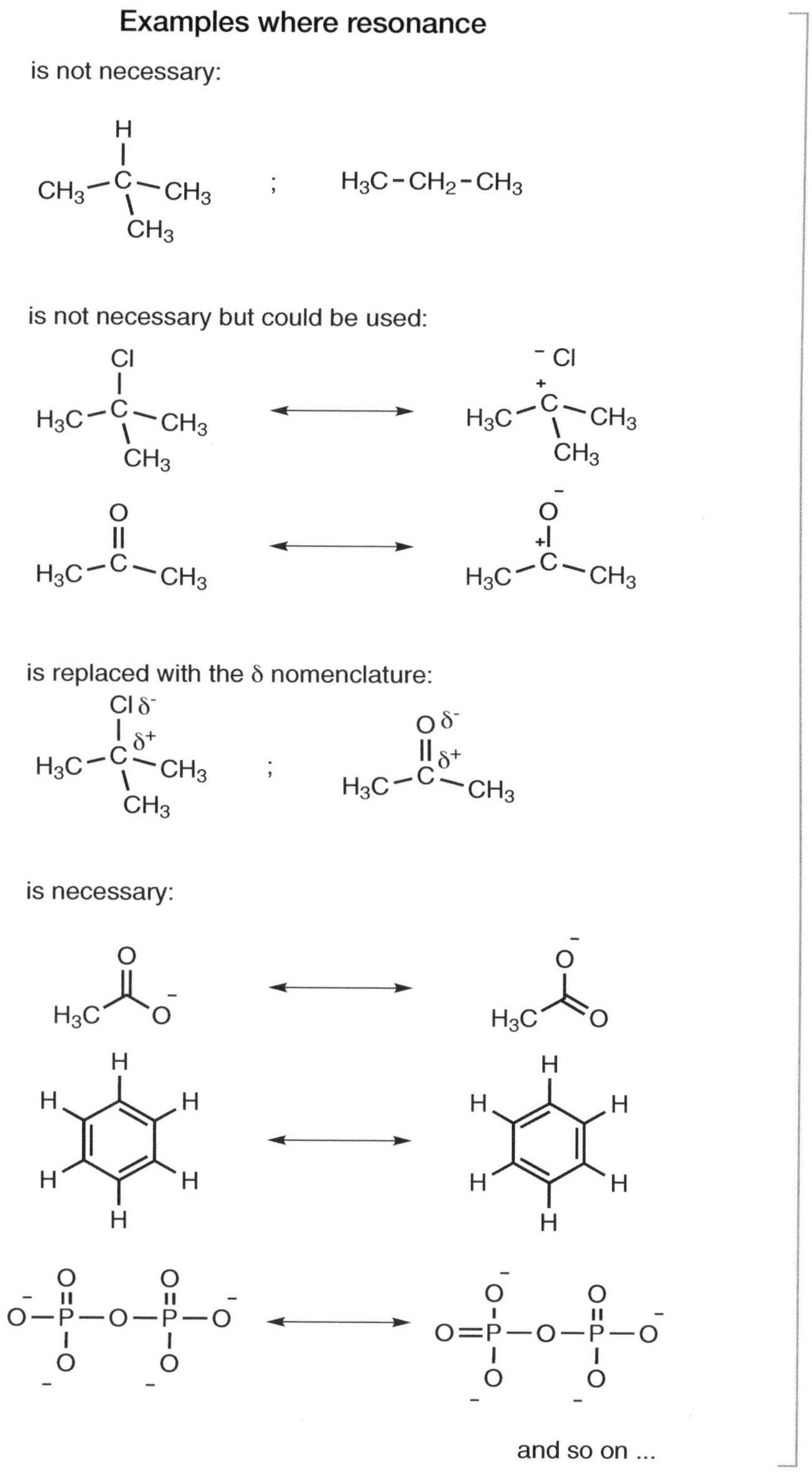

Examples where resonance

is not necessary:

is not necessary but could be used:

is replaced with the δ nomenclature:

is necessary:

and so on ...

FIGURE 5.7

Examples of molecular structures where resonance is, or is not, necessary.

PROBLEM 5.9

Use structural drawings to produce examples that show how the following concepts are connected: resonance; leaving groups; high energy bonds (weak bonds); low pK_a acids and their conjugate bases; heterolytic bond breaking.

PROBLEM 5.10

Use structural drawings to support the hypothesis that the low pK_a of both H_3PO_4 and H_2SO_4 arise from resonance stabilization of their conjugate bases.

PROBLEM 5.11

Although the Brønsted-Lowry acidity of a carboxylic acid, and of an alcohol, both involve loss of a proton from an OH group, the former functional group has a pK_a many orders of magnitude lower than the latter. Show how resonance theory is consistent with this large acidity difference.

PROBLEM 5.12

Although the tertiary butyl cation, $(CH_3)_3C^+$, is easily formed from $(CH_3)_3C$-Cl -in a solvent that favors ionization, a high dielectric solvent, this carbocation can only be produced from $(CH_3)_3C$-OH on addition of a Brønsted-Lowry acid such as H_2SO_4 or alternatively, by reaction of $(CH_3)_3C$-OH with a carboxylic acid to form an ester such as an acetate, $(CH_3)_3C$-O-C=O(CH_3). Explain these facts.

PROBLEM 5.13

In Figure 5.3, although only a single structure represents acetyl coenzyme A, there are other representations, resonance structures, for this molecule. What are they?

PROBLEM 5.14

All structural representations of molecules may be shown with differing arrangements of the electrons while keeping the atoms in the same position. How is one to know when a resonance structure is reasonable? Give several examples each of resonance structures versus equilibrium processes involving both chemical reactions and conformational changes.

A molecule may be thought of as atoms in a sea of electrons, as one might think of anchored ships in a vigorously moving ocean. How does this metaphoric view of a molecule fit and not fit with the resonance concept?

PROBLEM 5.15

In section 4.7 the term hyperconjugation was introduced (no-bond resonance) as applying to stabilization of a carbocation. How does this idea relate to resonance? Could hyperconjugation be reasonably applied to a hydrocarbon as in Figure 5.7?

5.5

Carbocations are the key to the syntheses of terpenes and steroids, but not without enzyme catalysis. Markounikov's rule is demonstrated in vivo.

THE STRUCTURE OF ISOPENTENYL DIPHOSPHATE (Figure 5.3) is not suited, by itself, for its role as a building block of the essential terpene and steroid molecules necessary for life.

The problem is the strength of the bond between the CH_2 and diphosphate groups. As noted in section 5.4, the CH_2 group is a poor site for the positive charge, and a carbocation would arise at the CH_2 group if the diphosphate group left with the two electrons that bind it to the CH_2 group, that is, to form $P_2O_7^{-4}$. A great leaving group does not have a good group to leave behind.

This problem, as we shall see, is easily solved by an isomerization, that is, by the transformation of isopentenyl diphosphate to its constitutional isomer dimethyl allyl diphosphate. As usual for chemical reactions in biological systems, the reaction is catalyzed by an enzyme. The enzyme is named following the reaction catalyzed, isopentenyl pyrophosphate isomerase with *ase* added (the words pyrophosphate and diphosphate are interchangeable).

Although understanding how the enzyme accomplishes the catalysis is beyond our reach right now, the folded state of the enzyme, exhibiting the backbone of the protein without showing the atoms, is reproduced in **Figure 5.8**. The structures of enzymes are experimentally determined by X-ray crystallography, which is of critical use to all of biochemistry and to the pharmaceutical industry as well, which uses enzyme structure in drug development.

The enzyme structure is not shown so much for a detailed analysis of the mechanism of the catalytic action of the enzyme but more to share with you the astonishment we all feel at the complexity and structures of these proteins. We will show other enzyme structures from time to time in this way without going into the exact mechanism of the enzyme action. In Chapter 7 we'll look carefully at the mechanism of the catalytic action of human pancreatic lipase, which aids in the catabolism of triglycerides. We'll come to understand that only a

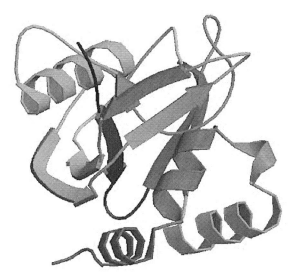

◀ **FIGURE 5.8**

Representation of the Folded Structure of the Enzyme Catalyzing the Conversion of Isopentenyl Diphosphate to Dimethyl Allyl Diphosphate

minute portion of the enzyme carries out the catalysis. Most of the enzyme serves only to bring the reacting molecules together in the precise way necessary for reaction.

Figure 5.9 exhibits one of the proposed mechanisms of the transformation of isopentenyl diphosphate to dimethylallyl diphosphate, the isomer in which the carbon bonded to the leaving group now allows, as we shall see below, resonance stabilization (section 5.4) of the positive charge. (By the way, notice that the "proper" names (nomenclature) (section 4.5) for these isomers are not being used –not unusual).

The precisely folded state of the enzyme shown in Figure 5.8 allows an isopentenyl diphosphate molecule to enter a site within the enzyme, the active site, with the double bond of the isopentenyl group perfectly aligned to accept a proton from the carboxylic acid group of the 116th amino acid in from the amino end of the protein, a glutamic acid residue. Figure 5.9 shows a representation of the mechanism. Quite amazing isn't it, all those hundreds of amino acids linked together into this long chain, this protein, fold into a shape just to make a perfect relationship between these two reactants and as we'll see another reactant to carry out the next step as well.

In section 4.10, in understanding the manner in which a proton adds to isobutene (reaction

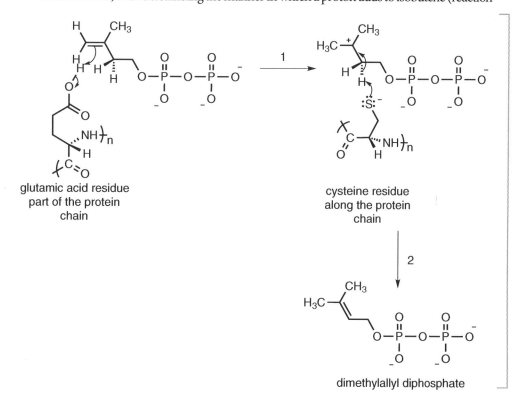

glutamic acid residue
part of the protein
chain

cysteine residue
along the protein
chain

dimethylallyl diphosphate

◀ **FIGURE 5.9**

Essential Mechanistic Steps In the Enzymatic Conversion of Isopentenyl Diphosphate to Dimethyl Allyl Diphosphate

1 in Figure 4.13) or how the produced carbocation adds to another isobutene, (reaction 2 in Figure 4.13) we understood that the end of the double bond to which the positive moiety adds, is controlled by the relative stability of the carbocation produced at the other end of the double bond – Markovnikov's rule. Carbocation stability is directly related to increased substitution (Figure 4.9) and therefore the incoming positive group adding to the double bond, be it a proton or a carbocation, will add to the least substituted end of the double bond.

To say it in another way, the π-electrons of the double bond are drawn to make the new bond with the incoming positive entity, therefore leaving the other end of the double bond, the more substituted end, to become positively charged. Markovnikov's rule, as we've discussed in section 4.10, predicted this reaction mode with double bonds long before the reasons for this mode of addition were understood.

Figure 5.9 shows Markovnikov's rule obeyed in the enzyme catalyzed reaction in step **1**. A proton, donated from the glutamic acid residue along the protein backbone, adds to the double bond of isopentenyl diphosphate, which is buried deep within the structure of the folded protein molecule represented in Figure 5.8.

The produced carbocation, shown in Figure 5.9, is precisely positioned for the next step. A cysteine residue, placed far away along the length of the chain from the glutamic acid residue, finds itself, because of the folded state of the protein (Figure 5.8), next to one of the hydrogen atoms on the CH_2 group adjacent to the carbocation site. Whereas the glutamic acid residue donates a proton, the cysteine residue, in its conjugate base form, is poised to accept a different proton in step **2**.

Here we find in these two steps leading to the dimethyl allyl diphosphate, two acid base reactions (sections 4.9 and 4.11). In step **1**, the double bond acts as a base accepting a proton from the carboxylic acid group of the glutamic acid residue while in step **2** the cysteine conjugate base accepts a proton from the $-CH_2$ group adjacent to the carbocation site.

The $-CH_2$ group therefore is a Brønsted-Lowry acid in this reaction. It may seem surprising that a hydrogen bonded to carbon could yield a proton, that is, could act as an acid, but this is not the usual carbon bound hydrogen. The adjacent positively charged carbon atom greatly weakens the adjacent carbon-hydrogen bond, which when broken forms not a negatively charged carbon but rather a carbon-carbon double bond, as shown in Figure 5.9.

The chemistry taking place in living systems is a subtle business and in this reaction of the cysteine anion with the CH_2 group there are two apparently equivalent hydrogen atoms available (Figure 5.9). Are they really equivalent? Does it matter which of the two hydrogen atoms on the $-CH_2$ group is abstracted, as a proton, by the negatively charged sulfur atom of the cysteine side chain ?

PROBLEM 5.16
How might the two enzyme catalyzed steps shown in Figure 5.9 be described as acid base reactions using the Lewis definition in place of the Brønsted-Lowry definition?

PROBLEM 5.17
Could steps **1** and **2** in the enzyme catalyzed reaction in Figure 5.9 be described in terms of the action of leaving groups and if so what would be the leaving group in each step?

PROBLEM 5.18
The consequence of the enzyme catalyzed reaction shown in Figure 5.9 is to convert a molecule that is not reactive in the steps to follow, to a molecule that is reactive. What is meant by this statement?

PROBLEM 5.19
How does resonance stabilization play a role in step **1** in Figure 5.9? If so, draw the resonance structures that define this role.

PROBLEM 5.20
Could the intermediate carbocation produced in step **1** in Figure 5.9 be stabilized by hyperconjugation that would help explain the loss of the proton in step **2**?

THE RELATIONSHIP OF THE TWO HYDROGEN ATOMS on the CH_2 group adjacent to the carbocation discussed above (Figure 5.9) does not depend on the carbocation. Their relationship is unchanged from the situation in isopentenyl diphosphate itself. The two hydrogen atoms are quite different. This difference can be understood by drawing two pictures in three dimensions of the CH_2 group (**Figure 5.10**). In one drawing we approach one of the hydrogen atoms with a symbol representing the enzyme, which we'll designate as $E_{(S)}$. E stands for enzyme and (S) stands for the fact that all the amino acids in the enzyme are of the (S) configuration (Figure 1.10).

Indeed, the enzyme itself is chiral not only from the (S) configuration of the amino acids but from the enzyme's overall shape as well, which is a chiral aspect of the structure that can not be designated by the Cahn Ingold Prelog (CIP) sequence rule. The point is that the enzyme is chiral with the (S) simply reminding us of this fact.

Let's look at the two drawings which appear in Figure 5.10. By approaching each of the hydrogen atoms in turn, we have made chiral the carbon they are attached to. A kind of chiral nomenclature can be used in this situation, which is adapted from the CIP method (section 1.12) and discussed below (**Figure 5.11**). Remembering what we learned about isomers in Chapter 1, ask the questions: are these two structural drawings in Figure 5.10 identical or isomeric? And if they are isomers, what kinds of isomers are they? If they are stereoisomers, what kind of stereoisomers are they? Are they enantiomers

5.6

Just as two molecules, which are constitutionally identical, can have a stereoisomeric relationship, two parts of a single molecule, which are constitutionally identical, can also have a stereoisomeric relationship.

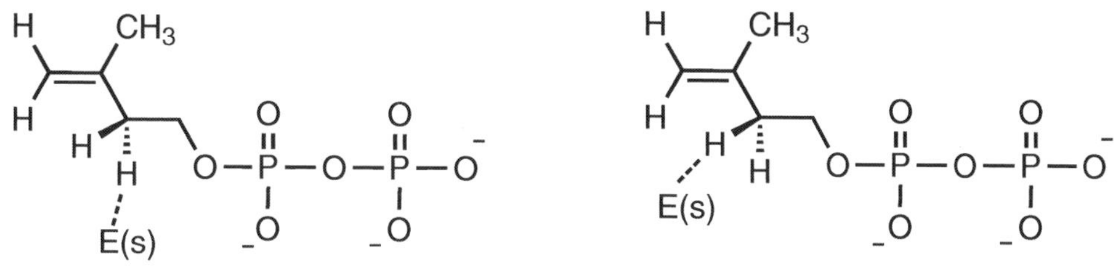

◀ **FIGURE 5.10**

Demonstration of Enantiotopic Hydrogen Atoms on a CH_2 Group of Isopentenyl Diphosphate.

or diastereomers? The answer one gets is that they certainly are stereoisomers, which means they are different but have identical bonding. Convince yourself that the two drawings are different, can not be superimposed on each other. Now convince yourself that these are not drawings of mirror images. For the two drawings to be mirror image related it would be necessary for there to be an $E_{(S)}$ for one structure and an $E_{(R)}$ on the other. The two structures can not be mirror image-related if both drawings contain $E_{(S)}$.

Because the two stereoisomeric drawings in Figure 5.10 are not mirror image related, these stereoisomers must be diastereomers. We learned that diastereomers differ in all

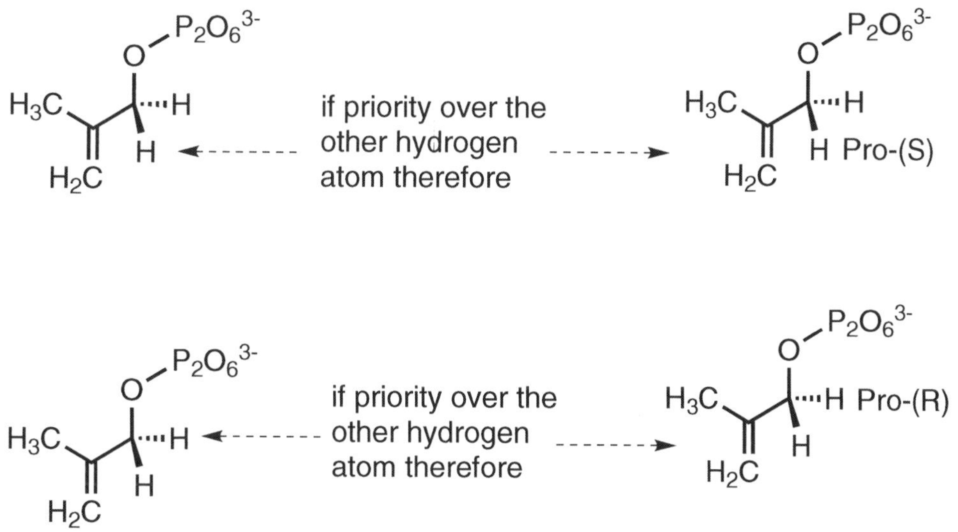

◀ **FIGURE 5.11**

Nomenclature of Enantiotopic Hydrogen Atoms

ways (section 1.8), in every aspect of their properties just as for two molecules of different structure. But the extent of that difference may vary from the very small to the very large depending on the details of the structure. The difference between two diastereomers can be profound, a life and death issue as we encountered in the difference between glucose and galactose in Chapter 3. What does all this mean in considering the difference in the interaction of the cysteine anion on the protein chain of the enzyme with these two hydrogen atoms (Figure 5.10)?

Enzymes have evolved to be exquisitely selective catalysts, which must mean that they fit perfectly, or nearly so, to the molecular changes in the reactions they catalyze. For example, as the isopentenyl diphosphate settles into the active site deep into the folded structure of the isomerase enzyme (Figure 5.8), a glutamic acid residue and a cysteine residue, which are located far apart along the backbone of the protein, must come close to the sites on the isopentenyl skeleton where they will act (Figure 5.9). The reactant fits like a key into a lock, the enzyme, as Emil Fischer, whom we learned about earlier for his work on sugars (section 3.11) put it more than one hundred years ago.

Such a fit means that the two hydrogen atoms on the CH_2 group can be expected to be enormously different in their relationship to the cysteine anion that will take one of them (Figure 5.9). And in fact using deuterium substitution for each of these hydrogen atoms in turn, chemists have learned that only one of the hydrogen atoms is taken by the cysteine in the isomerization process.

The other hydrogen atom on the CH_2 group does not take part in the reaction. You might think of lying on your back in bed next to a telephone. Your two arms, representing the two hydrogen atoms we are considering, are in an entirely different relationship to the telephone. The phone rings. You can pick it up without turning over only using one of your hands. The isopentenyl diphosphate carbocation rests in the enzyme pocket in only one way and turning over is not allowed. Only one of the hydrogen atoms can be reached by the cysteine residue on the protein chain.

Even when the enzyme is not present, even in a solvent in a glass flask, the two hydrogen atoms on the CH_2 group we have been focusing on in isopentenyl diphosphate differ from each other. In the presence of some chiral influence they could have distinguishable chemical shifts in the proton NMR spectrum and under ideal circumstances undergo spin-spin splitting to exhibit a doublet of doublets (sections 2.4 and 2.5).

If the two hydrogen atoms did not differ from each other in some manner, the enzyme would not be able to distinguish these hydrogen atoms, that is, to choose one over the other. Let's look further into how the two hydrogen atoms on the CH_2 group of isopentenyl diphosphate differ from each other.

Enantiomers can be distinguished by a chiral probe, be it a chiral molecule or as discussed in section 1.10, by chiral light. We have seen this stereochemical principle at work in the discussion around Figure 1.8. The reason that the enzyme can distinguish between the two hydrogen atoms on the CH_2 group we are focusing on in Figure 5.10 is that the enzyme is a chiral entity approaching two enantiomerically related hydrogen atoms. What we are discovering here is that just as two molecules can be related as mirror image isomers, as enantiomers, two structurally identical parts of a single molecule can just as well be related as mirror image entities. But here we don't use the word enantiomer, we call these structurally identical moieties within a single molecule, enantiotopic (pro-chiral), a term created by **Kurt Mislow** of Princeton University, who also gave us the term diastereotopic.

How can you tell if two structurally identical groups or atoms of a single molecule are enantiotopic? Draw the molecule in three dimensions, as already done in Figure 5.10, and give priority (as if its atomic number were higher) in turn to each of the structurally identical groups or atoms and determine if the atom they are connected to can be assigned (R) or (S) configuration. Because we are not dealing here with separate molecules but rather with parts of a single molecule, we use a different term.

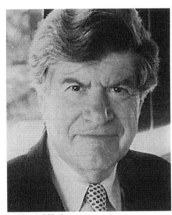

Kurt Mislow

Instead of (R) or (S) we use pro-R or pro-S. Figure 5.11 shows this procedure at work for the CH_2 group of interest in isopentenyl diphosphate.

The ability to distinguish perfectly, or nearly perfectly, enantiotopic groups in enzyme-catalyzed reactions is the best testimony to the highly evolved fit of enzymes to the reactions they catalyze and reveals some marvelous characteristics of biochemical reactions. One of the most amazing situations in which enantiotopic groups are distinguished is in the biochemistry of the citric acid cycle we are going to study in Chapter 8, and specifically in section 8.8.

PROBLEM 5 .21

Using the procedures discussed in the text associated with Figures 5.10 and 5.11, test the statement that in four coordinate carbon with two constitutionally identical groups and two different groups, R_2CXY, that the identical groups will be related enantiotopically. Assign pro-R and pro-S designations to the enantiotopic groupings in your test examples.

PROBLEM 5.22

Draw the structure of 1,3-propanediol, $HO-CH_2-CH_2-CH_2-OH$, showing all atoms in three dimensions. Using the principles in section 5.6, what are the stereochemical relationships between the following constitutionally identical parts of this molecule?

(a) the two CH_2-OH groups.

(b) the two hydrogen atoms on the central carbon atom.

(c) the two hydrogen atoms on either of the CH_2-OH groups.

If an enzyme catalyzed reaction was to take place, in which a CH_2-OH group on 1,3-propanediol was oxidized to an aldehyde group, CH=O, could the two hydrogen atoms be distinguished in this *in vivo* reaction? Could the *in vivo* reaction distinguish between the two CH_2-OH at the ends of the 1,3-propanediol?

PROBLEM 5.23

Explain the fact that enzymatic reduction by yeast of the deuterated aldehyde, 1-deuterioacetaldehyde, CH_3-C(D)=0, yields deuterated ethanol, CH_3CHD-OH, which is optically active. And why does the CH_3CHD-OH produced in this enzymatic reaction lose only the hydrogen atom and not the deuterium atom when enzymatically oxidized so that only CH_3-C(D)=0 is produced?

PROBLEM 5.24

A parallel to life (youth replaces age) is found as a consequence of the phenomenon of enantiotopicity in the citric acid cycle, a process which we will study in Chapter 8. Which constitutionally identical groupings of atoms in citric acid are enantiotopic? Are there other constitutionally identical groups within citric acid that are stereoisomerically related but are not enantiotopic? If so what are they and what is their stereoisomeric relationship? The structure of citric acid is: $HO_2C-CH_2-C(OH)(CO_2H)-CH_2-CO_2H$.

PROBLEM 5.25

If two constitutionally identical groups can be mirror image related, enantiotopic, then two constitutionally identical groups should also be possible to be related as diastereomers. Suggest a molecule to demonstrate a diastereomeric relationship between two constitutional identical groups within that molecule.

5.7

Why is a five carbon entity with the carbon skeleton of isoprene so well suited to produce such a wide variety of biologically important chemicals, the terpenes?

WE NOW HAVE TWO MOLECULES with double bonds, isopentenyl diphosphate and dimethylallyl diphosphate (**Figure 5.12**). The latter structural isomer is highly reactive. Loss of $P_2O_7^{-4}$ leaves behind not an unstable carbocation, $-CH_2^+$, but rather a resonance stabilized $-CH_2^+$ carbocation as shown in Figure 5.12.

Formation of the positively charged dimethyl allyl group causes a reaction with isopentenyl diphosphate which closely resembles reaction (2) in Figure 4.13 on the route to the high octane gasoline, 2,2,4-trimenthyl pentane. Again the Markovnikov rule is at work as seen in Figure 5.12. The addition of the carbocation to isopentenyl diphosphate occurs at the least substituted end of the double bond. (reaction 2, Figure 5.12)

Loss of a proton from the CH_2 group adjacent to the carbocation (carbon 7 of the eight carbon linear chain), just as in the synthesis of dimethyl allyl diphosphate in Figure 5.9 then yields geranyl diphosphate. Geranyl diphosphate is perfectly set up to form another resonance stabilized carbocation by loss of $P_2O_7^{-4}$ and therefore the ability to add another isopentenyl diphosphate to yield farnesyl diphosphate (Figure 5.12).

Virtually all chemical reactions taking place in vivo are enzyme catalyzed. The

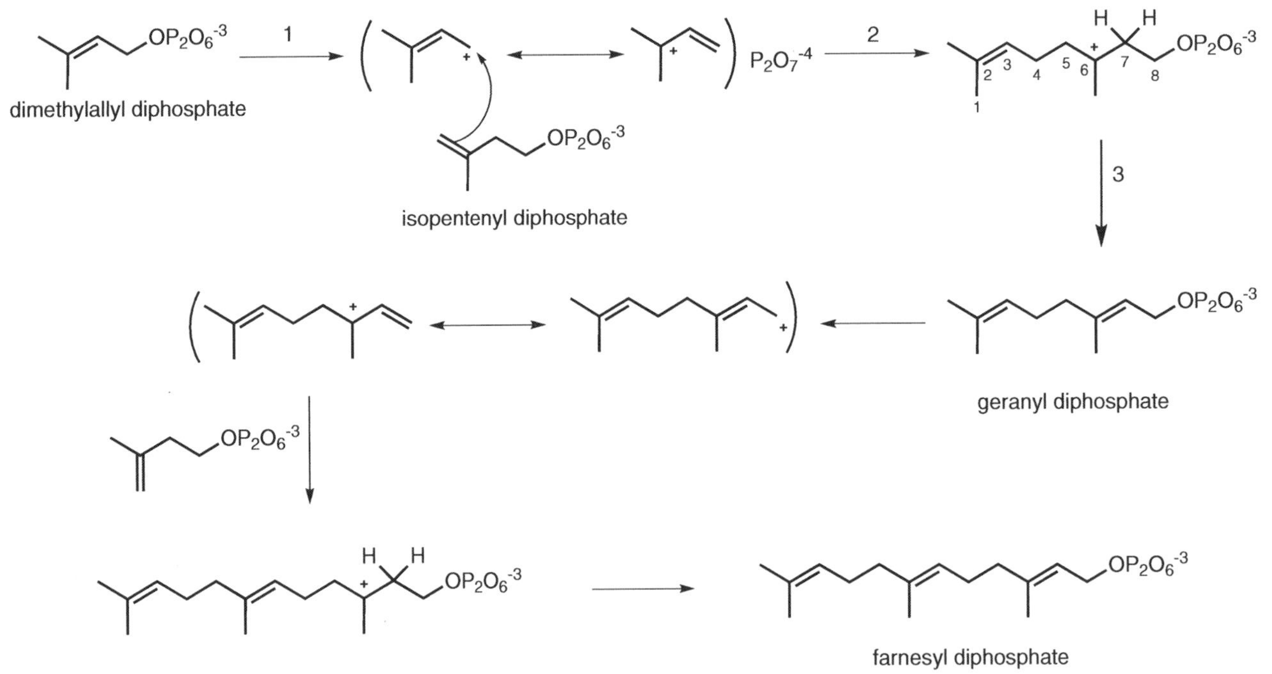

▲ FIGURE 5.12

Mechanism for the Conversion of Isopentenyl Diphosphate to Geranyl Diphosphate and Farnesyl Diphosphate

reactions shown in Figure 5.12 leading to both geranyl and farnesyl diphosphate are both catalyzed by farnesyl diphosphate synthase named, as usual, by the job being done by the catalyst. But we'll only focus on certain enzymes from time to time to make a special point. Our focus is on the organic chemistry of the substrates. We, nevertheless, stay aware that each enzyme catalytic process has its own fascinating details as we've seen in one example with the isomerase (section 5.5).

You might wonder where the names of the individual terpenes come from. The answer is usually from the plants from which they were first isolated. For example, the word geranyl derives from the genus Geranium from which terpenes with the carbon skeleton of geranyl diphosphate can be isolated. The Wikipedia web site for geranyl acetate in which the diphosphate group in geranyl diphosphate is replaced by acetate informs us that this terpene can be found in a cornucopia of flavors and fragrances: "*60 essential oils, including Ceylon citronella, palmarosa, lemon grass, petit grain, geranium, coriander, carrot and sassafras.*"

A similar search under farnesol, in which the diphosphate group is replaced by a hydroxyl group, revealed the source as a majestic oak tree of the genus Quercus robur

L., called in Italian La Farnia. This tree which grows to a very old age, even over one thousand years, and was once common to Europe and especially to the Po Valley in Italy, may have been the source of this sesquiterpene, that is, a terpene with fifteen carbon atoms. A distinguished family, the **Farnese** who were influential in Renaissance Italy and whose members included a Duke of Parma and a Pope is thought, perhaps, to have derived the family name from this oak tree. A Wikipedia site for farnesol revealed the interesting characteristics of this terpene:

"Farnesol is present in many essential oils such as citronella, neroli, cyclamen, lemon grass, tuberose, rose, musk, balsam and tolu. It is used in perfumery to emphasize the odors of sweet floral perfumes. Farnesol is also a natural pesticide for mites and is a pheromone for several other insects. In a 1994 report released by five top cigarette companies, farnesol was listed as one of 599 additives to cigarettes. It is a flavoring ingredient."

Given the geranyl and farnesyl carbon skeletons (Figure 5.12), nature has the capacity to synthesize all kinds of wonderful fragrances and flavors (section 5.2) that please us and terpenes that serve a multitude of biological functions, many of which are not yet elucidated. Just as one example of a well known terpene we should look at limonene, the main odor and flavor constituent of citrus fruits.

The mechanism of the enzymatic synthesis of limonene is shown in **Figure 5.13** where another chiral aspect of biological synthetic pathways is apparent. In the step that forms the six-membered ring, which is another example of a carbocation adding to a double bond, two enantiomers could be formed. Is it surprising that only one, the (R) enantiomer of limonene is found in citrus fruits?

The closing of the six-membered ring by making a bond between carbon 6 and carbon 1 in the structure in Figure 5.13 brings us back to section 1.13 where we learned that Pauling used quantum mechanical ideas to understand how rotation around single bonds can easily take place and how molecules can, therefore, rapidly change among a variety of shapes, that is, conformational isomers. There are single bonds in geranyl diphosphate between carbons 5 and 6, carbons 4 and 5, carbons 3 and 4, and carbons 3 and 2, which allow geranyl diphosphate to take a variety of conformations, which can rapidly interconvert.

Among these conformations is one, shown in Figure 5.13 (**2**) (represented by the resonance structures 2a/2b) which is almost, but not quite, suited to form the six-membered ring necessary for the synthesis of limonene. Structure **2** is indeed not quite suited to close the ring. The problem is that positively charged carbon 1 is pointed in the wrong direction to add to the double bond between carbons 6 and 7. Moving carbon 1 into the correct position to close the ring must involve rotation about a double bond.

Figure 5.13 shows the problem. Rotation around a carbon-carbon double bond, although not disturbing the σ-orbital overlap of the bond, must break the π-bond as the two ends of the double bond twist out of the same plane during the motion. Although the **p** orbitals can overlap in the initial state **2** and in the final state **3** reached after the conformational change exhibited in Figure 5.13, the p-orbitals must be perpendicular to each other in the intermediate state (look ahead to section 11.2), the midpoint in the conversion between **2** and **3**. The π-bond therefore breaks at the midpoint of the motion, making the motion not possible under normal circumstances.

The double bond between carbon atoms 2 and 3 in structure **2** (Figure 5.13) is not a normal double bond. This double bond is participating in resonance stabilization(2a/2b). The molecule may be considered as having overlap of the p orbitals at carbon atoms 1, 2 and 3 with a distributed positive charge. This p orbital overlap blocks the conformational motion necessary to form the carbocation necessary to close the ring, that is to form **3**.

What is hypothesized to occur to allow the necessary conformational motion is for the diphosphate group, $P_2O_7^{-4}$, to add reversibly to carbon 3 in the carbocation described by the resonance structures **2a/2b**. This structure (**i**) now has a true single bond between

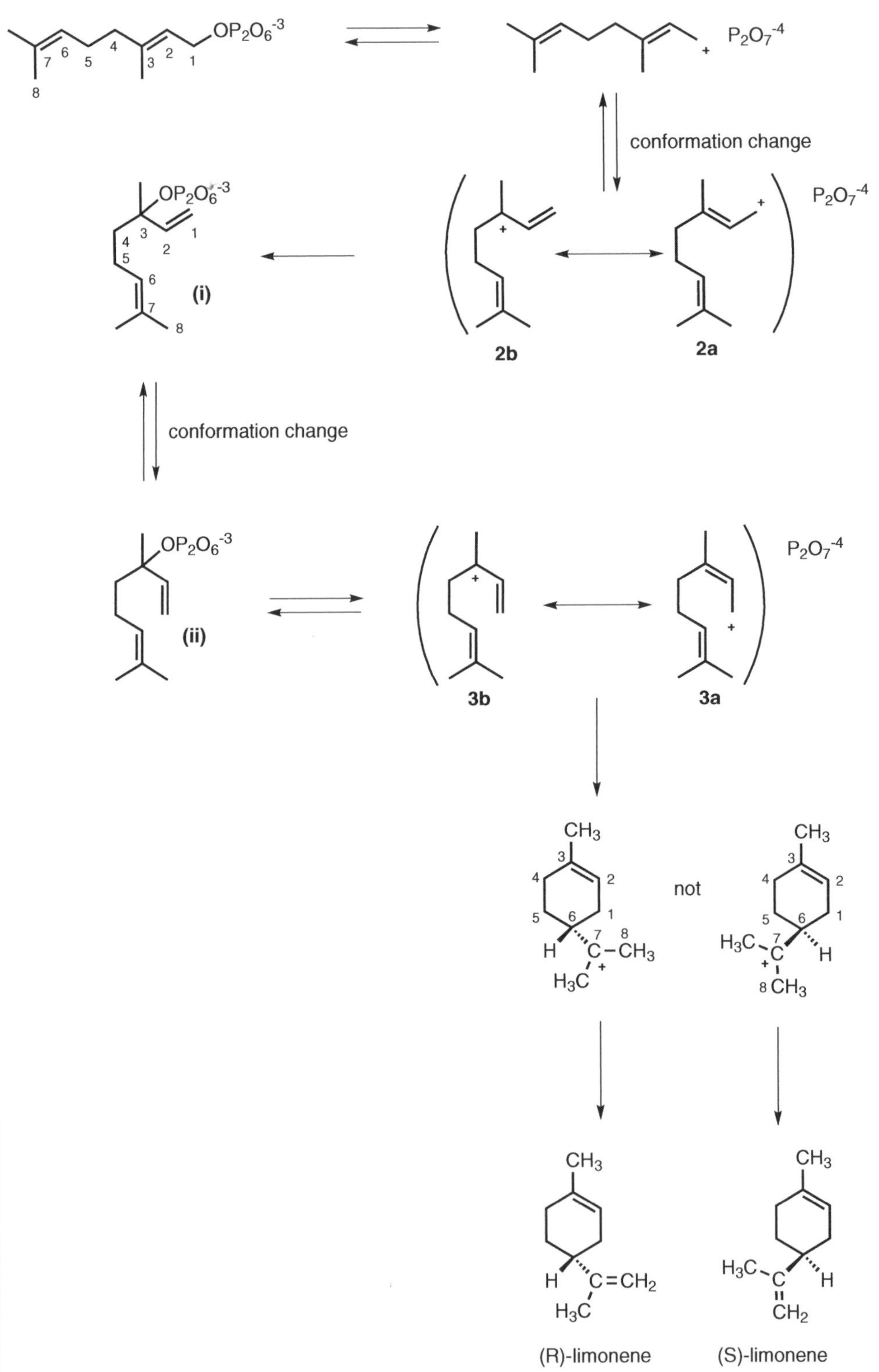

FIGURE 5.13

Mechanism for the Conversion of Geranyl Diphosphate to Limonene

carbon atoms 2 and 3 as shown in Figure 5.13, allowing the conformational motion necessary to bring carbon atom 1 in place (**ii**) to form the ring. Loss of the diphosphate group from carbon atom 3 then forms the carbocation **3a/3b**.

Now we are set up for the classic reaction of carbocations, addition to a double bond seen first in chapter 4 from production of high octane gasoline (Figure 4.13). But in **3** (represented by the resonance structures **3a/3b** in Figure 5.13) this addition is within the molecule, an intramolecular addition belonging therefore to the category of chemical reactions such as the ring closing of a sugar in Chapter 3 or the 1,2 shift of a carbocation discussed extensively in Chapter 4, or as we're about to discover in the in vivo synthesis of lanosterol (section 5.10).

There's a point of interest in nature's path to limonene that is worth noting. You might ask why the positively charged carbon does not add to the other end of the double bond between carbons **6** and **7** and form a seven-membered ring. The first answer is related to that discussed in Chapter 3 where we found out how chemists came to understand the special stabilities of six-membered rings. Nature, in forming limonene, chooses the most stable ring structure by adding the carbocation to the end of the double bond at carbon **6**. But, as in many biological processes, the results are often determined by multiple factors. Forming the carbocation as shown not only leads to the six-membered ring but also places the new positive charge on the most substituted end of that double bond - Markovnikov's rule is obeyed here.

This positive charge at carbon 7 then loses a proton to a base supplied by the enzyme to form a new double bond between carbons 7 and **8**. But why not form a double bond between carbons **6** and **7**? Such a double bond would place an sp^2 hybridized carbon atom in the ring disrupting the ideal chair conformation (Chapter 3). In addition, perhaps the constitutional isomer of limonene that would form does not have the attractive characteristic that limonene has. There's the related question of why (R)-limonene is formed and not (S)-limonene. A reasonable answer may be the unpleasant turpentine smell of this enantiomer, pointed out to me by Alexander Ritthaler, an excited sophomore chemical engineering student studying organic chemistry at the Polytechnic.

PROBLEM 5.26
What parallel is there between the chemistry that produces 2,2,4-trimethyl pentane and the chemistry that produces geranyl diphosphate?

PROBLEM 5.27
How does the formation of a single enantiomer of limonene demonstrate that there is an enantiotopic choice involved in the closing of the six-membered ring, that is, in the formation of limonene from the resonance-stabilized carbocation 3a/3b? Does this reaction require that the faces of the double bond, above and below the double bond, can be related as enantiotopic, just as constitutionally identical groups can be related enantiotopically?

PROBLEM 5.28
Use an orbital description of the carbocation produced from geranyl diphosphate in Figure 5.12 to explain why rotation around the carbon-2 carbon-3 bond would require more energy than rotation about any of the single bonds in that structure, but less energy than rotation around a double bond in the uncharged precursor.

PROBLEM 5.29
Show the molecule that would be formed if the ring closing reaction from intermediate carbocation 3 in Figure 5.13 made a bond between carbon 1 and carbon 7. How does the intermediate carbocation formed from this ring closing act to retard this closing?

PROBLEM 5.30
How is the rule of Markovnikov involved in Ruzicka's terpene rule?

PROBLEM 5.31
From Figure 5.12, which outlines the mechanism of the in vivo production of geranyl diphosphate and farnesyl diphosphate, show all examples of the roles of the following: leaving groups; Markovnikov's rule; resonance; enantiotopic groups; Ruzicka's terpene rule.

5.8

Nature chooses the terpene route to gain entry to the family of steroids.

ADDITION OF CARBOCATIONS TO DOUBLE BONDS and 1,2 shifts were the reactivity features of positively charged carbon which Guy Ourisson informed us were all that are necessary for formation of the higher terpenes (section 5.2). Ourisson could have gone further. These same characteristic reactions of carbocations are all that is necessary to convert a derivative of squalene to lanosterol (**Figure 5.14**), the precursor to cholesterol (Figure 5.2).

Squalene is an exception to the general rule that terpenes are named for plants. The Latin word for shark is Squalus and this triterpene (30 carbon atoms) was first isolated from shark liver by a Japanese scientist. Some sharks, especially deep water sharks, may have as much as 90% of the oils in their livers consisting of squalene, which is thought, because of the low density of this hydrocarbon, to aid the buoyancy and therefore swimming capability of the shark. It is claimed that fishermen from the Suruga Bay of Japan, in the shadow of Mount Fuji, traditionally drank an extract from the livers of deep sea sharks to maintain their health, a practice apparently common to Scandinavian fishermen as well. I found many web sites that claimed the ancient name of this substance was "samedawa," which in Japanese apparently translates to "cure all," and there are spas today that serve Japanese clients using this name. Apparently, the Chinese were aware of this as well. The respected healer, Li Ji Chin, in the Ming Dynasty, which dates from 1368, identified shark liver oil as an important source of health and longevity.

FIGURE 5.14 ▶

Isopentenyl Carbon Skeleton Related to Squalene and Lanosterol

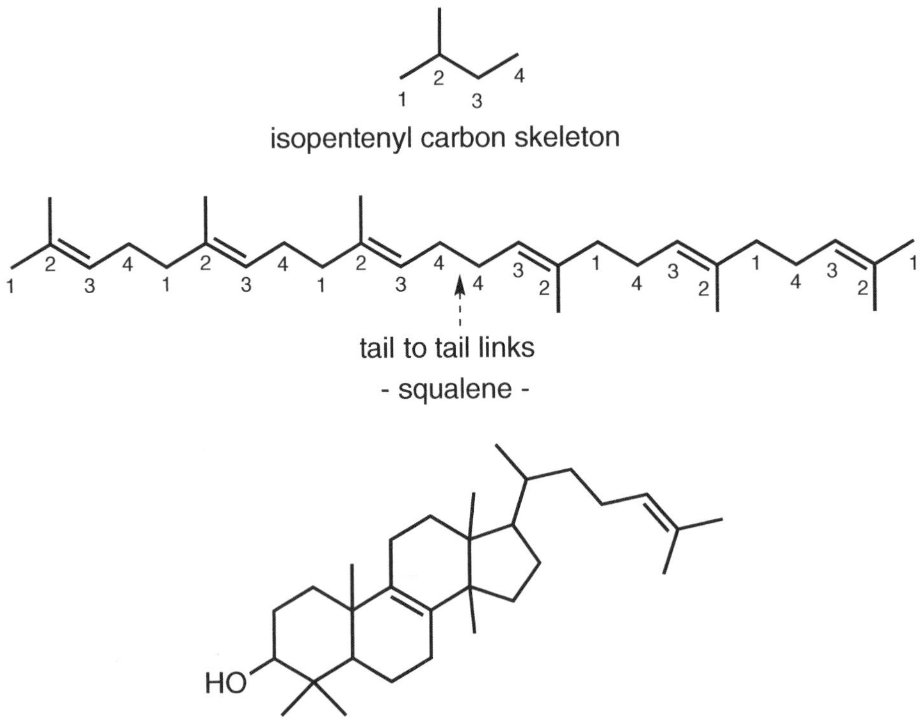

isopentenyl carbon skeleton

tail to tail links
- squalene -

lanosterol

The structure of squalene in Figure 5.14 compared to the structure of the terpenes in Figures 5.4-5.6 shows something new, which can be seen by following the numbered carbon atoms in the structure. The two fifteen carbon pieces, which are each derived from farnesyl diphosphate (Figure 5.12) are put together to form squalene in what are called tail-to-tail coupling, requiring an entirely different kind of mechanism than seen in Figure 5.12 for the synthesis of both the ten and fifteen carbon terpenes geranyl and farnesyl diphosphate. These terpenes are put together head to tail.

Chemists think of the carbon with the two methyl groups, what are called geminal methyl groups, as the head, with the methylene end, the CH_2 group, as the tail. Look again at the reaction between isopentenyl diphosphate and the carbocation produced from dimethyl allyl diphosphate in Figure 5.12. The head of the isopentenyl molecule adds to the tail of the dimethyl allyl molecule in forming the geranyl skeleton. Similarly, in the same figure, the head of the isopentenyl diphosphate adds to the carbocation at the tail of the geranyl carbon skeleton to form the farnesyl carbon skeleton.

When looking at the carbon skeleton of squalene it appears that the head-to-tail connection, carbon-4 to carbon-1, does occur from both ends of the squalene molecule up to the site where we see carbon-4 connected to carbon-4, in the middle of the structure (Figure 5.14). The isopentenyl carbon skeleton in squalene is numbered 1 to 4 with the methyl group on carbon-2, just as in the isoprene rule in Figure 5.6. This connection between carbon-4 and carbon-4 is the place where the two fifteen carbon pieces were connected tail to tail.

The complexity of this tail-to-tail mechanism in the biological synthesis of squalene is a bit advanced for our studies but what is not too advanced for us is the mechanism by which squalene is converted to what looks like a complicated molecule, lanosterol (Figure 5.14). Lanosterol is quite an important biological molecule - the precursor of cholesterol and, therefore, to all the steroids with their essential life functions.

How did chemists ever figure out how the connections were made in forming squalene and for that matter how squalene was converted to lanosterol and then lanosterol to cholesterol. To accomplish this, one has to follows where each carbon ends up starting from the precursors to isopentenyl diphosphate, as we'll come across later in the book. In fact, the isopentenyl diphosphate is made from the terminal two carbon moiety, $CH_3C=O$, within acetyl co-enzyme A (Figure 5.3), the acetyl group.

How is one to trace where these acetyl groups end up in cholesterol? The path from the two carbon piece to the larger molecules is impossible to figure out without tracing the source of each carbon atom. There is only one way to trace this path and that is to tag the carbon atoms. By putting a tag on the carbon atoms of the acetyl groups in acetyl coenzyme A and seeing where this tag ends up in squalene, in lanosterol, in cholesterol, one can understand how the building up of these molecules is accomplished by nature. However, how is it possible to distinguish one carbon atom from another? Isotopes are the answer.

Knowledge of the detailed mechanisms of the conversion of acetyl coenzyme A to isopentenyl diphosphate and on to geranyl diphosphate and farnesyl diphosphate (Figure 5.12), let alone to squalene and then on to lanosterol was not possible until isotopes of carbon and hydrogen became available for biochemical studies, a byproduct of the research that led to the atomic bomb.

Much of the early work in the United States in understanding how isotopes of uranium could be used to produce the bomb was conducted at Columbia University in New York City where **Harold C. Urey** had earlier initiated research in atomic structure. Urey was a student of G. N. Lewis, whom we have learned was so influential in the development of modern chemical ideas of bonding (section 4.11). Lewis encouraged Urey to continue his studies in Copenhagen with Niels Bohr who was responsible for the fundamental theoretical work on the atom. This is background to Urey's work at Columbia, which let him to the isolation in 1931 of an isotope of hydrogen, deuterium, for which he won a Nobel Prize in 1934.

Accurate measurements of atomic weights of the elements in early versions of mass spectrometers, devices that we learned about in Chapter 2 (section 2.1), showed that an isotope of hydrogen might exist and Urey realized that such an isotope being heavier than hydrogen must have a lower vapor pressure. He took four litres of liquid hydrogen and distilled it down to one cubic centimeter of the liquid allowing the proportion of deuterium to become high enough to prove its presence.

During the 1930s, Urey's group then separated isotopes of oxygen, carbon, nitrogen and sulfur, and began research that proved essential in elucidating the mechanisms of biosynthetic pathways. In 1933, directing isotope discoveries to biochemical questions

■ **Harold C. Urey**

■ **Rudolph Schoenheimer**

■ **Konrad Bloch**

■ **Feodor Lynen**

became a serious undertaking when the distinguished German biochemist, **Rudolph Schoenheimer** joined Urey's laboratory after being forced to leave Germany by Hitler's assent to power. Many German scientists of Jewish descent were forced to leave the country during these years, among them **Konrad Bloch**, who emigrated to the United States where he obtained his Ph.D. at Columbia University in 1938.

Bloch was invited to join Schoenheimer's group where he investigated how isotopes could be used to elucidate biological processes, research which proved able to solve otherwise intractable problems and became a focus of Bloch's life in science. Using isotopes and other methods, Bloch was able to trace the path by which squalene was converted to lanosterol, which then formed cholesterol, leading to his Nobel Prize in 1964.

Isotopes of carbon and hydrogen were very valuable because small molecules could be obtained with these isotopes, so that deuterium and ^{13}C or ^{14}C, could replace the normal atomic weight ^{1}H and ^{12}C. The position of these isotopes could then be located in the larger biological molecules allowing not only proof that the smaller molecules were used by nature for the synthesis but even the manner in which the bonds were made in the biological synthesis. The isotopes were appropriately called tracers.

Bloch shared the Noble Prize with **Feodor Lynen**, a German biochemist of almost identical age who had survived the war in Germany and carried out research in which he also devoted his scientific life to tracing the synthetic pathways in biological processes. Lynen was also an avid skier, an endeavor that led to a severe broken leg in a ski accident just as young men his age were being forced to join Hitler Youth and then the German army. His broken leg proved his good luck in escaping from participating in the military activities so many were forced into in those years in Germany prior to the Second World War.

We'll hear more about Lynen's work in Chapter 7 where we'll discuss fatty acid catabolism, that is, breakdown of fats, a process that involves not carbocations but rather carbanions, that is, negatively charged carbon. Lynen also was responsible for elucidating the role of a critically important building block and energy source in life, acetyl coenzyme A. We've seen the structure of acetyl coenzyme A in Figure 5.3 and we've noted this molecule as the source of the carbon atoms in isopentenyl diphosphate. Acetyl coenzyme A is therefore the ultimate source of all the carbon atoms in lanosterol and the steroids that are derived from lanosterol. We're going to discover many principles of organic chemistry when we study the biological synthesis of isopentenyl diphosphate in Chapter 8.

PROBLEM 5.32

Place the symbol, *, on carbon atom 4 of the isopentenyl skeleton in Figure 5.14. This star represents replacing the usual ^{12}C with ^{13}C, which then tags this particular carbon atom isotopically. Now trace the fate of the tagged carbon atom and show where it will end up in geranyl diphosphate, in farnesyl diphosphate, and in squalene.

Now tag the CH_3 group, carbon atom 1, in the isopentenyl skeleton ( ) in the same way, returning carbon atoms 4 to ^{12}C. Carry out the same trace of this isotopic substitution as you did above.

PROBLEM 5.33

Do you encounter a problem in carrying the tracing of the tagged carbon atoms from squalene to lanosterol, some disorder in their arrangement? What do you think could be the reason considering that every carbon atom in squalene is found in lanosterol?

OUR FOCUS NOW IS ON THE PATH from squalene to lanosterol and the critical intermediate on this path, the carbocation. There is a common principle involved in both the synthesis of limonene (Figure 5.13) and lanosterol (Figure 5.14). Both molecules are cyclic and both are enzymatically synthesized from acyclic precursors, geranyl diphosphate and squalene respectively. In the discussion of limonene synthesis (section 5.7) we were reminded of the principles of conformational analysis and the rotation about single bonds necessary to bring the atoms of the precursor monoterpene into proximity to form the limonene ring. These conformational changes in the ten carbon precursor to limonene are dwarfed by what is necessary for ring closing to form lanosterol. The problem was put succinctly in Bloch's own words in his Nobel Lecture, given on December 11, 1964:

*"The early proposals were mainly concerned with the problem of ring formation and quite uniformly they envisioned an origin of the tetracyclic steroidal ring system from an appropriately folded long-chain precursor. This intuition proved to be correct. All of the speculative schemes have in some way influenced the research that was to take place later, but none equaled in perspicacity L. Ruzicka's unifying hypothesis on the common origin of the terpenes and steroids and the suggestion by **Sir Robert Robinson** that cholesterol might arise by cyclization of the hydrocarbon squalene."*

Let's follow how this cyclization of squalene to lanosterol, the precursor of cholesterol takes place, the details of which were revealed by Bloch's work with isotopes playing the critical analytical role, while carbocations do the job. The cyclization is enzymatically catalyzed with oxidosqualene: lanosterol cyclase, the structure of which is shown in **Figure 5.15**. Again, as for isopentenyl diphosphate isomerase in Figure 5.8, the picture of this enzyme is included just to see the wonder of its folded structure without going into the details of its mechanistic action.

Figure 5.16 shows the structure of the precursor of lanosterol, which initiates the series of carbocation additions to double bonds, which form the lanosterol carbon skeleton. Here we are introduced to a new functional group (section 3.9), the epoxide (or oxirane) ring containing two carbon atoms and one oxygen atom. The bonds in the 30 carbon atom structure shown in Figure 5.16 are identical to squalene (Figure 5.14) except that the double bond at one end of the squalene chain, between carbon atoms 2 and 3 has been converted to an oxirane ring and the shape of the molecule has been changed by many rotations about single bonds, conformational motions (section

5.9

The conversion of the open chain 30 carbon molecule to a molecule with many fused rings requires the open chain to fold into a state bringing many atoms in close proximity and also requires the presence of a small strained molecule, which springs open to start the process.

■ Sir Robert Robinson

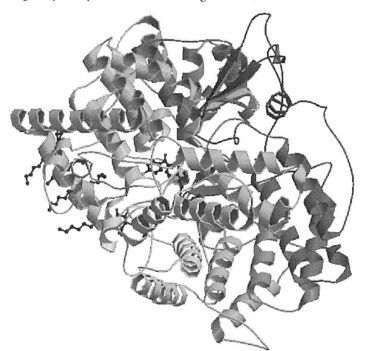

◄ FIGURE 5.15

Representation of the Folded Enzyme that Catalyzes the Formation of Lanosterol from Oxidosqualene

1.13), to prepare for the series of reactions that will lead to lanosterol.

In section 3.2 we learned about Baeyer's objections to the existence of cyclic molecules and specifically that a six-membered ring would be considerably strained, arising from the fact that the angle could not come close to the ideal tetrahedral angle. As we learned, this objection arose from Baeyer's assumption that the ring would be flat, in a single plane, and therefore with the geometry of a hexagon with 120° internal angles.

Well, six membered rings, as we learned in Chapter 3, are not flat and this solved the problem. But in an epoxide, there are only three atoms, which means the ring must exist in a plane. Because the internal angles must add up to 180° it is necessary for the angles at the two carbon atoms to be very far removed from the tetrahedral angle. In fact, if all the angles in the ring were the same, the deviation from the ideal angle would be nearly 50° at each carbon atom. Moreover, the internal angle at singly bonded oxygen is normally about 105 degrees.

FIGURE 5.16 ▶

Structure of Oxidosqualene

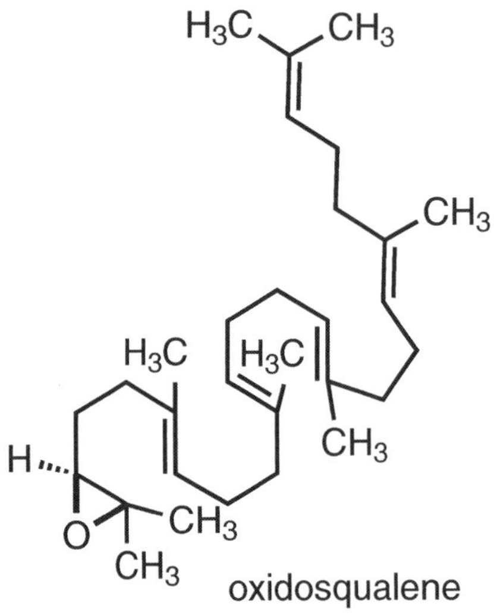

oxidosqualene

However, rings with three atoms do exist. Cyclopropane with three carbon atoms, C_3H_6 is a known molecule, a gas, with a boiling point of -33°, which crystallizes at -128°. There is no question of its existence and it has been known since first synthesized in 1881. For many years cyclopropane was used as an anesthetic. But there were warnings in the medical literature about its explosive potential.

It's interesting to see what a Wikipedia site says about cyclopropane under the heading of safety: "*Because of the strain in the carbon-carbon bonds of cyclopropane, the molecule has an enormous amount of potential energy. In pure form, it will break down to form linear hydrocarbons, including "normal", non-cyclic propene. This decomposition is potentially explosive,* especially if the cyclopropane is liquified, pressurized, or contained within tanks. Explosions of cyclopropane and oxygen are even more powerful, because the energy released by the formation of normal propane is compounded by the energy released via the oxidation of the carbon and hydrogen present. At room temperature, sufficient volumes of liquified cyclopropane will self-detonate. To guard against this, the liquid is shipped in cylinders filled with tungsten wool, which prevents high-speed collisions between molecules and vastly improves stability. Pipes to carry cyclopropane must likewise be of small diameter, or else filled with unreactive metal or glass wool, to prevent explosions. Even if these precautions are followed, cyclopropane is dangerous to handle and manufacture, and is no longer used for anesthesia.*"

Ethylene oxide, C_2H_4O, the simplest epoxide, that is, the fundamental molecule defining the functional group nature uses to initiate the cyclization of squalene (Figure 5.16) is also a long known compound first synthesized in the 1850s. It is an important industrial intermediate. In fact approximately 12% of the entire production of ethylene, which itself is produced in the billions of pounds, and is the major product of steam cracking of petroleum, a subject we'll take up in Chapter 9, is converted to ethylene oxide and then to other derived industrial products. These industrial products derived from ethylene oxide take advantage of the reactivity derived from the ring strain of ethylene oxide. There is an explosive danger in this molecule and industrial accidents involving ethylene oxide have occurred.

In all rings made of three atoms, the covalent bonds connecting the atoms defining the ring (the two carbon atoms and the oxygen atom in the epoxide ring in Figure 5.16) have unusual properties. These covalent bonds do not behave as σ-bonds usually do. Theoretical studies lead to a description of these bonds as bent bonds, in which the maximum electron density does not lie along the internuclear angle but rather extends outward away from the angle defining the ring, as shown in **Figure 5.17**.

In this way the formal σ-bonds (section 1.4), the bonds defined by the lines in the structural drawing linking the atoms in the ring move in the direction of the geometry of π-bonds therefore contributing to the bond weakness and reactivity of three membered

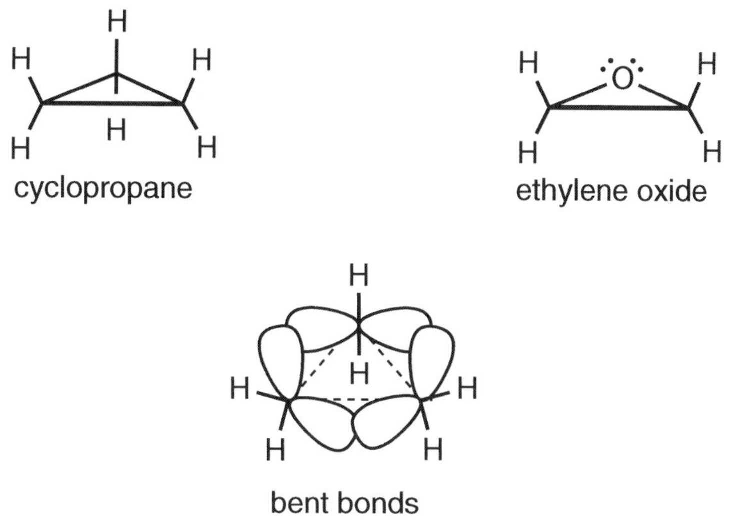

cyclopropane

ethylene oxide

bent bonds

◀ **FIGURE 5.17**

Orbital Picture of Bent Bonds for Cyclopropane and Ethylene Oxide

rings. The straight-on overlap of the orbitals making the bond in an unstrained molecule is not permitted in this three membered ring structure (Figure 5.17).

Another structural characteristic of three membered rings leading to their instability and reactivity is torsional strain arising from eclipsing of bonds introduced in section 1.13 and reinforced in the study of six membered rings in sections 3.2 to 3.5.

In three-membered rings, which have no conformational alternatives to existing in a plane, the necessity for all the carbon-hydrogen bonds to be eclipsed (Figure 5.17), adds torsional strain to the angle strain discussed above. Nature in using an epoxide ring has chosen a functional group that is ready to spring open to relieve all this strain. And springing open is precisely what must happen to start the series of reactions to lead to lanosterol.

The two pairs of nonbonding electrons on the oxygen atom in all epoxide rings cause the ring to act as a base and therefore easily undergo protonation from a Brønsted-Lowry acid. And the necessary proton donor is supplied by the enzyme that is responsible for folding the squalene epoxide into the proper conformational shape (Figure 5.16), that is, the enzyme (Figure 5.15) that is ushering the process to its conclusion. Such an acid base reaction opens the highly strained ring and initiates the first step in the conversion of squalene epoxide to lanosterol, as we'll discover in the next section.

PROBLEM 5.34

Use a Newman projection along any of the carbon-carbon bonds in cyclopropane (Figure 5.17) to test for torsional strain as one source of the instability of the three membered ring. Consider the two lone pairs of electrons on the oxygen atom in ethylene oxide. Would there be any strain associated with the two lone pairs of electrons?

PROBLEM 5.35

Sometimes the term "banana bond" is used to describe the carbon-carbon bonds in cyclopropane and ethylene oxide or the carbon-oxygen bonds in ethylene oxide. Is this justified?

PROBLEM 5.36

Why are the bonds in a three membered ring seen as intermediate between π and σ bonds? Draw the structures using the necessary orbitals to support this description.

PROBLEM 5.37

While the oxygen atom in H_2O makes an angle of about 105° with the two hydrogen atoms, the angle of the oxygen in ethylene oxide is far smaller causing a change in the hybridization of orbitals. Might this change have any effect on the basic nature of this oxygen atom, on the reaction with a proton source?

PROBLEM 5.38

Since the internuclear angle in a three membered ring is 60°, does that mean that the angle made by the H-C-H must be far larger than the usual tetrahedral angle? Because a larger angle is associated with hybridization for the bonding orbitals of more s character (sp 180°; sp^2 120°; sp^3 109.8°), does this mean that the C-H bonds in a three membered ring have more s character than they would in a tetrahedral structure that is not angle strained? Give reasons for your answers.

5.10

Given the proper conformation of oxidosqualene, the derived carbocation simply has to add to double bonds and carry out 1,2 shifts to produce lanosterol.

FIGURE 5.18 EXHIBITS THE INTERMEDIATE produced following the protonation step on the epoxide ring of the properly folded squalene oxide. The series of reactions that follow (1-5) all belong to the category of addition of a carbocation to a double bond. **Figure 5.19** continues the reaction path to lanosterol, but here the reactions (6-9) belong to the category of carbocation 1,2 shifts.

In this way we find, in the path from the protonated oxidosqualene to lanosterol, that there are necessary only the reactive characteristics of carbocations, the intermediate we were introduced to in the industrial chemistry leading to the syntheses of high octane gasoline (Chapter 4).

First there occur five reactions, all intramolecular, in which a carbocation adds to a double bond within the same molecule (Figure 5.18). From these reactions, the basic ring structure of lanosterol is formed. The next four reactions shown in Figure 5.19 then rearrange the structure to its final form by a series of carbocation 1,2 shifts. The last reaction in Figure 5.19 then expels a proton, which is necessary to rid lanosterol of its positive charge, and to place a double bond where it is required in the structure and finally, to complete the catalytic cycle, which began with the addition of a proton. It's stunning, and it is worth saying again that every reaction parallels reactions used industrially to convert petroleum fractions to high octane gasoline as we studied in Chapter 4.

The mechanism in Figures 5.18 and 5.19 shows only the bond making steps in the conversion of squalene epoxide to lanosterol. However, there is a great deal of stereochemistry involved in the formation of lanosterol, starting from the fact that a single enantiomer of the epoxide (Figure 5.16) is the starting point, and on to the fact that lanosterol is formed as one of a large number of possible diastereomers and as a single enantiomer of that diastereomer (**Figure 5 .20**).

These stereoisomeric choices in the cyclization then define the stereochemistry of cholesterol by a path we won't cover. We can not, also, follow here all the detailed stereochemical steps that intervene between this starting point and the final lanosterol structure. But we can see again, as discussed earlier, as in all enzyme catalyzed reactions, with rare exception, that when stereochemical choices are possible, only one path will be followed.

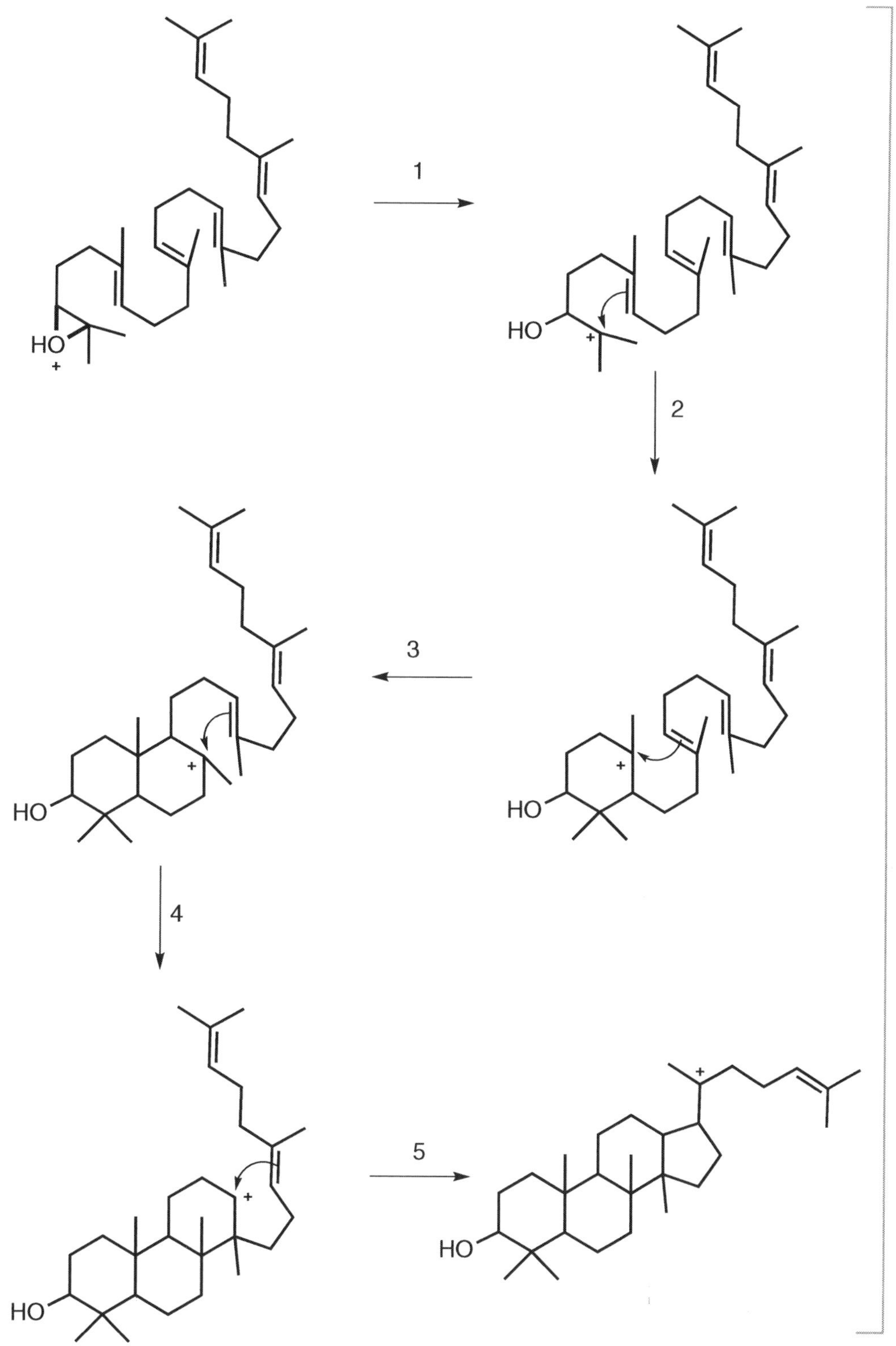

◀ **FIGURE 5.18**

Carbocation Addition to Double Bonds in the Enzyme Catalyzed Formation of the Four Fused Rings of Lanosterol

PROBLEM 5.39

Show how formation of lanosterol from squalene oxide demonstrates reactions of carbocations presented in Chapter 4. Do each of these reactions take place in a predictable manner taking account of the relative stability of carbocations?

PROBLEM 5.40

There is one addition of a carbocation to a double bond that does not obey Markovnikov's rule. Which is it and can you offer an explanation for how it might happen?

FIGURE 5.19 ▶

Carbocation 1,2-Shifts Leading to the Final Skeleton of Lanosterol

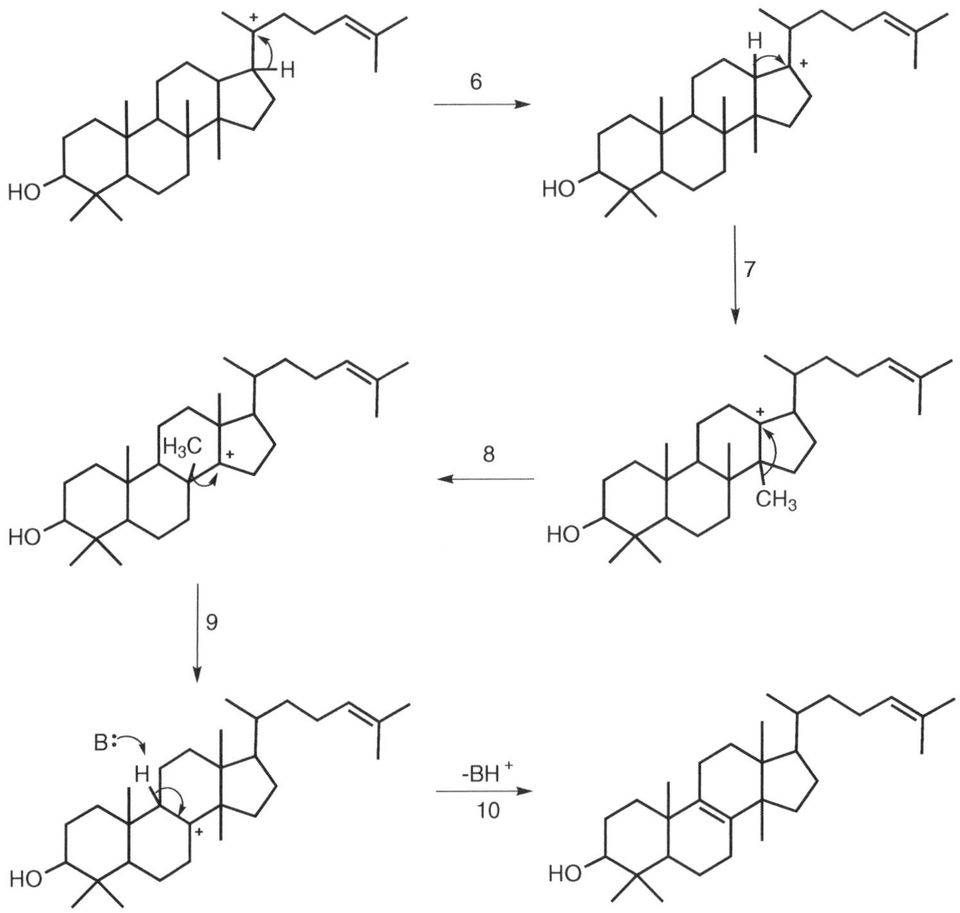

FIGURE 5.20 ▶

Stereochemical Representation of the Molecular Structure of Lanosterol

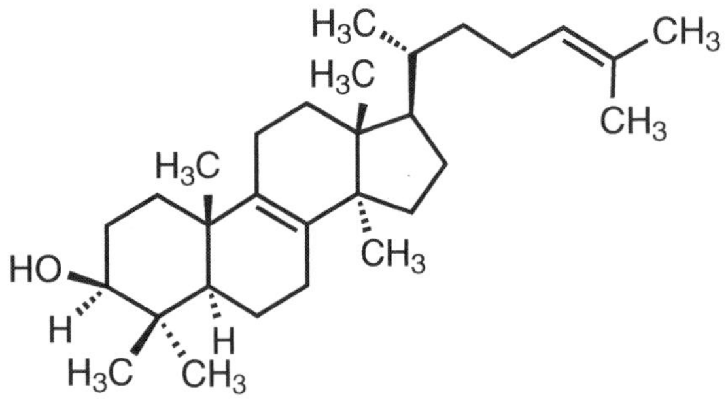

PROBLEM 5.41

Why is there a large entropic cost to the formation of lanosterol?

PROBLEM 5.42

Even without proton addition to the oxygen of the epoxide ring shown in Figure 5.16, the ring could spring open from the strain. However, the price would be to produce the conjugate bases of a very weak acid. What is meant by this?

PROBLEM 5.43

Compare the steroid structures in Figures 5.1 and 5.2 with that of lanosterol in Figure 5.20. Define the structural changes necessary to convert lanosterol to cholesterol and then from cholesterol to the other steroids shown – what atoms and bonds have to be changed? How does the stereochemistry of the steroids fit the stereochemistry of lanosterol?

CHAPTER FIVE SUMMARY of the Essential Material

THE CHAPTER INTRODUCES THE FAMILY OF TERPENES and their connection to lanosterol, the precursor of the steroids. The structures of two key biological molecules necessary for the synthesis of the steroids are presented, acetyl coenzyme A and isopentenyl diphosphate. We are then led to the work of Ruzicka and the isoprene rule, which demonstrates how the arrangement of the carbon atoms in the structure of isoprene can be found in all terpenes including even in natural rubber.

To understand the mechanism by which the terpenes are synthesized in vivo we had to study the nature of leaving groups and their connections to conjugate bases of strong acids. This took us into the concept of resonance, and understanding how the necessity to present the electron distribution of organic molecules using multiple structural representations is connected to stability in all systems including in many leaving groups.

We then studied how carbocations play the critical role in terpene synthesis and the realization that isopentenyl diphosphate, although featuring an excellent leaving group has that group attached to a carbon atom that forms an unstable carbocation. The transformation that nature takes to change the structure of isopentenyl isophosphate introduces the nature of stereochemical relationships within constitutionally identical parts of a molecule, enantiotopic and diastereotopic groups. We see how the exquisite stereospecificity of enzymes allows distinguishing such stereochemically distinct groups.

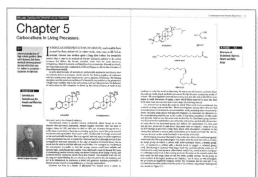

We come to the detailed mechanism of terpene formation and see how the essential step is carbocation addition to a double bond, a reaction we were introduced to in the industrial synthesis of high octane gasoline. This class of reaction is shown to be responsible for the ring formation of limonene and discussion of this mechanism brings up issues of resonance and rotation around single bonds and also the preference for six-membered rings when other ring sizes are possible. The importance of Markovnikov's rule is enforced again and again in this chapter.

We come then to the conversion from squalene to lanosterol and understand how conformational changes can be very complicated but greatly simplified by enzyme control. The mechanism of this conversion introduces the nature of three membered rings and the kinds of bonding and strain associated with these species and how nature makes use of this strain to initiate carbocation formation in the lanosterol precursor. Finally, we discover how the ring closing steps from squalene to lanosterol involve addition of carbocations to double bonds following Markovnikov's rule with, however, one exception and then how the final lanosterol structure is formed by the 1,2 shifts we discovered in catalytic cracking of petroleum, but taking place here in vivo. Finally, we come to appreciate how in a structure of so many stereochemical possibilities that only one is produced, testifying to the power of enzyme specificity. And presented in the chapter are two overall views of enzymes allowing us to see how their folded structures are so important even if only very few of many amino acids that make up these proteins are actually involved in the syntheses they catalyze.

Chapter 6

Aromatic: A Word That Came to Mean Something Other Than Odor in the Chemical Sciences

6.1

The Discovery of Benzene

P ARTINGTON (SECTION 4.9), IN HIS CLASSIC "A History of Chemistry," wrote about **Michael Faraday**: *"In his time Faraday was a model for scientific men. Of humble origin, he rose by his genius to the highest rank of scientific eminence, and his moral character and integrity were on the same level."*

Indeed, Faraday was of humble origins. He was born into a family that was not well off. His father was a village blacksmith and young Michael at the age of thirteen, one of four sons, had to end his formal schooling and become an apprentice to a bookbinder. But putting a brilliant and inquisitive young man in a shop full of books, in which he asked and was granted permission to read them, was all it took to set Faraday on a path of great accomplishment. His abilities and his intense interest in science led Faraday to the public lectures of one of the great scientists of that time, Humphrey Davy. Faraday not only attended the lectures but as well kept copious notes, which he wrote up in the form of a book that he sent to Davy. This led to an interview with the great man, which led Faraday to be offered a position with access to facilities to conduct experiments and therefore the beginning of a remarkable story, which is worth following by simply going to Google under the name Michael Faraday.

It was not long before Faraday's scientific accomplishments led him to became a fellow of the Royal Society (FRS) and a year later at the age of thirty three he read a paper before the society on June 16, 1825, titled *"On new compounds of carbon and hydrogen, and on certain other products obtained during the decomposition of oil by heat."*

Faraday, who disliked the distinction between chemistry and physics and thought of himself as a natural philosopher, had long been interested in the oil-gas increasingly used to light streets and homes in English cities in the latter part of the eighteenth and into the nineteenth century. David Gordon founded one of the companies producing oil gas, the Portable Gas Company, where Michael Faraday's older brother Robert was employed. Mr. Gordon had asked Robert if his brother might be interested in investigating a liquid that deposited on the portable gas valves, a liquid that interfered with the workings of the device. Let's follow Michael Faraday's own words:

"The object of the paper which I have the honour of submitting at this time to the attention of the Royal Society, is to describe particularly two new compounds of carbon and hydrogen, and generally, other products obtained during the decomposition of oil by heat. My attention was first called to the substances formed in oil at moderate and at high temperatures, in the year 1820; and since then I have endeavoured to lay hold of every opportunity for obtaining information on the subject. A particularly favourable one has been afforded me lately through the kindness of Mr. GORDON, who has furnished me with considerable quantities of a fluid obtained during the compression of oil gas, of which I had some years possessed small portions, sufficient to excite great interest, but not to satisfy it."

In the twenty six pages to follow this introduction, Faraday describes a remarkable series of experiments and analyses, which are astonishing for what was accomplished considering the chemical complexity of what he was studying and the tools available in the 1820s. In fact in 1968, R. Kaiser who worked for Badische Anilin- und Soda-Fabrik AG, a corporation derived from the company Mr. Gordon founded, published

■ Michael Faraday

a paper entitled: *""Bicarburet of Hydrogen." Reappraisal of the Discovery of Benzene in 1825 with the Analytical Methods of 1968."* Here Kaiser documents with modern techniques how complex a separation and analytical problem Faraday had overcome.

From his effort, involving multiple distillations and crystallizations, Faraday was able to isolate a liquid that he reports as crystallizing close to 32° F and which he analyzed as having the composition C_2H, hence the name bicarburet of hydrogen. This composition was based on an atomic weight of carbon assigned as 6 with hydrogen as 1 and therefore corresponds to $(CH)_{2n}$ in modern terms, where n could be any number.

He also measured the vapor density, which revealed that the "particles" of the substance, what we would call the molecular weight, was about 39. Correcting this for the assignment of hydrogen gas as H and therefore 1, as was believed in that time, instead of 2, as we know today for H_2, would double this number to close to the molecular weight of benzene, 78, corresponding to C_6H_6.

With this isolation of benzene, Faraday set in motion a molecular mystery whose solution took over 100 years and was not fully accepted by chemists for nearly the next 125 years. Let's follow this story and discover how benzene played a central role in the industrial processes leading to the most important plastics of our time. The story also leads to theoretical understanding of the special properties of benzene, which takes us to understanding the fundamental nature of the nucleotide bases in DNA and also certain coenzymes.

Faraday described his newly isolated compound in the following manner: *"Bicarburet of hydrogen appears in common circumstances as a colourless transparent liquid, having an odor resembling that of oil gas, and partaking also of that of almonds."*

Many compounds isolated in those early years of the nineteenth century from natural sources were pleasant to the smell, among them extracts from cinnamon bark, wintergreen leaves, vanilla beans and anise seeds, all reported as aromatic essential oils. These compounds were analyzed in the manner Faraday used for benzene and found to generally have a lower ratio of hydrogen to carbon than found for what came to be known as alkanes where the ratio of H/C is generally greater than 2. In those early years it is not surprising that benzene should be seen as belonging to that same class of aromatic molecules.

PROBLEM 6.1

Use web resources to discover the chemical structures of the molecules responsible for the properties noted in: "Many compounds isolated in those early years of the nineteenth century from natural sources were pleasant to the smell, among them extracts from cinnamon bark, wintergreen leaves, vanilla beans and anise seeds, all reported as aromatic essential oils." Do any obey the isoprene rule (section 5.2, Figure 5.6)?

FIGURE 1.6 EXHIBITS THE STRUCTURES OF SETS of molecules that are structural isomers. The formulas for the two isomeric sets in this figure are given as C_6H_{14} and $C_6H_{14}O$. In other words there are twice as many hydrogen atoms plus two more, for the number of carbon atoms, or that is, C_nH_{2n+2}. For six carbon atoms one has 12+2 hydrogen atoms. This formula corresponds to a general rule for organic molecules that contain only single bonds and as well in which no ring is present. The formula simply expresses the four coordinate nature of carbon seen first in the formula for methane, CH_4, which fits the same formula for n=1. All these molecules are called saturated.

Now try this formula for any of the structures of glucose in Chapters 1 and 3 and discover that two hydrogen atoms are missing. The formula for glucose is $C_6H_{12}O_6$. This formula fits if glucose is written in the cyclic form or in the open form (Figure 3.11).

For every ring in an organic molecule, as in the cyclic form of glucose, the formula

6.2

A Short Diversion about the Ratio of Hydrogen to Carbon in Various Organic Molecules.

must contain two fewer hydrogen atoms. You can discover the necessity for this reduction in the number of hydrogen atoms by considering n-butane, which has, as we've seen (section 4.5), four carbons in a linear array and, as expected, has the formula C_4H_{10}. Now link the two carbon atoms at the ends of this linear array to each other, which can only be accomplished by replacing two carbon-hydrogen bonds with the new carbon-carbon bond. Conclusion: for every ring, two hydrogen atoms are lost from the formula. If the two hydrogen atoms missing arise from the presence of a ring, the ring containing molecule is still called saturated.

However, if you try this single ring formula, C_nH_{2n}, on limonene (Figure 5.13) you get the wrong answer, not $C_{10}H_{20}$, but rather $C_{10}H_{16}$. The answer is that in addition to the ring in limonene there are two double bonds in this terpene.

Just as connecting the two carbon atoms at the ends of n-butane to each other to form a ring causes a loss of two hydrogen atoms from the formula, so doubly connecting two carbon atoms to each other, that is forming a double bond, also requires the removal of the two hydrogen atoms that would have been bonded to these carbon atoms. Any molecule that has two hydrogen atoms missing from its formula, arising from the presence of a carbon-carbon double bond, is called unsaturated.

For every two hydrogen atoms missing from the saturated formula (C_nH_{2n+2} to C_nH_{2n}), where the missing hydrogen atoms arise from a carbon-carbon double bond we assign the molecule's degree of unsaturation. Isobutylene, C_4H_8, has one degree of unsaturation. If a molecule has two carbon-carbon double bonds, as is the situation in limonene, we assign the molecule to the class of two degrees of unsaturation. In the situation of limonene there are six hydrogen atoms missing from the saturated formula ($C_{10}H_{16}$) but only four of these hydrogen atoms arise from double bonds, two double bonds. Two of the missing hydrogen atoms arise from the ring structure.

In summary, if you have no rings or double bonds the formula is: C_nH_{2n+2}; if one ring without a double bond, or one double bond without a ring, the formula is: C_nH_{2n}; if one ring and one double bond, or two double bonds and no ring, or two rings and no double bonds, the formula is: C_nH_{2n-2}; and if as in limonene there is one ring and two double bonds then the formula is: C_nH_{2n-4}. And so on you can derive a formula for any number of rings and double bonds. And as we've seen in the open form of glucose, the double bond causing a loss of two hydrogen atoms from the structure does not have to be between carbon and carbon, but can be between carbon and any other element, such as oxygen in the aldehyde group in the open form of glucose.

Now you can try this derivation of the formula yielding the ratio of carbon to hydrogen on many different molecules. Try it on the unsaturated cyclic molecule lanosterol (Figure 5.20). I guarantee it will work.

PROBLEM 6.2

Create a series of molecules all with ten carbon atoms and various numbers of hydrogen atoms (without formal charges and obeying the octet rule) so as to be fully saturated, C_nH_{2n+2}, and then sequentially, with two carbons less than fully saturated for each example down to the formula C_nH_{2n-18}. Is it possible?

PROBLEM 6.3

In section 6.2 one type of unsaturation was not mentioned, the triple bond. As for an alkene such as ethylene (ethene), C_2H_4, the smallest number of carbon atoms in a triple bonded molecule is also two, acetylene, C_2H_2. How would the formula for a molecule containing a triple bond be altered? Develop a hybridization orbital picture of a triple bond and predict the geometry of a triple bond.

PROBLEM 6.4

Draw as many structures as you can think of with the formula C_6H_6, which are not cyclic and for which there are no formal charges. How many degrees of unsaturation are revealed by this formula?

PROBLEM 6.5

Predict the number of hydrogen atoms in cholesterol (Figure 5.2) based on the number of carbon atoms and the structure. Now count the number of hydrogen atoms in the structure. Did you get the same answer?

PROBLEM 6.6

Is the formula for allene, $H_2C=C=CH_2$, consistent with the rules in section 6.2?

PROBLEM 6.7

Could you have created the ten carbon atom molecule with fewer hydrogen atoms than 2n-18 (referred to in Problem 6.2) if you had allowed the incorporation of one or more triple bonds?

PROBLEM 6.8

For each structure involved in answering problems 6.2 and 6.6, propose hybridization of orbitals consistent with the structure and then predict the geometry around each carbon atom in each structure.

A S POINTED OUT IN SECTION 6.1, molecules isolated from natural sources were often pleasantly odorous, especially those molecules with too few hydrogen atoms for the carbon atoms present, that is, less than 2n+2 hydrogen atoms in modern terms (where n equals the number of carbon atoms, section 6.2).

We have to add the phrase "in modern terms" because it was only a few years before these natural substances were first encountered that chemists were still arguing about the reality of atoms and **John Dalton**'s atomic theory. This was hardly a time when molecular formula existed on solid ground. Dalton first revealed his atomic theory in an informal manner in 1804 to the distinguished Scottish chemist, Thomas Thomson of the University of Glasgow.

It's fascinating to follow the interaction between the younger scientist, Thomas Thomson, then 31, and his distinguished elder, 38 year old John Dalton. As reported by Partington, Thomson wrote the following in his diary in 1804:

"Aug.26, Sunday, Called on Mr. Henry, and found him; dined with his father and drank tea in Mr. Henry junior's, in company with Mr. Dalton. Mr. Dalton had been lately occupied with experiments on the carburetted hydrogen. He finds three species. 1. Olefiant gas, composed of an atom of hydrogen and an atom of carbon. 2. Gas of marshes, composed of two atoms of hydrogen and one of carbon. 3. Oxide of carbon, composed of an atom of carbon and one of oxygen. He has suggested the following ingenious method of ascertaining the constituents of bodies."

Dalton's ingenious method was to draw symbols to represent the atoms he was hypothesizing. The symbols Dalton drew represented his hypothesis that molecules were composed of discrete entities with each entity having a specific weight.

Thomson, in his extensive writing in the years that followed, became a defender of Dalton and the atomic theory. In 1831 in discussing his conversation with Dalton, Thomson remembered Dalton telling him that the atomic theory first occurred to him during his investigations of olefiant gas and carburetted hydrogen gases, now understand to be ethylene, C_2H_4, and methane, CH_4. Although these ratios of carbon and hydrogen were not accurately determined at that time because the ratios of the atomic weights of hydrogen and carbon were not correctly known, Dalton nevertheless was able to see that the ratios between the two substances were multiples of each other. Let's follow Dalton's reasoning as recounted later by Thomson: *"If we reckon the*

6.3

When Faraday discovered benzene, the formula for a molecule was a key piece of information—really the most important, if not the only piece of information available.

■ **John Dalton**

carbon in each the same, then carburetted hydrogen gas contains exactly twice as much hydrogen as olefiant gas does."

In this way, using modern atomic symbols, we come up with CH for ethylene and CH_2 for methane. Now we understand that the hydrogen was not atomic weight one but rather two and so these ratios would change to CH_2 and CH_4 respectively. Multiply the former by 2 and we get the modern formulas for ethylene and, most important, we gain Dalton's insight that if the compounds were formed of atoms, then the ratios must be in multiple proportions since you could not have a fraction of an atom.

Let's follow the theory as Dalton hypothesized in his notebook on September 6, 1803 and reported by Partington in his Volume Three "A History of Chemistry."

"(i) Matter consists of small ultimate particles or atoms. (ii) Atoms are indivisible and cannot be created or destroyed. (iii) All atoms of a given element are identical and have the same invariable weight. (iv) Atoms of different elements have different weights. (v) The particle of a compound is formed from a fixed number of atoms of its component elements."

Dalton had arrived at his atomic theory after many years of the study of the atmosphere and how it responded to pressure and temperature and the study of the mixing of gases. Early on Dalton thought of the atmosphere as composed of ultimate particles but it took until 1803 that Dalton's notebooks showed symbols for atoms. They didn't have everything quite right in those early days, just shortly before Faraday entered the scientific world, but the stage was being set to reinforce the ancient statement that Newton had used again about science advancing by "dwarfs standing on the shoulders of giants."

Thomson started using symbols for the atoms in a quantitative manner shortly after a meeting with Dalton, discussed Dalton's atomic theory in a book he published a few years later and was a defender of Dalton's ideas against his critics. It is interesting to see how what we now call molecules were represented by Thomson: oxygen, carbon and hydrogen were designated by *w*, *c* and *h*; carbon dioxide was *2w+c*, methane was *c+h*; water was *w+h*; and particularly interesting was sugar, *5w+3c+4h*.

If benzene had been known at that time, it would have been realized that benzene has even less hydrogen than olefiant gas for the same weight of carbon - remarkably less, C_2H using the values for the elements at that time, which translates to CH using modern values for the atomic weights of the elements and with the modern formula for the molecule, six times that proportion, C_6H_6, as noted in section 6.1.

6.4
The stage was now set to propose a structure for benzene that would explain its properties.

AN IMPORTANT INSIGHT into the chemical properties of these odorous molecules with a reduced proportion of hydrogen to carbon was their reactivity. Modern chemistry has come to understand the reactive character of these "odorous" molecules as a consequence of understanding the weakly overlapping π bonds of alkenes (Figure 4.12). However, in alkenes the reactivity of the double bonds, responsible for their role in the biochemical processes that lead to the terpenes (Figure 5.12, 5.13) and the industrial preparation of high octane gasoline (Figure 4.13), differ greatly from the reactivity properties of overlapping π bonds in benzene.

In the initial report by Faraday on the discovery of bi-carburet of hydrogen, that is, benzene, and in the chemical investigations by others that followed his 1825 report to the Royal Society, benzene was found to be quite special beyond the ratio of carbon and hydrogen. Benzene was surprisingly unreactive.

In fact, Faraday investigated the reactivity characteristics of his newly discovered bi-carburet of hydrogen and reported, for example, an absence of reaction with Cl_2 in the dark. This is surprising considering that olefiant gas reacts readily with Cl_2 under identical conditions in the dark, to form $C_2H_4Cl_2$.

There is something very special about this bi-carburet of hydrogen. Why should a molecule with a formula CH, which is less hydrogen than in olefiant gas, be so unreactive? And moreover, as Faraday's molecule was further investigated by others over the following years, the reactivity it did exhibit differs fundamentally from the alkenes.

While alkenes underwent chemical reactions in which new groups added to the alkene, such as we've seen in Figures 4.13 and 5.12 and 5.13 referred to above, benzene resisted this kind of reaction. In this addition reaction to double bonds the two electrons in the π bond are converted to σ bonds. The product molecule after the addition is no longer an alkene.

In Faraday's molecule the incoming groups rather replaced the hydrogen atoms of the molecule, and then only with great difficulty – only using extreme conditions for the reaction, which differed greatly from the easy reactivity of the alkenes. The reactions of alkenes are termed by organic chemists, appropriately, addition reactions, which we've seen in the discussions of Markovnikov addition (section 4.10), while the reactions of aromatic molecules are designated as substitution reactions. We'll see more about these substitution reactions shortly.

In the forty years between 1825 and 1865, Faraday's discovery was confirmed by others and as well bi-carburet of hydrogen was synthesized by other routes. One of these paths produced bi-carburet of hydrogen from benzoic acid, leading to the name change of Faraday's discovery, which now became benzene.

Although benzoic acid may be familiar to many of us in its prosaic role as a preservative of pickles, this molecule, from which benzene derives its name has a storied history of its own. It is the main constituent of gum benzoin obtained from plants grown in Southeast Asia, a substance that has long use varying from incense in religious ceremonies to moderating evaporation in perfumes. Its name is reported to be derived from Arabic via Italian (*lubān jāwī* (لهب ان خ اوي) "frankincense from Java").

Nevertheless, by the 1850s the word aromatic had lost all its relationship to odor and was used only to describe benzene and those molecules derived from benzene, molecules that had a surprisingly low and unusual chemical reactivity. Another surprise about these molecules derived from benzene was the number of structural isomers obtained when benzene did undergo reaction. As we'll see shortly, there seemed to be fewer structural isomers than could be accounted for by the various arrangements of the atoms hypothesized to account for the properties of benzene.

No satisfactory explanation for the arrangement of the six carbon and six hydrogen atoms in benzene could be imagined to explain all the experimental facts until a remarkable paper was published by August Kekulé in 1865 in the Bulletin de la Societe Chimique in Paris. Kekulé, whom we have encountered before (section 1.11), as the chemist who advanced the idea of the tetravalence of carbon, in what he called "*substances grasses*," had turned his attention to the problem of benzene.

Do you remember the story of the unfortunate Archibald Couper told in section 1.11? This was the young man who worked for the powerful boss, Charles Adolph Wurtz, who had blocked Couper's paper from appearing so that Kekulé received all the credit for the four-coordinate nature of bonding of carbon. That happened in 1857. In 1865, Kekulé now still quite young, only 36 years old, again sponsored by Wurtz, startled the chemical world with his idea that benzene was a cyclic molecule. Although Kekulé was German, born in Darmstadt, he wrote this now famous paper in French. Let's see part of what Kekulé had to say in an English translation prepared by Partington:

"*In all aromatic compounds one and the same atomic groups is present, or they contain a common nucleus composed of six carbon atoms....Six carbon atoms so linked form an open chain with eight unsatisfied affinity units. If the further assumption is made that the two carbon atoms which terminate the chain are bound to one another by an affinity unit, a closed chain is obtained, which still contains six free affinity units.*

It is from this closed chain that the substances called 'aromatic' are derived." Kekulé goes on to write *"The theory requires that only one modification of monochlorobenzene and of pentachlorobenzene can exist, but several isomers (probably three) of di-, tri-, and tetrachlorobenzene."*

Although Kekulé had been attracted to chemistry by his teachers when he was nineteen years old, he was, at first, a student of architecture based on his talents in drawing and mathematics. These abilities apparently were at work in the visualization necessary for the development of his structural theoretical ideas, which could be called molecular architecture.

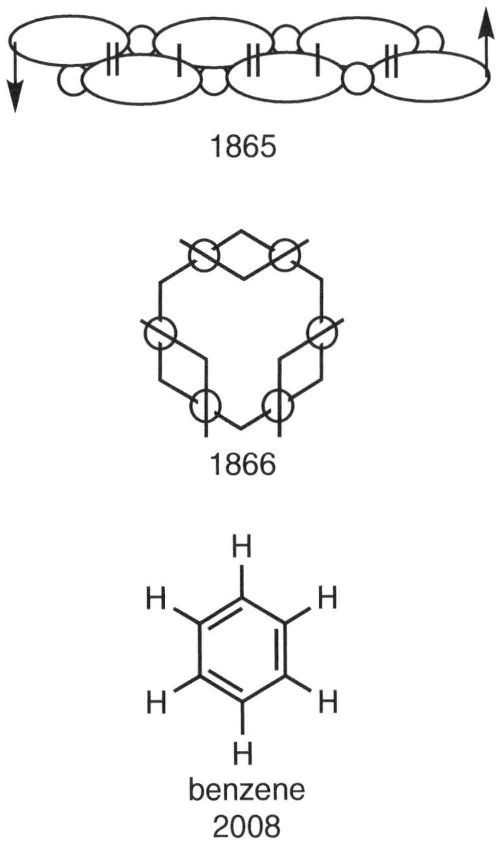

FIGURE 6.1 ▶

Evolution of the Structure of Benzene from 1865 to the Present

1865

1866

benzene
2008

Kekulé spends more than a page of his 1865 publication with drawings demonstrating his structural idea for benzene including drawings for many, if not all, of the derivatives of benzene that had been synthesized up to that time. The second drawing of this series, which Kekulé named "Chaine fermee," is redrawn in **Figure 6.1** showing the six carbon atoms with their oblong shape and the six affinity units represented by six circles. The up and down arrows represent the closing of the ring, which in this first publication on Kekulé's ideas for benzene did not propose the shape taken by this "closed chain." The double lines and single lines between the oblong carbon atoms, in modern terms, would represent double and single bonds. In another paper in 1865 he proposed a hexagon for the closed chain and by 1866, his final structure showing how the bonding arrangements could be made within the hexagon structure. Finally, in Figure 6.1 we show the structure of the benzene ring used today, evolved from Kekulé's proposal.

In Kekulé's structures drawn in Figure 6.1, only what Kekulé called the "common nucleus" is shown, that is only the six carbon atoms and how they are connected. In the 1865 structure the six small circles stand for, in Kekulé's words, "the six free affinity units," which allow the connection to the six hydrogen atoms to complete the formula for benzene, known by that time to be C_6H_6. The lines extending away from Kekulé's hexagon structure and similarly away from the modern structure also connect to hydrogen atoms in benzene itself. However, Kekulé allowed the possibility that these affinity units could allow connection to atoms other than hydrogen, which he called side chains, a word still used today. It's interesting to compare Kekulé's first presentation of these side chains with today's structures as shown in **Figure 6.2**.

The first experimental evidence that Kekulé's structure was correct was his prediction of the number of isomers, or what were called modifications, possible when one or more hydrogen atoms in benzene are substituted for by another atom.

3. benzine

4. benzine chloride

5. benzine bi-chloride

8. alcool phenique

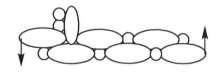

9. aniline

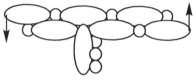

12. toluol

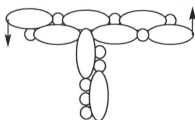

15. ethyle-benzine

18. alcool benzylique

As Kekulé put it in his 1865 paper, repeated from above: "…….*one modification of monochlorobenzene and of pentachlorobenzene can exist, but several isomers (probably three) of di-, tri-, and tetrachlorobenzene.*" **Figure 6.3** exhibits the structural isomers possible for these chlorobenzenes, using the modern structure for benzene, demonstrating the remarkable prescience of this 19th century chemist.

▲ **FIGURE 6.2**

Representations of the Molecular Structures of Benzene Derivatives from 1865 and the Present

monochlorobenzene

pentachlorobenzene

FIGURE 6.3 ▶

Substituted Benzene Nomenclature

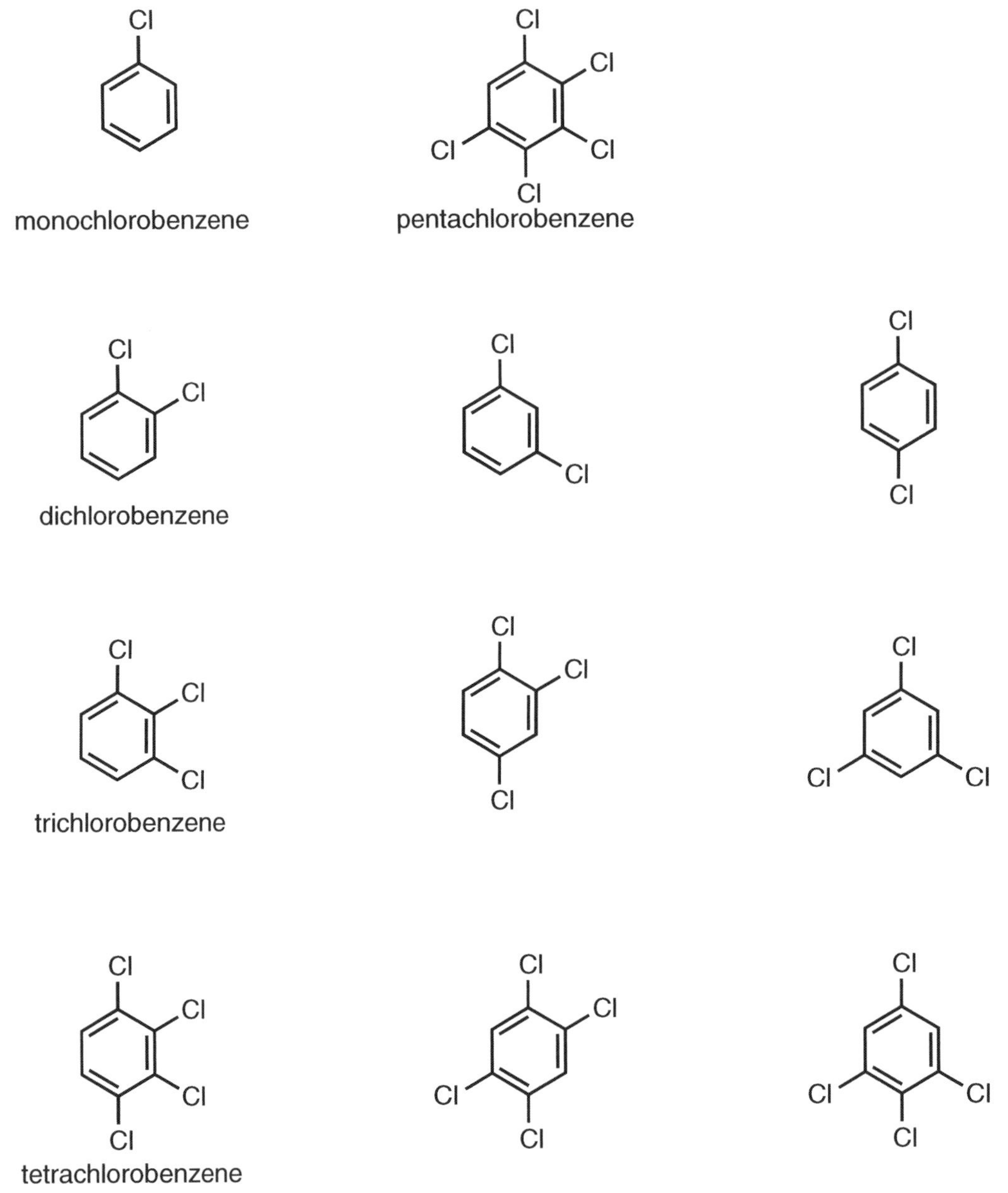

dichlorobenzene

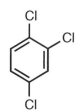

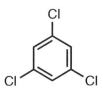

trichlorobenzene

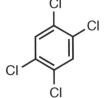

tetrachlorobenzene

PROBLEM 6.9

The structure for benzene proposed by Kekulé in 1866 was a six membered ring. If this ring contains three double bonds as we represent the structure now and two of the hydrogen atoms that are on adjacent carbon atoms are replaced by, let's say, chlorine, then how many different possibilities are there? What would these isomers be?

PROBLEM 6.10

What is meant by the difference between addition and substitution for a double bond?

PROBLEM 6.11

Why should the double bonds we draw in the structure of benzene (Figure 6.1) be so unreactive compared to a double bond in an alkene?

LATER IN THIS CHAPTER WE'LL COME ACROSS many substitution products of benzene, that is, molecules in which an atom or atomic arrangement replaces one or more of the original six hydrogen atoms on the benzene ring. We'll also find many sensible and many arbitrary names based on mode of discovery or some other aspect of the history of the molecule. However, there is a fundamental set of rules for naming disubstituted benzenes as exemplified by the three dichlorobenzene isomers in Figure 6.3. Going from left to right we name these molecules: ortho-dichlorobenzene, meta-dichlorobenzene and para-dichlorobenzene.

If there are more than two substituents on the benzene ring, that is, groups that have replaced more than two hydrogen atoms, then generally a number system is used. Going from left to right for the trichlorobenzenes we would name these molecules 1,2,3-trichlorobenzene, 1,2,4-trichlorobenzene and 1,3,5-trichlorobenzene. Similarly for tetrachlorobenzene the names are 1,2,3,4-tetrachlorobenzene, 1,2,4,5-tetrachlorobenzene and 1,2,3,5-tetrachlorobenzene.

As we encounter more benzene derivatives other aspects of the nomenclature of these molecules will be introduced as necessary.

6.5

Brief Stop for Benzene Nomenclature

PROBLEM 6.12

In Figure 6.3, the third tetrachlorobenzene could be named as 1,2,3,5 or 1,3,4,5. Which is preferable and why?

PROBLEM 6.13

Why are the names, ortho, meta and para only used for disubstituted benzenes?

Most new ideas meet with objections, but one particular objection to Kekulé's structural proposal for benzene required an addition to his original hypothesis.

Early on, some chemists questioned the prediction of the number of isomers for molecules derived from benzene, structural isomers according to our definitions of Chapter 1 (section 1.8). Inspection of the dichlorobenzenes in Figure 6.3 immediately shows the problem. In the ortho-dichlorobenzene, could not the two adjacent chlorine atoms be separated by a single bond within the hexagonal ring or alternatively by a double bond? These would be different isomers, but no experiment has ever been able to isolate more than three isomers of dichlorobenzene.

Similar considerations involving the fixed double and single bonds in all of the chlorinated benzenes in Figure 6.3 lead to predictions of more isomers than could be experimentally found and shown in the figure. At first excuses were made for the experimental difficulty of separating isomers with such similar structures. But Kekulé looked more deeply. Let's see his explanation in his own words, again translated from the original 1872 publication into English by Partington.

"The atoms in the systems which we call molecules must be considered to be constantly in motion......the movement must be of such a kind that all atoms forming the system retain the same relative arrangement....The same carbon atom is, therefore, during the first unit of time in double linkage with one of the adjoining carbon atoms, and in the second unit of time with the other....The ordinary formula of benzene represents only the contacts made during the first unit of time, or only one phase, and thus the view has arisen that the di-derivatives with the positions 1,2 and 1,6 must necessarily be different. If the above hypothesis, or a similar one, may be considered to be correct, it follows that this difference is only apparent, not real."

The two *"phases"* Kekulé was pointing to are represented in modern terms in **Figure 6.4**. Today we would call these molecular representations resonance structures and

6.6

Objections to Kekulé's hexagonal ring structure for benzene required an explanation that was equivalent to the concept of resonance.

replace Kekulé's *"atomic movements"* with electron distributions. In fact what Kekulé was writing in 1872 was his realization that a single structural representation of benzene was inadequate to portray the true structure of this molecule. This is the essence of Pauling's idea when he used the word resonance (section 5.4), which was based on a still earlier realization by G. N. Lewis (section 4.11) that a single representation of a molecule may be inadequate to fully represent the electron distributions in the bonds.

Pauling's idea of resonance was based on quantum mechanical considerations, which led him to the conclusion that the necessity for using resonance structures in molecular descriptions leads to an increased stability of the molecule over what would

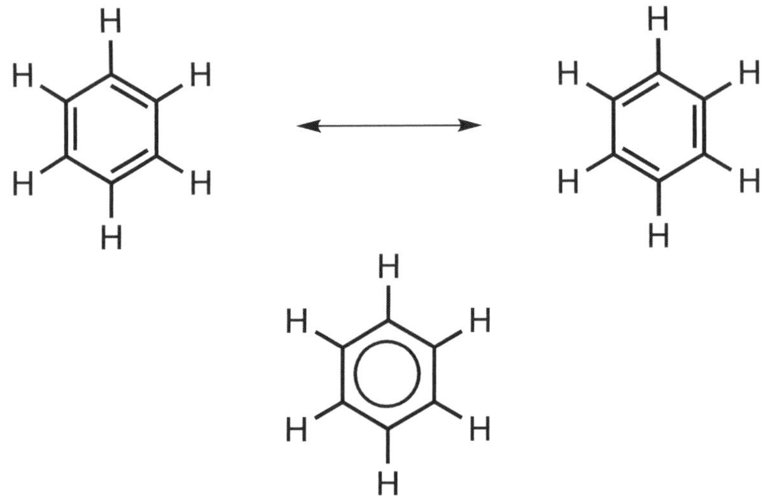

be expected from a single structure. We'll rewrite Pauling's words here from section 5.4: *"The molecule could be described as fluctuating rapidly between two electronic formulas, and achieving stability greater than that of either formula through "resonance energy" of this fluctuation."*

We now understand that the resonance structures are **not** fluctuating but rather simply multiple representations to portray a single molecular entity. Resonance structures are necessary in the situation when a single representation of the molecule, a single drawing, is not adequate to show the true structure of the molecule. However, Pauling's conclusion about greater stability is correct and we have already seen this idea at work in the conjugate bases of certain strong acids and in the phosphate based leaving groups so favored by nature (section 5.4).

But, as we shall see, Pauling's ideas were not adequate, in themselves, to understand the special properties of benzene. A complete understanding of the properties of benzene would not be found until a great German physicist, Erich Hückel, turned his attention to this molecule, as we'll look at in detail, shortly.

As introduced in section 5.4 the double headed arrow between two structures is the symbolism used to designate that these structures are not in equilibrium with each other, are not, in other words, exchanging or interconverting or *"fluctuating"* with each other. Rather this double headed arrow informs us that both structural representations are necessary to portray a truer picture of the molecule. In the situation of benzene (Figure 6.4), there are no localized double and single bonds but rather any two carbon atoms are bonded by some intermediate bonding state between a single and a double bond. Some representations of benzene remove the double bonds entirely and simply place a circle within the hexagon as also shown in Figure 6.4. But this kind of representation, as we'll see, gives up the ability to make predictions about the chemical behavior of benzene.

If the double and single bonds in benzene are not fixed in position, the criticism of Kekulé's isomeric predictions is removed. The two chlorine atoms in ortho-dichlorobenzene in Figure 6.3 do not reside on carbon atoms separated by a single or by a double bond but rather by a special kind of bond peculiar to benzene, neither single nor double.

But all of this discussion about the benzene structure was hypothesis. Is there really only one kind of bond connecting the carbon atoms in the benzene ring? If this were true, then the benzene ring would have to be a perfect hexagon with internal angles of 120° because all the connecting links would be identical. The test of this idea did not become possible until scientists discovered how X-rays could be diffracted by crystals of organic molecules, allowing the precise placement of the atoms within a molecule on the length scale of Angstrøms.

An X-ray analysis of derivatives of benzene had to wait until two reports, in 1929 and 1931, which analyzed the diffraction of X-rays from hexamethylbenzene and hexachlorobenzene, molecules that could form the kinds of crystals necessary to diffract X-rays to the resolution necessary. The results demonstrated that all the bonds between the carbon atoms in the benzene rings were indeed identical and that the ring was a perfect hexagon, that is, was flat, with internal angles of 120°. The benzene structure, although also a six-membered ring of carbon atoms, differed greatly from what was known about the six-membered rings in diamond, which were puckered as we have discussed (Figure 3.2). Kekulé had it right!

The X-ray work confirming Kekulé's structure for benzene was carried out by one of the first women allowed into the men's club of science in the early part of the 20th century, **Kathleen Lonsdale**. Her brilliance overcame the prejudice against women in science at that time. She was born in 1903 in Ireland, one of the Yardley children of County Kildare, whose family moved to England to avoid the political troubles in Ireland at that time. The very interesting story of her life can be found easily by using Google, but there are two items on the web I'd like to mention here. It is reported that when she married in 1927, she had the idea of leaving research and settling down to homemaking, but her husband dissuaded her by declaring he had not married a housekeeper. Kathleen and Thomas Lonsdale had three children and apparently a very happy marriage. Indeed, Thomas had certainly married someone who was more than a housekeeper. His wife was not only a distinguished scientist who made a brilliant career but, following on her Quaker beliefs, waged a campaign against war leading even to her imprisonment for one month as a punishment for her unwillingness in 1939, as World War II began, to participate in defense duties.

■ **Kathleen Lonsdale**

PROBLEM 6.14
Draw structures for all the isomers that would be present for each of the four multiply substituted benzenes shown in Figure 6.3 if the positions of the double bonds were fixed.

PROBLEM 6.15
How does the experimental evidence that the benzene ring is a perfect hexagon argue against isolated double and single bonds in this molecule?

PROBLEM 6.16
Considering that each carbon atom in a benzene ring is bonded to three other atoms, show how the necessary orbital hybridization makes isolated single and double bonds unreasonable. Which orbitals would participate in the hybridization if the benzene ring hexagon were drawn perpendicular to the page?

6.7

Hydrogenation of benzene yields a quantitative measure of the aromatic stability of benzene.

IT IS VERY DIFFICULT TO ADD TO THE "DOUBLE BONDS" in benzene, a resistance that is seen very clearly by comparing the conditions necessary to add hydrogen, H_2, to alkenes versus adding hydrogen to benzene. This difference can be expressed in quantitative form by the thermal changes occurring on catalyzed reaction of addition of hydrogen to alkenes and to benzene **(Figure 6.5)**. These two measures belong to the realm of kinetics and thermodynamics respectively, as we'll discuss a bit later.

Addition of H_2 to the double bond of a typical alkene such as cis-2-butene is exothermic by 28.6 kcal/mole (Figure 6.5), meaning that a large amount of heat is evolved arising from the fact that the bonds made in such a reaction are far stronger than the bonds broken. Let's apply some fundamental general chemistry to this reaction. The energies associated with breaking chemical bonds are widely available on the web by simply using Google under a heading of the general form, bond energies. Breaking an H-H bond requires an energy input of about 103 kcal/mole while breaking a π-bond costs about 63 kcal/mole. However, making a C-H bond releases about 99 kcal/mole and since two C-H bonds are made in the hydrogenation of an alkene double bond, one balances the gain of about 198 kcal/mole against the necessary energy input to break the π-bond and the H-H bond of 166 kcal/mole. That's where the heat release comes from.

If we consider the "double bonds" in benzene to be normal double bonds, such as those in cis-2-butene, then we could calculate that adding three molecules of H_2 to benzene to convert all the π-bonds to σ-bonds, that is, to convert C_6H_6 to C_6H_{12} (cyclohexane) there should be an exotherm of 3x28.6 kcal/mole. But the measured heat liberated is far smaller, 49.8 kcal/mole (Figure 6.5). The difference is 36 kcal/mole. What does this energy difference mean?

A π-bond is far weaker than a σ-bond, as we've seen in the properties of both the carbon-oxygen double bond and the carbon-carbon double bond (sections 3.13 and 4.10)), weak enough that even taking into account the necessity of breaking an H-H bond to convert an alkene to an alkane, the process is always exothermic, that is, heat is released. Put in another way, the total combined energy of an alkene molecule and a molecule of

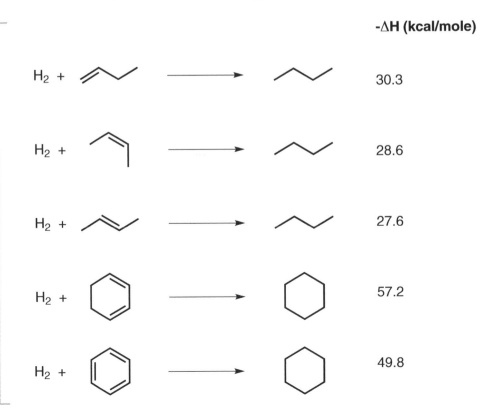

-ΔH (kcal/mole)

30.3

28.6

27.6

57.2

49.8

H_2 is higher than the energy of an alkane with the same number of carbon atoms. The greater is the difference in the stability of alkene/H_2 compared to the alkane (Figure 6.5), the larger the amount of heat released. This outcome means that there is something too stable about the "double bonds" in benzene compared to what is expected of the double bonds found in alkenes. Too little heat is released when benzene adds 3 H_2 molecules to form the alkane, cyclohexane. Quantitatively, benzene is "too stable" by 36 kcal/mole, as judged by the amount of heat released on conversion to cyclohexane (Figure 6.5).

The observation that too little heat is released in the hydrogenation of benzene is long known from measurements made in the last decade of the nineteenth century. The focus in those years on the importance of the quantitative measures of changes in heat associated with chemical reactions arose from the conviction that heat change would yield an insight into the forces holding molecules together. Credit is given to two major figures in chemistry of the nineteenth century, **Pierre Marcelin Berthelot** and **Julius Thomsen**, the former French, and the other Danish. Berthelot (1827-1907) is the more famous of the two and one of the best known chemists in the history of the science. He has been called by some the father of organic chemistry. He also had other accomplishments and in fact some streets in Paris bear his name following not so much on his scientific work but on his roles in public life, even gaining the position of Foreign Minister of France.

Thomsen (1826-1909) was responsible for the first ideas that led to the modern form of the periodic table in recognizing the special status of the noble gases, a status of zero valence, and as well for a discovery of great practical importance: Thomsen became wealthy by patenting a process to manufacture sodium carbonate, essential for the manufacture of soap, from a mineral found mostly in Greenland, which was controlled, at that time, by Denmark. Later this same mineral was used in the Hall process for the manufacture of aluminum: Al_2O_3 was electrolyzed in a solution of this molten mineral, cryolite, Na_3AlF_6.

Both Berthelot and Thomsen focused a great deal of attention on the heat changes noted in chemical reactions and carried out many experiments honing the quantitative accuracy of such measurements. But Thomsen is given credit for first pointing out that the heat released on adding 3 molecules of H_2 to benzene was far too small (Figure 6.5), casting therefore doubt on Kekulé's structure. How could double bonds, as Kekulé suggested were part of his cyclic structure, release such a small amount of heat when hydrogen is added to them?

Let's take a moment to look at another thermodynamic aspect of the hydrogenation reaction, the factors involved in the overall equilibrium of the reaction. To be favorable, the free energy change, ΔG, must be negative. A large negative enthalpic term, such as measured (Figure 6.5), favors a negative free energy change. However, on the hydrogenation of an alkene, or of benzene, the entropy change is also negative arising from more molecules combining to form fewer molecules. In other words, the $-\Delta S$, although leading to $-T(-\Delta S)$ and therefore $+T\Delta S$ will not overwhelm the $-\Delta H$ even the smaller $-\Delta H$ for hydrogenation of benzene. This result means that the free energy change, $\Delta G = -\Delta H - T(-\Delta S)$, will therefore be negative consistent with the hydrogenation taking place: $-\Delta H > -T(-\Delta S)$. The equilibrium constant will be large in the direction of hydrogenation for alkenes and for benzene..

As we understand from the most fundamental study of chemistry, equilibrium is the state that can be reached but the speed of reaching that state depends on factors that can be independent of the equilibrium constant. Hydrogenation of an alkene or of benzene, are perfect examples of the distinction between equilibrium constant and rate. You can wait a long time and discover that nothing happens when either an alkene or benzene is in the presence of an atmosphere of pure H_2.

However, in the presence of finely divided metals such as platinum or palladium prepared by reduction of the oxides of the metals with H_2 under one or at most a few atmospheres of pressure, an overwhelming and fast conversion occurs from the alkene to the alkane. Other metals prepared in a finely divided form, also at low pressure of H_2, also speed this reaction. In this manner, conversion of the two sp^2 carbon atoms at the ends of the double bond to sp^3 carbon atoms bound to hydrogen atoms, that is, reduction of the double bond occurs rapidly under mild conditions.

I remember having a well used Parr shaker, which described what it did, in my laboratory for over forty years (a considerably older version than the one shown below). It is a device that worked well at the low pressures that were adequate for hydrogenation of alkenes and related double bonds that were not part of an aromatic structure. Parr shakers, although hardly in their current form, have a long history dating from the early 1920s. **Paul Sabatier**, the Nobel Prize winning French chemist, discovered that

PARR SHAKER ▶

Parr Shaker, the first product to allow hydrogenation using elevated pressures and temperatures.

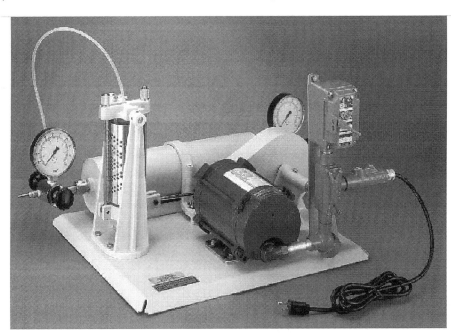

addition of metals to double bond containing molecules would speed hydrogenation and that shaking helped. Shaking makes sense, considering that hydrogenation is a heterogeneous process.

On the contrary, using similar catalysts, adding H_2 to benzene to produce the cyclic saturated cyclohexane ring, requires a high pressure apparatus, which chemists call a "bomb" considering that 2000 pounds per square inch pressure of H_2 must be used in addition to the fact that more active catalysts are necessary. As a young professor I remember we had a special organization, outside of the campus, where such difficult hydrogenations were conducted behind blast shields. We could have expected this, which is another example of the increased resistance of benzene, compared to "normal double bonds" (Figure 6.5), to convert the six π-electrons to σ-electrons, to convert C_6H_6 to C_6H_{12}.

■ **Paul Sabatier**

PROBLEM 6.17
π-Bonds are weaker than σ-bonds, accounting for the fact that energy is released when H_2 is added to a double bond – a less stable molecule is converted to a more stable molecule. Is this correct? If so explain why.

PROBLEM 6.18
What conclusions can you draw about structure and alkene stability from the fact that the three butene isomers in Figure 6.5, in order 1-butene, cis-2-butene and trans-2-butene, differ in the amount of energy released.

PROBLEM 6.19

What conclusion can you draw from the facts in Figure 6.5 that hydrogenation of the two double bonds in 1,3-cyclohexadiene releases close to twice the amount of energy as hydrogenation of one of the butene isomers, while hydrogenation of the three double bonds in benzene does not release three times the energy released by this butane isomer?

PROBLEM 6.20

All the enthalpy changes in Figure 6.5 are negative numbers. Considering also the entropy change associated with the hydrogenation of any of the molecules in Figure 6.5 what effect would increased temperature have on the equilibrium constant for the hydrogenations?

PROBLEM 6.21

How could cyclohexanol, $C_6H_{11}OH$, be synthesized from phenol (hydroxybenzene), C_6H_5OH? Why are derivatives of cyclohexane often synthesized from derivatives of benzene?

6.8

**Understanding Benzene:
Erich Hückel's Theory**

AS THE IDEAS OF QUANTUM MECHANICS in the 1920s began to be looked at to yield greater understanding of chemical phenomena there arose two approaches, one termed valence bond theory, which kept the fundamental localized bonds familiar to chemists, the kinds of bonds that G. N. Lewis had first shown to be so useful (section 4.11) and another approach termed molecular orbital theory, which attempted to understand molecular properties by developing a series of orbitals encompassing the entire molecule.

In the early 1930s, **Erich Hückel**, a German physicist, published a series of papers in which he discussed his theoretical work considering both the valence bond and the molecular orbital approach to the understanding of benzene. Although he decided that the molecular orbital approach offered the clearer potential to deal with the question of aromaticity, the valence bond approach, supported by Linus Pauling, which led to the concept of resonance, and which was more easily understood by organic chemists, was widely accepted. Therefore attempts were made to apply this theoretical idea to the benzene problem. The special properties of benzene discussed in detail above, were proposed to arise from the resonance necessary to describe the bonding in benzene: the two identical resonance structures (Figure 6.4) were proposed as accounting for the aromatic properties.

As Pauling had proposed (section 5.4) from considerations of valence bond theory, which had earlier been developed by Hückel, a molecule described as a hybrid of resonance structures was more stable than each of the structures. Benzene seemed to fit as it was certainly acting as a more stable molecule than the two double bonded resonance structures (Figure 6.4).

However, in one of those twists in how research progresses, a nitrogen containing natural product, pelletierine, a chemical obtained from pomegranate bark, was isolated in 1877, and was to play a role in understanding benzene. The name pelletierine honors the memory of a chemist who originated studies of nitrogen containing natural products. Its structure led to a result that made simple resonance ideas impossible as a way to justify the properties of benzene.

Pierre-Joseph Pelletier (1788-1842) was among the first to isolate chemicals from nature that were to prove important to medicine, and showed the way that chemists could advance human health by going beyond whole plant extracts, which had been used since as long as one can peer back into human behavior.

Pelletierine attracted the attention of **Richard Willstätter**, who won the Nobel Prize in chemistry in 1912 for his elucidation of many of the chemical properties and structure of the chlorophylls and the connection of chlorophyll to hemoglobin.

■ **Erich Hückel**

■ **Pierre-Joseph Palletier**

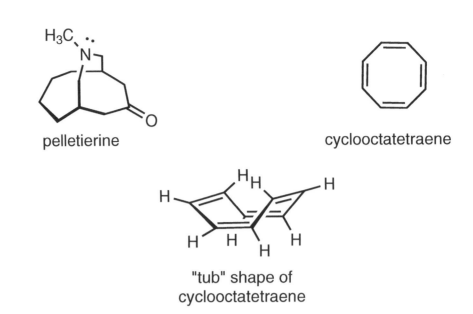

pelletierine

cyclooctatetraene

"tub" shape of
cyclooctatetraene

Richard Willstätter

Isidor Fankuchen

Willstätter was a highly beloved man of great principle who was driven out of Germany by his unwillingness to accept anti-Semitism, even when not directed against him. The final straw for him was the Nazis coming to power. Willstätter held strong views about other subjects including about the nature of a scientific education, which is of relevance to the philosophy of this book. Quoting from Partington: "*Willstätter thought the lecture courses should include the history of discoverers, substances, and theories, a purely factual treatment being more suited to a technical school than a university.*"

Chlorophyll had, in fact, first been isolated by Pelletier in the early 1800s, which perhaps played a role in Willstätter's interest in pelletierine. At any rate Willstätter in 1905 was able to convert pelletierine to a highly unsaturated molecule, cyclooctatetrene, that was expected to be an eight membered analog to benzene (Figure 6.6). However, the experimental facts proved otherwise. The four double bonds in cyclooctatetraene behaved as do the double bonds in alkenes, undergoing facile addition reactions such as adding four moles of HCl or four moles of Br_2 (see the discussion around Figures 6.13 and 11.7). The heat released on addition of four moles of H_2 to cyclooctatetraene is only about 2 kcal/mole less than the heat expected to be released from four isolated double bonds.

The key unanswered question of that time was the shape of the cyclooctatetraene. Would it be flat like benzene? But determining the shape of cyclooctatetraene turned out to be a difficult problem The first structural data derived from electron diffraction, the same technique used to determine the structure of cyclohexane (section 3.3) were not definitive. It was clear that only x-ray crystallography could yield the answer. But cyclooctatetraene had a melting point below the melting point of ice, -3° C; the molecule was a liquid at ambient temperature.

Isidor Fankuchen had developed a distinguished school of crystallography at the Polytechnic Institute of Brooklyn, the institution, now the Polytechnic Institute of New York University, from which this book is being written. He wanted to carry out crystallographic studies at low temperature, and therefore developed a low temperature technique well suited to solving the structure of this theoretically important molecule. In 1948 the results of the x-ray analysis were published demonstrating unequivocally that cyclooctatetraene was not flat like benzene but rather tub shaped (**Figure 6.6**) and as well that there are two kinds of bond distances, double and single bonds alternating in the structure. The molecule is the antithesis of benzene, antiaromatic as coined by **Ronald Breslow** of Columbia University for those molecules, which could appear to be aromatic based on their ring structure and p-electrons, but failed to exhibit aromatic character.

Poly, as Fankuchen's school was, and is still, fondly known, was a powerhouse of science and technology in the decades after the end of the Second World War, to some extent arising from distinguished faculty who were refugees from the European turmoil. One of the most important for Poly was **Herman Mark**, who initiated the first institute to investigate polymers in the United States and turned it into an enterprise that spread across both the academic and industrial globe. Mark, who was a pioneer in crystallographic studies in Europe, was a co-author with Fankuchen of the publication about cyclooctatetraene, which stimulates telling a story about this great scientist, which Linus Pauling often told and which I heard directly as true from Professor Mark himself.

Herman Mark was a distinguished scientist in Germany and the head of research at a division of I. G. Farben. When the Nazis came to power in Germany he was forced by the newly imposed racial laws to move to a professor's position in Austria, the country of his birth. What follows can be found on the web and from other sources.

"In early 1938 Mark began preparing to leave Austria. He clandestinely started to buy platinum wire, worth roughly $50,000, which he bent into coat hangers while his wife knitted covers so that the hangers could be taken out of the country. When Hitler's troops invaded Austria and declared the Anschluss (the political union of Germany and Austria), Mark was arrested. He was also stripped of his passport. He retrieved his passport by paying a bribe to an influential "friend" equal to a year's salary, and he obtained a visa to enter Canada where he had been offered a position at an industrial firm. At the end of April, Mark and his family mounted a Nazi flag on the radiator of their car, strapped ski equipment on the roof, and drove across the border, reaching Zurich the next day. From there, the family traveled to England and then to Canada."

Cyclooctatetraene offered a difficult problem to the resonance approach because identical resonance structures, as seen in benzene, could be realized for this eight member ring if the ring were flat. Could it be possible to understand why cyclooctatetraene acted like a normal alkene? A well accepted answer, at the time, can be found in a textbook published in 1959. In their 5th edition of "The Chemistry of Organic Compounds," by **James B. Conant** and Albert H. Blatt, it was put it this way:

"The explanation for the smaller amount of resonance stabilization and the greater reactivity of cyclooctatetraene is almost certainly to be found in the strain which would be present in a planar eight-membered ring."

Conant was both a distinguished organic chemist and administrator. He held various important posts during his career varying from President of Harvard University to Ambassador to West Germany.

While a perfect hexagon has internal angles of 120°, a perfect octagon has internal angles of 135° and therefore there was a deviation from the ideal angle of trivalent carbon of 15° for each of the eight carbon atoms. We now understand that such an angle strain is hardly insurmountable (look back at section 5.9) but this angle strain did offer a reason, although we know now, not a correct reason, for something that could not be otherwise understood.

The evidence was already in hand by the late 1940s that cyclooctatetraene easily added two electrons, via various chemical and electrochemical processes, to form a dianion, a double negatively charged ring structure. In 1960, **T. J. Katz**, a young professor at Columbia University, published a nuclear magnetic resonance (Chapter 2, section 2.3) study showing that the cyclooctatetraene dianion was flat. This conclusion was later confirmed by crystallographic x-ray studies, which showed that like benzene the bond lengths of the dianion of the eight membered ring are all equal while the internal angles were the expected 135° of the perfect octagon. The structure for this ten π-electron ring of carbon atoms with two negative charges fit perfectly for aromatic character.

The first reference in Katz's 1960 paper is to Erich Hückel's book published in 1938. In this book Hückel discussed work he had published in 1931-32 on the use of molecular orbital theory in a simplified form, where only the π-electrons of benzene were considered. This effort gave rise to a series of molecular orbitals (look ahead to section 12.21) in which the six π-electrons of benzene led to a filled shell, a molecular orbital equivalent of the filled atomic

■ **Ronald Breslow**

■ **Herman Mark**

■ **James B. Conant**

■ **T. J. Katz**

orbital shells of the noble gases. Hückel's analysis placed benzene as a noble molecule.

Like neon, one of the unreactive noble gases, benzene would also be loath to give up its special stability. And this stability came from an orbital situation that could not be described by localized double bonds but rather by molecular orbitals in which the 6 π electrons were delocalized. In fact, later measurements show that the π electrons of benzene form a continuous electron circuit around the ring giving rise to a magnetic field, which can be detected by nuclear magnetic resonance experiments by its effect on the chemical shift and spin-spin splitting of nearby atoms (sections 2.4, 2.5).

Although Hückel did not specifically state a rule himself, his work also demonstrated that other rings of carbon atoms led to molecular orbitals in which numbers of electrons corresponding to 4n + 2, where n could be any integer including zero, led to a similar "stability" arising from filled shells.

In Hückel's own words: *"Das Auftreten der Zahl 2 + 4 = 6 für eine abgeschlossene Elektronengruppe bei den Ringen hat denselben Grund wie das Aftreten der Zahl 2 + 6 = 8 für eine abgeschlossene Elektronengruppe ("Oktett") bei den Atomen."*

The molecular orbital diagrams derived from Hückel's work for a series of rings of different numbers of sp² carbon atoms existing in a plane, including benzene and cyclooctatetraene (imagined to be flat), are shown in **Figure 6.7**. Following Pauli's principle and Hund's rule, that is, the procedures for filling atomic orbitals, one readily sees the necessity of 4n + 2 electrons as necessary to form filled shells of the orbitals that contribute stabilizing properties to the molecule (see below).

For every ring shown for numbers of carbon atoms from 3 to 8 (Figure 6.7) the lowest orbital, to be filled first, is non-degenerate, that is, another orbital of equal energy does not exist. Therefore every ring size must begin with two electrons in this lowest orbital on the way to a filled shell. However, at higher energies, the orbitals exist in pairs of equal energy, degenerate states. Therefore a filled shell must add to the two electrons in the lowest energy orbital, four electrons multiplied by the number of degenerate pairs.

Electrons that occupy orbitals designated by a negative sign (Figure 6.7) will destabilize the molecule and detract from aromatic character so that the total number of electrons

FIGURE 6.7 ▼

Hückel Molecular Orbitals for Three to Ten Member-Rings that are Fully Unsaturated Leading to the 4n+2 Rule

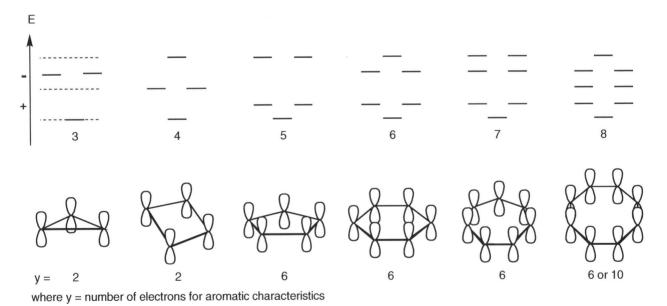

where y = number of electrons for aromatic characteristics

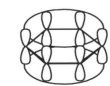

pictorial representation of
the π electrons in benzene

allowed for aromaticity is limited to the molecular orbitals with a positive or zero coefficient.

The numbers of electrons to satisfy the closed shell picture just described for each ring size for bonding and non-bonding molecular orbitals (positive or zero coefficients) are shown in Figure 6.7, designated as "y" and as can readily be seen, for the reasons just given, each number satisfies the equation 4n + 2 with n equal to zero or an integer.

However, only for benzene does the correct number, 6 (4n+2, n=1) correspond to both an uncharged state and the required planar ring with an internal angle of 120º, which corresponds to the required angle for sp² hybridization. The absence of a formal charge in each of the rings shown in Figure 6.7 requires that the number of electrons equal the number of p orbitals shown for each ring. Only in benzene is the number of electrons equal to the number of p orbitals. For one example, the three member ring with two electrons (4n+2, with n=0) would be aromatic but would have to carry a positive charge and angle strain.

The seven member ring with six electrons (4n+2 with n=1) would be aromatic but would have to carry a positive charge. In addition, the internal angles of a heptagon are 128.57º, causing therefore angle strain. Consider every other ring and find similar problems with attaining the aromatic state with sources of instability from charge and angle strain.

Among the cyclic molecules in Figure 6.7, one can justifiably state that benzene is a noble molecule, perfect in every way: no formal charge with the number of electrons, 6, equal to the number of p orbitals; and the structure of a hexagon with internal angles of 120º and therefore no angle strain. Benzene, as recognized by Hückel, is the molecular analog to the noble elements in the periodic table – a filled shell of bonding molecular orbitals finding a parallel to a filled shell of atomic orbitals. Perfect!

Chemists have confirmed the predictions of the molecular orbital pictures shown in Figure 6.7 with many experiments as shown in **Figure 6.8** for reactions 1, 3 and 5, which are also just a few of the many experimental findings supporting Hückel's theoretical work.

The early work of Hückel has been shown, as well, to be of exceptional use to the understanding of a variety of other molecular properties. In Robert S. Mulliken's

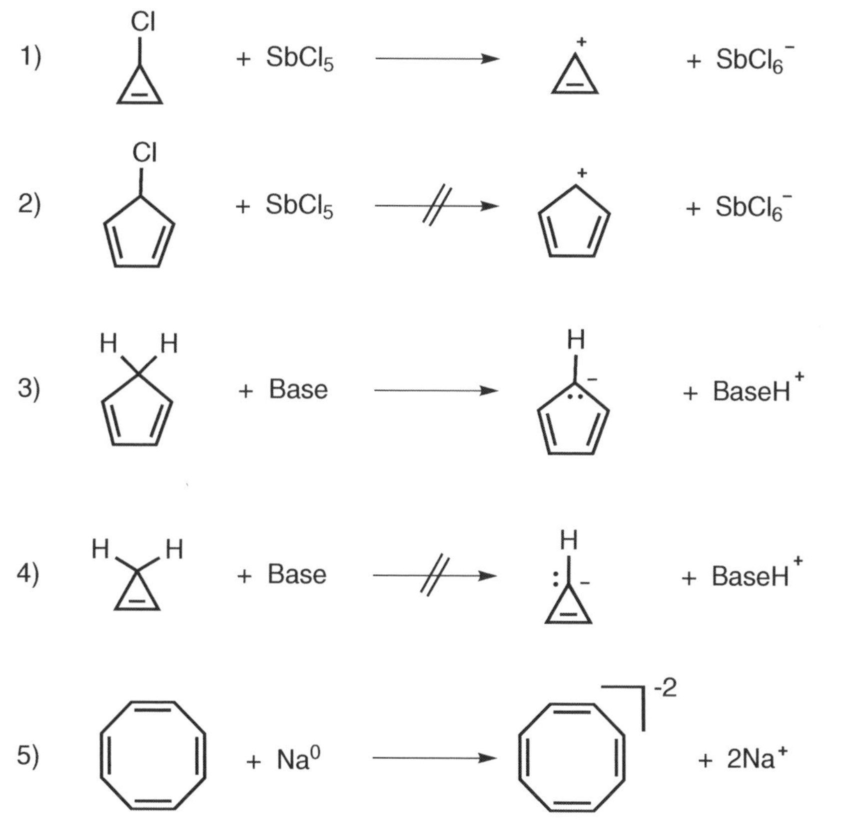

◀ **FIGURE 6.8**

Chemical reactions Of unsaturated cyclic molecules demonstrate aromatic and anti-aromatic characteristics.

Nobel Prize address in 1966, entitled "Spectroscopy and Molecular Orbitals," he writes: *"Hückel developed his very simple LCAO (linear combination of atomic orbitals) treatment for the π-electrons in unsaturated and organic molecules, a procedure which while rough and subject to some serious limitations, has been very useful to the organic chemists for some years."* In fact, the use of HMO theory (Hückel Molecular Orbital) was shortly to become even more useful to understanding chemistry in the years immediately following Mulliken's prize with the understanding that molecular orbital theory can explain a large class of reactions, which for many years defied explanation (section 12.21).

Hückel showed the value of quantum mechanical ideas applied to chemistry long before the computer technology existed to allow molecular orbital theory to be applied with its full potential.

In spite of these pioneering major contributions, to which must be added the theory with Debye on electrolytes in solution, Hückel was not successful in the world of science during his lifetime. **Jerome Berson** in his book "Chemical Creativity," based on extensive study of the man and his work, has offered a convincing analysis of the reasons for the delay in acceptance of Hückel's theoretical work. The delay corresponded with a very difficult career path for Hückel, exemplified by the fact that he even had trouble gaining a permanent position in German academia and, remarkably, did not gain full professorship until the year before he retired. Part of the problem was that he never made his work easily accessible to chemists, which was added to the disdain in which influential physicists of that time held a colleague whose work was interesting to organic chemists. It was not until 1953 that Hückel's mathematical expression of his theory was translated by others into graphical form showing how a certain number of electrons would lead to a filled shell of bonding molecular orbitals for each ring size, so that the source of the so-called magic number for aromaticity, 4n + 2, could be readily understood.

Apparently, the first, or one of the first times Hückel's ideas came to the attention of organic chemists arose from an unexpected source. **William Doering**, who was a nonagenarian emeritus professor of chemistry at Harvard University, was a graduate student at Harvard University in the late 1930s and decided to take a year off before continuing his studies. The year was 1939 and Doering took a bicycle trip through Germany where he describes passing through a University town (Munich he remembers) where he stopped in front of a bookstore and saw and purchased Hückel's now famous monograph: "Grundzüge der Theorie ungesättigter und aromatischer Verbindungen," published by Verlag Chemie in Berlin in 1938. Here Doering found, as he reported in a paper he published in 1951 in the Journal of the American Chemical Society, the source of, as Doering put it, the 2 + 4n rule.

Valence bond theory, in its truncated form as applied by Pauling and his former graduate student, G. W. Wheland, could not solve the mystery of the special properties of benzene. Molecular orbitals involving all the π-electrons of the molecule were necessary, but we shall see how valence bond theory, and the resonance idea (section 5.4) that arose from it allowed understanding of the properties of benzene when this molecule is forced to undergo a chemical reaction, electrophilic aromatic substitution. In this manner, application of resonance structures, which was incapable of explaining the aromatic character of benzene could yield insight into problems plaguing the chemical industry in their synthesis of essential precursors to commercially important polymers.

But before we delve into the reactivity properties of benzene let's take a short sojourn (section 6.9) to see how Hückel's theoretical work yields insight into certain molecules nature uses for critical functions.

Jerome Berson

William Doering

PROBLEM 6.22

Based on Hückel's theory of aromaticity, the 4n+2 rule, and the data in Figure 6.5, what enthalpic change would you predict for the hydrogenation of cyclooctatetraene to cyclooctane?

PROBLEM 6.23

Organic chemists have synthesized many molecules with larger rings than in cyclooctatetraene but also with alternating double bonds to test Hückel's theory of aromaticity. Noting that even-numbers of carbon atoms in a ring are necessary for alternating double bonds, draw the structures of these rings from 10 through 20 and predict which will tend toward planar conformations and which will undergo substitution rather than addition reactions.

PROBLEM 6.24

Explain how the chemical properties of benzene and its molecular orbital description make it reasonable to see benzene as the molecular analog of a noble gas.

PROBLEM 6.25

Although a five membered ring of carbon atoms (Figure 6.8) can fit into the 4n+2 rule with n = 1, the aromatic stabilization is less than found for benzene. What structural characteristics in this ring contribute to making it less aromatic than benzene?

PROBLEM 6.26

Draw the structure of a ring of seven carbon atoms with three alternating double bonds with the seventh carbon atom bearing a hydrogen atom and a bromine atom (C_7H_7Br). Why is the bond between carbon and bromine so easily broken to form the conjugate base of HBr?

PROBLEM 6.27

Draw the structure of a five membered ring analog (C_5H_5Br) of the seven membered ring drawn for Problem 6.26. Why is the bond between carbon and bromine so difficult to break to form the conjugate base of HBr?

PROBLEM 6.28

While the pK_a of cyclopentadiene (example 3 in Figure 6.8) is relatively low at +16, which is a moderate acid, the pK_a of the seven member ring analog of this molecule, cycloheptatriene, C_7H_8, is much higher with a pKa in the range of +36 – in other words, an exceptionally weak acid. Explain these experimental facts.

PROBLEM 6.29

Account for the three reactions in Figure 6.8 that occur easily and the two reactions that do not occur.

PROBLEM 6.30

Account for the experimental finding that a ring of four carbon atoms with two double bonds, cyclobutadiene, is very difficult to synthesize, and when successfully obtained is found to take the shape of a rectangle rather than a square. Would the two double bonds in this rectangle be shorter or longer than the two single bonds?

PROBLEM 6.31

Account for the numbers given for y in Figure 6.7 as the number of electrons for each ring size that would lead to aromatic character.

PROBLEM 6.32

What is the meaning of the + and – signs in Figure 6.7 and what importance do these signs have for the number of electrons necessary for aromatic character?

6.9 Applications of Hückel's Theory to Biologically Important Molecules

IN THE SUMMARY OF HIS THEORETICAL WORK (the book that William Doering saw in the bookstore in Munich (section 6.8),) Hückel proposed many molecules subject to aromatic characteristics in addition to benzene. The criteria for application of the theory were that the rings be planar and that only the π-electrons be considered. Within these structural restrictions there is a remarkable value to the 4n + 2 rule for aromaticity. Some of the molecules Hückel originally considered are related structurally to the aromatic characteristics of the nucleotide bases used in both deoxyribonucleic acids (DNA) and ribonucleic acids (RNA) shown in Figure 6.9. More examples of biologically important molecules in which aromatic characteristics play critical roles are many coenzymes and certain of the amino acids. Several coenzymes are shown in Figure 6.10.

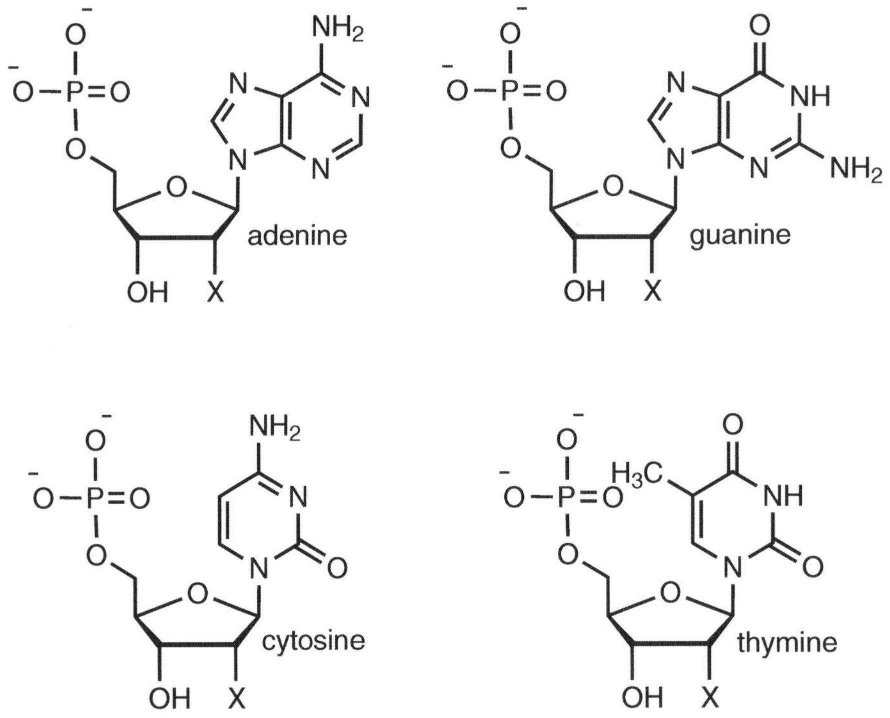

for adenine, guanine, cytosine: X = H for DNA
X = OH for RNA
for thymine: X = H for DNA

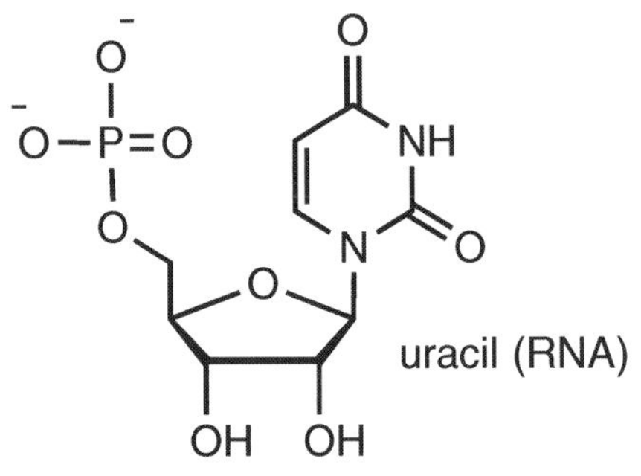

The proper name of these nucleotides are:
2-deoxyadenosine-5'-phosphate and so-on.

FIGURE 6.9 ▶ **Aromatic Rings in Nucleotides**

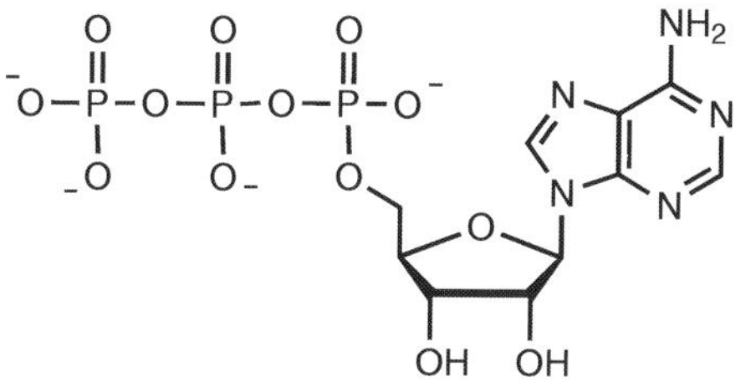

adenosine triphosphate, ATP

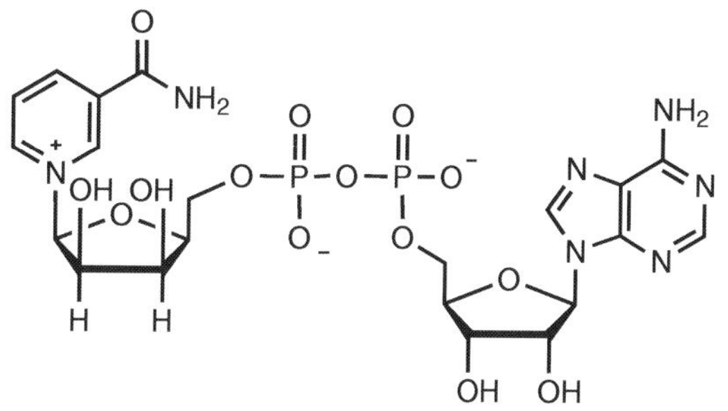

nicotinamide adenine dinucleotide, NAD⁺

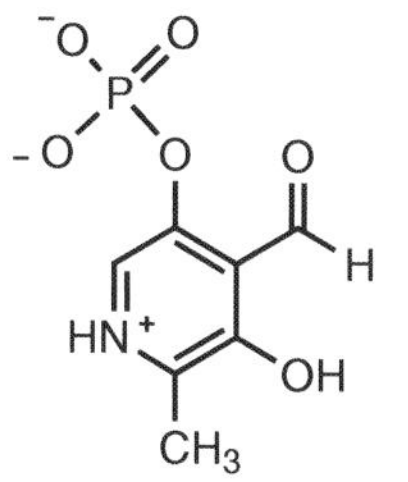

Aromatic Rings in Coenzymes

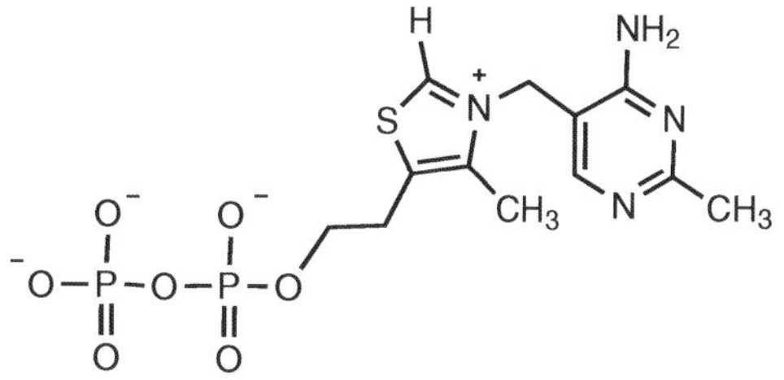

pyridoxal phosphate

thiamine diphosphate

PROBLEM 6.33

For each of the unsaturated rings in Figures 6.9 and 6.10 specify the source and number of the electrons contributing to aromatic character by drawing all the structures showing <u>all atoms and all nonbonding electrons</u>. How do the spatial dispositions of the nonbonding electrons in these molecules, which are specified by the orbital hybridizations, have an effect on their contribution to the aromaticity?

C
UMENE, IS AN IMPORTANT large volume industrial chemical because it is a precursor of two very important commercial polymers, polycarbonate and epoxy resin (**Figure 6.11**), which are themselves sold in the billions of pounds. While the applications of epoxy resins reside primarily in adhesives and coatings, polycarbonates are more widely used as what are called engineering plastics. The commercial uses of polycarbonates arise from their high impact resistance, their temperature resistance and their optical properties. One is amazed at how many common materials and devices contain this polymer as a major component: sunglass and eyeglass lenses, safety glasses, automotive headlamp lenses, compact discs, drinking bottles, iPod/Mp3 player cases, riot shields, visors, many signs and displays, computer and phone casings and other electronic devices: important stuff.

The reaction between propylene and benzene shown in Figure 6.11, is used to create billions of pounds of cumene world-wide, and is an example of one of the most iinvestigated reactions of organic chemistry, electrophilic aromatic substitution. However, byproducts, under the heading of multialkylation (**Figure 6.12**), have plagued this industrial process over the many years of its use. In our understanding of why these byproducts are formed and the means of overcoming the problem we will greatly deepen our understanding of the special properties of that "*common nucleus composed of six carbon atoms*," as Kekulé put it in 1865 (section 6.4), that is, of the special properties of benzene and all of its derivatives.

We've come across the concepts of "electrophilic" and "nucleophilic" before (section

6.10

Cumene, the common name for isopropyl benzene, is produced by the world chemical industry at the level of billions of pounds. The industrial process introduces us to electrophilic aromatic substitution, and the Friedel-Crafts reaction, and to a confrontation between industry's goals and organic chemistry's principles.

Overview of the Formation of Polycarbonate and Epoxy Resin from Petroleum

3.14) but what is electrophilic aromatic substitution? As we learned, chemists use the word electrophilic to mean "seeker or lover of electrons" after the Greek word philos. The opposite of electrophilic is nucleophilic, meaning seeker of the nucleus, which comes from the Latin nucleus, "little nut" or kernel. The nucleus of the atom is positive and, therefore, nucleophilic means seeker of the positive. In an electrophilic aromatic substitution reaction, a positive entity, an electrophile, usually a carbocation reacts with a source of electrons, an aromatic ring, which is the nucleophile. Electrophilic aromatic substitution is a classic example of a polar reaction.

We've come across this polar reactivity before in Chapter 4 where we discovered chemical reactions involving addition to the alkene double bond (section 4.10). Polar addition reactions were responsible for the formation of high octane gasoline (Figure 4.13), which took place when a proton and then when a carbocation, both electrophiles, reacted with the double bond of isobutene, a nucleophile, to produce carbocations. This mode of double bond reactivity in which a positive moiety, an electrophile, adds to a double bond, the nucleophile, is a recurring theme in the chemistry of alkenes, which we saw several times over in the formation of lanosterol from squalene oxide (Figure 5.18) and in all the reactions leading to the terpenes (section 5.7).

The addition of a hydrohalic acid, such as HCl, to double bonds, which involves the intermediacy of a carbocation produced when the electrophilic proton adds to the nucleophilic double bond, is a very well studied example of the interaction of an electrophile with a nucleophile. This reaction is shown for several alkenes, with the mechanism of this addition reaction, which is the same for all alkenes, shown for one of the alkenes in **Figure 6.13**, propylene.

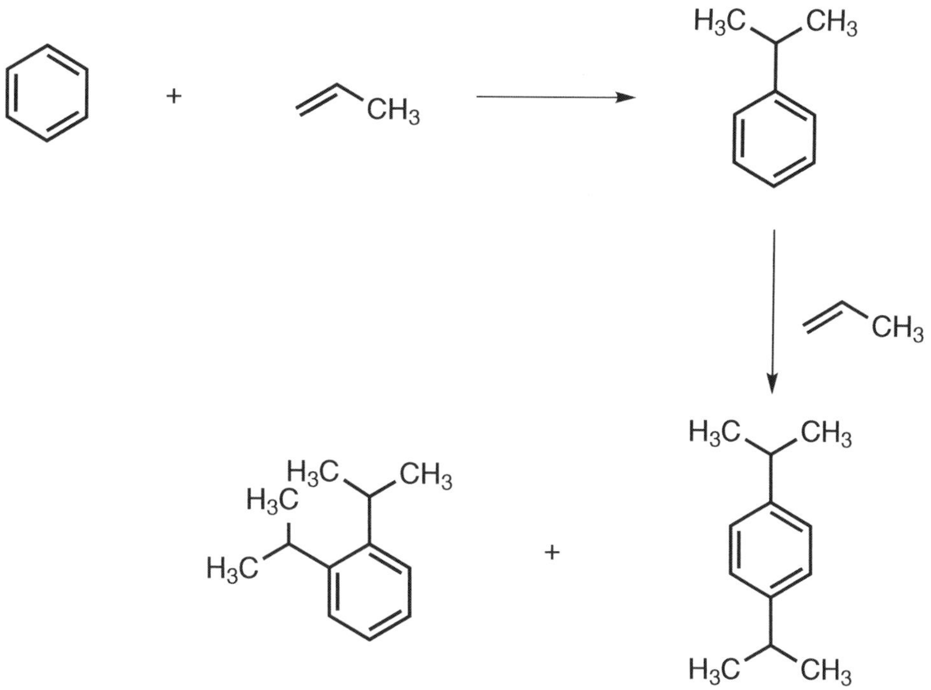

◄ **FIGURE 6.12**

Cumene and unwanted multialkylation products are formed from propylene and benzene.

The alkyl chlorides produced from the five alkenes shown in Figure 6.13 are consistent with the rule of Markovnikov discussed in section 4.10. Following this rule, in every example in Figure 6.13, the chloride anion, which acts as a nucleophile, resides on the carbon at what was the most substituted end of the double bond. The principle that more highly substituted carbon forms the most stable carbocation, which is at work in gaining high octane gasoline in catalytic cracking of naphtha fractions of petroleum (section 4.6) and in the route to 100 octane 2,2,4-trimethyl-pentane in the acid catalyzed reaction of isobutene with isobutene (Figure 4.13) is the source of the results in Figure 6.13. The proton adds to the end of the double bond that is least substituted as required to place the carbocation at the other end of the double bond, the most substituted site. Now we are prepared to turn our attention to benzene.

Considering the resonance structures in Figure 6.4, which show three double bonds, we might expect benzene, before we were exposed to Hückel (section 6.8), to behave parallel to the first step taken by an alkene, to act as a nucleophile. A partnering electrophile would be necessary and for the formation of cumene (Figures 6.11 and 6.12), which bears an isopropyl group on one of the six benzene carbon atoms the appropriate electrophile would be the isopropyl cation. We've seen the isopropyl cation as an intermediate in Figure 6.13 in the mechanism (reaction 1) for the synthesis of 2-chloropropane (isopropyl chloride) from propylene and HCl.

In the classic industrial synthesis, isopropyl cation is, in fact, produced when propylene is mixed with small amounts of HCl and aluminum trichloride, as catalysts. However, the results are quite different from the addition reaction that would be expected if the nucleophilic benzene were acting as a conventional double bond. Figure 6.14 exhibits the reactions taking place in this industrial process, an example of a reaction named after two chemists, Friedel and Crafts, who died in 1899 and 1917, respectively, but whose names live on to identify one of the most famous reactions in organic chemistry, a reaction they created in the nineteenth century.

Charles Friedel was a professor at the Sorbonne, in France, a Frenchman born in Strasbourg. **James Crafts** was an American born in Boston. In those years, late in the nineteenth century, things in the chemical sciences were quite different – the United States was a backwater in the sciences. The most advanced research and facilities were overseas and ambitious young chemists went to Europe to study. Crafts followed this path and therefore spent a great deal of time in Europe, including in the laboratory of

▪ **Charles Friedel**

▪ **James Crafts**

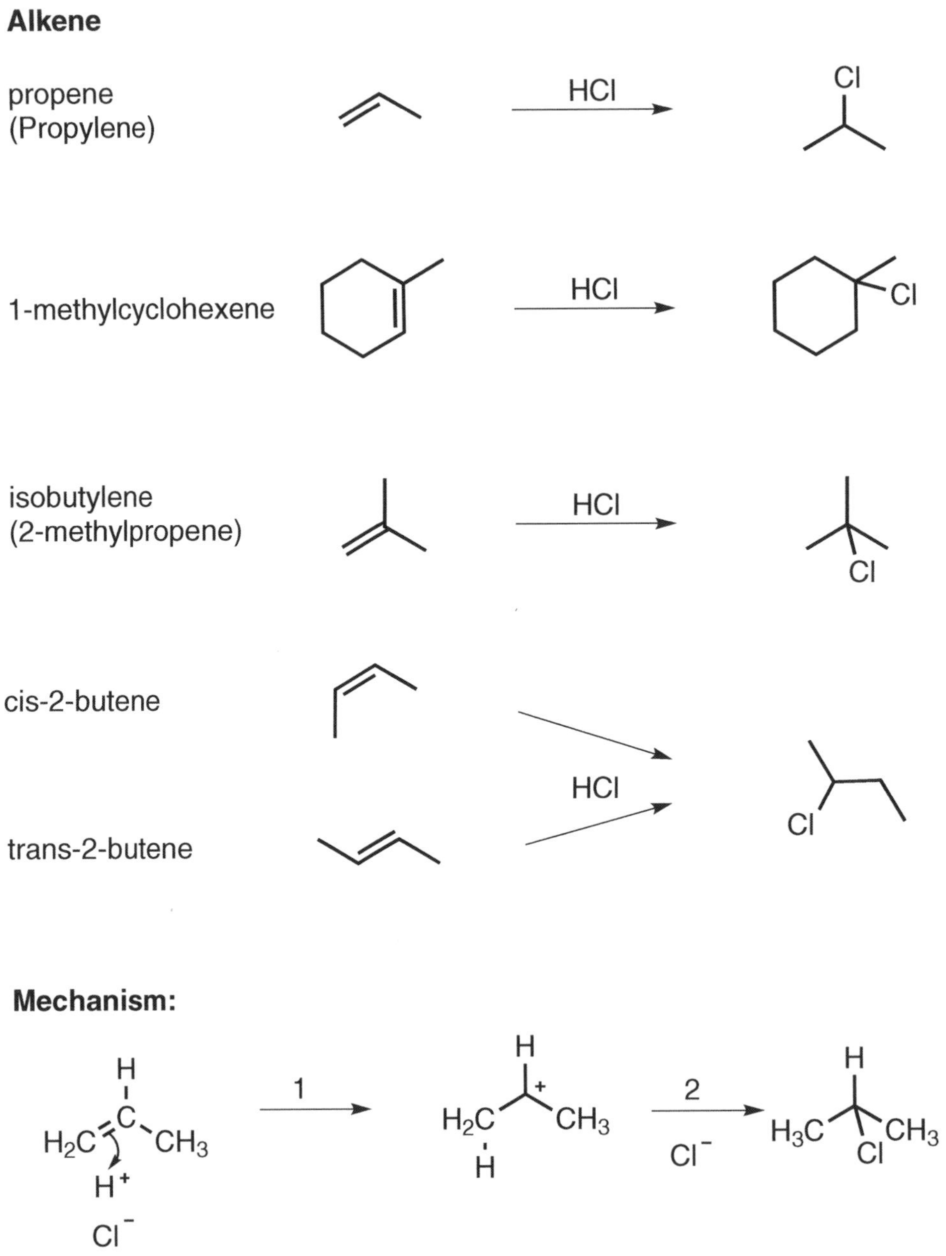
Alkene
propene
(Propylene)
HCl
Cl
1-methylcyclohexene
HCl
Cl
isobutylene
(2-methylpropene)
HCl
Cl
cis-2-butene
HCl
trans-2-butene
Cl
Mechanism:
H
H2C
C
CH3
H+
Cl-
1
H
+
H2C
CH3
H
2
Cl-
H
H3C
CH3
Cl

FIGURE 6.13 ▶

Addition of HCl to Various Alkenes Following Markovnikov's Rule

Professor Friedel. In 1877, the mentor and his student, working together, developed the use of aluminum trichloride to catalyze reactions with benzene.

Friedel was professor of mineralogy at the Sorbonne and then, later, professor of organic chemistry, possibly accounting for his combination of the chemistry of aluminum and the chemistry of carbon. The combination of these elements could be seen as an early version of what we today call organometallic chemistry, a field involving a large number of elements and offering powerful methods of synthesis of organic compounds found in industrial processes and in the in vivo chemistry essential to sustaining life.

Crafts, a gifted individual who excelled in many areas, returned to the United States, and was invited in 1897 to be president of the Massachusetts Institute of Technology (MIT), something he is reported to have seen as so much less interesting to him than scientific research that he resigned after less than three years to return to the laboratory.

In contrast to the reaction of HCl with propylene in Figure 6.13, where Cl^- is available to react with the isopropyl cation, Cl^- is not available in the Friedel-Crafts industrial reaction (**Figure 6.14**). The Cl^- is tied up with the aluminum trichloride to produce tetrachloro aluminate ion (reaction 1). This difference means that isopropyl cation produced by addition of H⁺ to propylene, reaction 2, needs to find another nucleophile other than Cl⁻ to satisfy the powerful electron-attracting properties of this positively charged intermediate.

The high reactivity of the isopropyl cation, arising from its extreme electron deficiency, leads to penetration of this carbocation into the aromatic core of the benzene ring (reaction 3), and this takes place in spite of the resistance to interruption of the

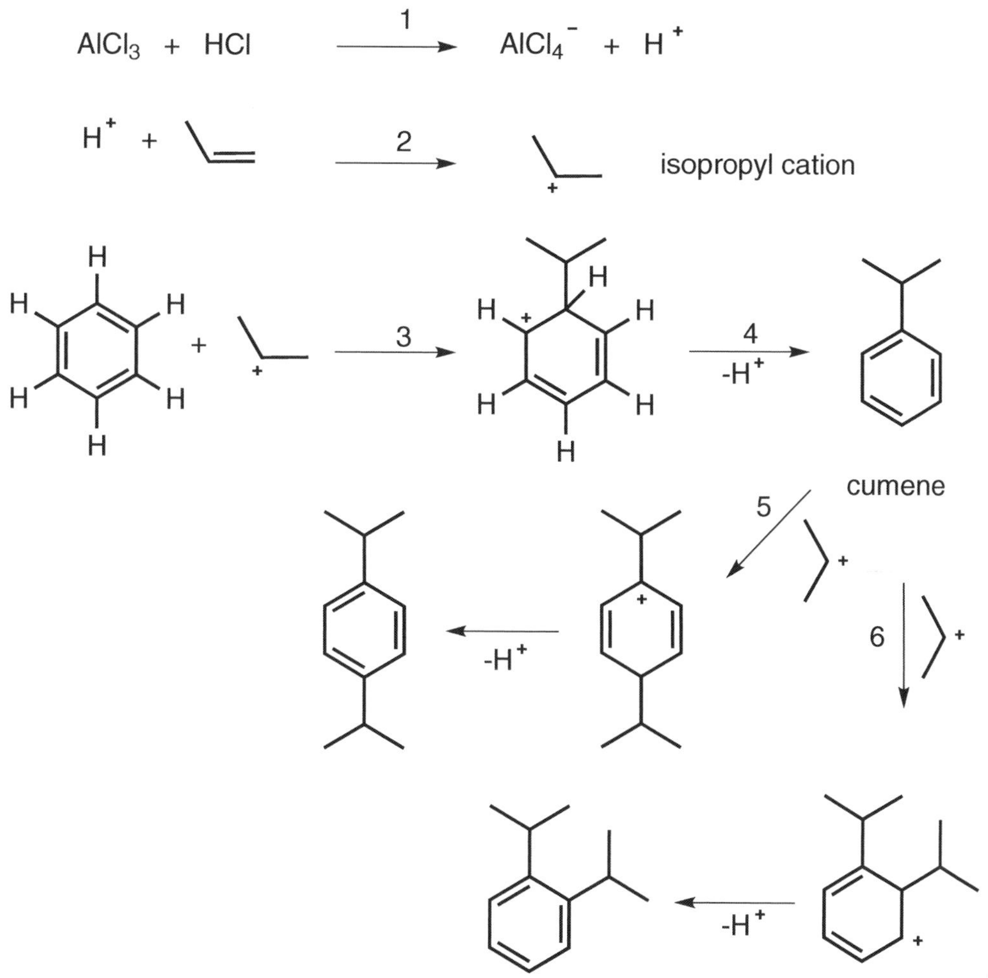

◄ FIGURE 6.14

Mechanism for the Formation of Isopropyl Benzene and Diisopropyl Benzene from Propylene and Benzene Catalyzed by Aluminum Trichloride and HCl

6 π-electron aromaticity. However, loss of the aromatic stabilization means reaction 3 (Figure 6.14) is a very difficult step, in fact the slowest step in the outlined reaction path.

We'll learn more shortly about the reasons that certain chemical reactions are faster or slower than others. But the reason reaction 3 in Figure 6.14 is slow is reasonable considering that breaking up the aromatic character of the ring must mean paying at least part of the 36 kcal/mole lowering of energy we learned about in section 6.7. As seems reasonable, the more energy a process takes, the more likely it will be slowed down. Shortly we'll understand the scientific basis for this intuition.

The carbocation produced in reaction 3 from addition of the isopropyl cation to the benzene ring, is an intermediate that can easily regain the 6 π-electron aromatic character. There are two choices, eject the just arrived isopropyl cation, that is, simply reverse step 3, or eject a proton from the carbon atom in the ring that the isopropyl group has become bonded to. The latter is shown in step 4 where the two electrons in the σ ortho bond that held the hydrogen atom to the ring are left behind to make up the necessary 6 π electrons for a return to aromatic character. In this manner, the isopropyl group substitutes for the hydrogen atom: the word substitution in electrophilic aromatic substitution. Cumene is produced.

However, all is not well. The cumene produced via this proton loss in step 4 turns out to be even more reactive with the available isopropyl cations in the reaction mixture than benzene, therefore leading via steps 5 and 6 to the production of undesired multialkylation byproducts (Figures 6.12 and 6.14).

Why is the resulting cumene more reactive to electrophilic aromatic substitution than benzene? And why do the additional isopropyl groups, the multialkylation products, form predominantly at the ortho and para position on the ring as shown in Figure 6.14?

The answer to this question involves an irony. It would have been expected that the original application of valence bond theory and the resonance approach, which had been discredited by not being able to explain the special stability of benzene (section 6.8) would be helpless to answer the question posed above. Although Hückel molecular orbital theory can yield the answer, the truncated valence bond theory used by Pauling with the resonance concept does resurrect itself, rising nicely to the occasion to yield the necessary insight.

To apply these resonance ideas it would be helpful to use a device that chemists use to describe the mechanisms (section 3.14) of chemical reactions, what are called reaction coordinate diagrams. Such diagrams begin and end with thermodynamic parameters, that is, the energies of the reactants and products of the reaction step focused on, but then attempt to trace the energy changes on moving between these starting and ending states. Because energy is connected to structure, a reaction coordinate diagram, in principle, has the potential to describe the motions of all the atoms and electrons in the reacting molecules as the chemical process proceeds from start to finish.

PROBLEM 6.34
All the reactions in Figure 6.13 may be considered acid-base reactions as well as nucleophile-electrophile reactions. Identify the reactants according to these designations. Are bases always nucleophilic and acids always electrophilic in all reactions in Figure 6.13? Answer this question also for the reactions in Figure 6.14.

PROBLEM 6.35
How does the mechanism shown for propylene in Figure 6.13, when applied to the reactions of cis and trans-2-butene with HCl (Figure 6.13) lead to the formation of identical products for the butene isomers? Does your answer to this question also yield the reason that both enantiomers of the chiral product, 2-chlorobutane are produced in the reactions of these diastereomeric 2-butenes?

PROBLEM 6.36
Redraw all the structures in Figure 6.14 showing all atoms. Use these structures, which show all the atoms, to present all the resonance structures possible for each positively charged intermediate. How is the particular proton determined, which is lost in the final step leading to the disubstituted product?

PROBLEM 6.37
If the reactions in Figure 6.14 were to follow the path of the reactions in Figure 6.13, what would be the products? Where do the two reaction paths, addition versus substitution, diverge?

PROBLEM 6.38
An alkene and a benzene ring both can act as nucleophiles in reactions with an electrophile, and both involve two steps, for example, 3 and 4 in Figure 6.14 and 1 and 2 in Figure 6.13. How do you believe steps 1 and 3 and then steps 2 and 4 in propylene and benzene respectively differ in their likely difficulty and how do you come to your conclusion? By difficulty is meant the amount of energy necessary to accomplish the reaction step which, as we'll see in section 6.11, is the energy of activation.

PROBLEM 6.39
In some older industrial procedures even unwanted trisubstituted alkylation products were produced. Predict the substitution pattern on the benzene ring of the trisubstituted isopropyl benzene from the para and from the ortho products in Figure 6.14. How did you make this prediction?

CHEMICAL REACTIONS BETWEEN molecules, that is, intermolecular reactions, require collisions between the reacting molecules. The number of molecular collisions increases as concentration increases because of the higher molecular density. Therefore the higher the concentration of the reacting molecules, the faster the chemical reactions will occur. The proportionality constant that connects the velocity of a chemical reaction to the concentrations of the reactants involved is called the rate constant, k. There is an entire division of chemistry called "chemical kinetics," which studies the interrelation between rate and concentration of reacting molecules. Such studies, which can become quite complicated are, nevertheless, exceptionally valuable in yielding information on the mechanism of chemical reactions. However complex chemical kinetic studies may be the fundamental necessary experimental measurements are always the same: how the concentrations of certain reactants and/or products vary with time. In this manner, the rate constant, k, of a chemical reaction can be determined.

6.11
Energy of Activation, Reaction Rate Constants, and Reaction Coordinate Diagrams

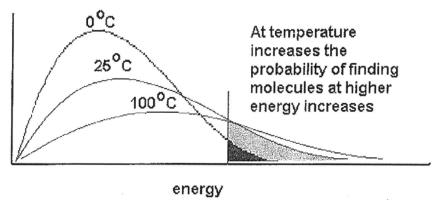

energy

◄ **FIGURE 6.15**

Representation of the Distribution of Energy of a Molecular Ensemble as a Function of Three Temperatures

Experimental work in the nineteenth century led to the understanding that knowing the overall starting materials and products of a chemical reaction, those chemicals that appear in the expression for the equilibrium constant, do not necessarily yield information that can be used to understand the speed with which the reaction takes place. In addition, in the waning years of the nineteenth century, experimentalists began to look closely at the effect of temperature on the velocity of chemical reactions and how this differed from the effect of temperature on equilibria. It was determined experimentally that the rate constant, k, of a chemical reaction was extraordinarily sensitive to temperature and entirely different from the effect of temperature on equilibrium. Why should this be? These disconnections between equilibrium descriptions of chemical reactions and the rate was the beginning of the development of the concept of mechanism (sections 3.14 and 3.15).

Many years before this increasing focus on what we call today chemical kinetics, **Leopold Pfaundler**, an Austrian chemist had an insight in understanding the effect of temperature on reaction velocity. Building on the understanding, by the 1850s, that the kinetic energies of molecules in a gas diverge from the average value and that only a small proportion have much greater energies than some critical value (as shown in **Figure 6.15** for what is called the Maxwell-Boltzmann distribution), Pfaundler proposed

▪ **Leopold Pfaundler**

that only the highest energy molecules in any ensemble would have the critical energy to undergo chemical change. This brings on stage a new player.

Svante Arrhenius was a child prodigy, who taught himself to read by the age of three and not long after became proficient enough in mathematics so that he started school at an advanced grade. His abilities were so extraordinary that the physics faculty at the University of Uppsala was not capable of supervising his progress and he had to move to the Academy of Sciences in Stockholm. His education led eventually to submission of a doctoral thesis, translated from the French as "*Investigations on the galvanic conductivity of electrolytes*," in which Arrhenius proposed that certain molecules in solution are dissociated into ions. His work was poorly understood at Uppsala leading to a poor grade for his thesis, in fact an almost failing grade. If not for the intervention of prominent scientists from within and outside Sweden, Arrhenius' life in science might have been blocked. His ideas were too original. But that originality was of great value as certainly demonstrated later by his receiving the Nobel Prize in Chemistry for work based on his poorly received thesis.

■ Svante Arrhenius

Arrhenius had many ideas. For one example, of interest to our twenty first century life, he published a paper in 1896 entitled: "*On the Influence of Carbonic Acid in the Air upon the Temperature of the Ground*." At that time the term carbonic acid was used for carbon dioxide and Arrhenius' paper was reporting on what we call today "global warming," which Arrhenius connected to CO_2 in the atmosphere. He pointed out the potential role of burning fossil fuels as the source of excess CO_2 - quite an advanced thinker, Professor Arrhenius.

Another interesting view that Arrhenius held was the idea that life on earth may have originated from spores blown across space by solar winds, the theory of panspermia or exogenesis.

For our purpose now, we focus on another of Arrhenius' ideas, on the effect of temperature on reaction rates. He took up Pfaundler's idea, noted above, that the Maxwell-Boltzmann distribution of kinetic energies for gases (Figure 6.15), if translated to molecules undergoing chemical reaction, would limit the number of molecules capable of undergoing reaction to those molecule with energies above a critical value. Arrhenius realized that Pfaundler's idea explained the strong temperature dependence of the rate of chemical reactions on temperature.

Boltzmann's work had predicted an exponential relationship between the proportion of molecules of a certain energy E as $e^{E/RT}$, with R the gas constant and T the absolute temperature. The number of molecules with a critical energy to be capable of reaction must therefore also exponentially depend on temperature.

Following this idea, Arrhenius created an empirical equation used to this day to connect energy of activation, E_a (the critical energy Pfaundler referred to) and absolute temperature, T, to rate constant, k. The result is known today as the "Arrhenius Equation."

$$k = Ae^{-E_a/RT} \quad \text{(Arrhenius Equation)}$$

We are now prepared to see how a reaction coordinate diagram (**Figure 6.16**) is valuable in understanding the mechanism of a chemical reaction, in particular here applied to steps 3 and 4 in the mechanism of the electrophilic aromatic substitution (Figure 6.14).

The reaction coordinate diagram in Figure 6.16 does not represent the precise path, nor should the absolute energy values along the ordinate be taken literally for the starting and ending states, but the diagram is a representation of the shape of the reaction path in the formation of cumene from isopropyl cation and benzene, outlined in Figure 6.14.

In general, a reaction coordinate diagram can be drawn for any step in the overall mechanism of a chemical reaction. Reaction coordinate diagrams may then be drawn for molecules and/or intermediates that don't appear in the overall equation for the reaction, that is, the equation from which thermodynamic calculations and therefore overall equilibrium constants are determined. A reaction coordinate diagram describes what is actually happening, what molecules and intermediates are coming into contact and/or changing structure along the path from starting materials to products.

In contrast, a thermodynamic analysis of the reaction in Figure 6.14 would include only benzene and propylene as the starting materials, and cumene and the ortho and para diisopropyl benzenes as the products in a balanced equation. The aluminum chloride and HCl are not consumed and therefore don't count in the overall accounting, nor do the intermediate positively charged species along the reaction path.

Let's see how to interpret the reaction coordinate diagram. Energy increases along the ordinate of the reaction coordinate diagram (Figure 6.16), so that the path that the continuously changing molecules (starting from the left of the graph) are following becomes more difficult. Few molecules attain the necessary energy to climb the path as temperature decreases.

From figure 6.15 we can understand the role of increasing temperature because a rising proportion of molecules attain the energy to reach any point on the path. The Arrhenius equation, shown above, presents this fact in quantitative terms. As T increases

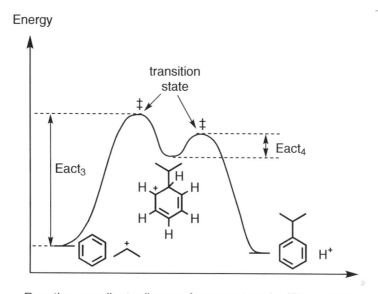

Reaction coordinate diagram for steps 3 and 4 (Figure 6.14)

◀ **FIGURE 6.16**

Reaction Coordinate Diagram for the Substitution Step In the Mechanism of Electrophilic Aromatic Substitution Leading to Cumene

the negative exponent decreases therefore causing k to increase. This happens because the A value in the Arrhenius equation is only weakly temperature dependent and is a large value. The exponential term subtracts from A, so that as the exponential term decreases, the value of k increases in an exponential manner.

Here a word about the A value helps to understand the nature of the reaction we are describing. The A value tends to the value of the rate constant k as the energy of activation, E_{act}, descreases and/or as the temperature increases, that is, as the exponential fraction tends toward zero and therefore $e^{-E/RT}$ tends toward unity. The rate constant, k, ideally, would equal A when energy no longer limits the speed of the reaction. The physical meaning of the A term is associated with the geometric restrictions on the reaction. Let's see what that means for the electrophilic aromatic substitution we are studying here.

No matter how much energy may be available for reaction (all molecules have adequate energy for reaction) the rate constant will be limited by the statistical probability of the isopropyl cation empty p-orbital approaching the face of the benzene ring rather than the edge of the ring. Only then can the electrons of the p-cloud interact with the electrophilic orbital of the cation. In general, in any chemical reaction, the more the geometry of the successful reaction path is restricted among many possibilities that can not lead to reaction, the smaller will be the value of A.

The most difficult path in Figure 6.16 describes the interaction between the isopropyl cation and the benzene molecule. This step is far more difficult, has a far higher energetic cost, which we call the energy of activation, than the following step in which the proton is expelled. In the former step, the aromatic character of the benzene ring is lost, while in the latter step the 6 π-electrons are returned to the cycle - aromaticity is regained.

Just as moving a crowd of people along a path with a barrier (in New York City one could think of a subway turnstile as the barrier) limits the persons per unit time to the speed with which the barrier is traversed, that is, no faster than the turnstile turns, so, similarly, the molecules passing along the reaction paths per unit time in any reaction, as for example in Figure 6.16, are limited by the slowest step along that path.

And the slowest step is that with the highest energy of activation, which is measured as the energy change from the lowest to the highest energy point along the path, the latter being the transition state, marked with the traditional symbol ‡. In the reaction described by the reaction coordinate diagram in Figure 6.16 there are two transition states and therefore two energies of activation. But only the activation energy for step 3 is rate limiting.

The transition state in the theory that stands behind the kind of diagram shown in Figure 6.16 is considered a molecular state that exists for even less time than the fleeting existence of an intermediate. In the ideal, the transition state exists for the lifetime of a bond vibration, in the range of nanoseconds. In (section 10.6) we'll note the work of Ahmed H. Zewail whose work addresses the possibility of direct observation of such rapid changes.

In the reaction coordinate diagram of an electrophilic aromatic substitution, the first transition state, that for step 3 (Figure 6.16), is followed by an intermediate sometimes named after George W. Wheland, an influential chemist who worked closely with Pauling and was influential in defending the theory of resonance as used by Pauling. Wheland came to Pauling's laboratory at Cal Tech after a doctoral degree with James Conant at Harvard. We quoted Conant's book (section 6.8) regarding the arguments offered to defend resonance theory against the absence of aromatic character in cyclooctatetraene. Pauling had great respect for Wheland, for whom no photograph could be found. Here is one 1970 quote from Pauling about Wheland: ""*The theory of quantum mechanical resonance of molecules among several valence-bond structures constituted a major addition to the classical structure of organic chemistry. This theory was developed in the period from 1931 on by a number of investigators including Slater, E. Huckel, G. W. Wheland and me.*" *Linus Pauling.*

The Wheland intermediate requires only the breaking of a bond between carbon and hydrogen to reform the 6π aromatic character of the ring, a step requiring a very small energy of activation, as shown in Figure 6.16, E_{act} (step 4)..

Study of the velocity of a chemical reaction as a function of the concentrations of reactants, as pointed out above, yields the reaction rate constant k at any temperature. In an ideal situation, the concentrations of reactants which affect the velocity of the reaction are those reactants that take part in the rate determining step.

Subsequent study of the temperature dependence of k and use of the Arrhenius equation yields, then, a method to obtain the energy of activation, E_a, for the rate determining step. According to the Arrhenius equation, plotting ln k versus 1/T should yield a straight line with slope ($-E_a$/RT) and intercept on the y-axis of ln A yielding a method of obtaining E_{act} of the rate determining step. In the reaction coordinate diagram in Figure 6.16, such experimental measurements would yield E_{act} (step **3**). The reaction coordinate diagram (Figure 6.16) then represents the physical meaning of E_a with representative drawings on this diagram of the structures, movements and changes of the reacting molecules along this path.

Using the device of a reaction coordinate diagram and understanding the role of energy of activation, we are now in a position to understand, in the next section, with the help of resonance structures, why multialkylation takes place in the Friedel-Crafts reaction for the industrial synthesis of cumene.

PROBLEM 6.40

The Arrhenius equation expresses the well known relationships between rate, temperature, and energy of activation. Carry out the following calculations to appreciate the large exponential effects on rate. Considering that the pre-exponential factors A and the energies of activation, E_a, are the same for two compared reactions, calculate the change in rate constant for an increase of 10° C. Now keep the pre-exponential factors and the temperature the same for both reactions and calculate the change in rate constant for a change in energy of activation from 20 to 25 kcal/mole.

PROBLEM 6.41

Bonds between carbon and deuterium are significantly stronger than bonds between carbon and hydrogen. Yet the rate of the aluminum chloride-catalyzed reaction between isopropyl chloride and benzene (Figure 6.14) is almost identical for reactions with benzene, C_6H_6, and deuterated benzene, C_6D_6. How is this experimental information consistent with the reaction coordinate diagram in Figure 6.16?

PROBLEM 6.42

Aluminum chloride-catalyzed reaction of isopropyl chloride with benzene yields in addition to multialkylation products, only one monoalkylation product, cumene. However, aluminum chloride-catalyzed reaction of n-butylchloride, $CH_3CH_2CH_2CH_2Cl$, yields two monoalkylation products, n-butylbenzene, $C_6H_5CH_2CH_2CH_2CH_3$, and larger amounts of 2-butylbenzene, $C_6H_5C(H)(CH_3)(CH_2CH_3)$. How does the mechanism of electrophilic aromatic substitution account for this rearrangement product?

PROBLEM 6.43

There is a variation of the Friedel-Crafts reaction that allows synthesis of n-butyl benzene and avoids the rearrangement product of Problem 6.42. Aluminum chloride catalyzes the reaction between butyryl chloride, $CH_3CH_2CH_2CO(Cl)$, and benzene. Butyryl chloride is derived from butyric acid in which the OH group of the carboxylic acid is replaced with a Cl atom (section 11.9). The first product is substitution of a hydrogen atom on the benzene ring to produce the ketone, $(C_6H_5)C=O(C_3H_7)$. Other catalysts then allow reaction of this ketone with the equivalent of two moles of H_2 to produce H_2O and convert the carbonyl carbon to a CH_2 group, therefore producing n-butyl benzene. The key intermediate carbocation responsible for this reaction has been proposed to be what is named an acylium ion, $C_3H_7C\equiv O^+ \leftrightarrow C_3H_7C^+=O$. Considering the structure of the acylium ion, offer an explanation as to why no rearrangement occurs. Draw structures of all the molecules noted above and propose a mechanism with all intermediates responsible for the reactions, showing all bonds and all electrons.

PROBLEM 6.44

Considering that an empty p orbital is associated with a carbocation of an sp^2 hybridized carbon, offer an explanation, based on hybridization of orbitals, for the fact that acylium ions, as produced in the aluminum chloride catalyzed reaction of carboxylic acid chlorides with benzene, do not undergo rearrangements while carbocations do undergo rearrangements as in problem 6.42.

FIGURE 6.17 EXHIBITS AN approximate representation, for the area in the reaction coordinate diagram around the Wheland intermediates, of the reaction of isopropyl cation with cumene compared to reaction with benzene, that is, steps 5 and 6 versus steps 3, in Figure 6.14.

6.12

Resonance Resurrected

As shown in Figure 6.17, the energy of activation for formation of the Wheland intermediate for step 3, the step that produces cumene (Figure 6.14), is significantly higher than the energy of activation for the formation of the Wheland intermediates for steps 5 or 6 of Figure 4.14, the steps that produce ortho diisopropyl benzene or para diisopropyl benzene from cumene and isopropyl cation.

The structures shown for the Wheland intermediates in Figures 6.14, 6.16 and

6.17 are not adequate to describe the electron distribution in these positively charged intermediates. Here is a situation where resonance is necessary for a complete representation (section 5.4).

The possible resonance structures for these Wheland intermediates for steps 3, 5 and 6 (Figure 6.14) are shown in Figure 6.18. In each of the three positively charged intermediates, three resonance structures are necessary to describe the overall electron distribution. A fourth could not be reasonably (section 5.4) drawn.

The three resonance structures (**Figure 6.18**) for each of the three intermediates (from steps 3, 5 and 6) demonstrate that the positive charge is distributed around the ring, a stabilizing parameter.

FIGURE 6.17 ▶

Comparison of Energies of Activation for Formation of Cumene Versus Diisopropyl Benzene Represented by Reaction Coordinate Diagrams

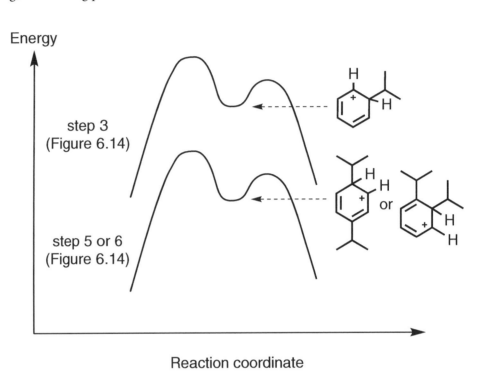

Distribution of charge in a molecule, a situation for which resonance structures are necessary, is generally associated with stability and we have seen this before in the lower pK_a of acids that form resonance stabilized conjugate bases such as $H_3C-CO_2^-$, HSO_4^- and $H_2PO_4^-$ and in the nature of leaving groups such as the biologically important $P_2O_7^{-4}$ (section 5.3 and Chapters 7 and 8).

In section 4.7 we explored the increased ease of formation of carbocations, which are more deeply imbedded in large molecules, that is, tertiary over secondary over primary and then CH_3^+, the most unstable. Whatever may be the theoretical basis of enhanced stability of more substituted carbocations, it is a fact that has a role to play in the multialkylation problem faced by the chemical industry.

In the resonance structures shown for step 3 (Figure 6.18) the positive charge is constrained to secondary carbon atoms, carbon atoms bound to two other carbon atoms and a hydrogen atom. However, in the resonance structures for steps 5 and 6 the positive charge finds its way to a tertiary carbon site, a carbon bound to three other carbon atoms. Therein one finds the answer to the multialkylation problem faced by the chemical industry.

The lower transition state energy and therefore the lower energy of activation seen in Figure 6.17 for steps 5 and 6 compared to step 3 arises from the stability of tertiary over secondary carbocations, which is quickly revealed, as noted above, by considering resonance stabilization. Application of resonance structures, although having failed, at first, to understand aromaticity (section 6.8), succeeds in explaining the enhanced reactivity of cumene over benzene to electrophilic aromatic substitution.

The difference in stability of more substituted carbocations, which is responsible for the path from isopentenyl diphosphate to the terpenes and steroids, and which serves industry in catalytic cracking producing high octane number branched hydrocarbons (section 4.7), and is the basis of Markovnikov's rule, works against industry in producing unwanted multialkylation byproducts in the production of cumene.

benzene

from Figure 6.14
- Intermediate -

step 3

step 5

step 6

meta substitution

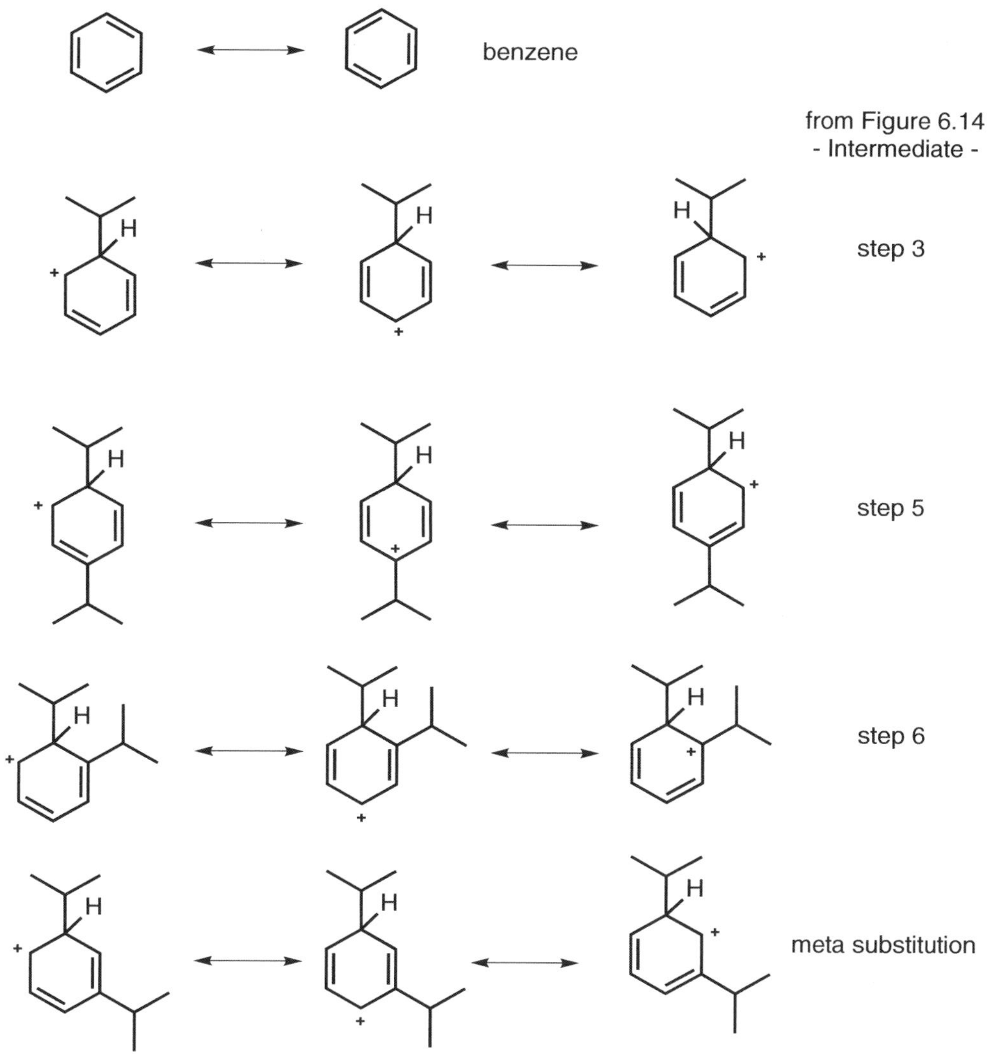

◄ FIGURE 6.18

Resonance Structures for the Intermediates Formed in the Electrophilic Aromatic Substitution of Isopropyl Cation with Benzene

Moreover, we now understand why the additional isopropyl group added to cumene does not appear at the meta position. As seen in the resonance structures in Figure 6.18, meta substitution of cumene, as for benzene, constrains the positive charge to secondary carbon sites. The energy of activation, therefore, for meta substitution of cumene, as for substitution of benzene, will be greater than for ortho or para substitution of cumene, causing the rate constant, k, for meta substitution, as for benzene substitution, to be lower.

It is hopeless for industrial production of cumene to try to overcome the basic characteristics of carbocation stability or resonance in repressing multialkylation in the production of cumene. Fighting with a fundamental principle of the science is a lost cause. But a way has been found based on conducting electrophilic aromatic substitution of benzene with propylene within the molecular size cavities of acidic zeolite pores, the same kinds of zeolites used in catalytic cracking of petroleum fractions. The interesting story of how this works, which we will not go into here, can be found in the book by Green and Wittcoff, *Organic Chemistry Principles and Industrial Practice*, pages 49-56 in Chapter 3.

Mark M. Green, Harold A. Wittcoff

Organic Chemistry Principles and Industrial Practice

WILEY-VCH

PROBLEM 6.45

How do the products produced in multialkylation of benzene correlate with carbocation stability?

PROBLEM 6.46

In the three steps labeled 5, 6, and "meta substitution" in Figure 6.18, substitute an NO_2 group for the isopropyl group already present on the ring. How would this change alter the relative stability of these three carbocation intermediates in this figure and why?

PROBLEM 6.47

Before the value of zeolites was discovered in reducing multialkylation in the electrophilic aromatic substitution of both cumene and ethylbenzene, industrial chemists and engineers conducted the reaction with a large excess of benzene. Why was this approach taken?

PROBLEM 6.48

By chance, industrial chemists and engineers discovered that reacting propylene with benzene in a zeolite cavity (section 4.2) led to a large reduction in the formation of multialkylation, that is, the formation of diisopropyl benzene (Figure 6.14). Considering the following facts, offer an explanation for the value of zeolites in reducing multialkylation: (1) although many diisopropyl benzene molecules are formed in the zeolite, the cavities in the zeolite are too small for the diisopropyl benzene to escape. However isopropyl benzene (cumene) is small enough to be able to leave the cavity; (2) the concentration of benzene in the zeolite, which is a small enough molecule to easily enter and leave the zeolite cavities, is very high; (3) the concentration of protons in the zeolite cavities is high; (3) the reaction paths 5 and 6 in Figure 6.14 are reversible.

6.13

Application of the Ideas of Resonance Stabilization of Wheland Intermediates in Electrophilic Aromatic Substitution

THE IDENTICAL PRINCIPLES used in understanding the results in Figures 6.12 and 6.14, resonance and carbocation stability, can be widely applied in understanding and making predictions about all electrophilic aromatic substitution reactions of benzene. Let's try this out on the results presented in **Figure 6.19**.

In Figure 6.19, variable monosubstituted benzenes are subjected to electrophilic aromatic substitution with the nitronium ion, a powerful electrophile, NO_2^+, which is produced by reacting HNO_3, nitric acid, with sulfuric acid, H_2SO_4. The sulfuric acid is the stronger acid (pK_a of -3 compared to -1.5 for nitric acid) and adds a proton to the nitric acid producing $H_2NO_3^+$, leading to loss of H_2O and therefore leaving behind NO_2^+.

Every reaction shown in Figure 6.19 follows the same general mechanistic path as for formation of cumene and its multialkylation (Figure 6.14) and the general reaction coordinate diagrams exhibited in Figures 6.16 and 6.17 also apply. Addition of the electrophile, NO_2^+ in each of the reactions in Figure 6.19, produces the positively charged Wheland intermediate in the slowest step, the step with the highest energy of activation. This intermediate then rapidly loses a proton to regain the aromatic 6 π electrons, producing the substitution product.

In each example in Figure 6.19 the Wheland intermediate is subject to resonance stabilization in a manner that is identical to that shown in Figure 6.18: Substitution at the ortho and para positions places positive charge at the site of the original substituent; Substitution at the meta position constrains the positive charge only to carbon atoms in the ring that bear a hydrogen atom.

The product results for reaction 1 (Figure 6.19) are consistent with the results for the multialkylation of cumene. The new substituting group resides predominantly at the ortho and para positions.

However, in reactions 2, 4 and 5 the incoming NO_2^+ group substitutes for a proton

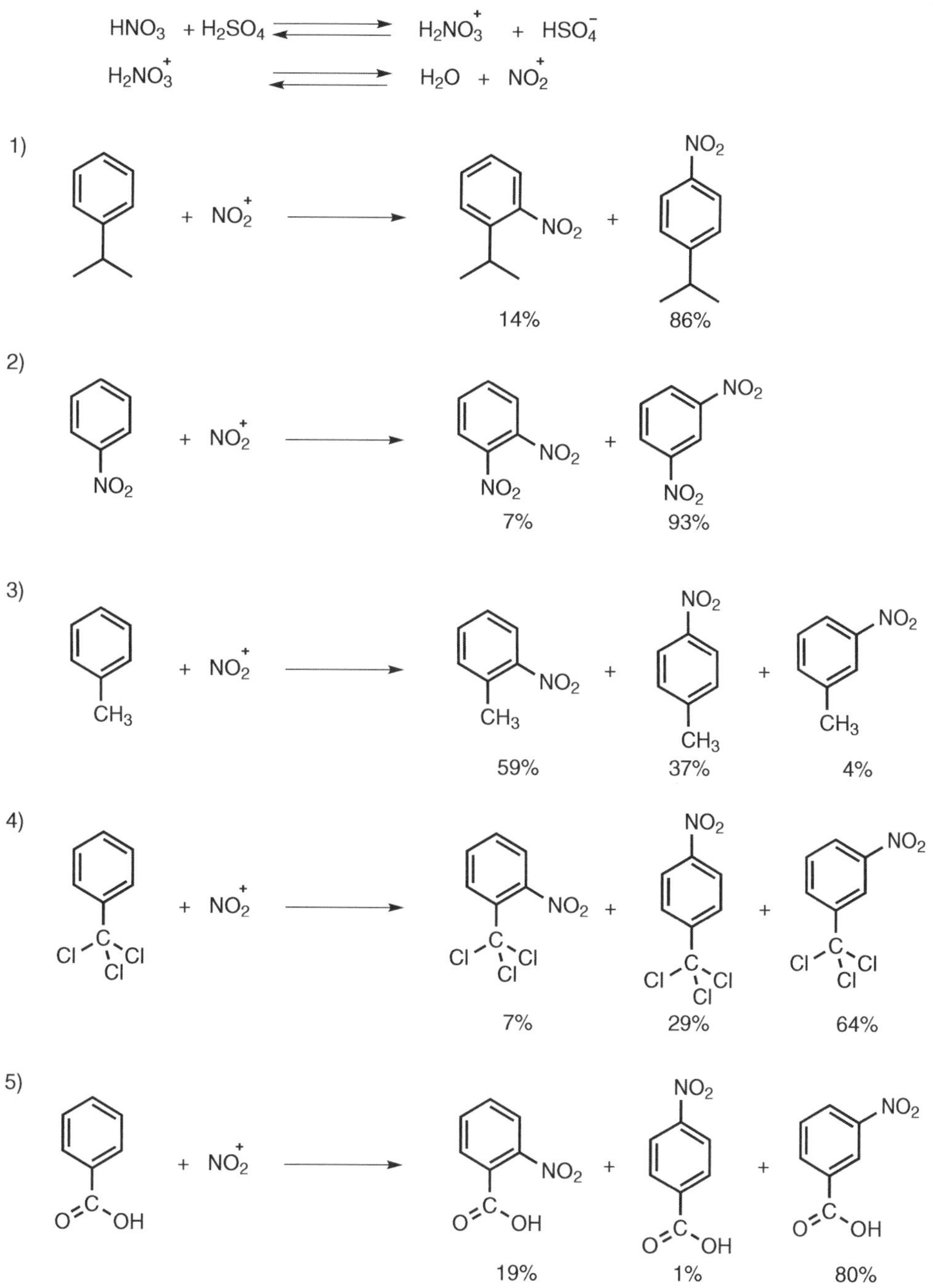

$$HNO_3 + H_2SO_4 \rightleftharpoons H_2NO_3^+ + HSO_4^-$$

$$H_2NO_3^+ \rightleftharpoons H_2O + NO_2^+$$

1) [benzene with isopropyl group] $+ NO_2^+ \longrightarrow$ [ortho-nitro isopropylbenzene] **14%** $+$ [para-nitro isopropylbenzene] **86%**

2) [nitrobenzene] $+ NO_2^+ \longrightarrow$ [ortho product] **7%** $+$ [meta product] **93%**

3) [toluene] $+ NO_2^+ \longrightarrow$ [ortho product] **59%** $+$ [para product] **37%** $+$ [meta product] **4%**

4) [benzene with CCl$_3$ group] $+ NO_2^+ \longrightarrow$ [ortho product] **7%** $+$ [para product] **29%** $+$ [meta product] **64%**

5) [benzoic acid] $+ NO_2^+ \longrightarrow$ [ortho product] **19%** $+$ [para product] **1%** $+$ [meta product] **80%**

◀ FIGURE 6.19

Electrophilic Aromatic Substitution of NO_2^+ on Various Substituted Benzenes Leading to Ortho/Para Versus Meta Substitution

not at the ortho and para positions but, rather, at the meta site on the ring. Why? The answer to this question rests in the nature of the initial substitution products on the rings in these three reactions; In reaction 2 this group is an NO_2 group, in reaction 4 a CCl_3 group and in reaction 5 a CO_2H group. These functional groups have something in common. The atoms attached to the benzene ring carbon are all electron deficient as shown in **Figure 6.20**.

It's not possible to satisfy the octet rule at any of the three atoms in a nitro group, $R-NO_2$, without a positive charge on nitrogen as seen in Figure 6.20. Turning now to the CCl_3 group, the electronegativity of chlorine compared to carbon (section 3.13) forces an electron deficiency at the carbon of the CCl_3 group. This electron deficiency is then transferred to the adjacent carbon atom on the benzene ring. Regarding the carboxylic acid group, the carbonyl carbon atom of this functional group is flanked by two electronegative oxygen atoms and as shown in Figure 6.20 gives rise to an electron deficiency at this carbon atom, which is bonded to the a benzene ring carbon atom.

The consequence of an electron deficient group bonded to a carbon atom on the benzene ring means that substitution of the newly incoming electrophile (NO_2^+ in Figure 6.19) at the ortho or para position in placing a positive charge at that carbon atom will be an <u>unfavorable</u> situation.

On the contrary, the consequence of an electron rich group bonded to a carbon atom on the benzene ring, which can stabilize an adjacent carbocation, means that substitution of the newly incoming electrophile at the ortho or para position in placing a positive charge at that carbon atom will be a <u>favorable</u> situation.

Therefore, reactions 1 and 3 favor ortho-para substitution while reactions 2, 4 and 5 disfavor ortho and para substitution and in disfavoring these ortho, para sites, meta substitution, although occurring slowly, nevertheless dominates the product distribution. In reaction 2 we see 93% of the substitution product is meta, in reaction 4, 64% is meta substitution, and in reaction 5, 80% is meta substitution. We'll leave for later more advanced inquiries in organic chemistry to delve into the details of the differing proportions of meta substitution for these reactions.

The difference in the ortho para ratio in 1 and 3 has a reasonable answer you will be given the opportunity to come up with in addressing problem 6.50 with the help of problem 6.51.

PROBLEM 6.49

For each of the five reactions in Figure 6.19, draw resonance structures, to offer an explanation for the relative favoring of ortho-para or meta substitution.

PROBLEM 6.50

Although ortho-para substitution is favored over meta substitution in both reactions 1 and 3 in Figure 6.19, there is far more ortho substitution in reaction 3 compared to reaction 1. Offer an explanation for this experimental fact.

PROBLEM 6.51

In the conformational equilibrium between axial and equatorial groups on a cyclohexane ring, although both methyl cyclohexane and isopropyl cyclohexane strongly favor occupying equatorial over axial positions, the isopropyl substituted cyclohexane favors the equatorial position more strongly. How does this experimental fact help to answer problem 6.50?

PROBLEM 6.52

Offer a reason for the fact that all substituted benzenes (NO_2, CCl_3, CO_2H) (Figure 6.19), which direct additional substitution to the meta position, react far more slowly than those substituted benzenes ($CH(CH_3)_2$, CH_3) which direct substitution to the ortho and para positions?

PROBLEM 6.53

TNT stands for 2,4,6-trinitrotoluene. Draw the structure of TNT. Could TNT be synthesized starting from nitrobenzene by electrophilic aromatic substitution using H_3C-Cl and $AlCl_3$ followed by additional reaction with nitric acid and sulfuric acid? Could TNT be synthesized starting from methyl benzene, that is, toluene followed by reacting this molecule with nitric acid and sulfuric acid? Discuss these two possible routes to this high explosive.

PROBLEM 6.54

While, as seen in Figure 6.19-reaction 3, toluene yields only 37% para substitution of the incoming nitro group, fluorobenzene and anisole, (F-C_6H_5 and CH_3O-C_6H_5), both yield more than 80% para nitro substitution. Both F and OCH_3 stabilize an adjacent carbocation as seen in the last two examples in Figure 10.18 (Chapter 10) and therefore direct the incoming NO_2^+ group to the ortho and para positions. However, while this resonance effect favors both ortho and para substitution, there is another effect by which these groups withdraw electrons, an inductive effect, that moves through the sigma rather than the pi bonds. How might this inductive effect offer an explanation for the far larger para substitution compared to toluene? How does the inductive effect depend on electronegativity?

CHAPTER SIX SUMMARY of the Essential Material

THIS CHAPTER INTRODUCES BENZENE and the concept of aromaticity. A molecule can be aromatic, a term we learn is associated with the history of benzene and the experimental observation of a low ratio of H to C. Ratio of carbon to hydrogen atoms brings up the subject of formula and how one can determine the degree of unsaturation or number of rings in a molecules by the ratio of C to H. We then get involved in the struggle to understand benzene's structure and the insight that led to the realization of its cyclic structure. We followed the necessity to invoke ideas of resonance to explain how all the three double and single bonds in the proposed structure were identical and why benzene can be shown experimentally to be a perfect hexagon. We are introduced to the nomenclature of substituted benzene and the numbers of isomers possible as different groups replace one or more of the six hydrogen atoms originally in the structure.

We discover how the special properties of benzene are revealed by experiments in which H_2 is added to convert benzene to cyclohexane and the reason that too little heat is released in this reaction compared to the heat released when alkenes are similarly reduced. The subject then turns to the failure of resonance ideas to explain the properties of benzene and the role that cyclooctatetraene plays in this understanding. The failure of resonance ideas to understand why cyclooctatetraene is tub shaped instead of flat brings up the contributions of Hückel and how molecular orbital theory solves the problem and also allows expansion of the class of molecules that

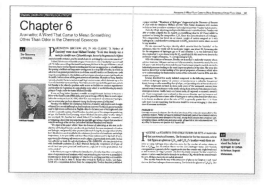

CHAPTER SIX SUMMARY continued

belong to the aromatic category. We show how molecules can be transformed to an aromatic state as a consequence of their reactive properties. We then study many biologically important molecules that are aromatic and discover how hybridization and the geometry of lone pairs of electrons makes the difference to the critical number of electrons necessary for aromatic character.

The second subject in the chapter then rests on understanding of the structure of benzene and the nature of aromaticity in presenting a problem to the chemical industry in synthesizing cumene, an important intermediate in the production of epoxy resin and polycarbonate. The question needing to be answered involves unwanted byproducts in cumene's production, products that place more isopropyl groups on the benzene ring than desired. Now we enter the world of electrophilic aromatic substitution and come to understand the difference between alkenes, which undergo addition reactions that consume their double bonds, and benzene undergoing substitution reactions that maintain the unsaturation and therefore the aromatic character of the ring. We study the mechanism of reaction of electrophiles to double bonds and then the reaction of electrophiles to benzene.

Mechanisms of chemical reactions can be expressed in terms of reaction coordinate diagrams, which in combination with ideas of the nineteenth century predict how and why reaction rates change with temperature. We understand how the Arrhenius equation works and the role of energy of activation and what rate determining steps and rate constants are. We discover what role a key intermediate, the Wheland intermediate, plays in electrophilic aromatic substitution and how, when applied to this intermediate, resonance ideas are resurrected. We understand why benzene undergoes unwanted multiple additions in reaction with isopropyl chloride and aluminum trichloride. From this understanding we see that the same factors of carbocation stability controlling the production of high octane gasoline discussed in Chapter 4, or the basis of Markovnikov addition to alkenes, are at work in electrophilic aromatic substitution. We understand why certain isomers of benzene are formed and others not, concerning questions of ortho, meta and para substitution. The answers are found in the nature of resonance structures. And, finally, we see how the ideas that answer the questions posed by industry in their quest for cumene allow predictions about the nature of many kinds of electrophilic aromatic substitution.

Chapter 7
Fatty Acid Catabolism and the Chemistry of the Carbonyl Group

7.1
The fatty acids in living organisms are saturated and unsaturated.

THE "FATS OF LIFE" BY CAROLINE POND is full of interesting information about its subject matter including a story I remember reading years ago in a popular textbook of organic chemistry written by two professors at New York University, which is that the hoofs of reindeer contain a higher proportion of unsaturated fatty acids than the upper body areas of this animal.

There are many different kinds of fatty acids, some of which are shown in **Figure 7.1**. These molecules, with variable length long hydrocarbon chains terminated by carboxylic acid groups, can be broken down into two classes, saturated and unsaturated. In fatty acids these terms take on special importance.

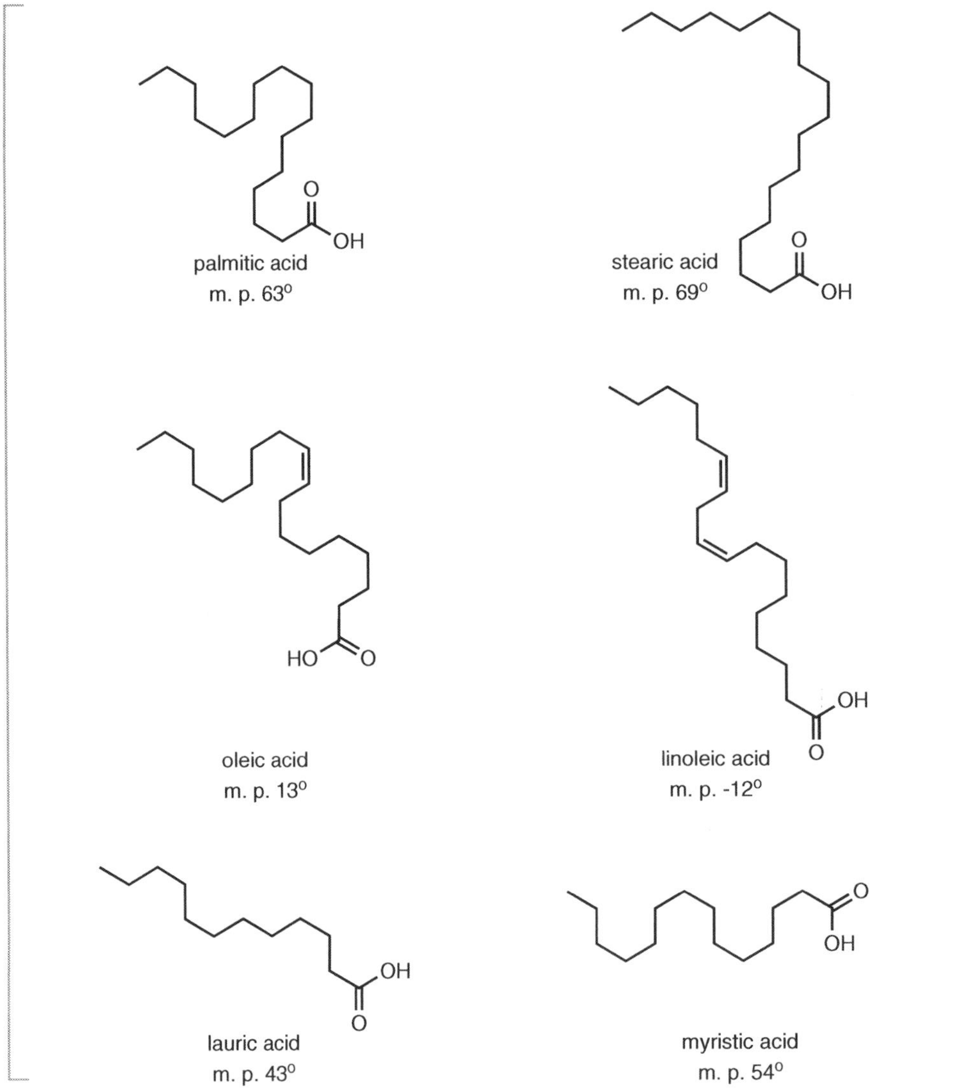

FIGURE 7.1 ▶

Structures and Melting Points of Various Saturated and Unsaturated Fatty Acids

palmitic acid
m. p. 63°

stearic acid
m. p. 69°

oleic acid
m. p. 13°

linoleic acid
m. p. -12°

lauric acid
m. p. 43°

myristic acid
m. p. 54°

In biological systems unsaturated refers to those molecules that contain one or more carbon-carbon multiple bonds (section 4.10, Figure 4.12), which refers in the most part to sp^2 carbon atom hybridized carbon-carbon double bonds in long chains made up mostly of sp^3 hybridized carbon atoms (section 1.4, Figure 1.2). We have hardly noted triple bonds in our studies, that is, sp hybridized carbon, and triple bonds, although not unknown, are rarely found in natural fatty acids, so, for now, we can restrict our definition of unsaturated fatty acids to those containing carbon-carbon double bonds, which, as we shall see, are essential in the role fatty acids play in living systems.

Caroline Pond, in her book, points out that the shingle-backed lizard, which lives in the deserts of western Australia, when fed a diet of unsaturated fatty acids likes to spend its time in cooler places compared to being fed a diet of saturated fatty acids after which it likes to hang around in warmer places. Other lizards apparently behave in a similar manner. That's pretty interesting.

As in the reindeer example mentioned above, differences in fatty acid composition are found routinely in different parts of warm blooded animals - more unsaturated in the appendages, more saturated in the inner body. Pond discusses the fact that neat's foot oil, which has been used as a lubricant since the middle ages, is derived from the fat in a cow's hoofs and is a liquid at room temperature while suet from the inner parts of the body tends to be solid. Pond also points out that marine plants have far more unsaturated fatty acids than terrestrial plants. Moreover fish have a higher proportion of unsaturated to saturated fatty acids than land dwelling animals and fish living in colder waters have a greater proportion of unsaturated fatty acids than fish living in warmer waters.

Before we discover how the differences between saturated and unsaturated fats serve life's functions, as seen in the examples just given, let's first understand something about how fats are found in living organisms.

PROBLEM 7.1
Given the fact that derivatives of fatty acids are critical components of cell membranes in living entities, and that unsaturated fatty acids melt at lower temperatures than saturated fatty acids of the same number of carbon atoms, can you offer a biological reason for the ratios of saturated to unsaturated fatty acids found in the animals and plants noted in this section?

PROBLEM 7.2
Do the proportions of fatty acids of different composition and degrees of unsaturation in Figure 7.3 (to be discussed below) support your answer to Problem 7.1?

PROBLEM 7.3
Redraw the structures in Figure 7.1 showing all atoms and all lone electron pairs. Assign hybridization and geometry to all carbon atoms that are not tetracoordinate. Assign E or Z configuration to the double bonds (section 11.2).

FATTY ACIDS IN VIVO ARE NOT FOUND as free carboxylic acids as shown for the structures of several fatty acids exhibited in Figure 7.1 but rather as esters of glycerol, that is, triglycerides. Now we will become familiar with carboxylic acids, alcohols and esters, important functional groups (section 3.9) widely found in organic molecules. The general structure of a triglyceride is shown in **Figure 7.2** in addition to the fatty acid composition of the triglycerides found in one vegetable oil, sesame oil. The structures of these four fatty acids are shown in Figure 7.1.

We've come across carboxylic acids (sections 4.9, 5.4) and also seen a fatty acid

7.2

Fatty Acids

before in Figure 5.2. They are an important functional group (section 3.9). As noted above, and as seen in Figure 7.1, fatty acids are structures with long hydrocarbon chains terminated by a carboxylic acid group. The functional group in a triglyceride, an ester, is formed by the combination of a carboxylic acid and a molecule containing hydroxyl groups, that is, an alcohol.

Glycerol, shown in **Figure 7.2**, contains three hydroxyl groups and therefore can form three ester groups with three fatty acids, which can all be the same or differ in any way. Esters could be thought of as anhydrides (meaning loss of water) of carboxylic acids and alcohols because the formation of an ester involves elimination of water, the HO⁻ group of the carboxylic acid and an H^+ from the alcohol hydroxyl group. In this chapter we'll study the properties of esters, as they are key to both the breakdown of fats, catabolism, and the build up of fats, anabolism (ancient Greek: *ballō* = I throw: *kata* = downward; *ana* = upward).

FIGURE 7.2 ▶

Structures of a Triglyceride and Glycerol

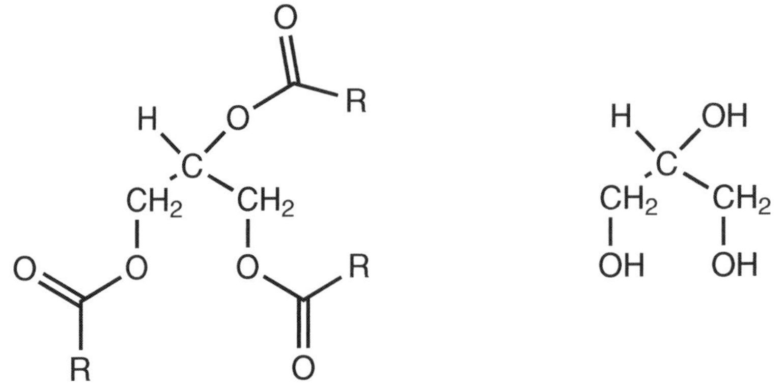

triglyceride
ester of glycerol

glycerol

R stands for fatty acids such as in sesame oil: 9% palmitic acid; 4% stearic acid; 41% oleic acid; 45% linoleic acid

There are two prominent biological roles of fatty acids and the esters they form with glycerol, one structural and the other an energy source. In their structural role, the fatty acids and their derivatives are important components of cell membranes, the feature of every cell that acts, among other roles, as the gateway for nutrients and exit out for expulsion of waste products of cellular metabolism. The hydrophobic nature of fatty acids, which derives from their long hydrocarbon chains, acts to separate the aqueous interior of the cell from the aqueous environment that surrounds the cell. It is the requirement that the membrane be a gateway, a passageway, which causes membranes to be composed of mixtures of saturated and unsaturated fatty acids.

As seen in Figure 7.1, unsaturated and saturated fatty acids of the same chain length, such as, oleic acid and stearic acid, melt differently, 13° C and 69° C, respectively. Here we find the answer to why the cell membranes in the reindeer's hoof contain a higher proportion of unsaturated triglycerides compared to the cell membranes in the warmer parts of the animal's body, and to why the cold blooded lizards prefer a warmer climate when fed saturated fatty acids. The melting point differences between saturated and unsaturated fats allow us to understand the proportions of the saturated to unsaturated fatty acids in the various animals and plants in **Figure 7.3**.

Cell membranes composed of triglycerides of increasing proportion of unsaturated fatty acids remain fluid, can therefore act as gateways, at lower temperatures than membranes composed of higher proportions of saturated fatty acids. Life exposed to lower temperatures turns to unsaturated fats to gain transport through cell membranes.

Fatty acid composition of some common edible fats and oils.

Percent by weight of total fatty acids.

Oil or Fat	Unsat./Sat. ratio	Saturated					Mono unsaturated	Poly unsaturated	
		Capric Acid	Lauric Acid	Myristic Acid	Palmitic Acid	Stearic Acid	Oleic Acid	Linoleic Acid (ω6)	Alpha Linolenic Acid (ω3)
		C10:0	C12:0	C14:0	C16:0	C18:0	C18:1	C18:2	C18:3
Almond Oil	9.7	-	-	-	7	2	69	17	-
Beef Tallow	0.9	-	-	3	24	19	43	3	1
Butterfat (cow)	0.5	3	3	11	27	12	29	2	1
Butterfat (goat)	0.5	7	3	9	25	12	27	3	1
Butterfat (human)	1.0	2	5	8	25	8	35	9	1
Canola Oil	15.7	-	-	-	4	2	62	22	10
Cocoa Butter	0.6	-	-	-	25	38	32	3	-
Cod Liver Oil	2.9	-	-	8	17	-	22	5	-
Coconut Oil	0.1	6	47	18	9	3	6	2	-
Corn Oil (Maize Oil)	6.7	-	-	-	11	2	28	58	1
Cottonseed Oil	2.8	-	-	1	22	3	19	54	1
Flaxseed Oil	9.0	-	-	-	3	7	21	16	53
Grape seed Oil	7.3	-	-	-	8	4	15	73	-
Illipe	0.6	-	-	-	17	45	35	1	-
Lard (Pork fat)	1.2	-	-	2	26	14	44	10	-
Olive Oil	4.6	-	-	-	13	3	71	10	1
Palm Oil	1.0	-	-	1	45	4	40	10	-
Palm Olein	1.3	-	-	1	37	4	46	11	-
Palm Kernel Oil	0.2	4	48	16	8	3	15	2	-
Peanut Oil	4.0	-	-	-	11	2	48	32	-
Safflower Oil*	10.1	-	-	-	7	2	13	78	-
Sesame Oil	6.6	-	-	-	9	4	41	45	-
Shea nut	1.1	-	1	-	4	39	44	5	-
Soybean Oil	5.7	-	-	-	11	4	24	54	7
Sunflower Oil*	7.3	-	-	-	7	5	19	68	1
Walnut Oil	5.3	-	-	-	11	5	28	51	5

* Not high-oleic variety.
Percentages may not add to 100% due to rounding and other constituents not listed.
Where percentages vary, average values are used.

◀ FIGURE 7.3

Table of the Proportions of the Various Fatty Acids from Different Animal And Plant Sources

We'll return to the reason for the lower melting points and therefore the higher fluidity of unsaturated compared to saturated fatty acids later in the book when we'll study natural rubber. (section 11.3) We'll also discover a related phenomenon about the differing kinds of polyethylene which the chemical industry can produce.

For now let's focus on how triglycerides are broken down to glycerol and fatty acids. These fatty acids then undergo catabolism to acetyl coenzyme A (Figure 5.3), which is an intermediate that is a source of energy and as well a biochemical building block. All these processes involve applications of fundamental principles of organic chemistry.

PROBLEM 7.4

Use the structures of the fatty acids in Figure 7.1 to understand the "structural code" used in Figure 7.3 and draw the structures of the fatty acid components of the triglycerides of many of the plants and animals in the table. In your drawing these structures do questions arise as to the proportion of each fatty acid within each triglyceride molecule? Are you able to answer the question in this problem unequivocally? Do questions of stereochemistry also arise?

PROBLEM 7.5

In the conversion of isopentenyl diphosphate to dimethyl allyl diphosphate in sections 5.5-5.6, we came across the concept of enantiotopic groups. Does this stereochemical designation apply to glycerol?

PROBLEM 7.6

Considering the shape around a cis versus a trans double bond and the fact that the favored conformation around the many CH_2 groups in a fatty acid chain take an anti conformation (section 1.13), and considering that in a membrane the chains are packed close together, offer a reason (hypothesize) why cis fatty acids make membranes that are more fluid, that is, have lower melting points, than trans fatty acids.

PROBLEM 7.7

Considering the composition of the triglycerides in sesame oil, Figure 7.2, and that there are three hydroxyl groups in glycerol available to form esters, how many different triglyceride structures are possible?

PROBLEM 7.8

Make a list of all the functional groups you have come across in the book so far or that you are aware of from elsewhere and draw their structures.

7.3
Saponification

IN CHAPTER 5, SECTION 5.1 we were introduced to Michel Chevreul, the beloved and long lived French chemist whose investigations of the water insoluble constituents of life led to the discovery of cholesterol and several fatty acids and also to the understanding that fatty acids occurred in nature as the esters of glycerol, a molecule he characterized but did not discover. Glycerol was discovered by **Carl Wilhelm Scheele** in the 1780s by heating fatty substances with litharge, a basic compound of lead. He called it oelsüss and described the substance as a sweet principle of oils and fats. Scheele's characterization of glycerol as a sweet principle reminded me that many years ago when I was a student, long before concerns about toxicity and environmental hazards came to the fore, there was tradition we heard about to taste new substances – place a trace on the tongue. Why do you think the Germans used the word carbonsäure for carboxylic acids?

It is difficult to mention Scheele's name without saying something more about this very interesting 18th century Swedish-German chemist. Scheele, who lived a relatively short life, from 1742 to 1786, maybe because of tasting too many chemicals, was not someone with enviable laboratory facilities. In spite of his first laboratory being described as a "cold and draughty wooden shed," Scheele discovered chlorine and oxygen and laid the foundation for photographic film in his discovery of the action of light on silver salts. Later in his career he was given a position where he was allowed to do research one day a week, by someone Partington called a "considerate master." Nevertheless, when Scheele turned his attention to what we call now organic chemistry he discovered many organic acids including tartaric, prussic, malic, lactic, uric and citric acids and for our current interests, a neutral molecule, as mentioned above, glycerol. I remember when I was a young research student, my mentor, Kurt

■ **Carl Wilhelm Scheele**

Mislow, telling me that the drive to carry out research by some is so overpowering that they will do it under any circumstances, not matter how difficult. I wonder if he was thinking of Scheele. What a chemist!

When Professor James Moore of Rensselaer Polytechnic Institute read the paragraph about Scheele he sent me the following note:

I know of this noted alchemist only because the Institute of Organic Chemistry at the U. of Mainz is on Johann-Joachim-Becher Weg. "The chemists are a strange class of mortals, impelled by an almost insane impulse to seek their pleasures amid smoke and vapour, soot and flame, poisons and poverty; yet among all these evils I seem to live so sweetly that may I die if I were to change places with the Persian king." Johann Joachim

saponification

Mechanism:

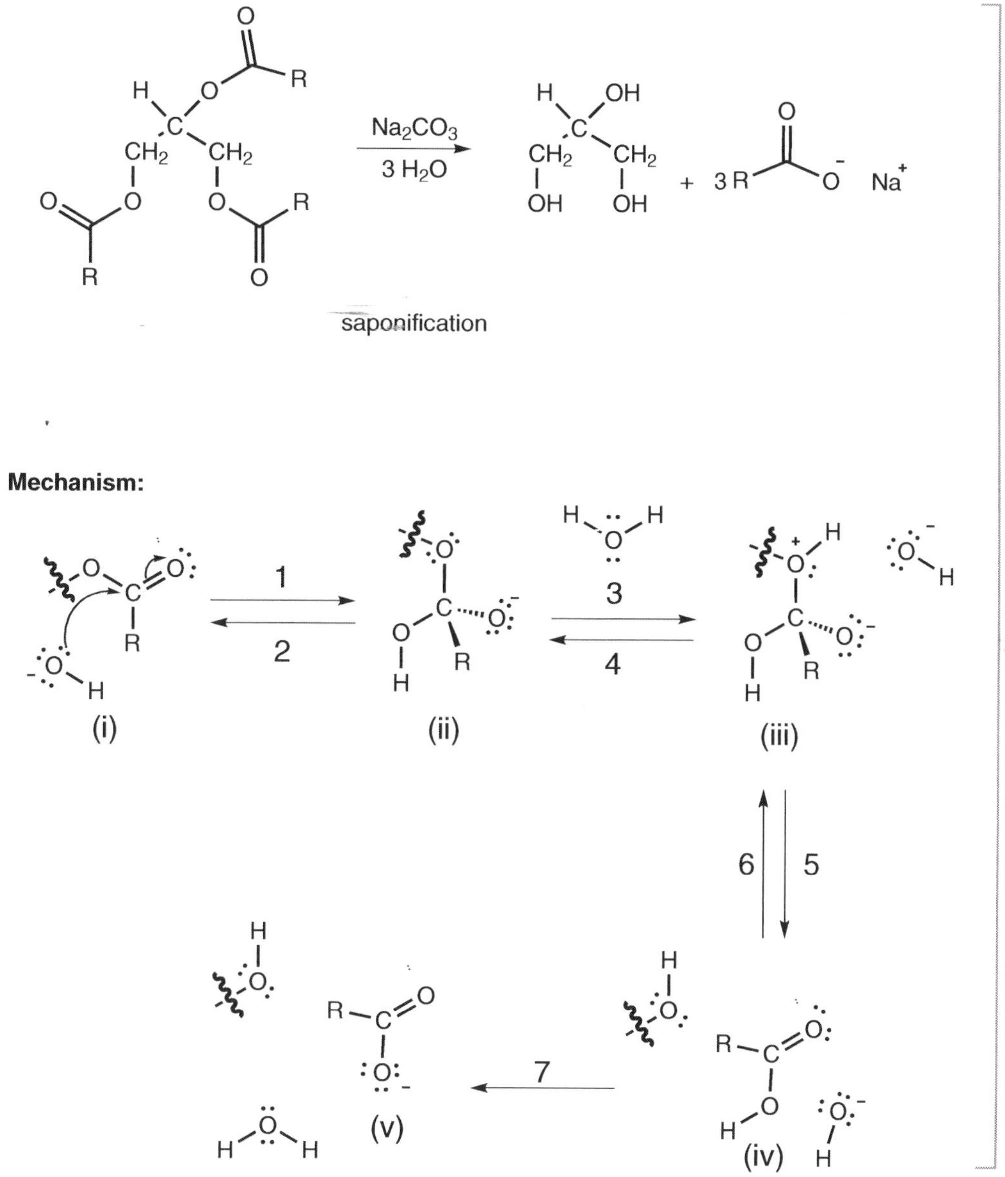

◄ FIGURE 7.4

In Vitro Basic Hydroly-sis (Saponification) of a Triglyceride

Becher, Physica subterranea (1667). Quoted in R. Oesper, The Human Side Scientists (1973), 11.

The history and production of soap is intertwined with the chemistry of the ester functional group, which links glycerol and the fatty acids (Figure 7.2). The earliest recorded history of the making and use of soap goes back nearly five thousand years, although its modern widespread use in bathing is much more recent (the last two hundred years or so). However, the fundamental chemistry for making soap has not changed over all these years. Fats or oils from plants or animals are subjected to a basic

substance in water. The earliest process certainly involved the accidental discovery that ashes from burning wood on mixing with animal fat and water from a cooking process produced a substance we call soap.

The basic substance used was alkali, a mixture of soda ash, that is, Na_2CO_3, and potash, K_2CO_3, which continued to be obtained from wood ash well into the 1700s. However, as the need for soap increased, as people bathed more frequently, the consequence was destruction of large swaths of European forests to get the wood to make the ash. Some potash could be obtained by burning sea weed and this was a source in England and Scotland with plenty of coastline but it was clear that some new source had to be found.

In the late 1700s **Nicolas Leblanc**, stimulated by a monetary prize from the French crown, a prize he was eventually denied because of the French Revolution, found a way to transform common salt, NaCl, to soda ash. Sulfuric acid was used to convert sodium chloride to sodium sulphate, Na_2SO_4, which was then reacted with limestone, $CaCO_3$, to produce the soap making chemical, soda ash (sodium carbonate). This history of soap brings us back to Carl Wilhelm Scheele, who along with all his other accomplishments noted above, was the one who discovered the essential reaction for the process in the conversion of common salt to sodium sulphate using sulfuric acid. And, in an interesting tale of the long and winding paths allowing Europeans to bathe in large numbers, we find that sulfuric acid was first produced by **Jabir ibn Hayyan**, an Arab-Persian alchemist who lived in the eighth century. In fact, the medieval Muslim world had advanced methods for making soap and a recipe found on the web from a manuscript of that time is: "*sesame oil, a sprinkle of potash, alkali and some lime - mix and boil - pour into a mold and leave to set to produce a hard soap.*" I guess you now know precisely the fatty acid salts in this eighth century soap.

The chemical process, saponification, which means "soap making" from the Latin word sapo for soap, is shown in **Figure 7.4** and is an example of basic hydrolysis of an ester. Hydrolysis is the perfect word to describe the change seen in Figure 7.4 - water, that is, hydro, causing the lysis, from the Greek for separation. In the process, the two parts of the ester bond are separated, the hydroxyl part from the carboxylic acid part. The mechanism for this hydrolysis, discussed in the section to follow, is also shown in Figure 7.4.

PROBLEM 7.9

For every structure in Figure 7.4 show all lone pairs of electrons and check on the accuracy of those already shown. Can you justify as correct all formal charges shown in the figure?

PROBLEM 7.10

Offer an explanation of why a fatty acid does not form as effective a soap as the salt of a fatty acid, that is, with an ionized end group, $-CO_2^-\ Na^+$.

PROBLEM 7.11

Imagine that an alcohol of structure ROH combines with a carboxylic acid of structure $R'CO_2H$ to form a molecule with the formula ROOCR'. Draw as many structures as you can for this formula evaluating formal charge and the octet rule. Is there more than one possibility without formal charge that obeys the octet rule?

PROBLEM 7.12

Amides are analogs of esters formed by replacing the alcohol with an amine, such as RNH_2. Propose a formula for the amide parallel to the formula for the ester in Problem 7.11 and then answer the same question posed in Problem 7.11.

PROBLEM 7.13

For the saponification of a triglyceride, use the mechanism in Figure 7.4 to trace the oxygen atom that was originally part of glycerol. Could you write a mechanism using a basic catalyst as in Figure 7.4 where this oxygen took another path?

PROBLEM 7.14

In the conversion of (i) to (ii) in Figure 7.4 do you see any analogy to the chemistry of the aldehyde group in glucose and galactose leading to ring closing of these sugars? On what structural characteristic, that is, on what common characteristic of the functional groups involved might this analogous chemical reactivity be based?

PROBLEM 7.15

In the mechanism for saponification in Figure 7.4, curved arrows are only used in the conversion from (i) to (ii). Use curved arrows to show the flow of electrons in all the other reactions making certain to show all electrons and formal charges.

PROBLEM 7.16

Using leaving group concepts (section 5.3) evaluate the relative probability of reaction 2 to reaction 5 in Figure 7.4. What does your answer reveal about the importance of reaction 3?

PROBLEM 7.17

Reaction 7 in Figure 7.4 is the only step that is shown as irreversible. Offer a reason for this irreversibility.

7.4

Similarities and Differences between Ketones and Aldehydes and Derivatives of Carboxylic Acids: Mechanism of Saponification

IN SECTION 3.13 and following sections of Chapter 3 we came to understand the role of the aldehyde functional group in both the formation of the cyclic structure of sugars and also as the source of galactosemia.

The carbonyl group, because of the weakness of the ϖ-bond and the large electronegativity difference between carbon and oxygen, is highly susceptible to attack at the carbon end of the double bond by electron rich moieties, that is, molecules or portions of molecules that are nucleophiles. We first came across the roles of nucleophiles and their electron poor counterparts, electrophiles, in section 3.14 in our study of the chemistry of glucose and then again in the study of the chemistry of benzene in section 6.10 and following sections. Nucleophiles and electrophiles are terms used to describe reactants in polar reactions.

All chemical reactions that involve interactions between positive and negative moieties come under the general heading of **polar reactions**, and the hydrolysis of an ester in converting it back to the functional groups from which it was formed (Figure 7.4), a carboxylic acid and an alcohol, is another example of this class of chemical reactions.

The saponification mechanism is initiated (reaction **1**, Figure 7.4) by attack of hydroxide ion, HO⁻, on the carbonyl group of the ester, opening the π bond. We've seen nucleophilic attack at carbonyl carbon before in the formation of a glucose ring by intramolecular reaction of the OH group on carbon-5 of glucose with the aldehyde carbon of the glucose. We've seen this type of chemistry again in the unfortunate reaction of the aldehyde group of the open galactose ring with protein-bound nucleophiles (section 3.14, Figures 3.15 and 3.16).

In fact all functional groups bearing a carbon-oxygen double bond, a carbonyl bond, C=O, are susceptible to nucleophilic attack in this manner. But there is a <u>big difference</u> for the outcome of this nucleophilic attack at carbonyl groups in aldehydes and ketones versus those in carboxylic acids or their derivatives. The difference has to do with the concept of leaving groups (section 5.3). Let's take a look back at the chemistry of the aldehyde groups in the sugars in Chapter 3 and the chemical reactions they undergo to understand this "big difference."

Figure 7.5 reproduces the forward reaction **5** from Figure 7.4. All chemistry of carbonyl groups is driven forward by the opening of the π bond, which breaks to deliver the two electrons of this bond to the electronegative oxygen and forms a new σ bond to the incoming nucleophile. However, the consequence of this breaking of the double bond to oxygen is formation of an intermediate in which the bond angles

Importance of the Leaving Group Structure in the Hydrolysis of a Glyceride Ester Bond

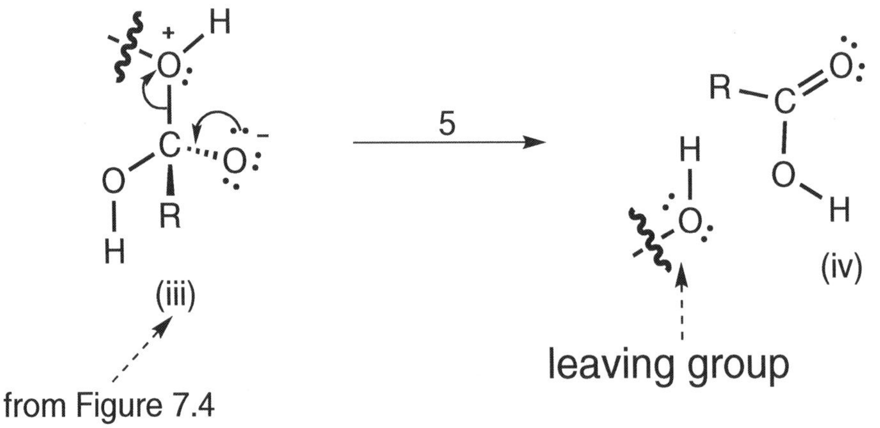

from Figure 7.4

leaving group

around carbon have been considerably reduced from what they were for the carbonyl group, from approximately 120° to approximately 109°, a more crowded arrangement. An exchange of a π bond for a stronger σ bond, an energy lowering transformation, is offset by an energy raising process in producing a more crowded bonding, the tetracoordinate intermediate (iii) in Figure 7.5. The counterbalance means that the carbon-oxygen double bond has a tendency to reform.

In the mechanism shown in Figure 7.5 (reproduced from step **5**, Figure 7.4) the reformation of the carbon-oxygen double bond from (iii) is shown to take a path that forms a carbonyl bond in an entirely new structure (iv). Here we come to the "big difference" between the carbon-oxygen double bonds in aldehydes or ketones versus carboxylic acids and their derivatives. The option to form a new carbonyl containing compound (step **5** leading to iv), may not be available from the tetracoordinate intermediate formed from all carbonyl compounds. Why not?

Forming a new carbonyl-containing molecule, which is a perfectly reasonable reaction path in the saponification mechanism shown in Figures 7.4 and 7.5, is not reasonably possible, is in fact virtually impossible, in the chemistry of the aldehyde group in glucose **(Figure 7.6)**. Whereas reaction **5** from Figures 7.4 and 7.5 occurs readily, reaction **1** in Figure 7.6 is impossible even if the final resulting product produced by reaction **1** is a perfectly stable structure, which it is. The product of reaction **1** (Figure 7.6) is an ester that would be formed by an intramolecular reaction between a hydroxyl group and a carboxylic acid group within the same molecule and is called a lactone, a functional group we will come across later in the book (section 12.12).

The difference between reaction **5**, which does take place (Figures 7.4 and 7.5), and reaction **1**, which does not take place (Figure 7.6) arises from the leaving groups (section 5.3) involved. Reaction **5** in reforming the carbonyl group of the carboxylic acid (iv) forces out one of the hydroxyl groups of glycerol. Reaction **1** (Figure 7.6) in reforming the carbonyl group and therefore leading to a cyclic ester (a lactone) **(b)** would have to force out a hydride ion, H:⁻.

As discussed in Chapter 4 (sections 4.9 and 4.11) the conjugate base of a Brønsted-Lowry strong acid must be a stable entity, such as Cl⁻ from hydrochloric acid, or $H_2PO_4^-$ from phosphoric acid or H_2O from H_3O^+, the hydronium ion. The latter ion is particularly relevant in the situation for reaction **5** in Figures 7.4 and 7.5. The glycerol hydroxyl group expelled in reaction **5** is the conjugate base of the protonated hydroxyl group shown in (iii), an analogously strong acid to the hydronium ion. Reaction **5** is driven forward by the ejection from the tetracoordinate intermediate, iii, of an excellent leaving group, the conjugate base of ROH_2^+.

If reaction **1** took place, it would eject from the tetracoordinate intermediate **(a, Figure 7.6)** a hydride ion, H:⁻, which is the conjugate base of H-H an impossibly weak acid. A hydride ion, H:⁻, is therefore an unlikely leaving group. Because of the unlikely leaving group reaction **1** in Figure 7.6 would not take place.

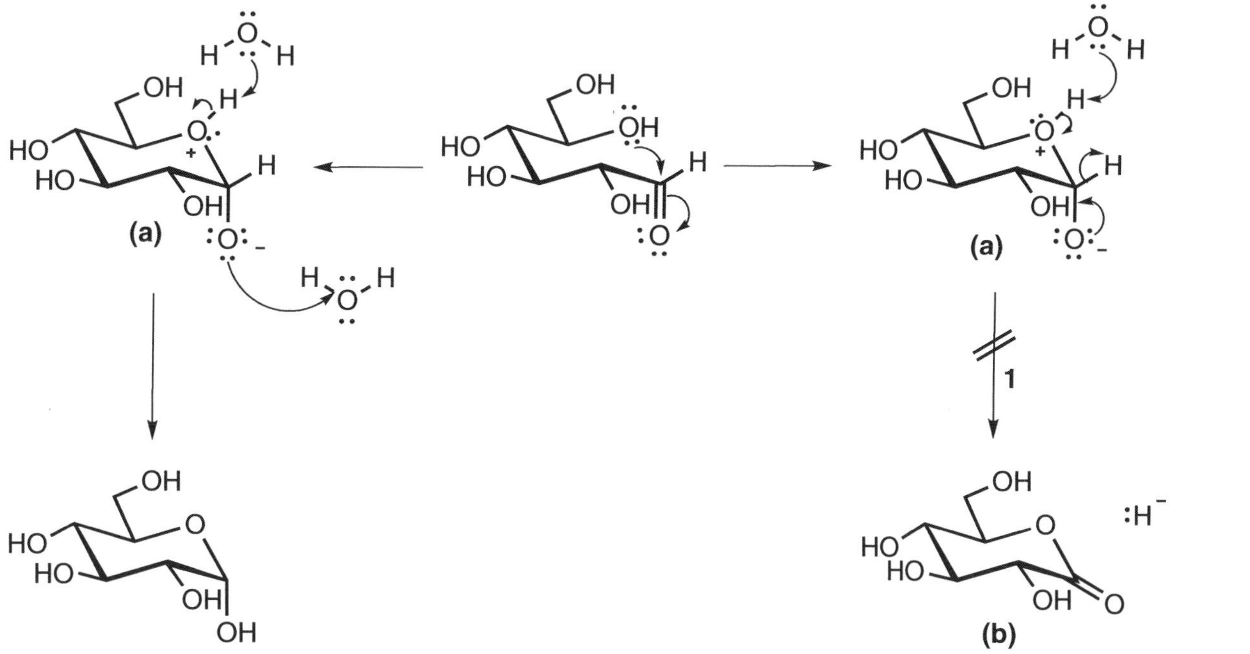

We can extend the discussion of the difference between the chemistry of (iii) in Figures 7.4 and 7.5 and that of (a) in Figure 7.6 to a general consequence of the four coordinate intermediates formed from aldehydes and ketones after nucleophilic attack. In these carbonyl functional groups, the flanking atoms to the carbonyl carbon are always carbon or hydrogen therefore requiring formation of a carbanion, R_3C: or H: to be expelled to reform the carbonyl group from a four-coordinate intermediate. **Figure 7.7** exhibits these ideas for a typical ketone, acetone, and a typical aldehyde, acetaldehyde with nucleophilic attack by methoxide ion (H_3CO:).

Now you might ask: If nucleophilic attack at the carbonyl carbon atom of aldehydes and ketones can take place and if the carbonyl group can not reform, as we've just learned, then what happens after this nucleophilic attack? You can answer your own question by looking at Chapter 3 at the ring closing reactions of the sugars and you'll be finding other examples later in the book (section 12.7).

In the special circumstances serving life's needs, the rules laid down above find an exception, which makes sense with regard to the leaving group ideas discussed above – in fact the exception, as they say, that proves the rule.

▲ FIGURE 7.6

Intramolecular nucleophilic attack shown from the open chain of glucose leads to addition rather than substitution because of an extremely poor leaving group.

▼ FIGURE 7.7

Nucleophilic attack at carbonyl carbon of ketones and aldehydes is shown not to lead to substitution because of extremely poor leaving groups.

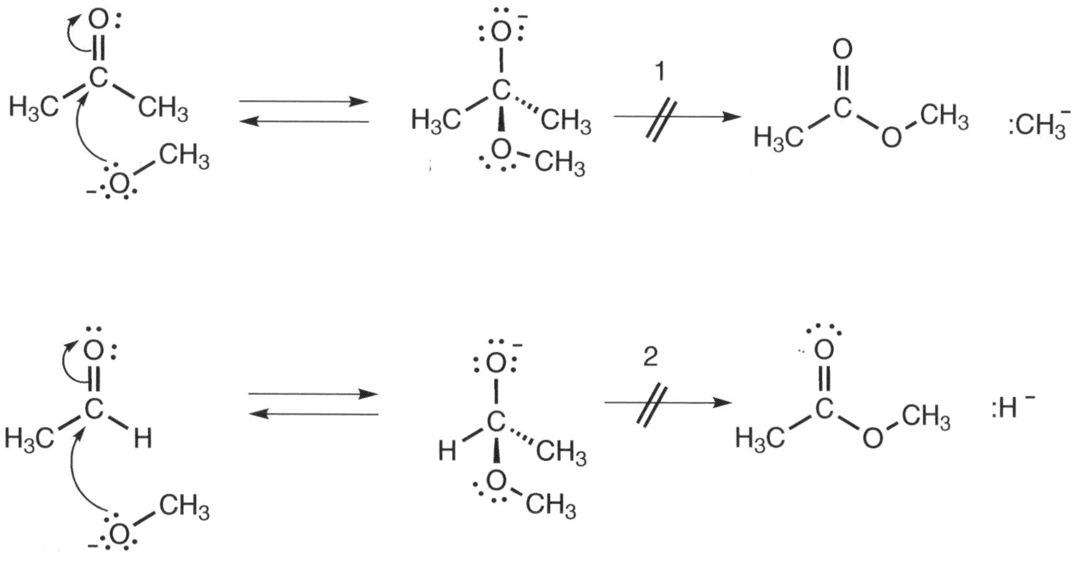

A carbanion can be expelled (not reaction **1** Figure 7.7) from a four coordinate intermediate arising from an attack at carbonyl carbon if the negative charge at carbon is stabilized. And, in fact, an essential step in the biochemical process by which fatty acids are broken down into two carbon pieces, catabolism, which we'll look into in this chapter, and the catabolism of sugars to be studied in the chapter to follow, will yield examples of this exception.

PROBLEM 7.18
Can you offer two reasons why aldehyde carbonyl groups are more susceptible to nucleophilic attack than ketone carbonyl groups? Now apply your ideas to predict the relative propensities to nucleophilic attack at the ketone carbonyl group by hydroxide ion on methyl ethyl ketone versus hydroxide ion on methyl tertiary butyl ketone.

PROBLEM 7.19
"Knockout drops" is the name for a drug that causes rapid unconsciousness in the victim. The formula is $Cl_3CCH(OH)_2$, chloral hydrate. Offer a structure for chloral hydrate and suggest why trichloroacetaldehyde, $Cl_3C-CH=O$, adds water to form chloral hydrate, which has little driving force to return to the carbon-oxygen double bond.

PROBLEM 7.20
An unusual reaction takes place between trichloroacetaldehyde and sodium hydroxide in water: $H(C=O)CCl_3 + OH^- \quad H(C=O)O^- + HCCl_3$. If one of the chlorine atoms in the starting aldehyde is replaced by hydrogen atoms the reaction does not take place under the same conditions. Here's an important piece of information. Chloroform, $HCCl_3$, is a moderately weak acid because of the stability of the trichloromethyl anion, $Cl_3C:^-$. Offer an explanation for why this reaction is an exception to the general rule for ketones.

PROBLEM 7.21
From your answer to problem 7.19 can you predict something about the properties of hexafluoroacetone $(F_3C)_2C=O$, in water?

PROBLEM 7.22
Draw the structure of the ester that would be formed from the reaction of phenol, C_6H_5OH (a benzene ring with one hydrogen replaced by OH) with acetic acid, H_3C-CO_2H and then another ester that would be formed by reaction of methanol with acetic acid. Based on the fact that phenol is a moderately weak acid (pK_a about 10), while methanol is a far weaker acid, offer an explanation for the higher reactivity for basic hydrolysis for the phenyl over the methyl ester. Show a possible mechanism for this hydrolysis based on the information in Figure 7.4.

PROBLEM 7.23
Suggest a carbon-carbon bond breaking reaction to reform the carbonyl group from the intermediate in Figure 7.6, which is just as improbable as the reaction shown not to take place.

7.5
Hydrolysis of the Triglyceride Ester Bonds: Enzyme Catalyzed Path

AS FOR ALL BIOCHEMICAL REACTIONS, hydrolysis of triglycerides is enzymatically catalyzed. In human beings this transformation takes place with an enzyme called pancreatic lipase. This enzyme has been very well studied, so that the mechanism by which the three amino acids that find themselves properly placed to carry out the hydrolysis has been determined. As understood from sections 5.5 and 5.9 it is the complex folding properties of the protein, which brings these three amino acids together in a "pocket." This pocket, that is, the active site, is designed to hold the molecule that will undergo chemical change.

The necessity for protein folding is demonstrated again for human pancreatic lipase by the very different positions along the lipase protein chain for these three amino acids: serine, the 152nd amino acid; aspartic acid, the 176th amino acid; and histidine, 263rd, which are all along the chain designated from the N-terminus of the protein. The protein contains 449 amino acids linked end to end in its structure. The folding of the lipase is shown in **Figure 7.8** and even if this is not the place to understand all the twists and turns in the structure shown in this figure, the structure certainly is, as were Figures 5.8 and 5.15, marvels to observe – to observe what nature has put to work for catalytic purposes.

It is interesting to see how the protein chain begins and ends. The sequence of amino acids begins with alanine, followed by aspartic acid, and then by glutamine and so on until the last three amino acids in the sequence, lysine, serine and finally, glycine. By convention, the beginning of the protein is designated as the end of the chain with a free amino group, NH$_2$, or some derivative of an amino group, while the end of the chain is terminated with a free carboxylic acid group, CO$_2$H, or some derivative of that group. The structures of the three amino acids (which we have drawn by hand) at each end of the protein chain are shown in **Figure 7.9**, although not shown in detail for their conformational states or their states of ionization.

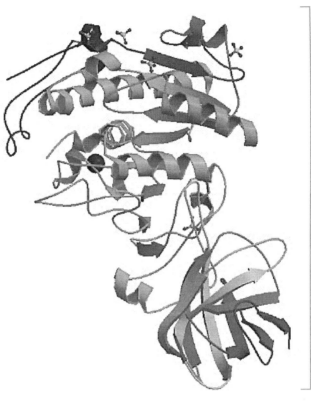

Representation of the Folded State of Human Pancreatic Lipase, the Enzyme that Catalyzes the Hydrolysis of the Ester Bonds of Triglycerides

Depending on pH, the various amino and carboxylic acid groups may appear as NH$_3^+$ or CO$_2^-$ or not ionized, NH$_2$ or CO$_2$H, respectively.

Figure 7.10 portrays a representation of the three amino acids that constitute the active site of the protein, that is, the amino acids participating in the hydrolysis of the ester bond between the glycerol hydroxyl group and the fatty acid carboxylic acid group. From left to right, there is the conjugate base of aspartic acid, histidine and serine. Figure 7.10 also includes the first steps of the mechanism of the hydrolysis.

In this first step, **1**, of the enzyme-catalyzed mechanism shown in Figure 7.10, we find the familiar curved arrows designating the movement of electrons and discover that the reaction path involves a coordinated process – several transfers of electrons and atoms take place in concert. The result of this cascade of reactions is that the

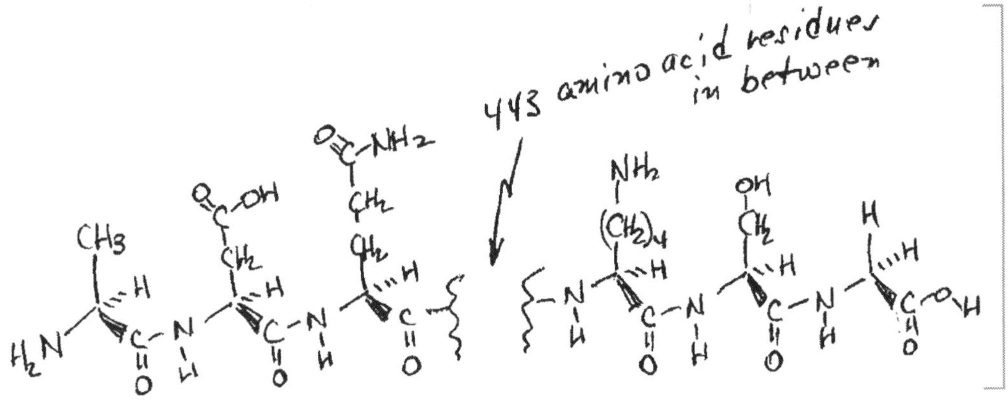

Three Amino Acids, Drawn by Hand, at Each of the Termini of the 443 Amino Acid Enzyme Shown In Figure 7.8

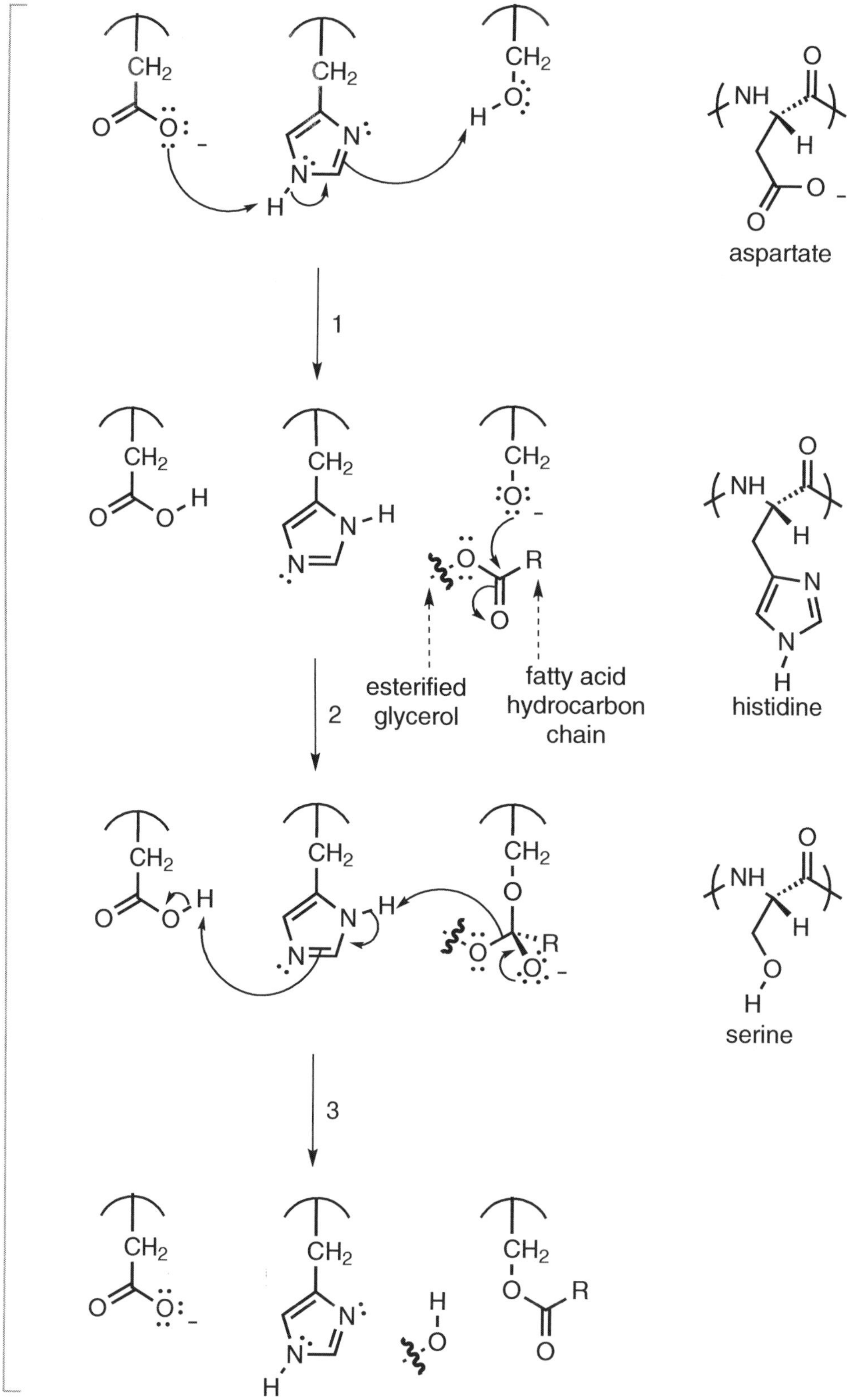

FIGURE 7.10 ▶

Enzyme catalyzed mechanism involving the three amino acids in the active site acting to transfer the fatty acid from the glyceride bond to an ester with a serine unit on the protein chain.

nitrogen containing histidine is acidic enough at one of the nitrogen atoms to donate a proton to the aspartate and also basic enough at the other nitrogen atom to remove the proton from the hydroxyl group of the serine residue. These proton transfers produce therefore, the conjugate base of serine, which is capable of nucleophilic attack at the carbonyl carbon of the ester group.

The next step, **2**, accomplishes the same result as reaction **1** in Figure 7.4, the π bond forming the carbonyl group is broken and a four-coordinate intermediate is formed.

Step **3** picks up the mechanism from the tetracoordinate intermediate where we see a demonstration of the beautiful advantage of the precise geometry arising from the precisely folded protein structure. In Figure 7.4, showing the in vitro hydrolysis of an ester, that is, saponification, most of the reactions were reversible. In the tetracoordinate intermediate shown in Figure 7.10, reformation of the carbonyl group could occur, in principle, via ejection of the serine oxygen atom, thereby reversing the prior step, **2**, or, on the other hand, by ejection of the glycerol oxygen atom, which advances the hydrolysis.

As shown in step **3**, the precise transfer of a proton from the histidine nitrogen atom to the oxygen atom of the glycerol moiety, a movement of atoms that arises from the precise folding of the protein, forces the reaction path to advance the hydrolysis. This kind of detailed control of the pathways of chemical reactions, seen routinely in enzymes and exemplified in Figure 7.10, is the envy of the most brilliant and successful synthetic organic chemists who find such control generally inaccessible to reactions in laboratories and in industrial processes. The precise control that goes on in the routine chemical processes within our bodies can not be easily reproduced under our control.

In Figure 7.10 in steps **1**, **2** and **3**, we've seen the ester bond between the fatty acid and the glycerol hydroxyl group, replaced by a new ester bond arising from the combination of the fatty acid with the hydroxyl group of the serine amino acid residue located at the 152nd position from the nitrogen end of the enzyme. Organic chemists, appropriately, call this kind of reaction a transesterification.

The fatty acid moiety is now buried deep within the lipase protein structure. Obviously, something else has to happen. The fatty acid has to be released from the protein and the protein has to be released to work on another triglyceride ester bond. As shown in **Figure 7.11**, a molecule of water comes to the rescue. And this water molecule does not originate from a pool of water. Again following the precision theme of the working of the enzyme, a single water molecule is let in to the reaction site and oriented in such a manner to do the necessary job.

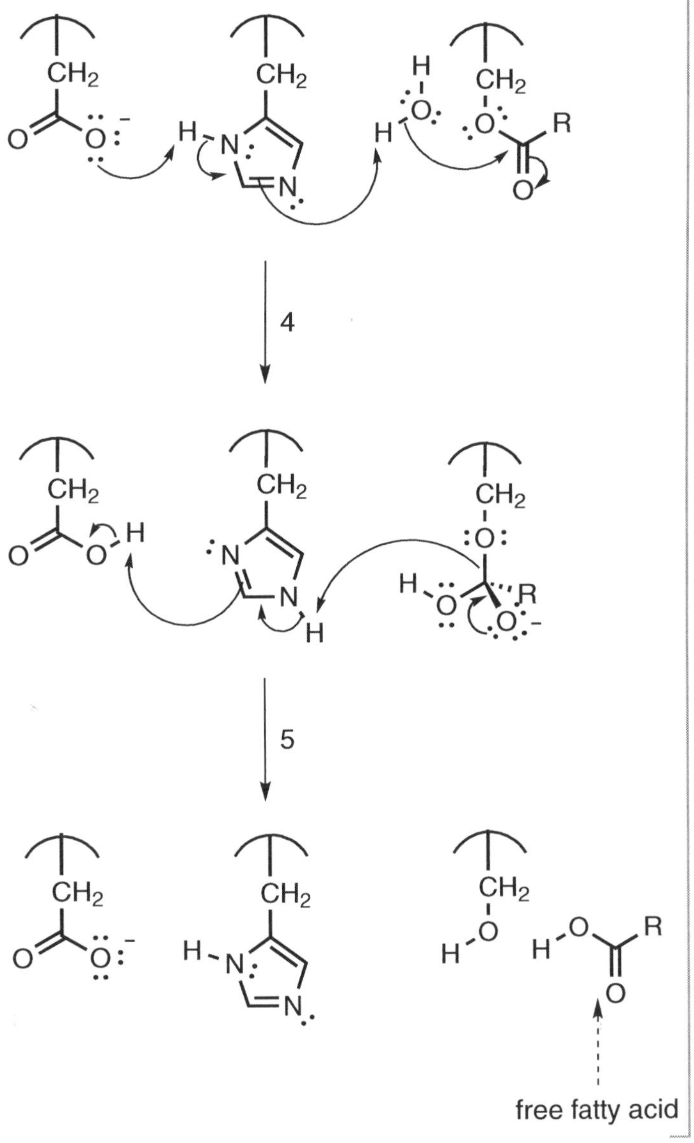

◄ **FIGURE 7.11**

Enzyme catalyzed mechanism involving the same amino acids as in figure 7.10, now converting the ester bond of the fatty acid with serine to the free fatty carboxylic acid.

free fatty acid

Just as one of the nitrogen atoms of the histidine took a proton from the serine hydroxyl group in initiating the reaction in Figure 7.10 (step **1**), so the histidine in Figure 7.11 takes a proton from a water molecule allowing nucleophilic attack from the resulting hydroxide ion, HO⁻, to produce a tetracoordinate intermediate (step **4**).

The step to follow, again demonstrates the specific reactivity arising from precise geometric placement of the reactive groups arising from the folding of the protein shown in Figure 7.8. In step **5**, if the hydrogen bound to nitrogen in histidine were able to reach the OH group, step **4** could be reversed. There is no intrinsic chemical reason other than a spatial relationship, for this nitrogen bound proton to add to the serine oxygen rather than the oxygen of the OH group. The spatial relationship, as just noted, does not allow approach of the proton to the OH group, but only to the serine oxygen. What follows from the delivery of the proton to the serine oxygen atom is that the reactive process proceeds as intended to release the fatty acid and return the enzyme active site to take another turn at the next glyceride ester bond presented to it. And this presentation, as you expect, will present the carbonyl group of the ester in precisely the correct spatial relationship to take step **1** (Figure 7.10) over again.

In this mechanistic portrayal of the working of the human lipase we see the essential characteristics of all enzyme-catalyzed reactions – reactive amino-acid-based groupings placed in precise spatial relationships around a substrate to accomplish the intended chemical process.

PROBLEM 7.24
Assign absolute configurations, (R) or (S), to the six amino acids at the ends of the lipase chain shown in Figure 7.9.

PROBLEM 7.25
Compare the steps in the saponification mechanism in Figure 7.4 with the parallel steps in Figures 7.10 and 7.11 showing where the action of the enzyme blocks a reaction that would reverse the overall hydrolysis of the ester bond.

PROBLEM 7.26
In the enzyme-catalyzed hydrolysis of the glyceride ester bond shown in Figures 7.10 and 7.11, a histidine unit on the protein chains plays a central role as both a Brønsted-Lowry base and an acid. Identify the reactions that justify this statement.

PROBLEM 7.27
In each reaction step in Figures 7.10 and 7.11 (**1-5**) identify all moieties that are acting as acids or bases and specify which role is played.

PROBLEM 7.28
Is the ring in the histidine unit in Figures 7.10 and 7.11 aromatic? If so, identify the electrons contributing to this aromaticity and determine if any reactions interfere with the aromatic stabilization.

PROBLEM 7.29
Consider the structures of the amino acids that are pointed to in Figures 7.9-7.11 when removed from the chain by hydrolysis of the linking amide bond. Identify all functional groups that could act as Brønsted-Lowry acids or bases. Now imagine glycine, lysine and glutamic acid dissolved in water. As you change the solution from acidic to basic what changes would you expect in the amino acid?

THERE IS A GREAT DEAL of complex, very interesting biochemistry involved in the fate of a fatty acid. But the essential organic chemistry, which is our focus, takes the fatty acid to form what is called a thioester, which is the stepping-off point before breaking up each fatty acid into many two carbon pieces, for example, nine molecules of acetyl coenzyme A from the eighteen carbon atoms of stearic acid. Let's look back to Chapter 5 (Figure 5.3) where we saw the structure of this important biological molecule, acetyl coenzyme A.

The first prerequisite step for breaking up the fatty acid is conversion to a thioester. Acetyl coenzyme A is also a thioester, that of acetic acid. An ester and a thioester differ by replacing an oxygen atom in an ester with a sulfur atom in a thioester. Taking account of the atoms, an ester is the combination of a carboxylic acid with a hydroxyl containing molecule, ROH, the focus of this chapter so far, while a thioester is the combination of a carboxylic acid with a thiol, RSH. Coenzyme A is a thiol.

The prerequisite for the breakdown of a fatty acid to yield life's energy and building blocks is for the fatty acid to be converted to a thioester of coenzyme A as shown in **Figure 7.12**. The steps in this conversion are, as for all biological chemical reactions, enzyme-catalyzed, but here we will focus on the chemical steps and the coenzymes involved, which reveal a great deal about how biological mechanisms use organic chemistry principles.

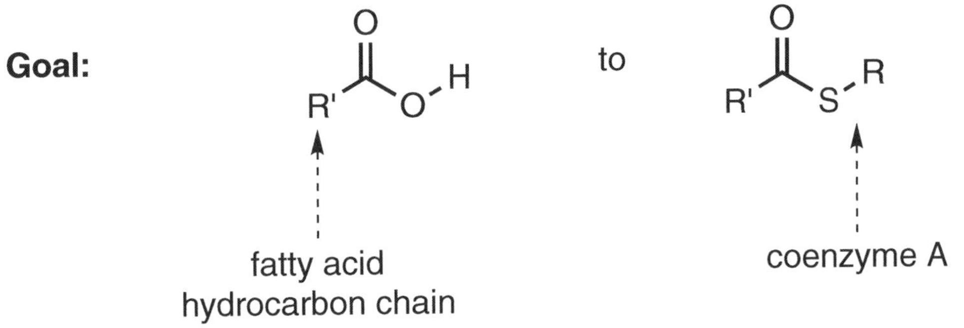

7.6

Biochemical Conversion of Fatty Acids to their Thioesters with Coenzyme A: The Key Role of Leaving Groups

Goal:

fatty acid
hydrocarbon chain

coenzyme A

◀ **FIGURE 7.12**

The biological process needs to form a thioester with coenzyme A from the free fatty carboxylic acid.

The overall transformation shown in Figure 7.12 replaces the OH group of the carboxylic acid functional group of the fatty acid with a thioester derived from coenzyme A. How is this to take place? To simplify our representations we'll replace the complex structure of coenzyme A with R and the hydrocarbon chain of the fatty acid with R'.

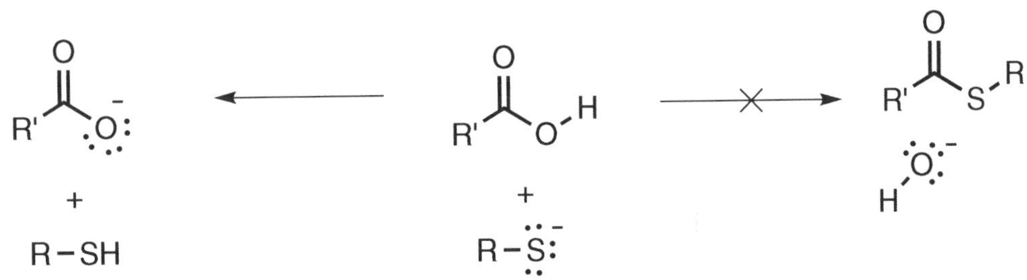

◀ **FIGURE 7.13**

Direct nucleophilic attack of the conjugate base of coenzyme A on the fatty carboxylic acid will not attain the goal of figure 7.12 Because of pK$_a$ values.

A thiol, RSH has a pKa (section 4.9) in the range of 10, which means that RS⁻ is readily accessible. It is the conjugate base of a moderate acid, and we could use RS⁻ as a nucleophile to attack the carbonyl group of the fatty carboxylic acid as shown in **Figure 7.13**. However this reaction could not work because the pK$_a$ of the carboxylic acid, in the range of 5, is far lower than that of RSH meaning that RS⁻ would abstract a proton from the carboxylic acid producing the carboxylate anion and RSH. All this is shown in Figure 7.13.

But let's say we had a "magic way" to allow the RS⁻ group to avoid removing the proton from the carboxylic acid and instead allowed RS⁻ to attack the carbonyl group of the carboxylic acid to produce the tetracoordinate intermediate as shown in **Figure 7.14**.

Can we predict the fate of this tetracoordinate intermediate, if we could, somehow, make it? The answer is not a good one for our intended purpose to form the thioester. The RS⁻ group will be ejected from the tetracoordinate intermediate shown in Figure 7.14 to reform the carbon-oxygen double bond. The competition between HO⁻ and RS⁻ for the best leaving group (section 5.3) is far in favor of the latter, the conjugate base of the stronger acid. RSH is a far stronger acid than HOH (between 5 and 6 orders of magnitude difference in K_a (section 4.9).

FIGURE 7.14 ▶

Direct nucleophilic attack of the conjugate base of coenzyme A on the fatty carboxylic acid will not attain the goal of figure 7.12 Because of leaving group differences.

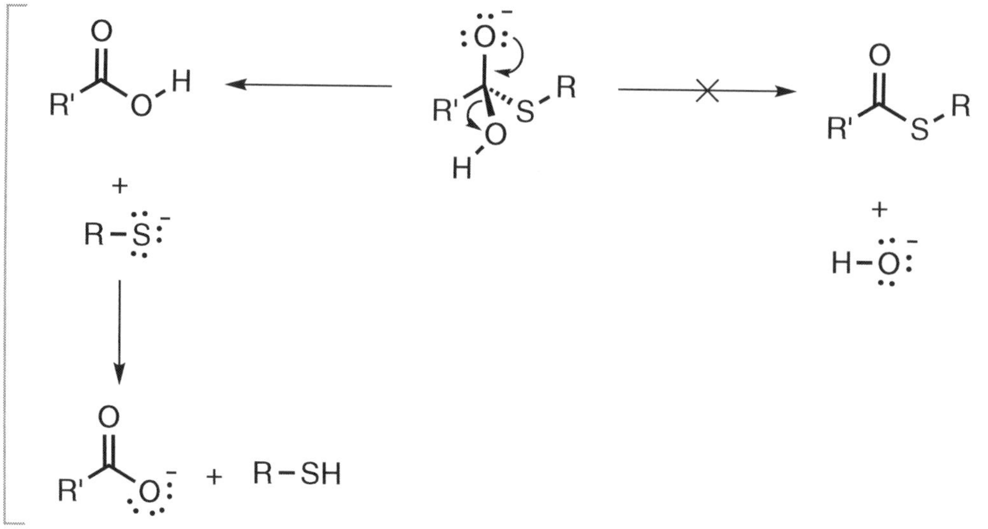

We observe in Figures 7.13 and 7.14 how different aspects of the properties of acids and bases, and the nature of leaving groups, work against nature's purpose if pursued as suggested in these figures. Whatever power nature may have, whatever enzyme precision may exist, there is no way to get around the principles of acid-base chemistry and the competitive nature of leaving groups (section 5.3).

But, naturally, there is another way: the OH group could be converted to a group that would be a better leaving group than RS⁻. One possibility is shown in **Figure 7.15**, where the tetracoordinate intermediate could eject phosphate instead of RS⁻. Phosphate is an excellent leaving group since it is the conjugate base of a very strong acid, phosphoric acid, which we learned about from the biochemistry leading to the terpenes (section 5.7).

FIGURE 7.15 ▶

An excellent leaving group at the acyl carbonyl group would allow the goal of figure 7.12 to be attained.

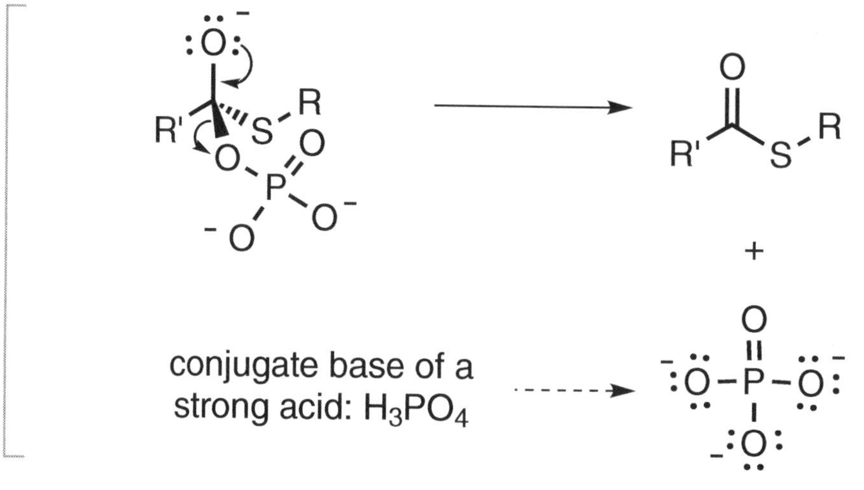

conjugate base of a strong acid: H_3PO_4

But how is one to convert the OH group of the carboxylic acid to the phosphate? The answer is to turn leaving group principles to our advantage as seen in **Figure 7.16**.

The OH group of the fatty carboxylic acid is not converted to a simple phosphate as shown in Figure 7.15 but rather to a more complex phosphate derived from adenosine triphosphate (ATP). ATP is a coenzyme (Figure 7.16), which plays one of the most critical roles in sustaining life, a principle that depends on leaving group chemistry, one example of which is seen in Figure 7.16. We'll hold until Chapter 8 (section 8.9) how these leaving group principles are used by ATP in a more general way.

The phosphorus atoms in the triphosphate moiety of ATP are parallel in their electrophilic characteristics to the carbon atoms of carbonyl-containing functional groups. The high electronegativity of oxygen compared to phosphorus and the weak π bond between oxygen and phosphorus parallel the relationship between carbon and oxygen in carbonyl groups and lead to the reaction shown in step **1** (Figure 7.16). Just as carbonyl

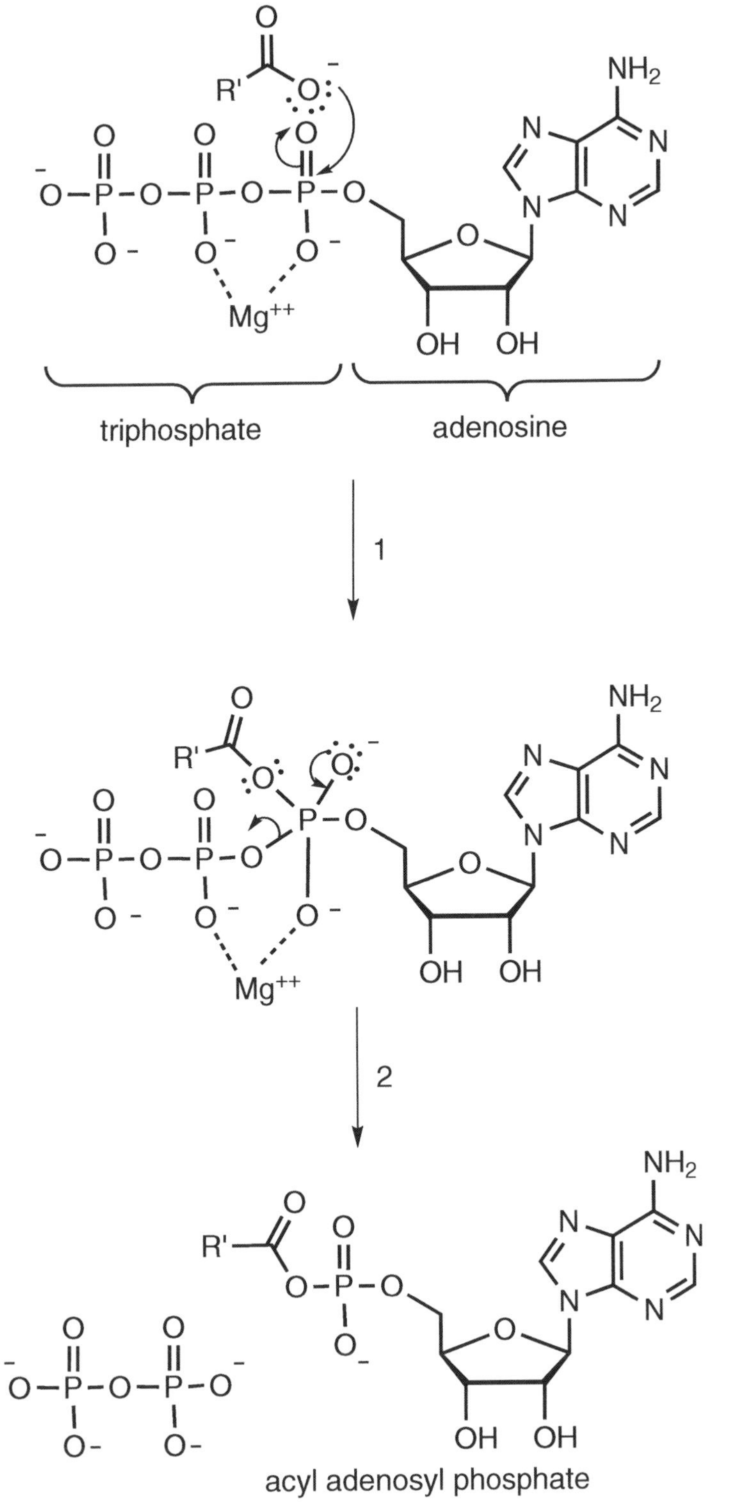

◄ FIGURE 7.16

Nucleophilic attack by fatty carboxylate anion on a phosphorus atom of ATP allows conversion to acyl adenosyl phosphate.

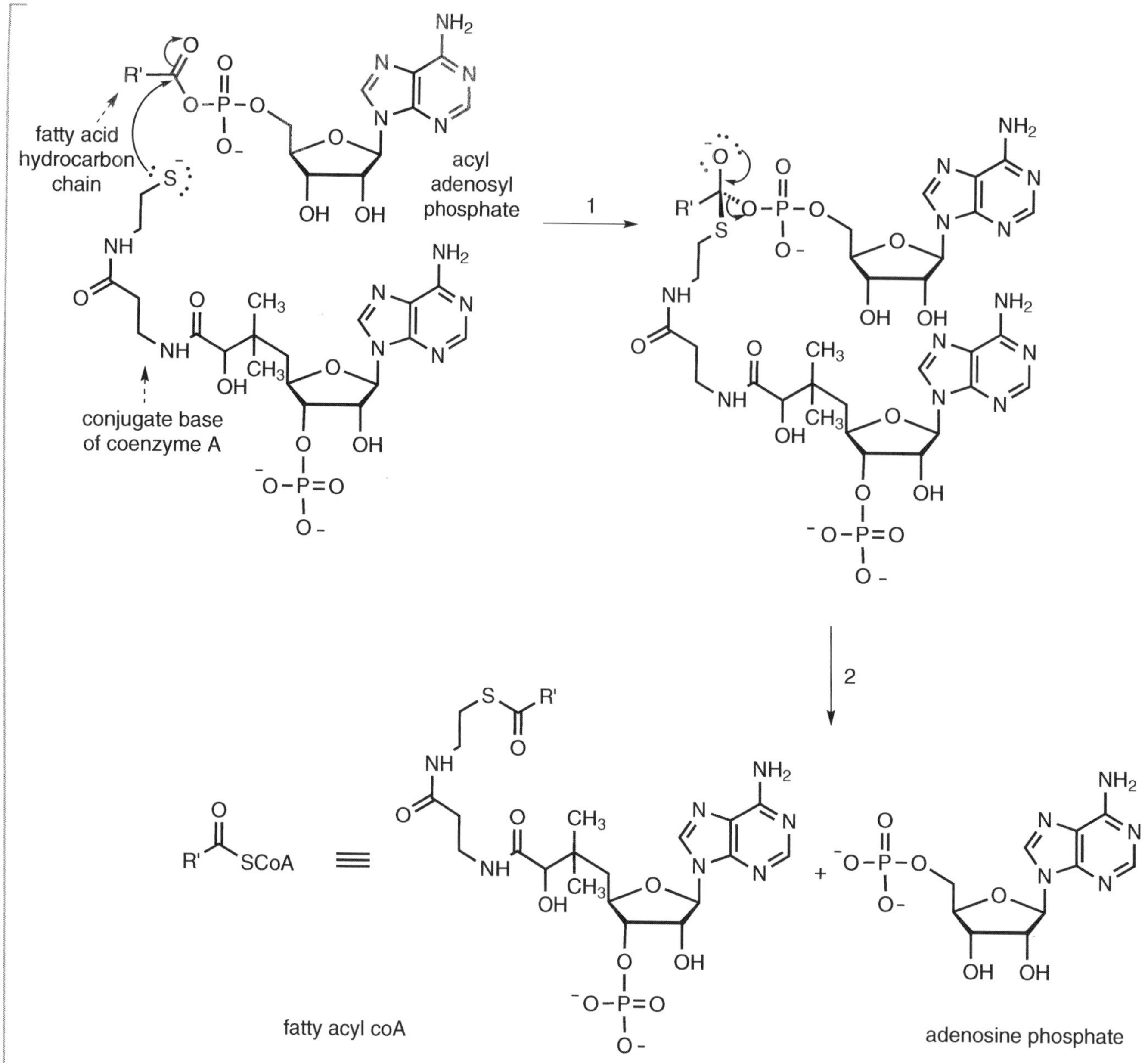

▲ FIGURE 7.17

Mechanism of the Acyl Adenosyl Phosphate Conversion to the Coenzyme A Thioester of the Fatty Acid

carbon following nucleophilic attack is changed from tricoordinate to tetracoordinate (section 7.4), so nucleophilic attack at a phosphorus atom in ATP also expands the coordination around the phosphorus atom, but for phosphorus, from tetracoordinate to pentacoordinate (Figure 7.16). And parallel to the situation at carbon, where the double bonded tricoordinate state and the tetracoordinate states can be interconverted, so can the tetracoordinate and pentacoordinate states of phosphorus be interconverted, which is precisely what happens in step **2** in Figure 7.16.

In a further parallel between the chemistry of carbonyl and phosphate, just as in the re-formation of the carbonyl group after nucleophilic attack at the carbonyl carbon atom requires ejection of one of the bound groups with the two electrons that connect it, this action is necessary in the pentacoordinate intermediate product of step **1** in Figure 7.16. What's possible for this ejection step?

Ejection of the oxygen singly bound to the CH_2 group of the adenosine moiety would produce the conjugate base $-O^-$ of an acid, $-OH$ with a pK_a in the range of 15-20. Not a great leaving group. The ejection of the carboxylate anion that formed the pentacoordinate intermediate in the first place, would be more reasonable. The pK_a of a carboxylic acid is near 5, a far stronger acid than an alcohol. But most reasonable and the group that is ejected in every event in this biochemical reaction, is the diphosphate group, as shown in Figure 7.16. This group, $P_2O_7^{-4}$ is the conjugate base of the very strong acid, $H_4P_2O_7$ with a large negative pK_a.

The fatty acid product produced in step **2** of Figure 7.16, acyl adenosyl phosphate, and the subsequent reaction with the conjugate base of coenzyme A are shown in Figure 7.17 demonstrating that the acyl adenosyl phosphate has all the properties to accomplish the reaction shown in Figure 7.15 to produce the desired coenzyme A derived fatty thioester.

Removal of a proton from the terminal sulfur atom of coenzyme A produces the negatively charged nucleophilic sulfur shown in **Figure 7.17**, which on attacking the carbonyl carbon of acyl adenosyl phosphate, produces the tetracoordinate intermediate shown as the product of step **1** (Figure 7.17). The return to the carbonyl group in step **2** (Figure 7.17) ejects the conjugate base of the stronger acid, also a derivative of phosphoric acid, adenosyl phosphate to produce the desired thio ester with coenzyme A.

Let's now look ahead to what happens next. Long before anything was known about the necessity of a thioester in the breakdown of fatty acids and before coenzyme A was discovered, a German chemist, **Franz Knoop**, carried out an experiment in 1904, which is shown in **Figure 7.18**.

■ **Franz Knoop**

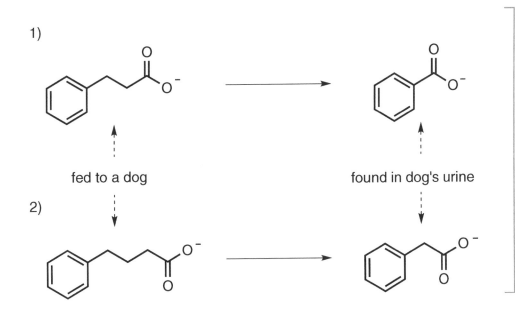

1)

fed to a dog

found in dog's urine

2)

The experiment is considered the first example of exploring the mechanism of a biochemical process by labeling. We've discovered the value of isotopes of carbon and hydrogen in the understanding of the mechanism for synthesis of the terpenes and steroids (section 5.8) and how this had to wait until after the Second World war, until after isotopes became available from the research that led to the atomic bomb. Knoop's work was no less a labeling experiment, although not with an isotope but rather by modification of a molecule that he fed to dogs.

As seen in Figure 7.18 when the fed molecule had an odd number of carbon atoms in the chain (two CH_2 groups and a CO_2^-) the product was entirely different than when the fed molecule had an even number of carbon atoms in the chain (three CH_2 groups and a CO_2^-). Knoop interpreted this to mean that the fed molecule was broken into two carbon pieces without the ability to break the bond between the benzene ring and the attached CH_2 group. This was taken to mean that fatty acids were catabolized also into two carbon entities, which we now know to be correct with the two carbon entities being acetyl CoA, the thioester of acetic acid and acetyl coenzyme A (Figure 5.3).

In the next two sections we'll follow the catabolism of the fatty acyl coenzyme A into two carbon moieties. This in vivo process is rich in demonstrating important principles of the science of organic chemistry.

PROBLEM 7.30

What does the far lower pK$_a$ of H$_4$P$_2$O$_7$ compared to ethyl mercaptan (CH$_3$CH$_2$SH) have to do with the reaction sequence in Figure 7.17?

PROBLEM 7.31

There are many derivatives of H$_4$P$_2$O7 in all of which the -OH group of the carboxylic acid is replaced, as for example: by -SR in a thioester; or by -NHR in an amide. Draw the structures of derivatives of carboxylic acids in which the –OH group has been replaced by: –O-(C=O)-R; –Cl; OPO$_3$$^{-2}$; -OCH$_3$; -OC$_6H_5$. Judge the relative reactivity of each of these carboxylic acid derivatives after the formation of a tetracoordinate intermediate from nucleophilic attack at the carbonyl carbon of the derivative by hydroxide ion, HO$^-$.

PROBLEM 7.32

Answer problem 7.30 for Figure 7.16 by comparing the pK$_a$ of acetic acid to phosphoric acid.

PROBLEM 7.33

If the conjugate base of coenzyme A, which attacks the carbonyl carbon in step **1** in Figure 7.17 had instead attacked the phosphorus atom producing a pentacoordinate intermediate at phosphorus instead of the tetracoordinate intermediate at carbon as shown, what could have been the product of this path. Use leaving group ideas to evaluate the breakdown of the compared intermediates proposed in this question.

PROBLEM 7.34

How did the experimental results in Figure 7.18 lead to the conclusion drawn about the nature of the catabolism of fatty acids?

7.7

Breaking a fatty acid down into two carbon pieces first requires introducing a double bond using an oxidizing coenzyme.

GIVEN A LONG HYDROCARBON CHAIN terminated by a thioester functional group, fatty acyl CoA, the end product of the chemical events in Figure 7.17, how does the biochemical process break this molecule down in an orderly manner? We've learned that the power of functional groups is their specific reactivity (section 3.9). Long chains of CH$_2$ groups, as in all saturated fatty acids (Figure 7.1) lack the specific reactivity necessary in the catabolic process. There is no chemical principle allowing any mechanism, biological or otherwise, to specifically break off two carbon moieties from the long hydrocarbon chain of the fatty acid. The many carbon-carbon bonds in the hydrocarbon chain are too similar and even if a bond breaking method were available, the bonds would break in a random manner. In fact, random breaking of carbon-carbon bonds is what happens in the industrial process that breaks petroleum derived hydrocarbons into smaller molecules, steam cracking, a subject we'll look into in Chapter 9.

What is needed for cleaving a two carbon atom moiety from the hydrocarbon fatty chain is a functional group within the chain to direct the reactive properties in a specific manner. The carbonyl group of the thioester bond plays this role. Let's see how that works.

The electron attracting power of the carbonyl functional group drives reasonable loss of protons from an adjacent C-H bond to a variety of bases and this is shown happening to the fatty acyl CoA in Figure 7.19. In this manner, the two electrons that made up the C-H bond are left behind, leaving the carbon atom with a negative charge, a **carbanion,** which is stabilized by interaction with the carbonyl group.

As shown in **Figure 7.19**, localizing the two electrons on the carbon adjacent to the carbonyl group, the α-carbon, does not truly represent the structure produced on loss of a proton. The negative charge is delocalized, which is shown by the necessity of drawing a resonance structure. It is this delocalization that is responsible for lowering the pK$_a$ to an

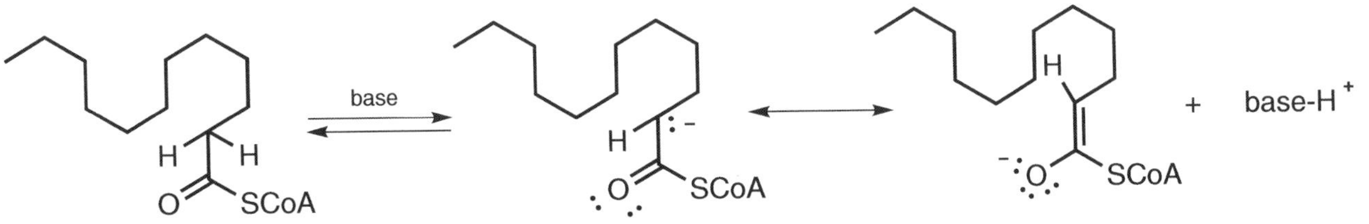

▲ **FIGURE 7.19**

Carbanion Stabilization on Carbon Adjacent to Carbonyl Via Resonance – The First Step in Enzyme Catalyzed Catabolism of Fatty Acids

extent so that a base is able to abstract a proton from the carbon atom α to the carbonyl carbon. In the absence of this adjacent carbonyl function the loss of a carbon- bound hydrogen atom as a proton would be virtually impossible, with a pK_a in the range of 50 instead of nearer to 20. As discussed in section 4.9, this circumstance means that the adjacent carbonyl group changes the equilibrium for proton loss in the range of 30 orders of magnitude, 10^{30}.

The loss of proton shown in Figure 7.19, in combination with another chemical step involving a coenzyme, discussed below, completes the intended chemical transformation to a fatty acid with a double bond adjacent to the thioester group, which is just the right functional group necessary for the first step on the path to catabolism of the fatty acid.

This chapter has introduced two coenzymes and showed how they are used for conversion of the fatty acid to the thioester (section 7.6). Coenzymes, which are also sometimes called cofactors, are molecules with particular reactive properties, as we've seen in both coenzyme A and in adenosine triphosphate. Coenzymes work in concert with enzymes but are not permanently bound to an enzyme. In that way, an enzyme can be specifically designed to work on some particular chemical reaction while the coenzyme has a characteristic that is valuable over a range of enzyme catalyzed reactions

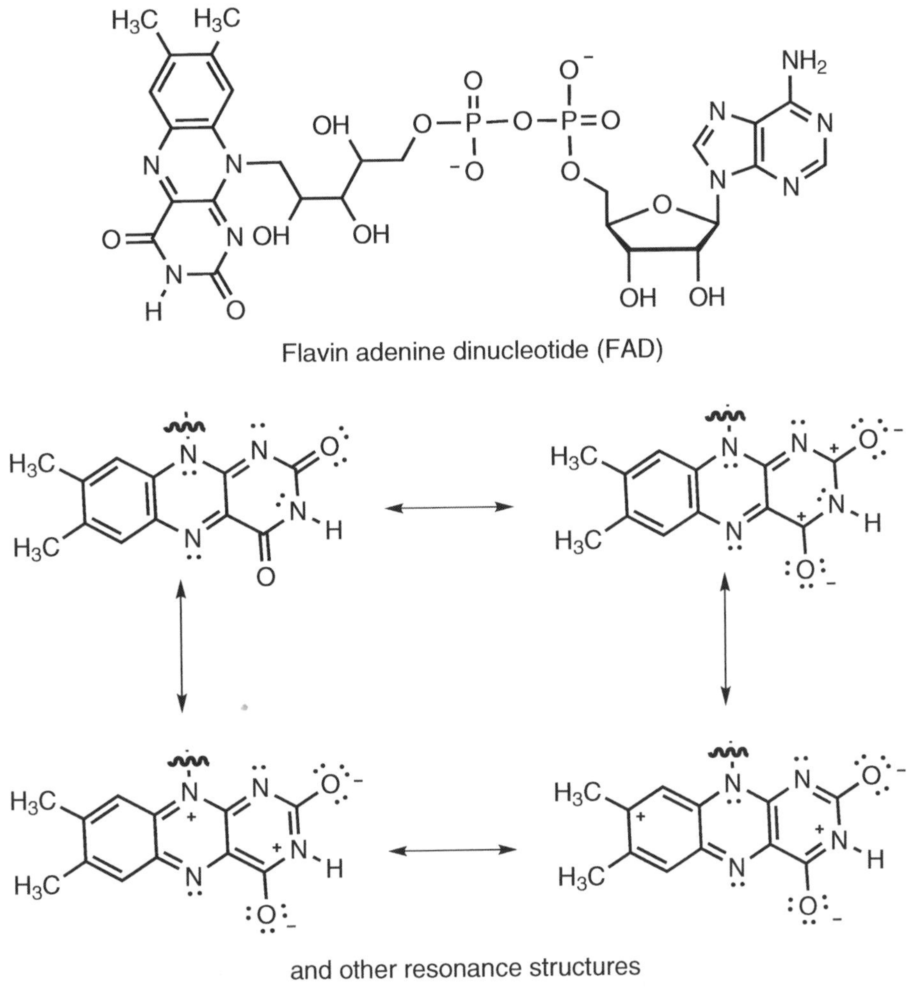

Flavin adenine dinucleotide (FAD)

◄ **FIGURE 7.20**

Aromatic ring current in flavin adenine dinucleotide (FAD) requires polarization of both carbonyl groups on one of the fused rings.

and other resonance structures

– a tool that does a certain kind of job and can be applied to a variety of tasks, that is, can work with a variety of enzymes - just as one tool, for example, a screw driver, can be useful to electrical, or wood working or plumbing jobs among others. Here's something interesting I found on a Wikipedia site on the web about coenzymes (cofactors).

Organic cofactors are often vitamins or are made from vitamins. Many contain the nucleotide adenosine monophosphate (AMP) as part of their structures, such as ATP, coenzyme A, FAD, and NAD+. This common structure may reflect a common evolutionary origin as part of ribozymes in an ancient RNA world. It has been suggested that the AMP part of the molecule can be considered a kind a "handle" by which the enzyme can "grasp" the coenzyme to switch it between different catalytic centers.

Our focus at this point is on another coenzyme, flavin adenine dinucleotide, **FAD**, the structure of which is shown in **Figure 7.20**.

By taking account of the appropriate lone pair electrons on nitrogen, FAD has 14 π electrons in orbitals perpendicular to the plane of the fused ring structure, which fits Hückel's 4n+2 rule, with n=3, (section 6.8). These 14 π electrons could form an aromatic ring current around the periphery of the fused three ring structure except for the orbitals on the carbon atoms of the two carbonyl groups on one of the fused rings. These p orbitals that must be involved in the aromatic ring current of the 14 π electrons are occupied with the π-bonds making up the carbonyl groups, a log jam, so-to-speak.

If the electron densities of the p orbitals on the two carbon atoms of the two carbonyl groups are transferred to the two oxygen atoms, the "log jam" would be broken as represented in three of the four resonance structures for FAD in Figure 7.20. The carbonyl groups in these three resonance structures are represented in a dipolar manner. The fourteen π electrons are able to move, in a manner of speaking, through available orbitals along the entire periphery of the three fused rings. The molecule exhibits an aromatic ring current just as in benzene (section 6.8).

The necessity for this charge displacement shown in the three out of four resonance structures in Figure 7.20 as a prerequisite for aromaticity, in addition to other factors, reduces the aromatic stabilization energy of FAD far below the stabilization energy of benzene (section 6.7). Benzene resists any chemical change that interferes with its aromatic character to a far greater extent than does FAD. This is nature's intent. It is this capacity for change in FAD, change between an oxidized and reduced state, **FADH$_2$**, which makes this molecule a valuable coenzyme as seen in **Figure 7.21**.

A bit further on in our study of the catabolism of fats we'll discover that another coenzyme, nicotinamide adenine dinucleotide, **NADH**, has a parallel characteristic,

FIGURE 7.21 ▶

FAD acts as an oxidizing agent by producing FADH$_2$.

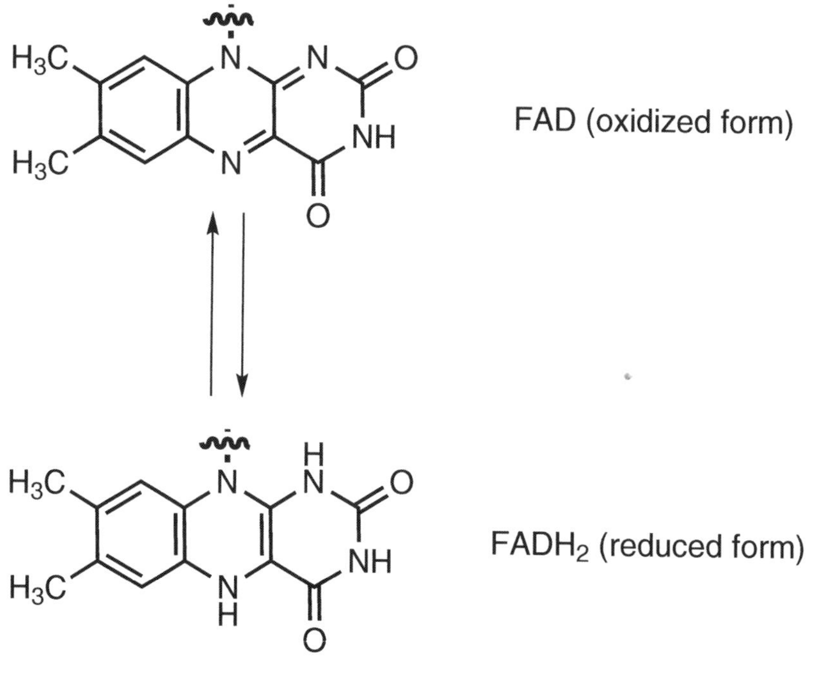

FAD (oxidized form)

FADH$_2$ (reduced form)

but in that situation (NAD⁺), the aromatic stabilization is reduced by the necessity for a positive charge. There follows, a capacity for change in both NAD^+ and FAD to accomplish a necessary coenzyme task.

Coenzymes are valuable when two states, such as an oxidized or reduced (section 3.8) form of the coenzyme are well balanced so that both are accessible. Benzene would make a poor coenzyme because it is powerfully resistant to change (Chapter 6). In FAD the two accessible states are, in fact, a reduced and an oxidized form. FAD, shown in Figures 7.21 is the oxidized state. The reduced state, $FADH_2$, differs from the oxidized state, FAD, from an electron accounting point of view, in the equivalent of the addition of two electrons and two protons, the equivalent of H_2.

$FADH_2$ also differs from FAD in that 16 rather than 14 π electrons can be counted around the periphery of the ring available for a ring current. But 16 electrons do not fit Hückel's 4n+2 rule so that there is no driving force to polarize the two carbonyl groups as was the situation described for FAD (Figure 7.20) and discussed just above. In this manner $FADH_2$ loses

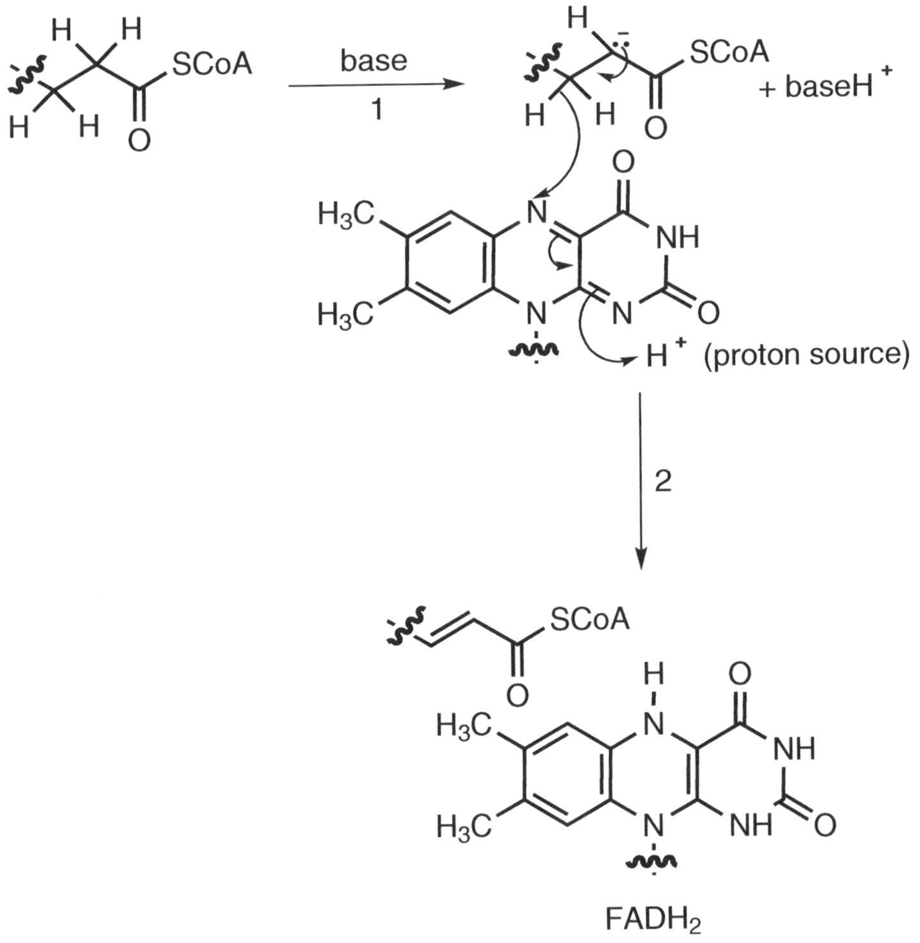

◀ **FIGURE 7.22**

Mechanism of the Conversion of the Carbanion Intermediate (Figure 7.19) to the Trans Double Bond Via Conversion of FAD to FADH₂

aromatic character over the entire three ring structure, but gains an uncharged state. These structural changes are a reasonable basis for the balance and therefore the ease of reversibility between FAD and $FADH_2$. FAD is driven to $FADH_2$ by the weak carbonyl π bond in FAD while $FADH_2$ is driven to FAD to recover the aromatic character in FAD.

The overall path from the coenzyme A derived thioester of the fatty acid all the way to acetyl CoA was an insight of Feodor Lynen to whom we were introduced in section 5.8 where we discussed Konrad Bloch with whom Lynen shared a Nobel Prize. Two chemists of nearly the same age stood on the stage in Stockholm together, one who was saved from being forced to join the Nazis before the Second World War by his anti-fascist views and a ski accident and the other who avoided death in the Holocaust by escaping from Germany at almost the same time.

Let's begin with one possible mechanism for producing the change in the fatty acyl CoA necessary for the initiation of the catabolic process. In this first step of the catabolism of fatty acids shown in **Figure 7.22** (step **1**) a proton is abstracted from the carbon atom α to the carbonyl group (Figure 7.19) forming a carbanion at this site. In step **2**, FAD can be seen at work in the acceptance of a hydride ion, H:⁻, from the β carbon of the fatty acid, with FAD converted to FADH₂ by addition also of a proton as shown in Figure 7.22 (step **2**).

The hydride ion has not been seen since studying the catalytic cracking of petroleum fractions (section 4.6) but in the situation we are focused on here, no carbocation is involved. Rather, the carbanion at the α-carbon supplies the two electrons for the π bond of the double bond. However, the double bond can only form if the β- carbon releases to FAD one of its two hydrogen atoms with the two electrons binding this hydrogen atom (Figure 7.22, step **2**). An unsaturated fatty acid with the double bond between α and β carbon atoms is produced.

Adding a negatively charged hydride ion to FAD would produce FADH⁻ with a negative charge, which is avoided by addition of a proton to produce FADH₂ (Figure 7.22, step **2**). FAD has been reduced by adding the elements of H₂ while the fatty acid has been oxidized by losing the elements of H₂ (Section 3.8).

PROBLEM 7.35

In the book up to this point there are many figures. Take the time to look at some of these figures and identify parts of molecules that fit the description of a functional group (section 3.9) and compare this list with the functional groups listed at the front of the book and also to your answer to problem 3.28.

PROBLEM 7.36

Resonance structures can be drawn for any molecule; H-H could be represented as H⁺ H⁻ However resonance representations are only reasonable contributors to the structure of a molecule when they themselves are reasonable structures. Demonstrate this principle by showing examples of resonance structures of the coenzymes discussed in this chapter that are and are not reasonable. What role does the octet rule and formal charge play in your answer?

PROBLEM 7.37

Account for all nonbonding electrons in the structure of FAD shown in Figure 7.20. Justify all formal charges and also determine the source of the 14 π electrons contributing to the aromatic stabilization of the three-fused-ring structure (the flavin). Do the same for the 10 π electrons contributing to the aromatic stabilization of the adenine two-ring-fused structure.

PROBLEM 7.38

(a) Why is it necessary for the π electrons of the two carbonyl groups in the structure of FAD to be transferred to the oxygen atoms of the carbonyl groups to allow aromatic stabilization?
(b) Is there any aromatic character in the structure of FADH₂? If so describe the orbital geometries of the electrons contributing to this aromaticity.

PROBLEM 7.39

In Figure 7.21 the oxidized and reduced states are FAD and FADH₂, respectively. Reduction, in this situation, therefore involves the net addition of H₂ and oxidation the removal of H₂. Look ahead and compare this hydrogen accounting to the oxidation/ reduction chemistry of alcohols and carbonyl compounds (Figure 7.25). Look back to compare the hydrogen accounting method to the oxidation/reduction chemistry of saturated and unsaturated hydrocarbons as in section 6.7.

PROBLEM 7.40

Objection was raised in the discussion around Figure 7.7 to the hydride ion, H⁻, as a leaving group, it being formally the conjugate base of a weak acid, H-H (pK_a about 35). However we have now seen hydride ions involved in carbocation chemistry in Figure 4.8 and now in Figure 7.22. What do you think about this apparent conflict?

PROBLEM 7.41
What C-H bond do the π electrons come from in the formation of the double bond (Figure 7.22) from the saturated fatty acyl coenzyme A?

PROBLEM 7.42
Identify all enantiotopic choices made in the formation of the α-β double bond from the saturated fatty acyl coenzyme A in Figure 7.22. There are four.

W E'VE SEEN THE REACTIVE CHARACTERISTICS of carbon-carbon double bonds arising from the weak π bond in the formation of the high octane gasoline, 2,2,4-trimethyl pentane (Figure 4.13), and in the formation of terpenes and lanosterol (Figures 5.12 and 5.18). If carbon-carbon double bonds are faced with electrophiles, such as carbocations or protons, when encapsulated by enzymes, or in highly acidic industrial environments, or in undergraduate organic chemistry laboratory demonstrations of Markovnikov's rule (section 5.10), it makes no difference. Their reactivity properties are identical in acting as nucleophiles when the π electrons respond to positively charged entities.

However, we see the π electrons of the double bond formed in Figure 7.22 in another role with no carbocations involved but a reactivity which also arises from the weakness of the π bond.

Step **1** in **Figure 7.23** is called a conjugate addition in contrast to the direction of addition of electrophiles directly to a carbonyl group. The driving force for the breaking of the π bond by nucleophilic addition as shown in Figure 7.23 is the flow of electrons shown in step **1** allowing a negatively charged carbon to arise adjacent to the carbonyl group, which is stabilized by delocalization, the same source of stabilization as shown in Figure 7.19.

The word conjugated has many meanings. In organic chemistry conjugated refers specifically to double bonds separated by single bonds. The two double bonds in 1,3 butadiene, $H_2C=CH-CH=CH_2$ are conjugated. An α, β unsaturated carbonyl compound, the fatty acid product of step **2** of Figure 7.22 has a conjugated carbon-carbon double bond which, in this molecule, is conjugated with a carbon-oxygen double bond.

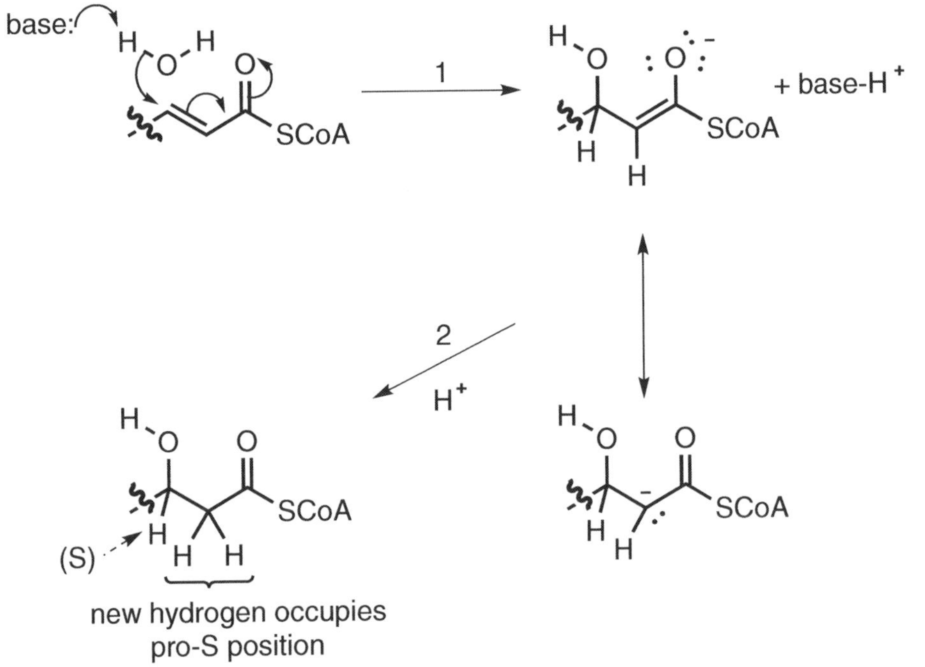

7.8

The Next Step in the Catabolism of the Fatty Acid: Conjugate Addition to a Double Bond

◀ **FIGURE 7.23**

Mechanism of the Addition of Water to the Double Bond Formed in Figure 7.22

Conjugate addition, which is the second step in the catabolic process that breaks down fatty acids and is the chemical reaction seen in Figure 7.23 (step **1**) is related to the Michael addition, after Arthur Michael who discovered conjugate addition of carbon nucleophiles in the 1880s. In the Michael reaction, which is widely used, as we'll discover in the synthesis of cholesterol studied in Chapter 12, (sections 12.4 and 12.5) the group adding is a carbon nucleophile although the conjugate nature of the addition is the same.

Arthur Michael was a distinguished organic chemist whose work overlapped the 19th and 20th centuries. Although he never received a college degree of any kind he held professor's positions in chemistry at both Tufts and Harvard Universities at different times in his life. He died in 1942 at the age of 89. It wasn't that Arthur Michael couldn't get into a college. He was, in fact, accepted at Harvard University as a young man. But illness stopped his attendance. Michael's New York State family was wealthy enough to be able to take an extended trip to Europe where young Michael recovered his health and was also able to enter the laboratories of famous chemists. One of these laboratories was headed by Wurtz, the mentor of Kekulé who we've heard so much about (sections 1.11 and 6.4).

▪ Arthur Michael

It was a smaller scientific world in those days and Arthur Michael eventually returned to the United States to a career in chemistry, (remember James Crafts who followed this path (section 6.10)) where his wealth allowed him to accumulate a distinguished collection of art and allowed him to work at home at certain times. It was not essential that he be paid. Arthur Michael's independent wealth, although unusual for a scientist these days, was more common for those interested in science in the 18th and 19th centuries when paid positions for scientists were far less common. Scientific work, I guess, was considered too much fun with an often uncertain practical, that is, money- earning importance. Well, that's interesting, but let's return to our focus on the chemistry.

It is hardly surprising that the β-hydroxyl substituted fatty thioester produced in step **2** (Figure 7.23) is of a single configuration, (S), and as well that the proton added to the carbanion produced in step **1** of this Figure occupies only the pro-S site (section 5.6). The stereospecificity seen here is another example of the connection between enzyme catalysis and stereochemical choice, which we've seen continuously and will continue to see throughout our studies of biochemical reactions. It's interesting to realize that the next step in the catabolic process, to be discussed in the next section, is oxidation of the hydroxyl group (removal of the elements of H_2) to a ketone, therefore erasing the stereochemical choice made in the addition of the HO⁻ group in step **1** (Figure 7.23).

Some might ask why the evolutionary process "bothers" to choose one configuration for the hydroxyl group or, for that matter, the pro-chiral site for the incoming proton considering that neither stereochemical choice will be "remembered" in the reactions to follow, as we'll see. Let's see what a great chemist had to say about this.

In 1975 the Nobel Prize was awarded to two chemists who carried out stereochemical investigations, one Vladimir Prelog, whom we've met around the discussion of the development of the (R) and (S) rules (section 1.12) and another, **John W. Cornforth**, who is worth quoting about the connection between enzymes and stereochemistry.

▪ John W. Cornforth

Cornforth, who was born in Australia, won half the prize in 1975 for his work on the stereochemistry of the biochemical path from the terpenes to lanosterol and cholesterol (sections 5.8-5.10). As did Lynen and Bloch, his experimental approach used the isotopes available after World War II. Cornforth, who accomplished so much in spite of being burdened by being deaf from his early teenage years, discovered many reactions in which a stereochemical choice was made in an enzyme catalyzed reaction where a subsequent reaction erased the choice, which as we shall see is the fate of the β hydroxyl compound produced in Figure 7.23. Cornforth realized from his work and that of others including, most notably Alexander Ogston, that the following must be true: "*We are thus forced to conclude that stereospecificity is inherent in the catalytic action of enzymes, and that enzymes would not be equally efficient as catalysts of non-stereospecific reactions.*"

In other words, for the enzyme to be an excellent catalyst, it must be geometrically well fit to the substrate as we've seen in the precise placements of protons in the lipase catalyzed hydrolysis of the ester bonds of triglycerides (section 7.5). A "byproduct" of that fit occurs when chiral choices are involved. Only one, (R) or (S), or pro-(R) or pro-(S), will be chosen as we are seeing in one of many examples in Figure 7.23, a choice that will be made even if that choice is erased in the chemical reactions that follow.

Cornforth's Nobel lecture is a treasure, which can be understood, in many essential ways, by a student of organic chemistry at the level of this book. I recommend it: http://nobelprize.org/nobel_prizes/chemistry/laureates/1975/cornforth-lecture.pdf

We'll come across Ogston, whom Cornforth mentions early in his Nobel lecture, later when we study the citric acid cycle in the next chapter, and discover that nature distinguishes between what could be called old and new, using stereochemistry in a way that causes philosophical musings about the meaning of life (section 8.8).

PROBLEM 7.43
Stabilization of negative charge at carbon by an adjacent carbonyl plays two essential roles in the transformation from fatty acyl CoA to β-hydroxyl fatty acyl CoA. What are they?

PROBLEM 7.44
Go over the biochemical processes in Chapter 5 and in this chapter up to this point and identify all reactions that involve stereochemical choices. Are any of these choices erased in reactions that immediately follow?

THE STEP THAT TAKES The hydroxyl-functionalized molecule product of step **2** in Figure 7.23 to the derived ketone requires an oxidation (section 3.8) and, therefore, an oxidizing agent. Nicotinamide adenine dinucleotide, NAD$^+$, which we have seen in Figure 6.10 and the structure of which is shown again in **Figure 7.24** is the necessary oxidizing agent. Also shown in Figure 7.24 is the structure of the reduced form of NAD$^+$, NADH, which is produced when NAD$^+$ acts as an oxidizing agent, as will be seen shortly.

7.9
Oxidation of β-Hydroxyl Fatty Acyl Coenzyme A Using an Enzyme and an Oxidizing Coenzyme

Nicotinamide, nicotine and nicotinic acid, which is also known as vitamin B-3 or niacin, all contain a pyridine ring, a six membered ring with five carbon atoms and one nitrogen atom, which contains 6 π electrons in orbitals perpendicular to the plane of the ring, an aromatic characteristic fitting the 4n+2 rule for n = 1 (section 6.8). Rings containing carbon and other atoms, especially nitrogen and oxygen, are called heterocycles.

Let's look more closely at NAD$^+$ and NADH in Figure 7.24. The unsaturated rings at the ends of the structure are adenine and nicotinamide. While important changes occur in the nicotinamide ring in the oxidized and reduced state of the coenzyme, the adenine ring remains unchanged, it simply being the handle (section 7.7). Both rings in the oxidized state of the coenzyme, NAD$^+$, are aromatic with 6 π electrons in the nicotinamide ring and 10 π electrons in adenine, both fitting the 4n+2 rule (section 6.9).

However in the reduced state of the coenzyme, NADH, the nicotinamide ring has added a hydrogen atom, with an extra electron (or from an accounting view, a proton with two electrons) allowing bonding this hydrogen to the para carbon of the ring. This result means that the equivalent of a hydride ion, H:$^-$, has been added to the nicotinamide ring. The consequence of this addition is that the ring is no longer aromatic while, at the same time, the ring is no longer positively charged at the nitrogen atom (Figure 7.24).

The exchange of aromatic character in NAD$^+$, which requires a positive charge on nitrogen, for the loss of the aromaticity in NADH, which is a ring without charge, can be seen as the source of the facile interconversion, that is, balance, between the nicotinamide adenine dinucleotide, NAD⁺, oxidized form

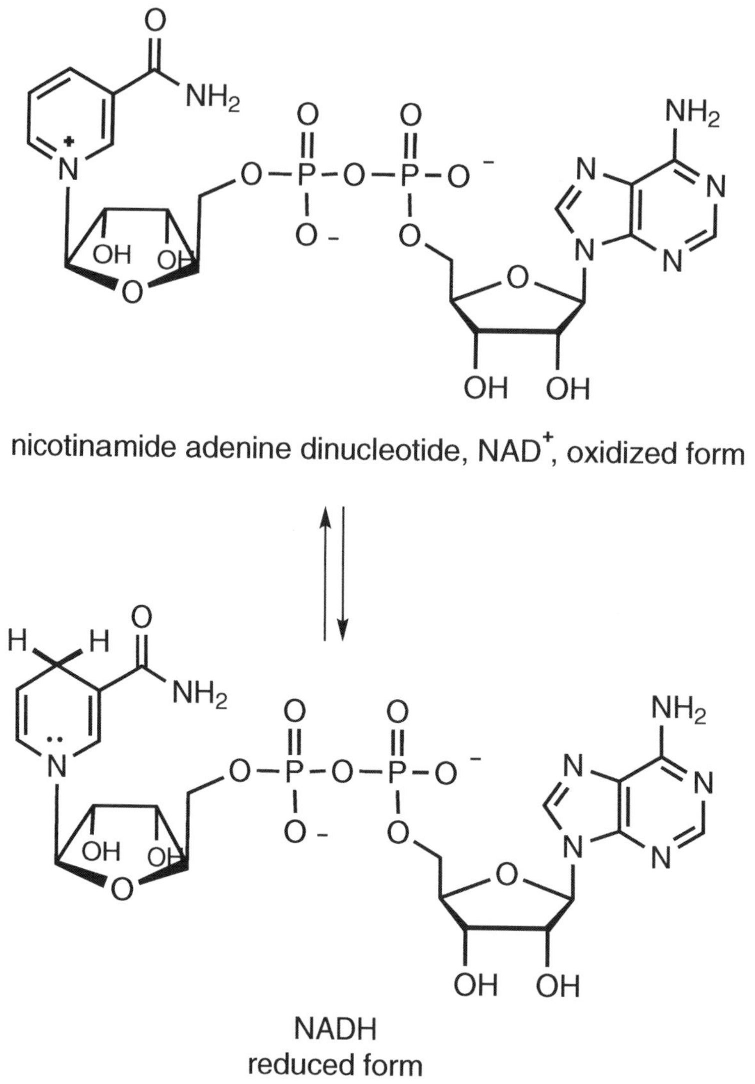

NADH
reduced form

oxidized and reduced states of this coenzyme and finds a parallel to the reason given earlier for the balance between the oxidized and reduced states of flavin adenine dinucleotide (section 7.7, Figure 7.21).

Nature uses the same "trick" in NAD⁺/NADH and FAD/FADH$_2$ to accomplish the necessity that these oxidizing/reducing coenzymes are balanced in their stability and therefore, easily interconvertible. FADH$_2$ will not "resist" the role of acting as a reducing agent in biochemical reactions and being converted to FAD because of the gain in aromatic character (Figure 7.21). Similarly, NADH will not resist acting as a reducing agent in biochemical reactions and therefore being converted to NAD⁺ because of the gain of aromatic character (Figure 7.24).

Equally, these two coenzyme systems in other biochemical reactions, can act as oxidizing agents converting FAD to FADH$_2$ and NAD⁺ to NADH. In the former, loss of the aromatic character means loss of formal charge and in the latter loss of aromatic character means loss of the necessity for the high energy dipolar orbital arrangement (section 7.7).

Figure 7.25 shows the mechanism for the enzyme catalyzed oxidation to the derived ketone of the β hydroxyl group, which was introduced to the fatty acyl coenzyme A (Figure 7.23). The enzyme catalyzing this oxidation is a dehydrogenase, named for the two hydrogen atoms that must be removed (the equivalent of H$_2$ and therefore an oxidation) to form the ketone. The coenzyme carrying out the oxidation is NAD⁺.

In Figures 7.8-7.11 we were introduced to pancreatic lipase, which carries out the hydrolysis of the ester bonds linking the fatty acid to glycerol. In the lipase, no coenzyme was involved so that all the participants in the hydrolysis were functional groups on the side chains of three amino acids, aspartate, histidine and serine, which are all part of the protein chain. In the mechanism of the dehydrogenase we focus on in **Figure 7.25**, there are also three participants in the reaction, but only two are amino acids that are part of the protein chain, glutamate and histidine. The third participant is the coenzyme NAD^+, which is not covalently bound to the protein, but fits nevertheless, as does the β-hydroxyl fatty acyl coenzyme A, into the active site of the enzyme.

As we realized from the distant spacing of the three amino acids along the contour of the enzyme protein chain in the lipase (Figures 7.8-7.11)), folding of the protein is necessary to bring the three active site amino acids into the close spatial relationship necessary to catalyze the hydrolysis. Such folding is necessary as well for

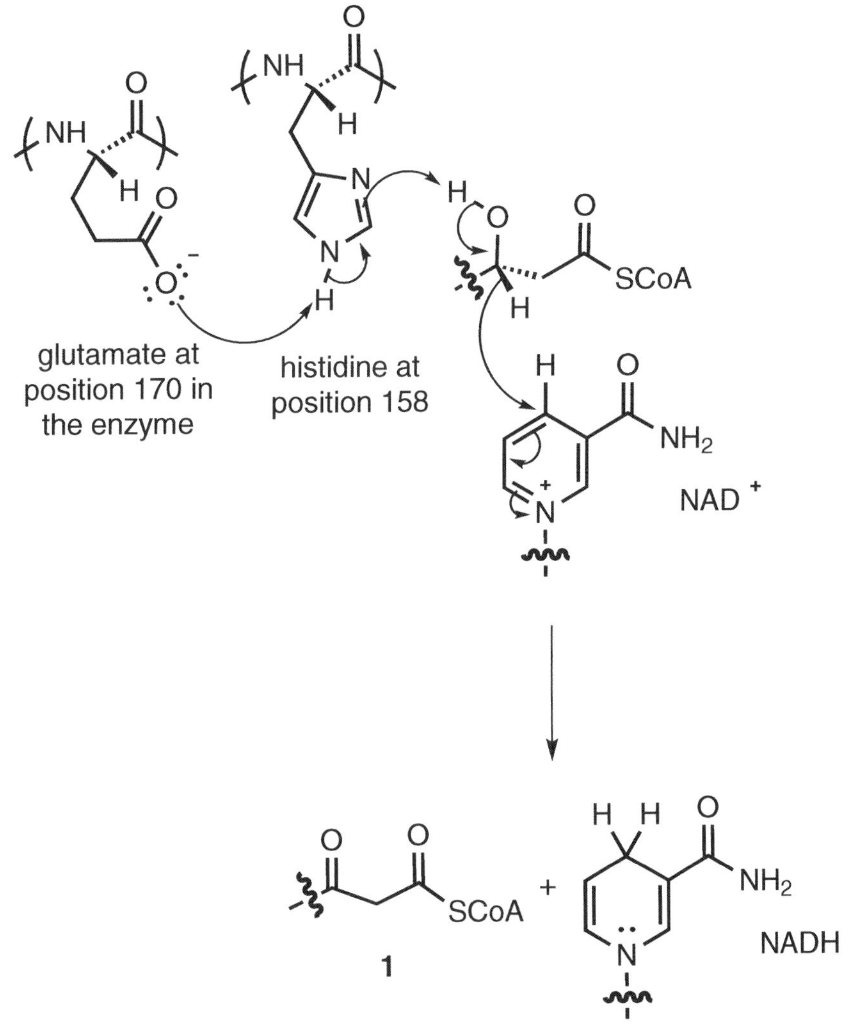

glutamate at position 170 in the enzyme

histidine at position 158

NAD^+

1

NADH

◄ FIGURE 7.25

Enzyme Catalyzed Mechanism for the Oxidation of β-Hydroxyl Group to Carbonyl on the Fatty Acyl Coenzyme A Via Conversion of NAD^+ to NADH

the dehydrogenase focused on here. The two amino acids participating in the reaction at the active site are located distantly along the contour of the chain from the nitrogen end, at position 170 for glutamate, and at position 158 for histidine.

In one of the many demonstrations of how nature uses the same methods again and again to accomplish very different tasks, the concerted steps shown in Figure 7.25 producing the β-keto fatty acyl CoA resemble those occurring in the first steps of the lipase catalyzed reaction.

In the lipase catalyzed hydrolysis, the carboxylate anion of an aspartate abstracts a proton from histidine. The histidine then abstracts a proton from the hydroxyl group of serine (Figure 7.10). In the dehydrogenase catalyzed oxidation, the carboxylate anion of a glutamate abstracts a proton from histidine. The histidine then abstracts a proton from the hydroxyl group located at the β-position of the fatty acid hydrocarbon chain (Figure 7.25).

The series of atomic and electron movements shown in Figure 7.25 can be expressed in acid/base terminology. The electrons available on oxygen in the Brønsted-Lowry base glutamate allow abstraction of a proton from histidine, which is acting as a Brønsted-Lowry acid. The electrons transferred within the histidine ring to the other nitrogen atom of the histidine then allow the histidine to act as a Brønsted-Lowry base to abstract the proton from the OH group on the fatty acid chain, which plays a Brønsted-Lowry acid role.

This is where the NAD⁺ enters to play its oxidizing role. A hydrogen atom with its two bonding electrons, H:⁻, a hydride ion, is transferred from the carbon bearing the hydroxyl group, which is called a carbinol carbon, to the para carbon atom of the nicotinamide ring. The electron pair on oxygen can now form the π bond using the newly empty orbital at the carbon atom from which the hydride ion was lost.

The two electrons associated with the hydride ion, which has added to the nicotinamide ring, displace the π electrons in the ring, which had formed the aromatic sextet of electrons in the nicotinamide ring, to form a lone pair of electrons on the nitrogen atom in the ring. This nitrogen atom therefore loses the positive charge as the ring simultaneously loses its aromatic character, the trade off discussed above as the basis of the reversible nature of the coenzyme (Figure 7.25).

The result of the curved arrows designating the electron motions and the movements of the atoms lead to the product acyl ketone **1** in Figure 7.25. The ketone carbonyl group, located β to the acyl carbonyl group as shown in the structure (**1**) is now set up for a reaction path that solves the problem of specifically breaking off a two carbon moiety from the fatty acid hydrocarbon chain. And this process is done while producing a new fatty acyl coenzyme A, a fatty thiol ester shorter by two carbons, which is poised to do the reaction paths shown in Figures 7.22, 7.23, 7.25 all over again. Let's see how that works in the next section.

PROBLEM 7.45

FAD and NAD⁺ shown in Figures 7.21 and 7.24 both act as oxidizing agents, which convert them to FADH₂ and NADH, respectively. Inorganic oxidizing agents carry out their function by electron transfer from the entity to be oxidized to the entity to be reduced, as in the reaction Fe + CuSO₄ = FeSO₄ + Cu, where the iron is oxidized and the copper is reduced. FAD oxidizes the fatty acid chain –CH₂-CH₂- to -CH=CH- as shown in Figure 7.22, while NAD⁺ oxidizes the hydroxyl group –C(OH)- to the carbonyl group –C(=O)- as shown in Figure 7.25. What parallels do you observe between the inorganic example given above and the two biological examples?

PROBLEM 7.46

Draw the structure of pyridine, which is an aromatic molecule (C₅H₅N) with a lone pair of electrons on nitrogen which does not contribute to the aromaticity of this heterocycle. The pyridine ring of NAD⁺ has no lone pair of electrons on nitrogen and is aromatic while the reduced form of this ring, NADH has a lone pair of electrons on nitrogen and is not aromatic. Justify these facts with electron counting and atomic orbital pictures.

PROBLEM 7.47

In Figure 7.25, the glutamate residue and the histidine residue are reactants in the active site. In the oxidizing reaction shown in this figure, are these residues oxidized or reduced or neither and how did you come to your answer?

PROBLEM 7.48

Is there any oxidation or reduction occurring in the reaction of an acid with a base, such as ammonia with hydrochloric acid or sodium hydroxide with acetic acid?

PROBLEM 7.49

(a) If the absolute configuration of the hydroxyl carbon atom in Figure 7.25 were (R) instead of (S), would the reaction still occur? (b) Now imagine the consequences of the following changes: (i) What would be the consequence if the chiral carbon atoms in the NAD⁺ were reversed? (ii) What would be the reactive consequence if the chirality of the NAD⁺ and the hydroxyl carbon atom were not changed but all the amino acids in the enzyme were reversed from (S) to (R)? (iii)

What would be the reactive consequence if every carbon atom in the enzyme, the coenzyme and the hydroxyl substrate were reversed in their absolute configurations? Offer reasons based on a structural consequence for your four conclusions Refer back to Chapter 1 (section 1.9).

PROBLEM 7.50

What aromatic characteristic is built into the structures of FAD/FADH$_2$ and NAD$^+$/NADH, which makes these coenzymes well suited as reversible oxidizing/reducing agents, which is a role benzene is not well suited for?

PROBLEM 7.51

What does the folding of the protein chain of the dehydrogenase enzyme catalyzing the oxidation of the β– hydroxyl acyl coenzyme A have to do with its catalytic function?

PROBLEM 7.52

Assign acid or base Brønsted-Lowry character to all participants in proton-transfer reactions for the dehydrogenase-catalyzed process in Figure 7.25 and for the lipase- catalyzed process in Figures 7.10 and 7.11.

W E'VE SEEN THE ELECTROPHILIC character of carbonyl functional groups in ketones, aldehydes and in all derivatives of carboxylic acids and learned that nucleophilic attack at the carbon atom of C=O is the signature characteristic of all carbonyl carbon. Nature takes this path, as seen in step **1** of **Figure 7.26**, as the opening step toward releasing two carbon atoms from the fatty acid chain, which we'll see in Figure 7.28. The nucleophile involved comes from a sulfur-containing-amino-acid along the enzyme chain, a cysteine unit at position 125 from the nitrogen end of the enzyme (see the discussion around Figure 7.8). The enzyme is appropriately named-ketoacyl-CoA thiolase.

The thiol functional group, -SH, on the side chain of the cysteine amino acids (Figure 7.26) in proteins must be ionized to its conjugate base –S$^-$ to act as a nucleophile for attack at the carbonyl carbon. The enzyme therefore supplies a base to abstract a proton from the cysteine SH group.

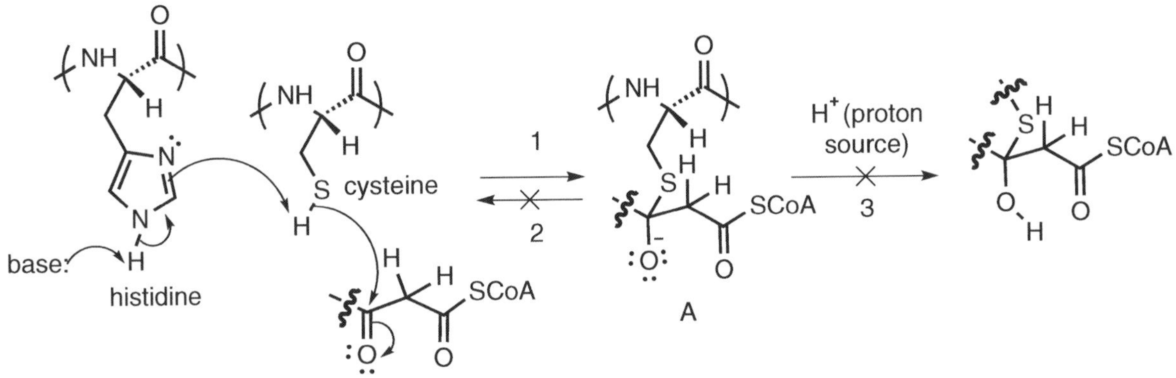

7.10

Cleaving a Two Carbon Fragment from the Fatty Acid Chain: the Retro-Claisen Reaction

As you might suspect from Figures 7.10, 7.11 and 7.25, it is a histidine unit called on to be that base, a histidine located along the protein backbone at position 375, the 375th amino acid from the nitrogen end of the chain, analogously to the enzyme described in Figure 7.8. And as we've been convinced in both the lipase and in the dehydrogenase, the two enzymes we've looked at earlier in this chapter, the histidine is waiting in the <u>exact</u> spatial relationship to the cysteine SH group to abstract the proton, as a consequence of the folding of the protein.

The result of the –S$^-$ nucleophilic attack at the carbonyl carbon atom shown in Figure 7.26 is the formation (step **1**) of the expected tetracoordinate intermediate, an

▲ FIGURE 7.26

Enzyme catalyzed nucleophilic addition of cysteine sulfur to the-β carbonyl group leads to a tetracoordinate intermediate.

intermediate that we've seen many times arising as a consequence of all nucleophilic attack at carbonyl groups.

The tetracoordinate intermediate (A) formed via step **1** in Figure 7.26 arose via nucleophilic addition to a ketone carbonyl group. In our study of saponification in sections 7.3 and 7.4 we discovered that the reactive consequences for tetracoordinate intermediates formed from ketone (or aldehyde) carbonyls are limited compared to the reactive consequences for tetracoordinate intermediates formed from acyl carbonyl. The difference between acyl carbonyl and carbonyl groups in ketones and aldehydes was understood as arising from leaving group possibilities, which was discussed in section 7.4 and exhibited in Figures 7.6 and 7.7.

If the rules arising from the discussion in section 7.4 were followed for the tetracoordinate intermediate shown in Figure 7.26, only two reactive possibilities would exist, reaction **2**, which reverses the nucleophilic addition, or reaction **3**, which simply adds a proton to the negatively charged oxygen atom formed as a consequence of the nucleophilic addition. However, the tetracoordinate intermediate formed in Figure 7.26, contrary to the rules for ketones, does not take one of the two paths described above (the cross X through these arrows). Although both are possible, the reverse of reaction **1** in Figure 7.26, reaction **2**, does not occur, nor does reaction **3** occur, which would produce a hydroxyl group.

Rather, the four-coordinate intermediate takes a path, which is not generally available to aldehydes and ketones as discussed in section 7.4 (Figures 7.6 and 7.7). The carbonyl of the thioester group, placed in a 1,3 relationship to the carbon of the tetracoordinate intermediate opens another possibility, a biochemical equivalent of the well known in vitro retro-Claisen reaction. Let's understand what this and the Claisen reaction are about, reactions associated with molecules in which two carbonyl groups are place in a 1,3 relationship, and then return to the fate of A in Figure 7.26

Figure 7.27 exhibits an example, (using methyl acetate) of a famous class of reactions in organic chemistry named after **Rainer Ludwig Claisen**, a famous German chemist who was active in the later 1800s and the early 1900s. Claisen was a student at the University of Bonn where he studied under Kekulé, who was professor at this University and whom we have heard so much about regarding the hypothesis of tetracoordinate carbon and the structure of benzene (sections 1.11 and 6.4) and who had a common mentor, Wurtz, with Michael (section 7.8). It was a small chemical world in the 1800s.

■ **Rainer Ludwig Claisen**

FIGURE 7.27 ▶

Examples of Claisen and Retro-Claisen Reactions

two choices for 2: return to
1 or retro-Claisen

The Claisen condensation in its many variations is one of the most important carbon-carbon bond forming reactions in organic chemistry and is widely used in the academic and industrial world for building complex structures from smaller molecules, especially in the pharmaceutical industry. We'll see more of this reaction in Chapter 8.

In Figure 7.27 in going from right to left (**4** reacts with **3** to yield **1**) we see how the Claisen reaction can make carbon-carbon bonds and we see how a resonance stabilized carbanion (**4**) attacks the carbonyl group of methyl acetate (**3**). The resulting tetracoordinate intermediate ejects a methoxide group to form a four-carbon backbone molecule (**1**) ultimately derived from two molecules, each with two carbon atoms.

Nature does use an equivalent of the Claisen reaction to make carbon-carbon σ bonds and in fact we will look closely in section 8.3 and following sections, at one example of this biological reaction in the formation of isopentenyl diphosphate, the intermediate responsible for the formation of terpenes and lanosterol (Chapter 5).

But here we are not making carbon-carbon bonds, but rather breaking them. Our interest here in Figure 7.27 is not in the reaction sequences from right to left, but rather from left to right, the reverse of the Claisen condensation, the retro-Claisen.

Now we'll discover the reason for the introduction of the β carbonyl group in fatty acid catabolism. A structural situation is created to set up a mechanism to break a two carbon moiety from the fatty acid chain, an in vivo parallel to the reverse of the reaction discovered by Claisen (Figure 7.27) so many years ago. However, let's see how this mechanism works in an enzymatically controlled biological system.

But before we move ahead into the enzymatically catalyzed path fated for the biological intermediate, **A** (Figure 7.26) let's say a word about Professor Claisen whom I so much appreciated when working in the lab, and who we can be reasonably certain would be astonished to know that he had uncovered what would one day be shown to be nature's way to both breakdown and build up carbon-carbon σ bonds.

Claisen, in addition to inventing chemical reactions, was a quite handy laboratory man and created a solution to a problem that plagues all those who work with distillable liquids– bumping. Many of the compounds Claisen worked with in developing the chemistry we are focused on here are liquids so that he had to use distillation as a key method of purification. Liquids are a problem.

As they come to boiling, as the temperature is raised in a distilling flask, the ideal observation, often unfortunately not happening, is gentle bubbles in the liquid and a slow rising of the vapor to be converted back to liquid in the condenser. What often happens instead is nothing, until to the experimenter's dread, a large portion, if not all of the distilling liquid, with all the impurities one had hoped to separate, rises suddenly and floods the condenser and the collection apparatus – bumping. Even as I write this, I remember my own dread as this occurred to my valuable liquid sample. This phenomenon can hopefully be interrupted in many ways, for example, putting glass wool into the distillation flask, but one of the best is to disturb the about-to-boil liquid with a stream of small bubbles.

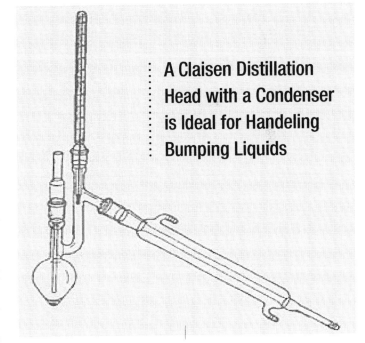

A Claisen Distillation Head with a Condenser is Ideal for Handeling Bumping Liquids

The picture shows Claisen's solution to the problem. In between the flask and the condenser is a distillation head to hold the thermometer and as well a capillary to allow a stream of air or whatever gas is desired into the distillation flask. The air, or often nitrogen, by attaching a balloon to the gas inlet, is drawn into the flask either because a vacuum is applied at the end of the condenser or simply because of the temperature difference between the condenser and the distilling flask.

Every laboratory of organic chemistry has a drawer with several Claisen distilling heads and many an organic chemist has praised Claisen as much for his distillation head as for his carbon-carbon bond forming reactions.

Suffice with the practical matters, let's return to our study of the carbon-carbon bond breaking step in the catabolism of fatty acids.

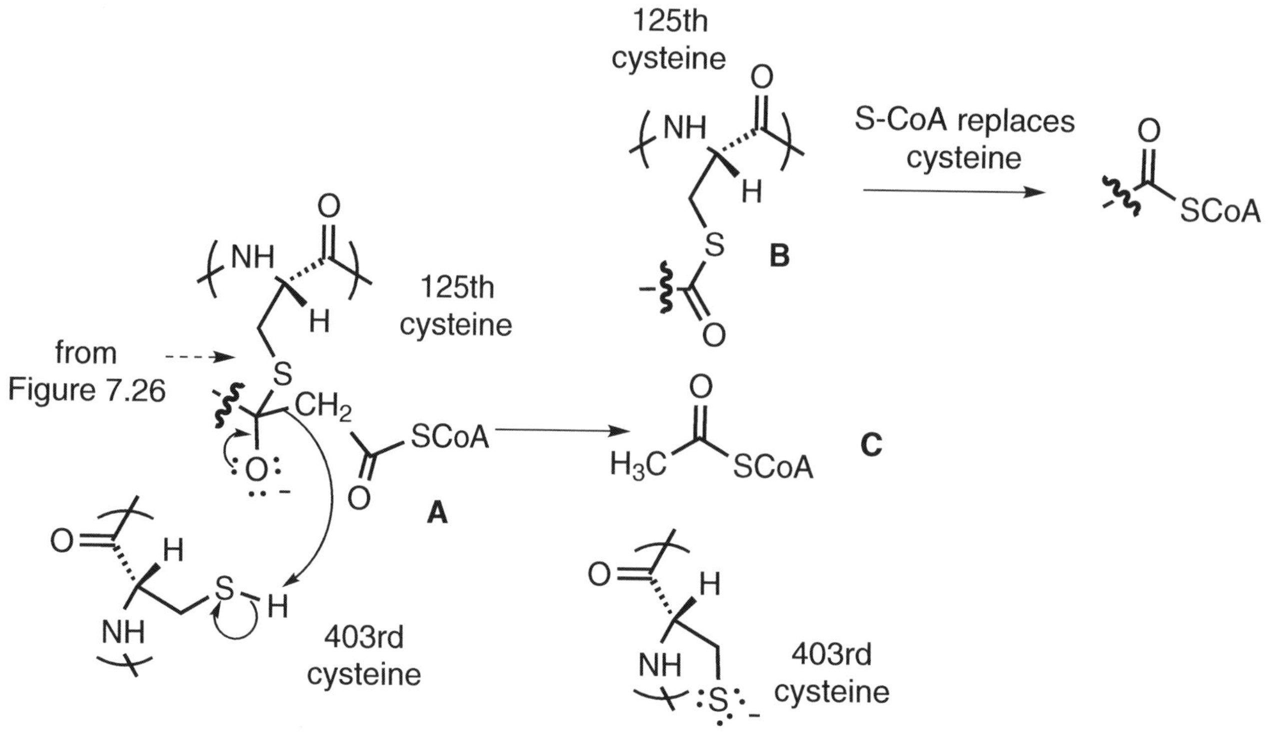

Mechanism of the Enzyme Catalyzed Retro-Claisen Reaction from the Tetra-coordinate Intermediate Formed in Figure 7.26

Figure 7.26 took us from the β keto fatty acyl CoA (**1**) to the tetracoordinate intermediate (**A**) arising from nucleophilic attack of the cysteine derived –S⁻ nucleophile. This tetracoordinate intermediate (**A**) is presented again in **Figure 7.28**.

What does not happen for **A** in Figure 7.28 is for the sulfur nucleophile to be ejected as in reaction **2** in Figure 7.26. What certainly happens, however, as shown in Figure 7.28, is that the tetracoordinate intermediate (A) instead undergoes carbon-carbon bond breaking, (analogous to formation of **3** and **4** from **1** in Figure 7.27), producing **B** and acetyl coenzymer A (**C**) (Figure 7.28).

In the steps not shown (Figure 7.28) **B** is converted to the thioester of the remaining fatty acid chain, now two carbons fewer. The process in summary accomplishes cleaving a two carbon entity away from the fatty acid, which then happens again and again until each two carbon entity of the chain is converted to a molecule of acetyl coenzyme A.

The movement of the reaction path forward to produce acetyl coenzyme A requires that the alternative reactions shown in Figure 7.26 (reactions **2** and **3**) do not take place. The wrong way, as we've seen in the control of the enzyme involved in pancreatic lipase (Figures 7.10 and 7.11), is blocked by the enzyme involved here, β-ketoacyl-CoA thiolase.

One of the functions of an enzyme is not only to lower the energy of activation of a reaction but, in doing so, to choose one reaction over another when more than one reaction is possible. The outline of how this is accomplished by β-ketoacyl-CoA thiolase is seen in Figure 7.28, by critical delivery of the proton from cysteine 403 to exactly the right site to aid the bond-breaking step to produce **C** and **B** from **A**.

PROBLEM 7.53

For many structures, which seem interesting to you, in each of the figures with molecular structures in this section, redraw the structure showing all atoms, including hydrogen atoms and as well showing all nonbonding electrons and therefore accounting for the presence or absence of formal charge and as well to test for the octet rule for every atom in the second row or above of the periodic table.

PROBLEM 7.54

Two perfectly reasonable reactions, **2** and **3**, do not occur in the β-ketoacyl-CoA thiolase. catalyzed reaction path in Figure 7.26. How does the absence of these two possible reactions demonstrate the power of enzyme catalysis and what does your answer have to do with the folded state of the enzyme?

PROBLEM 7.55

While the pK_a of a C-H bond α to a carbonyl group can be in the range of 15 to 20, the pK_a of a C-H bond in a saturated hydrocarbon is estimated to be in the range of at 60 to 80. How does this information apply to the function of 1,3-dicarbonyl compounds in the retro-Claisen reaction?

PROBLEM 7.56

What does your answer to problem 7.55 have to do with the concept of leaving groups?

PROBLEM 7.57

What does the pK_a of a C-H bond α to carbonyl have to do with the first two steps of the catabolic chemistry of the coenzyme A ester of a fatty acid?

PROBLEM 7.58

Propose a series of reactions based on the Claisen condensation (**3** to **1** in Figure 7.27) to prepare various linear β-keto carboxylic acids given that you have available methanol, and a series of carboxylic acid, RCO_2H, and acids and bases of differing strengths.

PROBLEM 7.59

Draw the structure of lauric acid (Figure 7.1) labeling random carbon atoms with a star (*) designating therefore the chosen carbon atoms as isotopes of carbon-12, that is, carbon -13. Consider the enzyme catalyzed catabolism of your labeled fatty acid and show the site of the starred carbon atoms in the six molecules of acetyl coenzyme A ultimately produced.

PROBLEM 7.60

If the result was strictly based on the relative leaving group propensity, which way would the reaction path go in the conflicting directions outlined in Figure 7.28? What does the pK_a of a thiol, RSH, versus that of a C-H bond adjacent to carbonyl have to do with your answer?

PROBLEM 7.61

Propose a mechanism for the trans thioesterification reaction B to acyl coenzyme A, shown in Figure 7.28.

PROBLEM 7.62

Offer structures to demonstrate the fact that while reduction of an aldehyde, R-CH(C=O), produces a primary alcohol, $R-CH_2-OH$, reduction of a ketone produces a secondary alcohol, R_2CH-OH. Create structures to justify the fact that tertiary alcohols, R_3C-OH, can not be oxidized while primary and secondary alcohols can be oxidized. Show structurally, why the oxidation of a secondary alcohol can yield only one product while the oxidation of a primary alcohol has two possible products.

CHAPTER 7 SUMMARY of the Essential Material

THE CHAPTER INTRODUCES NEW KINDS OF MOLECULES, fatty acids with their long hydrocarbon chains and bearing a functional group, the carboxylic acids and the derived esters formed by reaction with alcohols. This brings us in particular to the esters of carboxylic acids with glycerol and their structural role in nature and the differing characteristics of saturated and unsaturated fats.

The chapter then turns to the energy-yielding role of the long hydrocarbon chains of the fatty acids released in the saponification process. The mechanism of saponification, hydrolysis with aqueous base, introduces the chemistry of the ester functional group under nucleophilic attack. Other carbonyl containing compounds are also subject to nucleophilic attack as seen in the aldehyde group of sugars. But there is a fundamental difference between acyl carbonyl as in esters and the carbonyl groups of aldehydes and ketones, a difference that we learn is associated with the concepts of leaving group chemistry and therefore the nature of conjugate bases of strong and weak acids. Understanding these differences allows seeing how nucleophilic attack at acyl carbonyl can lead to substitution while that at aldehyde or ketone carbonyl leads to addition to the carbon-oxygen π bond.

With the background in carbonyl chemistry associated with the beginning of the chapter we are prepared to understand the in vivo hydrolysis of glycerides and the role of enzymes in controlling the spatial relationships of the reactive amino acids

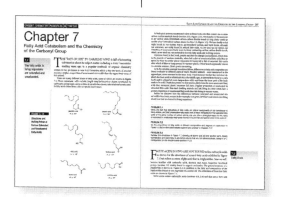

in the protein. It is seen that the fundamental mechanisms of saponification and in vivo enzyme catalyzed hydrolysis of glycerides, although different, are based on identical organic chemistry principles and that the ideas of leaving groups are key to this understanding.

Release of the fatty acid from the ester bond with glycerol sets the stage for conversion of the fatty acid to a biochemical-compatible form, which takes place in vivo by conversion to a thioester with coenzyme A. We learn about the fundamental nature of coenzymes and the roles they play in enzyme catalysis. Again leaving groups are central to the conversion of the free carboxylic acid to the thioester and here another coenzyme comes to play, adenosine triphosphate (ATP). We see that nucleophilic substitution at the phosphorus-oxygen double bond is a replay of that at the acyl carbon-oxygen double bond with ejection of a leaving group. The formation of the thioester of the fatty acid with coenzyme A then sets up the catabolic process, that is, the breaking down of the fatty thioester to acetyl coenzyme A.

The next stage in the in vivo breakdown of the fat takes advantage of the lowered pK_a of a C-H bond adjacent to a carbonyl, a feature of this structural characteristic that arises from resonance stabilization of the carbanion formed on carbon adjacent to a carbonyl. This stabilization factor is then essential to the formation of a conjugated double bond in the fatty thioester of coenzyme A, which is followed by conjugate addition of water to that double bond. The addition of water forms a secondary OH group at a site in the fatty chain that is followed by oxidation to the derived ketone These steps are enzyme catalyzed with the use of the coenzymes FAD and NAD⁺, respectively. We look at the workings of these coenzymes and discover that the nature of each of their oxidation and reduction powers arises from an aromatic character that can be easily lost and then gained again when they act as reducing agents, and the structural reasons for this balance, which serves nature so well.

The ketone arising from the steps outlined above is structured perfectly to undergo a famous reaction long known to organic chemists, the retro-Claisen reaction, which cleaves two carbon atoms from the carbonyl terminus of the fatty acid chain. The hydrocarbon chain, now shorter by two carbon atoms, can undergo another and another two carbon loss via the same sequence of steps until the chain is consumed. These reactions again rest on the stability of negatively charged carbon adjacent to carbonyl carbon, a consequence of resonance stabilization. Here we learn leaving groups can sometimes take on surprising forms while, at the same time, keeping the identical principles at work.

Chapter 8

Carbanions and Carbonyl Chemistry: Sugar Catabolism, Isopentenyl Diphosphate Synthesis and the Citric Acid Cycle

8.1

Nature's Problem with the Catabolism of Glucose and its Solution

WE'VE STUDIED MUCH ABOUT GLUCOSE in Chapters 1 and 3, but to serve life's metabolic functions, this six carbon carbohydrate must be broken down to smaller molecules, that is, be subject to catabolism. Consistent with life's efficiency, the catabolic process for glucose leads ultimately to the identical molecule produced in the catabolism of fatty acids, acetyl coenzyme A (Figures 5.3, 7.28). In this way the processes to follow, which are necessary to sustain life can be carried out on one molecule, acetyl coenzyme A, independent of the original food source.

No matter what the source of the acetyl coenzyme A, it serves identically as the carbon source for the citric acid cycle. We've already seen how the thioesters of fatty acids are broken down into acetyl coenzyme A (section 7.7 and following sections). Let's now follow the critical steps of the catabolic process which take glucose to the same two carbon entity. Demonstrating the beautiful efficiency of life's processes, the identical organic chemistry principles are used by nature to solve the problem.

The path to acetyl coenzyme A from glucose, which contains six carbon atoms, produces not three molecules of acetyl coenzyme A, but rather two molecules of acetyl coenzyme A and two molecules of carbon dioxide. For this process to occur, the six carbon glucose molecule must be first broken down into two three carbon molecules. A carbon-carbon bond breaking step is necessary and this broken bond must be between carbon atoms 3 and 4 of the glucose molecule to gain two, three carbon, moieties.

Let's look again at the structure of glucose to understand the problem of breaking the carbon-3 to carbon-4 bond (**Figure 8.1**). Certainly, closed forms of glucose intervene in the biochemical events but for demonstration of the fundamental principles of organic chemistry involved, the open form serves perfectly well. Moreover, it is the open form of the sugar, as seen below, which must be present for the critical steps in the process.

We've seen in the catabolism of fatty acids that the carbon-carbon bond breaking step is driven by the weakness of a bond α-β to a carbonyl group (section 7.10). A tetracoordinate intermediate at the β-position with a negatively charged oxygen atom reforms the carbonyl group, from which the tetracoordinate intermediate was derived, as the driving force to break the weak bond (Figures 7.27 and 7.28).

FIGURE 8.1 ▶

Molecular Structure of the Open Form of D-Glucose

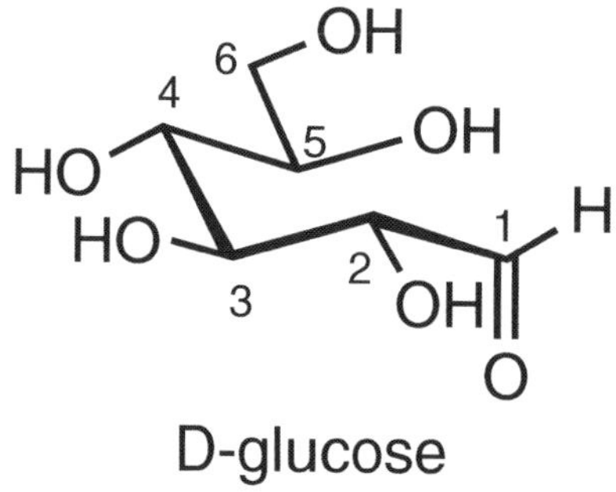

D-glucose

Consider an identical mechanism for breaking a carbon-carbon bond existing in glucose to that used in the fatty acid. But for glucose we substitute the aldehyde as the carbonyl group stabilizing the carbanion arising from the bond breaking event. Let's see what would happen.

The initiating event shown in **Figure 8.2** would be abstraction of a proton from the hydroxyl group on carbon-3. The intermediate formed, called by organic chemists an alkoxide, R-O⁻, **A**, could be compared to the tetracoordinate intermediate formed by reaction of a nucleophile with a ketone, precisely as we have observed in the catabolic mechanism for fatty acids. (section 7.10). After all, is there not a tetracoordinate carbon atom in structure **A** with a negatively charged oxygen atom bonded to it? An intermediate with these structural features will be formed from a wide range of carbonyl groups via a nucleophilic attack or, with the same result, as we are seeing here, from an alcohol via proton abstraction from the OH group (Figure 8.2).

If a π-bond were to be formed from this alkoxide ion, **A**, one of the four groups bonded to carbon-3 of the glucose chain would have to be ejected. And for the same reason as that for the carbon-carbon bond-breaking responsible for the reaction in Figure 7.28 in the fatty chain, the carbon-carbon bond breaking in the situation in glucose shown in Figure 8.2 (step 2) must be between carbon-2 and carbon-3.

As a consequence of alkoxide intermediate **A** in Figure 8.2 the two electrons in this bond between carbon atoms 2 and 3 would be delivered to carbon-2 with resonance

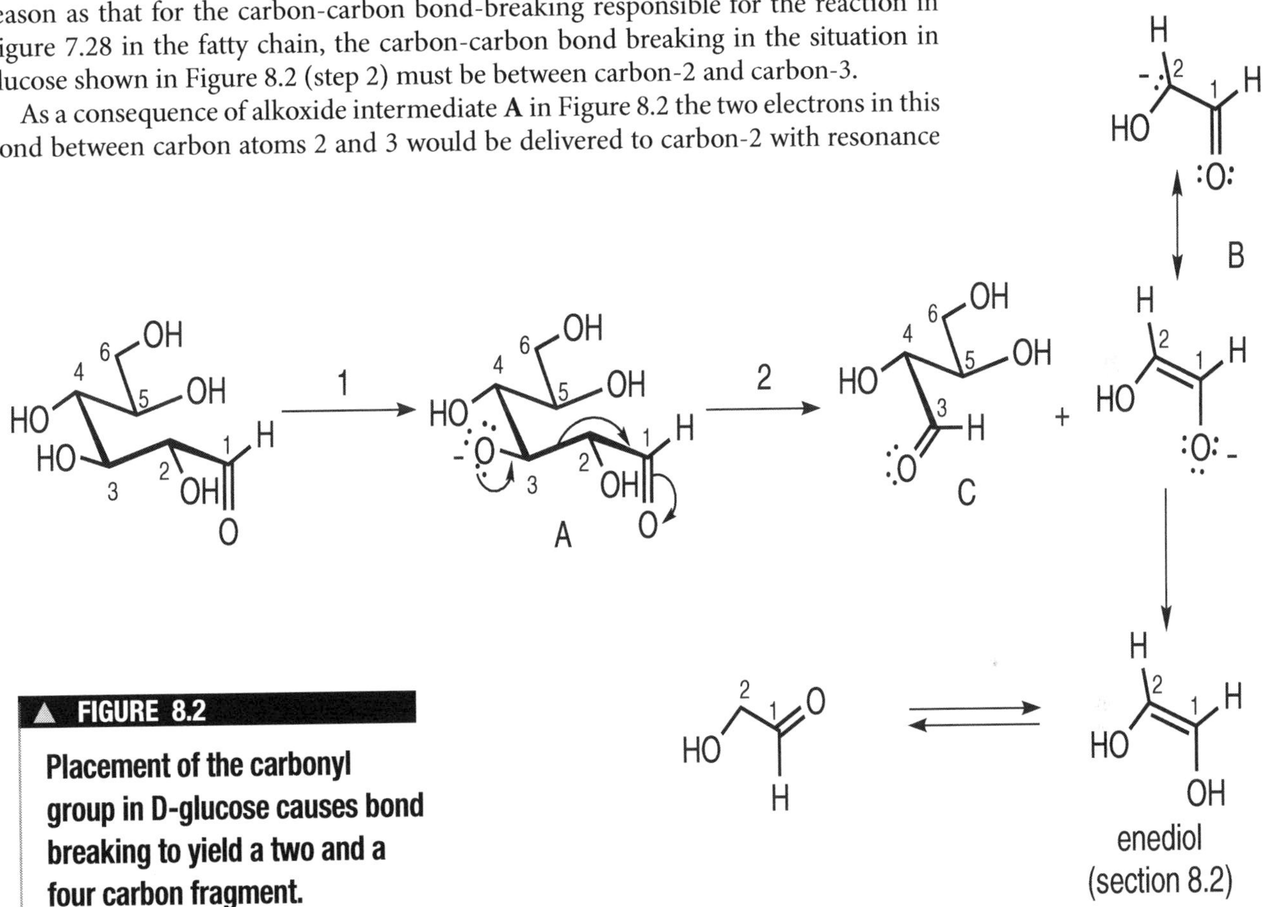

▲ FIGURE 8.2

Placement of the carbonyl group in D-glucose causes bond breaking to yield a two and a four carbon fragment.

stabilization by the adjacent aldehyde carbonyl group (B). However, instead of two three carbon moieties produced, the bond breaking occurring between carbon-2 and carbon-3 produces one two carbon and one four carbon molecule, **B** and **C**. Breaking down glucose in this manner would greatly complicate the chemistry that nature had to follow because instead of one kind of breakdown product there would be two to be dealt with.

The problem is that the carbonyl group is located on the anomeric carbon atom, carbon-1 (section 4.10). The weakened α-β carbon-carbon bond is therefore between carbon-2 and carbon-3 not where it is wanted, which is between carbon-3 and carbon-4 (Figure 8.2).

To accomplish a symmetrical breakdown of the six carbon atoms of glucose requires a change in the structure of the sugar, namely moving the carbon oxygen double bond from carbon-1 to carbon-2, from an aldehyde to a ketone carbonyl. Such a change would convert glucose to fructose. In fructose, the carbonyl group is located in a manner to allow breaking the carbon-carbon bond between C-3 and C-4 therefore breaking the sugar down into two three carbon moieties as will be shown below.

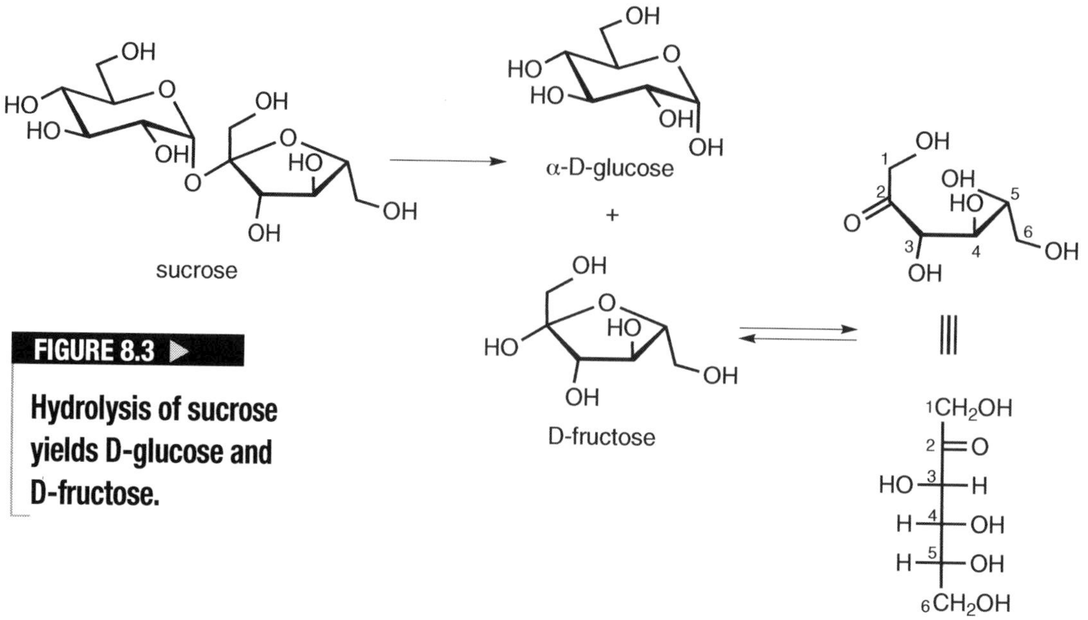

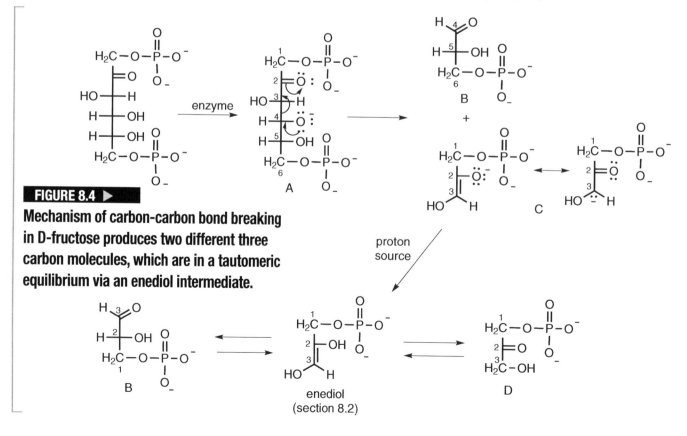

FIGURE 8.3 ▶

Hydrolysis of sucrose yields D-glucose and D-fructose.

In Chapter 3 (section 3.7) we were introduced to historical information on sugar and learned that commonly available sugar, sucrose, is a disaccharide composed of glucose and fructose. The structure of fructose is shown in **Figure 8.3** in the open and closed forms of this carbohydrate. The research effort described in Chapter 3 (sections 3.8 and following sections), which led to the structure of glucose similarly led to the structure of fructose so that is has long been known that the former molecule is an aldehyde and the latter a ketone, in their open forms.

Inspection of the open structure of D-fructose and application of the identical chemical reaction shown in Figure 8.2 (steps **1** and **2**), must lead to breaking the σ bond between carbon-3 and carbon-4, therefore breaking the fructose molecule into two molecules with identical numbers of carbon atoms, three each. **(Figure 8.4)** Although not changing the understanding of the underlying chemistry whatsoever, this catabolic step does not take place from D-fructose, but rather from a phosphorylated structure.

FIGURE 8.4 ▶

Mechanism of carbon-carbon bond breaking in D-fructose produces two different three carbon molecules, which are in a tautomeric equilibrium via an enediol intermediate.

Apparently, as seen in Figure 8.4, we have told the truth about fragmenting D-fructose into two molecules with three carbon atoms each, but have not told the truth about fragmenting D-fructose into two identical molecules. The alkoxide, **A**, derived by loss of a proton from the OH on carbon-4 of fructose does weaken, and cause breaking of the bond between carbon atoms 3 and 4 as shown by the path of the curved arrows. The weakening of this bond is enabled by the stabilization of the carbanion formed on carbon-3 as a consequence of the carbonyl group at carbon-2. Two different molecules are formed, **B**, and **C** the latter a resonance-stabilized carbanion, which picks up a proton to form **D** (Figure 8.4).

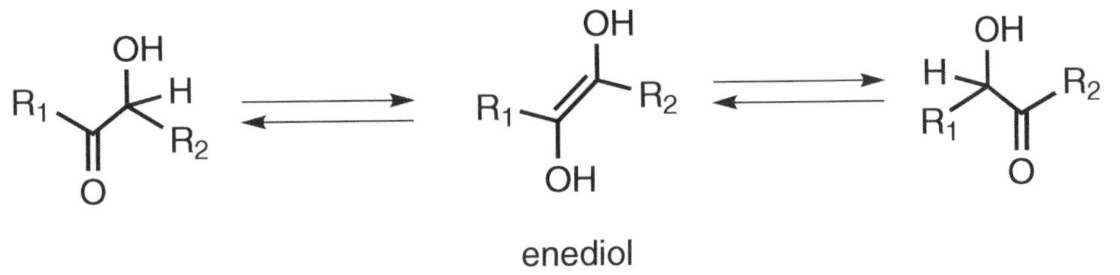

◄ **FIGURE 8.5**

General Structure to Demonstrate Tautomeric Equilibrium via an Enediol Intermediate

B and **D** (Figure 8.4) are clearly different but it turns out that both molecules are in rapid equilibrium with a common intermediate allowing therefore the conversion of one into the other. And because nature prefers **B** over **D**, this mechanism converts all of **D** into **B**. For the in vivo reactions that follow, the carbonyl group must be an aldehyde rather than a ketone.

The common intermediate allowing the conversion of **D** into **B** is called an enediol (Figure 8.4 and **Figure 8.5**). If two carbon atoms are singly bonded to each other and one is a carbonyl carbon and the other a tetracoordinate carbon with an OH group a structural situation exists allowing a rapid exchange to take place through an enediol. The carbon atom with the OH group can be transformed to the carbonyl carbon atom while the carbonyl carbon atom is transformed to a carbon atom with an OH group. This is the situation just discussed for **D** and **B** (Figure 8.4).

The situation just described as the basis for an enediol is, interestingly enough, also the situation for the structures of glucose and fructose. In glucose, carbon-1 is an aldehyde carbon while carbon-2 is carbon bearing a hydroxyl group. In fructose, carbon-1 bears a hydroxyl group while carbon-2 is a carbonyl carbon. Just as for **D**

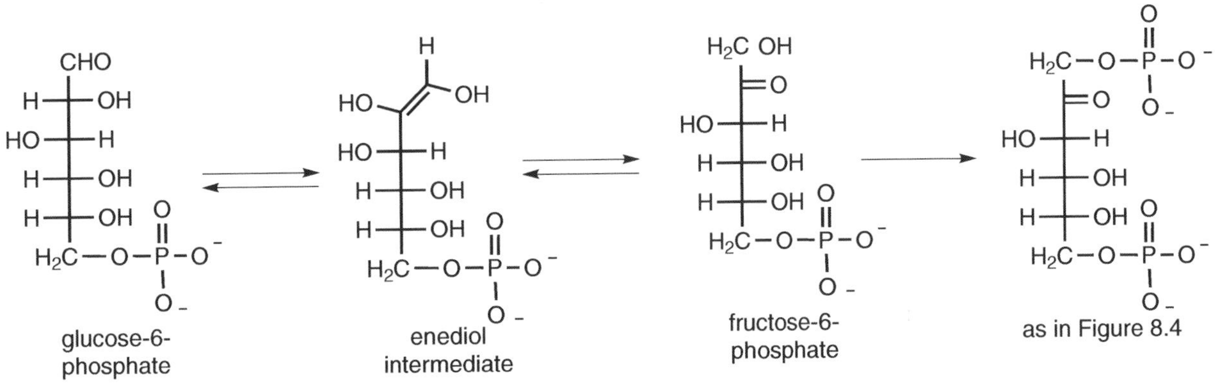

glucose-6-phosphate   enediol intermediate   fructose-6-phosphate   as in Figure 8.4

and **B**, glucose and fructose can be transformed into each other via an enediol (**Figure 8.6**). Here the reactions that follow require that the carbonyl carbon be the ketone rather than the aldehyde, that is, the fructose structure is favored to accomplish the necessary bond breaking step, as discussed above (Figure 8.4).

The biochemical process uses the enediol intermediate accessible to both glyceraldehyde 3-phosphate, **B**, and dihydroxyacetone phosphate, **D** (Figure 8.5), to change the ketone group in the latter to the aldehyde group in the former and then uses the same kind of intermediate to change the aldehyde group in glucose to the ketone group in fructose (Figure 8.6). Although the enediol can yield the aldehyde or the ketone in both situations,

▲ **FIGURE 8.6**

D-Glucose-6-phosphate and D-fructose-6-phosphate are tautomers equilibrating via an enediol intermediate.

these are enzyme moderated isomerizations and so the enzymes catalyzing these processes control which is favored, aldehyde in the two three-carbon catabolic products and ketone in the starting phosphorylated glucose-fructose exchange.

How beautiful – nature uses the enediol intermediate to convert a carbonyl group in the wrong place, carbon-1, in glucose, to the right place, carbon-2, in fructose to enable a carbon-carbon bond to be broken using a driving force that is identical to the one used in the carbon-carbon bond breaking in fats (section 7.10). And then the enediol intermediate allows the two different molecules produced from the fructose fragmentation to form a single structure.

The enediol (Figures 8.5 and 8.6) is a special case of a more general phenomenon known as tautomerism. In the next section, let's see why tautomerism is so well suited to nature's purposes.

PROBLEM 8.1

For many structures, which seem interesting to you, in each of the figures in this section, redraw the structure showing all atoms, including hydrogen atoms and showing all non-bonding electrons and therefore accounting for the presence or absence of formal charge and as well to test for the octet rule for every atom in the second row or above of the periodic table.

PROBLEM 8.2

Assign (R) or (S) nomenclature to all chiral carbon atoms in the structures of glucose and fructose in Figures 8.4 and 8.6.

PROBLEM 8.3

What is nature's problem with the catabolism of glucose? Why is it necessary to isomerize the glucose skeleton to the fructose skeleton?

PROBLEM 8.4

What is the common fundamental driving force for the carbon-carbon bond breaking steps in the catabolism of a fatty acid and of glucose? Does the fundamental nature of this step differ for glucose versus fructose.

PROBLEM 8.5

Compare the carbon-carbon bond breaking step in the catabolism of a fatty acid and that in the carbon-carbon bond breaking step of fructose. Is there a path that could be taken from the tetracoordinate intermediate in the fatty acid catabolism that is not possible in the parallel intermediate in the fructose situation?

PROBLEM 8.6

Why would formation of the alkoxide functional group from a hydroxyl group on carbon-3 or carbon-5 in phosphorylated fructose (Figure 8.4) not lead to carbon-carbon bond breaking?

PROBLEM 8.7

Propose possible sources for the protons necessary for the conversion from C to D in Figure 8.4, assuming that water is not involved.

PROBLEM 8.8

Is it correct that one of the resonance structures of C in Figure 8.4 picks up the proton?

PROBLEM 8.9

What structural characteristics must a molecule have in order to be able to form an enediol intermediate? Offer several examples of molecules that you imagine with these structural features.

PROBLEM 8.10

Research shows that the hydrogen atoms bound to the hydroxyl groups in the enediol (Figure 8.5) are not found in the equilibrated carbonyl compounds produced from the enediol. Propose an explanation for this experimental fact by showing mechanistic steps considering that the enediol equilibria take place in water.

PROBLEM 8.11

Use your proposal in answering question 8.10 to propose a detailed mechanism, showing all proton transfers, for the conversion of glucose phosphate to fructose phosphate in Figure 8.6.

PROBLEM 8.12

Answer question 8.10 for the structures shown in Figure 8.7 (to follow) with the interconversion shown also taking place in water.

PROBLEM 8.13

The structure of sucrose is shown in Figure 8.3. Create a mechanism in water solution in which D̄-α-glucose reacts with the open form of D-fructose, also shown in Figure 8.3, to form sucrose by eliminating a molecule of H_2O. Does nucleophilic attack at ketone carbonyl play a role in your mechanism?

PROBLEM 8.14

The open structure of D-fructose is shown in Figure 8.3. Create a mechanism to produce the closed form of D-fructose, also shown in Figure 8.3, in water solution. Does nucleophilic attack at ketone carbonyl play a role in your mechanism?

PROBLEM 8.15

There are five chiral centers in the closed form of D-glucose and four chiral centers in the closed form of D-fructose. Assign (R) or (S) configurations to each of the common chiral centers. How do these assignments demonstrate the connection between these sugars?

I N CHAPTER 1 (SECTION 1.13) the late nineteenth century struggle to understand how it was possible for molecules to change their shape was described and an important chemist was introduced who contributed to the picture that emerged, the picture that gave us the concept of conformation. That chemist was Johannes Wislicenus.

Now we discover that Wislicenus, at that time, was interested in other dynamic aspects of molecular structure. In section 3.14 we quoted from Carl Noller's classic text of organic chemistry about the concept of mechanism. There can be found in this book, as well, a to-the-point description of the nature and history of the concept of tautomerism: "... *tautomerism is the dynamic equilibrium existing between two spontaneously interconvertable isomers.*" Sounds like what we're seeing in Figures 8.5 and 8.6.

Noller informs us that Wislicenus and Claisen, the latter whom we just learned about in Chapter 7 (section 7.10), worked at nearly the same time in the late 1800s, investigating molecules that seemed to exist in more than one form. The word tautomerism had been coined just before this time and could have been applied to their discoveries, which are effective demonstrations of this dynamic aspect of molecular structure (**Figure 8.7**).

Noller describes how Wislicenus' compound (Figure 8.7) formed a liquid that when treated with ferric chloride turned red but this compound also formed a solid that did not give a color with the same reagent. However, the solid, on standing slowly changed into the liquid, which turned red.

Noller describes that Claisen had a similar experience with his compound (Figure 8.7). One form of Claisen's compound melted at 85-90° C and in alcohol solution gave a red color with ferric chloride. From a solution of this compound in alcohol, further crystallization took place to yield a compound melting between 109-112° C, which gave no color with ferric chloride but on standing slowly turned red.

The evidence described above and other experiments convinced both Wislicenus

8.2

Tautomerism: Enediols are a special case of the dynamic interconversion between enol and keto tautomers.

and Claisen that they were observing the isomerization processes (tautomerism) shown in Figure 8.7.

Several years later, a distinguished chemist, **Julius Wilhelm Brühl** (1850-1911), who was born in Warsaw but whose professional life was in Germany, gave names that are used to this day for these isomers, enol and keto forms. The term keto form is used even if the carbonyl group is an aldehyde and not a ketone. Brühl was a pioneer in the use of refractive index data for the identification of liquids and for estimation of compound purity. The refractive index method among others has

FIGURE 8.7 ▶

Two Classic Examples of Enol/Keto Equilibrium

Wislicenus

Claisen

⬛ **Julius Wilhelm Brühl**

been used to estimate the proportions of enol and keto forms in various molecules. **Figure 8.8** shows examples of several carbonyl containing molecules with estimates of the proportions of keto and enol forms.

I well remember as a graduate student nearly fifty years ago the old fashioned refractometer and being required to obtain the refractive index of all liquid samples, and the pleasure of seeing this number, out to three or four places after the decimal point, match up to those measured in other laboratories as a means to judge if one had the right compound and to test its purity. However, this requirement seems to have fallen out of favor in modern laboratories.

Inspection of the molecular structures in Figure 8.7 and Figure 8.8 shows clearly the applicability of the names enol and keto. A structure with a double bond, alk*ene*, and a hydroxyl group, alcoh*ol*, on one of the carbon atoms of that double bond is an enol. A compound with a ketone group (or an aldehyde group) is a keto form.

Now the question arises as to the conditions that allow the enol and keto forms to interconvert so rapidly. An early answer arose from studies of a compound that resembles closely one we have discussed, compound **1** in Figure 7.27, with the only difference being an ethyl instead of a methyl ester group. The structure of this carefully studied molecule is one of those shown in Figure 8.8.

Kurt Heinrich Meyer was one of the leading chemists of the early years of the twentieth century. He worked with two of the greats of the late 1800s, Baeyer and Willstätter, both of whom have been discussed in this book for their important contributions (sections 3.2 and 6.8). It was during those years before World War I that Meyer became interested in the enol/keto problem while he was in Baeyer's laboratory and discovered a way to add bromine to an enol/keto mixture so that only the enol would react by bromine adding to the double bond. In this manner Meyer effected a separation of the tautomers, although one was changed in the process.

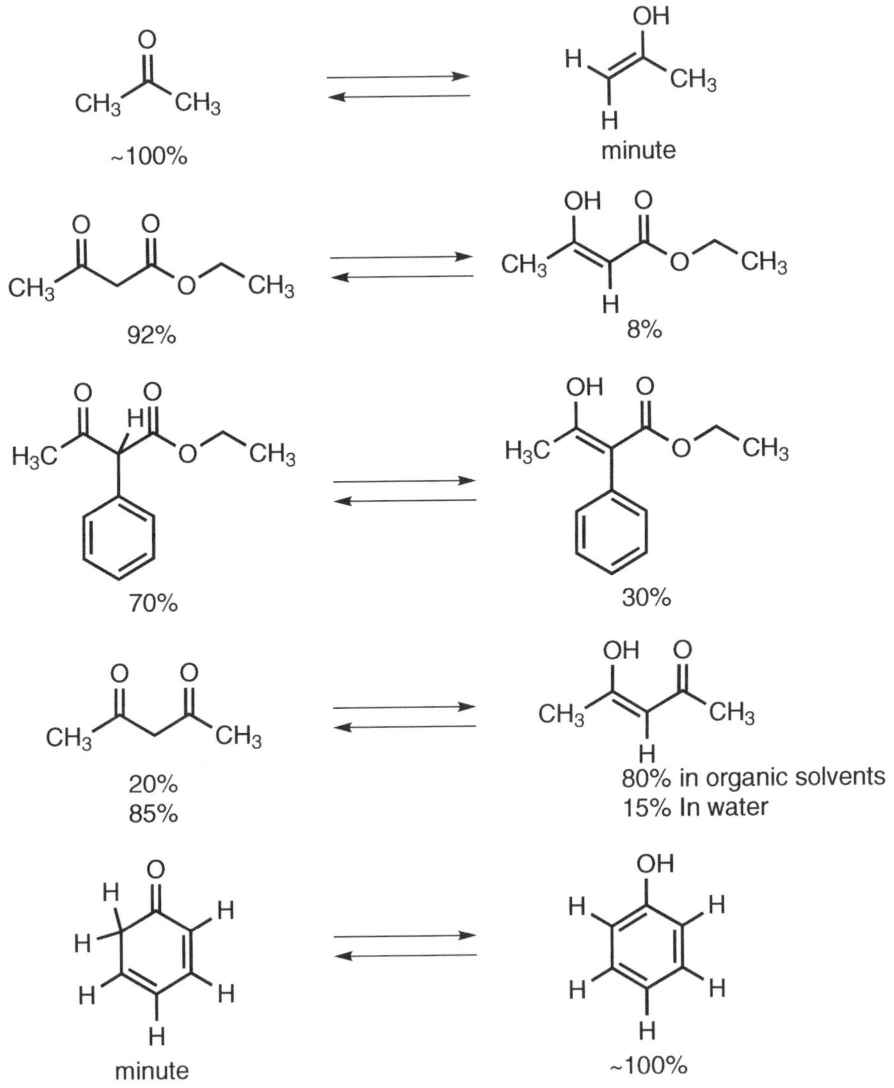

◄ **FIGURE 8.8**

Examples of how the Structures of Different Carbonyl Containing Compounds Affect the Equilibrium between Enol and Keto Tautomers

After the war, Meyer took up the problem again to attempt to separate an enol from a keto tautomer without using a difference in their reactivity. In 1920 he succeeded and in doing so confirmed, in a beautiful way, the importance of a proton source, an acid catalyst, to speed the transformation between the tautomeric forms. Using a specially treated distillation apparatus to greatly reduce the acidity of the silica surface, Meyer was able to separate the keto and enol forms of ethyl acetoacetate (the second example in Figure 8.8). Their slow interconversion in the virtual absence of a proton source allowed the tautomers to distill separately.

Meyer was a collaborator of Herman Mark, whom we learned about in section 6.8, as the chemist who fled the Nazis with the platinum coat hangers. Meyer, in the same year he separated the keto and enol forms of ethyl acetoacetate was invited to head the research laboratories of IG Farbenindustrie and given free reign to carry out fundamental work, even if it was not related to the company's commercial interests. Several years later he hired Herman Mark to join the company. These two chemists used their time together to lay the foundations for understanding the nature of polymers and especially biologically interesting polymers such as natural rubber and cellulose.

Unfortunately for both of these distinguished scientists, and certainly for IG Farbenindustrie, the Nazi racial laws forced both of them out of their positions. While Mark ended up in Canada and then the United States Meyer was able to accept a professor's position at the University of Geneva. During this period in Geneva, Meyer developed further his interests in using concepts of physical chemistry to understand organic chemical phenomena, a field we now call physical organic chemistry. He, as well, turned his work toward understanding the molecular basis of biology, which is an important focus of chemistry in the twenty first century – well ahead of his time in many ways, Professor Meyer.

The now well understood mechanism by which keto and enol forms interconvert is shown in aqueous solvents for ethyl acetoacetate (**Figure 8.9**). Although the structural difference between the keto and enol forms appears to simply be movement of a hydrogen atom, it is more complicated than simply transferring a hydrogen atom from one part of the molecule to another part of the same molecule. In the absence of an external proton source the transformation is exceptionally slow, allowing Meyer's laboratory to separate the two tautomers, under the conditions of a distillation, as noted above. However, with the aqueous proton source as a catalyst, shown as the hydronium ion, H_3O^+, the tautomerization is rapid (Figure 8.9).

Given this introduction to enol and keto forms of carbonyl containing molecules we can see how the phenomenon of tautomerism leads to the enediol intermediate allowing conversion of glucose-6-phosphate into fructose-6-phosphate (Figure 8.6) and the conversion of dihydroxyacetone phosphate, **D**, into glyceraldehyde-3-phosphate, **B** also through the enediol intermediate (Figure 8.4).

An enediol is simply the product of the enolization of a carbonyl group with an adjacent tetracoordinate carbon atom bearing a hydroxyl group, allowing therefore exchange of the

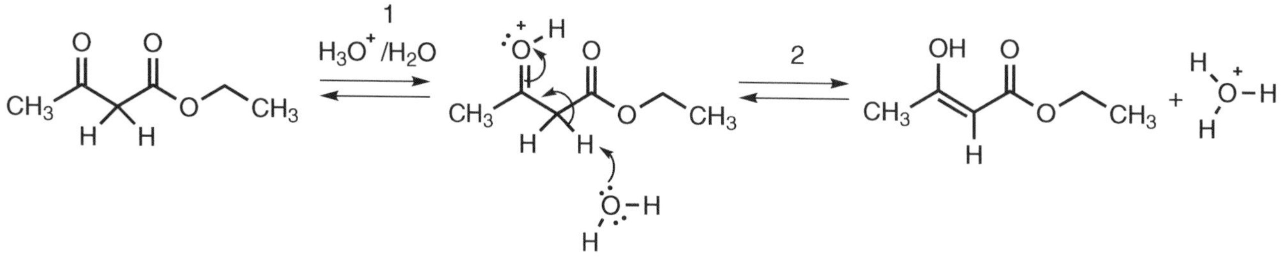

FIGURE 8.9 ▲

Water Mediated Mechanism of an Enol/Keto Interconversion

state of these two carbon atoms as used so effectively in the catabolism of sugars.

A discussion of enol and keto forms is an opportunity to reinforce the fundamental difference between equilibrium structures and resonance structures. Figure 8.9 presents the acid catalyzed mechanism for the interconversion of the enol and keto forms of ethylacetoacetate. Enol and keto forms can also be obtained via a base catalyzed process as shown in **Figure 8.10** for acetylacetone.

Imagine an interconverting enol and keto form of acetylacetone in water to which is added sodium hydroxide. Proton abstraction from either the keto tautomer or from the enol tautomer yields the identical negatively charged intermediate **C**, which is stabilized by resonance as shown in Figure 8.10. Now neutralize the base so that a proton can be added to **C**. The products will be **A** and **B**, which are different molecules with different

FIGURE 8.10 ▶

Base Catalyzed Interconversion Between Enol/Keto Tautomers Involves an Intermediate Described by Three Resonance Structures

C is one molecule represented by three resonance structures

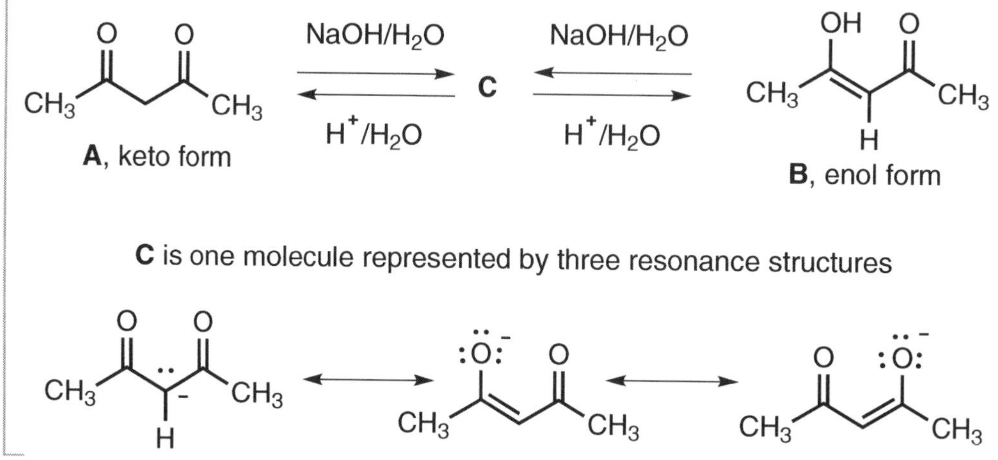

arrangements of their atoms. **A** and **B** are constitutional isomers (section 1.8).

The three representations of **C** shown in Figure 8.10 are simply three ways to represent the electron distribution of this molecule. A single structural representation of **C** using Lewis bonding nomenclature is impossible. However, if we were able to portray **C** using orbitals and electron densities, a single representation would suffice.

Here we see in Figure 8.10 a clear portrayal of the difference between isomers and resonance structures designated by two arrows pointing in opposite directions ($\rightleftarrows$) for isomers and a double headed arrow ($\leftrightarrow$) for resonance structures.

PROBLEM 8.16
What structural characteristics distinguish tautomeric isomers from conformational isomers?

PROBLEM 8.17
Why would addition of Br_2 to a solution of ethyl acetoacetate (Figure 8.9) cause the loss of the ketone carbonyl group?

PROBLEM 8.18
Keto forms are intrinsically more stable than enol forms. Offer explanations for the increased proportions of enol forms shown in the examples in Figure 8.8.

PROBLEM 8.19
For each of the five examples in Figure 8.8 write out a reasonable mechanism for the interconversions between keto and enol forms catalyzed by acid in water, that is, by H_3O^+.

PROBLEM 8.20
What fundamental principle demonstrated in Figure 8.10 determines the difference between isomers and resonance structures?

I N CHAPTER 5 WE LEARNED about the five carbon atom molecule, isopentenyl diphosphate, which is the source of the terpenes (section 5.7 and following sections). Where do the five carbon atoms of isopentenyl diphosphate come from? The answer is three molecules of acetyl coenzyme A, the end product of the catabolism of fatty acids, as discussed in Chapter 7.

The number of molecules of acetyl coenzyme A produced from each fatty acid equals the number of carbon atoms in the fatty acid divided by two. Each glucose molecule, on the other hand, yields two molecules of acetyl coenzyme A. You've seen all the steps in the production of acetyl coenzyme A from a fatty acid but not all the steps starting from glucose. The three carbon atom molecule, glyceraldehyde-3-phosphate, produced in the catabolism of glucose (section 8.1, Figure 8.4), goes on through a series of intermediates to produce pyruvic acid, which then loses one carbon atom as CO_2 to also yield acetyl coenzyme A. The formation of pyruvic acid and the mechanism of the loss of CO_2 is a bit complicated and best left for later in your studies of this science. The outline of these sources of acetyl coenzyme A is shown in **Figure 8.11**.

If three molecules of acetyl coenzyme A combine to form a five carbon molecule, then one carbon atom has to be lost. Let's see how this happens and come to realize that all the elements of organic chemical reactivity discussed in this chapter and Chapter 7 come together to allow understanding of this conversion. **Figure 8.12** shows the structure of isopentenyl diphosphate and the structures of three molecules of acetyl coenzyme A.

Each carbon atom in each of the three acetyl coenzyme A molecules is numbered. The numbers assigned to the isopentenyl diphosphate carbon atoms then show the relationship

8.3

We've seen how the reverse of the Claisen condensation in the catabolism of both fats and sugars causes breaking of carbon-carbon bonds. Let's see how nature uses the Claisen reaction in the other direction, to make carbon-carbon bonds.

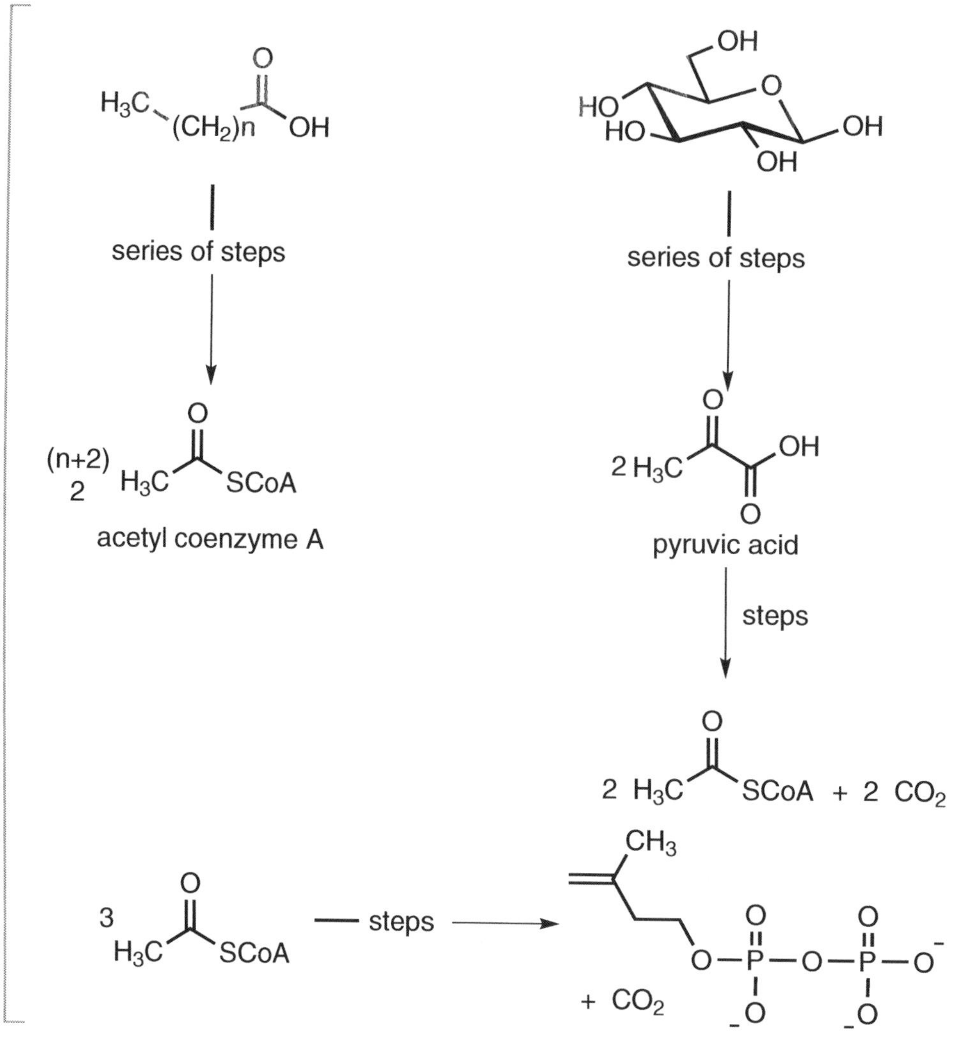

FIGURE 8.11 ▶

Overview of the Conversion of Fatty Acids and Sugars to Acetyl Coenzyme A

of each carbon atom to the precursor acetyl coenzyme A molecules. The source of the carbon atoms in isopentenyl diphosphate shown with numbers in Figure 8.12 arises from experimental evidence derived from labeling experiments using isotopes of carbon and hydrogen as we've discussed earlier for discovering the sources of the carbon and hydrogen atoms in the higher terpenes and lanosterol (section 5.8).

The numbers on the carbon atoms of isopentenyl diphosphate and the three molecules of acetyl coenzyme A in Figure 8.12 lead to an interesting puzzle. How are those three molecules of acetyl coenzyme A put together by the biological mechanism to lead to the connections shown by these numbers? It seems that carbon-1 and carbon-4 have to form a bond connecting two of the acetyl coenzyme A molecules. In addition carbon-4 and carbon-5 have also to form a bond to connect the third acetyl coenzyme A. And finally carbon-6 has to somehow become the carbon atom of a CO_2 molecule.

Carbon-1 and carbon-4 remind us of a mechanism for forming a carbon-carbon bond using the Claisen reaction as seen in Figure 7.27 in the reaction of **4** with **3**. If acetyl coenzyme A (i) reacted in a parallel manner with acetyl coenzyme (ii) as **4** reacts with **3** in Figure 7.27, the product would be a structure parallel to **1** in Figure 7.27.

This reaction is a Claisen reaction and makes the necessary connection between carbon-1 of (i) and carbon-4 of (ii) in producing **A** (**Figure 8.13**). The formation of the bond between carbon-1 and carbon-4 takes place by reactions **2 and 3** (Figure 8.13).

There is a detail in Figure 8.13, which you may have noticed. The thioester group of (ii) has been changed from the thioester of coenzyme A (SCoA) to that of cysteine. The necessity for this makes sense. It allows the Claisen condensation between the two thio esters to take place in a precise manner by being associated with the enzyme. One of the thioesters, bound to the cysteine sulfur, is held in place

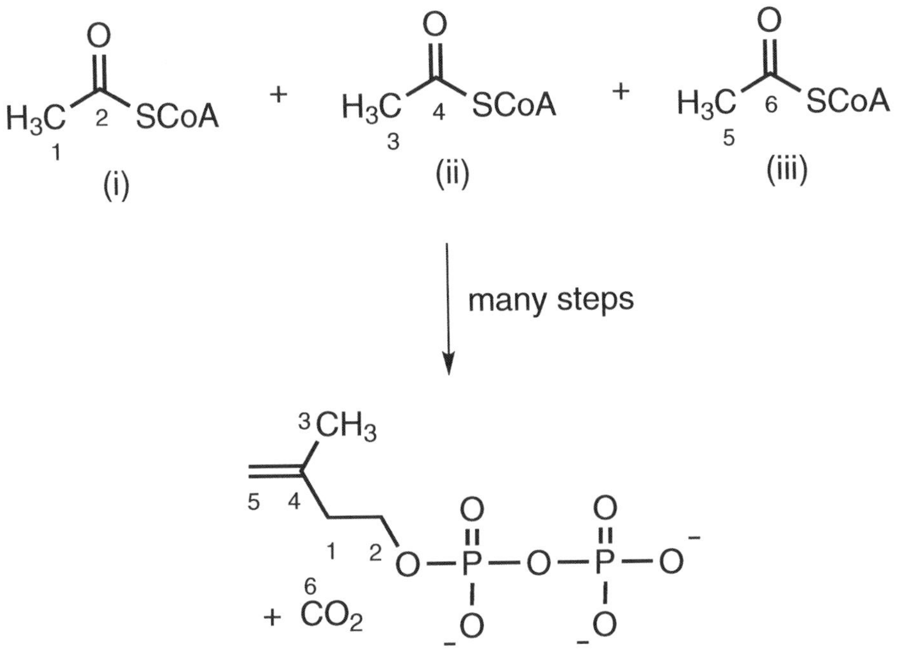

◄ **FIGURE 8.12**

Tracing the Carbon Atoms in the Conversion of Three Molecules of Acetyl Coenzyme A to One Isopentenyl Diphosphate

by a covalent bond to the protein backbone of the enzyme.

The product, **A**, of the Claisen reaction in Figure 8.13 is set up for another reaction in which the carbanion formed at carbon-5 from acetyl coenzyme A (iii) (Figure 8.12) forms an identical intermediate as that formed from (i) in Figure 8.13 but undergoes a different kind of reaction. The subsequent nucleophilic attack on **A** by the negatively charged intermediate formed from iii is not at the acyl carbonyl, which would have caused organic chemists to call this a Claisen reaction, but rather at the ketone carbonyl group of **A**, carbon-4. This is an example of what is known in organic chemistry as an Aldol condensation, a carbon-carbon bond forming reaction equally as important as the Claisen condensation. Let's learn more about the Aldol condensation before we go on with the synthetic path from A (Figure 8.13) to isopentenyl diphosphate (section 8.5).

▼ **FIGURE 8.13**

First step in the enzyme catalyzed synthesis of isopentenyl diphosphate is a Claisen condensation.

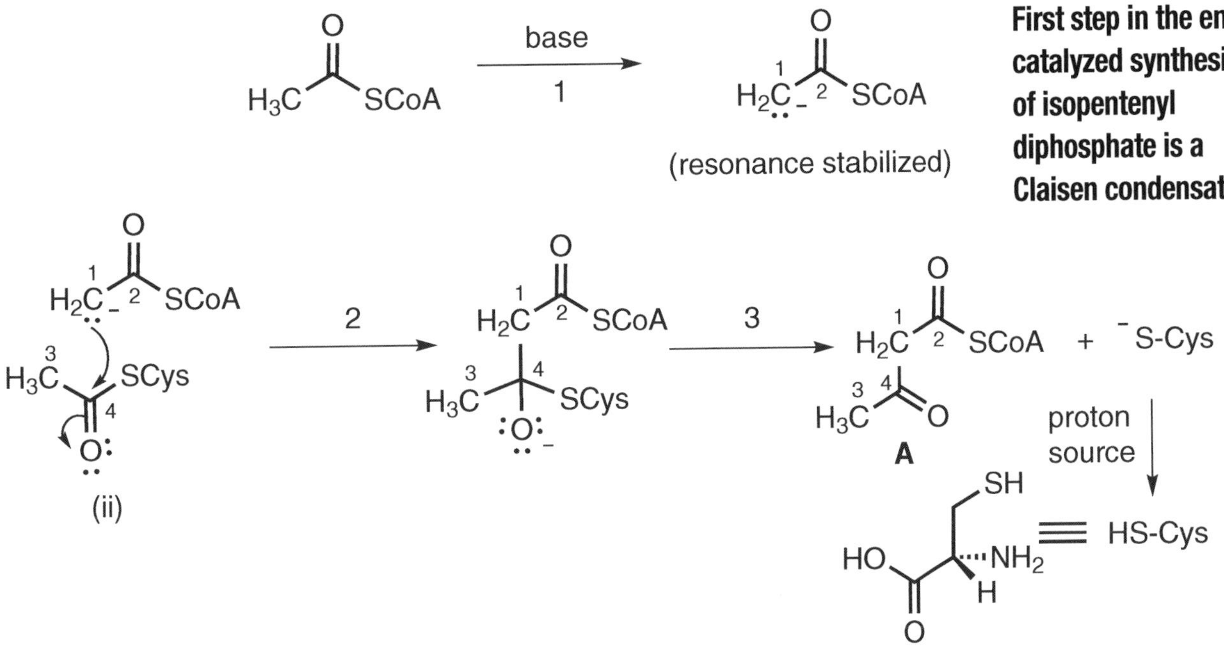

PROBLEM 8.21

Circle the critical bond breaking step for the catabolism of both fats and sugars and show how the mechanism for the bond breaking arises from the same driving force.

PROBLEM 8.22

Abstraction by base of a proton from C-5 with loss of the phosphate from C-6 in **B** in Figure 8.4 would lead to an enol, which would be well on the way to produce the pyruvic acid shown in Figure 8.11. Outline the process in detail. What kind of coenzyme would be necessary for the full transformation to the pyruvic acid?

PROBLEM 8.23

Looking ahead, show examples of steps in the formation of isopentenyl diphosphate that can be identified as Claisen condensations or as Aldol condensations and explain the difference.

8.4

The Aldol Condensation

THERE IS A CLASSICAL REACTION in organic chemistry called the Aldol condensation. Do you remember the story of Wurtz and Kekulé from the 1850s when the idea about tetracoordinate carbon was developed leading to the scooping of young Archibald Couper (section 1.11)? Do you remember the story of Kekulé's development of the idea of the structure of benzene in the 1860s (section 6.4), again sponsored by Wurtz? Well, here is another story about Kekulé in the next decade, in 1872.

In investigations of the chemical reactions associated with acetaldehyde, Kekulé, now working independently of Wurtz, showed that a "condensation" took place combining two molecules of acetaldehyde to yield what is called crotonaldehyde. At the same time, Kekulé's former mentor, Wurtz, working in competition with Kekulé, showed that the crotonaldehyde Kekulé discovered arose from an intermediate structure that Kekulé had not uncovered. Wurtz's compound contained both an <u>ald</u>ehyde and hydroxyl (alcoh<u>ol</u>) functional groups so he called it an aldol. The reaction, shown in **Figure 8.14**, investigated by these famous chemists in the early 1870s, was apparently investigated and published at least five years earlier by the Russian chemist, **Aleksandr Porfiryevich Borodin**, who was not at first cited by Kekulé. Borodin was apparently too much of a gentleman to force the issue of his priority and is more

FIGURE 8.14 ▶

Example and Mechanism of an Aldol Condensation

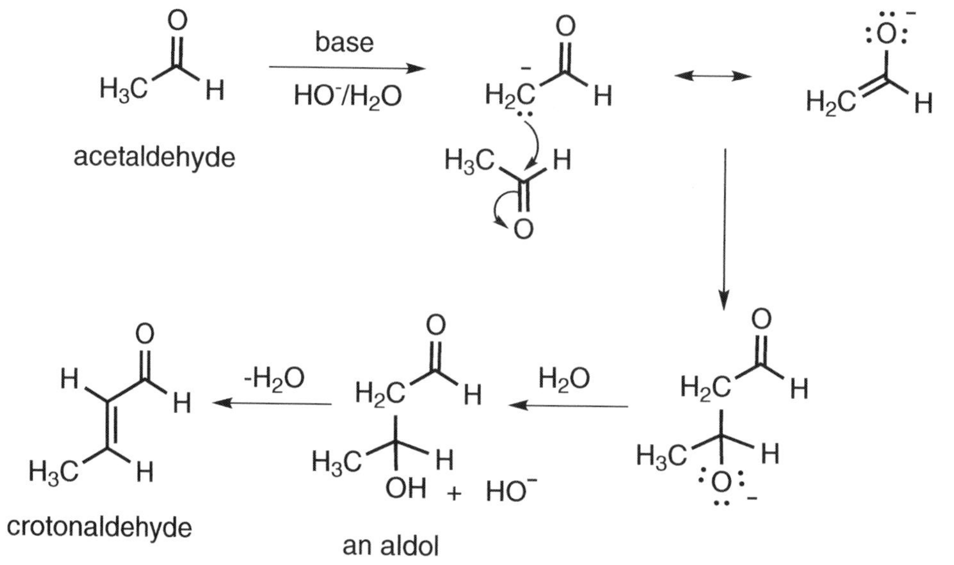

rememberd for his musical compositions than as the chemist who discovered one of the most important reactions in organic chemistry, the Aldol condensation.

Let's look at the reaction steps taken in Figure 8.14. The C-H bond adjacent to a carbonyl group is lost to a base – no surprise. The carbanion produced is resonance stabilized and acts as a nucleophile to attack a carbonyl group in another molecule. Because the carbonyl group is not acyl, but rather aldehyde, the product is an alkoxide ion, which takes a proton from a water molecule to yield an alcohol. This is Wurtz's aldol. Here, we come across an addition reaction at carbonyl carbon, the required fate of nucleophilic attack at either aldehydes or ketones in contrast to the substitution chemistry possible for acyl carbonyl (section 7.4).

In the Aldol reaction we see the similarity and contrast to the Claisen reaction. Both reactions involve attack at carbonyl carbon by resonance-stabilized carbanions but the carbonyl target in the Claisen reaction is acyl while that in the Aldol reaction is either an aldehyde or a ketone.

We'll delay discussion of the step that eliminates water from the aldol. The mechanism for formation of the double bond can take many paths but is essentially driven by the stability of a double bond adjacent to carbonyl and the large negative heat of formation of water. The double bond in crotonaldehyde is conjugated with the carbonyl group just as is the double bond formed in the first step of fatty acid catabolism in section 7.7 (Figure 7.22).

In any chemical structure, π-bonds that are conjugated with each other lend stability to the molecule and any chemical reaction that forms conjugated π bonds will be favored. Transformation from the aldol to crotonaldehyde forms two contiguous π bonds. We'll see more examples of this principle in chapters to follow.

Given the Aldol condensation (Figure 8.14) let's return to the biosynthesis of isopentenyl diphosphate and look at nature's path for further transformations of **A** (Figure 8.13) ,which involves an Aldol condensation (section 8.5).

▪ **Alexander Porfiryevich Borodin**

PROBLEM 8.24
What does the concept of leaving groups have to do with the difference between the Claisen condensation and the Aldol condensation?

THE INTERMEDIATE **A** on the path to isopentenyl diphosphate (Figure 8.13) has been formed from two of the required acetyl coenzyme A molecules, i and ii (Figure 8.12), although the thioester group of ii has been exchanged to allow connection to the enzyme. The third coenzyme A, iii, now must be reacted with **A** to make a bond between carbon-4 of **A** and carbon-5 of iii (Figure 8.12). This step is shown in **Figure 8.15**, and it takes place by nucleophilic attack of a carbanion, which is stabilized by resonance with an adjacent carbonyl group, on a ketone carbonyl group.

It may be a stretch considering that a ketone rather than an aldehyde is involved, but the essential features of the reaction shown in Figure 8.15 are the same as those of the Aldol condensation shown in Figure 8.14 and organic chemists call the reaction in Figure 8.15, aldol-like. It is.

For the same reason that ii in Figure 8.13 was bound to the enzyme backbone by forming a thioester with a cysteine residue, the electrophile **A** is attached to the enzyme catalyzing this reaction by substituting the coenzyme thioester with a thioester derived from cysteine (step **1** Figure 8.15). In another parallel to the Claisen-bond making step exhibited in Figure 8.13, a base converts acetyl coenzyme A (iii) to the carbanion. Again here as in Figure 8.13 the carbanion can be represented as shown in

8.5

Continuing on the Path to Isopentenyl Diphosphate

FIGURE 8.15 ▶

Second enzyme catalyzed step for addition of the third acetyl coenzyme A in the synthesis of isopentenyl diphosphate uses an Aldol condensation.

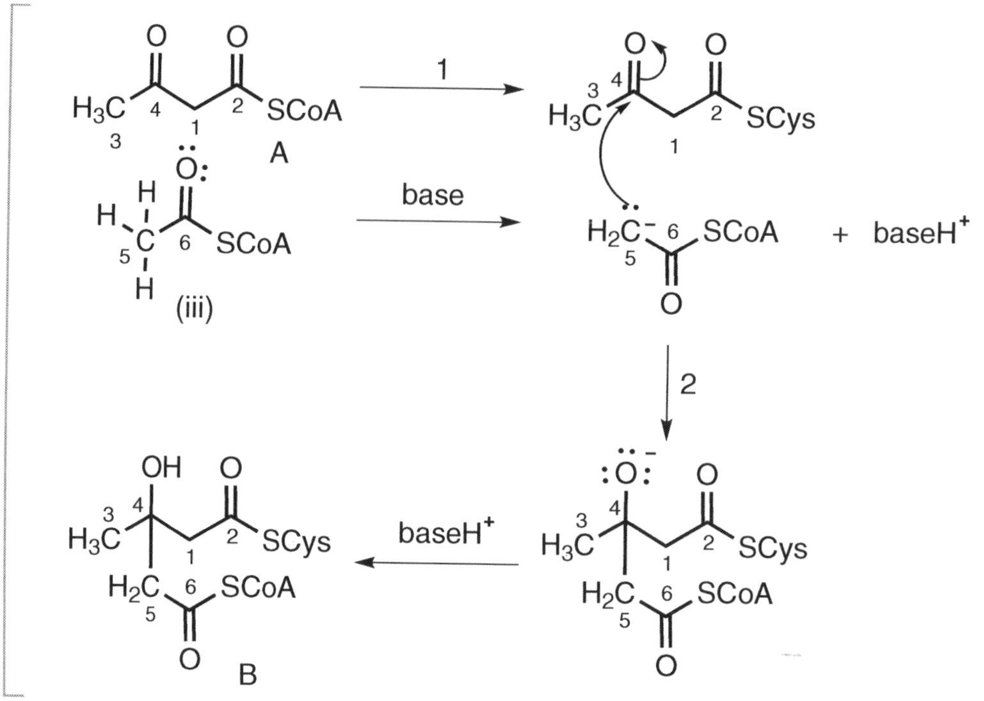

Figure 8.15 or as the enolate resonance structure, which is not shown. Either resonance structure could have been used to show reaction **2** in Figure 8.15.

The next steps for the conversion of **B** (Figure 8.15) to the precursor of isopentenyl diphosphate are shown in **Figure 8.16**. Without going into mechanistic detail the steps make sense. First (reaction **1**) coenzyme A is released by hydrolysis of the thioester, forming the carboxylate anion. This step (**1**) might involve an enzyme using serine as in the hydrolysis of triglycerides (section 7.5)

Secondly, a reduction mechanism follows, which could use NADH as the necessary coenzyme. This step (**2**) involves reduction at acyl carbonyl taking it first, as least formally, to the aldehyde and then from the aldehyde to the primary alcohol. In the third step (**3**) the two hydroxyl groups (OH) are converted to phosphate groups, which likely involves adenosine triphosphate, ATP (sections 7.6, 8.9, Figures 7.16, 8.22 and 8.23).

The final step requires the loss of carbon-6 as CO_2. Let's see how that happens (Figure 8.17) and then how it is related to carbon-carbon bond weakening in other reactions we have studied in Chapter 7 and in this chapter.

The bond breaking step (**Figure 8.17**) occurs because the electrons making up the

FIGURE 8.16 ▶

Hydrolysis and phosphorylation in the enzyme catalyzed path to isopentenyl diphosphate prior to the loss of carbon dioxide yields the final skeleton.

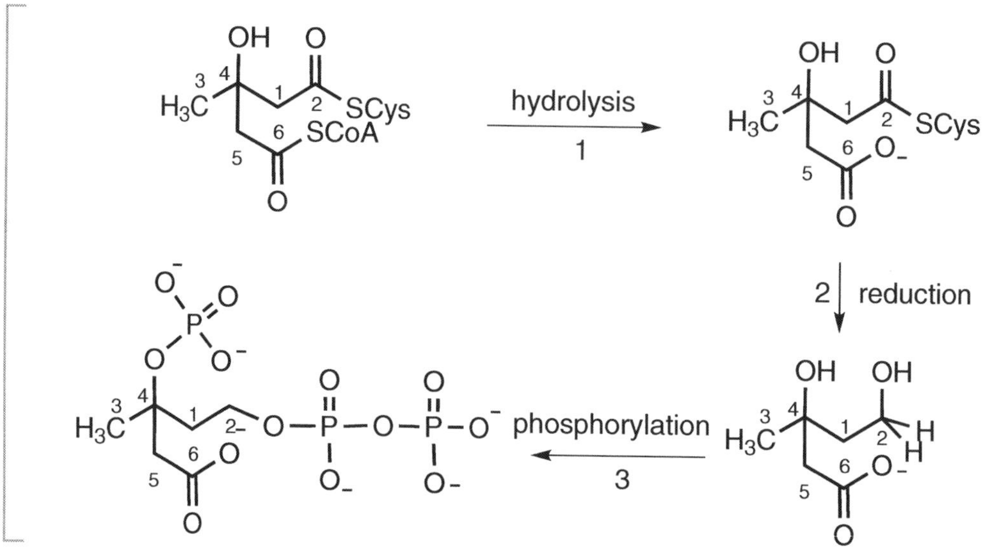

bond holding the CO_2 group, that is, the bond between carbon-6 and carbon-5, have "some place to go," so-to-speak. The electrons that made the bond between these carbon atoms are transferred into the orbital "space," which became available by loss of the phosphate, PO_4^{-3}, group from carbon-4. This is a reasonable reaction path arising from the fact that phosphate is an excellent leaving group (section 5.4), which means that the bond holding it to carbon is weak. As shown in Figure 8.17, all the

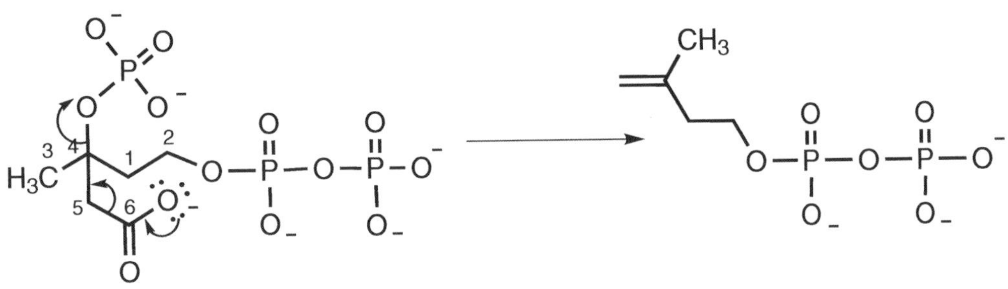

▲ **FIGURE 8.17**

Enzyme catalyzed loss of carbon dioxide is allowed by loss of phosphate to form isopentenyl diphosphate.

electrons flow in a concerted manner to break the two bonds simultaneously, carbon-carbon and carbon oxygen. Let's look at some parallel reactions in **Figure 8.18** that can be described by the principle of electrons having "some place to go," that is, into an orbital that is empty and available to form another kind of bond.

The bond breaking converting **A** to **B** and **C** in Figure 8.4 is reproduced in Figure 8.18 (**1**). This key reaction in the catabolism of fructose occurs, because the electrons making up the bond between carbon-3 and carbon-4 holding the two three-carbon moieties of phosphorylated fructose together have "some place to go," that is, into the orbital on carbon-2, which had participated in forming the carbonyl group π-bond. This is a reasonable reaction path arising from the fact that the π-bond is weak and with the high electronegativity of oxygen accepting the two electrons that had made up the carbonyl π-bond, carbon-2 becomes an excellent sink for the two electrons of the σ bond that are displaced. All the electrons flow in a concerted manner to break simultaneously the two bonds, σ carbon-carbon and π carbon-oxygen

A second example in Figure 8.18 is the bond breaking (**2**) between carbon-2 and carbon-3, the key reaction in the catabolism of a fatty acid (Figure 7.28). Here the electrons making up this carbon-carbon σ-bond in the fatty acid chain have "some place

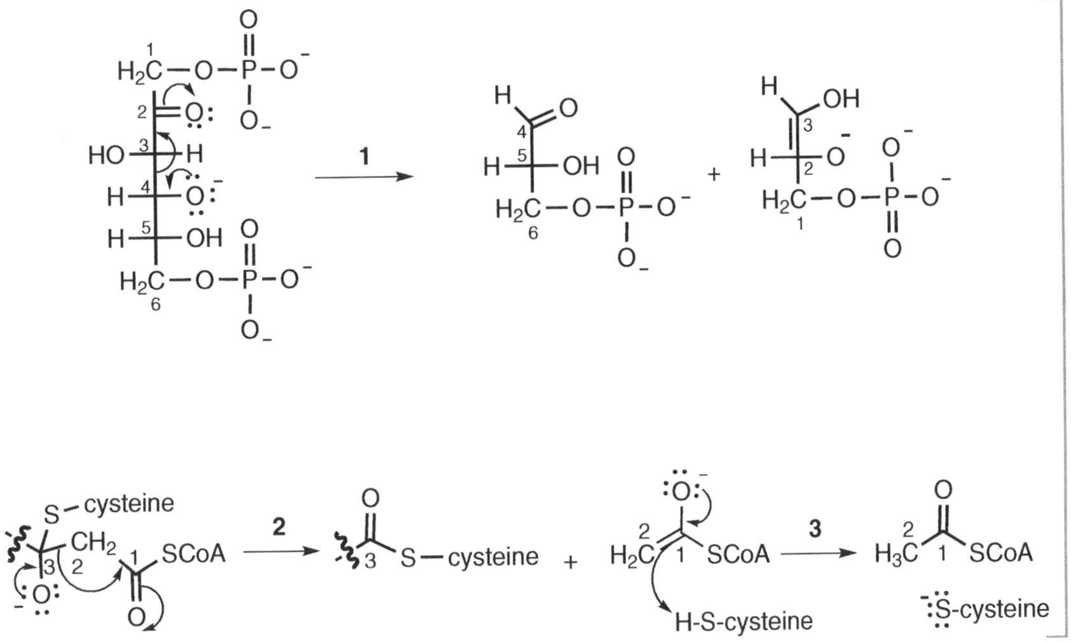

◀ **FIGURE 8.18**

Two Examples of Carbon-Carbon Bond Breaking Driven by Formation of Resonance Stabilized Carbanions

to go," that is, in the direction of forming an enolate. The two electrons making up the π-bond between carbon-1 and oxygen are now displaced to the oxygen atom, freeing therefore the p-orbital on carbon-1 to form the π-bond of the enolate between carbon atoms 1 and 2. As shown in 2 (Figure 8.18), all the electrons flow in a concerted manner to simultaneously break the two bonds, carbon-carbon and π-carbon-oxygen. The 403rd cysteine (Figure 7.28) rapidly transfers a proton to carbon-2 using two of the electrons in this newly formed enolate π-bond to form acetyl coenzyme A (3; Figure 8.18).

Looking at the three bond breaking steps in Figures 8.17 and 8.18 in the manner just discussed shows how these very different reactions along widely different biological paths: the synthesis of a terpene starting material; the catabolism of a carbohydrate; and the catabolism of a fat, are intimately related in sharing a common principle of organic chemistry, which could be expressed in an embarrassingly simple way by saying that electrons have to have "some place to go." Many reactions of organic chemistry can be understood by realizing this necessity, that is, for an available unoccupied orbital which can participate in a bonding interaction, a new π- or σ-bond. We're going to see this in detail in Chapter 12 (section 12.21) in a class of reactions called pericyclic.

PROBLEM 8.25

Show all the essential mechanistic steps, without involving the details of enzymes, in the conversion of three molecules of acetyl coenzyme A to isopentenyl diphosphate. Label the carbon atoms in the three molecules of acetyl coenzyme A as 1 through 6 and then show where these carbon atoms end up in the final product.

PROBLEM 8.26

Imagine you have isolated the aldol shown in Figure 8.14. Now dissolve this molecule in water with a small amount of acid so that the hydroxyl oxygen is protonated. Alternatively, dissolve this molecule in water with a small amount of base so that a proton is lost from the CH_2 group adjacent to the carbonyl group. Draw reasonable mechanistic paths for formation of crotonaldehyde from the aldol via both reaction scenarios outlined above.

PROBLEM 8.27

In Figure 8.15 an Aldol reaction is shown. Show the reactive path that would have been taken for the identical intermediates if a Claisen reaction were to take place.

PROBLEM 8.28

Using examples from other biochemical reactions studied, propose reasonable mechanistic paths and coenzymes that could affect steps 1 and 2 and 3 in Figure 8.16.

PROBLEM 8.29

Describe each of the three reactions shown in Figures 8.17 and 8.18 by identifying what could be described as the leaving group in each case. Considering that a leaving group can be described as the conjugate base of an acid, write the structure of the corresponding acid for each of the three leaving groups. Which would be the strongest and which the weakest acid?

PROBLEM 8.30

In the last paragraph of this section a sentence ends with: "….., which could be expressed in an embarrassingly simple way by saying that electrons have to have some place to go. Many reactions of organic chemistry can be understood by realizing this necessity." Go back over various reactions studied in the book and test the accuracy of the statement. Look ahead to Chapter 12, section 12.21 for a detailed discussion about occupied and unoccupied molecular orbitals.

PROBLEM 8.31

Do you find the comparisons made in Figure 8.17 and 8.18 justified?

VERY CELL OF THE TRILLIONS OF CELLS in your body finds the energy to sustain the cell's functions as an output of the mitochondria of the cell. The mitochondria have been called the cell's "power plants." They are the the the sites of the essential elements of respiration by which the food we eat in combination with the oxygen we breathe is converted to the energy we need to live.

In some cells there are astonishing numbers of these organelles. For example it is estimated that in a single liver cell there may be found up to 2,000 mitochondria constituting approximately one fifth of the total volume of the cell. One definition of an organelle is a specialized subunit within a cell that has a specific function, and is usually separately enclosed within its own lipid membrane. The word comes from the idea that an organelle is to a cell, as an organ is to the whole body of the animal, but smaller.

Mitochondria are fascinating organelles in their connection to evolution. It is thought that production of the necessary energy to sustain life was inefficient in the earliest versions of the kinds of cells that make up the bodies of all animals, including us, that is, eukaryotic cells. Over some period of time, perhaps more than a billion years ago, primordial mitochondria were incorporated into these early eukaryotic cells by a process called endosymbiosis.

The primordial mitochondria are thought to have been free standing early cells called prokaryotes with efficient biochemical mechanisms to convert the products of catabolism we have studied in this and the preceding chapter to the energy necessary for the life of the cell. In this way of thinking, the cells making up our body are not necessarily evolved only from linear changes arising from mutations. Evolution could also take place from what could be called networking in which different life forms found it advantageous to combine their functions. Such networking could involve the gathering of energy-yielding-nutrients by the primordial eukaryotic cells with the efficient energy yielding mechanisms of the mitochondrial prokaryotic cells.

What's going on inside these mitochondria? One answer is something we have already studied – the conversion of fatty acids to fatty acyl coenzyme A and on to acetyl coenzyme A (section 7.7 and following sections). Another mitochondrial process is the taking in of the pyruvate, the conjugate base of pyruvic acid, produced by the breakdown of glucose in the cytoplasm of the cell (Figure 8.11) and converting the pyruvate to acetyl coenzyme A. And then there is the central chemical process taking place within every mitochondrion known on earth, the citric acid cycle. In this cycle, every two carbon atoms that enter the cycle as acetyl coenzyme A replace two carbon atoms that were already part of the molecules in the cycle. The replaced carbon atoms are then converted to two molecules of carbon dioxide.

An oxidation therefore takes place (Sections 3.8, 7.7, 7.9) causing the production of reduced forms of the coenzymes we have studied in Chapter 7, NADH and FADH$_2$. These reduced coenzymes in a process called oxidative phosphorylation, reduce oxygen which sets up a mechanism by which adenosine triphosphate, ATP (section 8.9) is produced. In fact it is estimated that for each glucose molecule yielding two molecules of acetyl coenzyme A entering the citric acid cycle, there are produced over thirty molecules of ATP. By comparison only two ATP molecules are produced from a molecule of glucose without the mitochondrial biochemical processes. Clearly, a cell deriving its energy from an incorporated mitochondrial organelle will have a considerable advantage.

You'll have to look further into biochemistry to flesh out the discussion above and it is possible that such study is in the future for many of you. But for now let's discover the organic chemistry underlying the citric acid cycle and discover that it fits into the reactive principles already studied in this chapter and the preceding chapter. And then we'll discover that we already know why ATP is the source of energy necessary to sustain life, it is all about leaving groups (section 5.3).

The first description of the cycle and the origin of the name citric acid cycle occurred

■ **Hans A. Krebs**

FIGURE 8.19 ▶

The Transformation of the Molecules of the Citric Acid Cycle

oxaloacetate

H_3C SCoA
-HSCoA

1

citrate

2

-H₂O

cis-aconitate

+NAD⁺
-NADH   10

+H₂O   3

malate

H₂O   9

isocitrate

+NAD⁺
-NADH   4

fumarate

oxalosuccinate

+FAD
-FADH₂   8

-CO₂   5

succinate

7
-CoASH

succinyl CoA

6

α-ketoglutarate

in 1937, as published by **Hans A. Krebs**. Krebs, following on his own experiments and based on considerable prior work that can be traced back to the beginnings of the twentieth century, understood the outline of the cycle of chemical reactions taking place. In honor of his contribution, the cycle is often called the Krebs cycle (**Figure 8.19**).

Krebs received the Nobel Prize in 1953 for his elucidation of the cycle and also was knighted by the English government for his scientific contributions. I highly recommend reading the transcript of the lecture he gave. It is a beautiful demonstration of the calm brilliance of this man's scientific thought. You have the background, in combination with the discussion in this section, to understand the essential points and may be stimulated to learn more by what you perhaps won't easily follow. The web site is: http://nobelprize.org/nobel_prizes/medicine/laureates/1953/krebs-lecture.pdf

The advances in organic chemistry during the nineteenth century, some of which have been outlined up to now in the book, meant that the structures of many carboxylic acids were known including the dicarboxylic acids. I remember as a student learning the mnemonic: o my such good apple pie, which stands for the dicarboxylic acids: oxalic, malonic, succinic, glutaric, adipic, pimelic. The structures are all $HO_2C(CH_2)_nCO_2H$ with n varying from zero in oxalic acid to five in pimelic acid. All of these dicarboxylic acids were known to organic chemists by the first years of the twentieth century and chemists began testing these and other carboxylic acids on animal muscle tissue samples to gain some information on the molecular basis of muscle action. The most

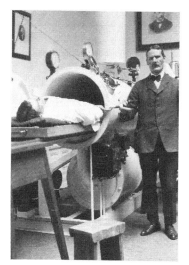

■ **Torsten Thunberg**

prominent among these scientists was **Torsten Thunberg**, professor of physiology at Lund in Sweden. Thunberg discovered what was called at the time "tissue respiration", a field he developed, which was also a focus of Hans Krebs' teacher in Germany, **Otto Heinrich Warburg**.

Thunberg took what we would call today a physical organic chemical approach, which involves varying structure as a test of the mechanism of a process. Krebs reported in his Nobel address that Thunberg between 1906 and 1920, tested the oxidation in muscle tissue of over 60 organic substances and discovered "the rapid oxidation of the salts of a number of acids, such as lactate, succinate, fumarate, malate, citrate and glutamate." The structures of these carboxylic acids are shown in **Figure 8.20** where one can see similarities in structure to the molecules in the citric acid cycle (Figure 8.19). It was realized in those years that the ability of muscle tissue to oxidize these carboxylic acids was related to the molecular events associated with the intake of oxygen in live animals or, in other words, to the consequences of breathing, of life itself.

If one follows the research in the years following Thunberg's seminal work, the investigation of the action of muscle tissue on organic molecules (especially such tissue from pigeons, where respiration is exceptionally rapid) was the main tool leading to understanding. Krebs was finally able to put a coherent picture together by the realization that the product of glucose catabolism, pyruvic acid (Figure 8.11) was the connection between glucose and the cycle (Figure 8.19). The biochemical process converts pyruvate to acetyl coenzyme A, which then enters the citric acid cycle as shown in step **1** in Figure 8.19

■ **Otto Heinrich Warburg**

lactic acid

succinic acid

fumaric acid

malic acid

citric acid

glutamic acid

◄ **FIGURE 8.20**

Carboxylic Acids Tested for Tissue Respiration

and the conjugate bases are named "...ate" as for example: lactate, succinate...etc.

But it took more than a decade for the full picture to develop, to understand that it was not the conjugate base of pyruvic acid, pyruvate, that was the entry point but rather acetyl coenzyme A, which is derived from pyruvate (Figure 8.11.). And acetyl coenzyme A derived from any source, including fatty acids, as we've seen in Chapter 7, and from other sources such as many amino acids, which we are not studying here, equivalently entered the cycle. It is now understood that many biochemical mechanisms lead to molecules that enter the Krebs cycle and that the cycle produces many molecules that are essential for the synthesis of molecules that life uses. If there be an essential element that defines life at the molecular level, it is the Krebs cycle. Here is what Krebs said about what he called the citric acid cycle in the lecture he gave on receiving the Nobel Prize in 1953:

"Before I conclude I would like to make an excursion into general biology, prompted by the remarkable fact that the reactions of the cycle have been found to occur in representatives of all forms of life, from unicellular bacteria and protozoa to the highest mammals. We have

long been familiar with the fact that the basic constituents of living matter, such as the amino acids and sugars, are essentially the same in all types of life. The study of the intermediary metabolism shows that the basic metabolic processes, in particular those providing energy, and those leading to the synthesis of cell constituents are also shared by all forms of life.

The existence of common features in different forms of life indicates some relationship between the different organisms, and according to the concept of evolution these relations stem from the circumstance that the higher organisms, in the course of millions of years, have gradually evolved from simpler ones. The concept of evolution postulates that living organisms have common roots, and in turn the existence of common features is powerful support for the concept of evolution. The presence of the same mechanism of energy production in all forms of life suggests two other inferences, firstly, that the mechanism of energy production has arisen very early in the evolutionary process, and secondly, that life, in its present forms, has arisen only once."

Hans Krebs died in 1981. He was born in Germany in 1900 where he studied medicine and chemistry and where his potential for research was realized early in his life. However, in June 1933, shortly after Hitler came to power he lost his position as a consequence of the racial laws. Luckily his potential for distinguished work was already known in England based on another cycle he elucidated in 1932, that which produces urea. Krebs left Germany for England, where he continued his work at Cambridge University and then the University of Sheffield, from where the citric acid cycle was published. In one biographical note the following appears: *"Although Krebs had renounced the Jewish faith twelve years earlier at the urging of his patriotic father, who believed wholeheartedly in the assimilation of all German Jews, this legal declaration proved insufficiently strong for the Nazis."* I found no information on the fate of the Krebs family in the horrible years in Germany following his emigration to England.

PROBLEM 8.32

With the exception of only step **6** in the citric acid cycle (Figure 8.19) every other step finds a parallel in biochemical reactions discussed in detail in this and the preceding chapter. Identify these parallels.

8.7

The Organic Chemistry of the Krebs Cycle

THE KREBS CYCLE OUTLINED IN FIGURE 8.19 does not contain all known details but does show the basic chemical changes. The reactions shown have been numbered in the figure. Lets look at these reactions in turn starting with the reaction of acetyl coenzyme A with the conjugate base of oxaloacetic acid, oxaloacetate, reaction **1**.

First a note of nomenclature: Why the name oxaloacetic acid? In the list of dicarboxylic acids shown above, the list memorized by this professor many years ago, using the mnemonic about apple pie, oxalic acid has the structure, $HO_2C\text{-}CO_2H$. Acetic acid has the structure $HO_2C\text{-}CH_3$. If one removes a hydrogen atom from the CH_3 group of acetic acid and an OH group from one of the carboxylic acid groups of oxalic acid to form a molecule of water, HOH, you get a combination of oxalic acid with acetic acid, oxaloacetic acid. Identical reasoning applies to the name oxalosuccinic acid, the conjugate base of which, oxalosuccinate is produced by reaction **4** of the citric acid cycle. Note however, as is all too usual, this method of naming does not follow the standard rules such as outlined in Chapter 4, section 4.5.

The ten steps shown for the citric acid cycle in Figure 8.19 are enzyme- catalyzed and the details of the catalytic processes are complicated and not appropriate for our discussion in this book. But what is appropriate is the realization that these reactions make sense based on fundamental principles of organic chemistry and, even more, that the principles of chemical reactivity discussed in this and the preceding chapter are found again in this cycle. Below, we'll discuss each step of the citric acid cycle in organic chemical terms. In every enzyme-catalyzed reaction in the citric acid cycle and, in fact, in every biochemical process, the principles of organic chemistry can be discovered.

Let's look at reaction **1**. It's an Aldol condensation just as we've seen before in one of the synthetic steps in the production of isopentenyl diphosphate (Figure 8.15). Remove a proton, H^+, from the methyl group of acetyl coenzyme A. Just as in Figure 8.15, the carbanion produced by removal of this proton is a powerful nucleophile and attacks the ketone carbonyl group of oxaloacetate to produce a hydroxyl group – classic aldol chemistry, although taking place on the carbonyl group of a ketone.

In addition, in step **1** the thioester group of acetyl coenzyme A is hydrolyzed to the conjugate base of the carboxylic acid. Let's look back at section 7.6 where we were describing nature's attempt to go in the opposite direction, the conversion of a fatty acid to the thioester with coenzyme A. In our study of how nature accomplishes this task, we discovered what would work in the opposite direction, that is, how to convert a thioester to the carboxylic acid (or equivalently, its conjugate base). Addition of the equivalent of HO^- to the thioester group leads to the intermediate seen in Figure 7.14, which then leads to the hydrolysis, which was the unwanted result in Figure 7.14 but is the desired result of reaction **1**.

How about reaction **2**? In section 8.4 where the Aldol reaction (condensation) is discussed we discovered that the commonly encountered path is loss of water to form the conjugated double bond from the alcohol that was produced via the nucleophilic attack at the carbonyl carbon (Figure 8.14). In the latter figure we formed crotonaldehyde, in Figure 8.19 as a consequence of water loss in step **2** we form cis-aconitate, the name given to the product of this step.

We've seen the addition of water to a conjugated double bond before in the catabolism of fatty acyl coenzyme A (Figure 7.23). Here we see it again in step **3**. Although the details differ, the basic idea is identical, a conjugate addition (section 7.8). The molecule produced is the conjugate base of a structural isomer of citric acid and therefore is called isocitrate.

One of Nature's purposes in the citric acid cycle is to eject two carbon atoms from the organism as CO_2 while producing reduced forms of the coenzymes NADH and $FADH_2$.

In section 8.5 (Figures 8.17 and 8.18) we saw the common reactivity principle behind breaking carbon-carbon bonds. The electrons in the bond to be broken have to have "someplace to go." In reaction **2** in Figure 8.18 the sink for the electrons that had formed the carbon-carbon bond to be broken is a negatively charged carbon α to a carbonyl group and, therefore, resonance-stabilized.

Going back now to the citric acid cycle, step **4** in producing a carbonyl group by the oxidation reaction shown, creates the necessary sink. The electrons holding the $-CO_2^-$ group in the middle of the isocitrate molecule had no place to go. In oxalosuccinate the situation is entirely changed. The electrons that hold the $-CO_2^-$ group have someplace to go by forming a resonance stabilized carbanion α to the newly created carbonyl group. And so step **5** takes place with loss of CO_2. The negative charge produced on loss of the carbon dioxide finds a proton and α-ketoglutarate is formed. Steps **4** and **5** are perfect for the citric acid cycle's goal – carbon dioxide is lost and a molecule of NAD^+ is converted to NADH.

The summary of the catabolic processes for both fats and sugars is presented in Figure 8.11. Each glucose molecule produces two molecules of pyruvic acid. The pyruvic acid then loses carbon dioxide to produce a molecule of acetyl coenzyme A. But we never discussed how this reaction takes place. It is a fascinating reaction and in fact requires a vitamin, thiamine, as a coenzyme. The absence of this vitamin, B-1, causes a disease called beri beri. You might like to look into it. There is much on the web and the history of the discovery of its cause, which involved the Japanese Navy.

There is a common structural feature in pyruvate and α-ketoglutarate. The $-CO_2^-$ group to be lost as CO_2 in the former to produce acetyl coenzyme A (Figure 8.11), and in the latter, to produce succinyl coenzyme A (via step **6** in the citric acid cycle), are bonded directly to a carbonyl group. In both structures there is no place for the electrons in the bond holding the CO_2 group to go and no easy reaction to alter this situation. Vitamin B-1 supplies a mechanistic path to overcome the problem but it is a bit complicated and we'll let it go as unnecessary for your studies now. However, the path created by thiamine for this problem does, as you suspect, give the electrons holding the carbon dioxide "someplace to go" in the thiamine.

The next step in the cycle, **7**, is also a bit beyond where we want to go in this book–note that H_2O is surprisingly not involved and we have also not shown all the coenzyme reactants involved. But rest assured we are not trying to hide reactions that defy organic chemistry principles. In steps **8, 9** and **10**, we are back on familiar ground.

In section 7.7 we discovered the necessity for incorporation of a double bond at a precise place in the fatty acid chain as the first step in the catabolic process and that the coenzyme called on for that task is FAD. The double bond produced in the fatty acid chain is conjugated with a carbonyl group just as is the double bond produced in step **8**, which uses the same coenzyme. Yes, the details are certainly different as is the enzyme, but the essential organic chemistry is identical. And step **9**, is another conjugate addition, just like step **3**.

Finally, step **10** is a perfectly routine oxidation of an alcohol to a ketone, analogous to step **4**, with NAD$^+$ converted to NADH (and also analogous to the oxidation step in the fatty acid catabolism in Figure 7.25). We've now returned full circle to oxaloacetate. Beautiful!

PROBLEM 8.33
By taking account of all bonding and nonbonding electrons in the molecules of the citric acid cycle in Figure 8.19, test for formal charge and the octet rule for all atoms in the second row of the periodic table.

PROBLEM 8.34
Write out the structures and names of all the dicarboxylic acids corresponding to the mnemonic about apple pie in the text (section 8.6), using both stick figures without showing carbon and hydrogen atoms and then with all these atoms shown.

PROBLEM 8.35
The functional group corresponding to the highest oxidation state accessible to a primary alcohol, can have a name ending in "-ic" or ending in "-ate." To which structures does this nomenclature refer? Apply the distinction of -ic versus -ate to the structures of the dicarboxylic acids you noted in your answer to problem 8.34.

PROBLEM 8.36
Considering that secondary hydroxyl groups, such as in lactic acid, or in malic acid, can be oxidized to ketones and that amino groups such as are encountered in glutamic acid can also be oxidized to ketones, what conclusion can you draw about Thunberg's experiments described in the text (section 8.6)?

PROBLEM 8.37
In most of the steps in the citric acid cycle (Figure 8.19) stereochemistry plays a role. Some of these stereochemical features are obvious such as the absolute configuration of the two chiral carbon atoms in isocitrate. Some are more subtle and in some steps stereochemistry is not involved. Identify the categories and the stereochemical features in the cycle, step by step.

PROBLEM 8.38
How does reaction 1 in Figure 8.19 demonstrate that the faces of the ketone group are related as enantiomers (enantiotopic)? Show this in a three dimensional drawing

PROBLEM 8.39
Assume that step **9** in Figure 8.19 takes place in a manner where the OH group and the H, which add to the double bond approach from opposite faces of the double bond. Show the mechanism of this process in a three-dimensional drawing using both a saw horse and a Newman projection. Identify the incoming H with a star (*) in the malate produced.

PROBLEM 8.40
Oxaloacetate has no stereoisomers, yet its synthesis in the citric acid cycle from malate takes place in a stereospecific manner. Explain this statement.

PROBLEM 8.41
The loss of CO_2 in step 5 (Figure 8.19) is a demonstration of what has been called in this book "electrons having some place to go." Show the reaction in structural detail.

PROBLEM 8.42
Even in vitro, in the absence of enzyme catalysis, oxalosuccinate would be unstable for loss of CO_2, while a-ketoglutarate is a stable molecule. Explain the difference.

PROBLEM 8.43
What comparable steps are there in the catabolic process for fatty acid degradation to steps **8**, **9** and **10** of the citric acid cycle?

I SSUES OF CHIRALITY ABOUND through the chemical reactions of the citric acid cycle shown in Figure 8.19 and considering that enzymes are involved at all stages, as we have discussed earlier, the reaction paths will involve only one of the enantiomers in molecules that are chiral (section 7.8). However, a problem arose in this regard only four years after Krebs' proposal of the cycle.

In 1941, as isotopes became available for biochemical research, which we've discussed with regard to the insights leading to elucidation of the mechanism of formation of lanosterol and cholesterol (section 5.8), use of isotopes of carbon revealed that in the formation of α-ketoglutarate from oxalosuccinate (step **5** in Figure 8.19), one of the carboxyl carbon atoms contained all of the carbon isotope that had been introduced from one of the reactants of step **1**. This seemed impossible because the oxaloacetate, after reacting with acetyl coenzyme A (step **1**) formed a symmetrical molecule, citrate.

Figure 8.21 (in which the numbered reaction steps correspond to those in Figure 8.19) shows what would be the fate of an isotope of carbon introduced in one of the reactants in step **1**. If the two CH_2-CO_2^- groups of citrate reacted identically in the steps to follow, as shown in Figure 8.21, then the isotope should appear with equal probability in both of the carboxyl carbon atoms in the α-ketoglutarate produced in step **5**.

But the experiment shows that the isotope does not appear equally in the two – $CH_2CO_2^-$ groups. This means that these two groups in citric acid had reacted differently. Why? Is citrate not a symmetrical molecule? The isotope experiment published in 1941 demonstrated that the isotope only appeared in the CO_2^- group α to the carbonyl group in α-ketoglutarate and therefore was lost as CO_2 in step 6 of the cycle. Citrate, therefore, could not be an intermediate in the citric acid cycle, causing, as you can imagine, considerable consternation for Krebs.

In fact Krebs had to wait until 1948 to be relieved of the anxiety arising from the isotope experiment. In that year there appeared in Nature a single page note by **Alexander George Ogston** pointing out, although in different terms, what we have learned about enantiotopic groups in section 5.6. Until that time no one had realized that structurally identical groups in a molecule could have a mirror image relationship and therefore be

▪ **Alexander Ogston**

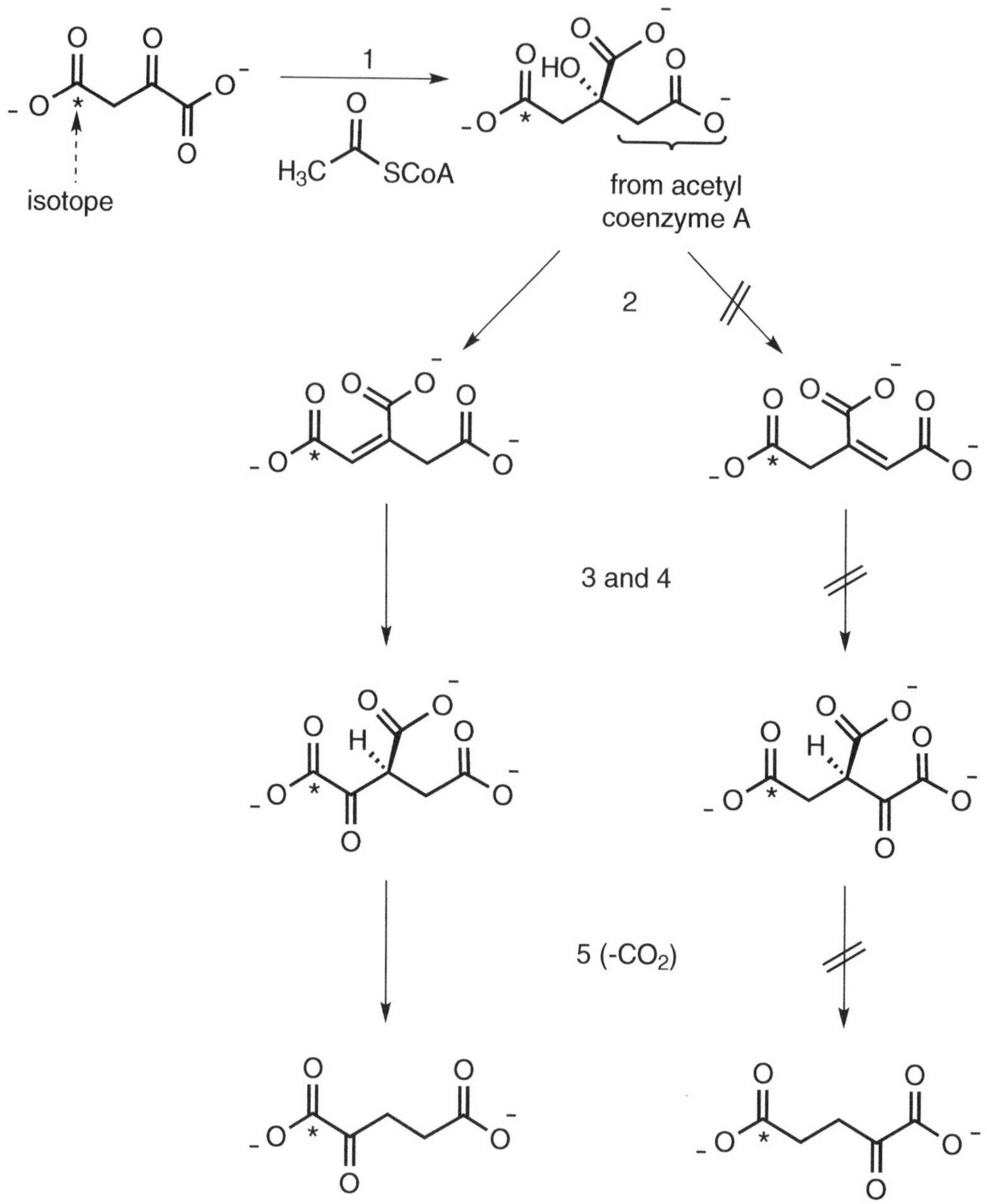

FIGURE 8.21 ▶

Results of the labeling experiment apparently demonstrated that Kreb's proposal of the citric acid cycle was impossible.

distinguished by a chiral reactant such as an enzyme. In a memoir published the year of Ogston's death in 1996 the following paragraph appeared:

"The Ogston concept of 'three point attachment,' conceived in two seconds and written the next day, was widely and rapidly accepted. Indeed shortly after publication of the idea, Krebs wrote excitedly to Sandy and subsequently spent much of his Harvey Lecture (1949) on the topic Sandy was always somewhat embarrassed by the importance others placed on an idea that had been conceived in such a brief time!"

There is something about the enzymatic inequality of the two structurally identical $H_2C\text{-}CO_2H$ groups in citric acid that has always fascinated me. It is not that the enzyme can distinguish these groups. That is a necessity as discussed in sections 5.6 and 7.8. The fascination is rather that the carbon atoms that enter the citric acid cycle from acetyl coenzyme A take one turn around the cycle while two carbon atoms in the oxaloacetate are ejected as two molecules of carbon dioxide, as revealed by the isotope experiment. Is that not an example of youth replacing age at the molecular level – or perhaps just chance? I prefer the former.

PROBLEM 8.44

Can you understand Figure 8.21? The essential idea is that the two $CH_2CO_2^-$ groups in citrate formed in reaction **1** are enantiotopic (section 5.6) and therefore the differently labeled molecules formed in step **2** are not formed equally because of the chirality of the catalyzing enzyme. In fact, one is formed to the exclusion of the other. Dwell on that a bit.

PROBLEM 8.45

Using an isotopic label (*) for the carbon atom of the methyl group in acetyl coenzyme A, show where this labeled carbon atom appears in the oxaloacetic acid produced after one round of the citric acid cycle if the pro-R CH_2-CO_2^- is involved in the formation of cis-aconitate or if the pro-S group is alternatively involved. Use your answer to assign which enantiotopic CH_2-CO_2^- is involved in the citric acid cycle.

PROBLEM 8.46

Citric acid is not a chiral molecule. Yet citric acid is full of chiral information in terms of constitutionally identical groups within the molecule that are related enantiotopically or diastereotopically. Describe the paired groups that can be described in these ways.

FRITZ LIPMANN WAS THE CO-WINNER of the Nobel Prize with Hans Krebs in 1953. Lipmann won the prize for his discovery of acetyl co-enzyme A, which is certainly an excellent fit with Krebs discovery of the citric acid cycle considering that acetyl coenzyme A is the connection between the catabolism of both fatty acids and glucose and the cycle (Figures 8.11 and 8.19). Lipmann had earlier, in 1941, introduced an important concept to biochemical thought – the idea that energy storage in life is associated with what he termed the high energy phosphate bonds in adenosine triphosphate. By high energy Lipmann meant that these bonds were weak and easily broken.

Many chemical reactions take place in living systems. I have seen, for one example, an estimate of about one thousand reactions taking place in the single cell bacterium, Escherichia Coli. Many of these reactions could not occur spontaneously at the concentrations necessary to sustain life because of a gain in free energy, that is, a positive change in free energy in going from reactants to products.

This is where ATP plays its role. The hydrolysis of ATP to adenosine diphosphate, ADP, and to adenosine monophosphate, AMP (shown in **Figure 8.22**) are very favorable reactions, with large negative changes in free energy. If this favorable change in free energy could be "coupled" with a reaction that has an unfavorable change in free energy, then the overall change might be favorable. Before we see what is meant by the word "couple," let's ask why the hydrolysis of ATP is a favorable reaction? Two factors play a role: the instability of ATP and the stability of the phosphate group released.

In **Figure 8.23** the bond breaking converting ATP to ADP releases the conjugate base, PO_4^{-3}, of a very strong acid, orthophosphoric acid, H_3PO_4, and, therefore, following on the discussion earlier in the book (section 5.3), an excellent leaving group is produced in the hydrolysis. The concept of the leaving group in organic chemistry and the connection to Brønsted-Lowry acids was the focus of section 5.3. Let's take the opportunity here to review some of these ideas.

A leaving group is most generally defined as a moiety in which bond breaking releases the electrons to the leaving group. Leaving groups therefore, generally, are likely to be negatively charged. In all strong Brønsted-Lowry acids, HA, A^- must be stable so that a high energy price is not paid for breaking the bond in the formation of the protonated base, for example in water: $HA + H_2O \rightarrow H_3O^+ + A^-$. The source of the stability of A^- is quite variable. HCl is a strong acid with the stability of Cl^- arising from a full shell electron configuration – analogous to the inert gas in the same row of the periodic table, argon.

If a chlorine atom is connected to a hydrogen atom, as in HCl, or connected to

Why is adenosine triphosphate, ATP, life's way of storing energy? In organic chemical terms we find an answer in the concept of leaving groups.

※ **Fritz Lipmann**

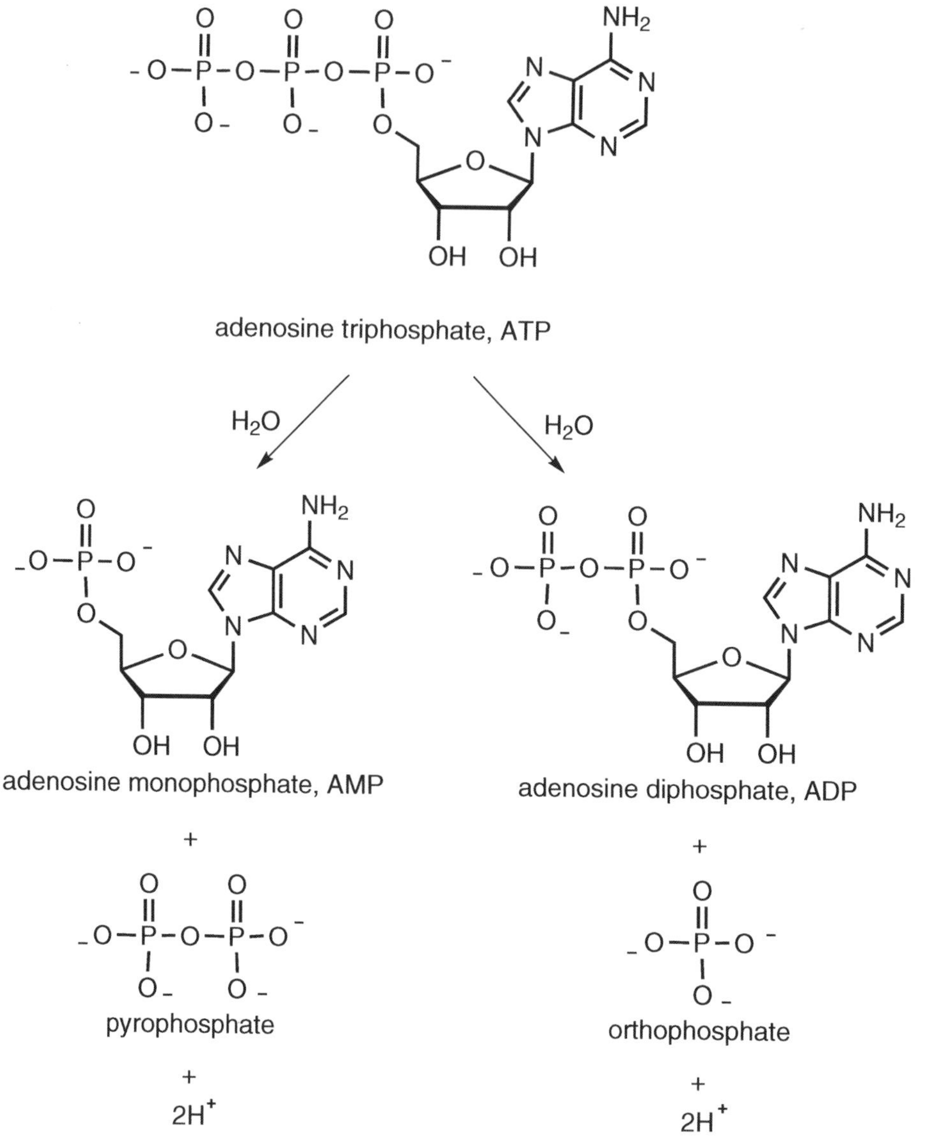

FIGURE 8.22 ▶

Adenosine triphosphate can be hydrolyzed to adenosine monophosphate or adenosine diphosphate.

a carbon atom, as in tertiary butyl chloride $(H_3C)_3C$-Cl, the weakness of the bond to chlorine (what Lipmann would call a high energy bond) is associated with the identical factor, this resemblance of Cl⁻ to argon. However, when Cl⁻ departs from H-Cl, it becomes the conjugate base of HCl, while departure from carbon makes Cl⁻ a leaving group.

The stability of PO_4^{-3} the conjugate base of phosphoric acid, H_3PO_4, arises from a different source, than that for Cl⁻, resonance stabilization. Just as shown in Figure 5.7 for $P_2O_7^{-4}$, the three negative charges on PO_4^{-3} are not constrained each to single oxygen atoms. However, ATP is not a Brønsted-Lowry acid and therefore orthophosphate is not, strictly speaking (see above) a conjugate base. These negatively charged groups are leaving groups. But the resonance stability is identical whatever the source of these phosphates, which means that the high acidity of the various phosphoric acids guarantees the easy hydrolysis of ATP.

The stability of the phosphate group produced on hydrolysis of ATP would be reason enough for the large negative change in free energy. But the ease of hydrolysis of ATP is further enhanced by a factor that destabilizes ATP, the repulsion arising from the large number of negative charges on the phosphate moiety.

In water solution, therefore, ATP is easily hydrolyzed. Even in the absence of water ATP is highly reactive to nucleophilic reactants. In Figure 7.16, the nucleophilic role is played by a fatty carboxylate. Although ADP is not produced in this reaction (Figure

7.16), the breaking of the phosphorus to oxygen bond in ATP is driven by the identical forces at work in the hydrolysis of ATP as shown in the mechanism in Figure 8.23. Both reactions (Figures 7.16 and 8.23) are driven forward by the resonance stabilization of the phosphate group, stabilized also by decrease of the negative charge density in the precursor ATP.

In Figure 8.16, step **3** is called a phosphorylation and produces a phosphate group from one of the OH groups and a diphosphate group from the other OH group. Where did these phosphorus containing groupings arising from step **3** (Figure 8.16) come from? Two molecules of ATP react in step **3**, with the two OH groups (one primary and the other secondary) acting as the nucleophiles to break the phosphorus oxygen bonds in the ATP molecules. In Figure 8.23 water plays this nucleophilic role and in Figure 8.16 the two hydroxyl groups (-OH) play this nucleophilic role.

What is nature's purpose in placing two excellent leaving groups on the intermediate produced by step **3** in Figure 8.16? One answer is found in step **1** of Figure 8.17, which shows the detailed electron movements underlying loss of carbon dioxide. Carbon

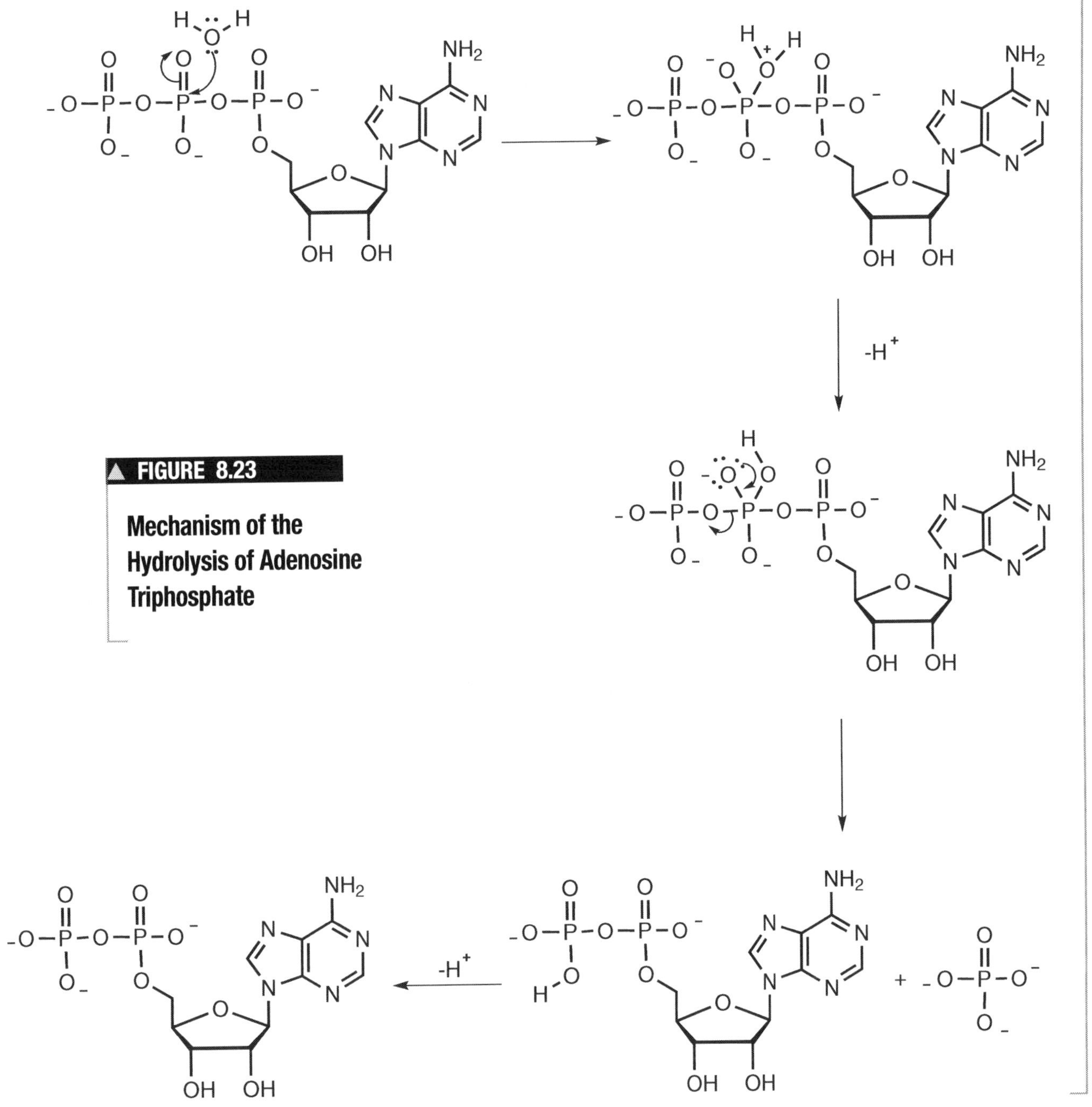

▲ **FIGURE 8.23**

Mechanism of the Hydrolysis of Adenosine Triphosphate

dioxide can be lost because of the ease of loss of the phosphate group. Nature needs an excellent leaving group in a precise structural position in order to accomplish loss of CO_2, the final step in the synthesis of isopentenyl diphosphate.

What about the diphosphate group formed from the other OH group in step **3** of Figure 8.16? Why is an excellent leaving group necessary at this site? For the answer we go back to Chapter 5 and to the in vivo synthetic steps involved in formation of the terpenes. In Figure 5.12, the formation of geranyl diphosphate requires first rearrangement of isopentenyl diphosphate, to dimethyl allyl diphosphate. Now the diphosphate group finds itself bonded to a carbon atom that is a resonance-stabilized carbocation when the $P_2O_7^{-4}$ group leaves.

Formation of this carbocation (Figure 5.12) is then followed by reaction with an isopentenyl diphosphate to form geranyl diphosphate. The geranyl diphosphate formed has the diphosphate group again in a position to be lost leaving behind a resonance stabilized carbocation, which reacts with another isopentenyl diphosphate to form farnesyl diphosphate.

These examples from terpene synthesis yield an understanding of the meaning of the word "coupled." If ATP is to be involved in the breaking of the bonds, the ATP molecule has to become part of the mechanistic path of the reaction.

The formation of the phosphate and diphosphate groups in step **3** of Figure 8.16 are favorable reactions because of the negative free energy associated with breaking the weak oxygen to phosphorus bonds of ATP. The products of this phosphorylation then give rise to the reactive properties that are necessary to form the higher terpenes, geranyl and farnesyl diphosphate. Another example is the excellent $P_2O_7^{-4}$ leaving group responsible for formation of the carbocation from dimethyl allyl diphosphate, the first step leading to limonene (Figure 5.13).

In the terpene syntheses described just above we see the role ATP plays in supplying the leaving groups necessary to drive these reactions, that is, to assure that these reactions occur with a large negative free energy change. Across the spectrum of in vivo reactions sustaining life this role of ATP is played continuously, that is, coupling the reactivity properties of ATP and the phosphorus-based leaving groups arising from ATP to allow reactions that would otherwise be impossible. It's all about leaving groups and nature's favorites are based on phosphorus.

In Lubert Stryer's classic text of biochemistry one learns fascinating aspects about ATP: an ATP molecule is used within one minute of its formation; a human being at rest consumes about 40 kg of ATP in 24 hours but while exercising, 0.5 kg per minute!!

PROBLEM 8.47

As we learned in section 6.7, about hydrogenation of alkene double bonds and of benzene, less stable double bonds yield more energy on addition of H_2. How would these reactions be described using Lipmann's term 'high energy bonds."

PROBLEM 8.48

Draw all reasonable resonance structures for both orthophosphate and diphosphate. Also draw any resonance structures for ATP, for ADP and for AMP. Judge the relative resonance stabilization in these negatively charged entities by tabulating the number of equivalent resonance structures that can be drawn for each one.

PROBLEM 8.49

Figure 8.23 shows a hydrolysis mechanism for production of ADP from ATP. Use this information to show a mechanism for the phosphorylations of the two hydroxyl groups in step **3** of Figure 8.16.

PROBLEM 8.50

ATP in vivo is often associated with Mg⁺⁺ counterions. Offer an explanation for the role of the magnesium ions and their placement in the structure of the ATP considering the discussion in the text about the stability of ATP. Would it be likely that a magnesium ion would be associated with the phosphate moiety lost in a hydrolysis step?

PROBLEM 8.51

How do the following examples demonstrate the meaning of the word "coupled" used to describe the manner in which ATP plays a role in favoring otherwise unfavorable in vivo reactions: Figure 8.17; Figure 7.16; Figure 7.17; Figure 5.13; Figure 5.12.

CHAPTER 8 SUMMARY of the Essential Material

WE BEGAN BY USING WHAT WAS LEARNED in Chapter 7 to realize that breaking glucose into two, three-carbon pieces using the retro-Claisen reaction was not possible because of the placement of the stabilizing carbonyl group as an aldehyde. We also saw an unusual introduction to the retro-Claisen reaction in which the initiating step was not nucleophilic attack at carbonyl carbon to form the alkoxide but rather forming the alkoxide via loss of a proton from a hydroxyl group. Nature solves the problem of the placement of the stabilizing carbonyl group by conversion of the glucose structure to the fructose structure by movement of the carbonyl to carbon-2 and, therefore, forming a ketone. Fructose and glucose make up the structure of sucrose so that this change does not introduce a strange sugar that is not a food source.

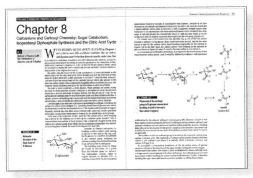

The subject now turns to enediols, the functional group whose properties are used by nature for both the interconversion between glucose and fructose and also the functional group that is used to convert the two three-carbon fragments formed by the retro-Claisen reaction from phosphorylated fructose into identical structures. The enediol discussion leads us into understanding the nature of enol and keto forms that are a consequence of the ease of proton loss from carbon-hydrogen bonds adjacent to carbonyl groups and we come to understand how the structural features of carbonyl containing molecules affect the formation of keto and enol forms. Here we come across the enolate, which brings up the difference between constitutional isomers and resonance structures and reinforces these ideas, the two-arrow versus the double-headed-arrow designations. This section ends with an overview of the conversions of both fatty acids and sugars to acetyl coenzyme A, the ultimate product of their catabolism.

We then turn our attention to the synthesis of a biochemical intermediate we have met in an earlier chapter, isopentenyl diphosphate, a molecule that is appropriate to study here because the in vivo process for its synthesis involves reactions that are variations of those studied in the catabolic process for both fats and sugars. The focus turns to the source of isopentenyl diphosphate and how three molecules of acetyl coenzyme A make up the five carbon atoms in this molecule so that one carbon atom has to be lost. Here we come across not the retro-Claisen condensation (reaction) but rather the Claisen condensation, which nature uses to form one of the key carbon-carbon bonds connecting two of the acetyl coenzyme A molecules together. In the next step the third acetyl coenzyme A molecule is linked to the first two by a new kind of carbon-carbon bond forming carbonyl based reaction, the Aldol condensation. We learn something about this

CHAPTER 8 SUMMARY continued

Aldol condensation and how it can lead to unsaturated carbonyl compounds in which the carbon-oxygen double bond and the carbon-carbon double bonds are conjugated.

We finally come to an intermediate structure on the path to isopentenyl diphosphate in which one carbon atom is lost as carbon dioxide, which only becomes possible because the electrons holding the carbon dioxide have "some place to go," an aphorism that summarizes the availability of a bonding orbital to accept the displaced electrons. Here we see the role of the leaving group again and come to realize that many of the reactions we've studied in Chapters 7 and 8 can be understood by expanding our understanding of leaving groups and the importance of weak bonds, that is, high energy bonds.

At this point, we begin to realize that many in vivo reactions can be understood to be variations on the same theme, played again and again, on different structures, a concept that is seen quite nicely in the citric acid cycle. In this cycle we discover we have seen many of the reactions before but in different form. We see how the coenzymes NAD^+ and FAD, play oxidizing roles seen before in the catabolism of fats. We learn about all kinds of new structures and the sometimes odd names they are known by sometimes easy to remember by devices based on apple pie. We see the problem that arose in accepting Krebs' proposal of the cycle because scientists did not understand a basic concept of stereochemistry we had studied in an earlier chapter – the fact that constitutionally identical groups within a molecule can be related as enantiomeric and therefore distinguished by enzymes.

Finally, the chapter closes with the realization that the role of adenosine triphosphate as a central player in life's chemistry is nothing more mysterious than the role leaving groups play in enhancing chemical reactivity.

Chapter 9

Investigating the Properties of Addition and
Condensation Polymers: Understanding
More about Free Radicals, Esters and Amides

9.1

**"If we knew what we
were doing, it wouldn't
be called research,
would it?"**

I DON'T KNOW IN WHAT YEAR Albert Einstein described research
with this question, but it does seem likely that it would not have affected
the decision in the 1920s by managers of chemical industries to carry
out what was called basic research – chemists being given the resources for investigations
not directly related to the existing commercial products of the industry.

Let's look at two materials, polyethylene and nylon, which arose from this basic
research and which made a great deal of money for two of the most powerful chemical
industries in the twentieth century, Imperial Chemical Industries (ICI) in England
and DuPont Corporation in the United States.

PROBLEM 9.1

Addition polymers such as polyethylene $(-CH_2-)_n$ have the same formulas, although not the
same molecular weight as the monomers from which they are made $(CH_2=CH_2)$. Are cellulose
and starch addition polymers and if not, why not?

9.2

**Polyethylene: The
Background Story**

P OLYETHYLENE in its various forms is produced world-wide in the
range of billions of pounds each year and finds it way into our lives in
many ways. The beginning of the story of polyethylene is told in **Herbert
Morawetz's** wonderful book "Polymers the Origins and Growth of a Science."

*"Just as DuPont decided in 1928 to hire Carothers to "work on problems of his own
selections," so at about the same time Imperial Chemical Industries (ICI) sent two of their
research scientists to Amsterdam to gain experience in A. Michels' laboratory in the study of
reactions under high pressure."*

DuPont's decision led to nylon, which we'll come to later in this chapter, while
ICI's decision led to polyethylene, which arose from a series of misunderstandings and
errors demonstrating a famous statement by a scientist we have already heard much
about, Louis Pasteur (section 1.10):

"In the fields of observation, chance favors only the mind that is prepared."

The reason that ICI decided to look into high pressure arose from academic
research that showed high pressure to cause unusual changes in chemical reactivity
and phase and because engineering advances in the early years of the twentieth
century in combination with new kinds of materials created during the first half of
that century allowed remarkable advances in the ability to attain high pressure. The
leader in this field was **Percy W. Bridgman**, who won the Nobel Prize in physics
in 1946 for his high pressure work, which he began around 1905. Bridgman's most
advanced high pressure apparatuses attained unbelievable pressures, in the range of
50,000 atmospheres! Imagine though that the pressure at the center of the earth is
estimated to be in the range of four million atmospheres and the sun ten thousand

■ **Percy Williams Bridgman**

times greater than that - plenty of room for advances in this field.

Professor Bridgman, who did not stray far from where he was born in Cambridge, Massachusetts in his education and then in his life's work as a faculty member at Harvard University, was a no-nonsense fellow who didn't dwell on the philosophical aspects of science as can be seen from the following quote from 1955, a point of view I certainly agree with:

"It seems to me that there is a good deal of ballyhoo about scientific method. I venture to think that the people who talk most about it are the people who do least about it. Scientific method is what working scientists do, not what other people or even they themselves may say about it."

The scientists who returned from Amsterdam to ICI decided to try a high pressure experiment. Ethylene ($H_2C=CH_2$) was becoming increasingly available as the major product of thermal cracking, which later evolved to become steam cracking of certain petroleum fractions (section 9.4). Cracking technologies for converting petroleum fractions into small molecule alkenes, which was an important boost for the chemical industry's need for a reliable source of starting chemicals for a wide range of products, were developed before the development of the catalytic cracking process discussed in Chapter 4 (section 4.2 and 4.6).

■ **Herbert Morawetz**

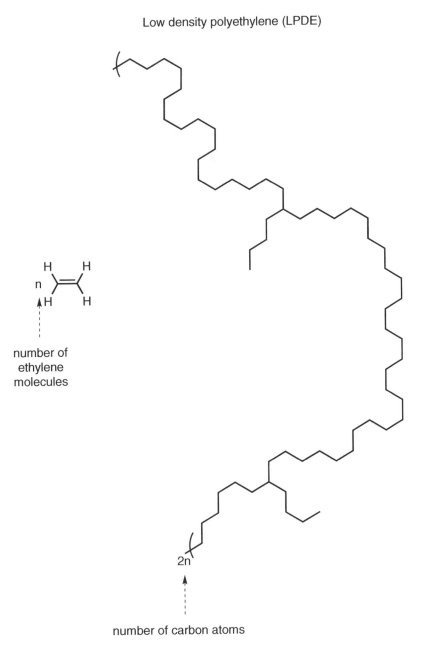

Low density polyethylene (LPDE)

n [ethylene structure]

number of
ethylene
molecules

2n

number of carbon atoms

◀ **FIGURE 9.1**

Structure of Low Density Polyethylene Showing the Branches

Starting in 1933, the ICI chemists tried mixing ethylene, the most abundant product of steam cracking, with various chemicals under high pressure, in the range of 1400 atmospheres, and high temperature. We now know that the reactions they were trying to accomplish are not possible, and, in fact, nothing came of their attempts except that when they opened the high pressure apparatus they found a small amount of a waxy solid, which analyzed as having the same composition as ethylene, two hydrogen atoms for each carbon atom, CH_2.

Realizing from the atomic composition that the waxy solid was a polymer of ethylene, the ICI chemists tried to improve its yield, by raising the pressure and probably the temperature as well. Their work was rewarded with an explosion, which closed down their research in this area for a couple of years. But eventually the experiments were begun again and through trial and error ICI was able, in 1939, to patent the production of polyethylene from ethylene producing a material for which they thought there might be a use but at the time had no idea of what that use might be.

But a use for polyethylene quickly arose as war came upon the world. Cables for radar had to be exceptionally well insulated against dielectric loss and the side with the best radar detection had a great advantage in "seeing" movement of the enemy from great distances. The polyethylene made by the ICI process, called now low density polyethylene (LDPE) (**Figure 9.1**) turned out to be an excellent material for this insulation.

PROBLEM 9.2

Consider chemical reactions in the gas phase in which the numbers of molecules on the left and right side of the equations differ. Would Le Chateliers rule give you a hint as to how these reactions would be affected by large increases in pressure?

PROBLEM 9.3

Did it make sense for the ICI chemists to try and increase the yield of polyethylene from ethylene by raising the pressure? Would lowering the pressure have helped or hurt their intended purpose?

PROBLEM 9.4

Consider the statement above: "Realizing from the atomic composition that the waxy solid was a polymer of ethylene,......" On what basis were the ICI chemists reasoning?

PROBLEM 9.5

Offer a reason for why polyethylene is called an addition polymer.

I N THE EARLY ICI ATTEMPTS at high pressure work, the apparatus was somewhat leaky and consequently small amounts of oxygen became mixed with the pressurized ethylene. Later it was understood that the oxygen was essential to the formation of polyethylene. Tightening up the system and avoiding all leakage of air, which was done to increase the yield, instead caused no polyethylene to be formed.

All introductory chemistry courses introduce the fact that O_2 is a ground state triplet, which means that the normal oxygen in the air we breathe is a diradical as shown in **Figure 9.2**. This figure also shows common reactivity patterns of O_2 in abstracting hydrogen atoms from hydrocarbons to form peroxides and carbon based

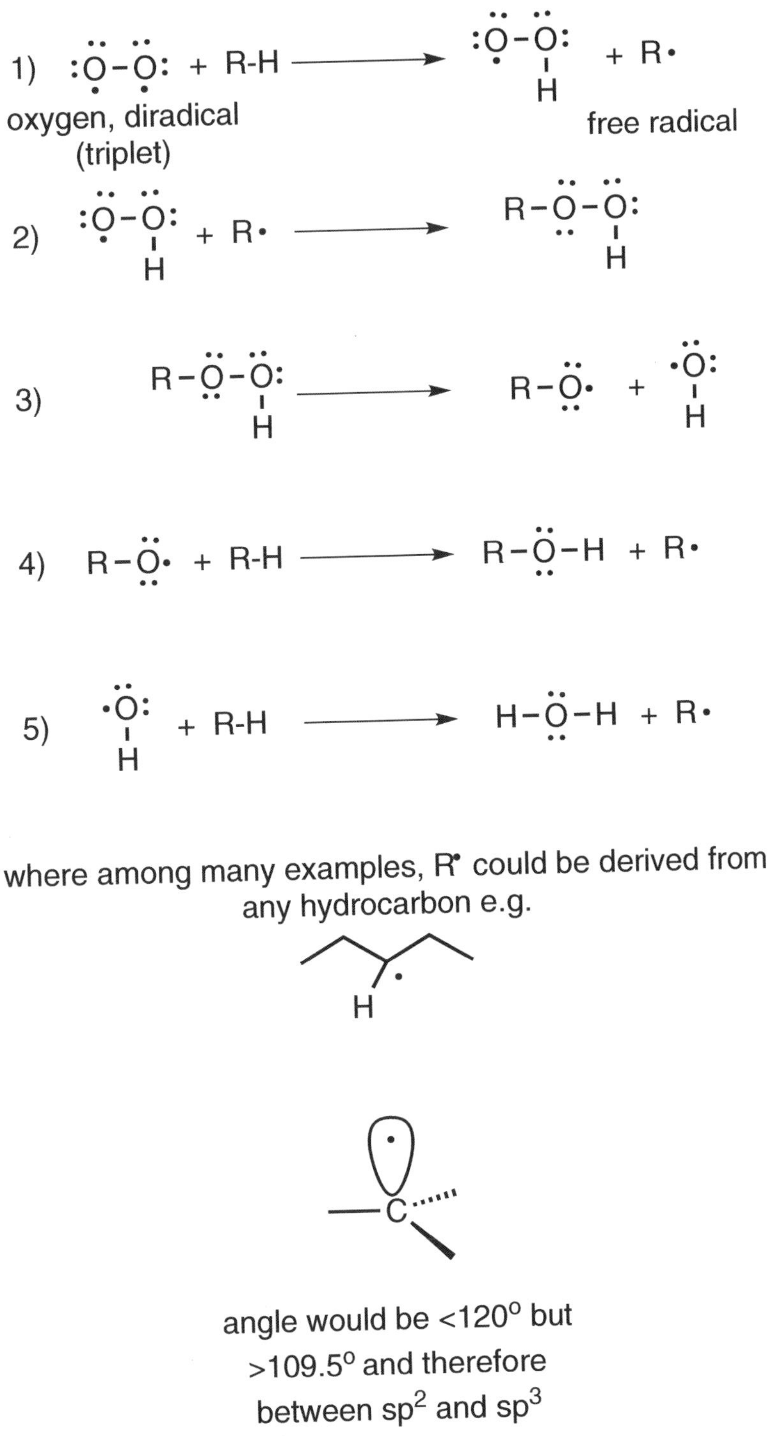

9.3

The Mechanistic Path to LDPE – Free Radicals

1) :Ö–Ö: + R-H ⟶ :Ö–Ö: + R·
                                    |
                                    H
oxygen, diradical   free radical
(triplet)

2) :Ö–Ö: + R· ⟶ R–Ö–Ö:
         |
        H |
                                      H

3) R–Ö–Ö: ⟶ R–Ö· + ·Ö:
     |
     H |
                                         H

4) R–Ö· + R-H ⟶ R–Ö–H + R·

5) ·Ö: + R-H ⟶ H–Ö–H + R·
    |
    H

◄ FIGURE 9.2

Examples of the Formation of Free Radicals and their Hybridized Orbital and Geometric Structure

where among many examples, R· could be derived from any hydrocarbon e.g.

angle would be <120° but >109.5° and therefore between sp² and sp³

radicals. We have not said much about radicals so far in the book (section 4.8, problem 4.31) but let's now learn more about the nature of these intermediates.

Two important reactive characteristics of radicals are shown in Figure 9.2, the abstraction of carbon bound hydrogen atoms, reactions **1**, **4** and **5** in this figure. In addition, there is the combination of two radicals to reform a bond, reaction **2** and the reverse of reaction **3**.

Reaction **3**, in the forward direction, which is the favored direction, is an exception to the general rule that at moderate temperatures covalent bonds do not break to form radicals. Bond breaking leading to two radicals is called homolytic cleavage in contrast to bond breaking that leads to a separation of charge (sections 4.7, 5.3), which is heterolytic cleavage. When chemical bonds are quite weak, even temperatures near to room temperature may be adequate to supply enough energy to homolytically break the bond. Peroxides, which are characterized by a single bond between two oxygen atoms, R-O-O-R, are such bonds and reaction **3** is an example of homolytic cleavage of this functional group.

Carbon based free radicals (R· in Figure 9.2) are less electron deficient than carbocations and as well differ from carbocations, which do not obey the octet rule (section 4.4). The difference is the empty p orbital in the carbocation, therefore causing the carbocation to be sp^2 hybridized so that the three atoms bonded to the positively charged carbon atom exist in a plane (Figure 4.6).

In a free radical the necessary placement of a single electron in what would be an empty orbital in a carbocation, causes some mixing of the 2s orbital into that p orbital leading to the hybridization picture presented in Figure 9.2. The hybridization is neither sp^2 nor sp^3 but somewhere in between. In the sp^3 hybridization the four hybrid orbitals contain 25% s and 75% p, while in the sp^2 hybridization each of the three hybrid orbitals contain 1/3 s and 2/3 p and there is one empty p orbital. The three atoms bonded to the central carbon atom in the free radical no longer exist in a plane associated with sp^2 hybridization nor do the three atoms bonded to the central carbon atom take a tetrahedral geometry. The three atoms bonded to the central carbon atom in the free radical are bent by several degrees from a plane as shown in Figure 9.2.

In Chapters 7 and 8 there was considerable focus on the reactive nature of carbanions, which together with carbocations and carbon-based free radicals make up the most important, but not all the reactive intermediates in organic chemistry.

Let's take a moment to step back to understand the hybridization and geometric pictures of these three reactive intermediates. There is an interesting change in angular geometry in moving from the empty p orbital in a carbocation (120° for the three bonds–a planar arrangement), to having an orbital that can hold an electron in the free radical about 115° for the three bonds to having an orbital that can hold two electrons in, a carbanion. The bond angle between the three groups bonded to the negatively charged carbon atom is now even smaller, in the range of 105°. You are given a chance to explain the relationships among these geometrical changes and how they fit with hybridization of atomic orbitals in the questions to follow in this section.

Not withstanding the differences between carbocations and free radicals, there are similarities. Just as carbocations are electrophilic, reactive with a source of electrons, a characteristic used to advantage in the industrial synthesis of the high octane gasoline, 2,2,4-trimethyl pentane (section 4.10), free radicals also react rapidly with double bonds. The consequence of this reactivity with ethylene allows understanding of the formation of polyethylene, as shown in **Figure 9.3**.

Figure 9.3 shows the consequences of the addition of a free radical to an ethylene molecule in the presence of a large number of ethylene molecules as in the high pressure apparatus used by ICI. The reactions shown in Figure 9.3 follow the sequence: initiation, propagation and termination, which is the definition of a chain mechanism, a mechanistic path we have seen before in section 4.10 for the production of 2,2,4-trimethyl pentane.

In section 4.4, the term "passing the buck" was used in the discussion about the 1,2 shift of carbocations responsible for the rearrangements taking place in catalytic

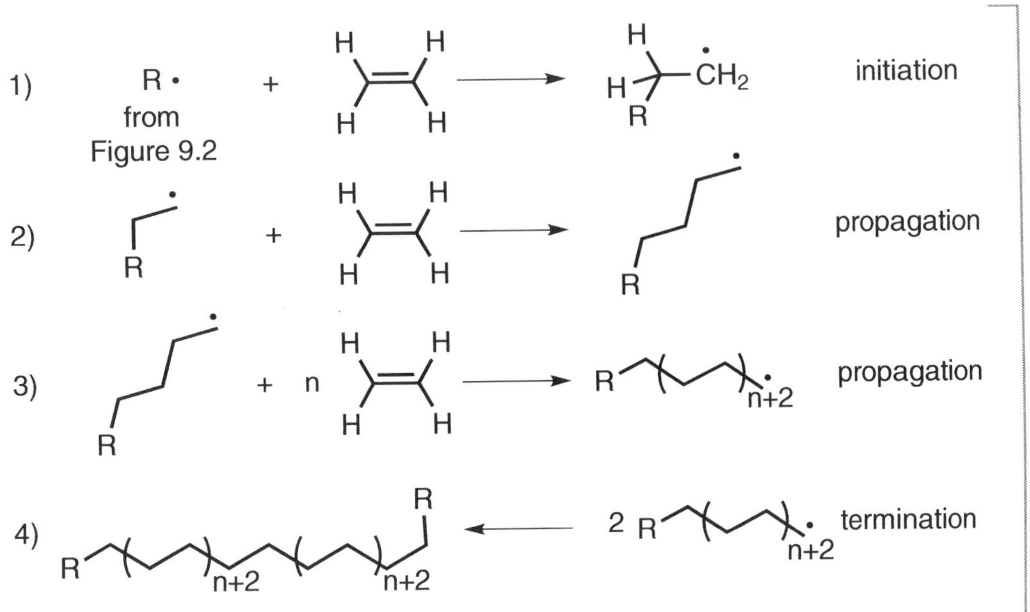

Chain Mechanism for the Free Radical Polymerization of Ethylene

cracking of petroleum fractions. "Passing the buck" is a characteristic of reactive intermediates responsible for many reactions in organic chemistry and we see this at work again here in Figure 9.3. The initiating radical, R·, in adding to the double bond of ethylene, passes the radical to one of the CH_2 groups of this ethylene molecule. The CH_2 group, which now bears the burden as the site of the radical, reaction **1** (Figure 9.3), passes the radical site onto the CH_2 group of the next ethylene molecule (reaction **2**) and so on in reaction **3**.

As long as these propagation reactions go on (reactions **2** and **3**) the polymer chain will grow longer and longer, that is, the molecular weight of the polymer will grow larger and larger, and therefore the degree of polymerization will become a larger number. One limit to the chain growth would be a reduction in the number of molecules of ethylene available for reaction. For this reason the molecular weight is related to the pressure, as is well known to the industry.

However, all chain mechanism are limited by termination reactions, that is, those reactions which eliminate free radicals which can "pass the buck," that can keep the propagation reactions going. Reaction **4** (Figure 9.3) is a prominent example of a termination reaction (certainly not the only one) in the free radical polymerization of ethylene. In reaction **4** two growing chains combine so that the free radical sites on the CH_2 groups at the chains' ends combine to form a covalent bond. There is no radical left to continue the propagation reactions.

The probability of two chains, with free radicals at their ends, to find each other (termination) is low as long as the population of ethylene molecules is high. The ethylene molecules capture the free radicals causing extension of the chains – the process of polymerization. Toward the end of the polymerization process, when few ethylene molecules are left, the termination reaction becomes competitive.

The chain mechanism shown in Figure 9.3 can not be the entire story because the mechanism does not account for the four carbon branches on the polyethylene chain shown in Figure 9.1. How are those structural features explained by the mechanism in Figure 9.3? The answer can be found in the fact that although abstraction of a carbon bound hydrogen atom occurs if the radical site is on oxygen as in reactions **1**, **4** and **5** (Figure 9.2) abstraction also occurs if the radical site is on carbon as: $R· + R'H → RH + R'·$.

Let's discover in **Figure 9.4** how a variation of the abstraction of a hydrogen atom by a carbon based free radical from another carbon atom is responsible for the four carbon atom branches on the polyethylene chain.

The ease of rotation about single bonds, conformational motion, was introduced

FIGURE 9.4 ▶

Backbiting Mechanism for the Formation of Branches in Polyethylene

1)

2)

3) → LDPE

in Chapter 1 (section 1.13). Imagine, therefore, the indescribably large number of conformations accessible to a long chain of CH_2 groups in polyethylene while the chain is growing or in its final state. Polymer science deals with this complexity using the power of statistical physics and some describe the array of conformations among the many chains in a sample of polyethylene as analogous to a plate of very long just cooked spaghetti. These rapid conformational motions along the growing polyethylene chain can occasionally bring the terminal $-CH_2$ radical site on the growing end of the chain in the vicinity of a carbon-bound hydrogen atom within the chain such as shown in the conformational motion in Figure 9.4.

Organic molecules love six membered rings, which was a focus of Chapter 3. This was a statement about structure. But six membered rings are also favored in many reactions of organic chemistry. Although we can't go into the details here, this fact brings up a reaction named after Derek Barton, whom we have met before in Chapter 3 (section 3.3). In the Barton reaction, a free radical is generated by converting an -OH group to a functional group with the structure $-ONO$. Loss of $NO^•$, results in formation of $-O^•$, which abstracts a hydrogen atom from a carbon atom in the same molecule so that the abstracted hydrogen atom, the oxygen and the intervening carbon atoms make up a six membered ring. In other words, the transition state of the reaction is constructed of four carbon atoms with the oxygen atom and the hydrogen atom. All this takes place in a molecule with a steroid structure and was the method by which an early version of the birth control pills were first synthesized in which **Carl Djerassi** played a large role.

There are many other examples in organic chemistry of chemical reactions taking place via six membered transition states including that shown in Figure 9.4 (reaction **1**), which transfers (passes the buck) the radical site from which the chain is growing, a CH_2 group, to a site five carbons within the chain. Ethylene molecules, following the same mechanistic path as shown in Figure 9.3, then grow from this new radical site, leaving the four carbon branch seen in the structure of LDPE in Figure 9.1. The site transfer is an intramolecular rearrangement (section 3.14), and happens in a random fashion so that the numbers and placements of the four carbon branches are not predictable, except on a statistical basis.

The creation of new catalysts for polymerization of ethylene, which blocked the formation of these branches, eventually put ICI out of the polyethylene business. ICI had not created the catalysts. The chemical industry has a name for what happened to ICI, "shut down economics." If you are in a product line, you had better be the inventor of new technologies that affect that product line. Not long ago, an astonishing example of shut down economics took place with the Kodak Corporation.

■ **Carl Djerassi**

PROBLEM 9.6

Following the rule that an atom with four different groups around it, such as the four hydrogen atoms around a carbon atom in methane, will be sp^3 hybridized, account for the hybridization of a carbon based free radical, $R_3C^{\cdot}$, while explaining why the angle between the R groups deviates from the tetrahedral angle.

PROBLEM 9.7

Using the distribution of energies as shown in Figure 6.15 explain why weaker bonds, such as the oxygen-oxygen bond in a peroxide, will break at lower temperatures than the temperatures necessary to break stronger bonds.

PROBLEM 9.8

Are you able to offer an explanation for why the bond angle for the three bonds in a carbanion is smaller than the tetrahedral angle as discussed in the text? How does your answer to this question relate to your answer to problem 9.6?

PROBLEM 9.9

How do the proportions of s and p orbitals in the hybridized orbitals used for bonding to the central carbon atom for a carbocation, a free radical and a carbanion vary? Answer this question also for the orbital that is not taking part in the bonding to the three pedant atoms in these three intermediates.

PROBLEM 9.10

Although addition of a radical, $R^{\cdot}$, to ethylene could not reveal any information about a free radical parallel to the Markovnikov rule (section 4.10), how do you think the rule might be adapted for free radicals based on the addition of $R^{\cdot}$ to propylene, $CH_3CH=CH_2$?

PROBLEM 9.11

Try to propose additional termination reactions to the chain growth in Figure 9.3.

PROBLEM 9.12

Does the mechanism responsible for formation of the branches in polyethylene (Figure 9.4) reveal that the stability of free radicals is parallel or opposite to the stabilities of carbocations?

PROBLEM 9.13

Using step **3** of Figure 9.3 as a model, draw structures with all atoms shown for the addition of several molecules of ethylene to the growing chain.

PROBLEM 9.14

Why would reaction **4** in Figure 9.3 increasingly occur as the degree of polymerization grew larger? And why would this increase in reaction **4** be reversed by adding more ethylene to the reactor?

PROBLEM 9.15

For reaction **1** in Figure 9.3 show the orbitals involved and geometry.

PROBLEM 9.16

Use Newman projections to test the idea that a small number of bond rotations can cause a large change in the overall shape of a polyethylene chain. Is your answer that this is reasonable, or not?

PROBLEM 9.17

Draw the structure of a chair conformation transition state for the six membered ring involved in step 1 of Figure 9.4.

PROBLEM 9.18

In the Barton reaction noted in this section, the functional group R-ONO plays the key role. Analyze this functional group by first counting electrons, then determine if it is possible to obey the octet rule and if avoiding a formal charge is possible. Carry this process out for the products of the bond breaking to produce R-O• and NO•.

PROBLEM 9.19

The Barton reaction used on a steroid structure can be found on the web. Look it up if you are interested in the history and the people involved in steroid chemistry. http://www.wiley-vch.de/templates/pdf/3527309837_c11.pdf

PROBLEM 9.20

What does the successful free radical polymerization of ethylene demonstrate about the strength of sp^2 carbon-hydrogen bonds ($=CH_2$)?

PROBLEM 9.21

Define and show at least one example each of the classes of free radical reactions involved in the production of LDPE

PROBLEM 9.22

Can you offer a reason as to how the branches produced by the free radical rearrangement shown in Figure 9.4 cause the low density property of LDPE?

9.4

An important reaction of free radicals is responsible for the large volume production of ethylene and other alkenes from the steam cracking of petroleum fractions.

THE CHEMICAL INDUSTRY needs starting materials with precisely predictable chemical reactivities. This means functional groups are necessary (section 3.9), and the simplest functional groups contain only carbon and hydrogen, which limits the family to triple bonds and double bonds. Early on, the favored starting material was acetylene. But acetylene, H-C≡C-H, is too unstable and unexpectedly explodes with devastating consequences. Ethylene has a valuable and predictable reactive nature and is much safer, so that when it was discovered that petroleum fractions could be heated to high temperatures to yield large quantities of small molecule alkenes (Figure 4.12), predominately ethylene, the chemical industry increasingly switched to the double bond and substantially abandoned the triple bond.

Some of the steps that initiate the chemical processes in steam cracking of petroleum fractions are identical to those initiating polymerization of ethylene, that is, involve oxygen (Figure 9.2). However, there are other sources of the initiating free radicals responsible for steam cracking arising from the exceptionally high temperature, in the range of 800° C, substantially higher than used for polymerization of ethylene. **Figure 9.5** outlines some of the important steps in steam cracking of a typical molecule found in the naphtha fraction of petroleum, n-decane.

As in the formation of 2,2,4-trimethyl pentane (section 4.10), and as seen in Figure 9.3 for formation of polyethylene, we encounter again in steam cracking, a chain mechanism. The detailed initiation, propagation and termination steps may differ but the overall mechanistic paths are parallel to each other.

The bond breaking propagation steps in Figure 9.5, reactions **2**, **3** and **4**, show another example of a class of reaction found in carbon-bound free radicals, β-cleavage. The name β-cleavage arises from the bond broken in this reaction, as shown in **Figure 9.6**, which also points out the overlapping electrons and orbitals

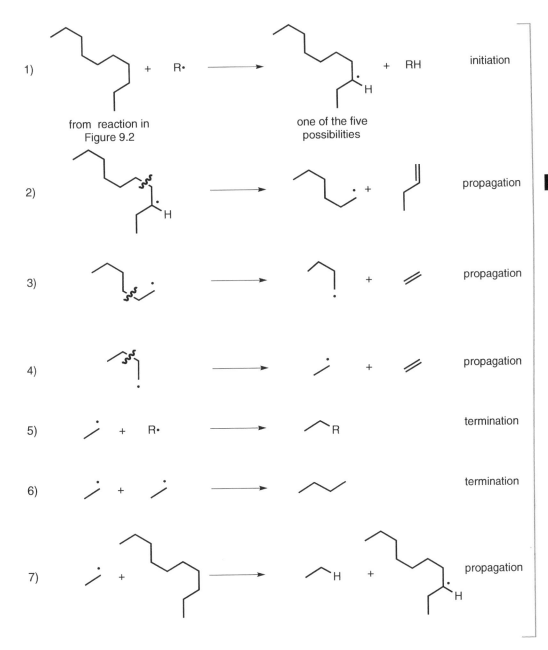

1) from reaction in Figure 9.2 → one of the five possibilities   initiation

2)   propagation

3)   propagation

4)   propagation

5)   termination

6)   termination

7)   propagation

◄ FIGURE 9.5

Free Radical Chain Mechanism in the Steam Cracking of Petroleum Fractions showing the β-Scission Step

leading to the formation of the π- bond resulting from this cleavage.

The β-cleavage steps shown in Figure 9.5 lead to two of the three most abundant alkenes formed in the steam cracking process, 1-butene (reaction **2**) and ethylene (reactions **3** and **4**). The orbital changes arising in β-cleavage shown in Figure 9.6 are chosen to demonstrate how the other most abundant alkene, propylene, would be formed as a consequence of the chance formation of the initiating radical at a different decane carbon atom.

Many other initiation, propagation and termination steps can be imagined in addition to those shown in Figure 9.5 and many other of these steps do, in fact, occur during the extremely short time, in the range of small fractions of a second, that the molecules in the petroleum fraction are exposed to the super heated steam.

Steam cracking is big business. The reactions shown in Figures 9.5 and 9.6 occur for a range of linear hydrocarbons in the naphtha fractions of petroleum and produce huge amounts of alkenes. For one example, within the last ten years a steam cracking unit constructed in the United States by the European companies BASF and Fina produces per year 1.7 billion kilograms of ethylene, propylene and the isomeric butenes, 1-butene and cis and trans 2- butene. Of this 1.7 billion kilograms,

FIGURE 9.6 ▶

Orbital Description of Free Radical β-Scission

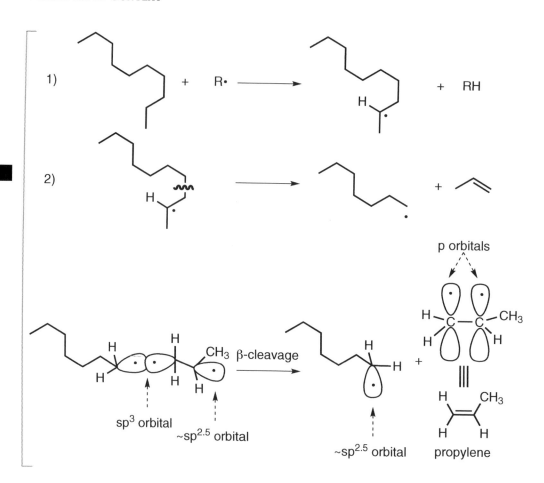

about 0.9 billion kilograms of ethylene are produced. Much of this ethylene is used to synthesize the various forms of polyethylene.

PROBLEM 9.23

While ethylene has one π bond and five σ bonds, acetylene has two π bonds and three σ bonds. How might this ratio of π to σ bonds affect the relative explosive nature of these molecules?

PROBLEM 9.24

Describe the differing sources of free radicals in steam cracking of petroleum fractions versus polymerization of ethylene?

PROBLEM 9.25

Trace all the orbital and geometry changes occurring in a β-scission reaction from ethyl radical, $H_3C\text{-}CH_2^{\bullet}$.

PROBLEM 9.26

Starting with n-heptane, $CH_3CH_2CH_2CH_2CH_2CH_2CH_3$, show how steam cracking could lead to the following alkenes: ethylene, propylene, 1-butene, cis and trans 2-butenes. Identify each step as initiation, propagation or termination.

PROBLEM 9.27

In your answer to Problem 9.26, show all reasonable termination steps.

PROBLEM 9.28

Can you offer an explanation for why steam cracking of petroleum is conducted at far higher temperatures than polymerization of ethylene?

AS NOTED IN SECTION 9.4, steam cracking requires exceptionally high temperatures. However, polymerization of ethylene must be conducted at far lower temperatures and, in fact, cooling equipment has to be involved to keep the temperature from getting out of control leading to what engineers call a "runaway" process. An out of control polymerization of ethylene has the potential to explode as the temperature may reach values in which ethylene can explosively decompose.

The different temperature characteristics necessary for steam cracking and ethylene polymerization offer a lesson in fundamental aspects of thermodynamics.

In Chapter 4 (section 4.10) we discussed the large difference in the strength of σ (sigma) versus π (pi) bonds. The difference arises from the linear overlap of the two sp³ orbitals in a sigma bond versus the weaker overlap of the two p orbitals in the π bond. Conversion, therefore, of σ to π bond, as is the situation in steam cracking (Figures 9.5 and 9.6) is an endothermic reaction. Energy must be added to the system to compensate for the formation of weaker bonds from stronger bonds, which translates to higher temperature favoring the cracking process.

Looking at steam cracking from another point of view, few molecules are converted to large numbers of small molecules, the large linear hydrocarbons in the naphtha fractions of petroleum converted to the alkene molecules. This increase in disorder translates to an increase in entropy in going from starting materials to products. Steam cracking is characterized by a positive, $+\Delta S$, change in entropy. Because the entropy term contributing to the free energy change is $-T\Delta S$, an increase in temperature associated with a reaction with a $+\Delta S$ will move ΔG to a larger negative number, therefore increasing the equilibrium constant, K, ($-\Delta G = -RT\ln K$). The product alkenes in steam cracking will be favored, therefore, by increased temperature considering both the entropy and the enthalpy change. Here we find in this simple thermodynamic analysis of steam cracking, one of the important contributing reasons for conducting steam cracking near 800° C.

In the polymerization of ethylene a π bond is converted to a σ bond – a weaker bond is converted to a stronger bond, precisely the opposite of the situation in steam cracking. The enthalpic change must therefore be exothermic since energy will be released. Removing energy, that is, lowering the temperature will favor the process.

The polymerization also differs in terms of entropy from steam cracking. Instead of many small molecules produced from a large molecule, rather very many small molecules, thousands in fact, are combined to form a very large molecule, a macromolecule – order from disorder, and therefore a negative change in entropy, $-\Delta S$. The entropy term therefore pushes the free energy change to more positive values, a less favorable reaction, a factor that is diminished as the temperature decreases.

Considering both the enthalpic and entropic changes in the polymerization of ethylene, an increase in temperature will therefore act to push ΔG toward more positive values and therefore disfavor the polymerization. In fact, in the synthesis of polymers, there is a temperature called the "ceiling temperature," above which no net polymerization will take place.

Therefore, for thermodynamic reasons and also to avoid the danger of an out of control reaction, heat must be continuously removed in the industrial reactors producing polyethylene – the temperature must be held below about 200° C.

The discussion above finds parallel in Chapter 6 (section 6.7) concerned with the hydrogenation of benzene and, as is the nature of thermodynamic arguments, can be applied to all chemical reactions.

9.5

Contrasting thermodynamic factors control polymerization of ethylene and steam cracking of the naphtha fraction of petroleum.

PROBLEM 9.29

What role do the thermodynamic parameters (enthalpy and entropy) play in the effect of temperature on steam cracking of petroleum fractions.

PROBLEM 9.30

With the realization that weak bonds are high energy bonds, as discussed in Chapter 8 (section 8.9), explain the fact that conversion of a σ bond to a π bond will be favored by an increase in temperature.

PROBLEM 9.31

How can the concept of ceiling temperature, discussed in this section, be understood from considerations of the enthalpic and entropic changes in polymerization?

PROBLEM 9.32

Is it possible to fit the concept of transition state into the discussions of this section?

PROBLEM 9.33

From thermodynamic considerations, does hydrogenation of benzene (section 6.7) resemble polymerization, or steam cracking, or neither?

9.6
Resonance works against the chemical industry again.

■ **Karl Ziegler**

I CAN'T FIND A RECORD of who might have tried to polymerize propylene ($H_2C=CH(CH_3)$) using the method ICI had developed for ethylene. Perhaps it was the ICI chemists. It would have made sense to try it. I do know that chemists at Monsanto Corporation in Texas in 1963 tried it and failed to obtain a polymer. Instead they observed the formation of a gooey liquid indicating a molecular weight far below that of a polymer. Long before that, in 1953, Italian chemists in the group of **Giulio Natta** using a catalyst developed in Germany by **Karl Ziegler** had successfully polymerized propylene to polypropylene. Let's hold that Nobel-Prize-winning story for the next section and first discover what the problem is with free radical polymerization of propylene, the basis of the attempt made by the Monsanto group.

No attempted polymerization of propylene using free radicals has ever been successful, in contrast to the success with ethylene (section 9.3). Oligomers, which are small polymers, do form, the beginning of polymerization, but the process is always aborted after several propylene molecules have added.

An analogous initiation step to the step occurring in the free radical polymerization of ethylene (Figure 9.3) begins the path toward what would be polymerization in propylene (step **1**, **Figure 9.7**). And the propagation steps, **2** and **3**, continue the usual chain mechanism. However, step **4**, terminates the growing chain before many propylene molecules have added. Why?

The problem is that the C-H bonds in the CH_3 group of propylene are weak bonds (high energy bonds (section 8.9)) as a consequence of the resonance stabilization of a free radical formed at that position when the C-H bond is homolytically broken. The weakness of this carbon-hydrogen bond offers an opportunity to the carbon radical at the growing end of an oligopropylene chain to abstract this hydrogen atom in competition with adding to the π bond of another propylene molecule as shown in step **4** Figure 9.7.

The ease of abstraction of a hydrogen atom from the methyl group of propylene arises from the resonance stabilization of the product allyl radical as shown in **Figure 9.8**.

Lowering the energy of activation of a reaction, **4** (Figure 9.7) as discussed here, makes this reaction more competitive with other possible reactive choices in the system, **2** and **3**. So instead of the reaction path following **2** and **3** to propagate the macromolecule, polypropylene, the reaction path follows **4**, generating an aborted chain.

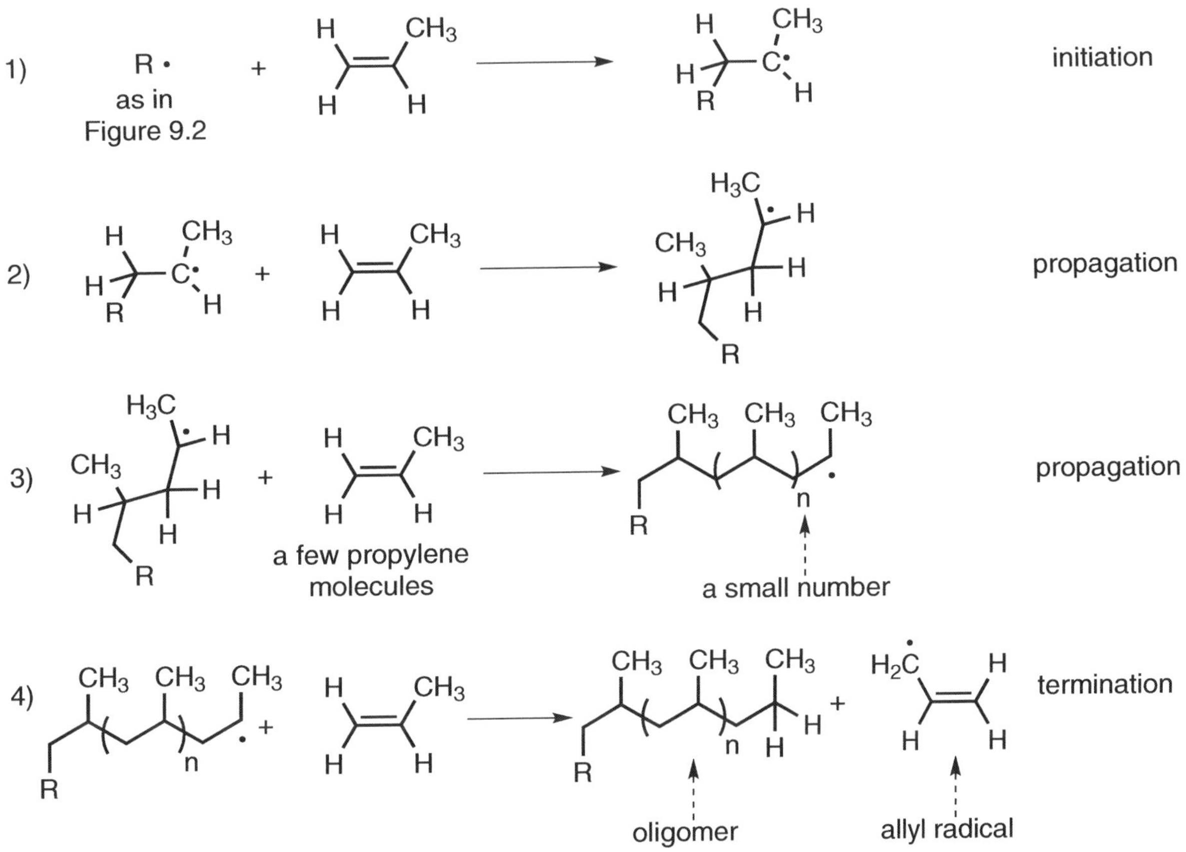

1) initiation

R •
as in
Figure 9.2

2) propagation

3) propagation

a few propylene
molecules

a small number

4) termination

oligomer

allyl radical

In Chapter 6, sections 6.12 and 6.13, in the discussion about the undesirable multialkylation of benzene plaguing the chemical industry, we learned how a resonance stabilized intermediate lowered the energy of activation of an unwanted reaction. Here again we see how a resonance stabilized intermediate takes a reaction path in an undesireable direction, step **4** in Figure 9.7.

But there is an additional part to the story of the aborted polymerization of propylene. Why does the free radical formed by loss of a hydrogen atom from the methyl group of a propylene molecule (designated allyl radical in Figure 9.7) not add to another propylene molecule π bond and start a new polymer chain growing? This possibility is shown in **Figure 9.9** giving a graphical answer to this question using a reaction coordinate diagram (section 6.11)

As shown in Figure 9.9, a resonance-stabilized free radical may be represented as of lower energy than a comparable radical that is not resonance stabilized, that is, R• versus the allyl radical in the reaction coordinate diagram. However, the addition of either radical to propylene will produce a new radical that is almost identical in energy. This means that the path for the allyl radical to add to a propylene π bond has a higher energy of activation than that for addition of R•. In fact, the energy of activation for addition of R• is accessible and so, as it has been put earlier, a passing of the buck occurs. However, the energy of activation of addition of the allyl radical to the π bond of a propylene molecule is far larger, translating to a smaller rate constant. Therefore the buck is not passed, so-to-speak, and a new polymer chain is not begun.

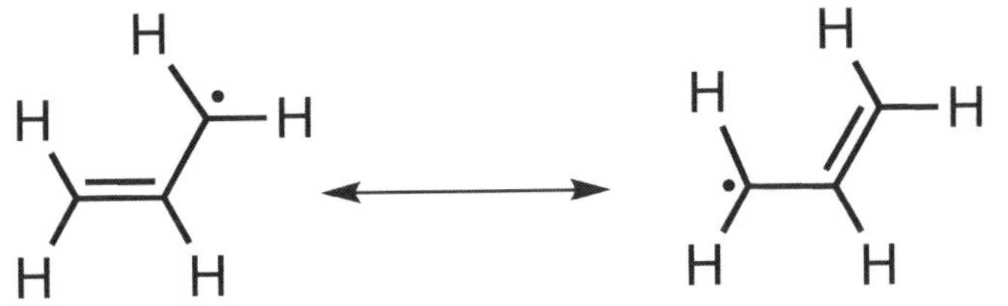

▲ **FIGURE 9.7**

Hydrogen Abstraction Mechanism Underlying the Aborted Free Radical Polymerization of Propylene

◄ **FIGURE 9.8**

Resonance Structures of Ally Radical

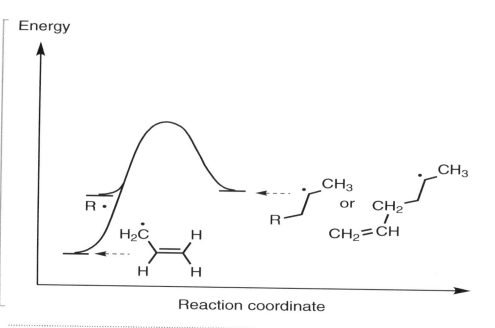

FIGURE 9.9 ▶

Reaction Coordinate Diagram Allows Understanding Why Allyl Radical Does Not Initiate Polymerization

---

PROBLEM 9.34

What is meant by the statement in the title of this section about resonance working against the chemical industry <u>again</u>?

PROBLEM 9.35

Does torsional motion play a role in the resonance stabilization of allyl radical ($CH_2=CH-CH_2$)?

PROBLEM 9.36

In Figure 9.9, two different initiation reactions are compared, which could lead to polypropylene. Offer an explanation for the fact that the starting energies of the two reaction paths differ greatly, while the transition states are almost identical. Why does this difference in the starting states but not in the transition states of the two paths cause one to initiate the polymerization and not the other?

PROBLEM 9.37

There are six C-H bonds in a propylene molecule. Why are the three C-H bonds of the methyl group so much weaker than the other C-H bonds in the molecule?

PROBLEM 9.38

The faces of the double bond of ethylene fundamentally differ from the faces of the double bond of propylene. Explain this fact.

9.7

A Short Story about a Nobel Prize

KARL ZIEGLER, A GERMAN CHEMIST internationally known for his chemical research before the Second World War, especially in areas associated with the overlap of organic chemistry and inorganic chemistry, that is, organometallic chemistry, was dedicated to a life of fundamental research. He did not want commercial interests to control his work. In 1943, when he was offered the directorship of what came to be known after the war as the Max-Planck-Institut für Kohlenforschung (Max-Planck-Institute for Carbon Research) in Mülheim/Ruhr, he feared that his freedom to conduct basic research would be limited by the interests of the German coal mining industry.

Giulio Natta, well known for his work on the structure of solid state substances by X-ray and other diffraction techniques, was happy to be associated with industrial advances and in 1938 became head of the Department of Industrial Chemistry at the Milan Polytechnic. In this position he worked both on basic research and on practical questions concerned with the development of synthetic rubber production in Italy. His efforts were closely associated with the large Italian chemical company, Montecatini.

In 1953, Natta learned that Ziegler's laboratory in Mülheim had discovered a catalyst based on aluminum compounds in combination with transition metal compounds, which was capable of polymerizing ethylene to give a polyethylene that was entirely different from the polymer formed by the ICI process we've learned about (section 9.2 and following sections). The Ziegler catalyst produced a polyethylene that melted higher and was far less deformable and using the new infrared spectroscopic methods becoming available in those years, was found to be without the four carbon branches that defined LDPE (Figure 9.1). The Ziegler catalyst produced a linear polyethylene that was substantially crystalline.

From the discussion in section 9.3 and from Figures 9.3 and 9.4 it is clear that this new linear polyethylene, which has come to be called HDPE, for high density polyethylene, could not involve free radicals in its synthesis.

Natta sent some of his co-workers to Mülheim to consult with the Ziegler laboratory about the new catalyst and then used the catalyst back in Milan to attempt to polymerize propylene. The experiment was a success in producing high molecular

■ **Giulio Natta**

isotactic

syndiotactic (alternating)

atactic (random)

◀ **FIGURE 9.10**

Tacticities of Polypropylene

weight polypropylene but a failure in what Natta had intended, which was to produce a new kind of rubber. He thought that the methyl groups that had to hang off the chain of polypropylene at every third carbon atom would lead to this property. The polypropylene produced had the opposite characteristic. It was hard and full of crystalline regions. Applying his knowledge of diffraction methods to this new material demonstrated that it had a characteristic found in biological polymers. Just as all the chiral centers in proteins are of a single configuration, and many proteins are substantially helical, so all the stereocenters in the polypropylene were of identical configuration and the polymer was helical.

This polymer came to be called isotactic with a structure shown in **Figure 9.10**, which also shows the structures of other forms of polypropylene that were produced by variations on the Ziegler catalyst. As well, many other alkenes were polymerized by the Ziegler catalyst and its variations, leading to many new kinds of materials. The research based on the results of the efforts of the German and Italian scientists had opened a new field in demonstrating that synthetic polymers could be made with the stereoregular properties normally thought to be associated only with biological polymers.

By 1963, when Ziegler and Natta stood on the stage at Stockholm to receive the Nobel Prize for their work, Natta, whose work was driven by industrial interests had done something of great importance to industry but had also opened a field of theoretical importance. Ziegler, whose interest was in the fundamental had also opened a field of industrial importance, which he pointed out in the beginning of his Nobel lecture by showing the many industrial plants opened around the world that produced polymers based on his catalyst.

In Chapter 10, when we learn about the connections between conformational analysis (section 1. 13) and elastomeric behavior of certain polymers, we'll come to understand why Natta thought that polypropylene would be a rubber and as well how the unsaturated fatty acids we discovered in Chapter 7, and low density polyethylene (LDPE), and *Hevea* rubber share a common characteristic, which is not shared by *gutta percha* and high density polyethylene (HDPE).

PROBLEM 9.39

"From the discussion in section 9.3 and from Figures 9.3 and 9.4 it is clear that this new linear polyethylene, which has come to be called HDPE, for high density polyethylene, could not involve free radicals in its synthesis." On what basis is this statement from the text true?

PROBLEM 9.40

Figure 9.10 shows three structures that are possible for polypropylene. Are these isomers and if so what kinds of isomers are these polymers?

PROBLEM 9.41

In Figure 9.10, two structural drawings appear for isotactic polypropylene with an equivalence sign between them. What is the relationship between these structural drawings? Does your answer have to take into account the equivalence or non-equivalence of the ends of the chains?

PROBLEM 9.42

Isotactic polypropylene (Figure 9.10) can only be formed if every propylene molecule adds to the growing chain end from the same face of the double bond. Use your set of models or three dimensional drawings to convince yourself if this statement is true or false.

PROBLEM 9.43

If the faces of a propylene molecule are related as enantiomers, that is, are enantiotopic (section 5.6), what does this inform you about the Ziegler catalyst?

PROBLEM 9.44

Could the catalytic sites on the Ziegler catalyst be related racemically without changing your answer to Problem 9.43?

PURITY HALL WAS THE BUILDING DUPONT named to house Wallace Carothers' research after he was induced to leave his young professor's position at Harvard to come to Wilmington, Delaware. The substantially higher salary was far less important than the promise that he would be given the resources to follow his interests independent of DuPont's commercial needs. The year was 1928 and he was told by DuPont to *"work on problems of his own selections."*

In a book of great insight about Wallace Carothers and his work, Matthew Hermes', "Enough for One Lifetime," traced the life of this great chemist and his invention of nylon. Here we learn how the professors at the University of Illinois, who importantly were consultants to DuPont, learned of the extraordinary capabilities of Carothers. Here is a quote from Hermes' book about the view of two giants of twentieth century organic chemistry, **Roger Adams** and **Carl Marvel**:

*"Among all of them, both faculty and students, Carothers read the scientific literature most assiduously, and among all of them he had the most retentive memory. Carothers clarified for them G. N. Lewis' octet theory....."*In the pages to follow Hermes goes on to paint a detailed picture of the extraordinary insights Carothers brought to the Illinois chemistry department and the great impression this young man made. It is little surprise that DuPont kept their eye on Carothers, leading to the creation of the position they made for him.

The problem that interested Carothers when he came to DuPont was the question of the ability to synthesize high polymers using linkages of well known functional groups such as the combination of a hydroxyl group with a carboxylic acid to form an ester, a functional group we have seen to be of great importance in Chapter 7.

Carothers hypothesized that if he had a molecule with two hydroxyl groups and another molecule with two carboxylic acid groups, and identical concentrations of both molecules, the esterification reaction would produce a polymer. The idea is shown in **Figure 9.11**.

In Chapter 6 (section 6.7) we learned of the great French chemist of the nineteenth century, Pierre Marcelin Berthelot, and discussed his interest in chemical bonds and the effect of heat. Berthelot had many interests and accomplishments. After all, he did have streets named after him in Paris. Another of his investigations, which he published in 1862, found that mixing equal numbers (moles) of molecules of alcohols and carboxylic acids did not lead to full conversion to the derived ester. When equilibrium was reached, when the proportions of the starting alcohol and carboxylic acid and the derived ester stopped changing, he found that much of the starting materials were still present. Moreover, Berthelot found that putting the pure ester in water led to the identical mixture obtained by starting with the alcohol and the carboxylic acid.

Berthelot's results predicted problems for Carothers' attempt to create a high polymer in which the units were linked together by ester bonds. Let's see why.

Carothers realized that to obtain a high polymer following the idea in Figure 9.11 there were critical prerequisites. The number of moles of the dicarboxylic acid and the diol had to be identical. For this to be attained, it was necessary for these starting materials to be absolutely pure so that when they were weighed no other substance contributed to this weight. In addition, Carothers realized that the bond holding the polymer chain together, the links in the chain, had to be stable. If a link came apart somewhere in the chain, the molecular weight of the chain would be immediately reduced and by a significant amount since the severed link could be anywhere in the chain. This is where Berthelot's results had great importance.

Avoiding breaking a link in the chain means avoiding hydrolysis of the ester bonds in Figure 9.11 that is, avoiding the reaction paths we studied in the hydrolysis of the ester bonds in triglycerides (sections 7.3 and 7.4). Now the problem: Mixing carboxylic acids and alcohols to form esters must release a molecule of water for each ester bond formed. The loss of a water molecule for each ester bond formed is seen in the stoichiometry of the reaction, which means just add up the atoms on each side of the equation in **Figure 9.12**.

9.8
We've followed the polyethylene thread that led from ICI's foray into basic research. Now let's follow the nylon fiber that unwound out of DuPont's move in the same direction: Polyesters first.

■ **Wallace Carothers**

■ **Roger Adams**

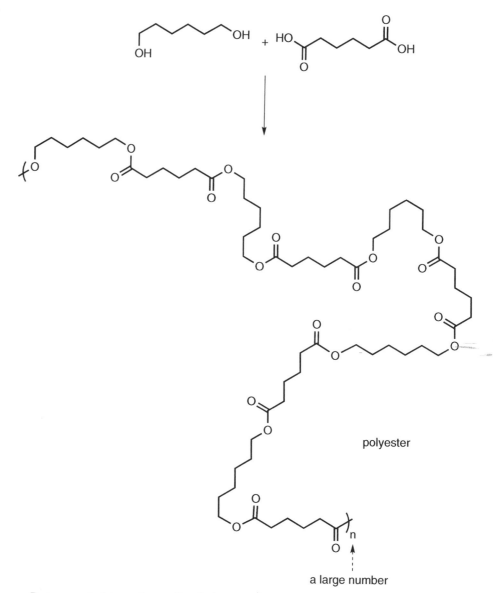

Formation of a Polyester by Reaction of 1,6-Hexanediol and Adipic Acid

polyester

a large number

▪ Carl Shipp Marvel

But, as noted just above, Berthelot, working approximately seventy years before Carothers joined DuPont, had discovered that mixing esters with water acted to return the ester to the starting alcohol and carboxylic acid. The water produced in the esterification reaction would act to reverse the esterification so that the polymerization would not occur to high molecular weights.

The reaction mechanism (**Figure 9.13**) for formation of an ester from a carboxylic acid and an alcohol shows that the absence of a good leaving group is the source of the incomplete formation of the ester. Instead, a combination of undirected proton transfers and nucleophilic attack at acyl carbon allow several paths to be followed without a clear direction. The series of reactions in Figure 9.13, and more that are not shown, can be compared to the enzyme controlled reactions in sections 7.5 and 7.6 where we see how nature exquisitely controls the proton transfers or provides a leaving group to drive the reacting system in the desired direction, either hydrolysis of the ester, or formation of the ester, although a thioester in the latter situation. In Figure 9.13 on the contrary the absence of this control allows each reaction to proceed in both directions setting up an equilibrium that can only be altered by removal of water.

In other words, to avoid hydrolysis as a consequence of the water released in the formation of the ester, the water must be removed as the ester bonds formed.

Luckily, around this time, work at the National Bureau of Standards had led to an invention called molecular distillation. At the high vacuum reached by diffusion

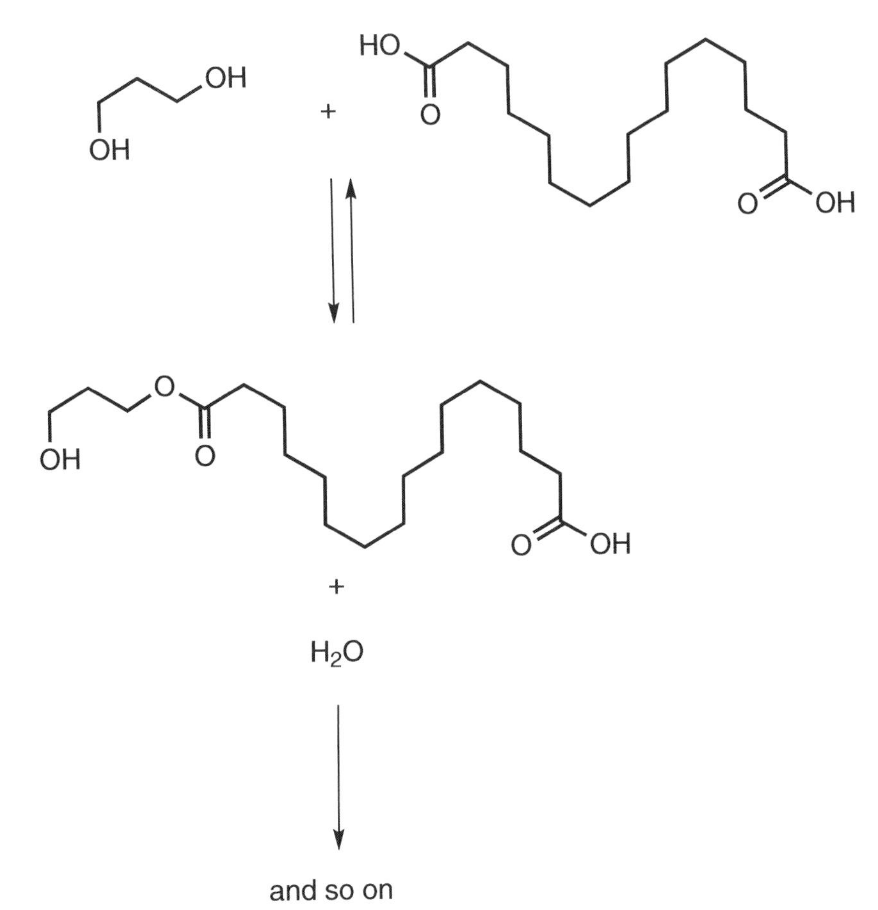

◀ **FIGURE 9.12**

Elimination of a Water Molecule for each Ester Bond in Forming a Polyester

pumps powered by the condensation of mercury vapor, the path taken by a water molecule before meeting another water molecule was quite long. Physical chemistry designates a mean free path to describe the distance a molecule must travel to meet another molecule. Such a path increases as vacuum increases. Water can not condense if the molecules don't meet each other. If a very cold surface is placed near to the esterification reaction which is producing the water molecule, then water molecules would hit the surface before meeting each other, and bulk water would not be produced. The ester bond would, therefore, not be hydrolyzed. The water molecules produced in the esterification (Figure 9.12) would be trapped on the cold surface.

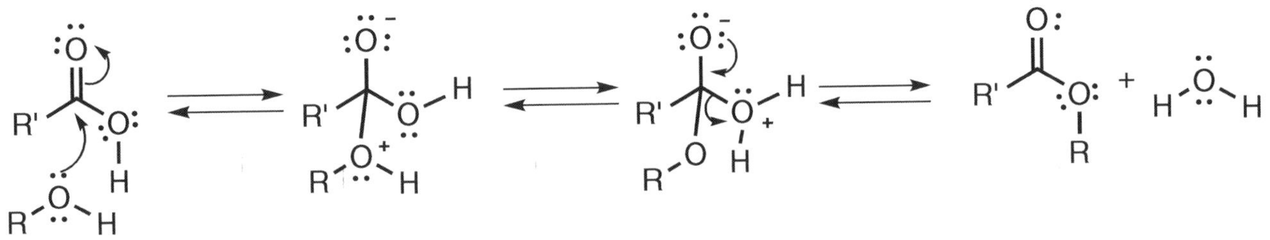

▲ **FIGURE 9.13**

Mechanism of Equilibrium Reactions Leading to the Formation of an Ester from an Alcohol and a Carboxylic Acid

In this way, and with other factors we'll not discuss here, the DuPont group in Purity Hall was able to synthesize a high molecular weight polyester, not the polyester shown in Figure 9.11, but rather one called polyester 3,16, a nomenclature they were to follow in the nylons. The number of carbon atoms in the dicarboxylic acid were designated by the second number, 16, and the number of carbon atoms in the diol by the first number, 3.

It was a marvelous time in Purity Hall as described in Hermes' book.

"*With Carothers downtown in Wilmington at the DuPont building ....Hill assembled his young cohorts and they ran through Purity Hall, tweezers in hand, drawing long lustrous, continuous filaments of their first artificial silk from the molten mass.*" And young they were. This was April 30, 1930 and Julian Hill had only received his Ph.D. from the Massachusetts Institute of Technology in 1928 and had only been at DuPont since 1929. What a beginning. Well, there is much to the story worth hearing, about organic chemistry and about life, told poignantly and beautifully in Hermes' book.

PROBLEM 9.45
Why are polyethylene and polypropylene named addition polymers whereas the polyester shown in Figure 9.11 is called a condensation polymer?

PROBLEM 9.46
Other proton transfer steps and intermediates are possible than those shown in the overall mechanism of esterification in Figure 9.13. Add these to the figure connecting each to the intermediates shown via the equilibrium arrows.

PROBLEM 9.47
Add all non-bonding electrons to the structures in Figure 9.13 and curved arrows to show the flow of electrons for each step.

PROBLEM 9.48
Identify all nucleophiles and electrophiles in the steps shown in Figure 9.13. Could any of the reactions shown be classified as acid-base reactions and if so to what category of acids or bases would they belong, Lewis or Brønsted-Lowry?

9.9
Nylon: But first let's take a look at proteins on which the nylons are modeled.

THE DISCOVERY OF A SYNTHETIC FIBER was unquestionably a commercially interesting finding to DuPont. However, although a wide variety of polyesters could be made by the technique which produced polyester 3,16, the unfortunate hydrolytic sensitivity of the ester bond discussed above, added to the fact that this class of polymers proved to be soluble in the cleaning fluids used for clothing and as well, melted at temperatures too near or below boiling water cancelled out the commercial interest as a fiber. And even if these problems were not enough to block commercial use of the polyesters as fibers, it would be an impossibly costly engineering task to construct a polymerization system with the vacuum equipment necessary to remove the water produced in the esterification.

Meanwhile, Carothers had turned his attention to another functional group, amides, which are far less susceptible to cleavage by hydrolysis. The absence of hydrolysis has to be the situation considering that the amino acid units in proteins, which exist in an aqueous environment, are linked by the amide group (Figures 5.8, 5.15 and 7.9).

According to Partington (section 4.9), Emil Fischer on whom we have focused much attention regarding the structure of glucose and its isomers (section 3.11), became interested in proteins after completing his work on the sugars and other biologically interesting molecules. Fischer, who left us with the often used metaphor for how enzymes function, as a lock and a key (section 5.6), was responsible for first clearly demonstrating the functional group by which proteins are linked. He also succeeded in synthesizing, in 1907, a polymer made up of eighteen amino acids. Fischer is credited with coining the term polypeptide, which he applied to this polymer.

As shown for several of the amino acid monomer units (Figures 7.8 and 7.9) for the enzyme that catalyzes the hydrolysis of the ester groups in triglycerides, the linking functional groups are amides, which are formed from carboxylic acids and amines, in an analogous reaction to the formation of esters.

Esters and amides of carboxylic acids are related functional groups in that both are derivatives of carboxylic acids. But there is a large difference in the properties of esters and amides in the equilibrium constants for their hydrolysis:

$$K_{ester} = [R'OH][RCO_2H]/[R'O_2CR][H_2O]$$

$$K_{amide} = [R'NH_2][RCO_2H]/[R'NHOCR][H_2O].$$

Whereas the equilibrium constant for ester hydrolysis is far larger than unity, that for amide hydrolysis is far smaller than unity so that the amide bond is inherently stable in water while the ester bond undergoes easy hydrolysis. The latter may be slow, if not catalyzed (section 7.3), but inevitably the bond will hydrolyze.

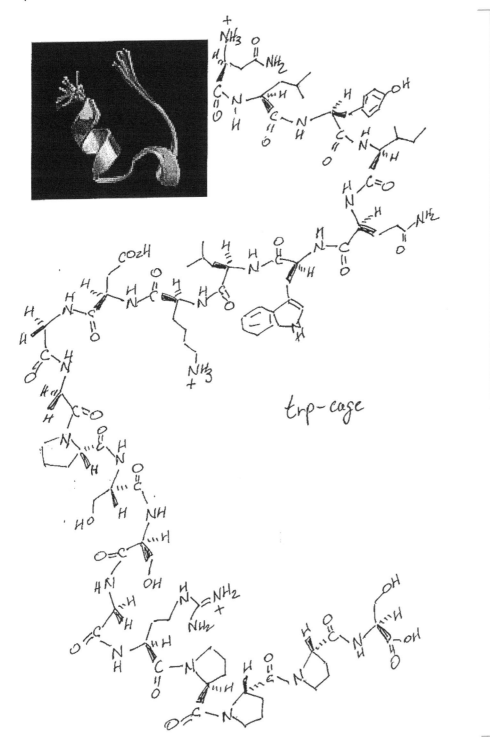

◄ FIGURE 9.14

Overall shape of Tryp-Cage Protein with the Constituent Amino Acids in order Drawn by Hand

As noted in section 9.8, the only way to drive the ester formation to completion, is to remove the water formed in the esterification. On the contrary the far smaller than unity equilibrium constant for amide formation allows the bond to both completely form in the presence of the water produced in the amidation and as well for the bond to be stable in the presence of water. When Carothers turned his attention to amides, he was addressing and solving the problem of water removal that had plagued the work on polyesters.

It seems opportune at this point, to show you a picture of a protein with most of the features of larger proteins but small enough to be completely described. This protein has been isolated by degradation of a somewhat larger peptide, one found in the saliva of Gila monsters. The parent peptide, named exendin-4, which has 36 amino acids, is a candidate for medical use in diabetes and is one of a large number of bioactive proteins that find their way into the victim by the bite of the Gila monster.

Trp-cage, which is derived by biochemists from exendin-4, has only 20 amino acids and has been the subject of considerable study because of the extreme rapidity, in the range of microseconds, which it folds into its final conformation (**Figure 9.14**). The mechanism by which proteins of complicated shape fold, such as those shown in Figures 5.8, 5.15 and 7.8 continues to be a critically important unsolved problem in biochemistry and having a small protein like trp-cage that folds rapidly offers a potential experimental tool to help solve this problem.

Take a close look at tryp-cage (Figure 9.14) before we get back to the story of the nylons. Tryp-cage, as for all proteins (Figures 7.8 and 7.9), and the nylons, share the identical backbone-linking-functional-group, amide, which undergoes hydrogen bonding (section 9.12). Hydrogen bonding plays a critical role in the fiber forming properties of the nylons and the folding of proteins.

PROBLEM 9.49

Create an analogous mechanism to that in Figure 9.13 by replacing ROH with RNH_2 so as to form an amide rather than an ester. Identify the equilibrium step in this mechanism, which in differing greatly from esterification would greatly reduce hydrolysis.

PROBLEM 9.50

Answer problems 9.47 and 9.48 related to the mechanism you created in problem 9.49.

PROBLEM 9.51

Write the chemical equation corresponding to the equilibrium constants for formation of amides and esters. Does the far larger value of K_{ester} over K_{amide} as discussed in this section help to answer problem 9.49?

PROBLEM 9.52

Identify all the amide groups in the protein shown in Figure 9.14.

PROBLEM 9.53

Show the amino acid structures, drawn in three dimensions, showing all the atoms, and the names of all the amino acids that would be produced if trp-cage were completely hydrolyzed. The web is rich in sites that show tables of amino acids and their names.

9.10

Nylon 6,6

THE SYNTHETIC WORK LEADING TO THE POLYESTERS (Figures 9.11 and 9.12) broke two barriers, as described by Herbert Morawetz in "Polymers, the Origins and Growth of a Science."

At the time in Europe when Carothers was carrying out his research there was still controversy about the possibilities that such large molecules as polymers could exist. There were schools of thought that the experimental properties, such as the high viscosity observed for what were claimed to be polymers, actually arose from colloids, that is, from associations of large numbers of small molecules. Although evidence was accumulating that macromolecules with covalently bonded structures existed, Carothers pointed out in the papers he published about the polyesters that there could be no question but that the polyesters were polymers. The ester functional groups necessary to gain the polymer properties was long known and beyond question.

In the synthetic work on polymers in Europe, the synthetic methods were based on free radicals such as described in sections 9.2 and 9.3 for polyethylene, and questions existed about the nature of the linking process. Mechanistic ideas about free radical chemistry were less developed in those years.

Carothers' work also introduced a new way to make polymers. As we've seen in the discussion about polyethylene, the formula for the alkene and the derived polymer is identical. Each ethylene molecule adds to the growing chain. Polyethylene and all polymers derived from alkenes are so called addition polymers and produced by what is called addition polymerization. Carothers was doing something entirely different. Each link in the chain to form the polyester was made by elimination of a water molecule so that the polyester and the monomers from which it is made had a different formula. This new method of polymerization was called condensation polymerization, which could be applied to amides as well as to esters.

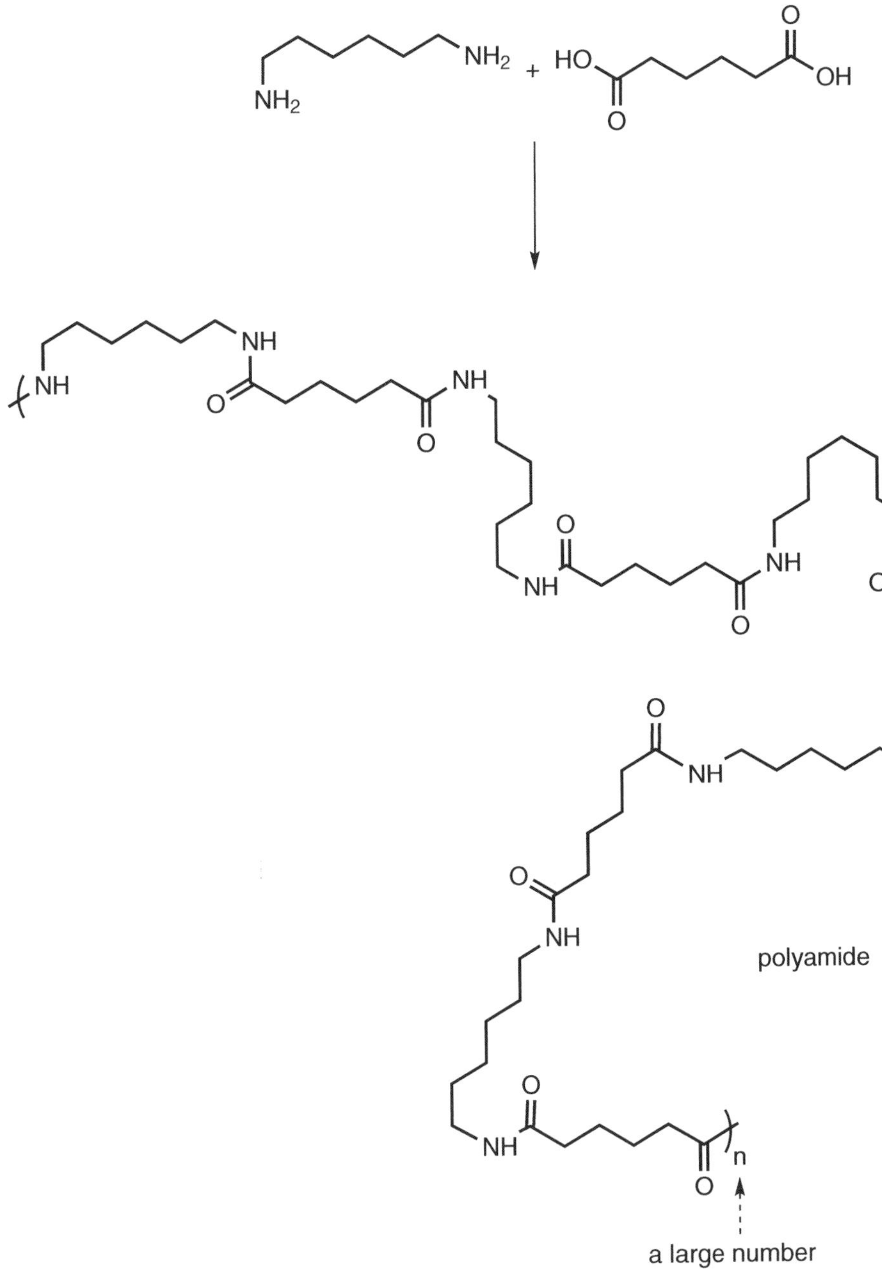

◀ FIGURE 9.15

Nylon 6,6 Formed from 1,6-Diaminohexane and Adipic Acid

When the group at Purity Hall turned their attention to amides as the linking functional group, a very interesting polyamide was produced as shown in **Figure 9.15** where the OH groups in Figure 9.11 were exchanged for NH groups. Figure 9.15 shows the structure of nylon 6,6.

In Green and Wittcoff's book noted in Chapter 6 (section 6.12) in the chapter entitled the Nylon Story, there is a section with the title: *"Carothers' Work at DuPont had Enormous Consequences for both DuPont and the Chemical Industry."* It certainly did as quoted from this section from Hermes' book: *"Nylon 6,6 made quite a stir when it first appeared in the form of women's stockings in a store in Wilmington, Delaware on May 15, 1940. There was bedlam; the 4000 pairs of nylon stockings were sold out in a few hours. At about $1.20 a pair, nearly five million pairs were sold throughout 1940. But within a year, nylon went from women's legs to the needs of the war.* Matthew Hermes further pointed out: *"The vanguard of the U.S. Army floated to earth in Normandy carried by and covered with nylon."* To further quote Hermes: *"DuPont sells billions of dollars worth of nylon each year now, and the profits from this single material*

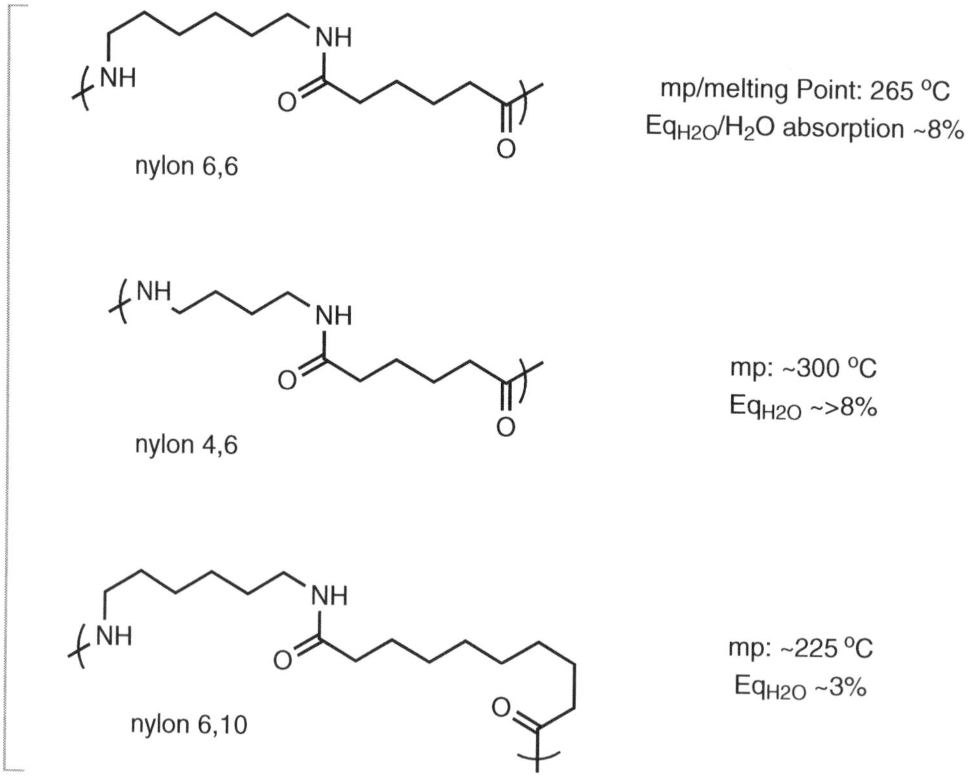

FIGURE 9.16 ▶

Water Absorption Equilibrium Values and Melting Points of Three Different Nylons

nylon 6,6

mp/melting Point: 265 °C
Eq_{H2O}/H_2O absorption ~8%

nylon 4,6

mp: ~300 °C
Eq_{H2O} ~>8%

nylon 6,10

mp: ~225 °C
Eq_{H2O} ~3%

have sustained the DuPont Company, the extended DuPont family and all the Company's stockholders for more than half a century."

There are many variations on the nylon theme, a few of which are seen in Figure 9.16. As can be seen in Figure 9.15 nylon is synthesized from a molecule with two NH_2 groups and a molecule with two CO_2H groups both of which contain six carbon atoms. Of all the nylons including those shown in **Figure 9.16** only nylon 6,6 is made from two six carbon molecules. One of the main reasons, but not the only reason, that DuPont chose nylon 6,6 as the company's commercial focus was the fact that both monomers, adipic acid (section 8.6) and 1,6-diaminohexane diamine (Figure 9.15) have the same number of carbon atoms as benzene, and therefore the possibility that nylon 6,6 could be produced from a starting material that was available in large quantities at a low price.

How DuPont used benzene for the purpose of nylon production and the problems associated with this use will be a subject for Chapter 10. That chapter will focus on the synthetic routes to the production of the small molecules necessary for the production of nylon 6,6, and the principles of organic chemistry one comes across in recounting that effort. For now let's look at how industry synthesizes nylon 6,6 from 1,6-diaminohexane (hexamethylene diamine) and adipic acid.

PROBLEM 9.54

In nucleophilic substitution at acyl carbon such as the formation of an ester or the saponification of an ester, a four coordinate intermediate is the first step. Draw a reaction coordinate diagram which describes hydrolysis of an ester in neutral water to form the alcohol and the carboxylic acid. Identify the rate determining step (section 6.11).

PROBLEM 9.55

Propose a reason, based on resonance, for the fact that twisting about the nitrogen to carbonyl carbon of an amide is more difficult than twisting about the carbon to oxygen bond of an ester. How might your answer relate to the fact that ammonia is a stronger base than water?

PROBLEM 9.56

Offer an explanation for the different water absorption and melting point properties of the nylons in figure 9.16.

PROBLEM 9.57

What synthetic first step, which we've seen earlier in the book, must be necessary to use benzene for the production of adipic acid?

PROBLEM 9.58

Propose monomers for the formation of the nylons 4,6 and 6,10 shown in Figure 9.16.

9.11

Hexamethylene diamine and adipic acid react together in the industrial process to produce nylon 6,6.

I N SAPONIFICATION, we looked at the breaking of an ester bond. Carried out in vitro (section 7.4), or within a cell (section 7.5), this process involves nucleophilic substitution at acyl carbon, that is, at a carbon atom that is double bonded to oxygen. In section 7.6 we looked into the formation of a thioester from a free fatty acid, a prerequisite to fatty acid catabolism, nature's way to derive the energy from the carbon-carbon bonds of the fatty acid hydrocarbon chain. Again, this process involves nucleophilic substitution at acyl carbon. In the formation of polyesters, first accomplished in Purity Hall by Carothers' group, nucleophilic substitution at acyl carbon is involved (section 9.8). In the formation of nylon 6,6, the subject we are taking up here, we are forming an amide from a carboxylic acid and an amine, which again is a process involving nucleophilic substitution at acyl carbon, a variation on all the themes noted in this paragraph.

In laboratory work at universities, where new reactions are discovered or used in new ways, we often don't maximize yield, because we are exploring. In industry where intense competition is the rule and where people's jobs, their way of living is at stake, where often large investments are involved, great care is taken to gain efficiency. In that sense industrial processes are like biochemical processes – life is involved.

Carothers knew that in condensation polymerization (section 9.8), the kind of polymerization that produces nylon 6,6 (Figure 9.15), precisely equal proportions of the two monomers adipic acid and hexamethylene diamine must be present to maximize the molecular weight, which is necessary to gain the best fiber forming properties. Certainly, one could weigh each of the monomers to make certain there were equal numbers of moles, but even the best weighing devices have an unacceptable uncertainty, which would be worse the larger the amounts weighed. And how would one know that some impurity such as a chain-terminating monoamine or

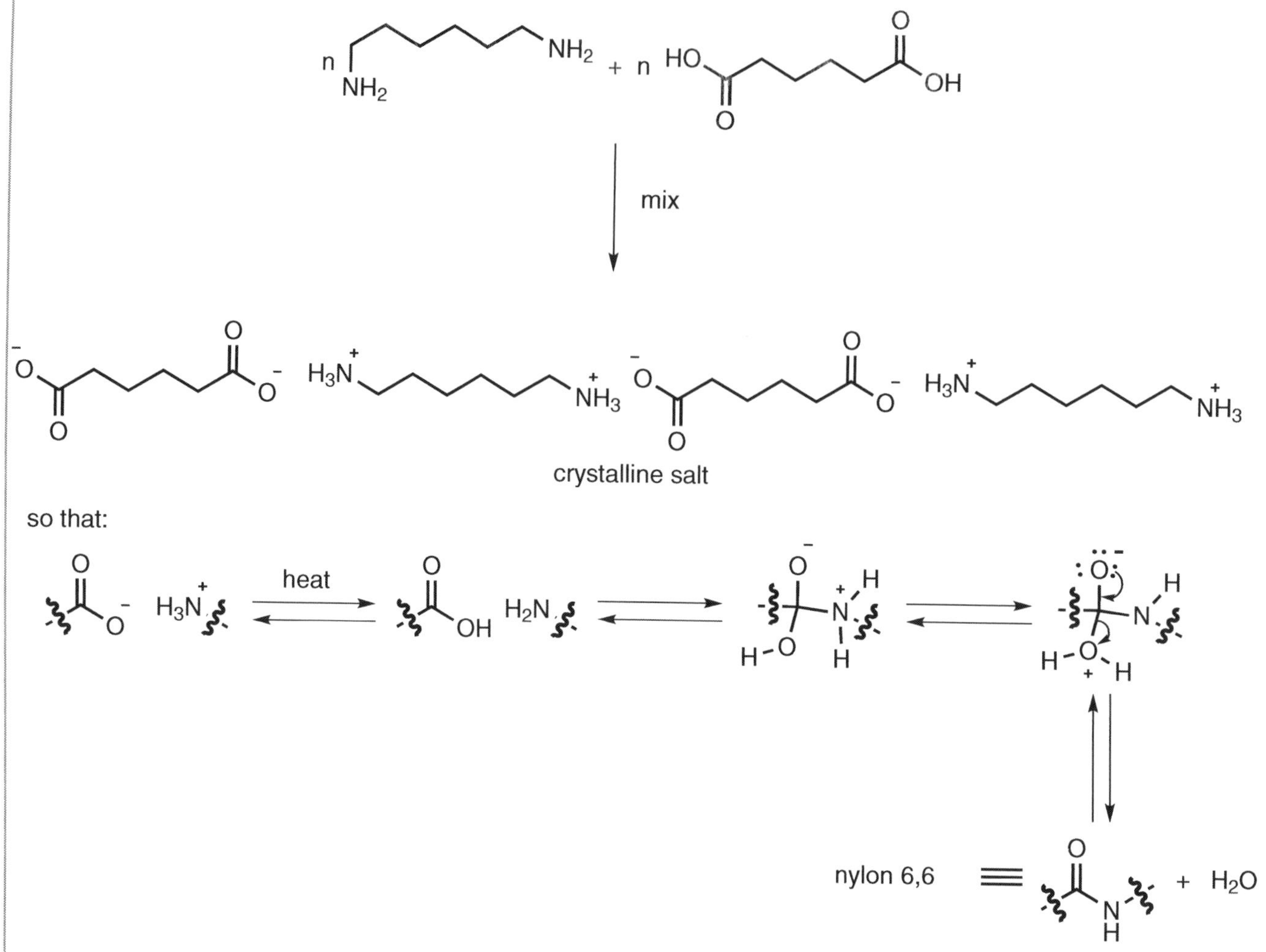

crystalline salt

so that:

nylon 6,6

▲ FIGURE 9.17

Mechanism for the Formation of nylon 6,6 by initial Formation of Crystals from 1,6-Diaminohexane and Adipic acid

monocarboxylic acid were not present and contributing to the weight. Here's where clever engineering came to the rescue (**Figure 9.17**).

In the industrial process the fact that the monomers to produce nylon 6,6 are an acid and a base means that adipic acid and hexamethylene diamine can first be mixed to form the salt as shown in Figure 9.17. Considerations of charge balance require that equal numbers of -NH_2 group and -CO_2H groups must be present to form the –NH_3^+ ^-O_2C- groups, which crystallize. The crystal therefore, in its pure state, that is its highest melting state, which can be achieved by repeated crystallization, must have exactly, to the molecule, equal numbers of adipic acid and hexamethylene chains. No weighing technique could ever accomplish such equivalence of molecular numbers in which both requirements are satisfied, stoichiometric balance and purity.

Moreover, as shown in Figure 9.17, heating the ammonium salt drives off H_2O to yield the amide group, linking the polymer chain together. Heating the crystalline salt of adipic acid and hexamethylene diamine will therefore yield nylon 6,6. And for the reasons that Carothers turned to the amide bond, as discussed in section 9.9, the water driven off in the amide-bond forming step will not reverse the step to reform the carboxylic acid and the amine.

PROBLEM 9.59

Show how the presence of a monocarboxylic acid or a monoamine would act to limit the molecular weight of a growing chain of nylon 6,6.

PROBLEM 9.60
What advantage does a crystalline step yield in the industrial production of nylon 6,6?

PROBLEM 9.61
A mechanism for the formation of the amide by heating the salt shown in Figure 9.17 would first lose a proton from the –NH$_3$$^+$ group. How might this step lead to a series of reactions that could form the amide? Is there any other first step toward the amide that you can imagine?

9.12
Why is nylon such an excellent fiber-forming substance? Because it mimics a property of silk–interchain hydrogen bonds.

IN SECTION 9.8 WE DESCRIBED HOW JULIAN HILL ran through Purity Hall drawing fibers of the polyester that was successfully formed by the laborious process of removing the water released from the esterification reaction. After this triumph, Hill would be anxious to repeat this experience. When it was clear that a high polymer was also formed from hexamethylene diamine and adipic acid, he melted some in a petri dish and stirred it with a glass rod. Imagine the excitement when Hill lifted the glass rod from the dish and began walking away with a fiber trailing after him and growing thinner and thinner as he walked. And not only did the fiber become thinner, it became stronger. As we've seen in section 9.10 this was a transforming moment for the DuPont Corporation.

In section 9.3, a sample of low density polyethylene was described as an entangled mass, like a plate of just cooked spaghetti. This was the situation for the nylon chains in that petri dish, in disarray and tangled one with another. As Hill drew the polymer threads outward on walking away from the molten mass, the polymer molecules lined up and came close together. This process of organization was helped greatly by the amide groups because as the chains took on a regular relationship with each other the amide groups in adjacent chains could form hydrogen bonds.

A regular arrangement of hydrogen bonds enhanced crystallization so that the nylon became very strong. If pulled on in one direction it would not yield because the chains were already stretched out and could be stretched no more. If pulled perpendicular to the stretched chain direction, hydrogen bonds leading to crystallinity would have to be overcome. These strength characteristics are precisely what are responsible for the properties of silk. Nylon is a mimic of silk as shown in **Figure 9.18**.

Protein folding, noted but hardly gone into in the book, essential for the properties of enzymes and for the biological properties of all proteins (sections 5.5, 5.9, 7.5), is an extremely complex phenomenon. What is certain about the forces controlling the manner in which proteins fold is that hydrogen bonding plays an important role. That is hardly saying enough about the importance of hydrogen bonding. Hydrogen bonding is responsible for the unique properties of water, why it is a liquid and not a gas, and why ice is less dense than liquid water, just for two examples. And the essential role water plays in biological phenomena is strongly associated with the hydrogen bonding properties of water. The double helix of DNA is held together by hydrogen bonding and so much more we can not go into here.

Hydrogen bonds belong to the category of what are called non-covalent interactions. Such interactions can be repulsive as we've seen in sections 3.3-3.5 where steric strain arises from overlapping van der Waals radii of groupings of atoms. But hydrogen bonds, which are driven by polarity differences, are always attractive. By polarity driven we mean that the attractive interaction in a hydrogen bond arises from the attraction between plus and minus. But the effect is not ionic and the energy of attraction is a small fraction of a covalent or an ionic bond, approximately in the range of 5 kcal/mole or that is, about 1/15th to 1/20th of a covalent bond. The attractive force arises when a hydrogen atom is covalently bonded to an electronegative atom,

FIGURE 9.18 ▶

Comparison of Hydrogen Bonding in the Structures of Silk Fibroin and Nylon 6

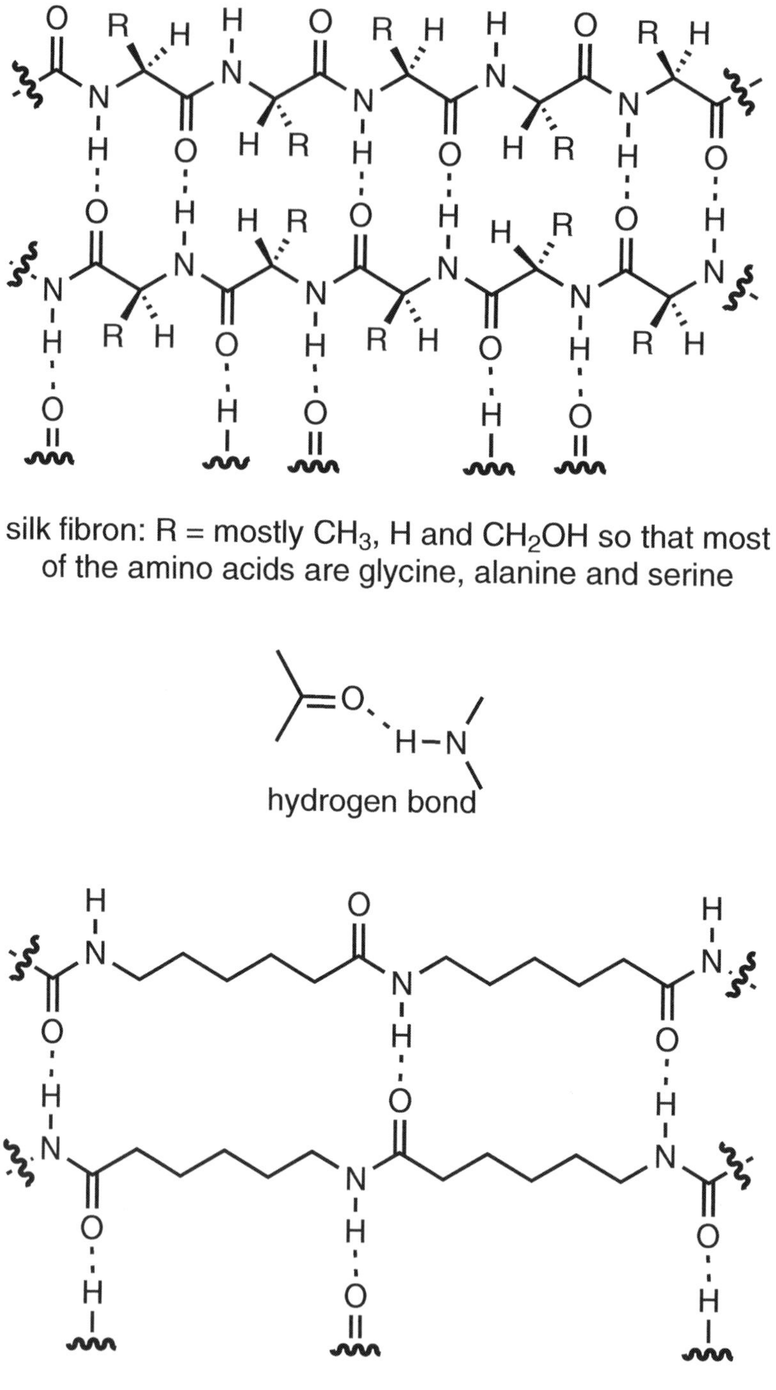

silk fibron: R = mostly CH_3, H and CH_2OH so that most of the amino acids are glycine, alanine and serine

hydrogen bond

typically oxygen or nitrogen and in the proximity of another electronegative atom bearing non-bonded electrons. As shown by the dashed line ($\_\ \_\ \_\ \_$) in Figure 9.18 hydrogen bonding plays a critical role in the properties of both silk and nylon.

PROBLEM 9.62

Methanol, H_3COH, boils at 65° C, a boiling point that is not attained by a hydrocarbon until reaching n-hexane, which boils at 69° C, while water boils at 100° C and acetic acid, CH_3CO_2H, boils at 118° C. Offer a structural explanation for these facts.

PROBLEM 9.63

Account for the hydrogen bonding shown in Figure 9.18 by showing the electrons involved. Does your answer suggest a geometrical necessity for forming the hydrogen bonds that requires alteration of the representations made in the figure?

PROBLEM 9.64

Note that nylon 6,6 is not portrayed in Figure 9.18, which instead shows what is known as nylon 6. Propose a monomer that could lead to nylon 6 and compare this monomer to the monomers that are needed to form nylon 6,6.

PROBLEM 9.65

Let's finish with a question concerned with the difference between addition and condensation polymerization. The sentence in the text: "As we've seen in the discussion about polyethylene, the formula for the polymer produced from an alkene is the same as the formula of the derived polymer." Why is this statement only approximately correct?

CHAPTER NINE SUMMARY of the Essential Material

WE BEGAN WITH THE STRUCTURE OF polyethylene and the randomly placed short branches off the main polymer chain and come to understand that polyethylene is an addition polymer and why it has this designation. The properties of free radicals are discussed with a comparison to carbocations and carbanions, the other major classes of chemical intermediates. Another example of a mechanism seen before, the chain mechanism, is involved in the synthesis of polyethylene, which we have designated LDPE. The chain mechanism combined with a property of free radicals allows understanding of the four carbon branches found in LDPE. We study the source of the small molecule that leads to LDPE, ethylene, and its source to the chemical industry by cracking of fractions of petroleum and look into the details of this particular kind of cracking and how another chemical property of free radicals is revealed, β-scission. In steam cracking we see another example of a chain mechanism. The comparison between the free radical polymerization of ethylene and the production of ethylene by steam cracking of petroleum fractions allows understanding of the thermodynamic differences between these processes where we see how entropy and enthalpy may play opposing roles.

The discussion then moves to the attempted free radical polymerization of propylene and the understanding of how resonance stabilization is the source of the failure to achieve a high polymer. This result leads to the value of reaction coordinate diagrams in understanding why one kind of free radical initiates polymerization while another does not. Finally we see that polymerization without the intervention of free radicals allows another kind of polyethylene to be made, HDPE, and also success in polymerizing propylene. The polypropylene produced by this method is found to have properties that normally are only associated with biological polymers with the implication that chirality plays a role in the polymerization.

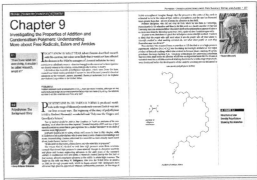

The chapter then changes direction in moving into the synthesis of condensation polymers and how such polymers differ fundamentally from addition polymers. We learn about the early work at DuPont and the problem with both the synthesis and the commercial potential of polyesters and how these problems are solved in the use of the amide functional group leading to nylon 6,6. This approach brings us back to the properties of carboxylic acids and their derivatives, esters and amides. We discover how the mechanisms of formation of amides and of esters have common features but differ in an essential manner. We learn about the parallel between the nylons and proteins in the linking functional group holding the chains together. The chapter ends with a discussion of hydrogen bonds and how they help understanding of the parallel properties of the nylons and silk.

We'll discover in the next chapter why DuPont chose nylon 6,6 over other possibilities with the general nylon structure, which comes from the ready availability of benzene and its six carbon atoms, which can be transformed into both adipic acid and hexamethylene diamine.

And not discussed is how much of the progress in the latter half of the 20th century was driven by the institute started by Herman F. Mark at the Polytechnic Institute of Brooklyn.

Chapter 10

The Industrial Road Toward Increasing Efficiency in the Synthesis of Hexamethylene Diamine with Stopovers at Kinetic Versus Thermodynamic Control of Chemical Reactions, Nucleophilic Substitutions, and with a Side Trip to Laboratory Reducing Agents

10.1

Benzene to Adipic Acid

COULD BENZENE, WHICH WE HAVE LEARNED is resistant to chemical change (section 6.7), be induced to undergo the considerable structural change necessary to form the starting materials to produce nylon 6,6?

H_2 can be added to benzene using highly active catalysts based on nickel or rhodium with high pressures of hydrogen and at high temperatures. From my own experience, adding hydrogen to benzene or derivatives of benzene is not done in the laboratory but is rather conducted in a special high temperature facility usually with blast walls and high pressure equipment. The extreme conditions require great care. You can imagine the danger for the large volumes involved in industrial production but the process is well designed to avoid disaster so that cyclohexane of high purity can be produced by addition of H_2 to benzene. This first step sets the stage for production of the starting materials to produce nylon 6,6.

FIGURE 10.1 ▶

Industrial Path from Benzene to Adipic Acid

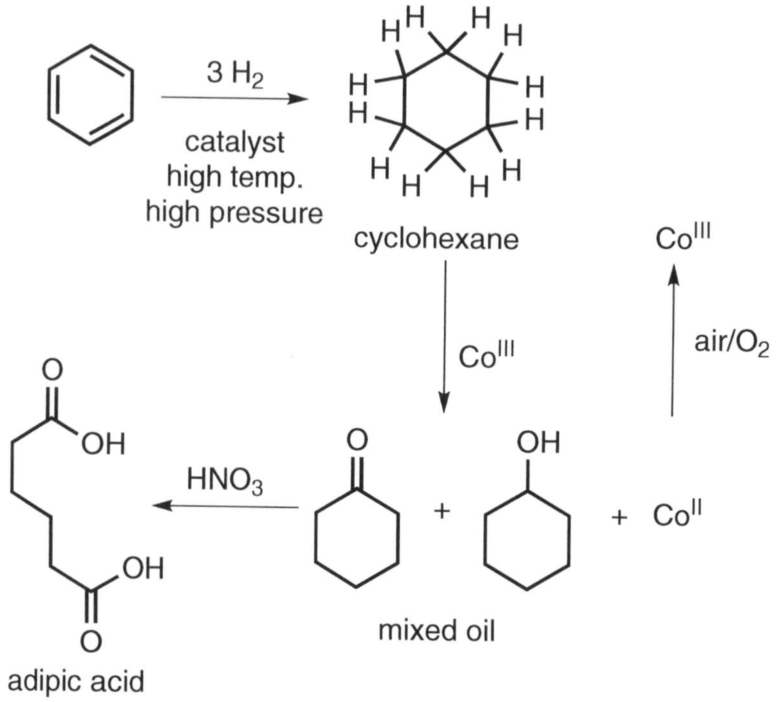

Although hydrocarbons can be oxidized to alcohols and carbonyl compounds using high oxidation states of cobalt (Co^{III}), these oxidation reactions are rarely used because mixtures of structural isomers are obtained. Consider how many constitutionally isomeric carbonyl compounds could be produced from a linear hydrocarbon chain such as n-hexane. However, this problem is averted in cyclohexane because of the symmetry of the molecule. Oxidation occurring at any of the six carbon atoms leads to the identical product. In the industrial process, shown in **Figure 10.1**, the oxidation reaction is conducted so that a mixture of cyclohexanone and cyclohexanol is obtained, so called "mixed oil."

Industrial processes in which a reactant is "free" are always desirable and this is the situation with the cobalt-based oxidation of cyclohexane. The Co^{III}, as a consequence of acting as an oxidizing agent is necessarily reduced to Co^{II}, an oxidation state of cobalt that is readily reoxidized to Co^{III} under the conditions of the industrial process by oxygen in the air. In metals, oxidation and reduction reactions are often easily understood as simply transfers of electrons in contrast to understanding the oxidation and reduction properties we've discussed for NAD$^+$/NADH and FAD/FADH$_2$ (section 7.9) or for oxidized states of carbon (section 3.8).

Oxidation by Co^{III} involves gaining an electron from the molecule to be oxidized, cyclohexane in this situation, therefore reducing the +3 state of the cobalt to +2. The oxygen in the air then oxidizes the +2 state of the cobalt by removing an electron, therefore returning the cobalt to the +3 state that is, to Co^{III}. The ultimate fate attained by the two atoms in O$_2$ is to form H$_2$O and to become bonded to the carbon atoms of the cyclohexane ring.

The portrayal of the oxidation and reduction states of the organic molecules and the cobalt in Figure 10.1 did not concern mechanism but what matters a great deal is that the Co^{II}/Co^{III} costs the industrial process nothing. Air is free assuming there is no pollution involved.

The mixture of the alcohol and the ketone does not require a separation step because the alcohol reacts with the next reactant, nitric acid, HNO$_3$, to produce the ketone, which then reacts further to produce the desired product, adipic acid (Figure 10.1).

In these reactions the nitric acid is an oxidizing agent, although a reagent that is not regenerated and therefore must be used stoichiometrically. The more the number of molecules of cyclohexane converted to adipic acid, the more the number of molecules of nitric acid that are consumed, not as in the cobalt-catalyzed oxidation of cyclohexane. Luckily, nitric acid is cheap, which contributes to the fact that the synthesis of adipic acid by this route has had an unusually long life without change in the chemical industry (since the 1930s), where constant change in processes and methods of synthesis is normally the rule.

The overall transformation from benzene to adipic acid is a reduction, benzene to cyclohexane, followed by oxidations, cyclohexane to adipic acid. The oxidation and reduction states of carbon are discerned as discussed in section 3.8 by the addition or elimination of hydrogen atoms. Benzene to cyclohexane adds three hydrogen molecules to benzene, a reduction. Oxidation of cyclohexane to adipic acid breaks a carbon-carbon bond and takes away two hydrogen molecules from the cyclohexane, the four hydrogen atoms bound to the carbon atoms that are converted to the two carboxylic acid groups.

Although reaction of nitric acid with cyclohexane is a commonly used experiment in undergraduate organic chemistry laboratories, the mechanism is a bit complex for us to handle here. But the next steps that convert adipic acid to hexamethylene diamine and then the reaction to produce nylon 6,6 are well suited to reinforce the chemistry of carboxylic acids and their derivatives, which we first encountered in the saponification and then catabolism of fats in Chapter 7.

PROBLEM 10.1

Can you imagine a single molecule that could yield a polymer with very similar properties to nylon 6,6. The structure of the polymer would be $-(NH(CH_2)_5C=O)_n$. Such a polymer exists and is called nylon 6. Carothers tried to make this polymer and failed thereby allowing German companies to gain an important foothold in the nylon business when their chemists succeeded in synthesizing nylon 6. The story is told in Chapter 5 of the Green and Wittcoff book (section 6.12).

PROBLEM 10.2

Why would the addition of the first molecule of H_2 to a benzene ring be far more difficult than addition of the remaining two molecules of H_2 necessary to complete the transformation to cyclohexane?

PROBLEM 10.3

This sentence appears in the text: "Consider how many constitutionally isomeric carbonyl compounds could be produced from a linear hydrocarbon chain such as n-hexane." Draw the structures of as many compounds with one carbonyl group that you can imagine as arising by oxidation of n-hexane. Name the functional group that each structure belongs to.

10.2

Nylon 6,6: Hexamethylene Diamine–the Classic Route from Adipic Acid

THERE ARE SIX CARBON ATOMS in adipic acid and six carbon atoms in hexamethylene diamine (Figure 9.15) and just as in biological systems, efficiency also drives industrial processes. In the industrial path to nylon 6,6 the diamine is obtained from the dicarboxylic acid so that no new carbon-carbon bonds have to be formed. In that way, all the carbon-carbon bonds in both monomers for production of nylon 6,6 have been derived from the carbon-carbon bonds in benzene. That's efficiency.

Ammonia (NH_3) is necessary for converting adipic acid to hexamethylene diamine (**Figure 10.2**). Nitric acid, as we've just seen (Figure 10.1) is necessary for production of adipic acid. Both these nitrogen containing molecules were not easily available until **Fritz Haber**, who won a Nobel Prize in 1919 for his work, invented a catalyst for converting ("fixing") atmospheric nitrogen to ammonia.

FIGURE 10.2 ▶

Mechanism and Path for the Conversion of Adipic Acid to 1,6-Diaminohexane (Hexamethylendiamine)

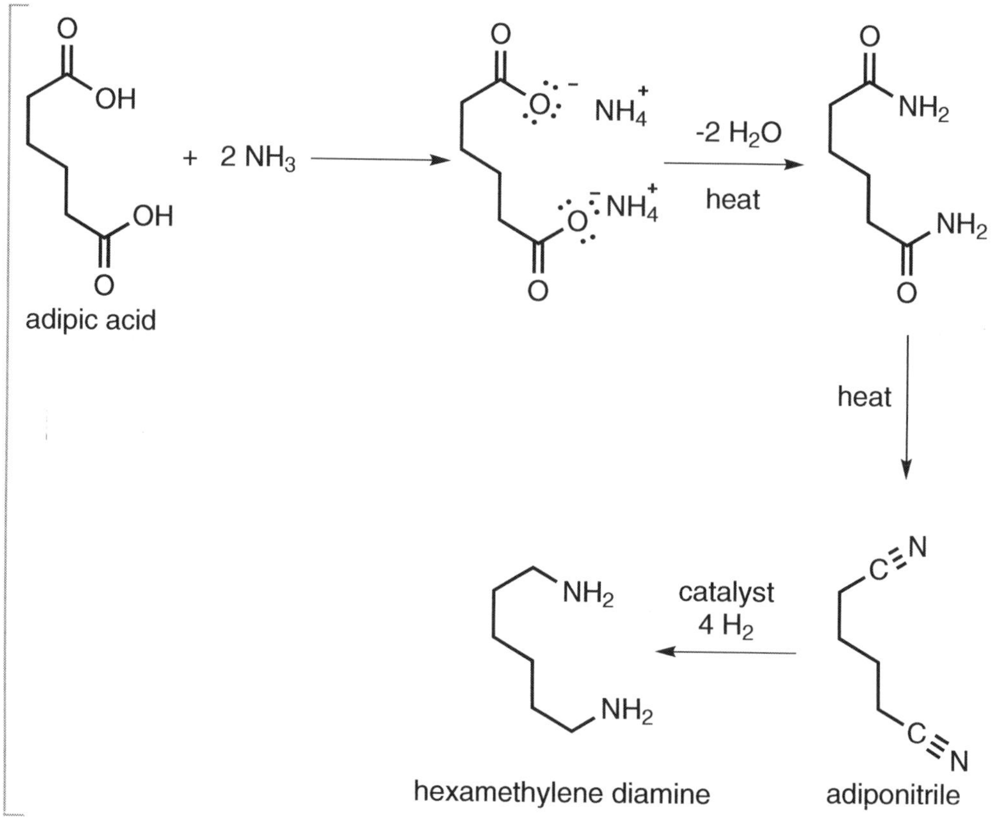

Putting aside momentarily the needs for production of nylon 6,6, ammonia is a critical necessity for the chemistry that sustains life, fertilizer for agricultural production of food. But nitric acid, which is produced by oxidation of ammonia, is a critical necessity for the chemistry that destroys life; explosives used in weapons can be formed from the same molecule used as a fertilizer and combined with nitric acid, ammonium nitrate. Haber's work therefore, as is sometimes the situation with discoveries in science, had positive and negative consequences.

Haber was a German patriot who is given credit for originating the use of poison gases in World War I, which is given credit for killing and maiming many soldiers on both sides of the battle and did little to change the outcome of the war. Haber's loyalty to the German war effort in World War I and his distinguished position in the world of science did him little good at the end of his life, which ended tragically when he was driven out of Germany in 1933 by the Nazi racial laws. He died in Switzerland in 1934.

The basic character of ammonia derives from the lone pair of electrons on nitrogen, which allows reaction with adipic acid to form the ammonium salt, a classic Brønsted-Lowry (sections 4.9) acid base reaction (Figures 9.17 and 10.2). Ammonium salts of carboxylic acids, on heating lose two molecules of H_2O, first to form the amide, the same functional group to be used to link the intended nylon structure (Figure 9.17) and then to a functional group we meet for the first time, nitrile (or cyano), $-C\equiv N$ (Figure 10.2).

The π-bonds in the triple bonds in the two nitrile groups in adiponitrile, formed by loss of water from the ammonium salt of adipic acid (Figure 10.2) can be reduced to amino, NH_2, groups. In the industrial process H_2 and a catalyst are used for this purpose. In a laboratory where handling of explosive gases is sometimes a problem, other reducing agents can be used. Let's take this opportunity to introduce you to two important reducing agents often used in laboratory research, which could be used to convert a nitrile group to an amino group.

■ **Fritz Haber**

PROBLEM 10.4
Boron trifluoride, BF_3, reacts with ammonia, NH_3, to form a salt in a Lewis acid-base reaction. Portray this reaction using the orbitals involved. G. N. Lewis saw no difference between Brønsted-Lowry acid-base reactions and those called Lewis acid-base reactions as long as the acid in the reaction of the Brønsted-Lowry type was the proton. For example, Lewis would not characterize the carboxylic acid in Figure 10.2 as the acid. Rather, the proton released from the carboxylic acid is the acid. Considering orbitals involved, in what way is a proton reacting with ammonia similar to the reaction of BF_3 with ammonia?

PROBLEM 10.5
Write a reasonable mechanism for the amide forming reaction in Figure 10.2.

T HE TRIPLE BOND IN A NITRILE GROUP has reactive properties that differ from a triple bond between carbon atoms, as in acetylene, $H-C\equiv C-H$. The difference arises from the electronegativity difference (section 2.4) between nitrogen and carbon, which is not encountered in carbon-carbon triple bonds. Nevertheless, the same weak p-p orbital arrangement is the source of the weakness of the two π bonds in any triple bond.

Lithium aluminum hydride and sodium borohydride (**Figure 10.3**) are common reducing agents in organic chemistry, the former the more powerful of the two and the one commonly used for the conversion of nitriles to amines. The mechanism of this reduction is thought to occur by addition of H:⁻ (hydride ion) to the carbon end of the triple bond with the polarity of the nitrile group driving the electrophilic character of the nitrile in this reaction.

10.3
A Side Trip to Laboratory Reducing Agents

FIGURE 10.3 ▶

Structures of Lithium Aluminum Hydride and Sodium Borohydride

$$Li\,^+AlH_4^-$$

lithium aluminum hydride
(LAH)

$$Na^+BH_4^-$$

sodium borohydride

Before we discuss the details of the reaction path it seems opportune to find an analogy between the two laboratory reducing agents, lithium aluminum hydride ($LiAlH_4$) and sodium borohydride ($NaBH_4$), which act via transfer of a hydride ion, and the two coenzymes, NADH and $FADH_2$ that play important reducing roles in biochemical reactions.

$LiAlH_4$ and $NaBH_4$ can act in only one direction, as reducing agents, as we'll see. On the contrary, the coenzyme couples NADH/NAD$^+$ and $FADH_2$/FAD, which we've seen involved in the catabolism of fatty acids (section 7.7), can act both as reducing agents and oxidizing agents.

The reduced forms of the coenzymes act as reducing agents by donation of a hydride ion (or its equivalent) to the substrate that is reduced. One example of this reducing

FIGURE 10.4 ▶

Mechanism for the transfer of two hydrides from two NADH coenzymes reduces an acyl coenzyme A thioester to the alcohol.

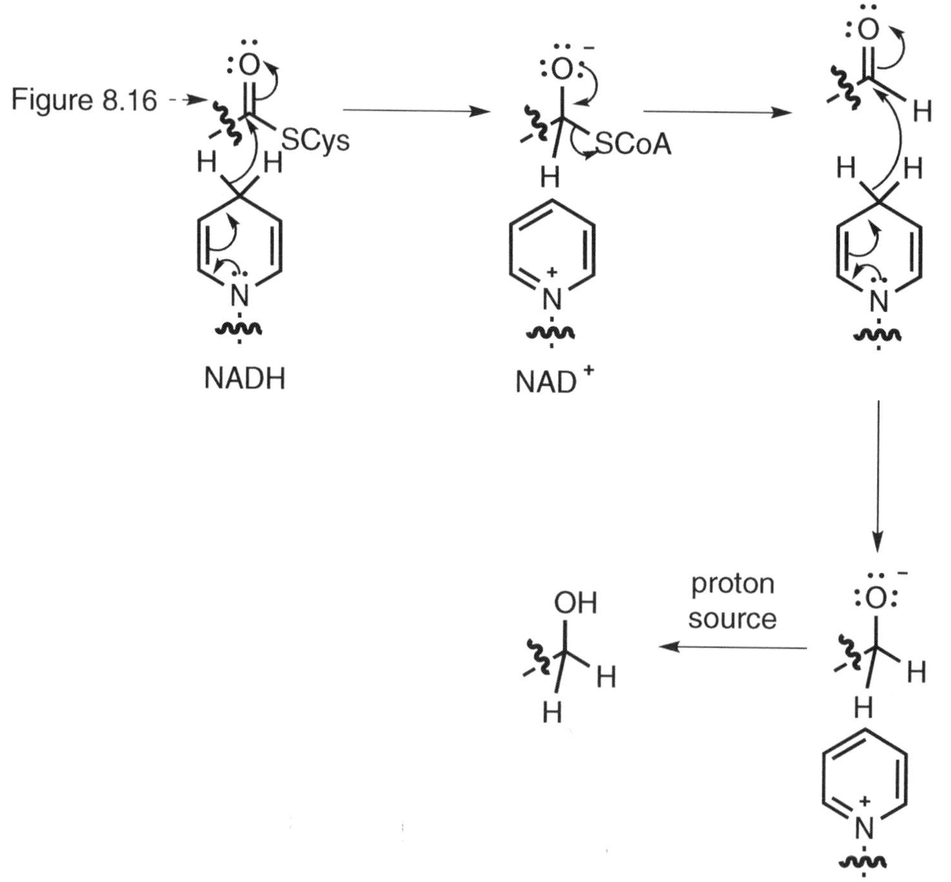

property is all that is necessary to see the hydride transfer analogy to lithium aluminum hydride and sodium borohydride. This process is found in the biochemical pathway to isopentenyl diphosphate. In step **2** of Figure 8.16 the thioester group is converted to a primary alcohol. The mechanism of this reduction, although not discussed around the presentation of that figure, is shown in **Figure 10.4**. The reducing agent is NADH.

Two molecules of NADH are utilized in the enzymatically controlled reactions converting the thioester first to the aldehyde and then to the alcohol. Two molecules of the oxidized form of the coenzyme, NAD$^+$, are produced. The reduction takes place

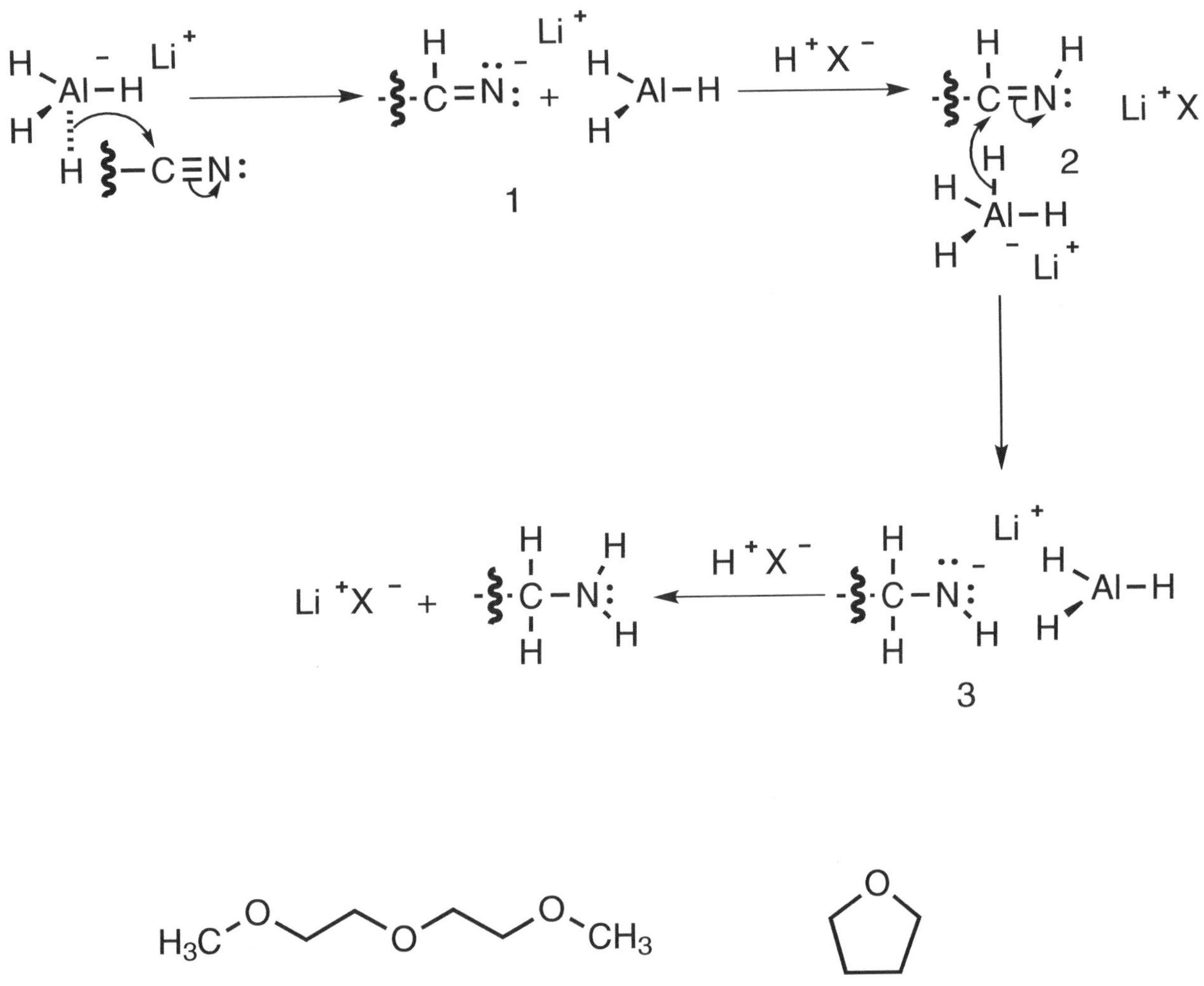

H₃C–O ⌢ O ⌢ O–CH₃

diglyme

tetrahydrofuran

by movement of a hydride ion from carbon on the nicotinamide ring to the oxidized carbon of the isopentenyl diphosphate intermediate (Figures 10.4, 8.16).

Now, let's return to the use of lithium aluminum hydride to reduce the nitrile to the amine. The details are wildly different but the end result, transfer of hydride is the same. In the enzymatic reaction, the reduced form NADH is neutral and the oxidized form NAD⁺ is charged, and the driving force for NADH to act as a reducing agent is to gain aromaticity (section 7.9). In the lithium aluminum hydride reduction, the reduced form, AlH₄⁻, is charged and the oxidized form, AlH₃, is neutral and the driving force for AlH₄⁻ to act as a reducing agent is the weak long bond between Al and H. Nevertheless, in both the enzymatic reaction, Figure 10.4, and the laboratory reaction, **Figure 10.5**, the differing driving forces lead to the same conclusion, transfer of a hydride ion to the substrate.

The solvent used for the reduction of nitriles with lithium aluminum hydride is usually diglyme or some other ether such as tetrahydrofuran. Here we are introduced to another functional group, ethers, which always contain an atomic arrangement in which two carbon groups are bonded to an oxygen atom, R-O-R. The structures of diglyme and tetrahydrofuran are shown in Figure 10.5.

Although almost all textbooks show the reaction, the detailed mechanism of the reduction of nitriles to amines with lithium aluminum hydride is rarely discussed. You can appreciate this from the complexities involved (Figure 10.5).

▲ **FIGURE 10.5**

Mechanism for the Transfer of Two Hydrides from Lithium Aluminum Hydride to a Nitrile to Produce the Primary Amine; and the Structures of the Necessary Ethereal Solvents

PROBLEM 10.6
Could the conversion of the thioester to the alcohol in Figure 10.4 use lithium aluminum hydride? Write a mechanism for the reaction.

PROBLEM 10.7
Analyze the bonding in a cyano group, -C≡N, using hybridization of the orbitals of the carbon and nitrogen atoms.

PROBLEM 10.8
Explain how the reaction of lithium aluminum hydride with a nitrile (cyano) group could be considered an acid-base reaction. Does the category of the acid-base reaction belong to the Lewis or Brønsted-Lowry type?

PROBLEM 10.9
Show the detailed movement of atoms and electrons including accounting of all the formal charges, for the chemical reactions in Figure 10.5.

10.4
Hexamethylene Diamine – An Attempt at a Better Route

A FAVORED ROUTE TO any commercial product produced by the chemical industry is one involving a minimum of steps separating the product from chemicals produced in petroleum cracking. The path to adipic acid passes this test while the path to hexamethylene diamine described in section 10.2 fails to pass. Industrial chemists have therefore long sought a replacement route to the diamine. But what could be the replacement? Six carbon atoms are needed.

Butadiene, which is produced in large amounts from steam cracking of the naphtha fraction of petroleum (section 9.4), has four carbon atoms and two double bonds meaning a wide range of reactivity. The DuPont chemists conceived of a path to hexamethylene diamine using these four carbon atoms with two more added from HCN (**Figure 10.6**).

HCN is a chemical that has been used in nasty ways. It was the active ingredient in Zyklon B, the killing chemical used in the Nazi death camps. From the story of the discovery of HCN, told below, one can hardly expect the horrible way this chemical has been used.

Early in the eighteenth century, about 1706, the first of the modern pigments, Prussian Blue (Berlin Blue), was synthesized in Germany. The extraordinary color of Prussian Blue was quickly appreciated for its value in painting as demonstrated by

The Great Wave of Kanagawa, color woodcut by artist Katsushika Hokusai, Year c. 1829–32

Hokusai's 1830 wood cut of Mount Fuji. Before Prussian Blue, the kind of blue necessary for such an image was impossible to obtain and those blues available tended to fade.

The story of the discovery of HCN brings us to Carl Wilhelm Scheele whom we heard about in section 7.3. According to Partington (section 4.9), Scheele was interested in Prussian Blue long before Hokusai used the pigment. As early as 1765 Scheele started work on the pigment and published his discovery of HCN in 1782-83. Prussian Blue is an inorganic compound of complex composition containing the cyanide ion, C≡N⁻ and related to ferrocyanide. Here is what Partington wrote in his *Volume Three* about

Scheele's work in this area, which makes an interesting contrast to current research in chemistry:

"*By distilling potassium ferrocyanide with diluted sulfuric acid he noticed a strong smell. He did not know the acid was poisonous; he says it had 'a peculiar, not unpleasant smell', and 'a taste which almost borders slightly on sweet and is somewhat heating in the mouth'.*" Partington goes on to write: "*It is difficult to understand how he escaped with his life.*"

It would be impossible to produce enough HCN for industrial purposes from Prussian

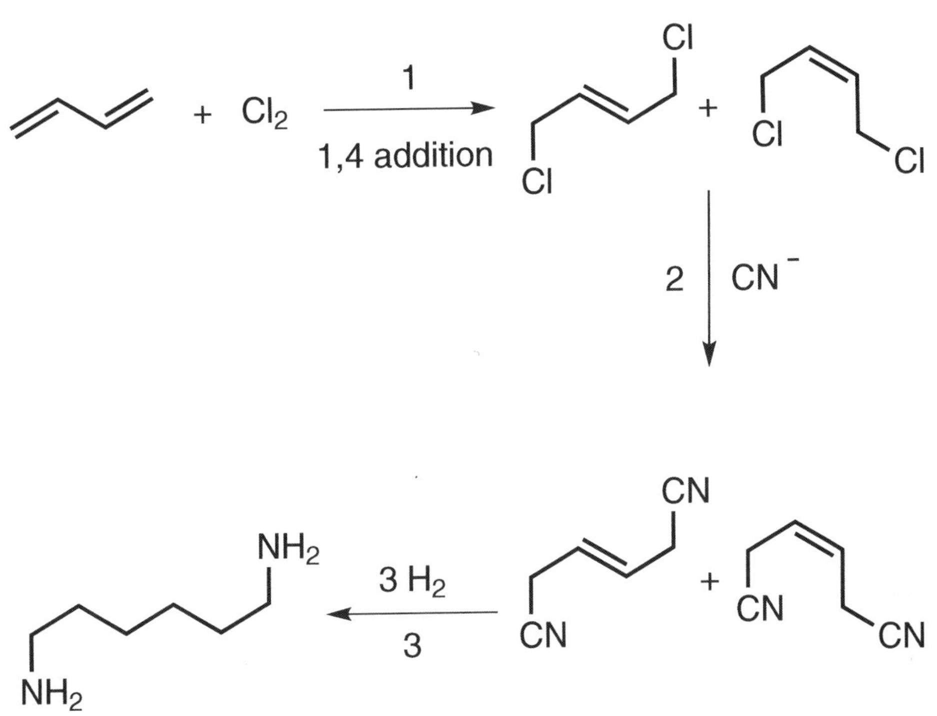

◀ **FIGURE 10.6**

Path for 1,4-Addition of Cl$_2$ to 1,3-Butadiene to Initiate the Production of 1,6-Diaminohexane

Blue or other similar inorganic sources. A process that does yield the large quantities necessary for industry and probably necessary for the Nazi death camps was invented in Germany in the early 1930s by Leonid Andrussow, a chemical engineer, after whom the process is named. At very high temperatures with a platinum catalyst, methane reacts with ammonia and oxygen as follows: $2\,CH_4 + 2\,NH_3 + 3\,O_2 \rightarrow 2\,HCN + 6\,H_2O$. With ammonia available from the work of Fritz Haber (section 10.2), and butadiene from petroleum (section 9.4), the materials for the production of large amounts of hexamethylene diamine via the route outlined in Figure 10.6 appeared to be in place.

However, for industrial production, there is a strong negative in the approach in Figure 10.6, which is the use of chlorine, Cl$_2$. All modern methods of production of chlorine involve electrolytic methods to oxidize chloride ion obtained from chloride salts such as sodium chloride. The chemical industry dislikes

but also:

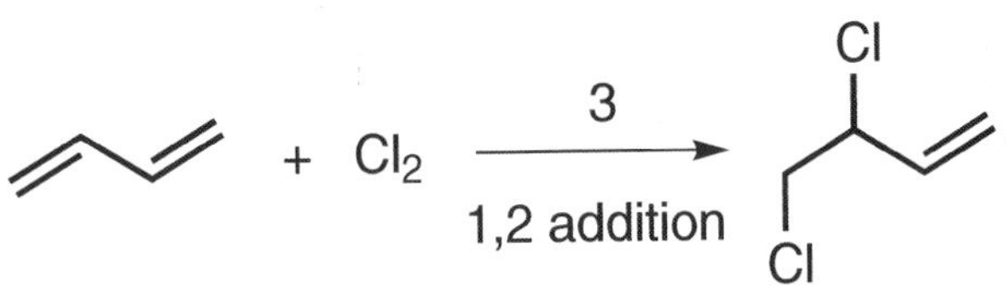

◀ **FIGURE 10.7**

1,2 Addition of Cl$_2$ to 1,3-Butadiene

paying for energy, and electricity is expensive. Nevertheless, DuPont was willing to put up with paying for Cl$_2$ until another problem with their scheme (Figure 10.6) arose that seemed to kill the idea.

Reaction **1** (Figure 10.6) produces not only the stereoisomers shown, which is a 1,4 addition of Cl$_2$ to the two double bonds in 1,3-butadiene but also the constitutional isomer, 1,2-dichlorobutane, which is a 1,2 addition or, that is, addition of both chlorine atoms to only one of the double bonds in the butadiene (**Figure 10.7**). Hexamethylene diamine could not be made, at least directly, from the 1,2 addition product. What is to be done with the 1,2-dichlorobutene? The conceived of process seemed to have a large problem in wasting a great deal of both chlorine and butadiene until the DuPont group realized that a discovery that came out of the invention of nylon could save the day.

FIGURE 10.8 ▶

Path from Vinylacetylene to Neoprene and Comparison to Natural Rubber

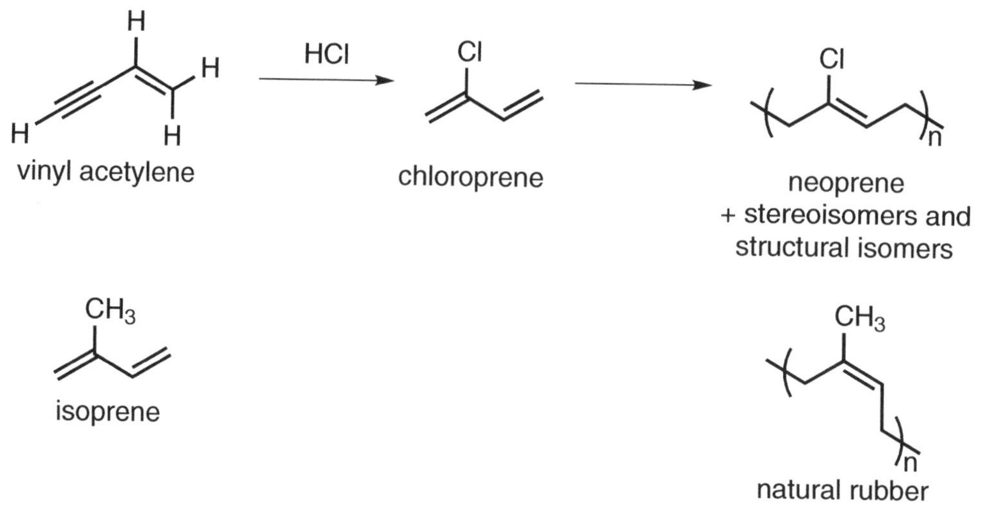

Carothers and his group in Purity Hall (Section 9.8) had discovered that vinyl acetylene, (a compound available to DuPont through connections to work at Notre Dame University), added HCl to form an analog of isoprene. Isoprene was known to be the structural building block of the terpenes and natural rubber (section 5.2). This new molecule, called chloroprene, when polymerized, yielded a rubbery polymer of commercial value, which came to be called neoprene (**Figure 10.8**). Although neoprene did not have the elastic qualities of natural rubber, which we'll discuss in Chapter 11, it did become a valuable commercial material for DuPont.

FIGURE 10.9 ▶

Path from 3,4-Dichloro-1-Butene to Chloroprene

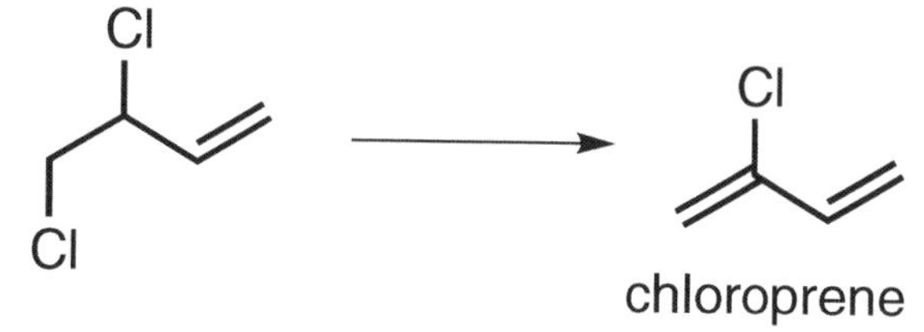

However, vinyl acetylene (Figure 10.8) is a very dangerous chemical. Like acetylene it is explosive and therefore not desirable for a large scale industrial process, blocking its use for producing chloroprene. The problem was solved by the previously unwanted 1,2 addition of Cl_2 to 1,3-butadiene (Figure 10.7) which produced the structural isomer that was not useful to synthesize hexamethylene diamine (Figure 10.6). Happily, the 1,2 addition product easily lost a molecule of HCl to form chloroprene (**Figure 10.9**). DuPont had in hand a safe route to a valuable elastomer.

Now, the addition of Cl_2 to 1,3 butadiene took on great importance to DuPont. While the 1,4 addition might lead to a new path to hexamethylene diamine, and therefore nylon 6,6, the 1,2 addition could offer a route to chloroprene and, therefore, to Neoprene rubber. The problem is, how to control the competing 1,2-versus 1,4-addition paths? What are the fundamental forces behind the competing paths and how might they be manipulated?

PROBLEM 10.10
In the 1,2 addition of Cl_2 to 1,3-butadiene two molecules are produced. What are they?

PROBLEM 10.11
Whereas chloroprene has no stereoisomers, the polymer derived from it, neoprene, has several stereoisomeric and constitutional isomeric possibilities. What are they and how are they related to each other?

PROBLEM 10.12
Figure 10.9 shows the reaction in which HCl is lost from 3,4 dichloro-1-butene to produce chloroprene. Loss of the other Cl as HCl would not produce chloroprene but this direction of loss of HCl is less likely because the first step in loss of HCl is loss of a proton to produce a carbanion. Explain the reasoning behind this understanding of the direction of HCl loss.

PROBLEM 10.13
Can the addition of Cl_2 or Br_2 to a double bond be considered an acid-base reaction and if so what kind of acid-base reaction and why? Analogously, could the reactions in Figures 10.10 and 10.11 be described as reactions between nucleophiles and electrophiles?

THE MECHANISM OF ADDITION OF HALOGENS such as chlorine or bromine, Cl_2 or Br_2, to double bonds has the halogen acting as an electrophile in seeking electrons from the double bond. The Cl-Cl bond is broken releasing chloride, Cl^-, and producing a positively charged intermediate called a halonium ion, in which positively charged chlorine bridges the two carbon atoms that formed the double bond. The released Cl^- then adds to one end of what was the double bond to form the dihalo addition product. This mechanism is shown in **Figure 10.10** for addition of Cl_2 to propylene.

In section 6.10 and particularly in Figure 6.13 we had seen electrophilic addition of HCl to double bonds in which the proton added first to form the positively charged intermediate with the following step, as in Figure 10.10, addition of the remaining Cl^-. In Figure 6.13 the addition followed a rule we ascribe to Markovnikov so that the most substituted carbon atom of the double bond takes on the positive charge and therefore is the site of the addition of Cl^-.

Although the product of addition of Cl_2 to a double bond can not reveal which carbon atom took the positive charge, the mechanistic forces favoring positive charge at the most substituted carbon end of the double bond are still at work. To make this point, we'll imagine that the two chlorine atoms in the attacking Cl_2 have been labeled. As seen in Figure 10.10 the Cl^- adds to the most substituted site, represented by resonance structures, the site of the most stable carbocation.

Now let's apply these ideas to the DuPont process in which Cl_2 is added to 1,3-butadiene, which adds a new element to the story arising from the additional double bond as shown in **Figure 10.11**.

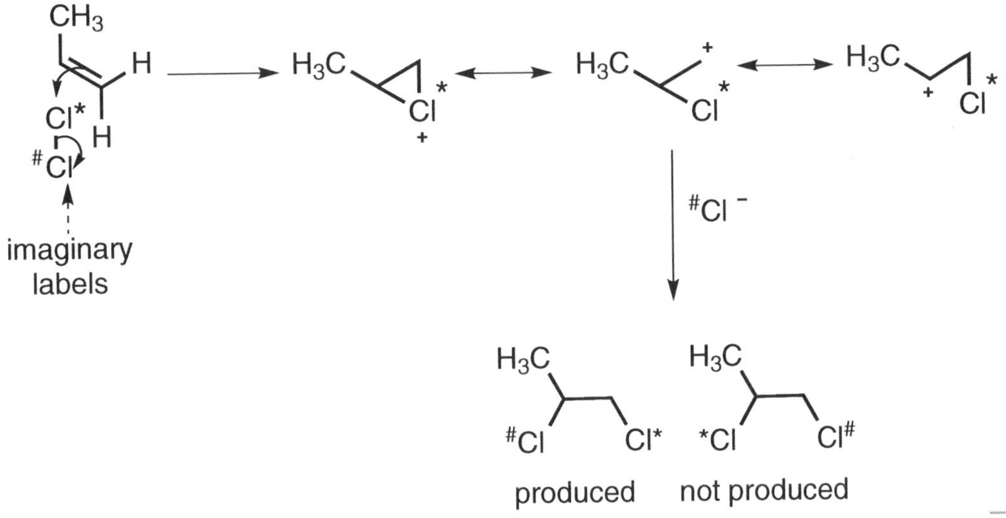

10.5
How industry overcomes a supposedly insurmountable problem arising from thermodynamic versus kinetic control in the addition of halogen to double bonds, to invent an elegant and commercially viable route to two commercial polymers, only to finally fail because of an unforeseen environmental consequence of their path.

◀ **FIGURE 10.10**

Mechanism for an imaginary experiment of the addition of labeled Cl_2 to propylene demonstrates the intermediacy of the halonium ion.

imaginary labels

produced   not produced

FIGURE 10.11 ▶

Resonance Structures Needed to Describe the Positively Charged Intermediate formed on Addition of Cl_2 to 1,3-Butadiene

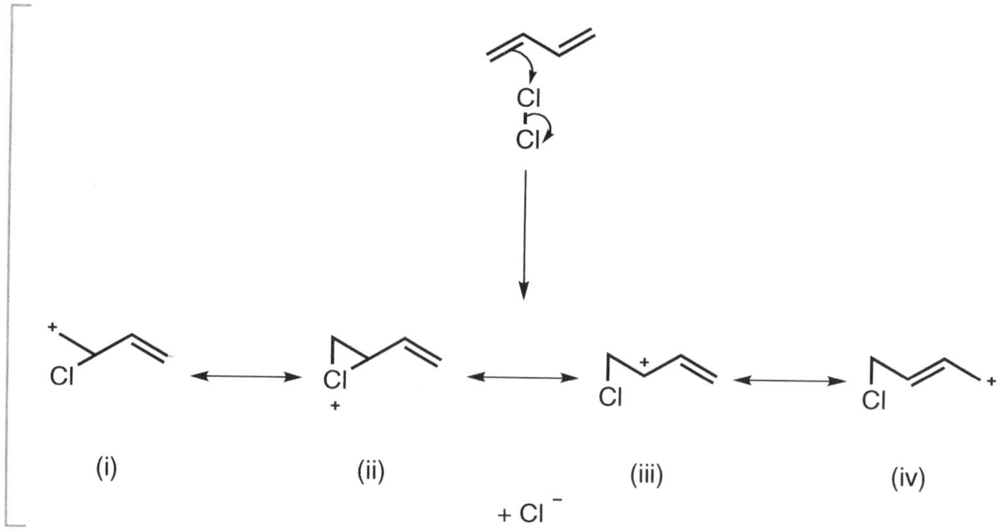

The resonance structures of the halonium ion shown in Figure 10.10 are clearly distinguished in favoring the more substituted carbon so that the left over Cl^- adds as shown. In Figure 10.11, resonance structure **iii** is preferred over **i** for the same reason. However, placing the carbocation as in **iii** puts the empty p orbital in conjugation with the p orbitals making up the other double bond, which means that another resonance structure must be considered, **iv**. The resonance structures **iii** and **iv** allow the positive charge to be delocalized over three atoms and we have learned that delocalization is most important in resonance stabilized structures (section 5.4).

If the structure of the intermediate produced by addition of chlorine to 1,3-butadiene were to be represented by **iii** and **iv**, it would be difficult to assign relative importance to these structures. Consideration of substitution of the positively charged carbon atom certainly favors **iii**. But what about the position of the double bond in **iii** versus **iv**.

To help to judge the difference in the position of the double bond, let's look back to section 6.7 and to Figure 6.5 with enthalpic data about adding H_2 to double bonds. The discussion in Chapter 6 was focused on the aromatic stability of benzene, but there is some information in Figure 6.5, which helps us to evaluate the difference in the double bond positions in **iii** and **iv** in Figure 10.11.

The first example in Figure 6.5 shows 30.3 kcal/mole of heat released on addition of H_2 to 1-butene. The second and third examples in this figure shows 28.6 kcal/mole of heat released on addition of H_2 to cis-2-butene and 27.6 kcal/mole heat released for trans-2-butene.

We will say more about cis and trans double bonds in Chapter 11 although we've already come across their important differences in two instances, the double bonds in fatty acids are overwhelmingly cis for critical structural reasons (section 7.2). And the difference between Neoprene rubber and natural rubber arises overwhelmingly from the trans and cis double bonds, respectively, in these unsaturated polymers (Figure 10.8).

The difference in the heat released on hydrogenation of cis- and trans-2-butene arises from the fact that trans double bonds are generally more stable than cis double bonds. In general, the larger the $-\Delta H$ for addition of H_2 to a double bond, the less stable the double bond. That's how we judged the extra stability of benzene compared to three isolated or even simply conjugated double bonds, as discussed in section 6.7.

And here is the clue to the difference between resonance structures **iii** and **iv** (Figure 10.11). Both cis and trans 2-butene release less heat on hydrogenation than does 1-butene, considerably less (Figure 6.5). Just as cis double bonds are less stable than the stereoisomeric trans double bonds, so less substituted double bonds are less stable than more substituted double bonds. From these considerations, resonance structure iv is favored over iii.

Consider the substitution of the carbocation sites: resonance structure **iii** is favored. Now consider the substitution of the double bonds in the compared resonance structures, which favors **iv**. In other words, the delocalization of the positive charge in the intermediate produced on addition of chlorine to 1,3-butadiene is balanced between **iii** and **iv** so that addition of the Cl⁻ occurs at both positions – the former yields the 1,2 addition product, while the latter yields the 1.4 addition product.

The above is what is called a kinetic argument because it has to do with the relative stabilities of intermediates and transition states (section 6.11). The energies of activation for 1,2 and 1,4 addition of Cl₂ to 1,3-butadiene are only slightly different, with the 1,2 path favored, so that both constitutional isomers are formed. However, the higher the temperature at which the reaction is run, the more 1,4 addition product is produced. Why?

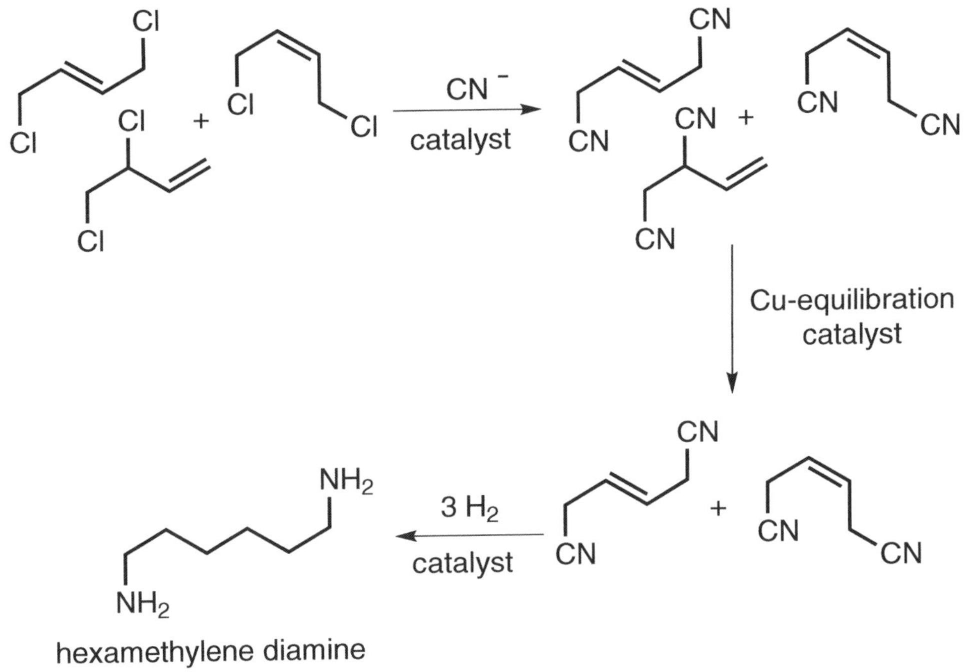

◀ **FIGURE 10.12**

Industrial Path for Formation of 1,6-Diaminohexane from the Mixture of the 1,4 and 1,2 Addition Products of Cl₂ to 1,3-Butadiene

The answer is found in the structures of the two constitutional isomers arising from 1,2-versus 1,4-addition of chlorine to 1,3-butadiene (Figures 10.6 and 10.7). We have just learned about the relative stabilities of differently substituted double bonds when judging the relative stability of the resonance structures in Figure 10.11. The stability difference as a function of substitution means that the final product of 1,4-addition must be more stable than the final product of 1,2-addition. If an equilibrium state were allowed to be reached, the 1,4 addition product would predominate. The reason that higher temperature produces more 1,4 addition product is that the addition of the two chlorine atoms to the four carbon skeleton of butadiene is reversible, allowing the equilibrium state to be reached. This situation is called thermodynamic control. Kinetic control, on the contrary, based on the relative stability of iii and iv favors 1,2 addition.

The DuPont process was designed to occur near 200° C allowing the equilibrium state to be reached. However, the stability of the 1,4-dichloro-2-butene (the 1,4 product) is not so much greater than the 3,4-dichloro-1-butene (the 1,2 product) as to exclude the latter from the equilibrium mixture. This result is a problem because the industrial process has to be able to form each one to the exclusion of the other to have separate paths to hexamethylene diamine and chloroprene.

Here's how industrial chemists and chemical engineers solved the separation problem. It was discovered that reaction of cyanide ion, CN⁻, as shown in **Figure 10.12**, with both the mixture of the 1,2 and 1,4 addition products, yielded a mixture of nitriles shown in Figure 10.12 and that a copper-based catalyst was found that converted the mixture overwhelmingly to the 1,4 addition product. Apparently, the stability of the

Henri Louis Le Chatelier

internal double bond in the dicyano isomers is enhanced compared to its stability in the dichloro isomers. Good piece of luck and we're apparently on the way to nylon.

In another lucky break, the 1,2 product, 3,4-dichloro-1-butene, boils at 123° C, while the cis- and trans- mixture of 1,4-dichloro-2-butene boils at 155° C. These boiling points are far enough apart to separate the 3,4-dichloro-1-butene by fractional distillation, leaving the 1,4-dichloro-2-butene in the pot where it could be isolated away from the 1,2 product.

Again, another lucky break; another copper catalyst was found that interconverted the isomeric 1,2 and 1,4 chlorine addition products, a catalyst that could be added to the distillation pot if the 1,2 product was desired. As the lower-boiling 3,4-dichloro-1-butene was removed in the distillation the equilibrating system sought to replenish this isomer. Even if the equilibrium state favored the 1,4-dichloro-2-butene, the lower boiling 3,4-dichloro-1-butene was continuously removed so that, eventually, all the 1,4 product was converted to the 1,2 addition product – a classic Le Chatelier displacement of equilibrium. We're on our way to neoprene rubber. Using this combination of equilibration catalysts and boiling point differences the DuPont chemists and chemical engineers were able to gain whatever product the commercial interests happened to be calling for at any particular time. Great!

Shakespeare wrote a play about 400 years ago with the title: "*All's Well that Ends Well*." I'm afraid I have to report to you that all the beautiful organic chemistry described in this section did not end well. A short time after DuPont started producing hexamethylene diamine or chloroprene via the addition of chlorine to butadiene, 1,4-dichloro-2-butene was discovered to be carcinogenic. DuPont dropped the process and had to turn the creative talents of it's chemists and engineers in other directions. The carcinogenic problem with 1,4-dichloro-2-butene arises from its activity as a site for nucleophilic substitution, the subject we take up in the next section.

It seems fair in mentioning Shakespear not to forget to mention **Henri Louis Le Chatelier**, whose principle worked so well for DuPont's solving the technical problem. He lived from 1850 to 1936 and was quite a notable scientist in France during those years, even with his name enscribed among a limited number on the Eiffel Tower and attaining the status of a Grand Office of the Legion of Honor of France. No less of an honor was the considerable space given to the principle and the life of Le Chatelier by Linus Pauling in his famous general chemistry book. Here's a web site for the curious among you: http://faculty.cua.edu/may/LeChatelier.pdf. Enjoy seeing what one person can accomplish.

PROBLEM 10.14
The fact that the 1,2-addition product is favored over the 1,4-addition product on reaction of Cl_2 with 1,3-butadiene reveals the relative stabilizing importance of double bond substitution versus carbocation substitution. What is revealed?

PROBLEM 10.15
Draw a reaction coordinate diagram that shows the conflict between kinetic versus thermodynamic control in the addition of Cl_2 to 1,3-butadiene. There are two conflicting carbocation intermediates produced, which differ in energy. What are they and how is it that the lower energy intermediate leads to the higher energy addition product.

PROBLEM 10.16
Answer the question posed in the text using a reaction coordinate diagram: "However, the higher the temperature at which the reaction is run, the more 1,4 addition produce is produced. Why?"

PROBLEM 10.17
Recount the story of how the separation of the 1,2 and 1,4 addition products of Cl_2 to 1,3-butadiene was accomplished by the DuPont corporation chemists and chemical engineers using physical and chemical procedures.

PROBLEM 10.18

Why was Scheele successful in producing HCN by adding sulfuric acid to a material with CN⁻ ions present? Could you produce H_2SO_4 by adding HCN to a material with sulfate ions present?

PROBLEM 10.19

How does the fact that 1,2 addition of chlorine to 1,3-butadiene is kinetically favored inform us that more substituted double bonds are not more important than more substituted carbocations?

PROBLEM 10.20

What is the basis of the following statement in the text: "Apparently, the stability of the internal double bond in the dicyano isomers is enhanced compared to its stability in the dichloro isomers."

THE REACTION OF CYANIDE ANION with the isomeric dichlorobutenes shown in Figure 10.12 is an example of nucleophilic substitution at saturated carbon adjacent to a double bond, allylic substitution. The cyanide ion, $C \equiv N^-$, is the nucleophile displacing Cl⁻, which is the leaving group. How does this reaction take place, what is its fundamental mechanism? The answer to this question forces us to look into one of the most studied reaction mechanisms in organic chemistry, which reveals many fundamental principles of the science allowing understanding of a wide range of phenomena ranging from the industrial to the biochemical. Let's begin by looking back to Chapter 5.

In the fundamental carbon-carbon bond-making that is prerequisite to the synthesis of the terpenes (section 5.7 (Figure 5.12)), the first step on the path is the reaction of isopentenyl diphosphate with a carbocation formed from dimethyl allyl diphosphate. In that reaction a carbon-carbon bond was made and a diphosphate group was displaced. Let's use this reaction to introduce us to the subject of this section but first let's be introduced again to an English chemist who played a critical role in the development of the ideas of nucleophilic substitution.

In section 1.12 we met **Christopher Ingold** who participated in generating the R, S nomenclature for chiral molecules. Claims to priority in issues of nomenclature have led to famous altercations among chemists with one of those involving Ingold and Sir Robert Robinson. If you need a break and as well an insight into the way some organic chemists related to each other behind the scenes, take a look at: https://webspace.yale.edu/chem125/125/history99/6Stereochemistry/CIP_Prelog/prelstory.html

Ingold contributed greatly to the way that organic chemists express their scientific ideas. He is given credit for the terms nucleophile and electrophile, which are so widely used throughout chemistry and in this book, and in addition the mechanistic descriptors, S_N1 and S_N2, which are the focus of this section. Ingold also independently proposed the concept we know as resonance but using the word mesomerism for which one can find the definition: "*The property of a compound having simultaneously the characteristics of two or more structural forms that differ only in the distribution of electrons. Such compounds are highly stable and cannot be properly represented by a single structural formula.*" Among some chemists the word mesomerism is seen as superior to resonance, since the latter suggests that the structures actually exist and are interconverting with each other, which we understand **not** to be the situation (section 5.4).

With regard to the focus of this section, one can find in a single paper published in 1935 by Christopher Ingold and his long term collaborator, Edward Hughes, in

10.6

The reactions of the isomeric dichlorobutenes with cyanide ion leads us to investigate one of the most studied reactions in organic chemistry, nucleophilic substitution at saturated carbon, which can take place at the extremes via the S_N1 or S_N2 mechanism.

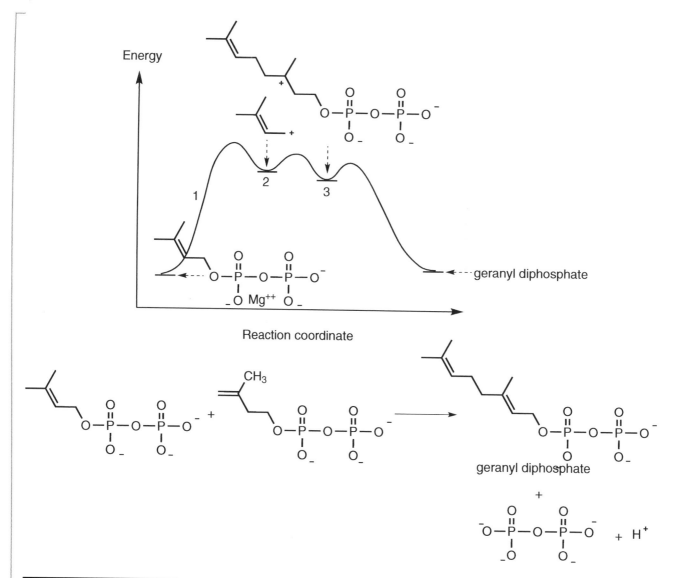

Reaction Coordinate Diagram Describes the Mechanism for the S$_N$1 Mechanism for Formation of Geranyl Diphosphate from Isopentenyl Diphosphate and Dimethyl Allyl Diphosphate

the Journal of the Chemical Society, most of what is written in modern textbooks of organic chemistry on the subject of nucleophilic substitution. The paper is entitled: "*Mechanism of Substitution at a Saturated Carbon Atom. Part IV. A Discussion of Constitutional and Solvent Effects on the Mechanism, Kinetics, Velocity and Orientation of Substitution,*"

In this 1935 paper, Hughes and Ingold defined the terms S$_N$1 and S$_N$2 and the conditions of the occurrence of these mechanisms using terms that are familiar to us. Let's look at the S$_N$1 reaction first.

Under conditions where: (1) a saturated carbon atom (sp^3 hybridized) in a molecule is bound to an excellent leaving group; (2) the carbon atom in question can form a stabilized carbocation (section 4.7); (3) conditions support charge separation (a reasonably high dielectric solvent); the leaving group may be displaced by an incoming nucleophile with the velocity of the reaction (section 6.11) not depending on the concentration of the incoming nucleophile.

Figure 10.13 is a representative reaction coordinate diagram, that is, a diagram in which the energy levels simply represent rather than even approximately fit, the actual energies of the reaction (section 6.11). The overall form of the diagram in Figure 10.13 fits the S$_N$1 mechanism proposed by Hughes and Ingold, which we have applied to the reaction of isopentenyl diphosphate with dimethyl allyl diphosphate (Figure 5.12).

In this reaction, as discussed in section 5.7, there is an excellent leaving group, diphosphate, bonded to a -CH$_2$ group, which when the bond to the diphosphate group is broken (step **1** in Figure 5.12) forms a resonance stabilized carbocation. The other player in this reaction, the nucleophile, is the double bond in isopentenyl diphosphate and in particular the π-electrons making up the double bond. The final step from intermediate **3** is simply the loss of a proton to form the geranyl diphosphate product.

As seen in the reaction coordinate diagram (Figure 10.13) the far largest energy of activation is associated with step **1**, the formation of the carbocation, intermediate 2. This step will, therefore, be the slowest step by far, the rate-determining step, subject to the identical general discussion for the nature of rate-determining steps around the mechanism and reaction coordinate diagram for electrophilic aromatic substitution (section 6.11, Figure 6.16).

In the mechanism presented in Figure 10.13, the rate of production of the product will not depend on the concentration of isopentenyl diphosphate, because the product, geranyl diphosphate, can not be produced until the ionization of the dimethyl allyl diphosphate takes place. The isopentenyl diphosphate does not take part in the formation of the carbocation. The rate equation (section 6.11) will therefore, ideally, be of the form: d[product]/d time = k[dimethyl allyl diphosphate]^1.

In other words the reaction rate depends only on the concentration of the dimethyl allyl diphosphate, which is expressed in the equation by the exponent 1; the reaction is first order in the concentration of the dimethylallyl diphosphate. This is the source of the 1 in the designation S_N1 with S for substitution and N for nucleophilic.

▪ **Christopher Ingold**

In their excellent book, "The Organic Chemistry of Biological Pathways," McMurry and Begley point out that based on the literature, there is uncertainty about the mechanism we have just ideally portrayed as an S_N1 reaction. The reaction may have some characteristics of what Hughes and Ingold described as an S_N2 mechanism. What is the S_N2 mechanism?

In large supermarkets with a focus on "natural foods" one can find multiple shelves with dietary supplements including many brands of SAMe, S-adenosyl- methionine (**Figure 10.14**). Some studies show health benefits for regularly taking S-adenosylmethionine including an effect on alleviating depression. This is hardly the place to go into those aspects of S-adenosylmethionine, or even the range of biochemical pathways this coenzyme takes part in, but S-adenosylmethionine does give us multiple opportunities to demonstrate the other mechanism that Hughes and Ingold described, the S_N2 mechanism.

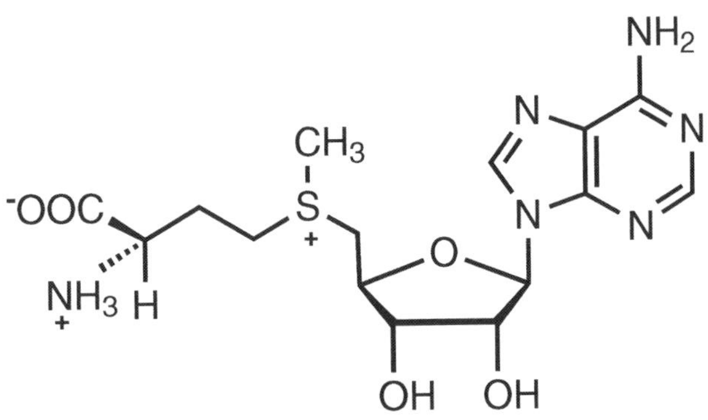

The first way that S-adenosylmethionine helps us understand the S_N2 reaction is in the formation of this coenzyme, which takes place by a reaction between adenosine triphosphate and methionine as shown in **Figure 10.15**. The reaction path producing S-adenosylmethionine fits all

S-adenosylmethionine

◄ FIGURE 10.14

Structure of S-Adenosylmethionine

the prerequisites for a nucleophilic displacement in that a lone pair of electrons on sulfur is available to form a bond to the CH_2 group, breaking the bond between the CH_2 group and the oxygen atom thereby releasing the triphosphate leaving group.

In contrast to the S_N1 mechanism, however, which defines the formation of geranyl diphosphate (Figure 10.13), the reaction forming S-adenosylmethionine takes place in a single step, which is represented by the reaction coordinate diagram shown in **Figure 10.16**. As a consequence of the mechanism presented in Figure 10.16, the

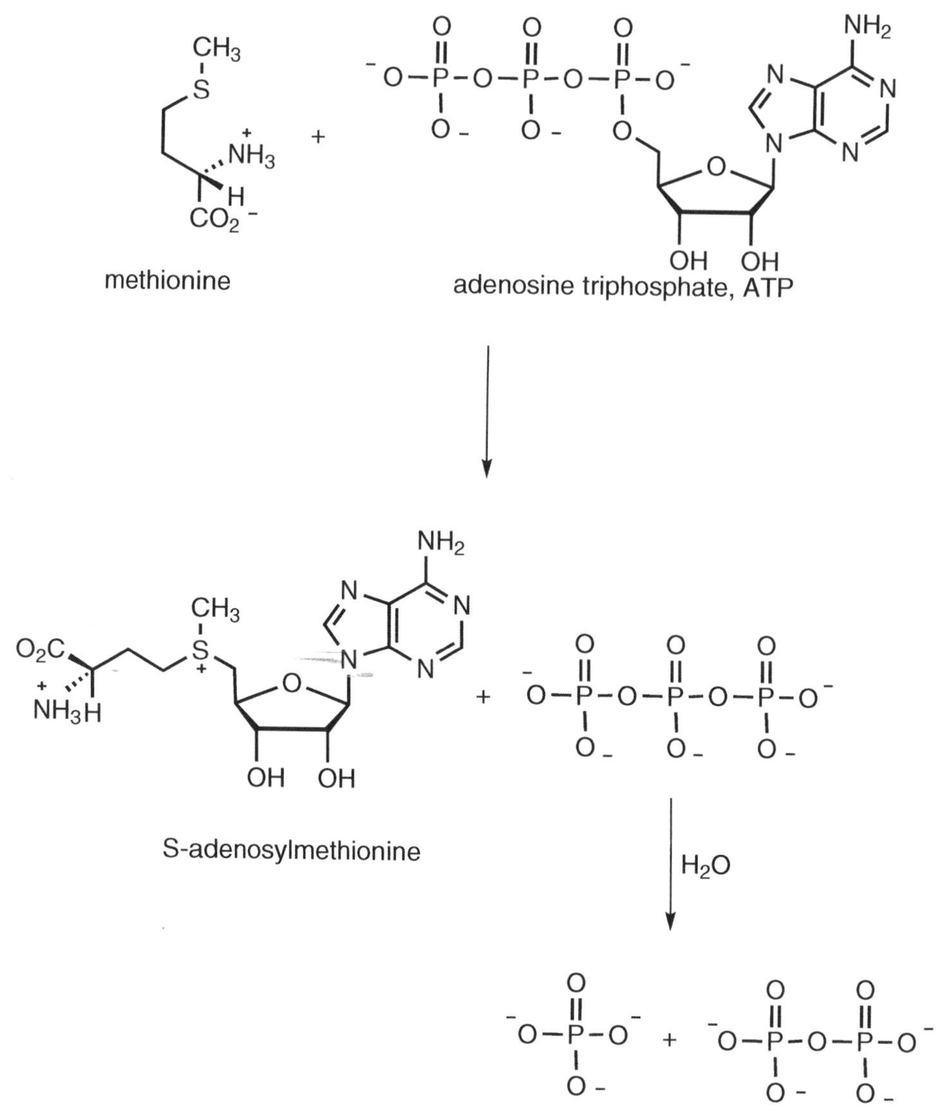

Biological Path from Methionine and Adenosine Triphosphate to S-Adenosylmethionine

rate of production of the product, S-adenosylmethionine, will depend on both the concentration of the starting methionine and on the adenosine triphosphate (ATP) concentration leading to the idealized rate expression:

d[SAMe]/d time = [methionine]^1[ATP]^1.

The reaction takes place in a single step that depends on the concentrations of both reactants, which is manifested by being first order in each of the reactants. The reaction is, therefore, a second order reaction, accounting for the 2 in the mechanistic descriptor, S_N2.

Why does the nucleophilic substitution reaction leading to geranyl diphosphate occur via the S_N1 mechanism as shown in Figure 10.13, while the nucleophilic substitution reaction leading to S-adenosylmethionine occurs via the S_N2 mechanism shown in Figure 10.16? There are many factors first pointed to in that 1935 paper by Hughes and Ingold and studied in great detail since that time, but the most important factor is the carbon atom at which the substitution takes place. Let's look at a diagram and an explanation offered in the Hughes and Ingold 1935 publication (Edward D. Hughes and Christopher K. Ingold, J. Chem. Soc., 1935, 244-255, DOI: 10.1039/JR9350000244 - Reproduced by permission of The Royal Society of Chemistry (**Figure 10.17**).

In Chapter 4, section 4.7, we learned that the success of catalytic cracking in producing the branched hydrocarbons necessary for high octane gasoline came from the intervention of carbocations and the fact that carbocation stability increases with

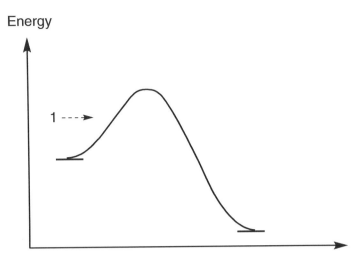

reaction 1

◀ **FIGURE 10.16**

Reaction Coordinate Diagram for the Formation of S-Adenosylmethionine via an S_N2 Mechanism

increased substitution at the positively charged carbon atom. This is the point that Hughes and Ingold are making (Figure 10.17) about "progressively increasing powers of electron release favoring the S_N1 mechanism." In recent years the identical idea about stability of carbocations is expressed by the concept of hyperconjugation or no-bond resonance as noted in section 4.7.

But there are other ways than increased substitution at the carbon undergoing the substitution that may favor the S_N1 mechanism. The carbocation can be stabilized also by resonance as for example in the situation in dimethyl allyl diphosphate where the CH_2 group that bears the positive charge is conjugated with a double bond (Figure 10.13, intermediate 2). In this situation the π electrons of the double bond have, to put it in the words of Hughes and Ingold, "powers of electron release." There are many

FIGURE 10.17 ▶

Key section of
Hughes and Ingold
publication describes
the characteristics of
both the S_N1 and S_N2
mechanisms.

Figure 10.17

of the *iso*propyl compound.   Under the conditions chosen for the hydrolysis of the secondary and tertiary compounds there is no disturbance due to olefin elimination.   The following relations are thus established for a series of groups, Alk, having progressively increasing powers of electron release :

Mechanism	....	S_N2		S_N1	
Group	....	CH_3	$CH_3{\rightarrow}CH_2$	$\begin{smallmatrix}CH_3\\CH_3\end{smallmatrix}{>}CH$	$\begin{smallmatrix}CH_3\\CH_3{\rightarrow}C\\CH_3\end{smallmatrix}$
Reaction order	....	2	2	1	1
Velocity	....	—— decrease ⟶		—— increase ⟶	

This is exactly the type of relationship which theory predicts : it is represented in Fig. 1 of Part III.

In attempting to generalise these conclusions, it should be noted that as they stand they apply primarily to strongly ionising solvents : in less strongly ionising media the mechanistic critical point should shift towards the right, so that the *iso*propyl group and even the *tert.*-butyl group could become included in the (S_N2) category. Furthermore, owing to the great loss of intensity accompanying the relay of polar effects through saturated hydrocarbon chains, the higher primary alkyl groups are likely to fall into the same class as ethyl, the higher secondary into the same class as *iso*propyl, and so on. The velocity relation in each class can readily be deduced when the mechanistic category is known. Hence, for normal alkyl groups we expect continuously diminishing differences, provided always that, when the differences really become small, they may show disturbances from factors other than that now considered.

examples of carbocations that are stabilized by nearby atomic moieties with powers of electron release as shown in just a few examples in **Figure 10.18**.

Hughes and Ingold included in their discussion of the factors controlling the competition between the two mechanisms, S_N1 and S_N2, the role of the solvent, the environment around the reaction with their sentence (Figure 10.17): *"in less strongly ionizing media.....,"* which would favor the S_N2 mechanism.

In summary therefore, to follow the ideas of Hughes and Ingold, the S_N1 reaction will be favored by nucleophilic substitution in strongly ionizing media when carbocations can be formed that are stabilized by electron release, whatever form the electron release takes–higher substitution or conjugated π- or nonbonding electrons. (Figure 10.18).

FIGURE 10.18 ▶

Various
Resonance
Structures for
Stabilization of
Carbocations

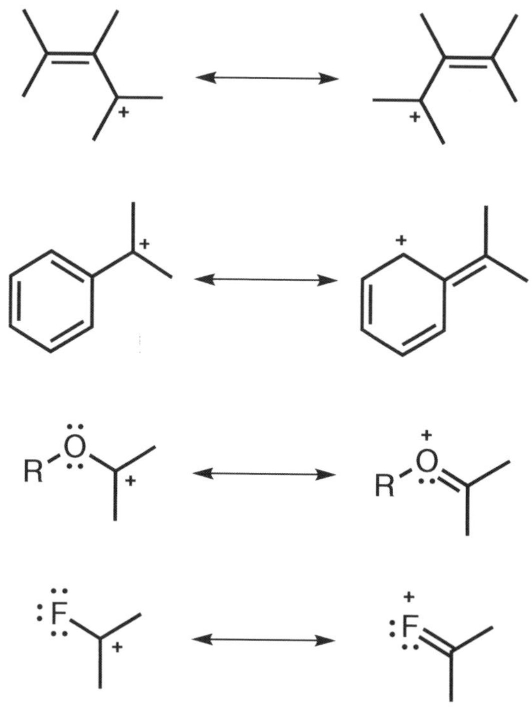

The control of the medium in the biologically-controlled reactions is under the control of the enzyme, so-to-speak, and somewhat beyond our studies here, but the structure of the molecule undergoing the substitution is grist for the Hughes and Ingold mill. Let's look again at the reaction of isopentenyl diphosphate with dimethylallyl diphosphate (Figure 10.13).

Taking only account of the site of nucleophilic substitution, a CH_2 group, would point to nucleophilic substitution via the S_N2 mechanism. However, taking account of the conjugation of a carbocation site on this CH_2 group with a double bond having "powers of electron release" (to put it in Hughes and Ingold's words) points to nucleophilic substitution via the S_N1 mechanism. Although the

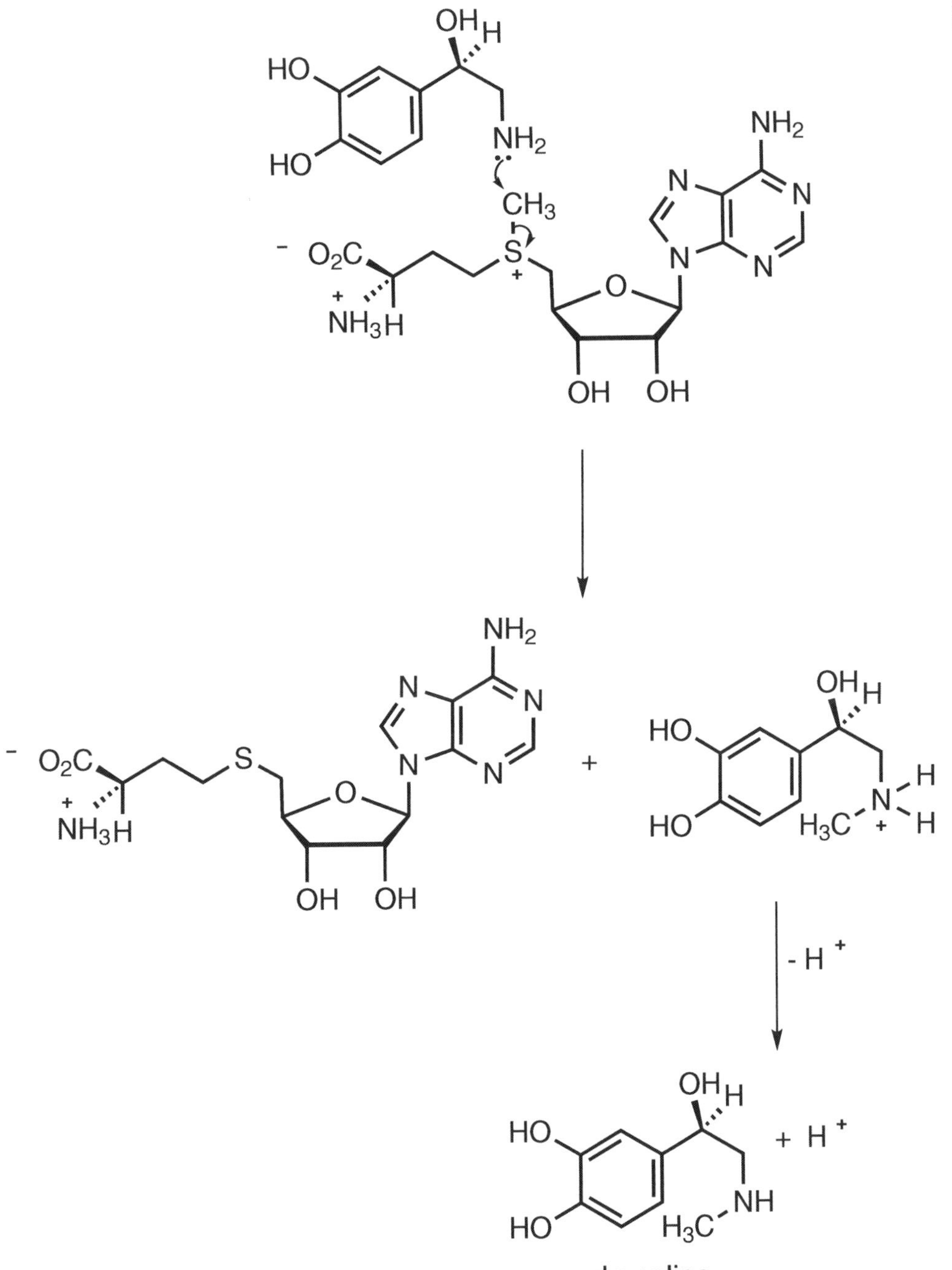

◀ **FIGURE 10.19**

S_N2 Mechanism by which Norepinephrine is Converted to Adrenaline (Epinephrine)

experiment necessary to discern the opposing mechanisms in this biological reaction has not been carried out, the vote seems in favor of the force of the electron release. We'll shortly see the kind of experiment that would be necessary.

But the CH_2 group at which substitution takes place in the formation of S-adenosylmethionine does not have access to electron releasing groups if a carbocation formed at this CH_2 group via an S_N1 mechanism. The formation of S-adenosylmethionine therefore takes place via the S_N2 mechanism as shown in Figure 10.16, as would have been predicted by Hughes and Ingold who saw substitution at either CH_3 or CH_2 as fitting this mechanism in which both reactants contribute to the velocity of the reaction.

Let's follow this line of reasoning in investigation of the biochemical function of S-adenosylmethionine. With a wide variety of substrates, S-adenosyl methionine transfers the sulfur bound methyl group to a nucleophile. A prominent example is the biochemical synthesis of the "fight or flight" hormone, epinephrine (adrenaline) formed from norepinephrine. Both are catecholamines and act as neurotransmitters in the body. The two hormones differ only in a single methyl group bound to nitrogen, as shown in **Figure 10.19**, which also shows the role of S-adenosylmethionine in the conversion of norepinephrine to epinephrine.

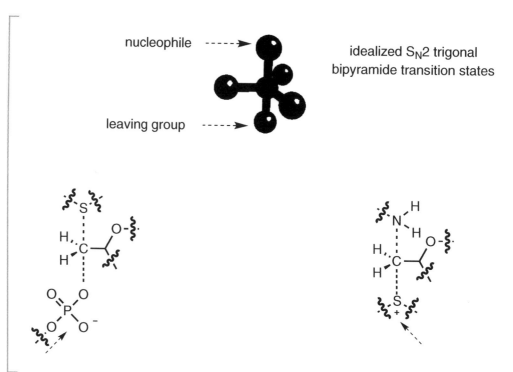

FIGURE 10.20 ▶

Representations of the Trigonal Bipyramide Transition State for the S$_N$2 Mechanism

The reaction shown in Figure 10.19 is a nucleophilic displacement reaction at a methyl group in which the amino group on the norepinephrine acts as the nucleophile and methionine sulfur is the leaving group. Following Hughes and Ingold (Figure 10.17), the fact that the site of substitution is at a CH$_3$ group requires that this biological methylation reaction takes place via an S$_N$2 mechanism. This mechanistic picture is shown using Ingold's curved arrow nomenclature in Figure 10.19.

The absence of stabilization of less substituted carbocations is certainly a reason for the system to avoid the S$_N$1 mechanism. However, is there another reason why less substituted carbon atoms favor the S$_N$2 over the S$_N$1 mechanism?

For a reaction to take place purely via the S$_N$2 mechanism the site of substitution must, at some place on the reaction path, become pentacoordinate, which is represented by the transition states shown in **Figure 10.20** for the reactions in Figures 10.16 and 10.19. This structure of these transition states must be a carbon atom associated with five groups, the leaving group, the incoming nucleophile and the three atoms that stay connected throughout. The incoming and leaving groups may be seen as partially bound to the carbon atom.

However we may represent the details of the transition state there is no doubt that it is crowded and therefore it makes sense that if the site of displacement occurs at a carbon atom that is not too crowded to begin with such as CH$_3$ and CH$_2$, the preferred mechanism will be S$_N$2.

The transition state in Figure 10.20 proposed for the S$_N$2 reactions shown in Figures 10.16 and 10.19 are states of such fleeting existence that they have not been able to be directly observed. Until recently there was no method to understand the intimate motions of the atoms in a chemical reaction near or at the transition state. How, then, is there any evidence for these transition states, which are hypothesized to exist on the time scale of a vibration or that is, in the range of femtoseconds, 10^{-15} of a second? How is one to know that the incoming nucleophiles, as shown in Figure

10.20, approach from the rear of the carbon atom from the view of the leaving group, so-called backside approach?

In recent years a field called femtochemistry has developed based on the work of **Ahmed H. Zewail** who was awarded the 1999 Nobel Prize in chemistry for the development of femtosecond spectroscopy. Extremely short pulses of lasers have been used to investigate the motions of atoms at energies near those of the transition state. Here are a few sentences from the statement of the committee which awarded the prize, and which reminds us of the history of the development of the ideas of chemical kinetics discussed earlier in Chapter 6 (section 6.11):

"The contribution for which Zewail is to receive the Nobel Prize means that we have reached the end of the road: no chemical reactions take place faster than this. With femtosecond spectroscopy we can for the first time observe in 'slow motion' what happens as the reaction barrier is crossed and hence also understand the mechanistic background to Arrhenius' formula for temperature dependence and to the formulae for which van't Hoff was awarded his Nobel Prize."

It is worth looking at the entire statement, which can be found at: http://nobelprize.org/nobel_prizes/chemistry/laureates/1999/chemback99.pdf

The answer to the question posed above about the transition state of the S_N2 reaction is increasingly addressed for more complex molecules by femtosecond spectroscopy. But another approach close to the hearts of organic chemists and widely applicable to complex molecules is to investigate the stereochemical consequence of the nucleophilic displacement as a probe of the transition state.

∎ **Ahmed Zewail**

PROBLEM 10.21
How do the energies of activation for the three steps in the reaction coordinate diagram in Figure 10.13 relate to the concept of the rate determining step?

PROBLEM 10.22
Designate all non-bonding electrons and account for all formal charges in the structures in Figures 10.14, 10.15, 10.16 and 10.19. Which electrons you have accounted for contribute to aromatic character?

PROBLEM 10.23
What factors cause S-adenosyl methionine (Figure 10.15) and its biological function to require the S_N2 mechanism?

PROBLEM 10.24
Why might a solvent of lower dielectric constant cause a nucleophilic substitution at saturated carbon to favor second order kinetics, that is, to favor the S_N2 mechanism?

PROBLEM 10.25
In what way may there be a mechanistic correlation between the production of the high octane gasoline, 2,2,4-trimethyl pentane (Figure 4.13), and the S_N1 mechanism of nucleophilic substitution?

PROBLEM 10.26
Make up a chart with increasing carbocation stability along one axis and solvent dielectric constant along the other axis and show how the gradual transformation from S_N1 to S_N2 mechanistic behavior will take place.

PROBLEM 10.27
How might enzyme catalysis influence the nucleophilic substitution mechanism between S_N1 and S_N2 in the formation of geranyl diphosphate without changing the solvent environment?

Paul Walden

10.7

Stereochemical Probes of Nucleophilic Displacement

THE FIRST STEREOCHEMICAL EXPERIMENT, which could lead to insight into the transition state of nucleophilic substitution, was carried out in the 1890s by **Pauls Valdens**, before transition state theory was developed. Valdens, a Baltic-German, was born in Latvia and educated there and in Leipzig and Munich. He moved to Germany after the Russian Revolution, became a professor at Rostock and Tübingen and changed his name to **Paul Walden**. Walden who died in 1957 at the age of 94 was highly regarded for his work but is best know for research he did in his 30s where he showed that an optically active molecule could be converted to its enantiomer by a series of nucleophilic substitution reactions (before the word nucleophile was coined). It was only many years later that Walden's observation, which came to be known as the "Walden Inversion," could be connected to transition state theory.

Dean Stanley Tarbell, a distinguished American organic chemist, who like Walden was interested in the history of organic chemistry, wrote an article in the Journal of Chemical Education in 1974 entitled: "The Chemical World of Paul Walden." Tarbell points out some interesting aspects of Walden's life:

"*...and he was evidently a talented political survivor, because he lived under many regimes (and outlived most of them), including the Romanov, Hohenzollern, Russian communist, Weimar Republic, Nazi, and West German governments. He was forced to move as a refugee at least four times; his home and personal library of 10,000 volumes were completely destroyed in the bombing of Rostock in 1942.*" We could edit the parenthetical phrase in Tarbell's writing: "(and outlived almost all of them with the single exception of the Soviet Union)."

In much work initially stimulated by Walden's efforts on interconversions of the enantiomers of malic acid, a dicarboxylic acid derived from apples and later to be shown as a key molecule in the citric acid cycle (sections 8.6 and 8.7), many nucleophilic substitution reactions have been studied with molecules that allowed observation of the stereochemical consequence of the reaction. As seen in **Figure 10.21**, these experiments yield a great deal of information about nucleophilic substitution, the kind of information not available from the reactions portrayed in section 10.6

The carbon atom sites of substitution in the formation of S-adenosylmethionine (Figure 10.16) and in the methyl transfer reaction in Figure 10.19 hold no stereochemical information. The products of these reactions therefore yield no information about the transition states. However, if the carbon atom site of substitution is a stereocenter, as shown for all the molecules undergoing nucleophilic substitution in Figure 10.21, then the stereochemical consequence of the reaction, which is revealed experimentally by the product produced, reveals detailed information about the transition state structure.

Three reactions are shown in Figure 10.21, nucleophilic substitution at primary,

Dean Stanley Tarbell

secondary and tertiary carbon atom sites. Because a primary carbon, a CH_2 group can not be the source of mirror image isomerization something special had to be done to explore the geometry of the transition state in a reaction such as **1** in Figure 10.21.

Over fifty years ago a young professor at Berkeley, **Andrew Streitwieser Jr.**, decided to apply the isotopic techniques that had proven so useful in elucidating biochemical mechanisms (section 5.8) to help to answer questions of mechanism in organic chemistry. In section 5.6 where the ideas of enantiotopic groups were first discussed, problem 5.23 introduced the possibility of producing a primary alcohol, ethanol, in a chiral form using deuterium substitution, CH_3CHDOH. This goal has been accomplished by reducing $CH_3CD=O$ enzymatically. Streitwieser used parallel enzymatic methods to prepare one enantiomer of 1-deuterio-1-butanol from which the substrate in reaction **1** in Figure 10.21 was synthesized.

Streitwieser obtained his Ph.D. in 1952 at the age of 25 and under the guidance of Bill Doering (who was a young professor at Columbia University) whom we met when we looked into the work of Erich Hückel (section 6.8) – small world – organic chemistry. The work Streitwieser did on elucidating the stereochemistry of nucleophilic substitution at primary carbon, such as reaction **1** in Figure 9.37, first demonstrated his creative abilities.

The transition state hypothesized in Figure 10.20 for the S_N2 reaction is very strongly supported by Streitwieser's results (**1** in Figure 10.21) because the backside attack hypothesized must lead to the nucleophile, the acetate group in **1**, displacing the leaving group with inversion of configuration. As Hughes and Ingold proposed (Figure 10.17), nucleophilic substitution at primary carbon is the ideal substrate for the S_N2 mechanism. Streitwieser's result demonstrated 100% inversion of configuration, within experimental error, a perfect S_N2 result.

■ **Andrew Streitwieser**

◀ **FIGURE 10.21**

Experiments that Reveal the Stereochemical Changes at the Substituted Carbon Atom Occurring in the S_N2 Versus the S_N1 Reaction

The extensive discussion in Hughes and Ingold's seminal 1935 paper (Figure 10.17) allowed the possibility that the S_N2 and S_N1 mechanisms could be mixed in a situation where one is not overwhelmingly favored. In fact the situation, although now well understood, is very complicated. The solvent can participate in loosening the bond to the leaving group with the solvent then displaced by the incoming nucleophiles in a variety of ion-paired states. This topic is a subject for advanced study but we can see the result of this complexity in reaction **2** in Figure 10.21 where a significant proportion of the reaction path leads to replacement of the leaving group with retention of configuration. Here no isotope is required because a secondary carbon atom site allows enantiomeric isomers.

Reaction **3** in Figure 10.21 would have been pointed to by Hughes and Ingold as ideal for the S_N1 reaction. The carbon atom site is highly substituted, tertiary, and at the same time a carbocation at this site would be resonance-stabilized by the conjugated benzene ring. It is, therefore, hardly a surprise based on understanding the fundamental forces at work, that the product of the nucleophilic displacement is formed near to the racemic state. The stabilized carbocation could exist long enough to free itself of all memory of the p-nitrobenzoate leaving group that had been attached to it, which it had thrown out, and become free of. The flat sp^2 hybridized carbocation (Figure 4.6), then without any chiral memory, that is, of the absolute configuration of the starting molecule, would be free to accept the incoming nucleophile, the acetate group in this situation, from either face, producing both enantiomers.

In summary we might say that those carbon atom sites of substitution that are poorly suited for the S_N1 reaction will be well suited for the S_N2 reaction and vice versa. However, there are intermediate situations in which the carbon atom site of substitution and/or the ionizing power of the medium are not ideal for either mechanism. There are many nucleophilic substitution reactions known in organic chemistry that fit into this category leading to experimental observations where the reaction kinetics are neither purely first order nor purely second order. For certain structural situations a mixed stereochemical result may signal the fact that the reaction path is neither entirely S_N1 nor S_N2.

These ideas can be applied to the substitution of the cyanide ions for the chloride ions in the abandoned industrial synthetic path toward hexamethylene diamine (Figures 10.6 and 10.12).

PROBLEM 10.31
Draw a reaction coordinate diagram to describe reaction **1** in Figure 10.21 including a structure for the transition state of this reaction.

PROBLEM 10.32
Answer the previous question for reaction **3** in Figure 10.21.

PROBLEM 10.33
Why is it possible to conclude for reaction **2** in Figure 10.21 that the reaction path is 76% S_N2 and 24% S_N1 mechanism?

PROBLEM 10.34
Use resonance structures and orbital representations to describe the concept of no-bond resonance (hyperconjugation, section 4.7) applied to tertiary butyl cation, $(CH_3)C^+$.

PROBLEM 10.35
How might the solvent in reaction **2** in Figure 10.21 act as a nucleophile via an S_N2 mechanism leading to retention of configuration from the starting material to the final product?

PROBLEM 10.36

Assign (R) or (S) nomenclature to the starting materials and products of the nucleophilic substitution reactions shown in Figure 10.21. Is it always necessary in an S_N2 reaction for (R) and (S) absolute configurations to switch?

PROBLEM 10.37

Apply the ideas of this section to the substitution of the cyanide ions for the chloride ions in the abandoned industrial synthetic path toward hexamethylene diamine (Figures 10.6 and 10.12). Would you predict S_N1 or S_N2 mechanisms, or mixtures of both mechanisms, and why?

CHAPTER TEN SUMMARY of the Essential Material

THE CHAPTER BEGINS WITH THE CHEMISTRY FOR CONVERSION of benzene to adipic acid, necessary for the industrial synthesis of nylon 6,6. Here we come across a practical application of addition of hydrogen to benzene followed by a catalytic oxidation and a stoichiometric oxidation. With adipic acid produced, the production of the other monomer for nylon 6,6 is attained by a series of reactions passing through the dinitrile, which can be reduced to hexamethylene diamine. Although the large scale of the industrial reduction requires use of hydrogen gas, reductions of nitriles on a laboratory scale use hydride donating reducing agents and in particular lithium aluminum hydride. We discover the method of action of this reducing agent and how it differs from the hydride reducing agents we've seen before, NADH and $FADH_2$ and understand how the biological agents are reversible in their action in contrast to the laboratory reducing agents.

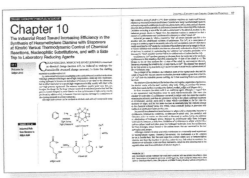

Industry sought a more efficient route to hexamethylene diamine, which led to the addition of chlorine to 1,3-butadiene with the idea to substitute nitrile groups for the chlorine in the produced dichlorobutenes to gain the necessary additional two carbon atoms. However, the addition of chlorine did not go as planned and isomeric structures were produced, which could not be avoided because they arose from a balance between kinetic and thermodynamic control, the source of which we studied in some detail. Catalysts were discovered that allowed equilibration of both the dichlorobutene isomers and the dinitrilebutene isomers. Using these catalysts in different ways in combination with fractional distillation allowed accomplishing separation of the isomers and therefore a process to produce the separated isomers. However, this valuable technology that lead to hexamethylene diamine and therefore to nylon 6,6 and also to chloroprene and therefore neoprene had to be abandoned because of toxic effects of the intermeiate involved.

The fundamental reaction used in the abandoned industrial approach to hexamethylene diamine involved a nucleophilic displacement leading us to look into the details of this class of reaction. In this way we discovered the classic mechanisms of nucleophilic displacement, S_N1 and S_N2 and the large amount of research that has revealed the structural basis of the competition between the two mechanisms. Here we used methods for the study of mechanism seen earlier in investigations of electrophilic aromatic substitution including the order of a reaction, rate-determining steps and reaction coordinate diagrams. Our study of this area led us back to biological systems where we saw both of these mechanisms taking place. We understood how transition state structures were hypothesized for both nucleophilic displacement mechanisms and how stereochemical experiments allowed testing of these hypotheses.

Chapter 11

Much can be learned about organic chemistry from the study of natural rubber and other elastomers

11.1
Two Different Trees

BOTH HEVEA BRASILIENSIS AND PALAQUIUM are tropical trees native to different parts of the world. The former, also called the rubber tree, was originally native to the Amazon rainforest while the latter, also known as Gutta-percha, is native to Southeast Asia, where in fact the rubber tree is now grown in large numbers on plantations. The Southeast Asian trees are the consequence of the (it is said) 70,000 seeds from Brazilian trees being stolen in the latter part of the nineteenth century and transplanted to new growing sites.

Although very few of these seeds sprouted, enough did that this thievery led to a great commercial activity spanning now three centuries, the rubber industry, and is said to have been the accomplishment of a hapless businessman, **Henry Wickham**, who never made any money. The plantations that arose from the stolen seeds, which eventually put the Brazilian rubber industry out of business, supply most of the natural rubber used in the world. The latex that is the source of the rubber is harvested from the trees in a labor- intensive process involving cutting into the tree in a precisely defined manner so as to not damage the tree's growth. The cut then allows the latex, a milky substance, to leak slowly into a cup placed below the cut. The latex is a protective substance manufactured by the tree to protect it when the bark is breached, and many plants are known to produce a milky exudate for this purpose.

Rubber is a word coined by **Joseph Priestley** in about 1770 when he discovered that the substance brought back from explorations in what is now South America was effective in rubbing out lead pencil marks from paper. This word displaced other words used for this elastic material, the most prominent being caoutchouc, derived, I have read, from the people who are native to the area in which the trees grow: caa meaning wood and o-cho meaning to run or to weep, therefore weeping wood. It is reported that caoutchouc was known long before Europeans discovered its properties, even to more than three thousand years ago. South American native people are known to have used the material extracted from the latex to form elastic balls with which they played life and death games and there are probably many more uses over these thousands of years that may forever remain unknown.

■ Henry Wickham

Priestley's word for the elastic substance derived from *Hevea brasiliens* was a minor accomplishment for this scientist who contributed greatly to setting the stage for the development of chemistry in the centuries that followed his work. This is not the place to lay out Priestley's accomplishments in chemistry but if there is any doubt, consider that the Priestley Medal is the highest honor conferred by the American Chemical Society and perhaps an even more important credit is that Partington in Volume Three of his "History of Chemistry" devotes an entire chapter to Priestley. If you are interested, the web is full of information about this great man. Let's now, however, return to that substance whose value is so much more than as an eraser of pencil marks.

The differing latex obtained from *Hevea brasiliensis* and from *Palaquium* are aqueous emulsions containing many substances including, of most interest to our studies and to commercial interests, microparticles of polymeric substances that differ in the two latexes. However, while the polymers differ, the formulae of the polymers in the latex from the two trees is identical. Therefore the different properties point to an isomeric relationship. As we've learned in Chapter 1, isomers with different properties may be constitutional isomers or diastereomers.

Isolation of the polymeric substance from *Hevea brasiliensis* yields an elastic material, what came to be known as rubber. The polymeric substance isolated from the latex of *Palaquium* know as *gutta percha* is an inelastic, hard substance that found early use as a tough coating for underwater cables and, in more modern times, as a tough coating for golf balls. It is used as well as a filling material for dental work, notably for root canal work.

◄ FIGURE 11.1

Comparison of the Structures of Havea Rubber and Gutta Percha

Hevea rubber   Gutta percha

Figure 11.1 shows the structure of a portion of the polymer chains of *Hevea* rubber and of *gutta percha*, which show that the difference between the two is not constitutional but rather diastereomeric. From a stereochemical point of view the differing polymer chains in Figure 11.1 find parallel to the different polymer chains in Figure 1.1. However, while the stereochemical difference between starch and cellulose arises from an axial versus an equatorial linking bond between the glucose units (section 3.16), the difference between *Hevea* rubber and *gutta percha* derive from a stereochemical property of certain double bonds, cis (Z) versus trans (E) configurations. Nevertheless, both pairs of isomers: cellulose and starch; and *Hevea* rubber and *gutta percha* are configurational diastereomers.

PROBLEM 11.1

Can you offer an explanation for the different properties of the two polymers in Figure 11.1 based on what you understand about unsaturated fatty acids and also the difference between LDPE and HDPE?

PROBLEM 11.2

Do the structures in Figure 11.1 obey the isoprene rule?

■ **Joseph Priestley**

IN THE CITRIC ACID CYCLE shown in Figure 8.19 one of the molecules of the cycle is fumarate, the conjugate base of fumaric acid, shown in Figure 8.20. If the identical groups on each end of the double bond, the two hydrogen atoms and the two carboxylic acid groups, were not on opposite sides of the double bond but instead were on the same side of the plane of the double bond, a different molecule is obtained, maleic acid. Fumaric and maleic acids are configurational diastereomers because rotation about the double bond, which would change each into the other, is not possible under ordinary conditions such as noted in Chapter 5 (section 5.7) concerned with the in vivo synthesis of limonene.

The rotation is restricted because the necessary motion has to traverse an energy barrier that is inaccessible at temperatures even considerably above room temperature. In the words of kinetic theory discussed in Chapter 6 (section 6.11) the energy of activation for interconversion between the cis and trans double bonds is too high. The reason is that the π bond must break for the interconversion to occur. One of the sp^2 hybridized carbon atoms at the end of the double bond must be twisted out of the plane, and if so the two p orbitals could no longer overlap. In the twisted intermediate structure shown in **Figure 11.2** the orbitals would be p_y and p_z.

fumaric acid π bond broken   maleic acid

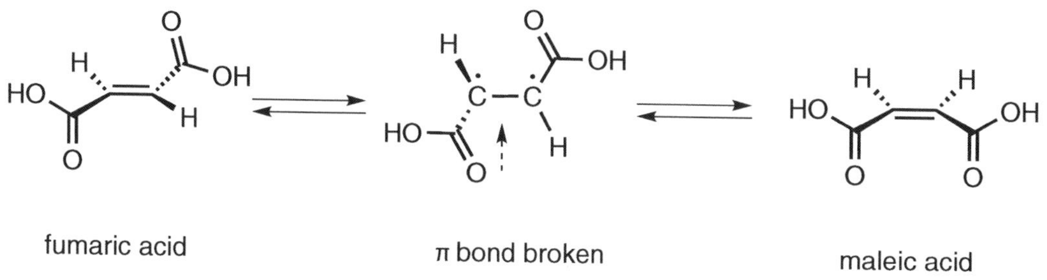

The structural difference between fumaric and maleic acids was first understood by van't Hoff in the same year, 1874, that he simultaneously with LeBel, introduced the concept that tetrahedral carbon could give rise to optical isomers, which finally explained the observation made by Pasteur so many years before (sections 1.10 and 1.11). Van't Hoff designated these unsaturated dicarboxylic acids, fumaric and maleic acids, as geometric isomers, a term that is no longer widely used. Instead a nomenclature introduced by Baeyer in 1890 is now widely used, cis and trans, from the Latin for on the same side and on the opposite side, respectively. Fumaric is therefore a trans alkene while maleic is a cis alkene.

Baeyer, whom we met in Chapter 3 (section 3.2), where we learned of his realization that rings might have to be strained, used the terms cis and trans not to describe the stereoisomeric possibilities in double bonds but, rather, to describe the configurational diastereomers he had synthesized in which two carboxylic acid groups reside on a cyclohexane ring, in particular for hexahydrophthalic acid (**Figure 11.3**). The synthetic path to these isomers is also shown in Figure 11.3.

In the ring compounds Baeyer studied, as for double bonds, but for a different reason, rotation to interconvert the cis and trans isomers is not possible without a prohibitive motion, But whereas in the double bonds this has to do with breaking a π-bond in a cyclic molecule the prohibitive motion involves breaking a σ-bond.

The situation for the double bonds in both *Hevea* rubber and *gutta percha* is ambiguous when using the cis and trans nomenclature. Although the former is designated cis and the latter trans, this nomenclature assumes that one considers only the main chain as on the same or opposite sides of the double bond respectively (Figure 11.1).

The most recent nomenclature takes care of this problem by using a variation of the Cahn-Ingold-Prelog rules (section 1.12). If the two groups on the same side of the

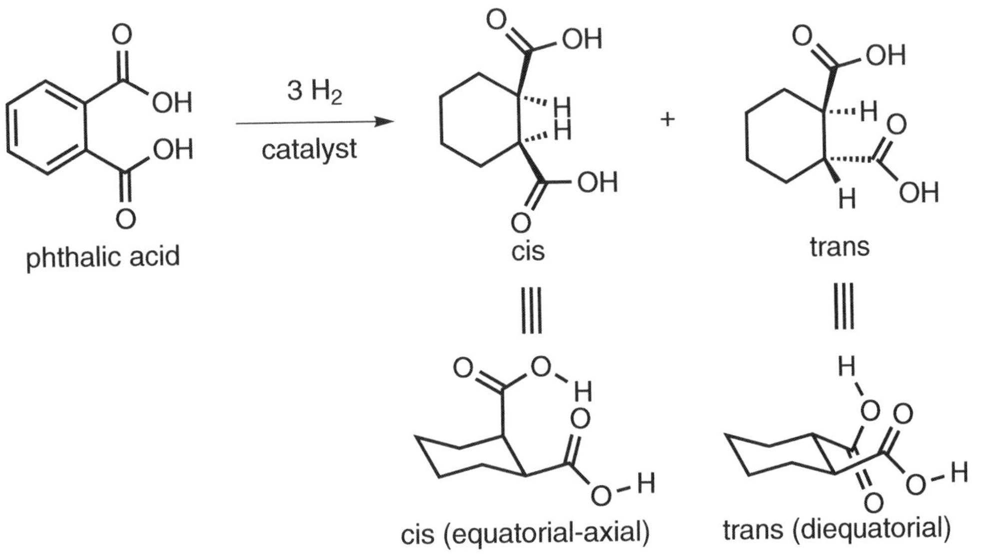

◀ **FIGURE 11.3**

Cis/Trans Isomers in Hexahydrophthalic Acid

double bond have the highest priority by these rules then the double bond is called Z for the German word for together, zusammen. If the highest priority groups are on the opposite side, the double bond is called E for entgegen, the German word for opposite. Using this nomenclature, fumaric acid would be E and maleic acid Z. For *Hevea* rubber the designation would be Z and for *gutta percha* E because a CH_2 group has a higher priority in the Cahn-Ingold-Prelog nomenclature than a CH_3 group.

On the other hand there are compounds with double bonds where there is no need for a stereochemical nomenclature; there is not a stereoisomer. Isobutylene (section 4.10) the alkene that first introduced us to this functional group has such a double bond as is the situation for ethylene and propylene studied in Chapter 9

Configurational diastereoisomerism arising from double bonds is exceptionally important in the role of fatty acids. We've noted in Chapter 7 (sections 7.1, 7.2) that unsaturated fatty acids, those fatty acids with double bonds in the hydrocarbon chains, are overwhelmingly cis and that cis-unsaturated fatty acids have lower melting points than saturated fatty acids. Let's find out how nature's choice of the cis double bond in unsaturated fatty acids arises from a reason related to nature's choice of a cis double bond in a material that has to be flexible, an elastomer, that is, natural rubber.

PROBLEM 11.3
How is energy of activation related to temperature in terms of the possibility of cis-trans isomerization of a double bond? What figure in Chapter 6 helps to answer this question?

PROBLEM 11.4
Use orbital diagrams for the structures in Figure 11.2 to demonstrate what is meant by the breaking of the double bond.

PROBLEM 11.5
Can flipping of the cyclohexane ring cause cis-trans isomerization in Figure 11.3?

PROBLEM 11.6
Why might strong base act to interconvert the cis and trans isomers in Figure 11.3?

PROBLEM 11.7
Is there any relationship between the cis and trans isomers in Figure 11.3 and the stereoisomers of tartaric acid studied in Chapter 1? Could (R) and (S) nomenclature be used to describe the stereoisomers in figure 11.3.

PROBLEM 11.8

Draw a planar representation of a cyclohexane ring with CH_3 groups on: (a) carbon atoms 1 and 2; (b) carbon atoms 1 and 3; (c) carbon atoms 1 and 4. In each of the three relationships (a-c) place the CH_3 groups cis to each other and then trans to each other. Convert these drawings to chair forms and determine the most stable conformation for each. Are any of these conformations chiral and will flipping of the chair forms have any effect on these chiral properties?

11.3

Why should the difference between a cis and trans double bond make the difference between an inelastic and an elastic material?

THE ELASTIC PROPERTIES OF RUBBER were long a scientific mystery. Over two hundred years ago scientists noticed that stretching rubber releases heat. I've tried this experiment by stretching a thick rubber tube while it was held at my lips, which are a part of the body sensitive to temperature and, sure enough, I felt the warmth. And, if this is not mysterious enough, there is something even more difficult to understand: natural rubber gets stiffer, that is more difficult to stretch when it is warmer.

When an automobile tire rotates a part of the tire in contact with the road flattens out, stretches and gets warm. Because the stretching and contraction of the tire is not perfectly reversible, the tire gets hotter. This heating process is of great importance to tire life and to the possibility of failure of the tire. In fact, the failure of tires on certain automobiles, which caused deaths when the vehicle turned over, arose from the fact that the tires were underinflated. The lower the pressure of air in the tire, the more the tire flattens out when it comes in contact with the pavement and therefore the more heat is released.

Natural rubber is the best of the elastomers in the reversible nature of this heating-cooling cycle. There exist many commercial elastomers, some of which we'll discuss in this chapter, but none are as useful for a range of purposes as natural rubber, which is, therefore, used for the most demanding task, that of an airplane tire. A tremendous amount of heat is released from the compression of the tire when an airplane lands and the tire's performance and, especially, the dissipation of the heat is critically important to the safety of those within.

The properties of natural rubber and other elastomers are a consequence of the properties of the polymer chains. When natural rubber is stretched, the polymer chains take on conformational states that allow the chains to stretch, that is, the distance from one end of the chain to another becomes greater. When the force causing the elastomer to extend, to stretch, is removed the material regains its original form and the polymer chains regain their original unstretched state. What we see in the material property is what is going on at the macromolecular level as seen in **Figure 11.4**.

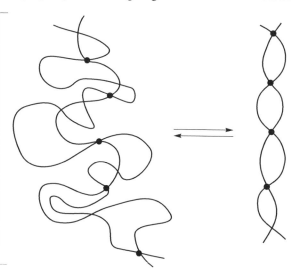

FIGURE 11.4 ▶

Representations of the Stretched and Coiled Forms of a Crosslinked Elastomer

The individual chains of both *Hevea* rubber and *gutta percha* are both capable of the conformational change described above, which is necessary for elastomeric properties. Why, then, the difference in their properties? The answer is found in the shape of the chains. The trans double bonds (*gutta percha*) allow different chains to approach closely to each other making it very difficult for the chains to undergo conformational motion because units of many chains would have to undergo the conformational motion in concert with each other. One

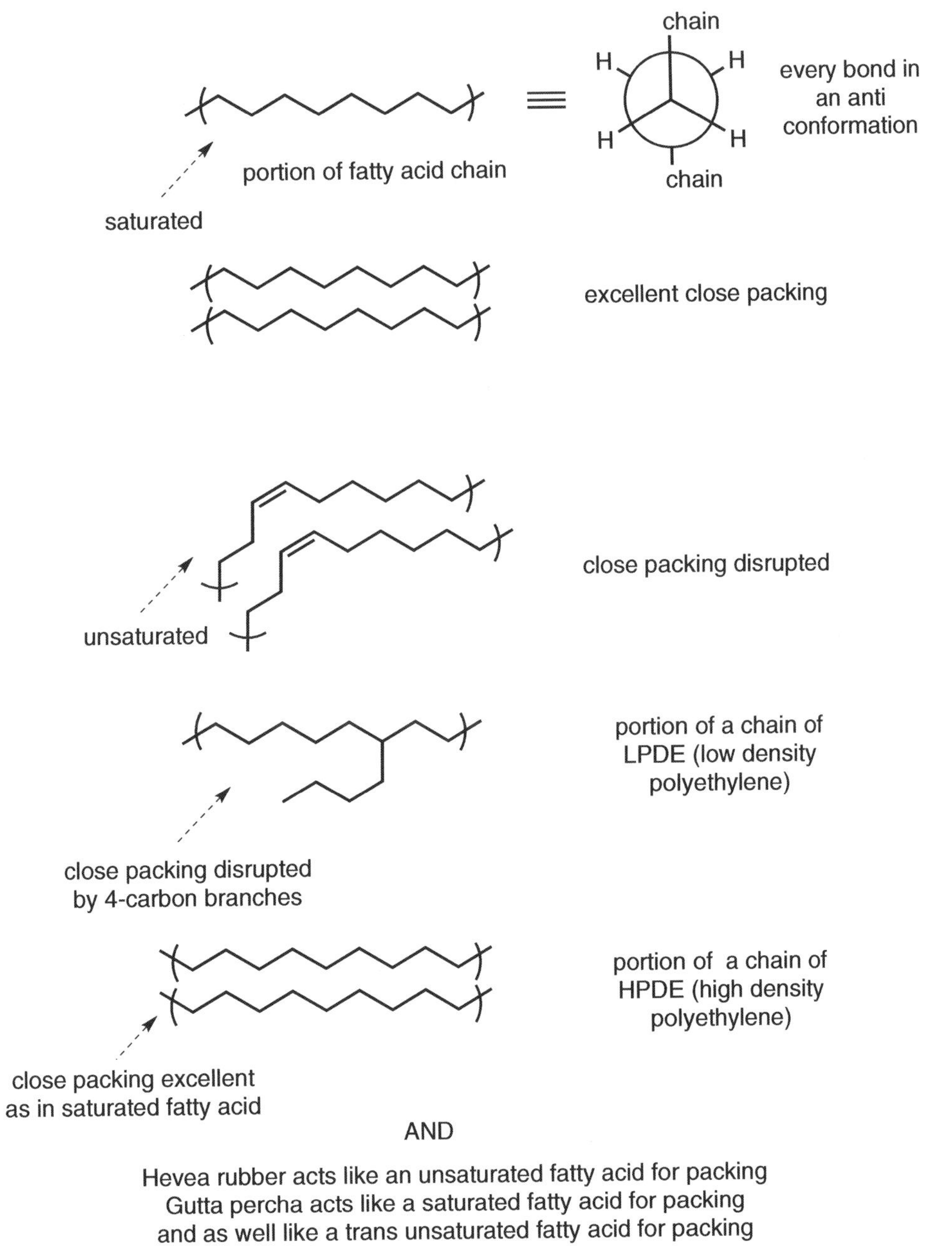

◀ **FIGURE 11.5**

Examples of the Effect of Double Bonds and Branching on Close Packing of Hydrocarbon Chains

portion of fatty acid chain

saturated

every bond in an anti conformation

excellent close packing

unsaturated

close packing disrupted

close packing disrupted by 4-carbon branches

portion of a chain of LPDE (low density polyethylene)

close packing excellent as in saturated fatty acid

portion of a chain of HPDE (high density polyethylene)

AND

Hevea rubber acts like an unsaturated fatty acid for packing
Gutta percha acts like a saturated fatty acid for packing
and as well like a trans unsaturated fatty acid for packing

chain can not move without the involvement of other chains. In addition, the close packing allowed by the extended regular shapes of the *gutta percha* chains allow small crystalline regions to form, which restrict even more tightly any chain motion.

Polymer chain packing, that is, how regular and close to each other are the individual polymer chains in the undissolved or unmelted state, is greatly dependent on the shapes of the individual chains to be packed together. This makes sense. These concepts extend to the packing properties of fatty acid chains. The cis-double bonds (Figures 7.1 and 7.3) in unsaturated fatty acids cause kinks in the direction of the chain of methylene (CH_2) groups around the double bond causing difficulty for the fatty acid chains to pack together closely. This situation gives rise to a fluid character for these unsaturated fatty acids, which is seen in a melting point that is lower than the melting point for saturated fatty acids where the chains can pack closely together. The close- packing of saturated fatty acids causes restriction in both conformational motion and movement of the chains relative to each other, accounting for the higher melting point.

In the discussion above we fulfilled the promise in Chapter 7 (section 7.2) to explain the different properties of saturated and unsaturated fatty acids. However, if the unsaturated fatty acids were not cis but instead trans, the packing would not be impeded. The overall chain shape with a trans double bond resembles that of a saturated chain (Figure 7.5).

These ideas can be extended to the differences in density between low and high density polyethylene, LDPE and HDPE. In section 9.7 we learned that the polyethylene, HDPE, produced by the Ziegler catalyst was of far higher density than the polyethylene, LDPE, produced by the ICI free radical process. The absence of the four-carbon branches in HDPE allows the polymer chains to pack more closely together allowing a higher degree of crystallinity and making HDPE more opaque and denser than LDPE.

In summary, therefore, as expressed in **Figure 11.5**, the cis double bonds in *Hevea* rubber as do the cis double bonds in unsaturated fatty acids as do the branches in LDPE keep the polymer chains apart so that conformational motions are not restricted by intermolecular forces.

Now let's turn our attention to why the temperature of *Hevea* rubber, that is, natural rubber, goes up when the rubber is stretched. To understand this characteristic of natural rubber we have first to understand the nature of the conformational changes that occur during the stretching process.

PROBLEM 11.9
What superior property of natural rubber might make it so valuable for airplane tires?

PROBLEM 11.10
Which of the following would be candidates for elastomeric properties and why: LDPE; HDPE; saturated fatty acid; cis-unsaturated fatty acid; isotactic polypropylene; atactic polypropylene?

PROBLEM 11.11
Hypothesize how anti, gauche or eclipsed conformations about the carbon-carbon bond in a long hydrocarbon chain could be correlated to the coiled and stretched states shown in Figure 11.4.

11.4

Why does rubber get hotter when stretched and why does rubber get stiffer at higher temperatures? The answer increases our knowledge of thermodynamics.

IT TAKES WORK TO STRETCH A PIECE OF RUBBER and thermodynamics teaches that if work is carried out, the product of that work must be of a higher energy equivalent to the work done.
If the product is not of higher energy then the work appears as heat. When rubber is stretched what happens to the work done?

The change that occurs within the individual polymer chains, as discussed in section 11.3, is from the random coil state to an extended shape. This change is shown in a representative manner in Figure 11.4. The question then arises as to what is happening to allow this change in polymer chain shape. Taking the change down to the bond motions involved shows, surprisingly, that not much conformational change is necessary.

In the structure of *Hevea* rubber shown in Figure 11.1 there are three σ bonds in each of the identical units that make up the chain around which torsional motions may occur. The conformational states that arise via these motions are more complicated than those encountered in the torsional motions about the central carbon-carbon bond in n-butane (section 3.5) but it can be shown that changes in the torsional motions about only few of these bonds is adequate to cause large changes in the shape of the chain, changes that resemble the differences between the coil and extended shapes represented in Figure 11.4.

As a consequence of the small number of torsional changes necessary, the work done to attain the stretched state of the rubber does not lead to a significantly higher energy state. As noted just above, therefore, the work must appear as heat, the heat of the tire against the pavement, the heat on your lip as you stretch the rubber band.

Thermodynamic considerations also offer an understanding of the stiffening of rubber as temperature increases. The stretched state of the rubber, as shown in Figure 11.4, is more orderly than the coiled state. This means that the stretched state is of lower entropy than the coiled state.

Consider the simple thermodynamic relationship between free energy, enthalpy and entropy: $\Delta G = \Delta H - T\Delta S$. If the change from the coiled to the stretched state of the rubber causes an entropy loss, then ΔS will be negative causing the term $-T(-\Delta S)$ to be positive. The free energy change, ΔG, will therefore become more positive as temperature increases causing the left side of the equation: coiled = stretch + heat to be shifted to the left. The rubber will be more difficult to stretch; it will be stiffer.

Thermodynamic ideas offer another way of understanding the stiffening of rubber as temperature increases. The transformation from coil to stretched state is exothermic, heat is released. If one adds heat by increasing the temperature the equilibrium is shifted to the left. In all exothermic equilibrium processes, increase of temperature causes a shift of the equilibrium to the left, to the coiled state in this situation, a stiffening of the rubber.

PROBLEM 11.12
What did Carnot and Joule in the 1840s have to do with finally understanding the scientific basis of the changes seen in natural rubber as a function of temperature?

11.5
Crosslinking of rubber is necessary.

ELASTIC BEHAVIOR REQUIRES THAT THE MATERIAL to be stretched has to return to its original state when the stretching force is removed. If the polymer chains can slide pass each other, that is, if the center of mass of each polymer chain changes when one pulls on the material, the material will stretch but will not return to the original state when the stretching force is removed. How is one to stop this sliding motion but still allow the conformational changes necessary for the elastic behavior? This influence is the role of the black dots in the coiled and stretched states shown in Figure 11.4. But what are these black dots, which are holding the chains from slipping past each other? What are these black dots that are linking the chains to each other, that are crosslinking the chains?

One of the Europeans whose name is long associated with bringing knowledge of a natural elastic material back to Europe was **Charles Marie de la Condamine**. He is given credit for informing Europeans of the name used for the tree from which the material was taken, *Hévé*. He also is given credit for the use of the word latex, derived from the Spanish word for milk, to describe the juice of the tree, and the word caoutchouc noted in section 11.1. Francois Fresmeau and Charles Marie de la Condamine, both Frenchmen, are credited independently in the middle years of the eighteenth century with connecting Europe to what was seen at the time as this strange material.

Both Frenchmen tried to discover ways to stabilize the elastic material in the latex to possibly develop commercial uses, an advance not to happen for at least a hundred years after they left the scene. However, there were reports that the native peoples who used caoutchouc exposed it to smoke and to the juice of citric fruits before using it for their games (section 11.1) or for other purposes such as making elastic bottles that could be used for what we might call a syringe – a fluid within the bottle could be expressed through a small opening by squeezing. With the hindsight derived from our knowledge of chemistry and of the structure of rubber (Figure 11.1) one could imagine that these kinds of treatments, (both smoke and citric juices are acidic) might cause chemical reactions involving the double bonds

■ **Charles Marie de la Condamine**

Charles Goodyear

Charles Macintosh

Nathaniel Hayward

Thomas Hancock

in which covalent bond formed randomly along the chains connecting the strands to each other, the black dots in Figure 11.4.

In fact, it is the reactivity of the double bonds in rubber, a reactivity that finds parallel to the reactivity of double bonds leading to high octane gasoline (Chapter 4) and to the terpenes (Chapter 5), which supply the chemistry to form the necessary crosslinks. And crosslinks allowed the transformation from a curious substance that constantly disappointed those interested in its commercial potential into an article of commerce whose value created great wealth.

The Goodyear Company is named for a man, **Charles Goodyear**, a Connecticut Yankee, who lived from 1800 to 1860 and who is primarily responsible for discovering the method by which the elastic material from a tree was transformed into a material the world could not live without. This is quite a remarkable legacy and especially remarkable because of the fact that Goodyear was constantly in debt and even several times in debtors' prison and who died with a large debt without a penny to his name.

Goodyear is reported to have had little interest in money and made terrible business decisions to be able to get back to his first love, experimentation. In his experiments Goodyear tried to find a way to stabilize rubber so that it would not melt to a sticky gum in hot weather. Others were having the same problem including in England where **Charles Macintosh** was making waterproof coats using rubber. I found the following statement on the web: "*Early coats had problems with smell, stiffness, and a tendency to melt in hot weather.....*"

Although Macintosh finally solved his problem with a process involving the use of sulfur, the original connection between rubber and sulfur is often attributed to an American, **Nathaniel Hayward**, who is reported to have dusted rubber with sulfur powder and exposed the mixture to the sunlight as a way to remove the stickiness. Hayward patented his discovery in the United States and turned it over to Goodyear with whom he worked. At some point it was discovered, it is said accidentally, that heating was far more effective than sunlight for transforming the sulfur-rubber mixture to a useful material. Nevertheless, the credit went to England where a patent was filed by **Thomas Hancock** who worked with Macintosh. Hancock came from a wealthy family long involved in the industrial revolution and with an interest in rubber. There is controversy about the chain of events but it is said that Hancock was given a sample of rubber with excellent properties, which did not have English patent protection. The sample was apparently a product of Goodyear's work in the United States.

What's going on when rubber is treated with sulfur and heated? The answer is that covalent bonds are formed between different chains. These are crosslinks, which are made of the sulfur atoms supplied by the added elemental sulfur. If too much sulfur is added, so many crosslinks are generated that the elasticity of the material is lost – molecular movement is highly retarded. But if just the right amount of sulfur is added, the sulfur crosslinks occur in a manner to stop the chains from slipping by each other in the stretched state but still allow the chains to change reversibly from the coil to the stretched state, that is, for the substance to act like rubber (Figure 11.4).

PROBLEM 11.13
Based on your knowledge of the reactive properties of double bonds, propose crosslinking reactions that could occur in rubber when exposed to the protons, the carbocations and the free radicals in a combination of smoke and citris juices.

T O ANSWER THIS QUESTION, we first have to understand the chemical properties of sulfur, which on a molecular level exists in several forms, prominent among them as an eight membered ring of sulfur atoms as shown in **Figure 11.6**.

In Chapter 10 in the discussion of the industrial routes to hexamethylene diamine to produce nylon 6,6, the mechanism of addition of Cl_2 to a carbon-carbon double bond was introduced (sections 10.4 and 10.5). Here we learned that the force of carbocation stability controls the addition of a halogen such as chlorine to a carbon-carbon double bond as well as the addition of a hydrohalide such as HCl (section 6.10, Figure 6.13). The mechanism of the addition of bromine, Br_2, follows the same mechanistic picture and for our particular purpose here, so does the addition of sulfur. The parallels are shown in **Figure 11.7** for the addition of Cl_2, Br_2, and S_8 to the unsymmetrical alkene, 2-methyl-2-butene, as a model for the double bond in *Hevea* rubber.

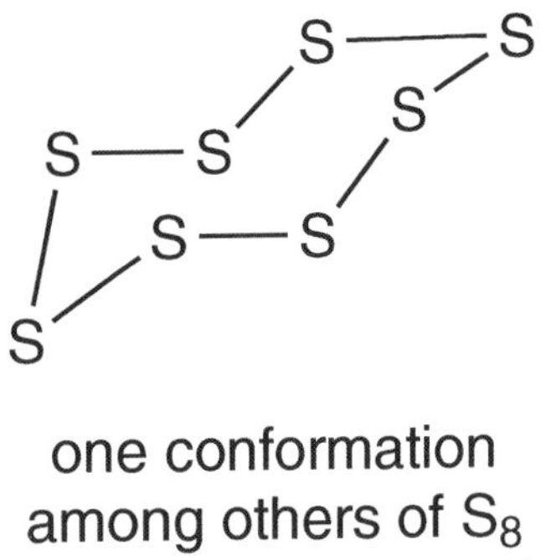

The three reaction paths in Figure 11.7 begin identically with the π bond of the double bond acting as a source of electrons to break the Cl-Cl bond, the Br-Br bond, and the S-S bond in a heterolytic manner so as to produce charged moieties as shown. For the halogens, this first step is known to produce what are called halonium ions in which the halogen bridges the carbon atoms of the breached double bond to produce a three membered ring, as shown in Figure 11.7. A similar intermediate is thought to be produced in the analogous sulfur addition as shown in Figure 11.7. However, in contrast to the reactions with Cl_2 and Br_2 the produced negatively charged entity remains connected via the six intervening sulfur atoms of the original S_8 reactant.

Closing of large rings (equal to or more than about ten atoms in the ring) does not readily take place simply because of the small probability of the ends coming close enough to react. When many atoms are involved in the chain that intends to form the

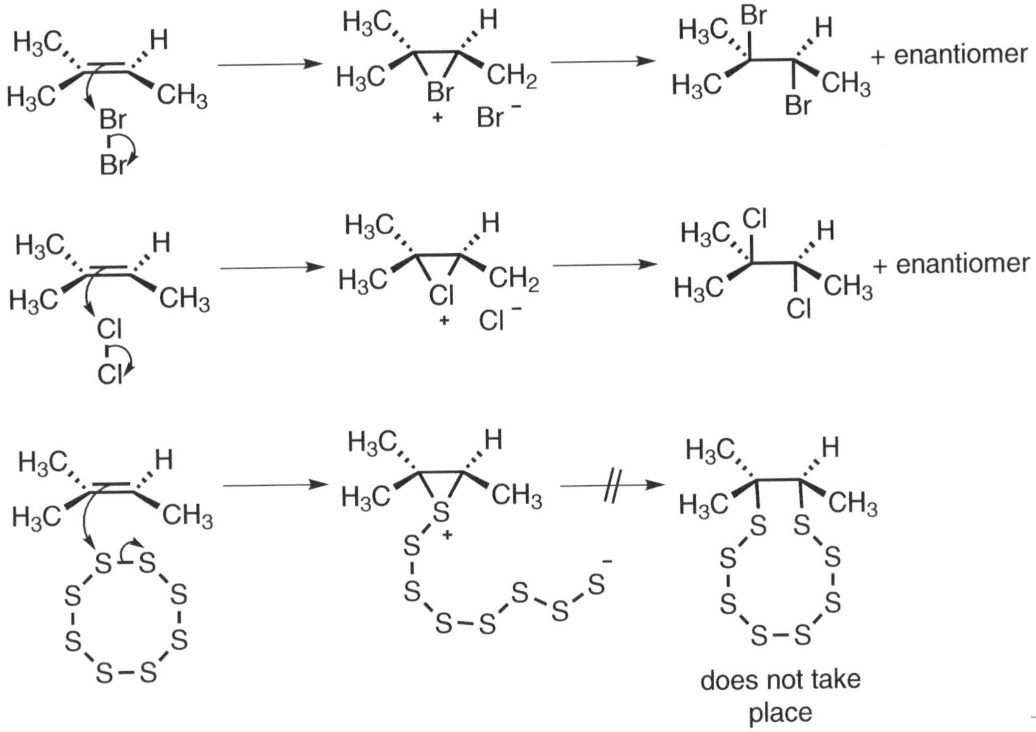

one conformation among others of S_8

◄ FIGURE 11.6

A Conformational Isomer of a Ring of Eight Sulfur Atoms

◄ FIGURE 11.7

Comparison of the Mechanism of Addition of Cl_2, Br_2, and S_8 to an Unsymmetrical Alkene

does not take place

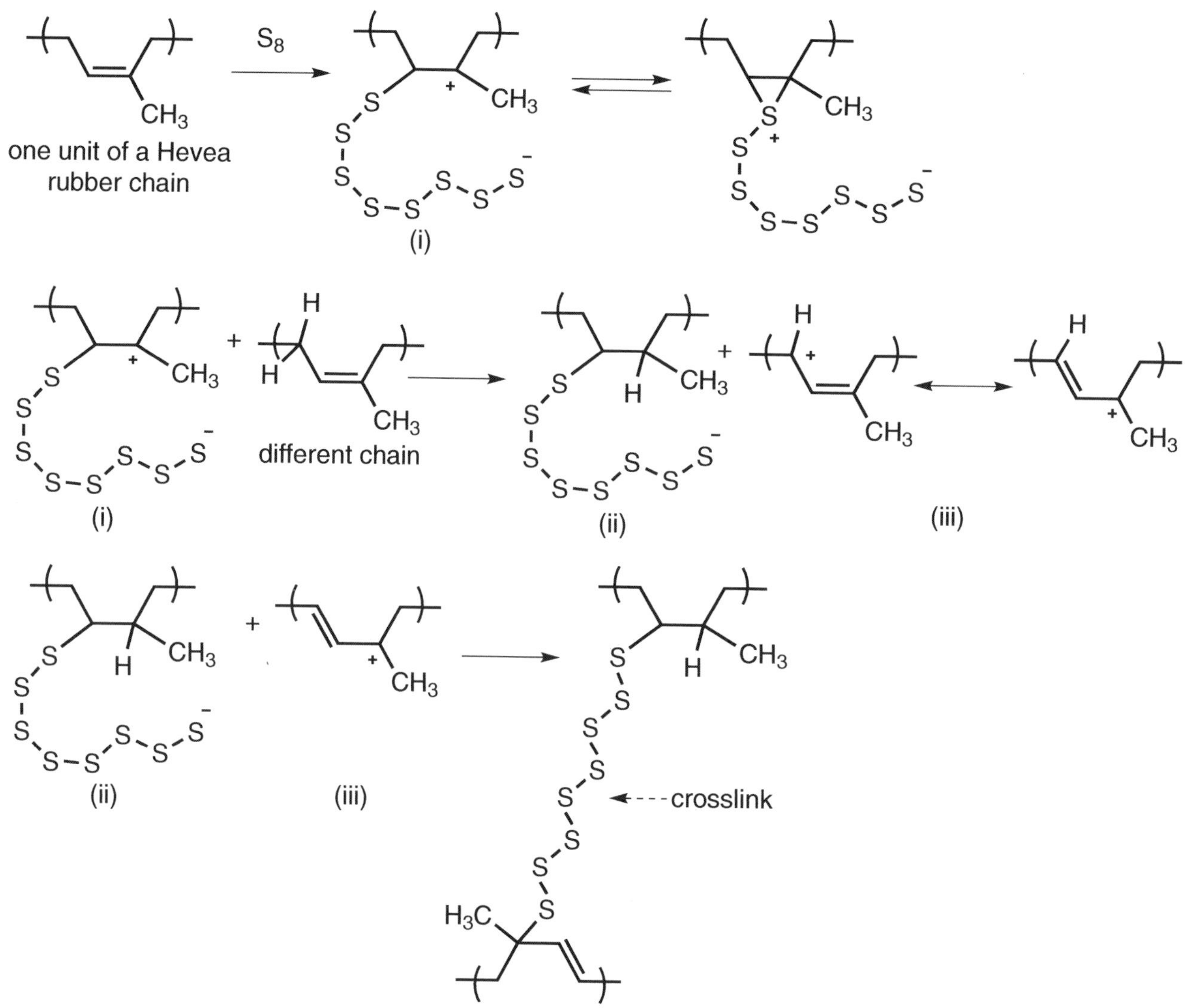

FIGURE 11.8 ▲

A Crosslinking Vulcanization Mechanism for Natural Rubber

ring, the number of conformations becomes very large. For this reason it is difficult for the chain of atoms to become restricted to the few conformations or even the single conformation necessary to close the ring, that is to make the necessary bond such as for one example the ring of ten atoms, eight sulfur atoms and two carbon atoms, that would form if the negatively charged sulfur atom in Figure 11.7 were to follow the path taken by the negatively charged halogen atoms in that figure. In the next chapter, section 12.13, we'll look into more detail about the closing of rings of various sizes.

The restriction of the negative end of the sulfur chain adding to the carbocation site, for the reason discussed above, opens other reaction possibilities for the carbocation. Other chains with their commonly occurring double bonds are nearby to the chain with the double bond that has reacted with S_8. The sulfur is being added to an amorphous solid material so that the chains are entangled with each other, as described by the "plate of cooked spagehetti" in describing addition polymers in Chapter 9 (section 9.3).

A reasonable mechanistic picture of the crosslinking step for the structure of *Hevea* rubber is shown in **Figure 11.8**. The sulfur added to the double bond forms the three membered ring which occasionally opens and closes alternately exposing each carbon atom to the full positive charge as shown in structure (**i**).

For the reasons discussed just above the negative sulfur atom is too flexible and far removed to rapidly add to the carbocation, allowing time for other reactions to take place. One of these reactions involves a hydride transfer from another *Hevea*

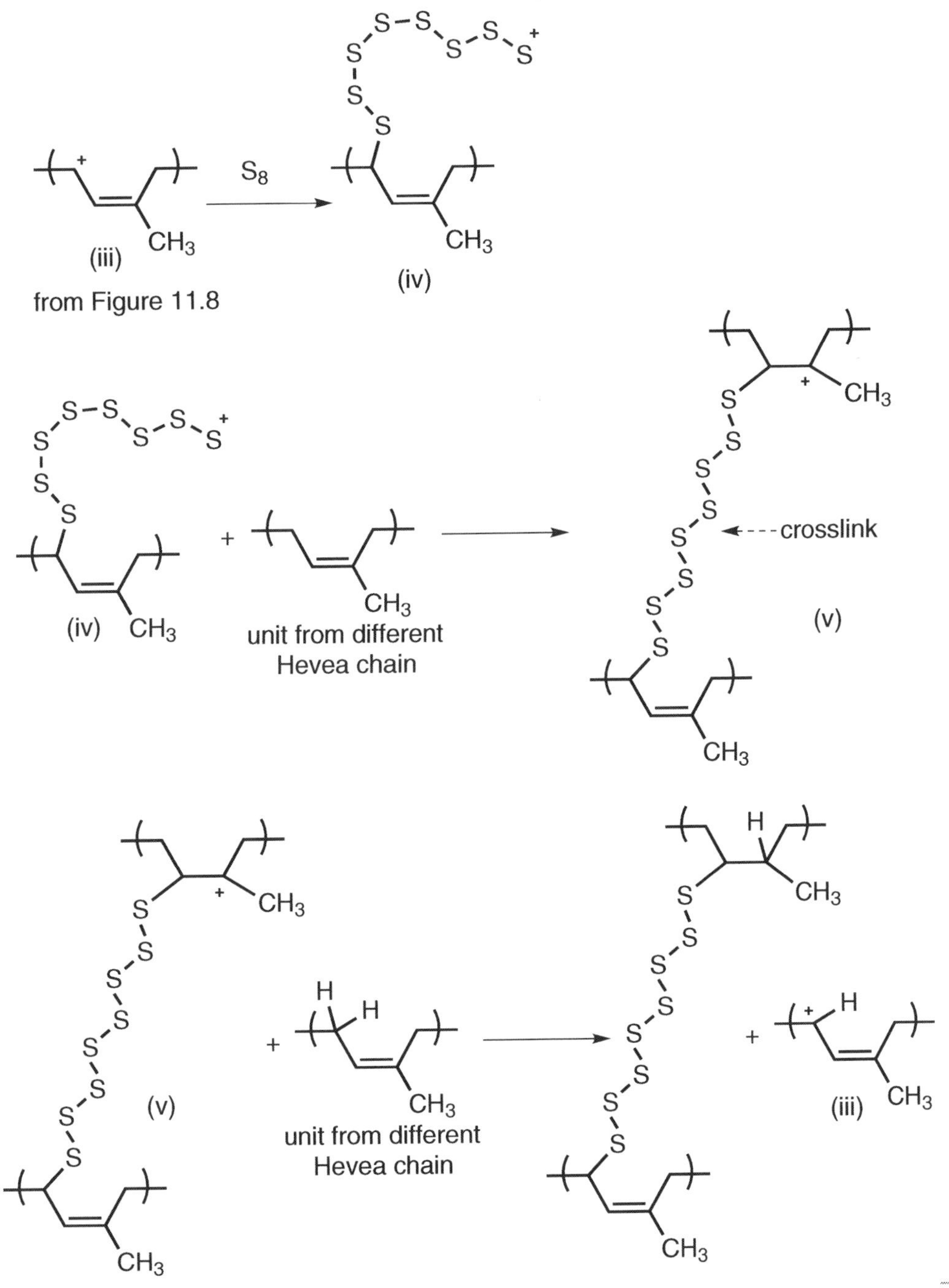

◀ **FIGURE 11.9**

A Crosslinking Chain Mechanism Variation on the Mechanism in Figure 11.8

chain, which leads to (**ii**) and (**iii**), the latter a carbocation that is more stable than the carbocation originally formed. We've seen these intermolecular hydride transfers to carbocations before as in the synthesis of 2,2,4-trimethyl pentane (Figure 4.13) and as well in the chemistry involved in catalytic cracking (section 4.6).

The new carbocation (**iii**) on the nearby chain is not only tertiary but also resonance stabilized (Figure 11.8).

Now we have a chain of eight sulfur atoms connected to one chain with a negative charge on the terminal sulfur atom and additionally another nearby chain with a positively charged carbon. The two (**ii** and **iii**) meet and a crosslink is formed. Mission accomplished but as we'll see below, there is an even better way for the crosslinking mechanism to unfold.

The usual tools for studying mechanism in organic chemistry, such as described in Chapter 6 (section 6.11) regarding electrophilic aromatic substitution, are not easily available for study of vulcanization of rubber arising primarily from the fact that one of the reactants is a poorly defined amorphous array of polymer chains. Adding to the problem is that the other reactant, sulfur, exists in a variety of states. S_8 as noted above is only one of several isomeric states for elemental sulfur. And a very serious impediment to understanding what is happening is that the final crosslinked material is a mixture that is extremely difficult to analyze at the molecular level.

The picture drawn in Figure 11.8 is therefore what we call an informed speculation, believed to be close to the truth. Further informed speculation leads to the distinct possibility of a chain mechanism, the kind of mechanism we have so far seen in other forms in both carbocations and in free radicals (sections 4.10, 9.3 and 9.4). A possible chain mechanism for vulcanization of rubber is presented in **Figure 11.9**. The first steps correspond to the first two steps in Figure 11.8 in which S_8 adds to a double bond followed by the hydride transfer to form the resonance stabilized carbocation intermediate (**iii**). The steps in Figure 11.8 leading to (**iii**) are initiation steps, with initiation defined identically to the definition in the earlier studies of chain mechanisms noted just above.

Sulfur-sulfur bonds are highly polarizable, which means that positive and negative dipoles can easily form. In the first step in Figure 11.9, S_8 reacts with the carbocation (**iii**) so that the last sulfur atom on the newly opened sulfur chain is charged not negatively, but positively. This positive sulfur is an electrophile and the electrons it seeks can be found in the double bond of a different *Hevea* chain producing crosslinked carbocation (**v**).

The step forming (**v**) is a crosslinking step that differs from the crosslinking step that formed by the reaction of (**ii**) and (**iii**) in Figure 11.8. The crosslinked structure (**v**) contains a positively charged carbon so that it is still reactive. A hydride ion, H^-, is abstracted from a unit of an adjacent chain to neutralize the charge in (**v**) while forming another (**iii**), which then reacts with another S_8 to continue the chain reaction. While the reaction of (**ii**) and (**iii**) in Figure 11.8 is a termination step of the chain mechanism, the reaction of (**v**) with a *Hevea* chain as shown in Figure 11.9 is a propagation step for the chain mechanism.

PROBLEM 11.14

Practice your ability to handle structural details by putting in all atomic symbols and lone electron pairs and searching for formal charge in Figures 11.8 and 11.9.

PROBLEM 11.15

The structure of S_8 is drawn as a puckered ring in Figure 11.6. Considering that there are no atomic substituents on the ring of sulfur atoms, what kinds of strain are being addressed by the puckering of the ring?

PROBLEM 11.16

In Figure 11.7, what would be the stereochemical consequence if the electrophilic halogen attacked the double bond from above instead of from below, as shown now?

PROBLEM 11.17

Can the difficulty of closing the sulfur ring in Figure 11.8 be expressed using entropic considerations? Would lower temperature help or hinder the ring closing?

PROBLEM 11.18

For the chain mechanism for vulcanization of rubber outlined in this section, assign initiation, propagation and termination steps. Now look at one or more of the chain mechanisms in earlier chapters and compare and contrast the probability of the parallel steps in those chain mechanisms.

W E NOW UNDERSTAND WHAT IS NECESSARY to form an elastomer and we see the commercial success of natural rubber and as well the limitation on its production derived from the manual labor involved in growing and tapping the trees. This limitation offers reason enough to attempt to make synthetic variations of natural rubber. But another important way to possibly improve on one characteristic of natural rubber is to find a way to make an elastomer without double bonds. The reason for this is that the many double bonds in natural rubber that have not taken part in the vulcanization process are highly reactive with a component of the air of cities that has greatly increased with pollution, ozone, O_3.

Ozone reacts with the remaining double bonds in the rubber chain to break both the π and σ bonds therefore reducing the chain length of the *Hevea* rubber polymer. Because only about one in one hundred of the double bonds on each chain react with the sulfur in the vulcanization process, rubber is a highly unsaturated material and therefore subject to severe degradation by ozone. An important goal for any synthetic rubber therefore is to avoid the presence of double bonds and this goal was addressed by DuPont chemists and chemical engineers.

A synthetic elastomer must be flexible and yet of a shape that does not allow close approach of the chains. In other words any synthetic elastomer must possess the essential characteristics found in natural rubber. And there must be the possibility of crosslinking. An early attempt to fit the bill avoiding the presence of double bonds was made by DuPont chemists using low density polyethylene, LDPE, (section 9.2) as a template polymer.

We've seen that LDPE should work as the basis of an elastomer (Figure 11.5) since the branches retard close approach of the different chains, accounting for the low density. In addition, a chain of all sigma bonds will allow the conformational changes necessary to respond to the stretching force. The problem is how to introduce the crosslinks. Without double bonds, vulcanization is impossible.

The approach DuPont took to the problem of introducing crosslinks used a property of free radicals. Three types of reactions of free radicals have been presented so far in the book. In Chapter 9 we first came across the addition of free radicals to double bonds, the class of reaction responsible for the polymerization reaction leading to the formation of LDPE. The short carbon branches in these polymers arise from another type of free radical reaction, abstraction of hydrogen atoms from C-H bonds (Figure 9.4) and finally the third type of free radical reaction we have encountered occurs as the important step in steam cracking of petroleum fractions, that is, β-scission, the cleavage of C-C and C-H bonds responsible for breaking the petroleum molecules down to smaller molecules with double bonds, the alkenes (section 9.4).

The route the DuPont chemists and chemical engineers took to incorporating crosslinks in LDPE involves a reaction of free radicals related to the reaction that forms the short carbon branches in LDPE, abstraction of hydrogen atoms from C-H bonds. As we're also about to discover a chain mechanism is involved, a class of mechanism which is most favored by industry for its efficiency and economy. However, the crosslinking process uses a free radical reaction, which although abstracting a hydrogen atom from a C-H bond, does it in a manner (Figure 11.10) we have not seen before. To see how this works we need first to discuss the bond energies in LDPE and in Cl_2 and in HCl.

The energy necessary to homolytically break the bond holding the two chlorine atoms together in Cl_2 is 58 kcal/mole, quite a bit smaller than the energy necessary to break a carbon-hydrogen bond. The bond energy between sp^3-hybridized carbon and hydrogen is in the range of 90 kcal/mole for a tertiary C-H bond (R_3C-H) and 95 kcal/mole for a secondary C-H bond (-CH_2-), the two kinds of C-H bonds along the backbone of the LDPE chain. The other important energy to consider in the DuPont process to produce Hypalon is the bond energy holding hydrogen and

11.7

Synthetic elastomers: Hypalon–crosslinking without double bonds requires introducing a functional group to a polyethylene chain.

chlorine together in HCl, 103 kcal/mole.

Homolytic bond dissociation energies, that is, the energy necessary to break a bond so that the two atoms forming the bond each receive one of the electrons that made up the bond, are quite different from the bond dissociation energies encountered when a bond is broken heterolytically, that is when one partner in the bond receives both electrons (Figure 4.9) as in the discussion of leaving groups (section 5.3).

We used the energies necessary to form carbocations (Figure 4.9) to help understand the chemistry in catalytic cracking, the synthesis of high octane gasoline, the Markovnikov rule, the formation of terpenes, the characteristics of electrophilic aromatic substitution and the competition between the competing mechanisms of nucleophilic substitution. We can use the bond energies for homolytic breaking of bonds similarly to help understand the chemistry to accomplish the chemical steps to incorporate a functional group along the backbone of LDPE (**Figure 11.10**). It is this functional group, sulfonyl chloride, reaction **4** in Figure 11.10, which, as we'll

FIGURE 11.10 ▶

Chain mechanism for the Industrial Production of Sulfonyl Chloride Substituted Polyethylene as a Precursor for the Production of Hypalon

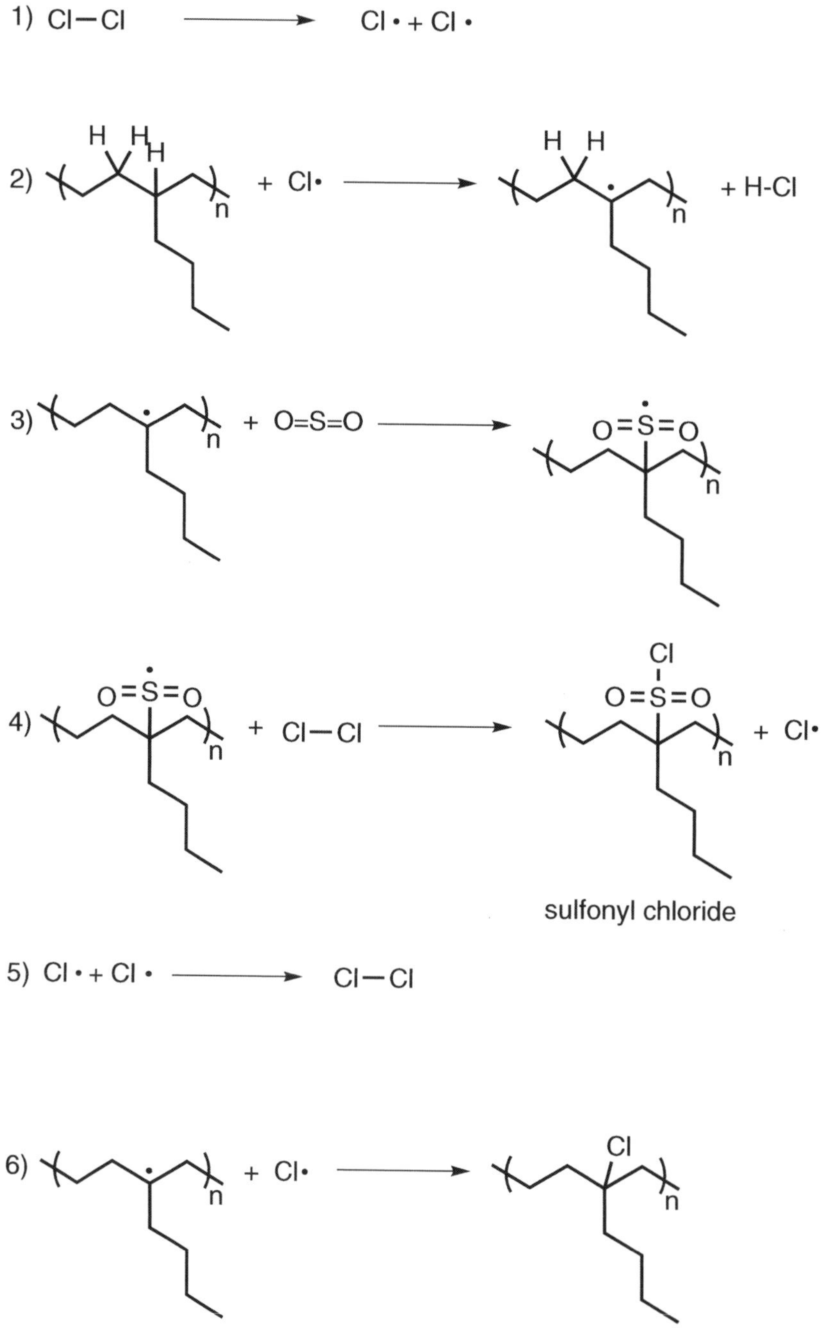

see in the section to follow allows the crosslinking to take place.

The small bond energy for the Cl-Cl bond makes the formation of two chlorine atoms, 2 Cl· from Cl₂ readily understandable with moderate inputs of energy by heat or light, energies that would be far too small to break C-C or C-H bonds. Homolytic cleavage of Cl₂ is the first step in the DuPont process (reaction **1** Figure 11.10).

Just as the bond dissociation energy informs us of the energy necessary to break a bond homolytically, this is also the energy released on forming this bond. A more stable bond with larger bond dissociation energy costs more energy to break the bond, and releases more energy when forming the bond. In general we can expect that chemical reactions will resist breaking stable bonds while tending to form stable bonds.

We've seen an example of this reactivity principle in the formation of LDPE where a weak bond is broken, the π-bond of the ethylene molecule, and a stronger bond is made, the σ-bond between the carbon atoms in the polymer chain (section 9.3). The energy released in the transformation of a weaker to a stronger bond is the source of the exotherm in polymerization (section 9.5).

Let's now test these ideas on the process created at DuPont. The comparative strength of the H-Cl bond (103 kcal/mole) versus that of the C-H bond noted above (90 to 95 kcal/mole) makes the second step (reaction **2**) in the process reasonable: a weaker bond is replaced by a stronger bond. The C-H bond-breaking in reaction **2** is most reasonable for the weaker of the C-H bonds along the backbone of the chain (90 kcal/mole), the tertiary C-H bond, in contrast to the secondary C-H bond in the CH₂ group (95 kcal/mole) (Figure 11.10). Therefore for reaction with the tertiary C-H bond, the exotherm of reaction **2** is 13 kcal/mole. The exotherm for abstraction of a hydrogen atom from the CH₂ group would alternatively be 8 kcal/mole. Now we have created a radical site at a carbon atom on the LDPE chain.

Sulfur dioxide, known for at least hundreds of years, is a cheap readily available chemical, which is a pollutant produced by many human activities, especially the burning of fossil fuels, but which can be made in large quantities by the reaction of elemental sulfur with oxygen. SO₂ has a characteristic that proved useful to the DuPont researchers. SO₂ reacts with carbon based radicals to form a strong carbon-sulfur bond and in doing so passes the radical site to the sulfur atom, as shown in reaction **3** of the DuPont process (Figure 11.10).

The energy of a bond between sulfur and chlorine is larger than that in Cl₂ so it is not surprising that the fourth step of the DuPont process (reaction **4**) breaks the Cl-Cl bond and forms an S-Cl bond releasing a chlorine atom, Cl·. Reaction **4** is exothermic just as reaction **2**.

Reaction **4** is the critical step, which defines the mechanism as a chain reaction. The released Cl· can react with another LDPE C-H bond forming another molecule of HCl, and a carbon based radical on the polymer chain, therefore restarting the reaction sequence (reactions **2** to **4**) and continuing the chain of reactions (Figure 11.10).

We can add the chain mechanism shown in Figure 11.10 to other chain mechanisms important to industry that we have come across. All chain reactions must end when one of the reactants is consumed, but a common end of the chain of reactions occurs in inevitable termination steps. The fewer of these termination steps compared to propagation steps the more product will be formed for each initiation step. In the mechanism shown in Figure 11.10 the initiation steps are reactions **1** and **2**. The propagation steps are reactions **3** and **4**, while two possible termination steps are reactions **5** and **6** (Figure 11.10), reactions that remove radicals from the reacting system. Whenever a termination reaction takes place then the chain of reactions has to be restarted by generating another chlorine atom by homolytic cleavage of Cl₂, reaction **1**.

PROBLEM 11.19

Show the details that support the following sentence from the section above: "We used the stability of carbocations of variable structure (Figure 4.9) to help understand the chemistry in catalytic cracking, the synthesis of high octane gasoline, the Markovnikov rule, the formation of terpenes, the characteristics of electrophilic aromatic substitution and the competition between the competing mechanisms of nucleophilic substitution."

PROBLEM 11.20

Try to see ahead before reading the next section and propose how the sulfonyl chloride produced in Figure 11.10 could be crosslinked realizing that a sulfonyl chloride is analogous in its reactivity to a carboxylic acid chloride.

PROBLEM 11.21

Express the bond energy ideas in this section in terms of the concept of high and low energy bonds commonly used in biochemical discussions such as in Chapters 7 and 8.

11.8

Crosslinking of Hypalon: the Parallel and Different Reactive Character of Carboxylic Acid Chlorides and Sulfonyl Chlorides

THE–SO$_2$CL GROUPS, SULFONYL CHLORIDES, pendant to the LDPE chain that are formed as the product of the free radical chain mechanism shown in Figure 11.10 are derivatives of sulfonic acids, -SO$_3$H, as shown in **Figure 11.11**. Sulfonic acids can be seen as analogs to carboxylic acids so that sulfonyl chlorides are analogs of carboxylic acid chlorides. Here we have an opportunity to reinforce two ideas, the concept of leaving groups introduced in section 5.3 and the concept of nucleophilic substitution at the carbonyl group of carboxylic acid derivatives discussed in Chapter 7 (section 7.3-7.6).

Let's review a bit. Chloride, Cl$^-$, is the conjugate base of HCl. As a consequence of the relative stability of Cl$^-$, HCl is a strong acid with a negative pK$_a$ (section 4.9). The stability of Cl$^-$, as discussed in section 4.9, derives from an identical electron configuration as one of the noble gases. All strong Lowry-Brønsted acids are characterized by the structure HX, in which X$^-$ is stabilized by some factor. However, the most common reason for this stabilization is resonance (section 5.4), as is the situation for sulphate or phosphate the conjugate bases of strong acids (section 4.9).

FIGURE 11.11 ▶

Comparison between Carboxylic and Sulfonic Acids and the Derived Acid Chlorides

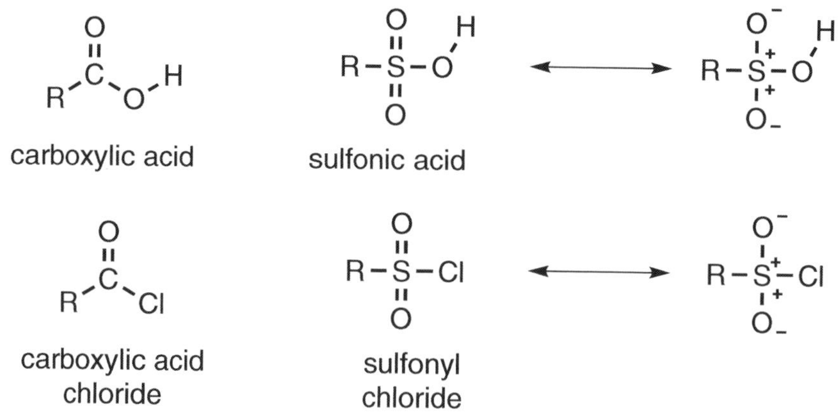

carboxylic acid   sulfonic acid

carboxylic acid chloride   sulfonyl chloride

When X is bound to some atom other than hydrogen, then the stability of X$^-$ causes a weakness in that bond so that heterolytic cleavage becomes likely. We've seen this in dimethyl allyl diphosphate (section 5.5), and in the various alkyl chlorides studied to evaluate the relative stability of carbocations (section 4.7). The stability of X$^-$ (Cl$^-$) is seen again in the chemistry that DuPont developed as the economical route to hexamethylene diamine, which failed for toxicity reasons (section 10.5). Then we call X$^-$ a leaving group. Whether a conjugate base or a leaving group, the stability of X$^-$ (whatever the source of the stability) is the critical factor in determining the strength of an acid, HX, or the weakness (high energy) of a bond R-X, or, as is our focus here, the reactivity of a sulfonyl choride, a molecule with a reactivity parallel to a carboxylic acid chloride.

Let's divert out attention momentarily from crosslinking of Hypalon to review a related subject, the susceptibility of carbonyl compounds, and especially acyl carbonyl, to nucleophilic attack. It is, after all, nucleophilic attack involving the sulfonyl chloride that leads to the crosslinking of Hypalon, although the detailed mechanism of the substitution reactions at the compared atoms, carbon and sulfur, differ.

PROBLEM 11.22
Nucleophilic attack at acyl carbonyl forms a tetracoordinate intermediate while nucleophilic attack at sulfonyl sulfur might form a pentacoordinate intermediate or transition state. Compare the geometries of these states.

PROBLEM 11.23
Taking account of all bonding and non-bonding electrons and using curved arrows, develop a reaction coordinate diagram to show the route from a carboxylic acid chloride to the ester formed by reaction with methanol (H_3C-OH).

11.9
A Review of Nucleophilic Attack at Carbonyl and Sulfonyl and the Role of Leaving Groups

WE'VE SEEN THAT CARBONYL GROUPS are all subject to nucleophilic attack but that the consequence of that attack differs when the carbonyl group is bound to a leaving group. These ideas were discussed in sections 7.3-7.6. Let's go over this subject again. It's an important subject, a central theme of organic chemistry. The reactivity of carbonyl in its structural variations is arguably the most important reactivity concept in organic chemistry, whether from the view of industrial chemistry or biochemistry.

From what has just been written, an acid chloride, and by analogy a sulfonyl chloride should have the same reactive characteristics as the carboxylic acid derivatives discussed in Chapter 7, but here via loss of the chloride anion as the leaving group. In chapter 7 we were focusing on both in vitro and in vivo hydrolysis of esters and on the formation of thioesters. Let's take a moment to review that material from Chapter 7. In section 7.3 we saw that saponification of triglycerides in vitro required hydrolysis of the ester bond, which connected the fatty acid and glycerol moieties. The reaction is base catalyzed by OH⁻ and the critical bond breaking step is reaction **5** in Figure 7.4. The leaving group for reaction **5** is a hydroxyl group of glycerol, which would be the conjugate base of this hydroxyl group bearing a proton and therefore a positive charge. If **ii** in Figure 7.4 were to go directly to **iv**, the leaving group would be not one of the hydroxyl groups of glycerol, R-OH, but rather R-O⁻. R-OH is the conjugate base of R-OH₂⁺, a strong acid, while R-O⁻ is the conjugate base of R-OH, a very weak acid. The stability of the group that has to leave to accomplish the saponification therefore must follow the reaction path **1, 3, 5**.

In section 7.4, we see reinforcement of the ideas in the above paragraph carried to an extreme in the reactive properties of aldehydes. Reaction **1** in Figure 7.7 is impossible because the necessary leaving group, H_3C:⁻, is the conjugate base of an impossibly weak acid, H_4C. Aldehydes and ketones, in which the carbonyl group is flanked by H or C atoms, do not offer a reasonable leaving group therefore blocking reactions **1** and **2** in Figure 7.7. However, in the exception that proves the rule when the carbon atom adjacent to a carbonyl group forms a stable carbanion, it may act as a leaving group as was seen in the breakdown of fats and sugars in Chapters 7 and 8 and also in the retro-Claisen condensation in Chapter 7.

Nature offers further confirmation of these ideas about carbonyl chemistry and leaving groups. In section 7.5 in following nature's path to hydrolysis of the ester bonds of triglycerides we see that the enzyme must obey the necessities of organic chemical reactivity to get the job done. In Figure 7.10, the critical bond-breaking reaction **3** can not take place without histidine delivering a proton to the oxygen that will leave the oxygen of the glycerol moiety, just as water delivers a proton to **ii** in reaction **3** of Figure 7.4. In both reactions, in vitro and in vivo, the leaving group must be R-OH rather than R-O⁻. It is hardly a surprise to see this course followed in Figure 7.11 where histidine delivers a proton to the oxygen of the serine side group (reaction 5) so that the leaving group is $-CH_2-OH$ and not $-CH_2-O^-$.

Further reinforcement of these ideas is found in section 7.6. The formation of the thioester of the fatty carboxylic acid released in Figure 7.11 must be converted to the thioester with coenzyme A (section 7.6) to enter the enzymatically-driven path of catabolism. And this process must involve excellent leaving groups based on phosphates.

In section 7.6 we observe the convoluted road that must be followed to accomplish the transformation from R-C(=O)OH where R here stands for the fatty acid hydrocarbon chain to R-(C=O)SR*, where R* stands for the coenzyme A moiety. All of these reactions occur, at significant cost to the energy sources in the living system, to assure that a good leaving group is involved. First the acyl adenosyl phosphate ester has to be formed via the use of adenosine triphosphate (ATP). Here the leaving group formed is $P_2O_7^{-4}$, the conjugate base of the strong acid, pyrophosphoric acid, $H_4P_2O_7$. Moreover, to complete the formation of the thioester, R-(C=O)SR*, another phosphorus-based leaving group is necessary, as seen in Figure 7.17 in reaction **2**.

With this short review, we are better prepared to understand the reactive properties of carboxylic acid chlorides and sulfonyl chlorides (Figure 11.11). We could predict that nucleophilic attack at either the carbonyl carbon of a carboxylic acid chloride or at the sulfur atom of a sulfonyl chloride would lead to loss of chloride anion with substitution by the nucleophile (**Figure 11.12**). Let's use a nucleophile based on nitrogen and, in particular, ammonia, NH_3.

The reaction path arising from nucleophilic attack by ammonia (Figure 11.12) for both the carboxylic acid chloride and for the sulfonyl chloride is driven to conclusion by the loss of the most stable leaving group on the substituted atom, which in both situations is the chloride ion, Cl⁻. The precise mechanistic details may differ. For the

FIGURE 11.12 ▶

Comparison between the Mechanisms for Nucleophilic Displacement on a Carboxylic Acid Chloride versus a Sulfonyl Chloride

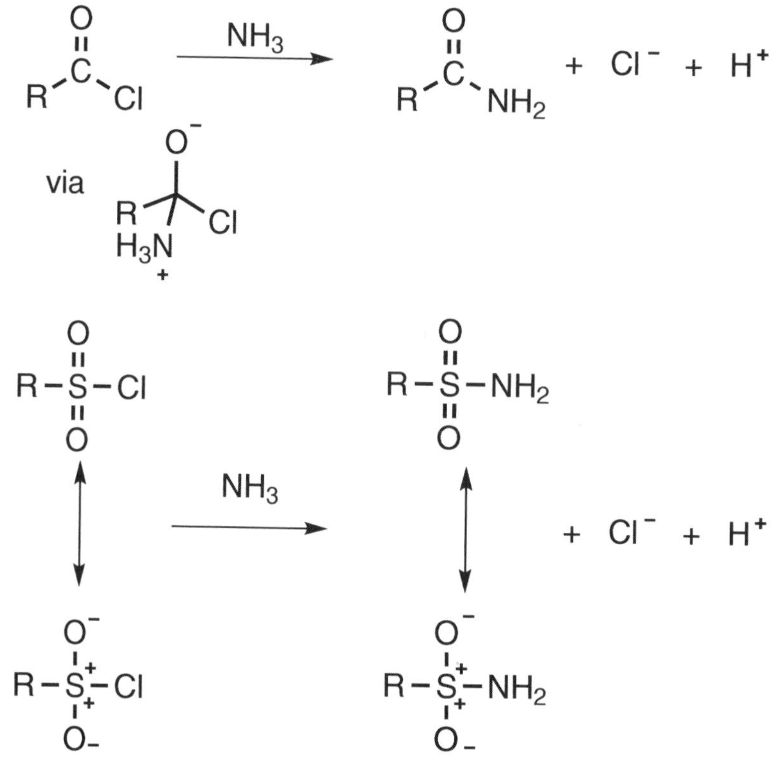

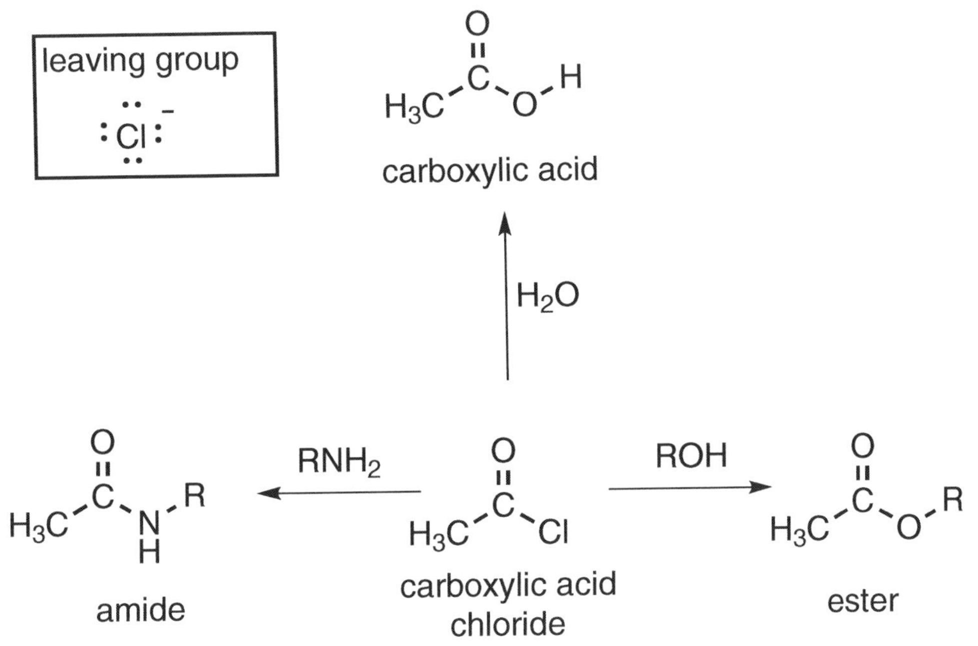

◄ **FIGURE 11.13**

Acyl nucleophilic substitution reactions show the importance of the chloride (Cl⁻) leaving group in reactions of carboxylic acid chlorides.

carboxylic acid chloride, the reaction occurs via the opening and then closing of the carbon-oxygen double bond. For the sulfonyl chloride, direct displacement of chloride is likely. However, both reactions are driven by an identical factor, loss of an excellent leaving group Cl⁻.

Carboxylic acid chlorides, sometimes simply called acid chlorides, are widely used in organic chemistry research laboratories as intermediates in the syntheses of derivatives of carboxylic acids such as esters and amides, as shown in **Figure 11.13** for one carboxylic acid chloride, acetyl chloride, the acid chloride derivative of acetic acid.

The reactions shown in Figure 11.13 are rapid at ambient temperature, which arises from the reactivity of the acid chloride with nucleophiles. This reactivity derives from the electron deficiency of the carbonyl carbon atom in the structure, a reactivity that arises from the two electron demanding atoms it is bound to, chlorine and oxygen — both of high electronegativity and to this factor is added the excellent leaving group.

It is not surprising, therefore, that another carboxylic acid derivative, which although somewhat less reactive than carboxylic acid chlorides, is also used in organic chemistry, carboxylic acid anhydrides, which are formally the combination of two carboxylic acids via elimination of water. As seen in **Figure 11.14**, the leaving group for nucleophilic displacement at the acyl carbon of these anhydrides is the conjugate base of an acid of moderate strength, a carboxylate anion.

Although we are now prepared to understand the role of the sulfonyl chloride in the crosslinking of Hypalon, let's take advantage of our insight into the chemistry of acyl carbon and the importance of leaving groups to take a look at a common industrial path to polycarbonates, which uses phosgene, $COCl_2$. Phosgene is the ultimate acid chloride, the acid chloride of carbonic acid, H_2CO_3. The outline of the chemistry of the process was shown in Figure 6.11 with the mechanism of the process, demonstrating the principles discussed in this section, shown in **Figure 11.15**.

Considering the reasons for the high reactivity of carboxylic acid chlorides discussed above, one can understand why phosgene is so much more reactive, with two chlorine atoms flanking the carbonyl carbon atom. Indeed, phosgene is so reactive that it was picked as a poison gas in World War I. Our lungs are full of nucleophilic functional groups capable of reacting with phosgene, including the moisture that is present.

FIGURE 11.14 ▶

Acyl nucleophilic substitution reactions show the importance of the carboxylate leaving group in reactions of anhydrides of carboxylic acids.

leaving group

carboxylic acid

H_2O

amide RNH_2 carboxylic acid anhydride ROH ester

- H_2O

2

■ **John Davy**

■ **Humphrey Davy**

Phosgene reacting with water, as for any carboxylic acid chloride, as shown in Figure 11.13 for acetyl chloride, will form HCl, a substance with a deadly effect on the tissues necessary for us to take a breath.

Considering the reactivity of this molecule, it is unlikely that phosgene appears naturally on earth except perhaps in trace amounts for short periods of time. Another substance hardly appearing in nature is another poison gas used in World War I, chlorine, Cl_2. In fact, the first synthesis of phosgene derives from chlorine, an element that was discovered by the great chemist we heard about in our study of saponification, Carl Wilhelm Scheele (Section 7.3).

It took about thirty five years after the discovery of chlorine for **John Davy**, the younger brother of the giant of 19th century chemistry, **Humphrey Davy**, to discover that shining light on a mixture of carbon monoxide and chlorine gave rise to a new substance. Humphrey Davy was the man who mentored Michael Faraday, as discussed in section 6.1. John Davy characterized his new discovery in various ways, but could hardly imagine it would be used in warfare or the synthesis of one of the most important plastics industry produces.

Here are John Davy's words published in 1812 and spoken before the Royal Society on February 6, 1812: "*The chlorine and carbonic oxide are, it is evident from these last facts, united by strong attractions; and as the properties of the substance as a peculiar compound are well characterized, it will be necessary to designate it by some simple name. I venture to propose that of phosgene, or phosgene gas; from φωσ, light, and γινομαι, to produce, which signifies formed by light; and as yet no other mode of producing it has been discovered.*"

John Davy was an excellent chemist, which means always thinking of control experiments to test your theory. If your experiment causes you to conclude, as Davy stated, that chlorine and carbonic oxide (Cl_2 and CO) are united by a strong force, this conclusion would be tested by comparing the reaction of Cl_2 with a molecule of related composition, CO_2, carbonic acid. Here is the result he obtained. "*I have exposed mixtures consisting of different proportions of chlorine and carbonic acid to light, but have obtained no new compound.*"

It's interesting from the point we are at now, to see what general conclusions John Davy made from theses investigations reported to the Royal Society in 1812.

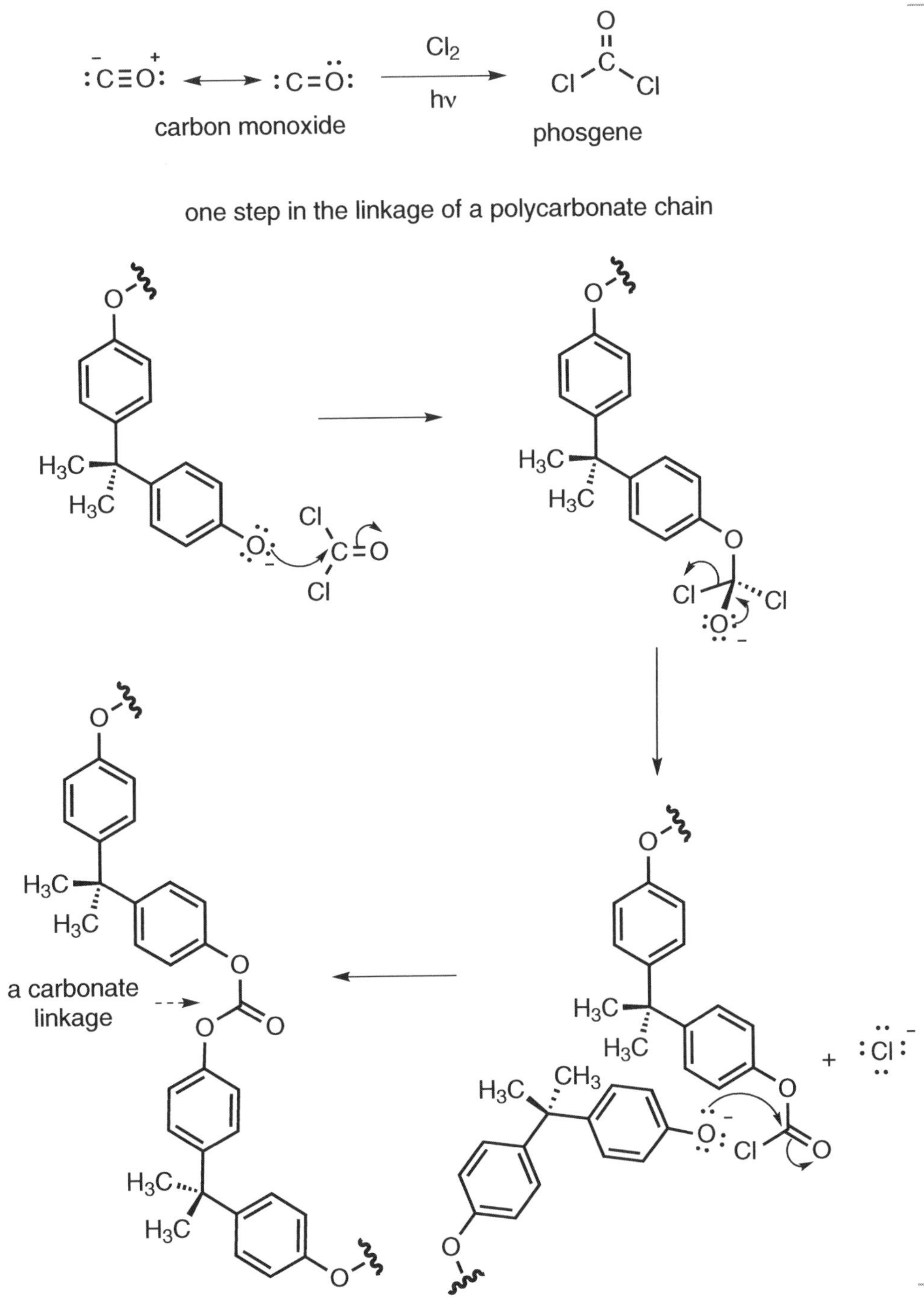

carbon monoxide

phosgene

one step in the linkage of a polycarbonate chain

a carbonate linkage

◀ **FIGURE 11.15**

Mechanism of polycarbonate formation demonstrates the importance of the chloride (Cl⁻) leaving group after nucleophilic attack on phosgene.

"The proportions in which bodies combine appear to be determined by fixed laws, which are exemplified in a variety of instances, and particularly in the present compound...... This relation of proportions is one of the most beautiful parts of chemical philosophy, and that which promises fairest, when prosecuted, of raising chemistry to the state and certainty of a mathematical science."

PROBLEM 11.24

Expand on the chemical details involved with the sentence in this section that begins with the words: "However, in the exception that proves the rule...."

PROBLEM 11.25

Account for the fact that reaction of ammonia, NH_3 (to produce the amide) is faster with an ester derived from p-nitrophenol ($HO-C_6H_4-NO_2$) and a carboxylic acid, than with an ester derived from the same carboxylic acid with methanol or with phenol (C_6H_5OH).

PROBLEM 11.26

Predict the products of the reaction of phosgene ($COCl_2$) with an excess of the following molecules: methanol; ammonia; water; phenol (C_6H_5OH); trimethyl amine ($(CH_3)_3N$).

PROBLEM 11.27

Consider a molecule called a mixed anhydride, an anhydride made from two different carboxylic acids such as from acetic acid and p-nitrobenzoic acid ($HO_2C\text{-}C_6H_4\text{-}NO_2$). Draw the structure of this anhydride and hypothesize about the products of reaction of this anhydride with ammonia, NH_3.

11.10

Sulfonamide: Crosslinking of Hypalon And Sulfa Drugs

LET'S NOW RETURN TO THE FORMATION of the crosslinks in Hypalon (**Figure 11.16**), which again demonstrates the principles discussed just above but now applied to the reaction of a sulfonyl chloride.

The two amino groups on ethylene diamine, one of the molecules used industrially for the crosslinking step, $H_2NCH_2CH_2NH_2$, allow a linkage between two sulfonyl groups along one polyethylene chain or between two sulfonyl groups on different chains. The latter process is favored by the industrial process, therefore forming the desired link between chains, a crosslink. A new functional group is formed as shown in Figure 11.16, a sulfonamide.

It's the way of chemistry that chemicals are neutral to their use, which we decide for our various purposes. The sulfonamide functional group forming the crosslink is the same functional group responsible for the action of the first effective weapon against bacteria, the sulfa drugs, which came to be as part of one of the marvelous stories of chemical discovery, leading to a Nobel Prize for the German chemist **Gerhard Domagk**.

The story brings us back to the dyes and Adolph von Baeyer whose Nobel Prize in 1905 was given for his discovery of how to turn tars into dyes, therefore giving great impetus to the chemical industry in general and in Germany in particular (section

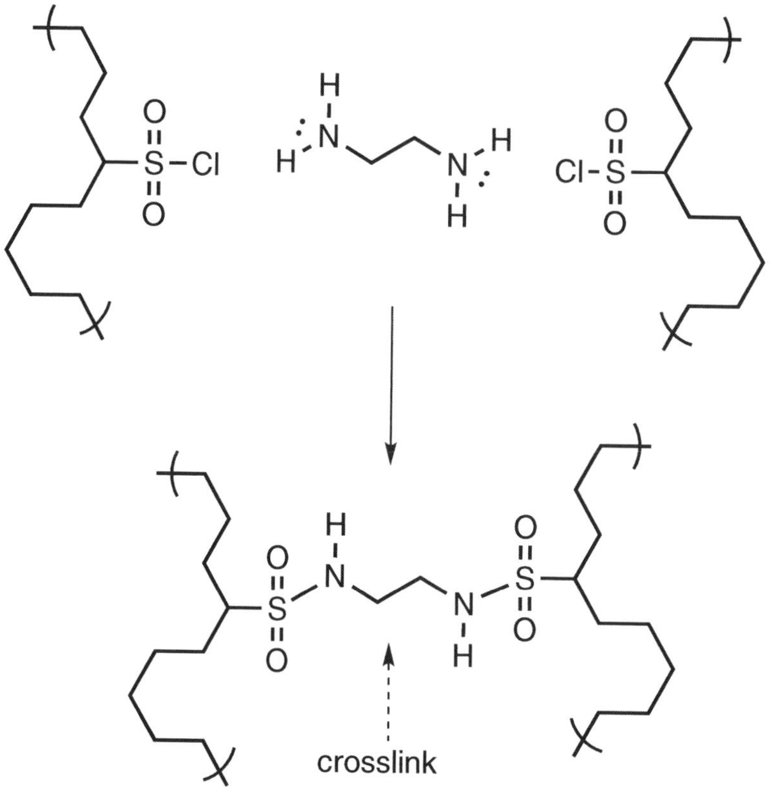

FIGURE 11.16 ▶

Crosslinking uses nucleophilic attack on a sulfonyl chloride for industrial production of hypalon.

3.2). In an article written by David M. Kiefer in the June 2001 issue of the American Chemical Society's "Today's Chemist at Work," the story of the discovery of the first sulfa drug is told (http://pubs.acs.org/subscribe/journals/tcaw/10/i06/html/06chemch.html).

Kiefer points out that before Prontosil, the first sulfa drug, which was introduced in the late 1930s, about 100,000 people died each year of pneumonia in the United States alone. These deaths were the tip of the iceberg from bacterial diseases including 2000 mothers a year who died in childbirth from puerperal sepsis (childbed fever). Gonorrhea affected 12 million Americans.

Domagk was a veteran of World War I who had been wounded and spent the war as a medic where he became interested in infected wounds. After the war he obtained a medical degree, which led him, eventually, to a position with I.G. FarbenIndustrie. The company had decided to determine if any of the large number of molecules they had synthesized as dyes could be useful in medicine. There was already a history of testing of industrial intermediates and other chemicals for their bactericidal properties when Domagk began his work at the company.

Domagk noticed that the azo dyes with the N=N functional groups (**Figure 11.17**) were held strongly to proteins in fibers and leathers. This gave Domagk the idea that such compounds could prove to be biologically active. The azo dye shown in Figure 11.17, (4-[(2,4-diaminophenyl)azo] benzenesulfonamide) proved the value of his idea and became the first sulfa drug. By the early 1940s deaths from bacterial diseases were a small fraction of what were seen before the sulfa drugs.

But there was an irony in the story. In 1936, shortly after Domagk's work on Prontosil became known, French chemists discovered that the azo compound was broken down in vivo to the simpler sulfonamide shown in **Figure 11.18**, with this molecule acting as the antibiotic. A clue came from the observation that Prontosil did not work in vitro. Adding Prontosil to a bacterial culture in a Petri dish was ineffective at controlling bacterial growth. Enzymes in the body were necessary for Prontosil to be effective, enzymes that broke Prontosil down to para-amino benzene sulfonamide.

The sulfonamide in Figure 11.18 had been first synthesized by a graduate student at the University of Vienna in 1908 and, without a patent in force, could be made and distributed by anyone therefore destroying the commercial potential of the discovery of Prontosil. In addition, Domagk's Nobel Prize gave him no monetary award. The Nazi government in Germany did not allow him to receive the prize when it was awarded in 1939, and after the war, when he received the prize he only received the certificate and the medal, the money was no longer available.

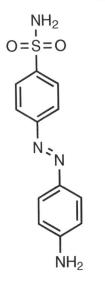

prontosil

◀ **FIGURE 11.17**

Prontosil, the first sulfur drug, features a sulfonamide functional group.

◀ **FIGURE 11.18**

The Active Drug Released from Prontosil in Vivo

PROBLEM 11.28
How might the industrial process for Hypalon have minimized the formation of intramolecular sulfonamide formation? Draw the structure of such links that are internal within a single polymer chain.

PROBLEM 11.29
Try to come up with a reasonable hybridization at sulfur in the sulfonamide functional group.

PROBLEM 11.30
Both Prontosil and para-amino benzene sulfonamide are not fully described by the structures shown in Figures 11.17 and 11.18. What is necessary for a full description of these molecules?

■ **Gerhard Domagk**

11.11 Industrial tradition rejects a perfectly good elastomer: more about free radicals.

NATURAL RUBBER AND HYPALON have in common that the element sulfur is involved in the crosslinking process for both. The way that sulfur is used, however, as we've seen in comparing section 11.6 and 11.10 is entirely different. But, nevertheless, the word vulcanization is used to describe both processes. Let's see how industry responded to an elastomer that could be crosslinked without using sulfur.

As we've learned (section 11.3), one of the two critical factors in elastomeric behavior is polymer chain flexibility. Flexibility can be blocked by structural characteristics of the chain, as for example in polymers in which there are only few bonds with rotational freedom. Polymers with a large proportion of aromatic character fit that situation making them what could be called anti-elastomers. Polycarbonate (Figure 11.15) is an example of an anti-elastomer.

Elastomeric behavior can also be blocked in inherently flexible chains because of intermolecular interactions in which the packing of the chains restricts the necessary conformational motions. Gutta percha is a perfect example as is high density polyethylene, HDPE, (section 11.3).

We have learned about a polymer that could work perfectly as an elastomer, low density polyethylene, LDPE. In addition to the fact that the chain backbone, with all the single bonds, is inherently flexible, the short carbon branches (Figure 11.5) preclude close chain packing. Both these structural factors allow the necessary conformational flexibility for elastomeric behavior. However, the site of the branches is random with the number of branches determined by the statistical probability of the back-biting mechanism (Figure 9.4). Industrial chemists and chemical engineers wanted to gain the flexibility of LDPE but with a structure that could be reliably predicted so that the properties of the intended elastomers could be controlled to maintain quality.

In section 9.6 we learned why propylene could not be polymerized using a free radical mechanism. The hydrogen atom on the methyl group in propylene could be abstracted by the free radical at the chain end, terminating the polymerization after only a few propylene units could add – an oligomer was formed. In section 9.7 we learned about the catalyst discovered by Karl Ziegler, which he used to synthesize a polymer of ethylene without the short carbon branches. This linear chain without branches came to be called high density polyethylene (HDPE). The branches in the free-radical-produced polyethylene (LDPE) (section 9.3) could not form in the absence of a free radical at the growing chain end (Figure 9.4).

However the formation of HDPE in combination with the ability to polymerize propylene, pointed the way to gain the flexibility found in LDPE but with branches that could be controlled.

FIGURE 11.19 ▶

An Elastomer Synthesized as a Copolymer of Ethylene and Proylene

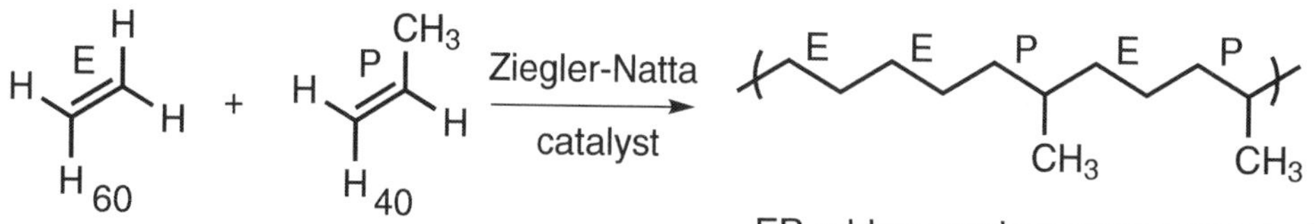

EP rubber-random arrangement of ethylene and propylene derived units.

If Ziegler's catalyst could be used to polymerize ethylene and to polymerize propylene, then why not use it to produce a polymer made of both these alkenes. This was the idea of the DuPont researchers – a great idea because the proportions could be varied to produce as many methyl groups along the backbone of the chain as desired – for example a ratio of 60:40 ethylene to propylene as shown in **Figure 11.19**. In this manner a commercially important elastomer was invented. And this polymer, as for Hypalon, would be an elastomer without double bonds and therefore without the

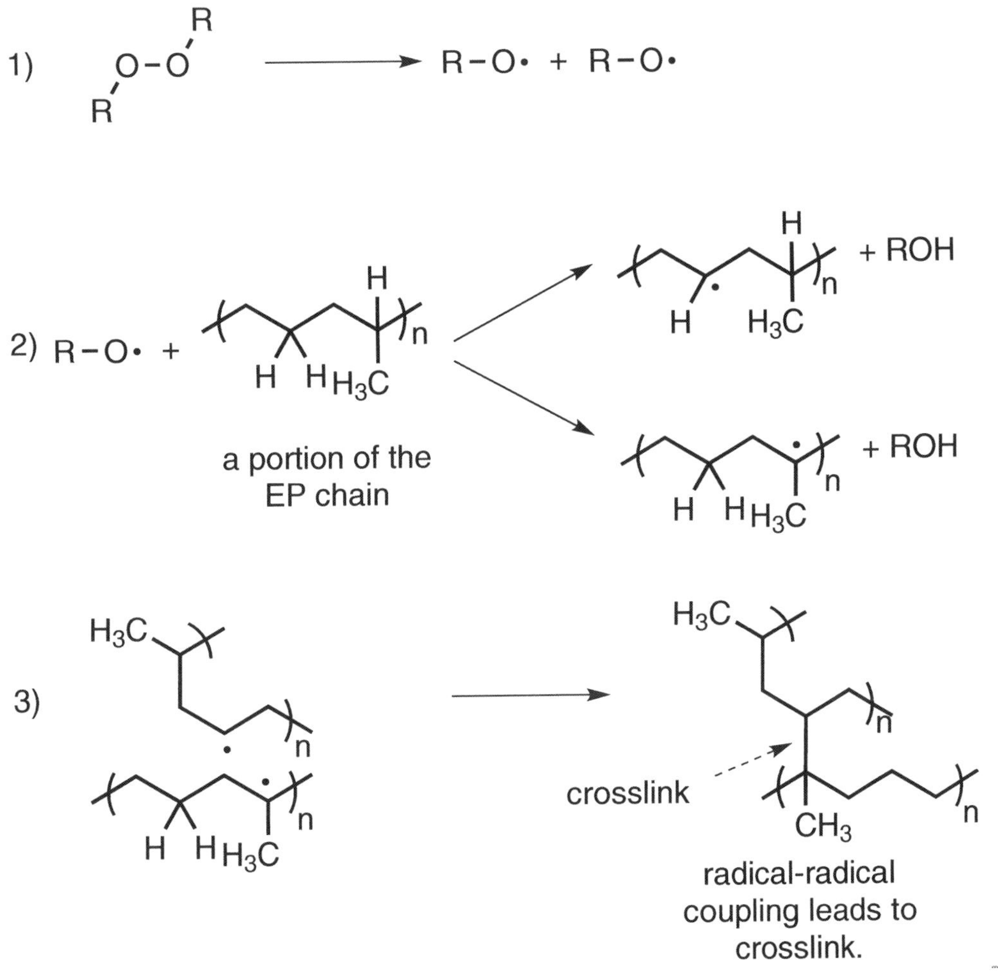

◄ **FIGURE 11.20**

Free Radical Mechanism Involving Hydrogen Atom Abstraction for Crosslinking of the Elastomer made from Ethylene and Propylene

disadvantage of decomposition from ozone in the atmosphere (section 11.7). Avoiding unsaturation in the elastomer was a driving force for the industrial innovation.

The polymer produced has the necessary flexibility but before it could be a commercial elastomer it had to be crosslinked. The crosslinking process used for Hypalon is expensive. Several chemical steps are involved not even mentioning the use of chlorine, which is always avoided by the chemical industry when possible. Chlorine is not only dangerous but causes profit to be reduced by having to pay for electrical power to produce it from sodium chloride or other chloride salts.

As shown in Figure 11.20 DuPont conceived of a way to crosslink the copolymer of ethylene and propylene that would avoid the expense associated with the crosslinking in Hypalon. Abstraction of hydrogen atoms along the EP backbone by reacting the copolymer with oxygen based radicals derived from peroxides (reaction **2** in Figure 11.20) led to crosslinks derived from radical-radical coupling between different chains (reaction **3**). The commercial idea was for DuPont to manufacture the ethylene-propylene copolymer (Figure 11.19) and sell it to the rubber manufacturers along with the technology to crosslink the polymer to the final product (outlined in Figure 11.20), the elastomer.

However, there was great resistance in the rubber industry to this kind of procedure. The rubber companies were set up for crosslinking processes using sulfur (vulcanization) and the free radical process developed by DuPont was too foreign to their experience and their equipment. They insisted on having a more familiar crosslinking process. Before we see how DuPont chemists and chemical engineers cleverly solved the problem, let's look more carefully at the process they had to abandon (**Figure 11.20**).

FIGURE 11.21 ▶

Industrial Method Introduces Double Bonds in Ethylene-Propylene Rubber to Allow Crosslinking by Vulcanization

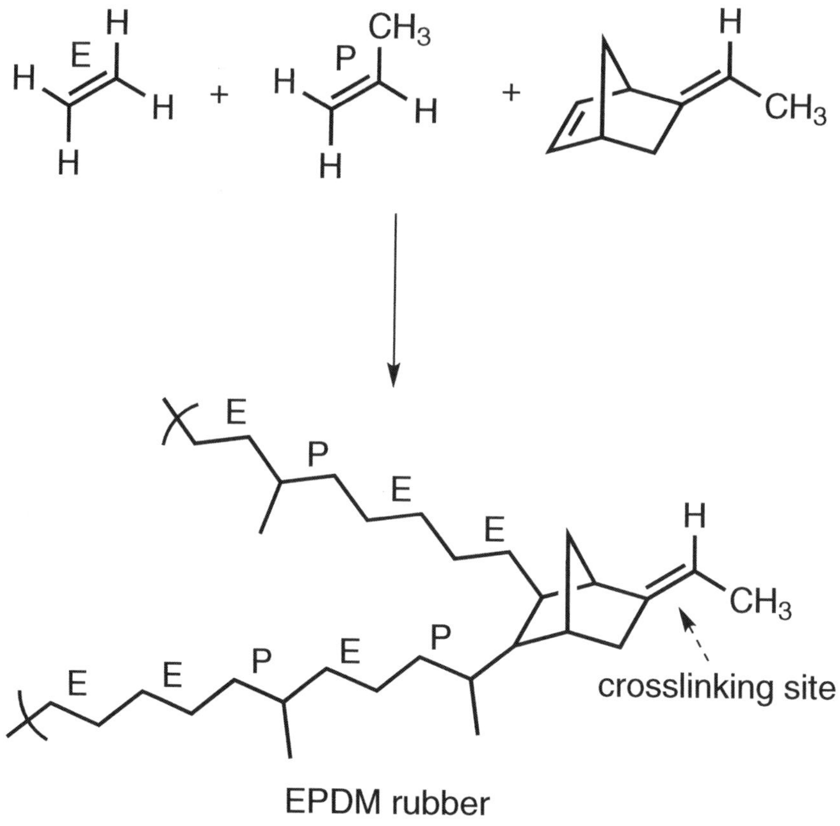

EPDM rubber

(random arrangement of the ethylene (E), propylene (P), and diene derived units.)

There is an interesting difference between the free radical mechanism in Figure 11.20 and the mechanisms discussed in Chapter 9 for the free radical polymerization of ethylene or the free radical chemistry involved in steam cracking. These are chain mechanisms in which the initiation reaction stimulates many reactions to follow (section 4.10) with the chain finally ended by a termination step. In these processes the termination steps are undesirable.

In Figure 11.20 the termination step, the crosslinking arising from the coupling of the radical sites on different chains, is the point of the process. Each coupling step (reaction **3**) requires one initiation step (reaction **1**). One crosslinking in Figure 11.20 consumes one molecule of the initiating peroxide—not a chain mechanism.

To make EP rubber fit the rubber industry, crosslinking had to involve sulfur vulcanization and therefore double bonds in the chains. DuPont solved the problem by adding another alkene to the mix with ethylene and propylene. Don't worry about the name, 5-ethylidenenorbornene (a diene) the important point is that the new additive had two double bonds, only one of which, the less substituted double bond, reacted in the polymerization process to become part of the backbone. In this way the number of double bonds in the polymer could be controlled by the amount of the diene added and the number of crosslinks, as for natural rubber, could be controlled by the amount of sulfur added. All of this was familiar to the natural rubber industry making the adapted DuPont product called EPDM, for ethylene propylene diene monomer, a success (**Figure 11.21**).

A big advantage of EPDM rubber is that the number of double bonds can be kept to a minimum therefore yielding weather and ozone resistance. This product is therefore used for roofing and also for parts of automobiles subject to extreme conditions such as radiators, heater hoses, bumpers, weather strips and seals. Good stuff.

PROBLEM 11.31

Step **3** in Figure 11.20 resembles a chain termination step in the free radical polymerization of ethylene, a step to be avoided. How might the industrial process for EP rubber be organized to enhance this step.

PROBLEM 11.32

Considering that propylene can copolymerize with ethylene in the Ziegler-Natta catalyzed process, but that only one of the double bonds in the diene shown in Figure 11.21 can participate, propose what other kinds of copolymers could be made using this catalyst and mixtures of alkene monomers. What do the observations about alkene copolymerization inform you about the nature of the catalyst?

PROBLEM 11.33

In step **3** of Figure 11.20 one crosslinked system is shown. Are there other radical reactions that would not lead to crosslinks?

11.12

Elastomers Without Covalent Crosslinks – the Glassy State

W E ALL KNOW THAT A CRYSTAL of an organic molecule is a solid with a precise melting point and with an organized arrangement of the molecular units that make up the crystal. Most organic molecules crystallize when the temperature is lowered to their melting point, although sometimes with difficulty. Every practicing organic chemist has developed the kinds of skills necessary for crystallization. To some it appears as a magical skill. I remember a story I heard about Emil Fischer, the great chemist we learned about in Chapter 3. Sugars, such as those discussed in Chapters 1 and 3, are often very difficult to crystallize and on cooling instead "oil out" as we say in the business of chemistry. And once a sample has formed an oil instead of the beautiful crystal we have been waiting for, experience tells us that crystallization from that substance in that container is a doomed expectation – alas.

In Fischer's laboratory where many sugars were synthesized in the experiments necessary to figure out the stereochemistry of D-glucose (section 3.11), crystallization was a critical step for purification and identification. It is said that as students struggled to crystallize their samples, that the great professor would often come by to offer encouragement and like magic, shortly after he left, the samples, which are often called mother liquors, would deliver the crystal. The values of the melting points for many sugars are exhibited in Figure 3.12.

Some say that Fischer carried tiny crystals of every variety of sugar in his beard, which no amount of washing would ever clean out. As Fischer stood over the laboratory bench and rubbed his beard pensively, these crystallites, which presumably found there way into Fischer's beard from his long exposure to the laboratory, were said to seed the crystallization of the students' samples. There could be some truth in that idea. I remember that one of the most effective ways to gain crystallization of a sample was to drop in a tiny amount of the same substance in crystalline form, or even that of another related crystalline substance.

However, there are some kinds of chemical substances that no amount of skill or intelligence or magic will ever cause to crystallize. Among such substances, polymers are prominent examples. There are several reasons for the difficulty and even impossibility of certain polymers to form crystals. Low density polyethylene is an excellent model for this behavior arising from the disorder in its structure with the randomly placed short carbon branches along the chain backbone (section 9.3). The regularity necessary for the structure of the crystal does not exist in the structure of

LDPE. Polycarbonate, as well, (section 11.15) has a structure with too many different kinds of moieties to gain the regular relationships necessary to form a crystal.

Another very important reason for the inability of polymers to crystallize is that they are highly viscous and exist with entangled relationships among the different polymer chains—like a plate of cooked spaghetti (section 9.3). As the temperature drops coming closer and closer to where crystallization could be possible, the movement of the chains, no matter how slow the cooling, is too slow for the chains to unentangle themselves and find the regular intermolecular positions necessary to form crystals.

So what is it that happens when a sample of a molten polymer is cooled and crystallization is not possible? Invariably, a glass is formed, a solid substance with the disorderly and transparent properties but without the flow properties of a liquid. There are other answers to why a glass forms. One view ascribes the formation of the glassy state to a thermodynamic necessity. A liquid has a higher heat capacity than a crystalline solid, which means that with falling temperature the entropy of the liquid would fall more steeply than the entropy of the crystal that would be formed below that temperature. In thermodynamic terms this is seen as a catastrophe, which has

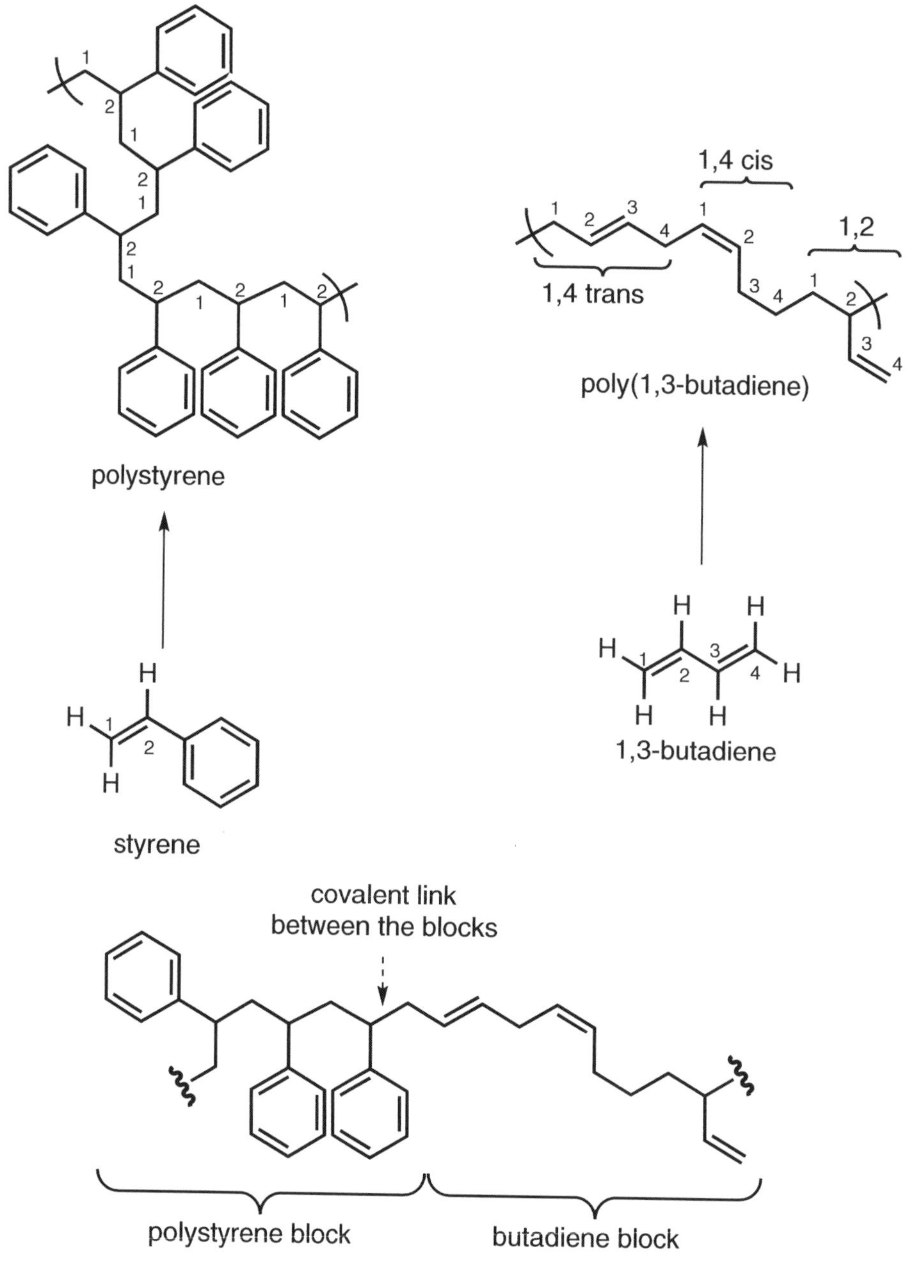

FIGURE 11.22 ▶

Representation of the Structure of a Block Copolymer Synthesized from 1,3-Butadiene and Styrene

come to be known as Kauzmann's paradox with the prediction that such a situation must force the crystal to form. But if the crystal formation is impossible, then the substance must form a glass, which in its restricted motions has the low heat capacity of a crystal.

Walter Kauzmann, who died at the age of 92 in 2009, was a gentle, brilliant giant of science who I was privileged to have on my doctoral committee. Little did I know then of his contributions to science. Not only did he offer insight into the glassy state of matter but of certainly equal, if not greater, importance was his suggestion that proteins fold to bury hydrophobic residues and to expose hydrophilic ones, a fundamental understanding of protein folding that has stood the test of time and is the basis of their complex shapes, as we've seen (sections 5.5, 5.9, 7.5).

Let's return to glass formation. There is a huge literature on the formation and properties of glass-forming polymers, not in the least because of their industrial importance. Many polymers become glassy as temperature is lowered beyond a certain value, which is called the glass transition temperature, T_g. There are many sources of information on T_g values for various polymers including an informative web site, http://en.wikipedia.org/wiki/Glasstransition#Polymers, which includes discussion of the basis of glass formation outlined above.

One prominent use that the chemical industry has made of the property of polymers to form glasses is in the invention of thermoplastic elastomers, which depend on another characteristic of polymers, their general immiscibility. Polymers don't mix well. Two structurally different polymers will generally phase-separate. The reason for this property is that the molecular weight of polymers is so high that in a sample there are far fewer molecules than in a sample of a small molecule. This situation leads to very low entropies of mixing.

■ **Walter Kauzmann**

In the absence of some attractive interaction between the structurally different polymers (hydrogen bonding (section 9.12) is one possibility) different polymers will not be miscible with each other. For example, the polymer made from styrene and the polymer from 1,3-butadiene, the structures of which are shown in **Figure 11.22**, are not miscible. When cast from a solvent the two polymers will separate into different parts of the resulting film.

Polystyrene is irregular in a different manner from the way that poly(1,3-butadiene) is irregular. In polystyrene all the units are linked in the same manner (1,2) while the pendant aromatic rings are configurationally irregular as in atactic polypropylene (Figure 9.10). Poly(1,3-butadiene) has an irregular backbone structure arising from the fact that the 1,3-butadiene units can polymerize parallel to what we learned about the addition of chlorine to this molecule (sections 10.4 and 10.5), that is, 1,4 cis and trans addition or 1,2 addition as shown in Figure 11.22. These differing additions of the 1,3-butadiene units to the growing chain occur in a random manner along the chain giving rise to a structure where lack of uniformity makes any possibility of crystallization highly unlikely.

Using a technique called "living polymerization," discovered by **Michael Szwarc** of Syracuse University College of Forestry, Shell Oil in the 1960s was able to synthesize a polymer with blocks of both polystyrene and poly(1,3-butadiene) (Figure 11.22). In a block copolymer there is a long chain of one kind of unit covalently linked to a long chain of another kind of unit as shown in Figure 11.22. Now let's understand how this block structure composed of styrene and 1,3-butadiene derived units can lead to an elastomer with very special commercially-interesting properties.

■ **Michael Szwarc**

Because of the pendant stiff aromatic rings that occur every three carbon atoms along the chain, polystyrene is far less flexible than poly(1,3-butadiene). The conformational motions along the chain backbone in polystyrene are impeded by the bulk of the aromatic rings while the conformational motions along the backbone of poly(1,3-butadiene) are hardly restricted. In fact the conformational flexibility of a poly(1,3-butadiene) chain is not dissimilar from a chain of *Hevea* rubber (section 11.1). These structural characteristics have a direct effect on the temperature at which these widely different polymers form the glassy state. The glass transition temperature of polystyrene is near 100°C while that of poly(1,3-butadiene) is close to -70°C!

Imagine that the block copolymer (Figure 11.22) is heated to well above 100°C. Both blocks are above their glass transition temperatures allowing chain movement and flow. For the reasons discussed above concerning polymer immiscibility, the two blocks with the differing structures will attempt to separate. But this separation is not entirely possible because the blocks are connected by covalent bonds. Polystyrene and poly(1,3-butadiene) blocks on different chains will form separate, but connected, regions in the material.

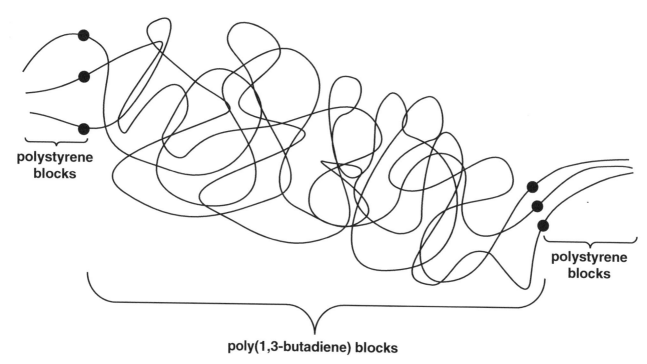

polystyrene blocks

polystyrene blocks

poly(1,3-butadiene) blocks

representation of a portion (3 chains) of Kraton

● **representation of the crosslinked glassy regions**

FIGURE 11.23 ▲

Representation of the Low and High Glass Transition Temperature Blocks shown in Figure 11.22, the Latter Acting as the Crosslinks

If now the temperature is brought to anywhere lower than 100°C but higher than negative 70°C, the chains within the poly(1,3-butadiene) picture will have the flexibility necessary for elastomeric behavior while the chains within the polystyrene region will form a rigid glassy mass. The glassy regions, in that way, form the necessary crosslinks. The block copolymer just described is a commercial material named Kraton in which the flexible poly(1,3-butadiene) blocks within each block copolymer chain have about 1300 units while the glass-forming polystyrene blocks are only about 140 units long. **Figure 11.23** is a representation of a portion of Kraton.

Kraton is widely marketed elastomer sold to the extent of hundreds of millions of pounds every year. Kraton and other block copolymers of this type, with widely variable glass transition temperatures for the blocks, are called thermoplastic elastomers because their elastic characteristics are temperature dependent – elastomeric at temperatures below the T_g of one block (the crosslinking block) and above the T_g of the other (the flexible block). At temperatures above the T_g of both blocks, the material can even flow, a characteristic allowing molding, which is of great commercial value.

PROBLEM 11.34

Why does Kauzmann's paradox being called a catastrophe have to do with thermodynamics?

PROBLEM 11.35

What is it about the structures of the gulose and idose (Figures 3.12 and 3.13) that make them so difficult to crystallize causing them to form glassy states?

PROBLEM 11.36

Why should the flexibility of the local segments of a polymer chain, that is, the speed with which conformational motions can take place, be a determining factor in the glass transition temperature?

PROBLEM 11.37

Looking along any of the single bonds in a polystyrene chain, draw a Newman projection and use this information to investigate why conformational change in polystyrene is slow while understanding why various conformations are possible.

PROBLEM 11.38

Describe the conformational properties of the block copolymer in Figure 11.23 at the following temperatures: 150° C; 30° C; and -100° C.

FIGURE 11.24 EXHIBITS THE ESSENTIAL FEATURES of the structure of Spandex (Lycra™, Elastane™) a thermoplastic elastomer invented by DuPont in the 1950s, which works by the same principle as Kraton in that the polymer consists of blocks that do not mix. But Spandex differs from Kraton in that the block responsible for the chains not slipping by each other, the crosslinking block, is not glass-forming but rather finds the necessary temperature-dependent attractive interactions in crystallization and hydrogen bonding. In Spandex, therefore, if the temperature is below the melting point of the crystal-forming- blocks and above the glass transition temperature of

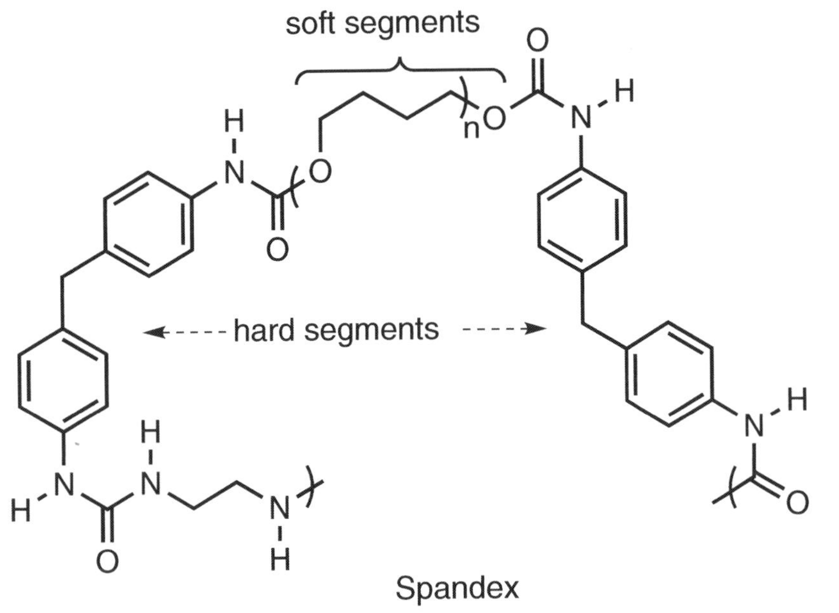

11.13

A thermoplastic elastomer that is not based on a glassy state: Spandex.

◄ FIGURE 11.24

Structures of the Soft and Hard Segments of the Spandex Polymer Chain

Spandex

the flexible blocks the material will act as an elastomer.

And what a remarkable elastomer Spandex is accounting for its fame as a "hug your body elasticity," which is responsible for its continued wide use as a fiber now fifty years after its invention. In the structure shown in Figure 11.24 there can be up to fifty recurring groups of $-O-CH_2-CH_2-CH_2-CH_2-$ moieties in the flexible segment of the chain allowing stretching of the fiber to a range of 500 to 600% over and over again while still returning to its original shape. If the segments responsible for the crosslinking, the hard segments, did not exert a powerful restraining force on slippage of the chains by each other, the original shape could not be regained.

Inspection of the blocks responsible for crosslinking (Figure 11.24) shows a functional group with N-H and C=O groups. The hydrogen bonding between these entities on different chains is the same phenomenon as that encountered in strengthening the fibers formed by nylon (section 9.12) while the regular placement of the stiff aromatic rings enhance the crystallization necessary for the crosslinking.

FIGURE 11.25 ▶

Mechanism of the Industrial Synthesis of the Flexible Block of Spandex from Tetrahydrofuran via a series of S$_N$2 Reactions

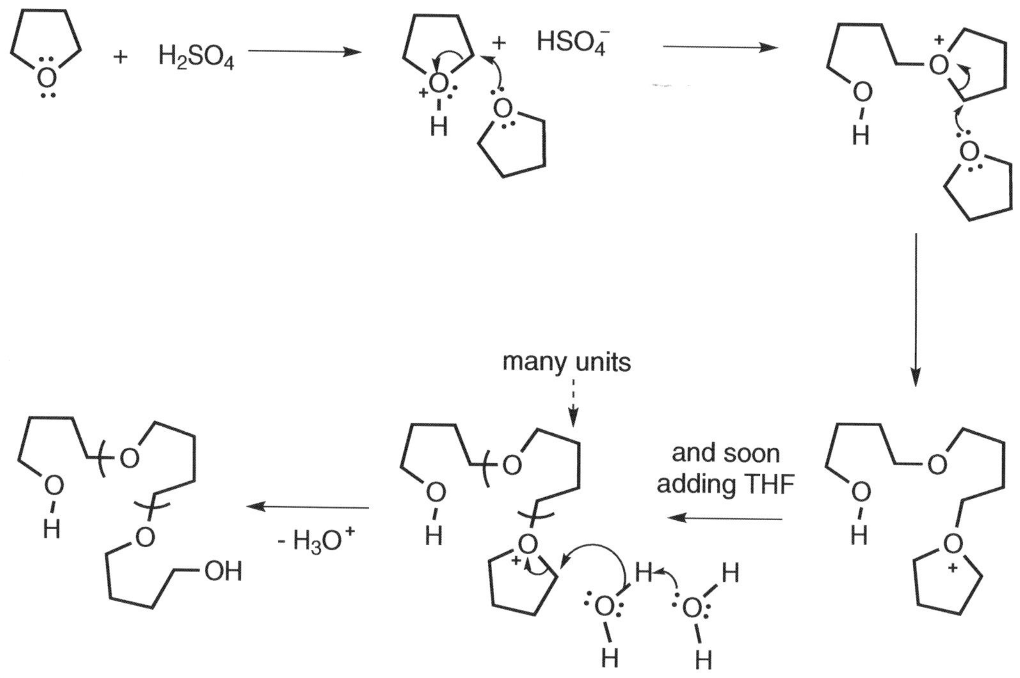

One of the starting materials for the industrial synthesis of Spandex is tetrahydrofuran (THF), an industrial chemical that has wide use and is produced yearly in the range of half a billion pounds. To produce the $-O-CH_2-CH_2-CH_2-CH_2-$ flexible segment the THF is treated with strong acid so that some of the molecules form a salt with the THF acting as a base (**Figure 11.25**).

The bond between the protonated oxygen and the adjacent CH$_2$ group in the THF ring is weakened allowing the lone pair of electrons on a THF molecule that has not acted as a base to act as a nucleophile in opening the salt-forming-THF-ring via an S$_N$2 reaction (section 10.6). In this manner the positive charge is now transferred from its original site to the oxygen of the ring that acted as the nucleophile. This process occurs again and again to form the polyether with the terminal hydroxyl group incorporated by reaction with water as shown in the last step of Figure 11.25.

The second key ingredient in the industrial synthesis of Spandex is methylene diphenyl diisocyanate, industrially known as MDI. In MDI we are introduced to a new functional group, -N=C=O, isocyanate. You might imagine, considering the two π-bonds and the large difference in electronegativity between the atoms that form the isocyanate group that a high reactivity is encountered. Indeed, isocyanates react quickly with nucleophilic functional groups such as OH and NH$_2$ as shown in **Figure 11.26**. Let's stop for a moment at the chemistry of isocyanates before continuing with Spandex.

In the reaction of nucleophiles with isocyanates the initial step is the familiar opening up of the C=O π-bond of the isocyanate group as shown in Figure 11.26.

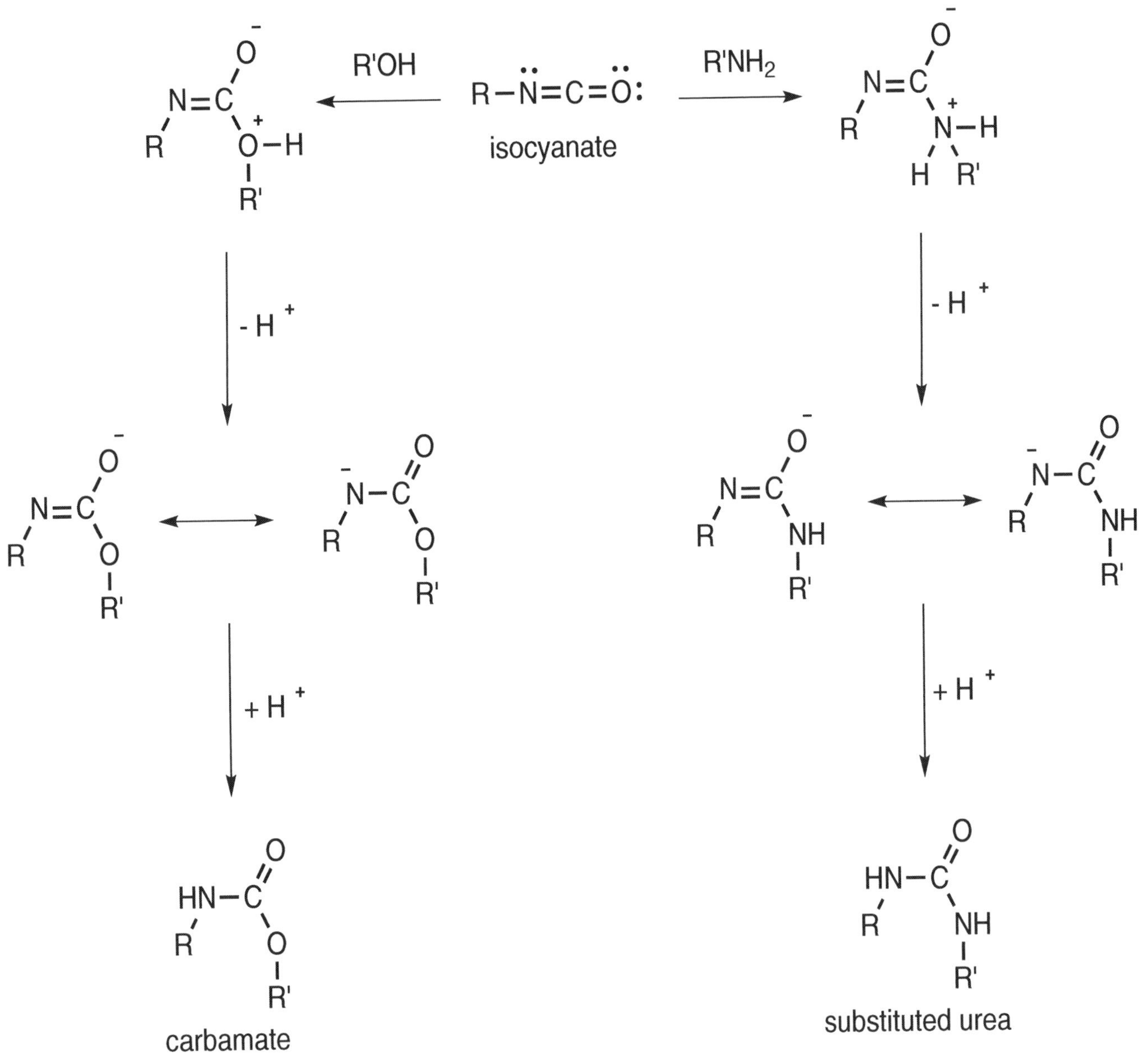

carbamate

substituted urea

The strange looking product that follows then loses a proton to the medium to yield a resonance-stabilized intermediate that adds a proton from the medium to the nitrogen to form either an ester of carbamic acid, called a carbamate or a substituted urea. As we'll see in a moment, these reactive characteristics of the isocyanate group allow the wedding of the flexible polyether block to the hydrogen bonding crystalline block.

A third large volume industrial chemical, produced annually in the billion pound range, ethylene diamine, which we've seen in the crosslinking of Hypalon (Figure 11.16), completes what is necessary to produce Spandex. In the structure of Spandex in Figure 11.24 you can find both the carbamate group and the substituted urea we've just seen in Figure 11.26.

▲ FIGURE 11.26

Mechanism of the Conversion of an Isocyanate to a Carbamate and to a Urea

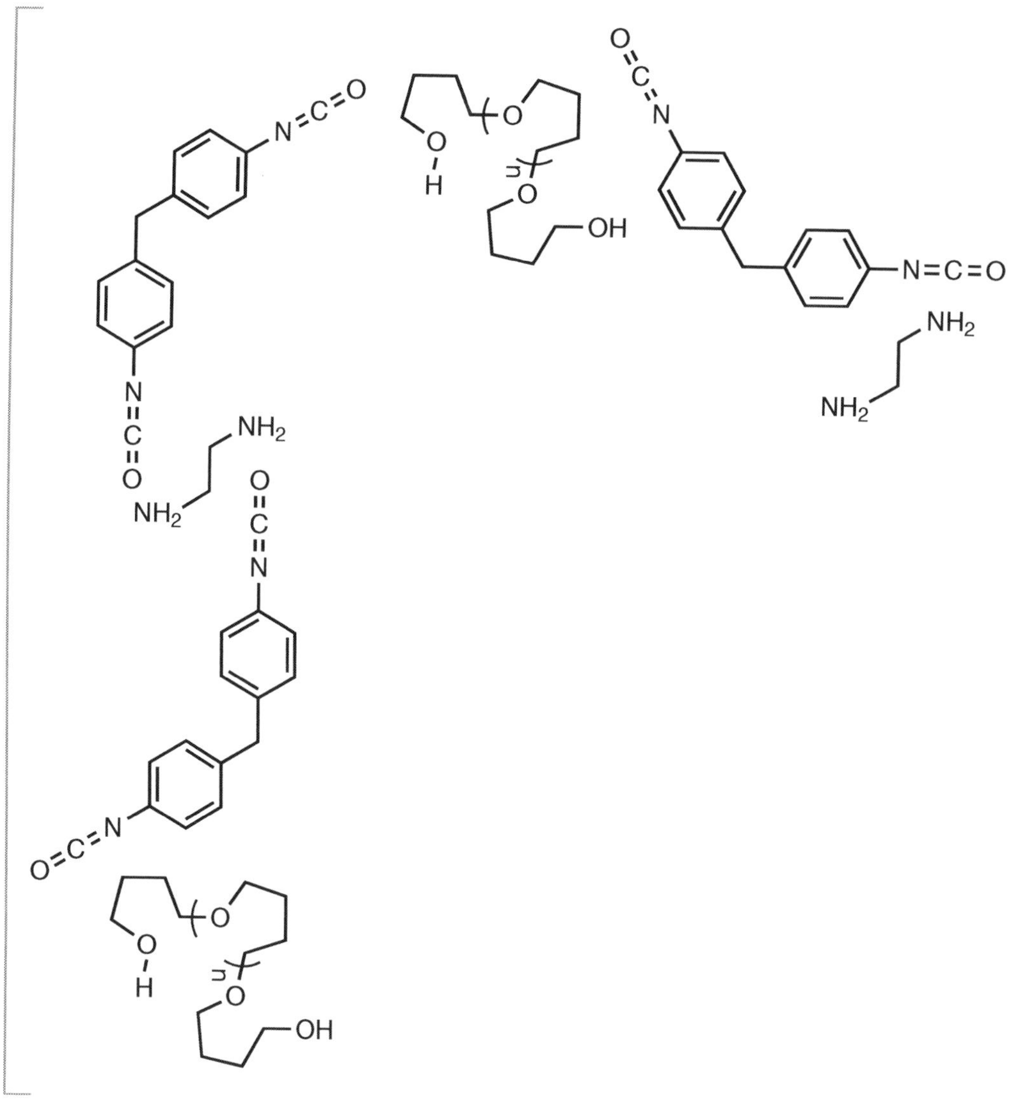

◄ **FIGURE 11.27**

Overall Industrial Path to the Production of Spandex from its Components

The carbamate is a product of the reaction of one of the isocyanate groups of methylene diphenyl diisocyanate with OH groups of the poly(tetramethylene glycol) product of the reactions in Figure 11.25. The substituted urea is the product of thereaction of the NH$_2$ groups of ethylene diamine with the other isocyanate groups of methylene diphenyl diisocyanate. These molecules arranged to form a segment of a Spandex chain, are shown in **Figure 11.27**.

PROBLEM 11.39

Describe the structural source of the fundamentally different conformational properties that cause the hard and soft segments of Spandex to differ from each other.

PROBLEM 11.40

Although Spandex and Kraton are both thermoplastic elastomers, their use temperatures depend on different structural factors. Describe the difference and show what happens at the upper and lower temperature limits of the elastic properties of Spandex.

PROBLEM 11.41

Tetrahydrofuran can be easily synthesized from an aromatic molecule. Describe the precursor to tetrahydrofuran and the source of its aromatic character.

PROBLEM 11.42
What are the high energy bonds in the synthetic path in Figure 11.25 and how are they created?

PROBLEM 11.43
How are the leaving groups in Figure 11.25 related to the leaving groups in the hydrolysis of esters, for example, in the saponification of triglycerides?

PROBLEM 11.44
Does Figure 11.25 describe a polymerization and if so, what kind of polymerization is it, addition or condensation?

PROBLEM 11.45
Describe the isocyanate functional group in terms of orbital hybridization. Why does your description require a linear geometry for this functional group?

PROBLEM 11.46
Show all lone pair electrons and justify all formal charges for the intermediates shown in Figure 11.26.

PROBLEM 11.47
Hydrolysis of a carbamate to a carbamic acid leads to rapid loss of CO_2. Do you find this a reasonable reactivity for the carbamic acid and why?

PROBLEM 11.48
Complete the formation of Spandex from the reactants in Figure 11.27.

CHAPTER ELEVEN SUMMARY of the Essential Material

HERE WE LEARN ABOUT NATURAL RUBBER and how the cis-double bonds are essential for its elastic properties, which is nicely shown by comparison to its diastereomer with trans double bonds. We look into the phenomenon of cis and trans and find this nomenclature extends beyond double bonds and also discover another way of naming double bonds based on German words. When cis and trans describe stereochemical relationships that are not due to double bonds we are lead back into the nature of six-membered rings and, by implication, other rings as well, although the latter are not discussed here. We discover why the cis double bond causes elastic properties and find out the relationship to both LDPE and unsaturated fatty acids and also why cis and trans double bonds are constrained by their bonding orbitals not to interconvert with each other.

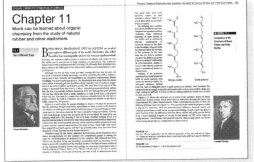

The chapter then spends some time looking into the thermodynamic understanding of the peculiar temperature dependent properties of natural rubber such as why rubber gets stiffer when hotter.

An important theme about elasticity is the necessity of crosslinking and we find out how sulfur allows crosslinking by its chemical properties so that sulfur reacts with the double bonds in natural rubber in a manner that is both similar but dissimilar from the reactions of double bonds with halogens. Here we come across another chain reaction such as we've seen before in earlier chapters.

The double bonds in natural rubber are not all crosslink sites leading to the possibility of breakdown of the rubber polymer chains by ozone. This deficiency caused industry to look for other kinds of elastomers in which double bonds are reduced in number or even eliminated. Here we come across Hypalon in which the rubber property is based on the conformational properties of polyethylene and the crosslinking, interestingly enough, arises from the reactive properties of a functional group, the sulfonamides, that in another form led to an effective battle against bacterial-caused diseases. Our study of this functional group and the crosslinking reaction leads to a review of the chemistry of the acyl functional group and how this chemistry finds a parallel reactivity in the properties of sulfur-oxygen double bonds.

In industry's search for elastomers without double bonds, an invention leads to crosslinking involving reactivity of free radicals that we have seen earlier to be responsible for the short carbon branches in LDPE. But this advance is rejected by the rubber industry which is wedded to the use of double bonds and sulfur. We learn how industry solved that problem using the phenomenon of copolymerization and the fact that not all double bonds are reactive to the catalyst derived from the work of Ziegler and Natta studied in Chapter 9.

Finally we come to elastomers in which covalent crosslinking is not necessary but is replaced with crosslinking arising from the fact that polymers form stiff glassy states and the fact that polymers of different structures don't mix with each other. We study the structural source of the glass-forming properties of polymers and the relationship of that property to conformational ideas. Here we come to block copolymers and thermoplastic elastomers and a commercial product called Kraton. From these ideas arises another kind of crosslinking that is not covalent, a crosslinking that uses small scale crystallization and hydrogen bonding such as found in the strength properties of both nylons and silk as we've studied in an earlier chapter. This development is the invention of Spandex, which still uses block copolymerization as in Kraton but not the glassy state to crosslink. The synthesis of Spandex introduces a new functional group, the isocyanate and we discover the source of its reactivity with nucleophiles leading to both substituted ureas and carbamates. The synthetic route to Spandex also returns us to the ideas of nucleophilic substitution in the formation of the conformationally mobile block that is based on the polymer formed via the reactive properties of tetrahydrofuran when subjected to acid catalysis.

Chapter 12
Synthesis–Part One

IN SEVERAL OF THE PRECEDING CHAPTERS our focus has been on structure and mechanism, the underpinning of organic chemistry. But we have touched on the synthesis of organic molecules in the course of our considerations, both small molecules, and macromolecules, the former mostly via biochemical processes. Synthesis carried out by organic chemists in their laboratories and in industry deserves more of our attention. The importance of synthesis does not arise only from the fundamental necessity of organic chemists to have access to a certain molecular structure, although that is certainly a major goal of synthesis, nor does the goal necessarily involve a commercial venture. Rather, as expressed by **R. B. Woodward**, perhaps the greatest of all organic chemists and someone whom we have met before (section 1.13), the goal of synthesis can be a way to measure the health of the science. And, in taking that measure, the deficiencies one encounters often lead to the necessity to create new kinds of chemical reactions and new insights into existing chemistry. In this way the power of the science is increased by undertaking the task of synthesis. Here are Woodward's words concerning his focus on chemicals produced by nature, so-called natural products:

"The synthesis of substances occurring in Nature, perhaps in greater measure than activities in any other area of organic chemistry, provides a measure of the condition and power of the science. For synthetic objectives are seldom if ever taken by chance, nor will the most painstaking, or inspired, purely observational activities suffice."

Woodward expressed this view of synthesis in his writings in 1956 and is quoted prominently in the wonderful book "Selected Organic Syntheses" by **Ian Fleming**, published in 1973 by John Wiley and Sons. I never doubted that Fleming's book is a treasure and have carefully guarded my copy over the years only to discover that others have certainly agreed with my assessment of the book. I just discovered that the book, although out-of-print, is still widely available thirty seven years after its publication. In this chapter we'll take examples from Fleming's book adapted to the level of this text.

Every organic chemist, to differing degrees, knows the methods necessary for the synthesis of organic molecules. And, according to the skills accessible to the chemists, the complexity of the target molecules will vary over a wide range. Here we are discussing not the simple transformation of one functional group into another, such as, for one example, the oxidation of a primary alcohol to an aldehyde or in the Chapter just completed, the conversion of an isocyanate to a carbamate. Although this may be complicated in certain circumstances, we are interested here in a larger goal, multistep synthesis. It is such a synthetic task involving many steps from the starting materials to the target that is the focus of this chapter.

A foundation necessary for all syntheses is knowledge of chemical reactions, and the more reactions one knows how to use, the more power one will have over the synthetic task. Studying synthetic pathways conducted by experts in synthesis will therefore introduce one to a variety of chemical reactions conducted in differing circumstances. This is the approach we'll take here. We'll look at how two experts carried out two multistep syntheses and from this complexity we'll discover the principles they used and the various chemical reactions they brought to the task.

■ Robert Burns Woodward

What's interesting is that experts in synthesis, who know myriads of chemical reactions and are expert in their limitations and how they are used, work backwards in planning a synthetic path. An expert in synthesis does not look at some simple starting molecules and then imagine how they could be transformed to the complex molecular target. Starting from the complex synthetic target, the expert imagines a path back to available starting materials. Let's follow the path of these two experts and then, later in the chapter, try their method ourselves.

Some of the reactions we'll encounter will be familiar from earlier chapters some will be new to you and worth learning about, and some may be new to you but not dealt with in the text. But overall you'll learn a great deal of chemistry and also learn how chemists work on synthetic problems. Let's get going by following the work of the master of the field, R. B. Woodward, and his synthesis of cholesterol.

■ **Ian Fleming**

PROBLEM 12.1

Although hardly of the complexity to be taken up in this chapter we have seen multiple steps to be required starting from readily available starting materials for the synthesis of polycarbonate and also of Spandex. Go back over these polymer structures and try to imagine other routes to their syntheses than those shown.

W HO WAS R. B. WOODWARD? We first met him in a story told by Melvin Newman about Woodward meeting Louis Armstrong (section 1.13) and now we are hearing about him again as the master of multistep syntheses of organic molecules.

12.2
R. B. Woodward

The best insight into Woodward and his work that I could find was written by another notable of the Harvard chemistry department, Frank H. Westheimer, in a book published in 2001, dedicated to Woodward and his work, which was edited by Otto Benfey and Peter Morris. There are some remarkable facts about Woodward. After graduating from Quincy, Massachusetts High School in 1933, recognized as a most brilliant and hard working student, Woodward entered the Massachusetts Institute of Technology (MIT) on a scholarship. He lost that scholarship after the fall semester of his sophomore year for poor grades. It seems that Woodward in many ways was unwilling to follow rules that were necessary to attain high grades at MIT. He was only seventeen.

It took Woodward a couple of years to earn enough money to return to MIT without the scholarship where Westheimer reports that he signed up for 186 credit hours of courses in one year meaning that Woodward was not able to attend all the classes he had signed up for. However, he was able to attend enough exams in enough courses to fulfill all the course requirements for both B.S. and Ph.D. He completed his doctoral degree in 1937 at age 22.

I've read elsewhere that Woodward's exceptional brilliance, perhaps genius, was recognized by the faculty at MIT who gave him unprecedented leeway in moving through the requirements of the school. Westheimer, among almost all organic chemists who know of Woodward's work, gave him credit for taking an intellectual approach to organic synthetic problems. Using the mechanistic theory available in those years, Woodward understood how complex molecules could be constructed and then set out with precise attention to detail and his own extremely hard work, and the hard work of those in his laboratory, to accomplish the construction of remarkably complex molecules.

Westheimer pointed out that much of the organic chemical approach to synthesis before Woodward took an empirical approach and rejected considerations of mechanism. As Westheimer stated it, "Then Woodward stepped into this intellectual

■ **Frank H. Westheimer**

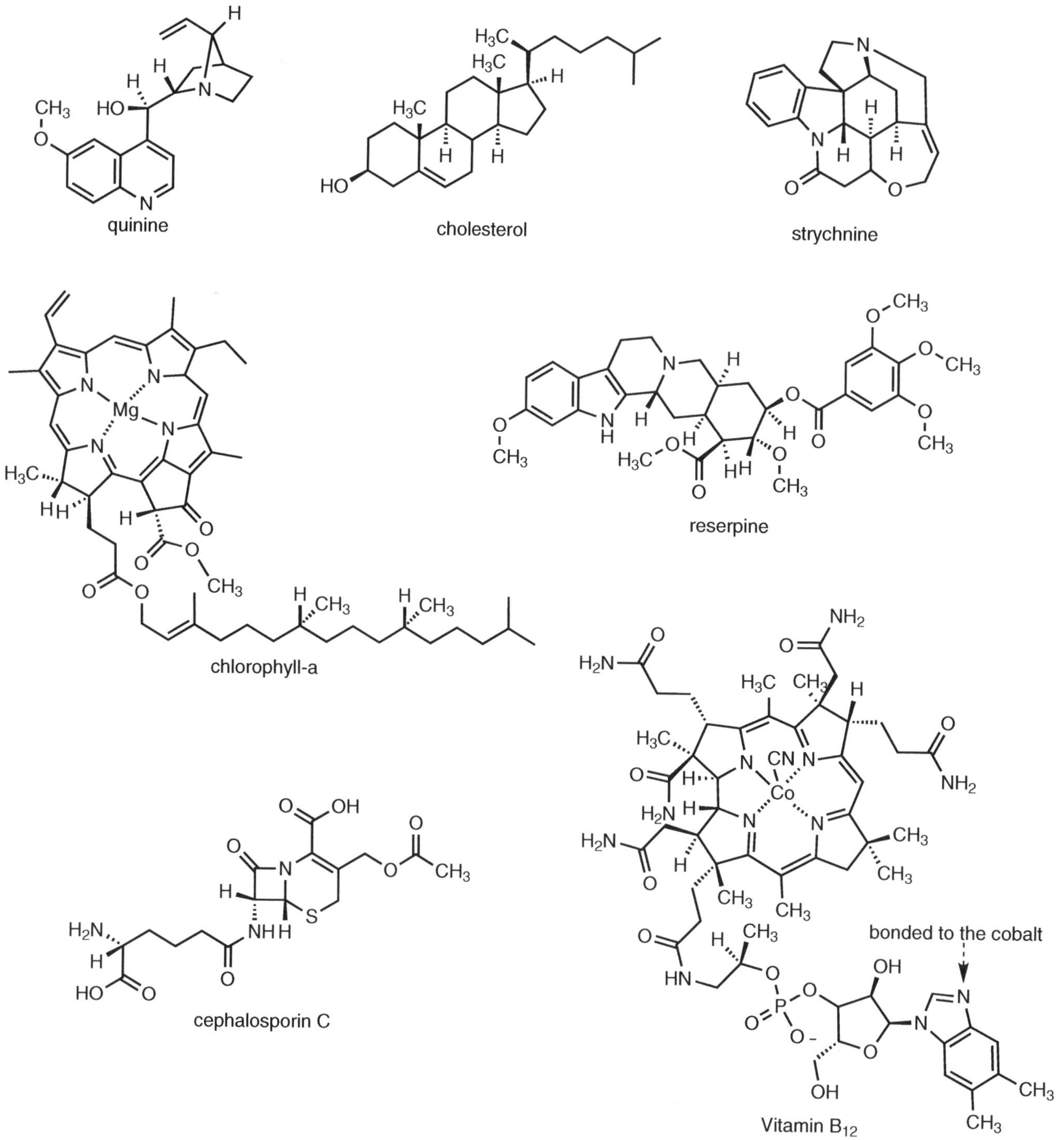

quinine

cholesterol

strychnine

chlorophyll-a

reserpine

cephalosporin C

bonded to the cobalt

Vitamin B$_{12}$

▲ FIGURE 12.1

Various molecules synthesized by Woodward's groups over the years.

vacuum and showed a better way." And a better way was certainly necessary to succeed at synthesis of the complex structures that were his goal. **Figure 12.1** shows the structures of the most prominent of the molecules Woodward's laboratory synthesized over decades of his research activities. Westheimer states that in the years from 1937 to Woodward's death in 1979 "he changed the way in which organic chemists thought about and performed synthesis." In fact, one can go further in saying that Woodward changed the way organic chemistry as a science was approached by chemists.

Woodward was quite the character, well known for his penchant for blue, as noted in the anecdote told about him in section 1.13, and as well by the many practical jokes he was noted for and his sense of the dramatic. His students once surprised him in a manner that served Woodward's dramatic flair by carrying him around Harvard Square in a sedan chair on the occasion of Woodward's 60th birthday. He was well described on a Wikipedia web site in the following way:

"His lectures were legendary and frequently used to last for three or four hours. [His longest known lecture defined the unit of time known as the "Woodward", and thereafter his other lectures were deemed to be so many "milli-Woodwards" long!] In many of these, he eschewed the use of slides and used to draw beautiful structures by using multicolored chalk. As a result, it was always easy to take good notes at a Woodward lecture. Typically, to begin a lecture, Woodward would arrive and lay out two large white handkerchiefs on the countertop. Upon one would be four or five colors of chalk (new pieces), neatly sorted by color, in a long row. Upon the other handkerchief would be placed an equally impressive row of cigarettes.

The previous cigarette would be used to light the next one. His famous Thursday seminars at Harvard often lasted well into the night. He had a fixation with blue, and all his suits, his car, and even his parking space were coloured in blue. In one of his laboratories, his students hung a large black and white photograph of the master from the ceiling, complete with a large blue "tie" appended. There it hung for some years (early 1970s), until scorched in a minor laboratory fire. He detested exercise, could get along with only a few hours of sleep every night, was a heavy smoker and enjoyed Scotch whiskey and a martini or two."

■ **Kurt Alder**

I well remember a lecture by Woodward when I was a postdoctoral fellow at Stanford. It was either late in 1966 or early 1967 and the auditorium was a jam packed group of excited chemists when he entered in his regal manner and turned to the blackboard. I remember my astonishment at the beauty of the colored chalk drawing of vitamin B12 (Figure 12.1) he produced on the black board. As he was about to turn to the audience to begin, a shout came from the back of the room. "Can't see it Bob." I was mortified.

Woodward never turned around but simply took the eraser and very calmly erased what I saw as a masterpiece of a structural drawing. He then picked up his colored chalk and drew again the structure just as magnificently as he had the first time but now about 50% larger. I mumbled to myself. This guy's a genius. I believe the lecture that followed, in the usual way for Woodward, lasted the better part of three hours.

Woodward's struggles in synthesis were rewarded not only by the attainment of the synthetic objective but as well by the many insights gained into the fundamental principles underlying organic chemistry. The most prominent of these insights was attained with the collaboration of a young colleague, Roald Hoffmann, leading to the Woodward-Hoffmann Rules. These rules allowed an understanding of how the ideas initiated by Erich Hückel to understand aromaticity and benzene (section 6.8) would also yield an understanding of many chemical reactions that for decades had not been understood, and which had been termed by many as "no-mechanism reactions." We'll look into these reactions later in section 12.21.

One of these so-called no-mechanism reactions, one of the simplest and most useful reactions known in organic chemistry, had been discovered by **Otto Diels** and **Kurt Alder** in 1928, which has come to be known as the Diels-Alder reaction. The discovery arose out of the doctoral thesis work of Alder when he studied with Diels, who was professor of chemistry at the University of Kiel. Professor and student won the Nobel Prize in 1950 for the discovery of this reaction.

■ **Otto Diels**

The Diels-Alder reaction is a beautifully simple way to make a six-membered ring, a ring size that you have already seen to be one of the most common in organic chemistry, making therefore this reaction exceptionally important in the synthesis of cyclic molecules. I saw on a Wikipedia site the term "Mona-Lisa of reactions in organic chemistry" used to describe the Diels-Alder reaction, I guess because the reaction is so perfect and easy to carry out – a smile of a reaction. Indeed, let's follow Woodward's synthesis of cholesterol in which the Diels-Alder reaction plays a central role as it does in the synthesis of the prostaglandins by E. J. Corey discussed in Part II of this chapter.

PROBLEM 12.2

Use the molecules in Figure 12.1 to practice your skill at interpreting structures by redrawing each one putting in all the atoms and the lone pairs of electrons and also checking for formal charges and the octet rule. Also identify in each structure any chiral atoms.

PROBLEM 12.3

In which of the structures in Figure 12.1 do you find conjugated double bonds and/or conjugated carbonyl groups.

PROBLEM 12.4

In which of the structures do you find aromaticity and, if so, assign the electrons contributing to the aromatic character.

12.3

Cholesterol: The First Step

WOODWARD HAD A LONG INTEREST in the reaction discovered by Diels and Alder, noted in the last section, and even published, in 1942 in the Journal of the American Chemical Society, a speculation on its mechanism, suggesting that the process was initiated by the transfer of an electron from one reactant to the other. Ironically, it was the insights leading to the Woodward-Hoffmann Rules, twenty three years later, which explained that electron transfer is not occurring and that orbital overlap is the key to the nature of this reaction. Nevertheless, Woodward's prescience is found by his including, parenthetically, the phrase "and possibly further by interparticular overlapping of the orbitals...." in his 1942 communication. Woodward never lost interest in the Diels-Alder reaction which, in fact, was to become the first step on his synthetic path to cholesterol.

Figure 12.2 includes examples of reactants that do and reactants that do not undergo the Diels-Alder reaction, although all the examples involve a conjugated diene (two double bonds separated by a single bond) reacting with an alkene or an alkyne. The lessons learned from the information in Figure 12.2 are known to all synthetic organic chemists: a successful Diels-Alder reaction requires that the diene be in a cis-like conformation about the double bond (s-cis conformation); the more enforced is the cis-like conformation over other conformational possibilities (s-trans conformation) the more readily does the reaction occur; the most reactive dienes are those with an electron withdrawing groups next to the two conjugated double bonds; all Diels-Alder reactions occur so that the two new σ-bonds are formed cis to each other.

These characteristics of the Diels-Alder reaction just described were entirely inexplicable to chemists until the Woodward-Hoffmann Rules were published. For our purpose in demonstrating the complex nature of multistep synthesis and learning new kinds of chemical reactions in the two parts of Chapter 12, it is not necessary to understand how molecular orbitals drive the characteristics of the Diels-Alder reaction or for that matter other "no-mechanism" reactions. Let's leave that discussion for section 12.21 until after we've seen how Woodward, and then Corey, used the Diels-Alder reaction for their synthetic purposes.

The Diels-Alder reaction is stereospecific as seen in two of the examples (**3** and **4**) in Figure 12.2. When the ene is cis-substituted (**3**) the cyclic product is cis-substituted – when the ene is trans-substituted (**4**) the cyclic product is trans substituted. This fact was the first problem to face Woodward in the synthesis of cholesterol as we'll see shortly.

The electron demanding characteristics desirable in the diene and the s-cis

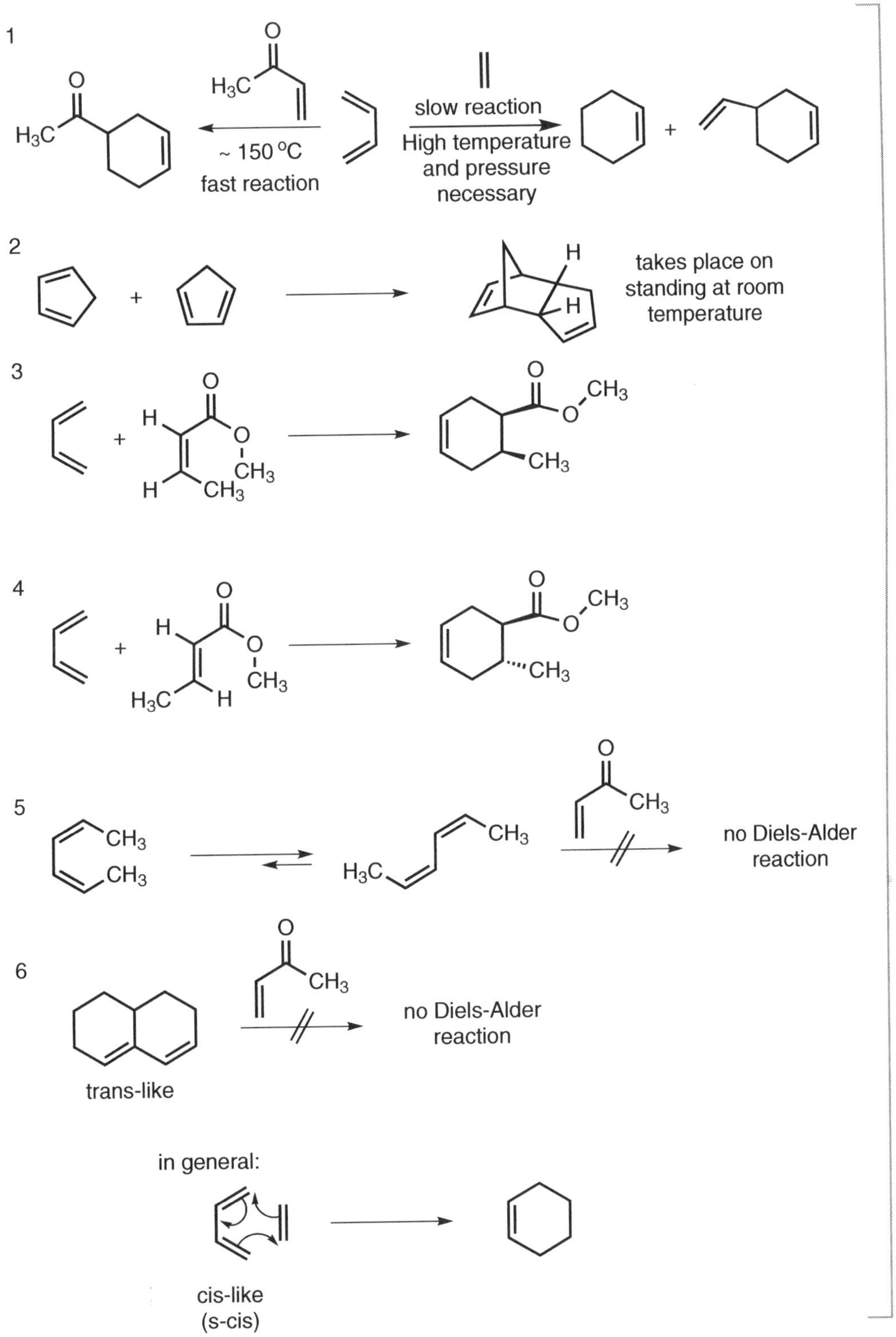

FIGURE 12.2

The characteristics of the Diels-Alder reaction are revealed.

relationship of the diene are precisely in line with the choice Woodward made with the quinone reactant in the first step (**1**) of his synthesis (**Figure 12.3**) so that the desired Diels-Alder reaction easily occurred. However, the stereochemistry at the junction of the two rings formed in step **1** of the synthesis (Figure 12.3) is cis, consistent with the properties of the Diels-Alder reaction (Figure 12.2). In his synthetic plan, Woodward planned that the junction between the two rings formed in step **1** would be transformed to the junction between the five and six-member rings in cholesterol. However, this junction is not cis but trans (Figure 12.3). Woodward expected this result and knew how to deal with it.

The work on the synthesis of cholesterol was ongoing around the time that Derek

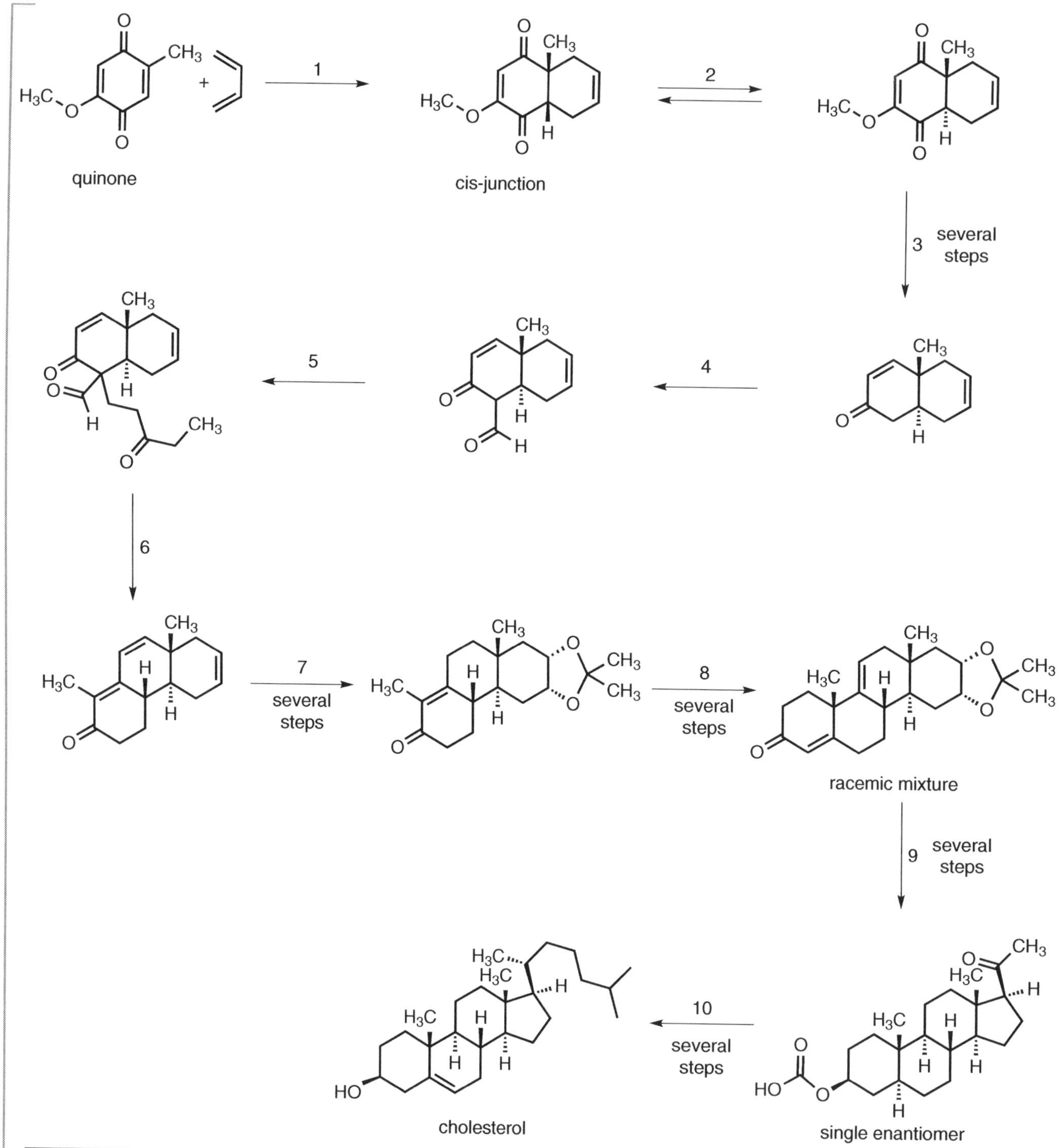

▲ FIGURE 12.3

▲ FIGURE 12.3

Woodward's Overall Path to Cholesterol

Barton was visiting Harvard and focusing his thoughts on the conformational properties of six-membered rings (section 3.3). From Barton's insights into conformation, it was realized that two fused cis cyclohexane rings are substituted with the adjacent bonds connecting the rings as axial and equatorial while trans-fused rings are diequatorial (look ahead to Figure 12.25). Woodward could therefore hope that, given the opportunity for interconversion between the cis- and trans-junctions, what is called epimerization, the trans-junction could predominate.

Organic chemists use the word "epimerization" to describe the kind of process Woodward was looking for, the interconversion between the two diastereomers shown in **Figure 12.4**. As we know from our studies of isomers in Chapter 1, the cis (**1**) and trans (**2**) ring juncture molecules shown in Figure 12.4 are diastereomers. Whatever the words used, the chemical structure for the Diels-Alder product

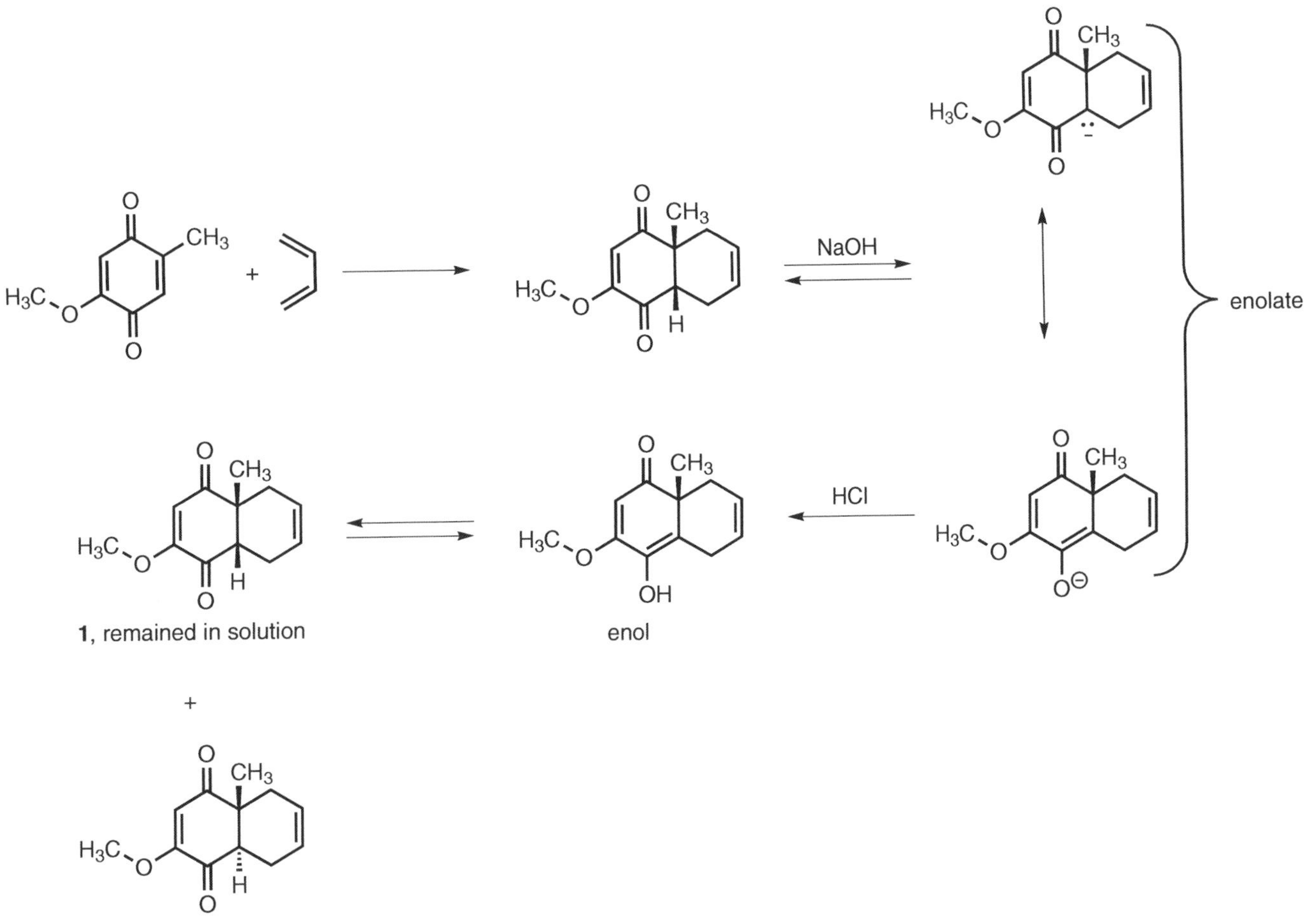

▲ **FIGURE 12.4**

The first steps to cholesterol involve a Diels-Alder reaction and an epimerization.

allowed epimerization, the change in configuration of a single stereocenter.

Adjacent to the hydrogen that needed to change its stereochemical position sits a carbonyl group, meaning that an enolate (base-catalyzed) (section 8.2) could form reversibly as shown in Figure 12.4. Luckily, when acid was added to the solution, the trans isomer (**2** in Figure 12.4) was induced to crystallize. The remaining unwanted cis isomer **1** was again subjected to the base-catalyzed formation of the enolate. By repeating this procedure the entire sample was transformed to the desired trans diastereomer. We've seen this kind of solution (although involving an entirely different chemistry) for a separation problem before, via equilibration in section 10.5, when DuPont was faced with separating the 1,2- and 1,4-addition products of the addition of Cl_2 to butadiene.

PROBLEM 12.5
Use your set of models or make up three dimensional drawings to demonstrate that in two trans fused cyclohexane rings, trans-decalin, the bonds emanating from each ring into the other ring take equatorial positions.

PROBLEM 12.6
Offer a hypothesis for the favoring of the s-trans over the s-cis conformation in example **5** in Figure 12.2.

PROBLEM 12.7
Take two white file cards and cut one out as a s-cis diene and the other as an alkene with cis substituents and use these cards to hypothesize a transition state for the Diels-Alder reaction

that fits the conclusions drawn from the experimental results in Figure 12.2. Now make up new file cards with an s-trans diene and try this again.

PROBLEM 12.8

Two products of the Diels-Alder reaction are shown for reaction 1 in Figure 12.2. How is each one formed?

PROBLEM 12.9

Use curved arrows to imagine the flow of electrons for each of the successful Diels-Alder reactions in Figure 12.2. Look ahead to section 12.21.

12.4

Cholesterol: Adding the Third Fused Ring

THE MOLECULE WITH TRANS-FUSED six-membered rings shown in Figure 12.4 (structure **2**), which arose from the Diels-Alder reaction shown in the first steps in Figure 12.3, was reacted with a reagent we've come across, lithium aluminum hydride, LiAlH$_4$ (section 10.3). In this reduction (**Figure 12.5**) the two carbonyl groups in the structures of Figure 12.4 are converted to secondary alcohols (**i**). Woodward was then able to convert this molecule (**i**) to structure (**ii**).

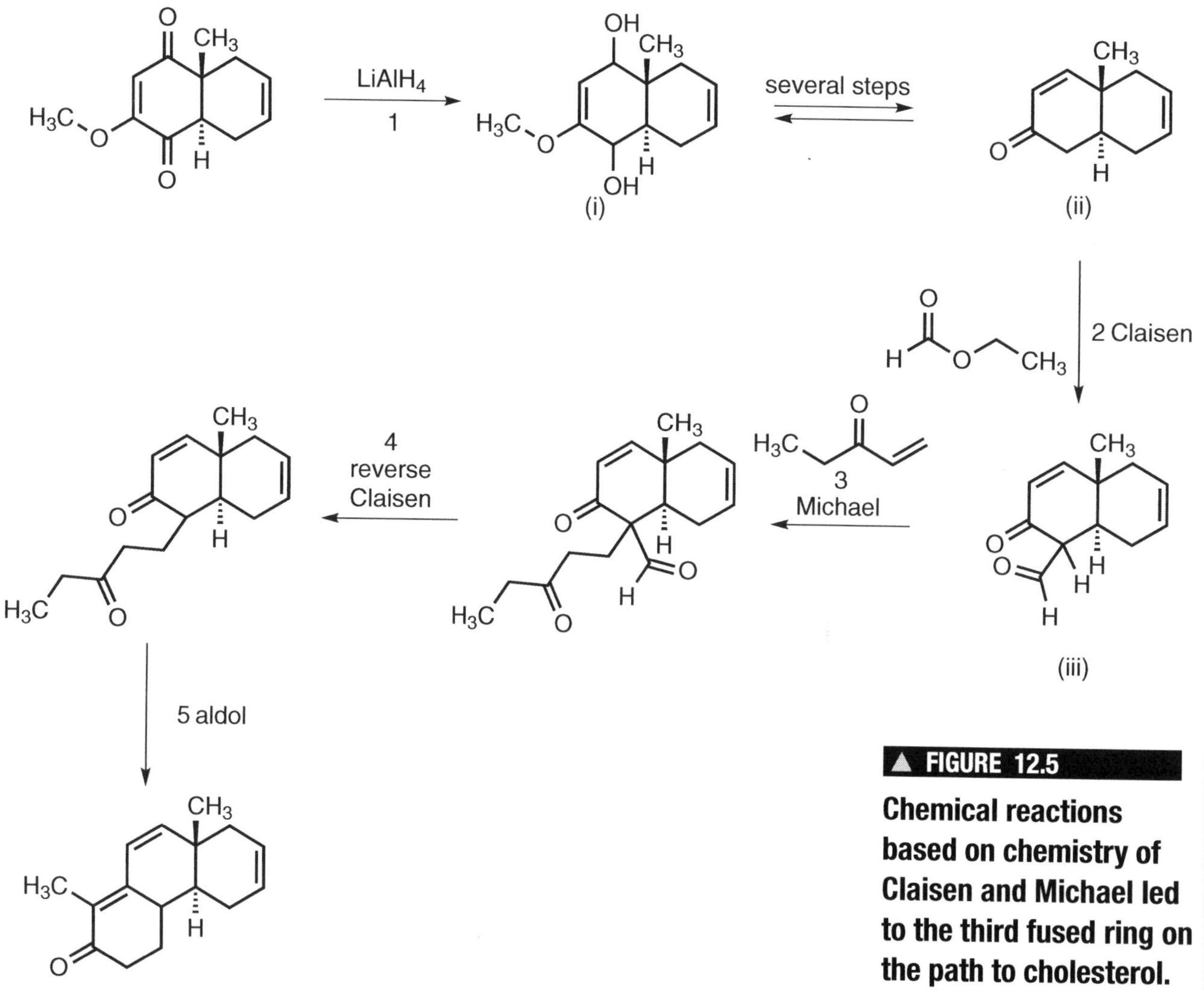

▲ FIGURE 12.5

Chemical reactions based on chemistry of Claisen and Michael led to the third fused ring on the path to cholesterol.

Using chemistry based on carbanions in ways that are familiar to us, a Claisen reaction (section 7.10 and 8.3), a Michael addition (conjugate addition) (section 7.8), a reverse Claisen reaction (section 7.10) and an Aldol condensation (section 8.4), **ii** was first converted to **iii** and then finally via step **5** on to the molecule with three fused rings.

These series of reactions were created at Oxford University in 1935 by an English chemist who created this synthetic path specifically as a path to the steroid skeleton and who was later knighted and won a Nobel Prize in 1947 for the advances he made in the synthetic chemistry of dyes and alkaloids. Sir Robert Robinson, whom we met before regarding his hypothesis of the biological formation of cholesterol (section 5.9), was a giant in the world of organic chemistry from early in his career when in 1917 he carried out the startling synthesis of a precursor to cocaine, until his death in 1975. As we shall see, Woodward made good use of what has come to be known as the Robinson ring annelation.

These series of carbonyl based carbanion reactions led decisively to forming a third fused ring as shown in Figure 12.5. Interestingly, as shown earlier in this book, the reactions that Woodward made use of in his laboratory (in vitro), noted just above, take place, almost since life originated, in the world of biochemistry, in vivo. Chemists had spent much of the nineteenth and twentieth centuries discovering chemical reactions that were unknowingly going on within their bodies and sustaining their lives. At the time that Woodward was working on the synthesis of cholesterol, making use of these reactions, this connection was not evident.

Let's look into the details of these reactions, **2 – 5**, (Figure 12.5), in Figures 12.6-12.8, which will reinforce our understanding of the in vitro use of these reactions, which we've been introduced to in vivo in Chapters 7 and 8.

The methylene (CH_2) group in (**i**) in **Figure 12.6**, adjacent to the ketone carbonyl, is capable of donating a proton to a base of appropriate strength—a Brønsted-Lowry acid-base reaction (**1**). Woodward used the conjugate base of methanol (CH_3OH), $Na^+ \; CH_3O^-$. The carbanion formed, which is shown as the enolate (**ii**) rather than the carbanion resonance structure, could then act as a nucleophile, a characteristic we've seen so often in both Chapters 7 and 8. Woodward made certain that the focus of this carbanion was the electrophilic carbon of the carbonyl group of the ethyl ester

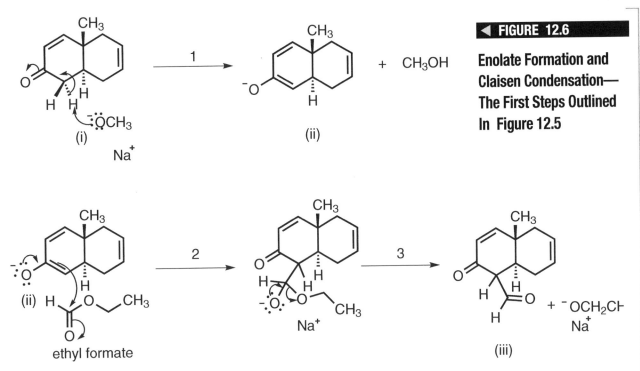

◀ FIGURE 12.6

Enolate Formation and Claisen Condensation— The First Steps Outlined In Figure 12.5

Claisen condensation

of formic acid, ethyl formate as shown in Figure 12.6, therefore conducting a Claisen condensation (**2**, **3**) (sections 7.10 and 8.3) leading to the molecule he had in mind (**iii**).

A single hydrogen atom remains on the carbon atom that was the site of the product of reactions **2** and **3** (**iii** Figure 12.6). Surrounded by two carbonyl groups, a Brønsted-Lowry acid-base reaction easily occurs as shown in Figure 12.7 (**1**) setting up another nucleophilic carbanion (two resonance structures shown). Woodward has now prepared the system for the Michael addition (section 7.8) he had in mind, reaction of this carbanion with ethyl vinyl ketone as shown in **Figure 12.7** step **2**, which was followed by addition of a proton, to yield the Michael adduct.

The Michael adduct shown in Figure 12.7 is now set up for reactions that will both occur on addition of the base that Woodward used, potassium hydroxide, KOH: a retro-Claisen to eliminate the aldehyde group; and an intramolecular Aldol condensation to form the third fused ring (**Figure 12.8**) . The mechanistic path with all the familiar features of these reactions of carbanions with carbonyl groups we have seen in Chapters 7 and 8 are shown in Figure 12.8.

The reverse of the Claisen condensation, called the retro-Claisen (section 7.10) is initiated in the usual manner by attack of the hydroxide anion to form the tetracoordinate intermediate (**1**) in Figure 12.8. The intermediate (**1**) collapses then to break the carbon-carbon bond to the ring carbon atom releasing what was the aldehyde group (shown as two resonance structures) so that a couple of proton transfers takes us to the product of the reverse Claisen, **2**.

The reactive system remains highly basic and although there are three carbon-hydrogen bonds accessible in **2** to act as proton sources in a Brønsted-Lowry acid-base reaction with hydroxide ion, only the one shown can lead to a productive result, which is attack by the resulting carbanion at the carbonyl group to form a six-membered ring (**3**), an intramolecular example of the Aldol condensation (section 8.4).

Here we find in structure **1** an example (Figure 12.8) of the different routes that can be taken by an alkoxide formed from a carbonyl group bonded to a leaving group (a rather complex one in this situation) and another example of an alkoxide formed from a ketone carbonyl (3). The alkoxide formed by attack at the carbonyl group in

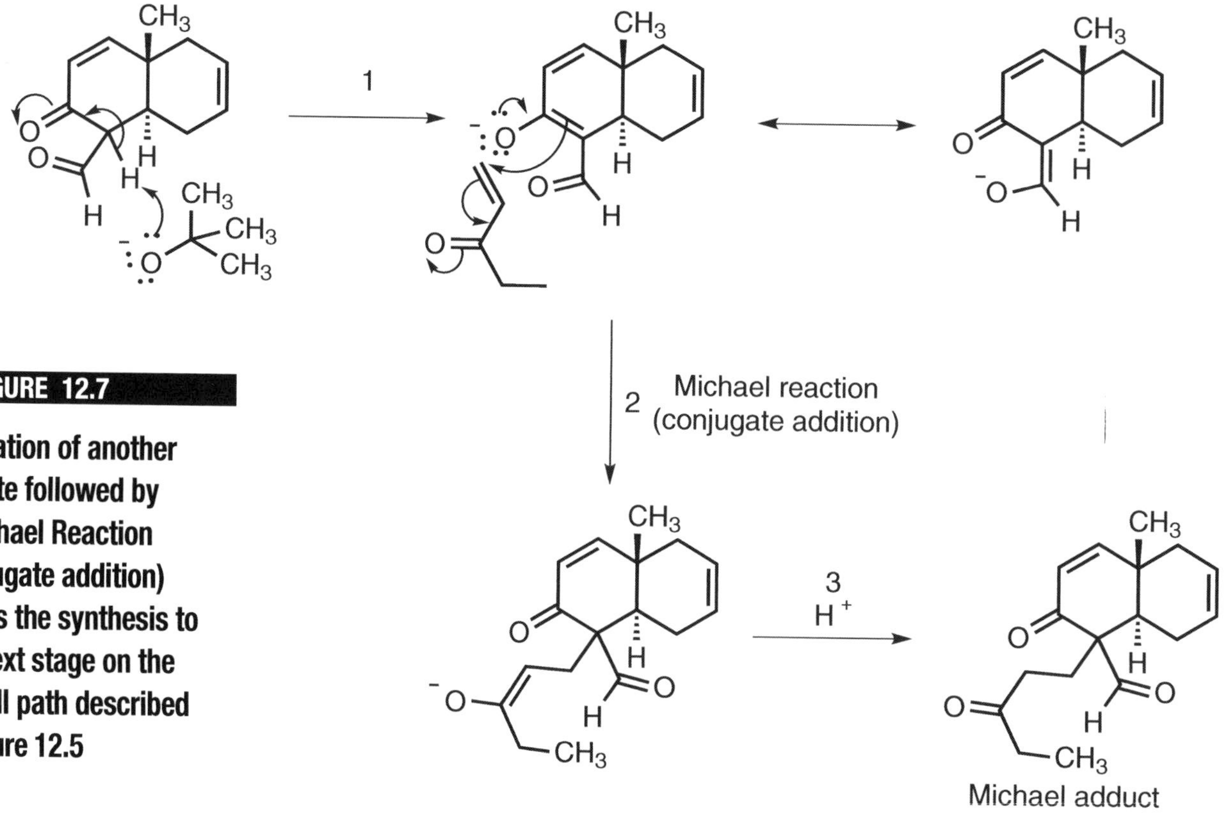

▶ FIGURE 12.7

Formation of another enolate followed by a Michael Reaction (conjugate addition) moves the synthesis to the next stage on the overall path described in figure 12.5

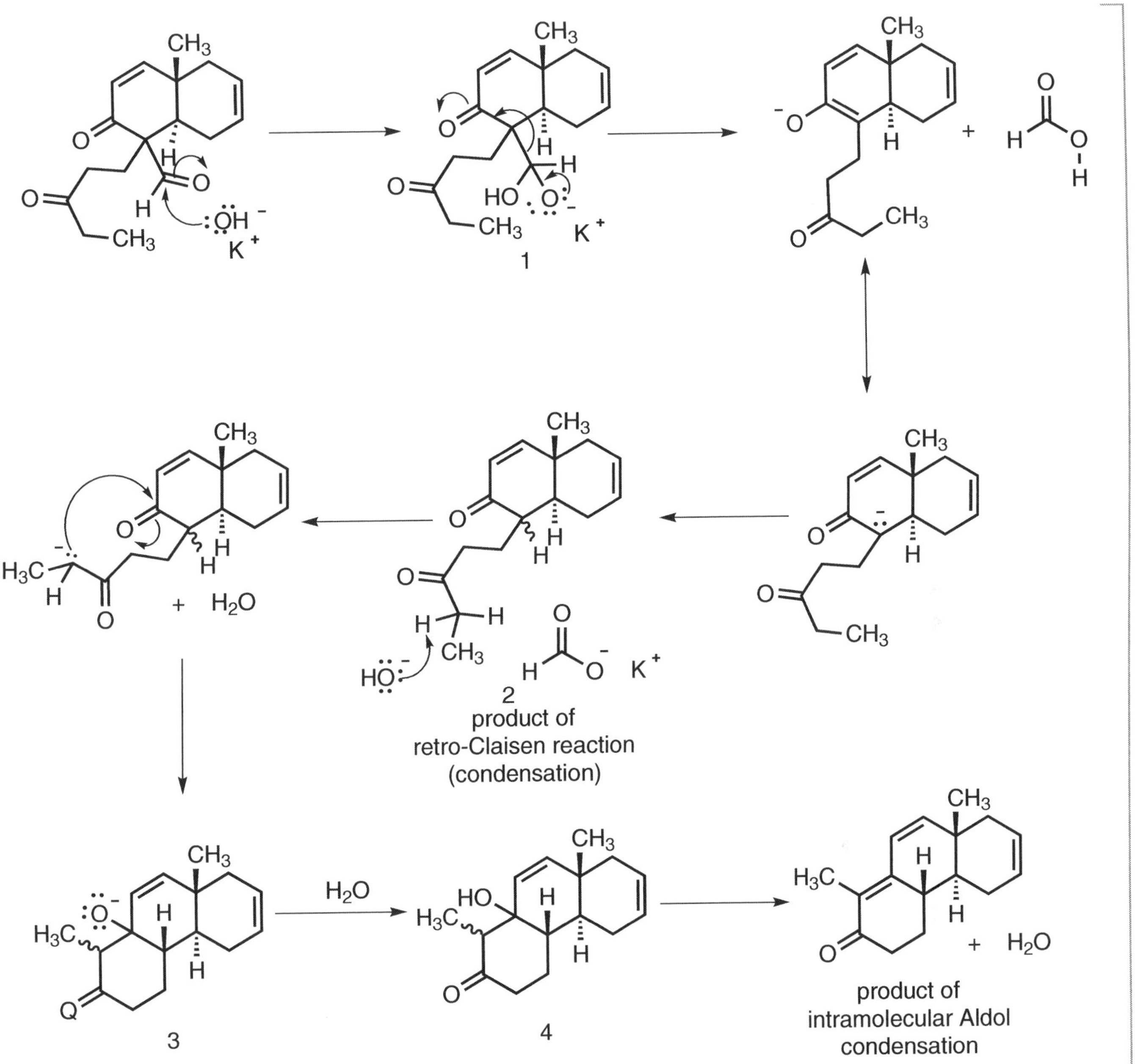

▲ FIGURE 12.8

A Retro-Claisen reaction followed by an intramolecular Aldol condensation summarizes the essential steps in completing the formation of the third fused ring described in figure 12.5.

1 can reform the carbonyl group by ejecting the resonance-stabilized carbanion. On the contrary, the alkoxide in **3**, formed by attack at a ketone carbonyl group does not have this option and therefore simply accepts its fate and picks up a proton forming a hydroxyl group, **4**. Loss of water follows to yield the product Woodward wanted to attain, the three-fused six-membered rings (Figure 12.8).

Note that the uncertain configuration, axial or equatorial of the methyl group in **3** and **4** (wiggly bond) did not matter because it was bound to a carbon atom in the ring that was destined to be bound to an sp²-hybridized carbon atom, as seen in the final intramolecular Aldol product (Figure 12.8).

PROBLEM 12.10
In Figure 12.6 which shows the detailed mechanism of the Claisen condensation using ethyl formate, step **3** takes place in the forward direction from the tetracoordinate intermediate. What does the possible competition of the reverse step reveal about the relative acidity of two different Brønsted-Lowry acids?

PROBLEM 12.11

Use structural three dimensional drawings to show all the stereoisomers produced (if any) as a result of each of the reactions in Figure 12.7.

PROBLEM 12.12

While the carbon-hydrogen bond in structure **2** in Figure 12.8 is important because it has a consequence in the stereochemistry of the three fused ring final product of this figure, the wiggly carbon to methyl group bond in structure **3**, also designating a mixture of configurations at this stereocenter, is not important because the carbon atom in the final structure in this figure becomes part of a double bond. Focusing on the mixture of stereoisomers in structure **2**, what conclusion might you draw about the fact that the closed ring leads to a single stereoisomer of the final product in Figure 12.8, although racemic.

PROBLEM 12.13

For the Aldol reaction in Figure 12.8 offer a reason why the loss of water has a more powerful driving force than for the loss of water from the model Aldol reaction in section 8.4.

12.5

Cholesterol: Setting the Stage for Adding the Fourth Fused Ring

Hugo Schiff

AS SEEN IN THE OVERVIEW OF WOODWARD'S synthesis of cholesterol in Figure 12.3, the critical steps in building the cholesterol skeleton are the formation of the fused ring system, the four fused rings in the final structure. The first step in the process depended on the Diels-Alder reaction, as we've seen in section 12.3. The next ring fusion step involved use of classic carbonyl-based reactions, Claisen, retro-Claisen and Aldol as discussed in section 12.4.

For incorporation of the fourth ring in the cholesterol skeleton and for adding the side chain on the five member ring in the final structure, Woodward again turned to chemistry based on carbanions and carbonyl groups but also to organometallic reactions, that is, reactions in which a metallic element is intimately involved. In particular Woodward made good use of the chemistry of carbon bound to magnesium, a reaction named after Victor Grignard, the Grignard reaction.

Let's jump ahead several steps from the structure of the final Aldol product in Figure 12.8 to an intermediate structure in which one of the double bonds was reduced and the other double bond functionalized as shown in structure (**1**) in **Figure 12.9**.

Look back at Figure 12.3 to the structure of cholesterol. Observe that the carbon atoms that attach the ring Woodward intends to fuse next, the fourth ring, have to be bonded to the sp^2 carbon atom bearing the methyl group and to the sp^2 carbonyl carbon atom in (**1**) (Figure 12.9). If a carbanion stabilized by the carbonyl group in (**1**) is to be involved in the necessary chemistry, a complex situation arises.

Consider the five portrayed hydrogen atoms in structure (**1**), which are on carbon atoms that are adjacent to sp^2 hybridized carbon atoms (three structurally different) and the carbanion structures (all resonance-stabilized) that could arise via a Brønsted-Lowry acid-base reaction of each of these three structurally different (A, B and C) proton sources (Figure 12.9).

The mixed good news for Woodward is that two of the three possibilities, B and C, have a resonance structure with a negative charge at the carbon bearing the methyl group, which is where he wants it. Happily, although this is a poor site for stabilization of negatively charged carbon it will therefore be most reactive once formed.

The bad news is that the most favorable site for proton abstraction, leading to resonance structures A, does not place a negative charge in a synthetically useful position. Organic chemists call such problems lack of regiospecificity, meaning that more than one site on a molecule can undergo reaction with the reagents used.

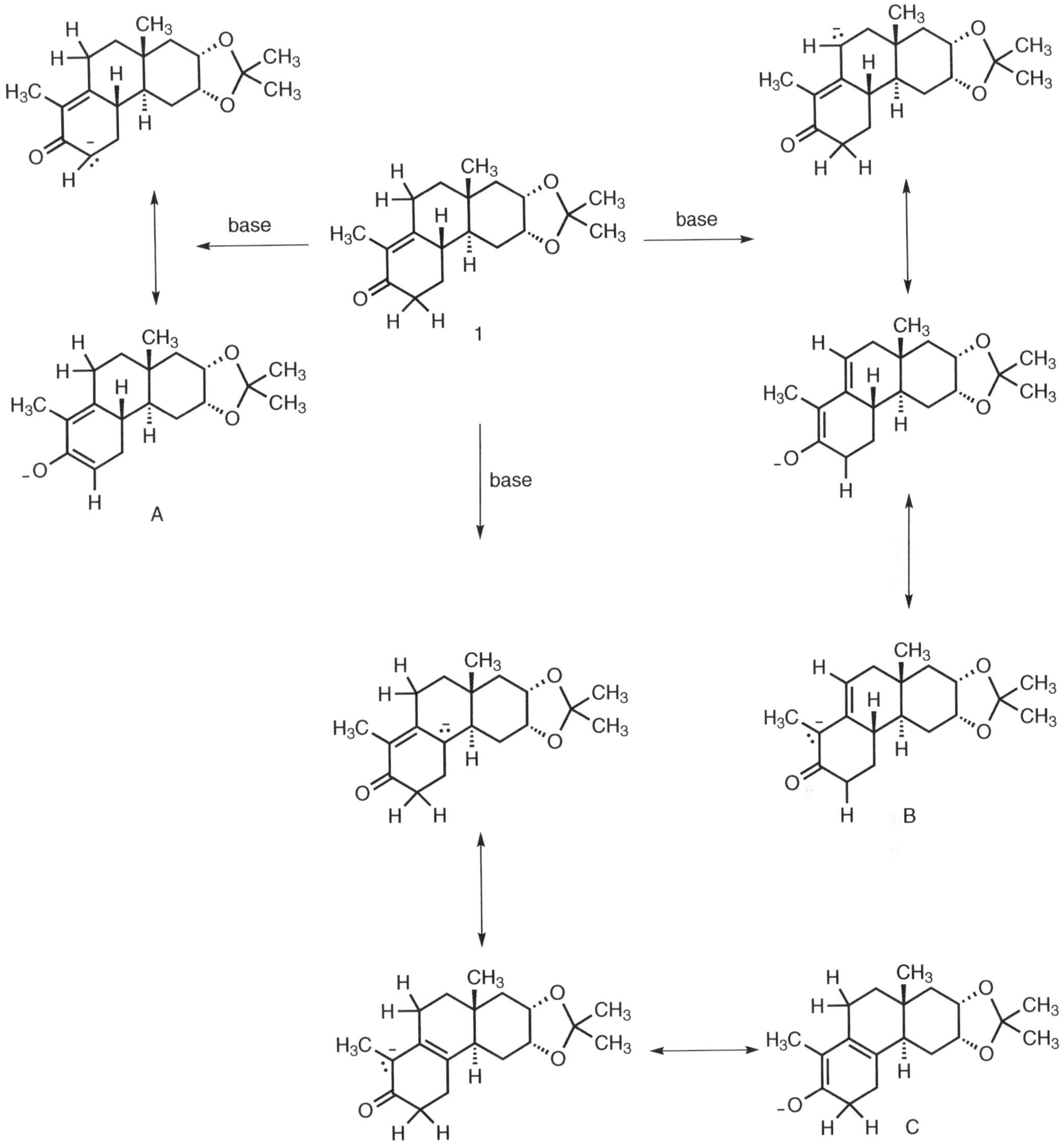

▲ FIGURE 12.9

The route to cholesterol has to involve discrimination among the three different enolate structures shown.

Woodward had somehow to protect the system from choice A (Figure 12.9). He did it not by somehow blocking proton abstraction leading to A but rather by making use of it. The scheme that Woodward used introduces us to a new class of functional groups based on the interaction of amines with carbonyl groups. For this strategy Woodward reached back to the work of a nineteenth century chemist, **Hugo Schiff**.

Schiff was a German chemist who studied for his doctoral degree under Friedrich Wöhler. In section 1.8 when the idea of structural isomerism was introduced we learned of Wöhler's nailing the coffin shut on the idea that a vital force was necessary to synthesize a molecule found in life with his synthesis of urea from a non-life connected source. That was in 1828 when Wöhler was 28 years old. It was 30 years later that Schiff, who was not born until 1834 took his degree with the aged famous Wöhler at the University of Göttingen. But Schiff had to leave Germany almost immediately after his degree. He

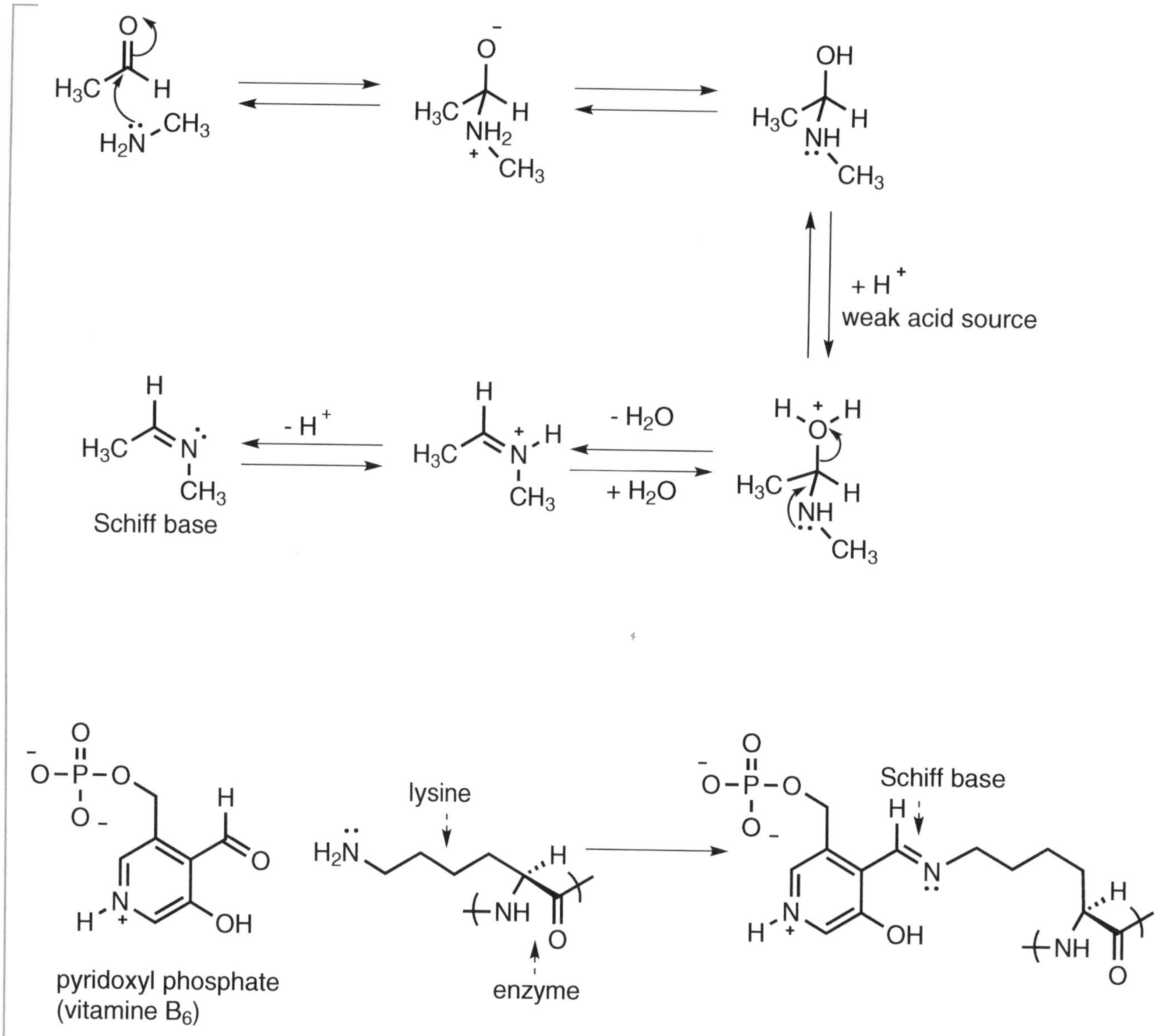

▲ FIGURE 12.10

Acetaldehyde and methyl amine and also vitamin B6 (pyridoxyl phosphate) demonstrate the formation of Schiff bases, imines.

was a trouble maker in the political sphere picking up the ideas of Karl Marx and the Communist Manifesto, which was published in 1848. Schiff did not, however, give up his interest in chemistry, eventually taking his considerable scientific talents to Switzerland and then to Italy where he remained as a Professor of Chemistry in Florence.

Schiff was an important influence on the development of chemistry in Italy, co-founding the Gazzetta Chimica Italiana, which for many years was the most important chemistry journal in Italy. He also maintained his interest in socialism, cofounding Avanti!, a newspaper that played a significant role in Italian politics over the years, and to this day. There is an institute named after Schiff at the University of Florence dedicated to the history of chemistry, a subject of no small interest to this text. Schiff was clearly not a man of narrow interests although our narrow focus now is on a functional group, the eneamine, which is related to one of Schiff's accomplishment, termed the Schiff base.

Schiff bases, as shown in **Figure 12.10**, are formed between aldehydes or ketones and primary amines, RNH_2 so that H_2O is lost forming a double bond between the carbonyl carbon atom and the nitrogen atom of the amine. The reaction is catalyzed by weak acid and takes place readily. The mechanism of the formation of Schiff bases for a typical example, between acetaldehyde and methyl amine, shown in Figure 12.10, follows the expected path, the classic attack of a nucleophile, nitrogen based in this situation, on the electrophilic carbonyl carbon atom. Addition of a proton from the usual weak acid catalysts used, followed by loss of water and loss of a proton produces the Schiff base (Figure 12.10). As shown in the figure the reactions are all easily reversible.

Schiff bases play an important role in biological processes. For one example, vitamin B6, necessary for the metabolism of amino acids, functions via initially forming a Schiff base between the NH_2 group of the side chain of a lysine along a protein chain of an aminotransferase enzyme and the aldehyde function of pyridoxyl phosphate (Figure 12.10).

There's a variation on the formation of a Schiff base, the eneamine, which is formed under the same conditions, but where the amine is not primary with two hydrogen atoms on the nitrogen but rather secondary, with only one hydrogen atom on the nitrogen, R_2NH. This substitution blocks the formation of the Schiff base but opens up the possibility of forming the related eneamine functional group (**Figure 12.11**). Eneamines have proven a valuable addition to the tools available for organic synthesis, as pioneered by **Gilbert Stork**, now emeritus professor of chemistry at Columbia University. Woodward, however, made use of an eneamine in a different manner than the Stork-eneamine reaction, to solve the problem outlined in Figure 12.9. And its use by Woodward introduces us to the concept of the protecting group, a very useful friend to synthetic problems, as we'll see.

Woodward's approach was to form the favored carbanion (A, Figure 12.9) followed by a Claisen condensation, analogously to the reaction in Figure 12.6. The resulting aldehyde was then reacted with phenyl methyl amine under mild acid catalysis

■ **Gilbert Stork**

(**Figure 12.12**). The resulting eneamine, which is formed by the identical mechanism as that exhibited in Figure 12.11, then restricts further reactivity at this site under the conditions necessary to form the enolates B and C in Figure 12.9. The eneamine, in this situation, is a protecting group.

Enolates B and C (Figure 12.9) are both represented by three resonance structures in which one resonance structure of each (B and C) places the negative charge on the carbon bearing the methyl group. In that way, it did not matter to Woodward if B or C enolate were formed. It would have been a great deal of trouble, best using deuterium labeling, a technique not well developed at that time, to tell which C-H bond was the

▲ **FIGURE 12.11**

Acetaldehyde and dimethyl amine illustrate the formation of an eneamine.

Woodward uses a Claisen Reaction followed by reaction with an amine to produce an eneamine protecting group at one of the enolate sites noted in figure 12.9.

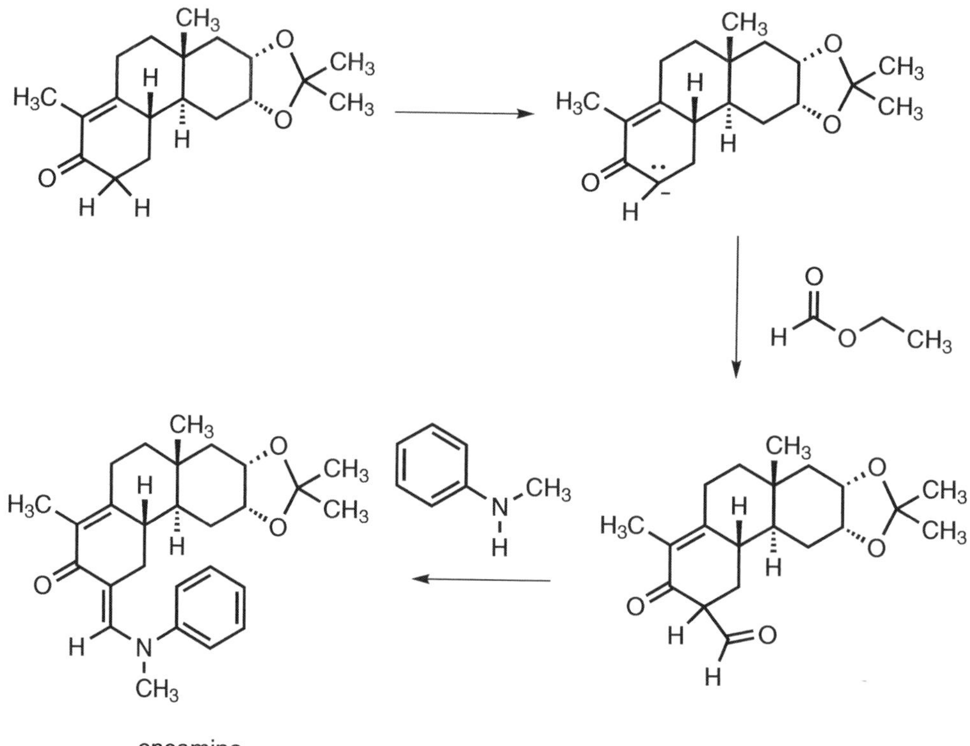

eneamine

source of the proton lost. At any rate, it was hardly worth the trouble.

Woodward knew from his experience with enolates, and the literature, that the most reactive site for the negative charge, the most nucleophilic site, would very likely be the one he wanted no matter if the enolate formed was B or C (Figure 12.9). He was now set up for the next step, a Michael addition to the desired enolate structure, with the site A protected (Figure 12.12) against reaction.

The Michael addition, introduced in section 7.8 as a variation of the general reaction class called conjugate addition, and the subsequent steps, are shown in **Figure 12.13**. These steps were followed by the use of stronger base, which then removed the eneamine protecting group releasing methyl phenyl amine and the conjugate base of formic acid as shown in Figure 12.13. We won't go into the mechanism of this "deprotection" step, that is, a reaction that removes a protecting group but give it a try with the first step hydroxide ion (OH⁻) from a strongly basic solution adding as a nucleophile to the carbon atom of the eneamine.

The reaction of HO⁻ with the eneamine protected ketone, the first step in Figure 12.13, produces the desired enolate with the most nucleophilic site for the negative charge as shown on the carbon atom Woodward needed, the carbon bearing the methyl group rather than the carbon atom bearing the hydrogen atom in the other resonance structure. Again here Woodward made use of a Michael addition (section 7.8) as he did in forming the second fused ring in Figure 12.7. In the latter situation vinyl ethyl ketone supplied the electrophilic double bond. As in all Michael additions, the necessary double bond has to allow stabilization of a negative charge after reacting with the attacking nucleophile. In vinyl ethyl ketone the stabilization arises from the adjacent carbonyl group shown as forming the enolate in Figure 12.7.

The reactant in Figure 12.13 is another molecule that can undergo Michael addition, acrylonitrile. We first came across the nitrile functional group (-C≡N) in the synthesis of hexamethylene diamine (sections 10.2, 10.3 and 10.4). There we learned about HCN, a moderate acid (pK$_a$ about 9), which means that the nitrile anion, ⁻:C≡N, has the necessary stability to be the conjugate base of this acid. The chemistry in sections just noted also demonstrates that ⁻:C≡N is a good nucleophile.

To these characteristics of the nitrile group now has to be added another. A

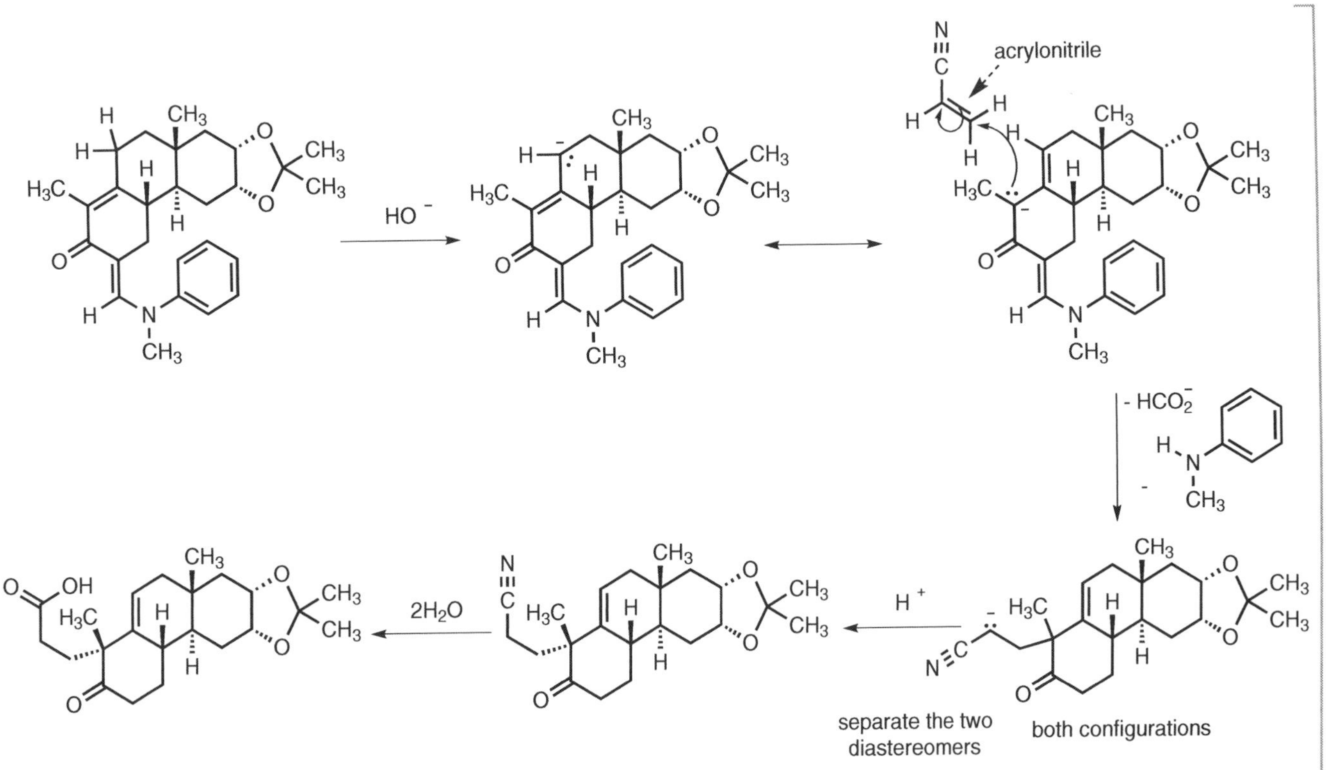

An enolate at the site not protected (Figure 12.12) undergoes a Michael reaction leading to the structure on the way to form the fourth fused ring.

carbanion adjacent to a nitrile group is stabilized analogously to the stability of a carbanion adjacent to a carbonyl group. This stabilization is what we are seeing in the structure produced on reaction of the enolate with acrylonitirle in Figure 12.13 and is the reason that acrylonitrile, CH_2=CH-C≡N, is an excellent participant in Michael additions, which Woodward was well aware of.

Unfortunately, the Michael adduct does not produce an overwhelming amount of a single diastereomer (Figure 12.13), but rather is a mixture. Diastereomers can be separated in many ways considering their different properties, as was done in this situation. One of the diastereomers, shown in Figure 12.13, was the molecule necessary to proceed along the synthetic path to cholesterol.

In the discussion on the synthesis of hexamethylene diamine (Chapter 10 section 10.3) note above, we saw that the nitrile functional group could be reduced, via H_2 and a catalyst, or via lithium aluminum hydride, to the amine, that is -C≡N to $-CH_2-NH_2$. The identical reducing conditions can convert a carboxylic acid to a primary alcohol, that is $-CO_2H$ to $-CH_2OH$. The nitrile and carboxylic acid functional groupsof carbon are both in the highest oxidation state accessible to carbon, other than CO_2. This circumstance means that the conversion between nitrile and carboxylic acid can take place by addition of water or removal of water catalyzed by acid or base. The former is shown for the nitrile group in the product of the Michael reaction producing the carboxylic acid, as shown in the final product in Figure 12.13.

We've been introduced to the ester functional group (section 7.3) and seen that activation of the carboxylic acid is the step normally taken to form the ester (section 7.6). One form of activation is conversion of the carboxylic acid to the anhydride (Figure 11.14). The anhydride of the carboxylic acid (Figure 12.13) was the functional group Woodward wanted, which was accomplished by reaction of the final product of Figure 12.13 with acetic anhydride (reaction 1, **Figure 12.14**) yielding what is called a mixed anhydride (**i**).

The ketone carbonyl group of structure (**i**), as we've seen for ketones (section 8.2), exists as an equilibrium mixture with the enol (**ii**, Figure 12.14). Although the enol in **ii** is formed to a very small extent compared to the tautomeric ketone, **i**, the OH group of the enol is highly reactive with the mixed anhydride to form the cyclic ester. Cyclic esters are called

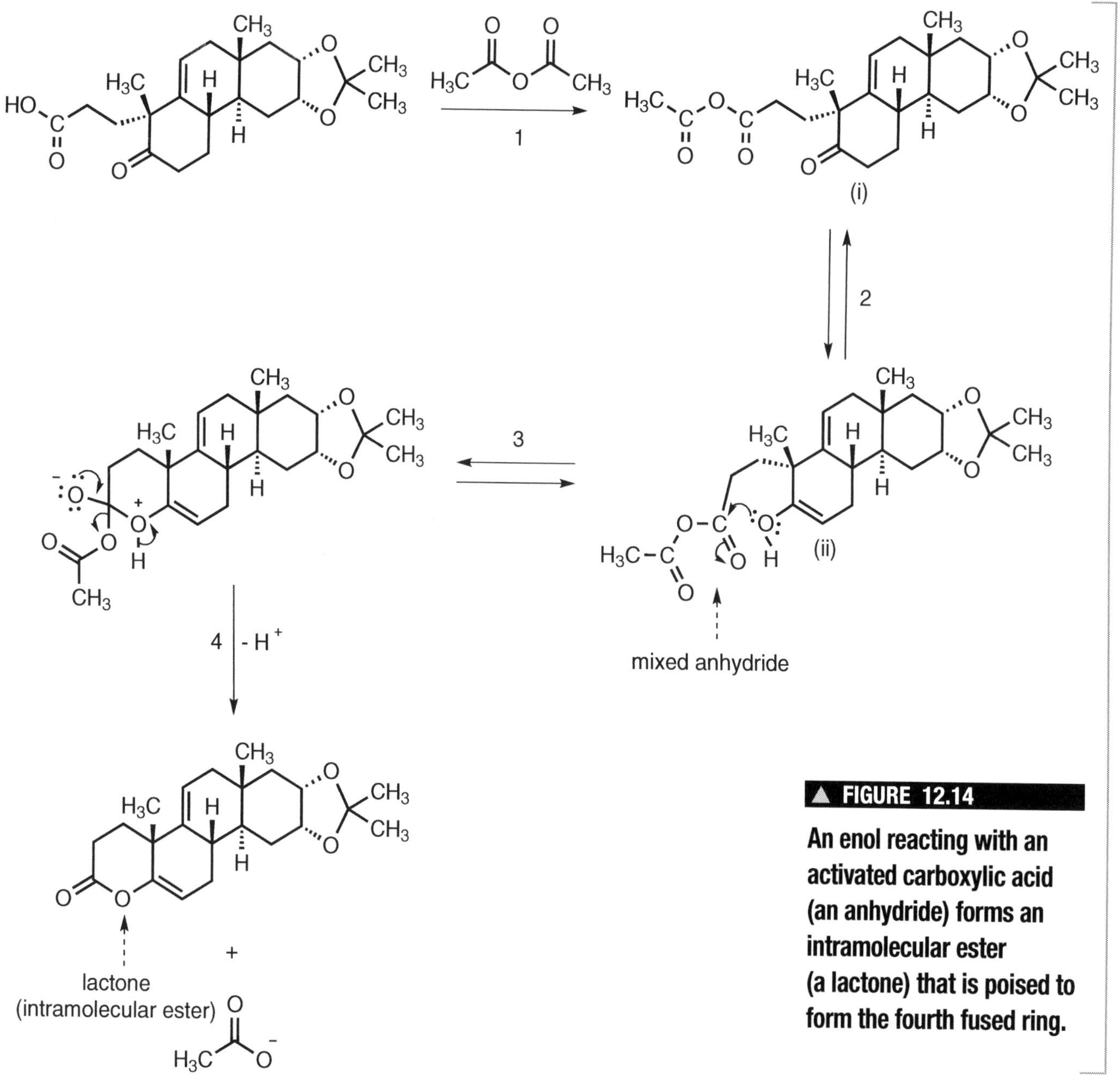

▲ FIGURE 12.14

An enol reacting with an activated carboxylic acid (an anhydride) forms an intramolecular ester (a lactone) that is poised to form the fourth fused ring.

lactones by organic chemists. As the lactone forms and removes an enol from the enol/keto equilibrium mixture, more keto form is converted to enol to compensate for the loss. In this way, the entire sample of **i** is converted to the lactone (Figure 12.14).

The results in Figures 12.13 and 12.14 were important breakthroughs for Woodward. He was now certain that cholesterol could be attained by making good use of a reagent named for a famous French chemist, **Victor Grignard**, who first disclosed the reagent in 1900 and won the Nobel Prize for his discovery in 1912.

PROBLEM 12.14

Redraw structure **1** in Figure 12.9 and circle each hydrogen atom lost to form the enolate structures A, B or C.

PROBLEM 12.15

What structural feature of **1** in Figure 12.9 distinguishes how proton loss to form A differs from the other possibilities?

PROBLEM 12.16

Offer a reason for the nucleophilic character of the CH$_2$ group in the eneamine in Figure 12.11.

PROBLEM 12.17

Why would water have to be removed from the reacting system in the syntheses of both Schiff bases and eneamines (Figure 12.10)? Look back to the synthesis of the polyesters in Chapter 9.

PROBLEM 12.18

The formation of the eneamine in Figure 12.12 does not occur from the carbonyl group on the ring but rather from the aldehyde carbonyl group. Offer an explanation for this selection.

PROBLEM 12.19

In Figure 12.13, two resonance structures are shown for the enolate. If I tell you that the negative charge at the carbon bearing the methyl group is the less stable site how does that help you to understand why that site should be most reactive? Do considerations of energy of activation and reaction coordinate diagrams help?

PROBLEM 12.20

The step in Figure 12.13 showing the Michael addition of acrylonitrile is complicated in also showing the loss of the eneamine protecting group. Offer a mechanism for the loss of the protecting group with a first step that could have hydroxide attacking the double bonded carbon atom bearing the methyl phenyl amine group. That would bring you back to the enol of the aldehyde that was the precursor to the eneamine. Now keep going. It all has to do with stability of the carbanion adjacent to the carbonyl group.

PROBLEM 12.21

Suggest a mechanism using aqueous base, OH⁻, to convert the nitrile group from the Michael addition in Figure 12.13 to the carboxylic acid group

PROBLEM 12.22

What does Figure 12.14 have to do with Le Chatelier's principle (section 10.5) ?

PROBLEM 12.23

Could Woodward have used the acid chloride instead of the mixed anhydride (refer to section 11.9) to accomplish the same objective in Figure 12.14? Would the objective have been accomplished without activating the carboxylic acid group? If the latter had been tried why would it have been necessary to remove water during the reaction?

» **Victor Grignard**

12.6

Woodward uses a Grignard reagent to form the fourth fused ring.

THE FINAL PRODUCT FORMED in Figure 12.14, the lactone, was a perfect set up for a sequence of reactions likely to yield the desired fourth fused ring. The first step involved reaction of the lactone (Figure 12.14) with the Grignard reagent derived from the reaction of methyl bromide with magnesium turnings (metallic magnesium).

Making Grignard reagents, as for many organic chemists, was for me, as a graduate student, one of my favorite reactions and considering how useful this combination of the metallic with the organic is, I had many opportunities to enjoy myself in the lab. I well remember now, so many years later, taking great pains to make certain the diethyl ether, (H$_3$C-CH$_2$)$_2$O, was free of water by using freshly cut sodium wire. When no bubbles of hydrogen gas evolved on adding the next piece of sodium, this proved the anhydrous state had been reached. There was paranoia about water. One prayed for a day with low humidity in those days before all labs were air conditioned.

After setting up the apparatus, simply a flask with a magnetic stirrer and a water condenser, with an equilibrating funnel topped with an inlet so that the whole system could be enveloped in a nitrogen atmosphere, and the proper amount of magnesium turnings were put in, one brought out the Bunsen burner. A mild flame, yellow not blue, was played over the entire glass apparatus and one watched with pleasure as the fog of moisture lifted from the surface of the glass. You didn't want to heat it too much. You'd get quite a show, and maybe more than you counted on, from the magnesium flare. (I won't tell you the story from long ago of what we graduate students sometimes did in the late of the night when Prof. was not around). When the system was cool to the touch, the proper amount of diethyl ether was added with the dissolved organic halide, which was to form the Grignard reagent.

One then hoped and stood helplessly. Would it go? If a white powder started to form, curses went through the lab and people who were not your friends came with unwanted advice you did not need. You well knew what the problem was. It was almost always the same – the system was wet. But with proper skill and care, and some tricks, such as, for one example, adding a crystal of iodine, a few bubbles started to rise from the magnesium turnings as the exothermic reaction between the organic halide and the metal caused the ether to boil. And then joy. The boiling became rapid. The ether visually condensed in the water condenser and a moderate flow of ether returned to the flask. And, best of all the ether in the flask would turn a dark color. You had it! In the experiment we are considering now in Woodward's laboratory, you had H_3CMgBr. I feel sorry about the loss of this experience for chemists now who are able to buy methyl magnesium bromide in ether in anhydrous sealed bottles.

What was this stuff and what did it do? Well, the answer is complicated and depends on the nature of R, but in general one could consider the Grignard reagent as formed of R^- and $MgBr^+$, a powerful reactant for carbonyl groups including esters and, as we'll see, particularly useful for the lactone (Figure 12.14) that Woodward confronted on his path to cholesterol. Some common reactions of Grignard reagents, including the reaction with esters (or lactones) are shown in **Figure 12.15 (2)**. There are many other uses of Grignard reagents but let's limit ourselves here to those that show the essential nucleophilic character of the R group as bonded to magnesium. Much research has shown the intermediate character of the bond between R and magnesium, which depends on many factors, not least of which is the nature of R and the halide used. The nucleophilic character is certainly seen in the examples involving carbonyl groups (**1** and **2**), including carbon dioxide (**3**) (Figure 12.15), and is also demonstrated by the reaction of Grignard reagents, made from a variety of halides, with ethylene oxide (**4**) as also shown in Figure 12.15.

In the particular example of the ester, **2**, in Figure 12.15, the result is determined by the ratio of the number of moles of Grignard reagent to the moles of ester. If two equivalents of Grignard reagent are used, then the result is a tertiary alcohol because the intermediate ketone formed, as shown in the example, will react again. However, if the reaction is conducted in such a manner to terminate the process after one mole of Grignard reagent has reacted, then the result will be the ketone. This path is precisely the one Woodward took as shown in **Figure 12.16**.

The first step, reaction 1, in Figure 12.16 is the familiar formation of the tetracoordinate intermediate following nucleophilic attack on a carbonyl group. In the second step (2), the tetracoordinate intermediate collapses to the trigonal state because there is a leaving group (sections 7.4-7.6), an unusual leaving group but one that can do the job as shown. Hydrolysis with a weak acid (reaction 3), usually ammonium chloride and water, yields the molecule Woodward wanted, the diketone (Figure 12.16).

Now Woodward was ready to forge ahead. Treatment with potassium hydroxide (KOH) of the diketone can yield three different enolates, but only the carbanion formed by proton loss from the methyl group (reaction **4**) can form the low energy six-membered ring (reaction **5**) and so the entire reaction path goes in that direction as shown in Figure 12.16.

The ring-closing reaction, just as in formation of the third fused ring (Figure 12.8), is an Aldol condensation (section 8.4). As commonly occurring in Aldol condensations, the resulting hydroxyl group is lost as water in the direction to form the most stable double

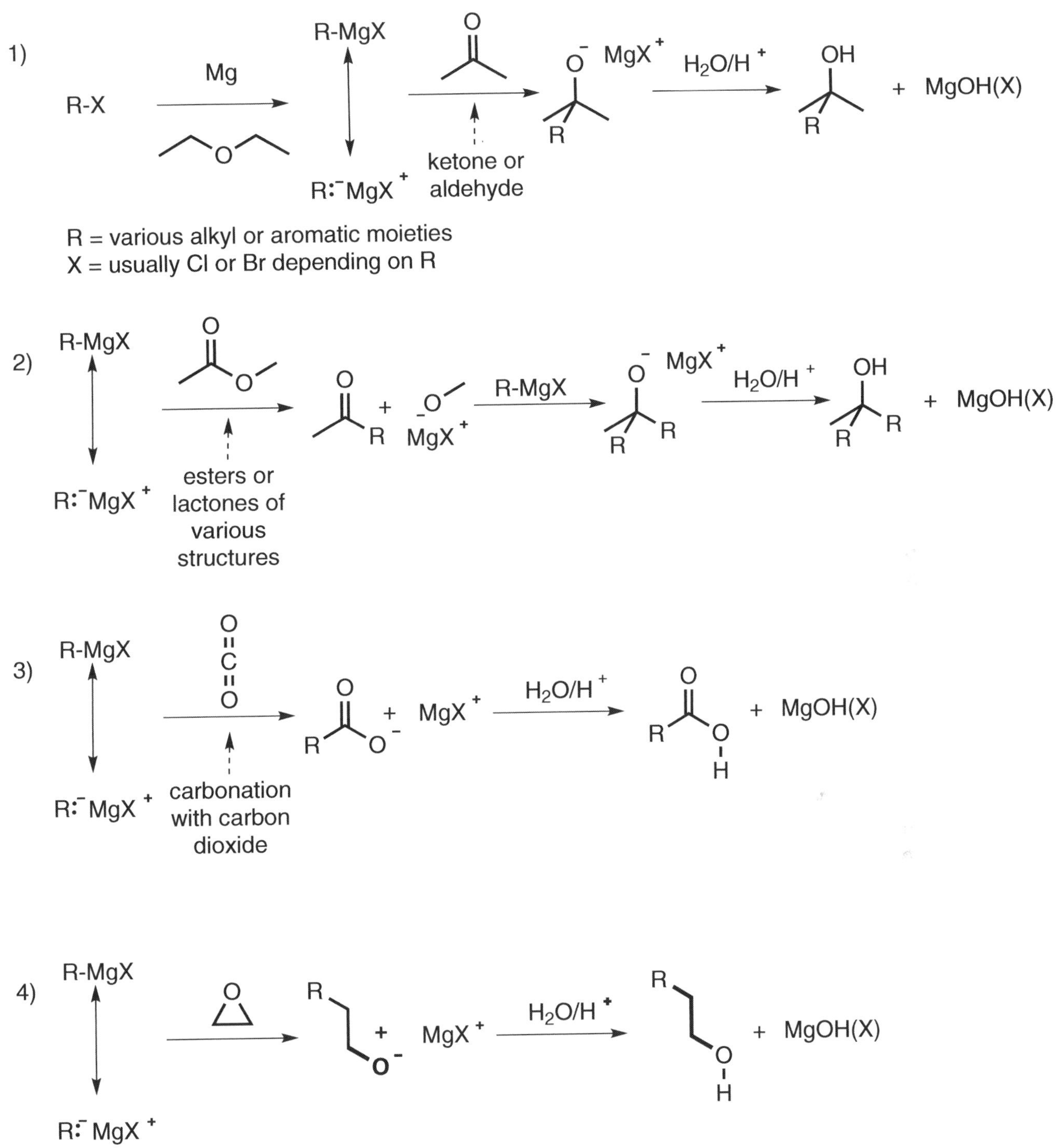

1)

R = various alkyl or aromatic moieties
X = usually Cl or Br depending on R

2)

3)

4)

bond, the double bond conjugated with the remaining carbonyl group. Beautiful—the four fused rings that will form the skeleton of cholesterol were in place.

Let's not look in detail at every step between the product formed by the series of reactions in Figure 12.16 and cholesterol. But if you feel you might enjoy delving more deeply, get a copy of Fleming's book mentioned in section 12.1. There is, however, one step in the remaining path to cholesterol that again demonstrates the importance of the Aldol condensation, a reaction that Woodward found so useful.

To introduce this Aldol condensation, we first have finally to address a feature of the synthetic path that you might have been wondering about. What is that six-membered ring containing two oxygen atoms that we have been carrying along since Figure 12.9? To begin to answer this question, we best return to the chemistry of the sugars in Chapter 3, which we'll up in the next section.

▲ FIGURE 12.15

Four fundamental reaction classes define the chemical characteristics of the Grignard reaction.

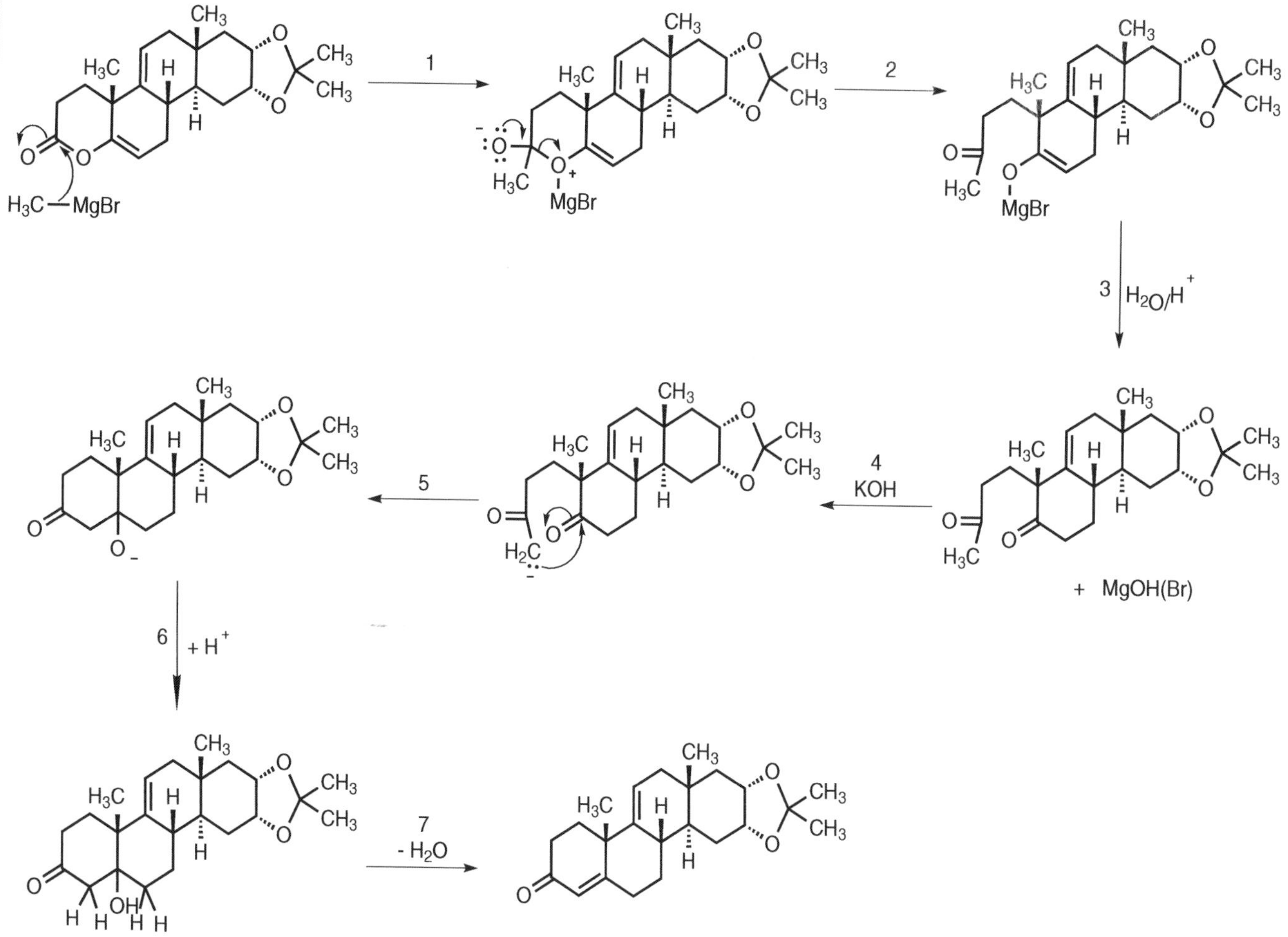

▲ FIGURE 12.16

Grignard reaction on the lactone formed in figure 12.14 followed by an Aldol condensation leads to the fourth fused ring necessary for cholesterol synthesis.

PROBLEM 12.24
Based on the story of my experience in the laboratory (section 12.6) with making Grignard reagents, what is meant by the advice about use of Grignard reagents that no active hydrogen source can be present? Is water an active hydrogen source and if water and methyl magnesium bromide were to come into contact what would be produced?

PROBLEM 12.25
For each of the reactions shown in Figure 12.15 show the movement of electrons in the mechanism using the curved arrow formalism.

PROBLEM 12.26
What is the high energy bond produced in reaction 1 in Figure 12.16 and how is this bond related to the properties of the conjugate base of an acid?

PROBLEM 12.27
Instead of reaction **5** in Figure 12.16, another Aldol reaction could have taken place but did not because of a characteristic of ring closing reactions to be covered in some detail later in this chapter but already alluded to in an earlier chapter about vulcanization. Outline the unwanted reaction that might have taken place.

PROBLEM 12.28
What structural characteristic drives loss of water from all Aldol condensations?

PROBLEM 12.29
Compare the structure of the final product of Figure 12.16 and cholesterol and identify the structural changes that must still be made.

HERE WAS A CLASS OF FUNCTIONAL GROUP that appeared in Chapters 1 and 3. In every structure of the closed form of glucose and the other diastereomers appearing in many figures in Chapters 1 and 3, and in the structures of the disaccharides, and also in cellulose and starch, there can be found at least one carbon atom in each structure with two oxygen atoms bonded to it. In the five membered ring with the two oxygen atoms shown many times in the intermediates on the path to cholesterol in the previous sections in this chapter there is a carbon atom with two oxygen atoms bonded to it. This structural characteristic identifies a class of functional groups designated as an acetal or a ketal.

Figure 12.17 shows the general structures of the various kinds of acetals and ketals (1 and 2) in general terms using R and R' and also for the specific hemiacetals and acetals formed by glucose (3) and its disaccharide, α-maltose (4) and finally (5) for the ketal Woodward used.

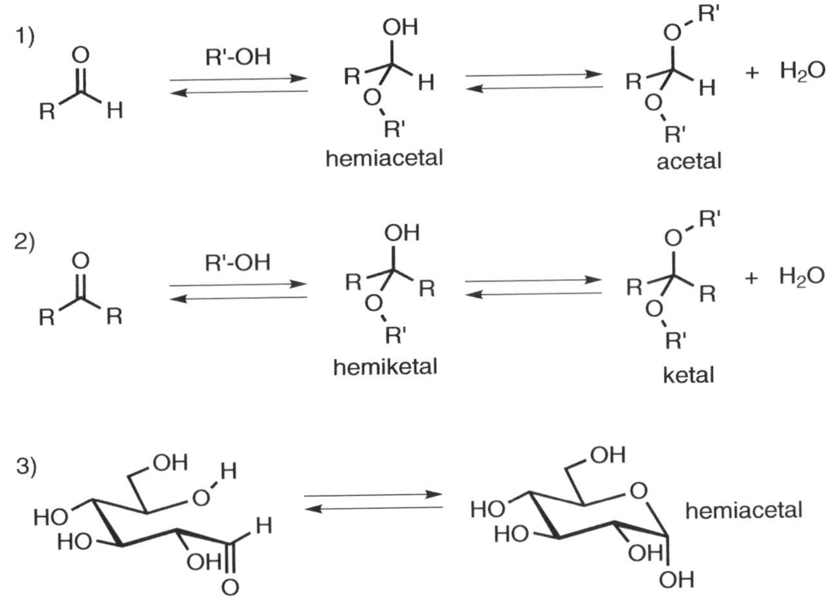

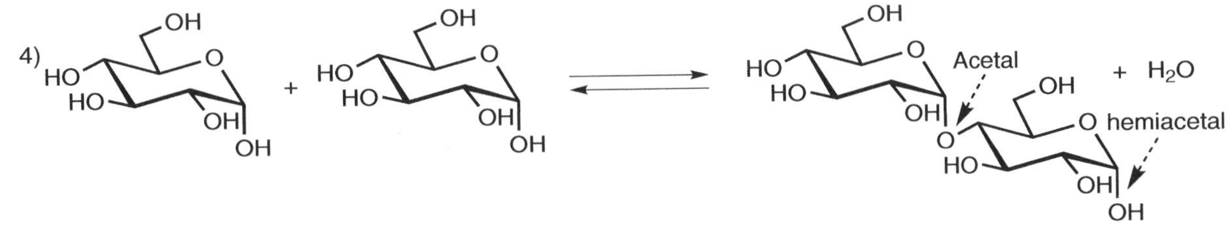

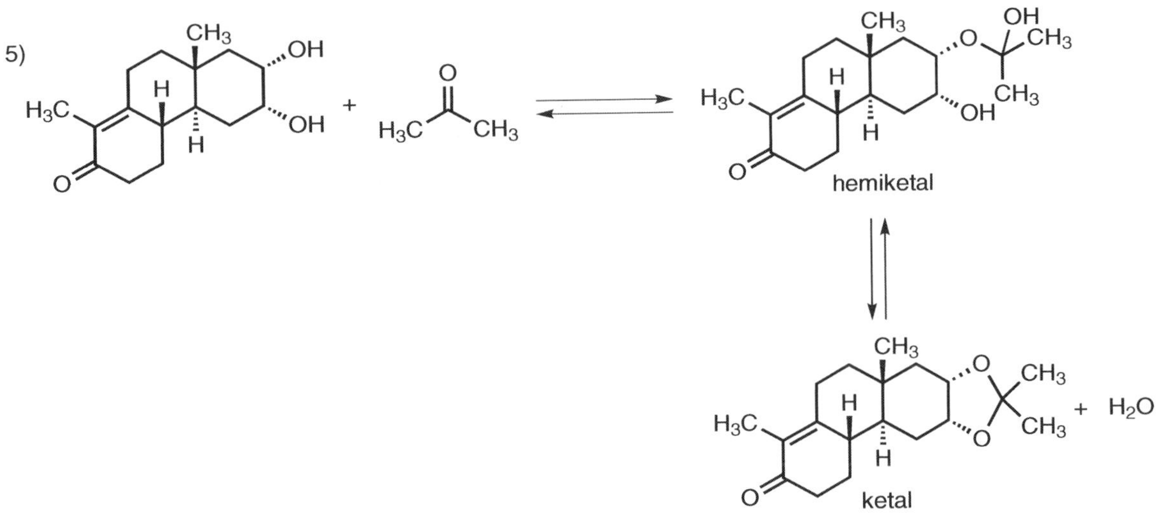

12.7

A diversion from the synthesis of cholesterol leads to understanding how Woodward used a ketal to protect a double bond.

◀ **FIGURE 12.17**

A series of chemical reactions varying from simple aldehydes and ketones to sugars and to a Woodward intermediate on the path to cholesterol demonstrate the equilibrating nature of hemiacetals and acetals and also hemiketals and ketals.

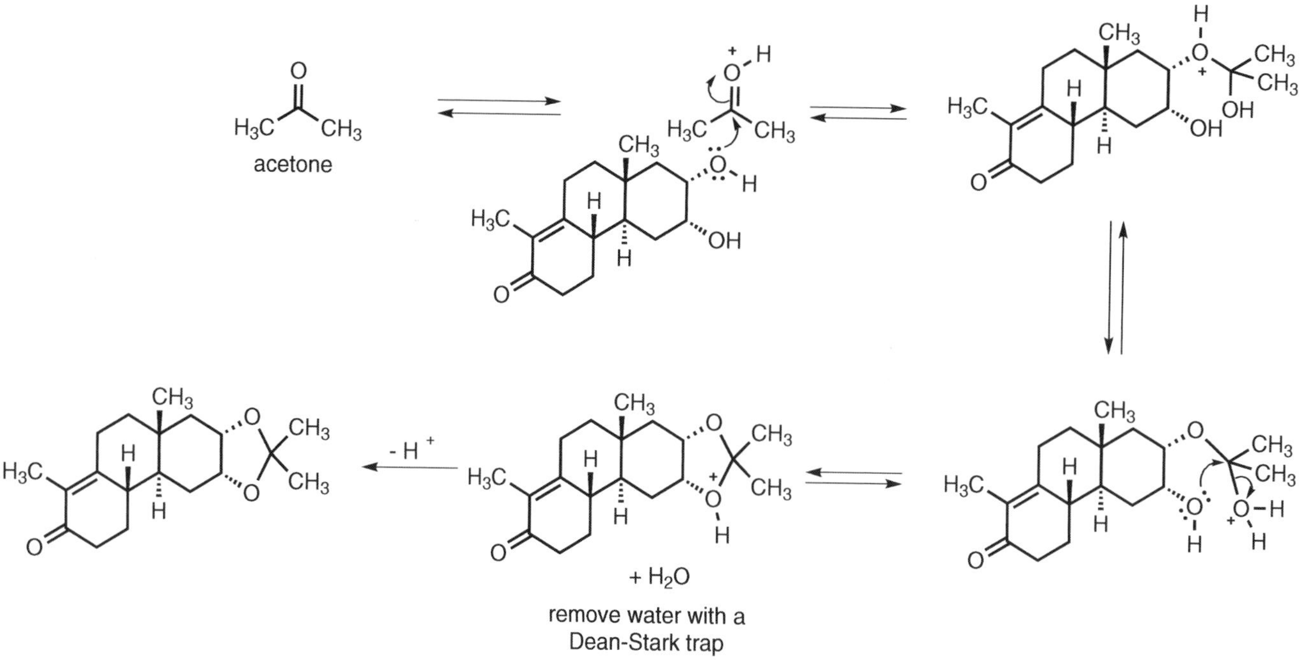

acetone

+ H₂O
remove water with a
Dean-Stark trap

▲ **FIGURE 12.18**

The lessons of figure 12.17 are applied to Woodward's intermediate to show the equilibrating structures possible.

The syntheses of acetals and ketals are catalyzed by acid. However, these functional groups exist in a rapid equilibrium with their precursor alcohols and carbonyl compounds so that the isolation of the ketal in reasonable quantities requires shifting the equilibrium continuously to the point where water is produced and then removing the water (**Figure 12.18**). In the laboratory, this is accomplished by using a device called a Dean-Stark trap, which is shown on the right. This apparatus was created in 1920 to measure the water content of petroleum. I well remember the pleasure of putting this clever apparatus together for the synthesis of ketals, a laboratory job that was certainly undertaken by Woodward's students.

For the formation of the ketal Woodward used (Figure 12.18) the flask would have been charged with the solvent benzene containing the proper proportions of acetone and the diol with a catalytic amount of acid. The mixture would have been brought to a boil in the pot (2). As the ketal formed, the water produced would form an azeotrope with the solvent benzene (a precise mixture of water and benzene that boiled at a constant temperature as if it were a pure substance) with this vapor finding its way to the water cooled (6,7) condenser (5) where the water and benzene separated. The liquid mixture would run down into the burette (8). Benzene is lighter than water so that it would run over back in (3) and be returned to the pot (2). The water could be removed using the stopcock (9). The amount of water produced (10) measures the extent of formation of the ketal. Beautiful!

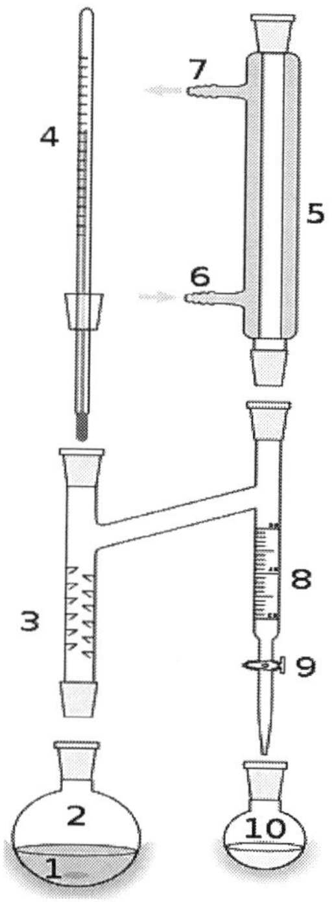

DEAN-STARK TRAP, an apparatus created in 1920 to measure the water content of petroleum

PROBLEM 12.30
What role does the acid catalyst play in the mechanism for formation of acetals and ketals and why is a low concentration of a mild rather than a strong acid used?

PROBLEM 12.31
The reversibility of formation of hemiacetals and hemiketals but not acetals and ketals in water solution in the absence of an acid catalyst is demonstrated by the different properties of glucose versus cellulose or starch. Offer an explanation for the reversibility, under these conditions, of one class of functional group and not the other.

PROBLEM 12.32
How does mutarotation of glucose relate to the chemistry of the mechanism for formation of hemiacetals, as shown in Figure 12.17?

12.8

The End Game

I N GENERAL, ADDING EXCESS WATER and a catalytic amount of acid to a ketal reforms the carbonyl compound and the alcohols from which the ketal was formed. Subjecting the final product formed in Figure 12.16 to this procedure releases acetone and produces the diol shown in **Figure 12.19**. In this manner, the diol from which the ketal was originally synthesized (Figure 12.18) is converted to the diol shown in Figure 12.19. There are several steps between the diol produced in reaction **1** (Figure 12.19) and cholesterol, which are not complicated, but for our purposes let's look carefully at only two of them, which are, again, uses of the Aldol and Grignard reactions.

When adjacent carbon atoms each carry a hydroxyl group, a vicinal arrangement (in contrast to a geminal arrangement in which the two identical groups reside on the same carbon atom) an easy oxidation can be carried out that breaks the bond connecting the two carbon atoms. An excellent reagent for this process is periodic

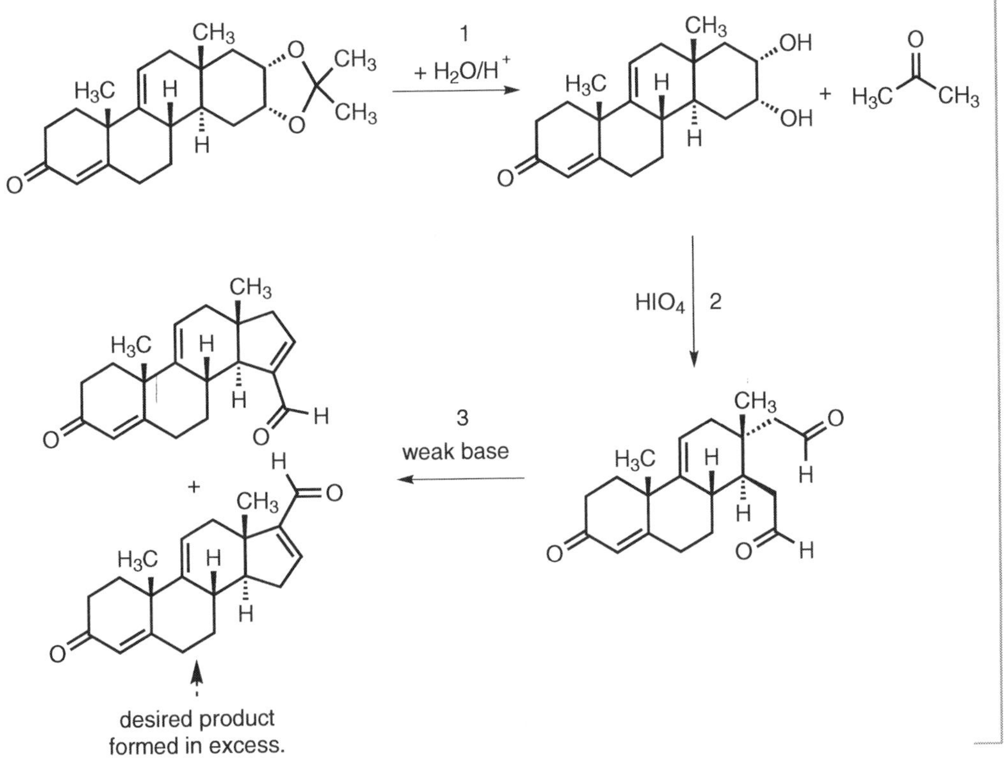

◀ **FIGURE 12.19**

The overall path to the necessary five membered ring in the structure of cholesterol is attained by removal of the acetone ketal followed by oxidation to aldehydes, which undergo Aldol condensations.

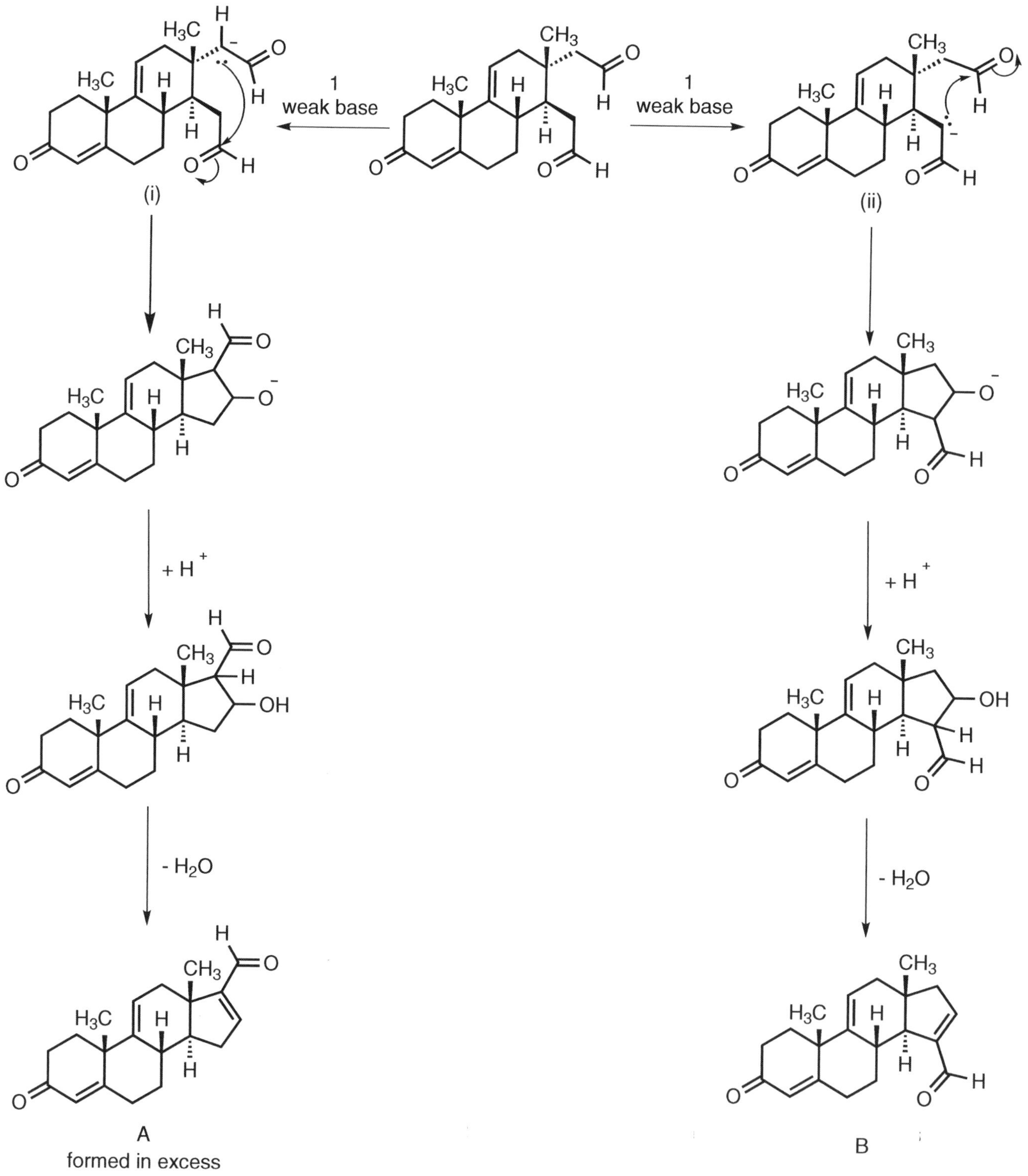

▲ **FIGURE 12.20**

The detailed nature of the two competing Aldol condensations are illustrated. Only one of these condensations forms the necessary five membered ring structure

acid (HIO_4), which was used by Woodward as shown in reaction **2** in Figure 12.19.

Breaking this six-membered ring was necessary because, in cholesterol, this site in the structure is not a six-member but rather a five-membered ring. On treatment with a weak base as shown in Figure 12.19 (reaction **3**) the two aldehyde groups undergo an Aldol reaction to form two constitutional isomers arising from the two possible carbanions formed from the dialdehyde. One of these constitutional isomers is the desired five-membered ring isomer. The usual mechanism (section 8.4), which applies to the formation of both products of this reaction is outlined in **Figure 12.20** where it can be seen that the first step (**1**) in producing the two enolates **i** and **ii** lead,

several steps

◀ **FIGURE 12.21**

The structure of a synthetic racemic intermediate is resolved with digitonin.

via this racemic intermediate

which was resolved with digitonin

inevitably, to the constitutional isomers **A** and **B**. Happily from the view of yield in the overall synthesis of cholesterol, isomer **A**, which has the aldehyde group placed properly for addition of the alkyl group pendant to the five membered ring in cholesterol, is formed in excess. How Woodward might have predicted this result is not clear. Sometimes in synthetic work you count on intuition and luck. Although not outlined here, several straightforward steps converted the desired five-membered product from Figure 12.20, which was fortunately formed in excess, to the ketone in **Figure 12.21**. On the way to this ketone the intermediate shown in Figure 12.21 was used to obtain the desired enantiomer that would lead to the natural enantiomer of cholesterol. Up to now in the synthesis, although the molecules were chiral, both enantiomers were present in equal proportions, the racemic state. How the resolution was accomplished is a story worth telling.

A plant that has been popular for centuries for stately gardens is the foxglove, *Digitalis purpurea*. A very complex molecule, digitonin can be isolated from this plant, with the useful property of precipitating molecules with steroid type structures.

Foxglove Plant

Considering the extensive chiral structure of digitonin (see above) it is hardly a surprise that the complexes formed are highly chirally specific. Indeed, when Woodward combined the racemic mixture of the intermediate structure shown in Figure 12.21 with digitonin, only the enantiomer that would lead to natural cholesterol was precipitated so that the forward synthetic path of the Harvard group was then able to focus only on this enantiomer, rather than the racemic mixtures necessarily used up to this point. Here is a classic demonstration of

Digitonin Structure

A **Grignard** reaction followed by elimination of water yields a mixture of double bonds, which are then hydrogenated, and followed then by a simple ester hydrolysis which yields the goal, cholesterol.

the principle of chiral separation discussed in section 1.9 (Figure 1.8).

The structure of the ketone produced via several steps from the aldehyde shown in Figure 12.21 no longer represents a racemic mixture but rather the structure of the single enantiomer, which the Woodward group worked with in the remaining steps in the synthesis. Up until now, all the chiral molecules in the figures showing the structures on the path to cholesterol, although showing single enantiomers were in fact racemic mixtures. This enantiomerically pure ketone (Figure 12.21) was then reacted with the Grignard reagent formed from 1-bromo-4-methyl-pentane (isohexyl bromide) (**Figure 12.22**).

Woodward got rid of the unwanted hydroxyl group produced from the Grignard reaction (reaction **1**, Figure 12.22) by converting the -OH group to an acetate group, $-OC=O(CH_3)$, which on heating eliminated acetic acid to form the double bonded mixture of stereoisomers and constitutional isomers (**i, ii, iii**).

Woodward was not concerned with obtaining a mixture of alkenes because his intention was to saturate the double bond formed, as was easily accomplished via platinum catalyzed addition of H_2, probably in a Parr shaker (section 6.7) (reaction **2**, Figure 12.22). All three constitutional alkene isomers shown would give rise to the same result, a mixture of saturated diastereomers arising from the two stereocenters formed in the hydrogenation. The desired isomer (**iv**) was readily separated from the others by various means, as is the situation for diastereomers in general (section 1.7). The final structure of cholesterol was now clearly in sight although we won't go into the few steps that remained (Figure 12.22).

Woodward's transformation of the alcohol produced by the Grignard reaction via conversion to the acetate followed by formation of the mixture of alkenes (Figure 12.22) should be put in context. Let's supply this context by using the opportunity to pay more attention to elimination reactions than we have, attention that is certainly deserved. (section 12.9).

PROBLEM 12.33
Develop a reasonable mechanism for reaction **1** in Figure 12.19.

PROBLEM 12.34
Going over Woodward's synthetic route up to Figure 12.19 can you find a reason why he felt he had to protect the double bond by conversion to a vicinal diol which was then protected as a ketal with acetone? Why did he carry this ketal through so many of the reactions up to this point?

PROBLEM 12.35
In Figure 12.20, in the final step of the Aldol condensations forming the five membered rings of the two constitutional isomers, water is lost to form the double bond under acidic conditions. What role might the proton play in this loss of water?

PROBLEM 12.36
In Figure 12.21 the key intermediate which was resolved with digitonin is shown. Let's try to take a step into synthesis. Think of an oxidation reaction and also use of a protecting group and also a Grignard reaction with limitations on the amount of CH_3MgBr and a Parr shaker with a tank of hydrogen gas. Can you come up with some, if not all, of the "several steps" leading to the product shown?

PROBLEM 12.37
Do you think that resolution using digitonin involved a physical or chemical interaction with the racemic molecule? Is one or the other necessary for success in the resolution?

PROBLEM 12.38
Let's say that you have available any products of cracking of petroleum (sections 4.10 and 9.4) and any reducing agent you would like and any solvents as well. In addition, you have a reagent, PBr_3, which can be used to transform an alcohol to a bromide. Finally, you have in hand a method to add water to a double bond in an anti-Markovnikov manner, a method we will be covering shortly, called hydroboration. Can you come up with a synthetic route to isohexyl bromide, and therefore the Grignard reagent used in step **1** of Figure 12.22?

W E'VE SEEN THE SYNTHESIS OF ALKENES most prominently in the catabolic process for fatty acids (section 7.7), and in the Aldol condensations (section 8.4). Chemistry has, however, many methods of forming double bonds. We'll leave out here industry's highly effective method of synthesizing alkenes by steam cracking of petroleum fractions (section 9.4), a procedure that, even if transferable to a laboratory scale, would not be suitable because of the uncontrolled nature of the reaction in yielding many different products.

In laboratory work, alkenes are synthesized in general via the intermediacy of a carbon-carbon bond with a C-H on one of the adjacent carbon atoms and a leaving group (X) on the other carbon atom. We've seen plenty of discussion of leaving groups, beginning with the understanding arising from the extensive discussion in Chapter 5 (section 5.3), that a leaving group is a generally negatively charged entity (X⁻), which is the conjugate base of an acid, HX. Double bonds are synthesized, therefore, by chemists in laboratories from the structural situation H-C-C-X. Such reactions are called elimination reactions, elimination of HX forms the double bond, the antithesis of addition reactions in which HX adds to a double bond.

We've seen plenty of chemical reactions that are designated addition reactions, beginning with the addition of carbocations to double bonds in section 4.10. Alternatively, we've seen substitution chemistry in electrophilic aromatic chemistry beginning with section 6.10. Nucleophilic substitution at acyl carbon for both saponification and biological hydrolysis of glycerides has been discussed in sections 7.3 to 7.5 while nucleophilic substitution at saturated carbon via the S_N1 and S_N2 reactions was the focus of section 10.6.

On the contrary there's been only little discussion of the other important class of reactions, those termed "elimination." The word has been used as in the formal elimination of water in the formation of esters from fatty acids and glycerol (sections 7.2 and 7.3) and in the formation of polyesters and nylons in Chapter 9 (section 9.8 and 9.10) and in this chapter in the elimination of water in the formation of ketals and acetals (section 12.7). Stimulated by Woodward's formation of the double bond after the Grignard reaction exhibited in Figure 12.22, let's give elimination chemistry a bit more attention

In the structural situation H-C-C-X, the prerequisite for elimination to form the double bond between these carbon atoms is reaction with a base to accept the proton that must be eliminated to balance the loss of X with the two electrons bonding it to carbon. Organic chemists well understand that a moiety that acts as a base can often, equally well, act as a nucleophile and vice versa. This exchange of reactive roles immediately causes a problem. A species intended to accept H⁺, such as HO⁻ might competitively act as a nucleophile and substitute for X⁻. Instead of obtaining the desired double bond, the system might deliver to us an OH group replacing X.

In fact, formation of double bonds is a very complex chemically reacting system for another reason. Imagine that the structural system in hand is $H-C(R)_2-C(R)X-C(R)_2-H$ in which the R groups might be different from each other and where loss of the hydrogen atom from one or the other carbon atom leads to different double bond structures, that is, constitutional isomers. How is one to control the structures of the elimination reactions of such a system? In other words, how is one to control which double bond will form? Although Woodward had no interest in controlling the double bond formation as shown in Figure 12.22 this is not the usual situation. And if that were not problem enough, it is possible that loss of either of the two hydrogen atoms on the same carbon atom adjacent to C-X such as in $H_2C(R)-C(R)X-C(R)_2-H$ could yield what are called geometric isomers, cis (Z) versus trans (E) diastereomers (section 11.2).

Setting out therefore on the synthesis of a double bond in a complex structure can be a mine field of complexity. Organic chemists have learned how to maneuver through this minefield, to some extent, but far from entirely, by varying bases and solvents and temperature and the nature of X and have learned what kinds of molecules give what

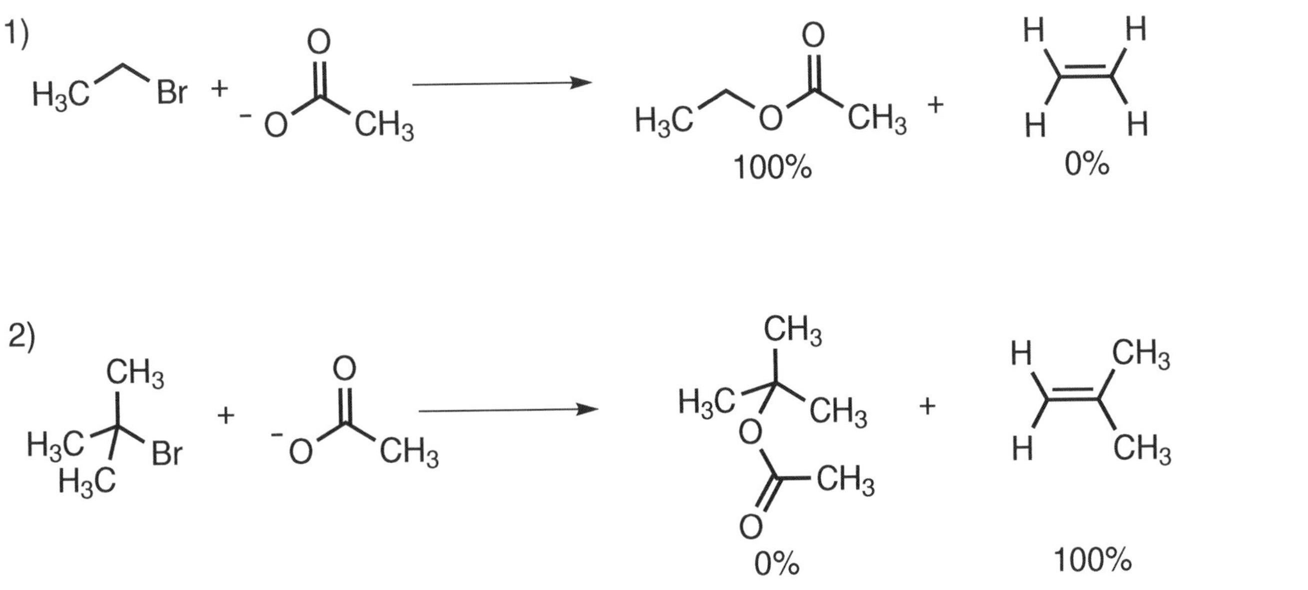

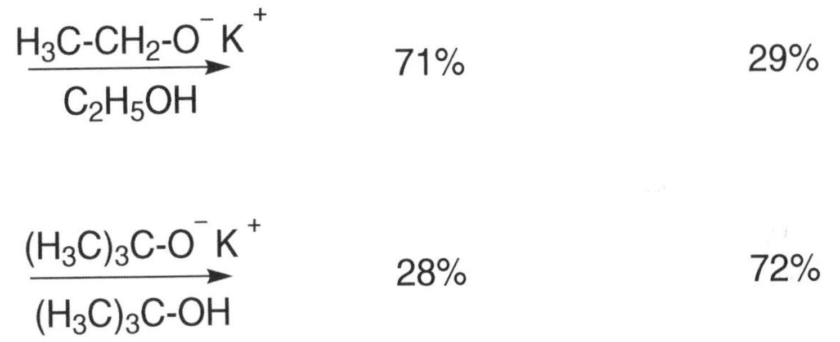

1)

100%   0%

2)

0%   100%

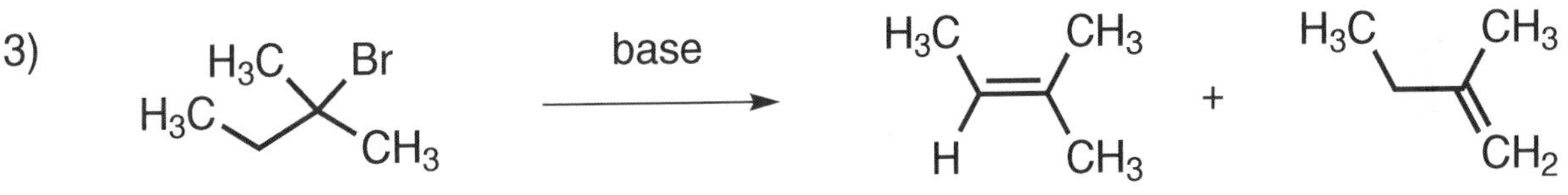

3)

△ FIGURE 12.23

Six differing reactions leading to double bonds yield insight into the complexity of elimination reactions and their overlap with nucleophilic substitution reactions.

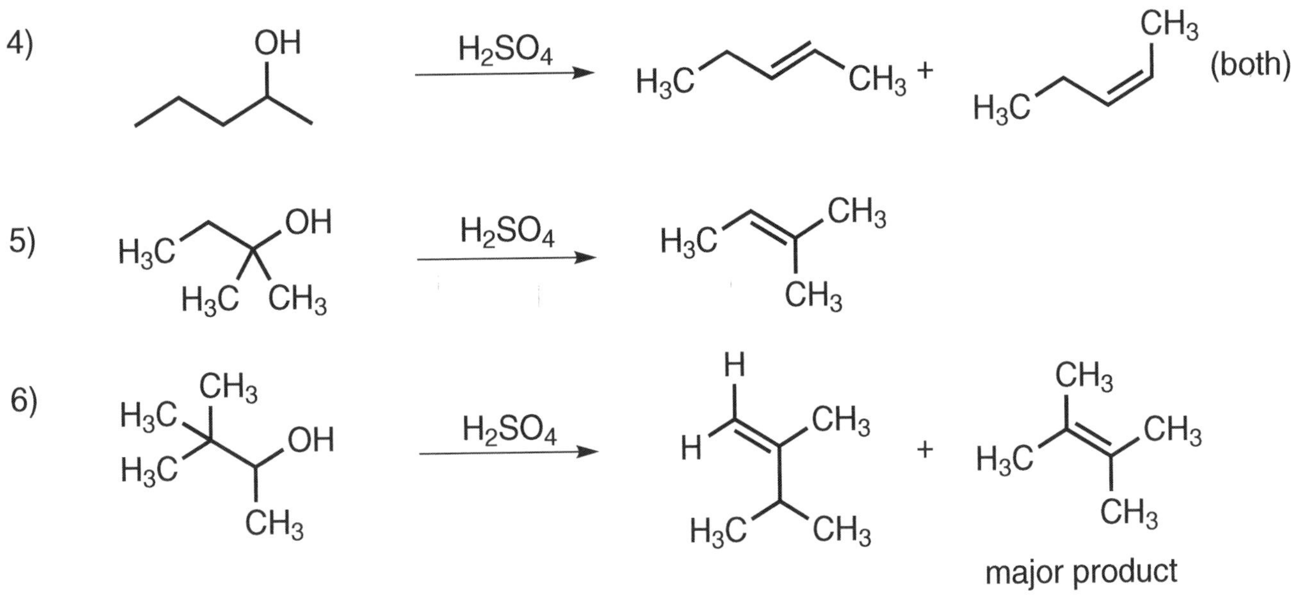

$H_3C\text{-}CH_2\text{-}O^- K^+$ / C_2H_5OH 71%   29%

$(H_3C)_3C\text{-}O^- K^+$ / $(H_3C)_3C\text{-}OH$ 28%   72%

4)

H_2SO_4 (both)

5)

H_2SO_4

6)

H_2SO_4

major product

kinds of results, a subject we will not go into here, except for exhibiting examples of the complexities involved and a short discussion below (**Figure 12.23**).

Muse over the varying results of the elimination reactions in Figure 12.23 and try to propose mechanistic reasons for the results. Good luck. At this stage in your studies, you do have the background in hand to come up with some reasonable answers. Excellent sources to help to understand the elimination reactions that produce double bonds can be found in Streitwieser and Heathcock's text, "Introduction to Organic Chemistry, 3rd edition" page 242 ff and in Carey and Sundberg, "Advanced Organic Chemistry, Part A, 3rd edition" page 368 ff.

In your effort, imagine two extreme mechanisms, E-1, an analogy to S_N1, where the elimination process is initiated by ionization of a leaving group forming therefore a carbocation. After formation of the carbocation, a proton on an adjacent carbon atom would be lost to a solvent molecule or to a base. The other extreme mechanism would be designated E-2, an analogy to S_N2, in which a base/nucleophile abstracts a proton from the carbon atom adjacent to the carbon atom bearing the leaving group, with loss of the proton occuring in concert with loss of the leaving group. The reaction coordinate diagram for the E-1 mechanism would resemble Figure 10.13 while the diagram for the E-2 mechanism would resemble Figure 10.16.

A third mechanism occurring in elimination reactions is termed E1cB, which is commonly encountered in elimination of water as the final step in the Aldol reaction (section 8.4). Here the carbonyl group increases the ease of proton abstraction of the proton from the adjacent C-H bond, which is then followed by loss of OH⁻ to form the double bond.

In general, however, organic chemists often avoid synthesizing alkenes in which multiple products can be formed and instead use reaction with more specific outcomes.

We're going to see some examples of how the synthesis of the alkene functional group can be controlled in the section that follows, in E. J. Corey's synthesis of the prostaglandins.

Let's return now to Woodward's triumphal arrival at the target of his quest. A few uncomplicated steps from the hydrogenated product (iv in Figure 12.22) and a crystalline white powder was in hand, identical to a sample of natural cholesterol that could have been handed down from Chevreul's work (section 5.1). The synthetic sample passed the ultimate test for identity and purity, a mixed melting point. Mixing the product produced in the Harvard laboratory with natural cholesterol gave no change in the melting point, softens at 149° C to a clear liquid at 150° C. That's it. The crystalline materials were identical. Head for the pub.

PROBLEM 12.39
Elimination via the E-2 mechanism is known to greatly favor an anti relationship between the leaving group and the eliminated hydrogen. Consider two diastereomers of the following structure: C_6H_5-C(CH_3)(H)-C(Br)(H)(CH_3). Use Newman projections to predict the alkene isomer that would be formed from each diastereomer considering the necessity for an anti relationship between the H and the Br.

PROBLEM 12.40
Suggest a transformation of the OH group in the product of reaction 1 in Figure 12.22 to form the alkenes i, ii, and iii. Show the H atom involved in the elimination to form each alkene.

PROBLEM 12.41
In Figure 12.23, assuming that the conditions are low dielectric constant for the solvent and, therefore, favoring S_N2 substitution, offer an explanation for reactions **1** and **2**.

PROBLEM 12.42
Can you think of a good reason for the results of the two reactions under **3** in Figure 12.23 using ideas of steric hindrance?

PROBLEM 12.43

In **4**, **5** and **6** of Figure 12.23 the mechanism for the elimination reaction differs fundamentally from the first three reactions with a focus on the leaving group. What role does the sulfuric acid play in this difference?

PROBLEM 12.44

In Woodward's synthesis of cholesterol he used the Aldol condensation several times. Show how the E1cB mechanism could be involved in these reactions when a double bond is formed.

Synthesis—Part Two

TWO NOBEL PRIZES for synthetic organic chemistry were awarded in the latter part of the 20th century, Robert Burns Woodward, 1965, whom we have met both in Chapter 1 and the first part of this chapter, and **Elias James Corey** 1990. Both Nobel laureates came from Massachusetts, and both were brought up by mothers widowed shortly after their births. They both obtained their bachelors and doctoral degrees at MIT where they became intrigued by synthetic organic chemistry. Both went to the University of Illinois after their doctoral degrees, Woodward for a short postdoctoral stay and Corey for a longer time as an assistant professor. Both carried out their Nobel Prize winning work at Harvard. Both were known for their intense dedication and their astonishingly creative work. The elder of the two, aware of the remarkable abilities of the younger, was instrumental in bringing Corey to Harvard from the University of Illinois. We've just seen R. B. Woodward's synthesis of cholesterol. Now let's look at how E. J. Corey overcame an entirely different class of multiple problems on the route to the synthesis of another kind of biologically important molecule.

There follows a quote (reportedly translated from the Swedish) from the presentation speech for the Nobel Prize Corey won in 1990 which brings our attention to the particular synthesis we'll study, a synthesis which demonstrates the unique character of Corey's abilities:*"Corey's most important syntheses are concerned with prostaglandins and related compounds. These often very instable compounds are responsible for multifarious and vital regulatory functions of significance in reproduction, blood coagulation and normal and pathological processes in the immune system. Their importance is witnessed by the awarding of the 1982 Nobel Prize in Physiology or Medicine to Professors Sune Bergström, Bengt Samuelsson and Sir John Vane for their discovery of prostaglandins and closely related biologically active compounds".*

Here are Ian Fleming's words (section 12.1), of Corey's ability in synthesis: *"One of the outstanding features of the brilliant work of E. J. Corey has been the number of new reactions he had devised. Most of these have been developed with a specific aim in mind, usually the synthesis of a natural product."*

Let's look at Corey's synthetic path to the prostaglandins and discover the source of the praise for this great organic chemist, and also learn about the chemical reactions Corey made use of in the synthesis of the prostaglandins. We'll begin by outlining the general reaction path Corey followed (**Figure 12.24**) followed by detailed study of several of the critical steps on that path.

12.10

Elias J. Corey

■ **Elias J. Corey**

12.11 Prostaglandin, the Beginning Steps

IN EXAMPLE 3 OF FIGURE 6.8 we discovered that cyclopentadiene, 1 (Figure 12.24) is acidic enough to lose a proton from its CH_2 group to form a cyclic array of parallel p-orbitals to gain a Hückel number of aromatic electrons. Although the lone pair of electrons with its attending negative charge is not powerfully nucleophilic arising from the contribution of these two electrons to the 6π electron Hückel number, nevertheless a reactive electrophile can induce reaction.

The powerful electrophile Corey used was $Cl\text{-}CH_2\text{-}OCH_3$. This molecule, chloromethyl methyl ether, features an excellent leaving group. Cl^-, bonded to a CH_2 group that would form a stable carbocation arising via resonance stabilization by the lone pair of electrons on the adjacent oxygen atom ($H_2C{=}O^+\text{-}CH_3$). This combination of characteristics enhancing reactivity via both the S_N1 and S_N2 reactions (section 10.6) makes for a molecule that is highly reactive with nucleophiles, including in vivo.

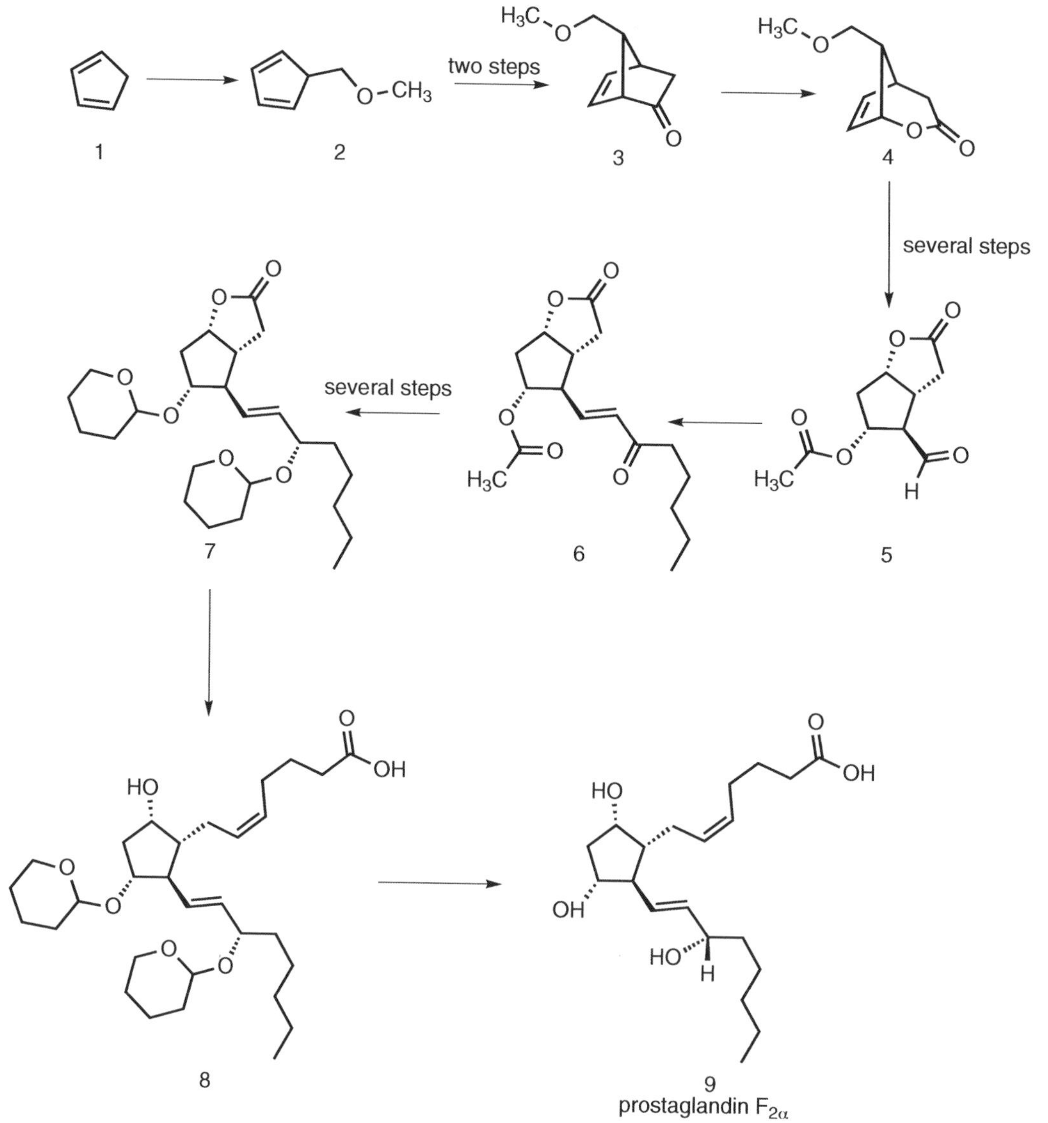

FIGURE 12.24 ▶ **Corey's Overall Path from Simple Starting Materials to Prostaglandin $F_{2\alpha}$**

It's not surprising that chloromethly methyl ether is a known carcinogen for reasons that are not dissimilar to the carcinogenic property leading DuPont to abandon the route to hexamethylene diamine (section 10.5). However, on the small scale use for laboratory synthesis, exposure to the chemical can be controlled in ways that are not reasonable, on an industrial scale.

The first reaction in Figure 12.24 shows the use of chloromethyl methyl ether to add

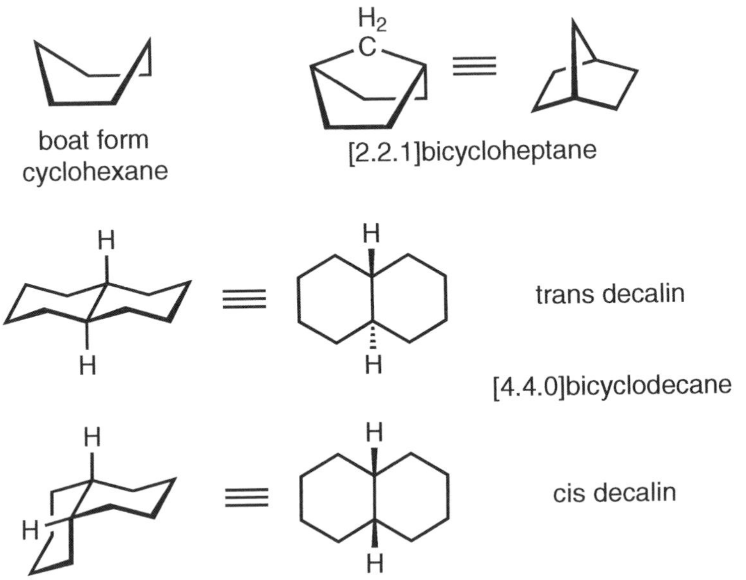

boat form
cyclohexane

[2.2.1]bicycloheptane

trans decalin

[4.4.0]bicyclodecane

cis decalin

◀ FIGURE 12.25

Three Possible Structural Variations for Fused Rings

the ether function to the cyclopentadiene, Corey's first step toward the prostaglandins.

The next two intermediates **3** and **4**, shown in Figure 12.24, are termed bicyclic molecules, a designation that organic chemists use for molecules with fused rings and also rings with what is called a spiro connection, which we'll not discuss here. In Woodward's synthesis of cholesterol we saw bicyclic molecules in which the two rings are connected to adjacent carbon atoms. Bicyclic molecules were introduced in Chapter 4 (section 4.3). **Figure 12.25** shows two classes of bicyclic molecules, one in which, as in the synthesis of cholesterol, the two fused rings are joined by adjacent carbon atoms, the basic structure being decalin, and one in which the rings are joined to carbon atoms related 1,4 to each other as in **3** and **4** in Figure 12.24.

The carbon structure of **3** (Figure 12.24) can be compared to a conformation of cyclohexane you were asked to consider in problem 3.10 in Chapter 3, the boat form. If the "bow and stern" of the boat form, as shown in Figure 12.25, are linked by a single CH$_2$ group, then we get the skeleton of **3** (Figure 12.24) . But how is one to synthesize **3** from **2**. The answer is the same reaction that Woodward used early in the synthesis of cholesterol, the Diels-Alder reactions (sections 12.2 and 12.3, Figure 12.3).The particular Diels-Alder reaction Corey chose is shown in **Figure 12.26**, which follows

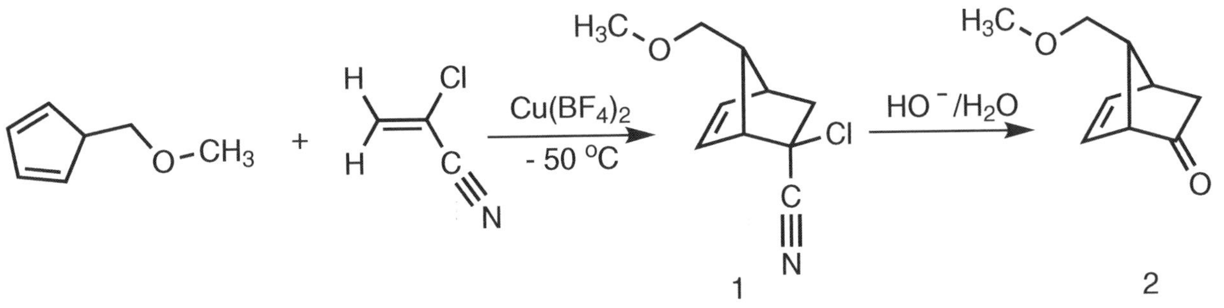

1

2

the classic pattern for this class of reaction (Figure 12.3). The reaction Corey used is, however, not as straightforward as shown in Figure 12.26. Although the configuration of the stereocenter with the Cl and C≡N groups is of no importance because it will be converted to a carbonyl group in the next step, as also shown in Figure 12.26, the configuration of the carbon atom bearing the CH$_2$-OCH$_3$ group is critically important because it determines a stereocenter configuration in the prostaglandin, as we'll see.

Corey determined that at low temperatures, in the range of -50° C, the product he wanted (**1** in Figure 12.26) would be obtained. But running a Diels-Alder reaction at

▲ FIGURE 12.26

Corey begins with a Diels-Alder reaction under special conditions, which yields a ketone by an unusual route.

such a low temperature was not usually possible but might become possible by greatly activating the electron demanding character of the ene, the $H_2C=C(C\equiv N)(Cl)$. This is the reason for the unusual use of a catalyst, the cupric fluoroborate, shown in Figure 12.26. The positive cupric ion (Cu^{+2}) complexes with the nitrogen of the nitrile group increasing the electron demand of the double bond to an extent that the Diels-Alder reaction readily took place at this extremely low temperature. The CH_2-OCH_3 took the necessary configuration as shown.

There is a general lesson in altering the conditions of the Diels-Alder reaction shown in Figure 12.26, a lesson that all synthetic organic chemists are very well aware of. A synthetic path can be well designed but each step may work or not work depending on the conditions used, which can vary not only the details of the molecular components of the synthetic path but the intimate details of how the reaction is run. I remember hearing a story about the synthetic laboratories at Harvard and in particular about Woodward's laboratories, that many students would be trying out the same general reaction Woodward had designed but varying the conditions, solvent, temperature, etc. to determine the best yield and even to make the reaction occur. It's sort of like cooking, isn't it? You can give different chefs the identical ingredients designed to make a delicious outcome but the details of how these ingredients are handled makes all the difference to the outcome. There's quite a bit of a "green thumb" involved in synthetic organic chemistry.

PROBLEM 12.45
A cyclic structure "spiro" can arise from a single carbon atom being the only juncture point between two rings. Explain how the chirality of a spiro structure arises without four different groups around any carbon atom in the structure. Suggest other structure with this kind of chirality, called axial chirality.

PROBLEM 12.46
Identify the axial and equatorial groups emanating from each ring into the fused ring for both cis and trans decalin.

PROBLEM 12.47
Draw reactions coordinate diagrams for reaction of chloromethyl methyl ether, ($Cl-CH_2-OCH_3$) with cyclopentadienyl anion (reaction 3 in Figure 6.8, Chapter 6) via the S_N1 and S_N2 mechanisms.

PROBLEM 12.48
In Figure 12.26 the product of the Diels-Alder reaction, **1**, is converted to the ketone with aqueous base. Offer a mechanism for this reaction realizing that both Cl and the CN can act as excellent leaving groups in this transformation.

12.12
Two remarkable rearrangements: The Baeyer-Villiger reaction forms a lactone, which is then rearranged to another lactone.

WE FIRST MET Friedrich Adolf von Baeyer back in Chapter 3 (section 3.2) and have come across this great chemist of the nineteenth century several times since. Late in his career he studied the effect of organic peroxides with a student, Victor Villiger, and at sixty five years of age Baeyer published the discovery of a reaction that proved critical to the path Corey choose for synthesizing prostaglandins, a reaction that has come to be known as the Baeyer-Villiger reaction. It is this reaction that took Corey from the ketone to the cyclic ester, the lactone, from **3** to **4** (Figure 12.24). The basic mechanism of the Baeyer-Villiger reaction applied to the formation of **4** from **3** is shown in **Figure 12.27**.

The Baeyer-Villiger reaction, which is generally acid catalyzed, requires a peracid. Corey used one that is commonly used in organic synthetic work, metachloro-perbenzoic acid. As seen in Figure 12.27, proton addition (**1**) to the

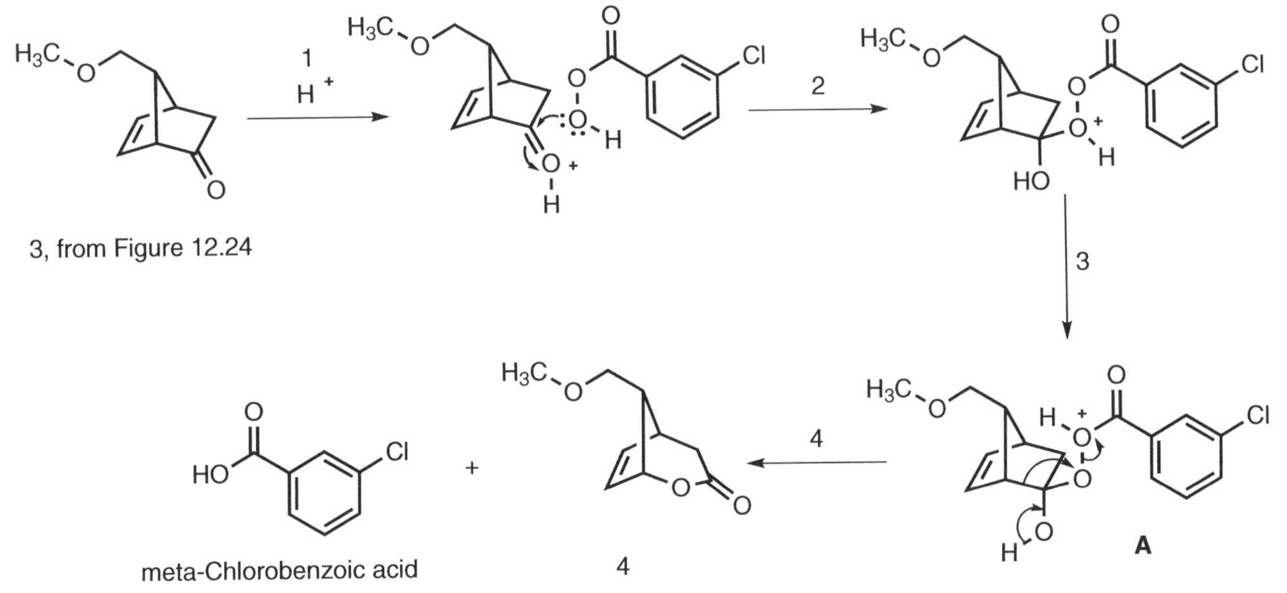

3, from Figure 12.24

meta-Chlorobenzoic acid + 4

oxygen atom of the starting ketone stimulates nucleophilic attack by the terminal OH group of the peracid (**2**) on the carbonyl group. Proton transfers (**3**) lead to the intermediate that undergoes the rearrangement (**A**). In **A**, the oxygen-oxygen bond (the peroxide bond) is greatly weakened causing the transfer of one of the carbon-carbon bonds flanking the tetracoordinate intermediate leading to the lactone, **4**, and metachlorobenzoic acid. The starting metachloroperbenzoic acid has been reduced in the course of the reaction to the metachlorobenzoic acid. The ketone, in turn, has been oxidized to the lactone.

On first appearance, the lactone, **4** (Figures 12.24 and 12.27) produced by the Baeyer-Villiger reaction does not appear to have a structure related at all to the prostaglandin structure. But Corey had designed **4** for precisely its relationship to the prostaglandin structure. Let's follow his view. Hydrolyzing the lactone (sections 7.3 and 7.4) to the carboxylic acid and the hydroxyl group yields **1** (**Figure 12.28**), which we can draw as a five member ring, bringing us closer to seeing what Corey was planning.

The next step required generating a new lactone structure, as seen in **5** in Figure 12.24, which had to involve one of the carbon atoms of the double bond in **1** (Figure 12.28). This was accomplished by a reaction that was remarkably effective in accomplishing Corey's objective. Potassium triiodide, KI_3, is a source of I^+, a cation that can add to double bonds.

We should appreciate what the reaction with KI_3 accomplished. The product of this reaction, **2** (Figure 12.28), has five chiral stereocenters, all of which are obtained in the correct relative configuration for the path to the prostaglandins that follows. And this did not take place by chance but rather by prediction, a brilliant accomplishment that stands at the heart of the success of the entire synthesis.

We've seen the positive state of halogens add to double bonds earlier to form halonium ions (section 10.5, Figure 10.10 and 10.11). The reaction we are looking at here is analogous except that instead of the negatively charged halogen adding to the halonium ion, the OH of the carboxylic acid group intervenes to form the lactone Corey wanted, **2** (Figure 12.28). The formation of the ring brings up a fundamental principle, which is discussed in detail in the next section.

Several simple steps, which we will not go into here substituted hydrogen for iodine, transformed the hydroxyl group in **2** to the acetate ester, the methyl ether group to the hydroxyl group and oxidized this hydroxyl group to the aldehyde to yield **5** (Figure 12.28, the same structural number as in Figure 12.24).

▲ **FIGURE 12. 27**

The ketone shown in figure 12.26 undergoes an oxidation with a peracid to a lactone via the Baeyer-Villiger reaction.

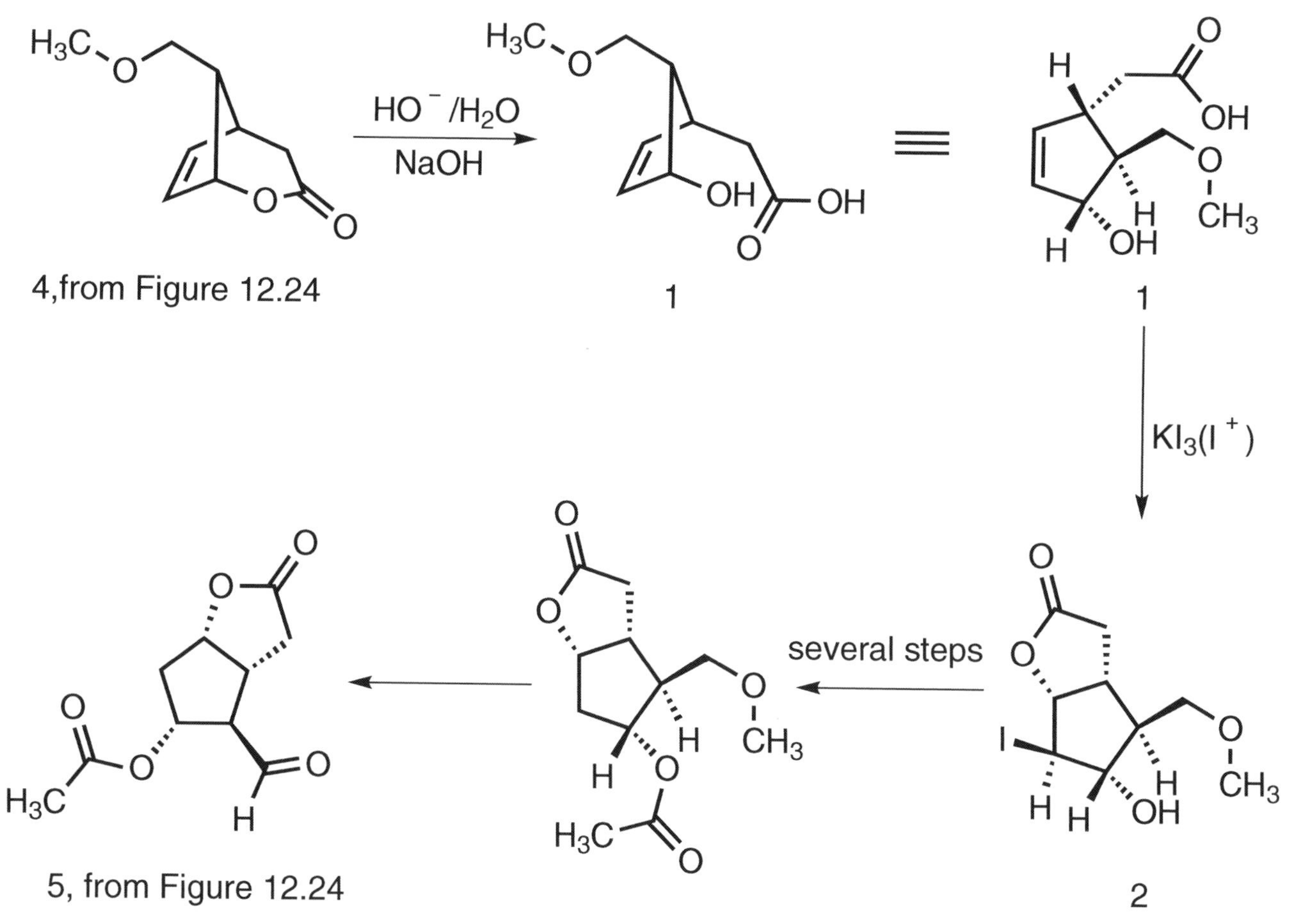

4, from Figure 12.24

1

1

$KI_3(I^+)$

several steps

5, from Figure 12.24

2

▲ FIGURE 12.28

Without showing a mechanism, the overall transformation from the product of the Baeyer-Villiger reaction to a critical intermediate is shown in the synthesis of prostaglandin F$_{2\alpha}$.

PROBLEM 12.49

In the intermediate in Figure 12.27, which undergoes reaction **4**, another rearrangement of a formally identically placed carbon-carbon bond could take place. What product would this have led to? And why was Corey certain the reaction would go the way he wanted based on the history of the Baeyer-Villiger reaction, which demonstrated that the more substituted carbon-carbon bond was favored to rearrange?

PROBLEM 12.50

There are two tetracoordinate intermediates shown in Figure 12.27. Using ideas of electrophilicity and electron flow shown by the curved arrows explain the necessity for the intermediate that undergoes the rearrangement.

PROBLEM 12.51

Looking ahead to the next section, draw three dimensional drawings with stereochemical features of the halonium ion formed on addition of I$^+$ to the double bond of **1** in Figure 12.28. Consider the reactive consequences if the I$^+$ had added from the top or the bottom of the double bond, or in the opening of the halonium ion to form a carbocation, from one side or the other side of the double bond – four possibilities in all. Which is necessary for the reaction to proceed in the direction Corey wanted as shown in Figure 12.29 (b)?

PROBLEM 12.52

The last line of Corey's 1969 Journal of the American Chemical Society paper (pages 5675-5677) states: "preliminary experiments demonstrate that the hydroxy acid is easily resolved." This is **1** in Figure 12.28. How would you attempt to carry out this resolution of **1** into the two enantiomers? And why is this intermediate, **1**, on the synthetic path so well suited for resolution?

THE CHEMISTRY FOLLOWING THE TREATMENT of **1** with KI$_3$ (Figure 12.28) teaches us about the issues facing organic chemists in the formation of rings. As the positively charged iodine atom, I$^+$, approaches the double bond from the side opposite the carboxylic acid group and the hydroxyl group, these two nucleophiles are poised within the structure of **1** to capture the positively charged carbon atom that arises (**Figure 12.29** (A) and (B)). Let's look at the consequences of each one acting as the capturing nucleophile.

The hydroxyl group certainly is nucleophilic enough to do this job but the consequence would be the formation of either a three (**i**) or four membered (**ii**) ring appended to the five membered ring (Figure 12.29 (A)). The carboxylic acid group, which does capture the positive charge generated by the attack of the I$^+$ on the double bond of **1** (Figure 12.29 (B)) forms an appended five membered ring (**iii**), which is the product Corey intended (**2** in Figure 12.28). But why does it also not form the appended six membered ring (**iv**) (Figure 12.29 (B))?

The data in **Figure 12.30**, although gathered many years after Corey's synthesis of the prostaglandins, give, us the answer to this question by revealing the strain energies of the various carbocycles shown and also the rate of formation of these rings, which we can use as a model for the choices shown in Figure 12.29. Let's find out why Corey

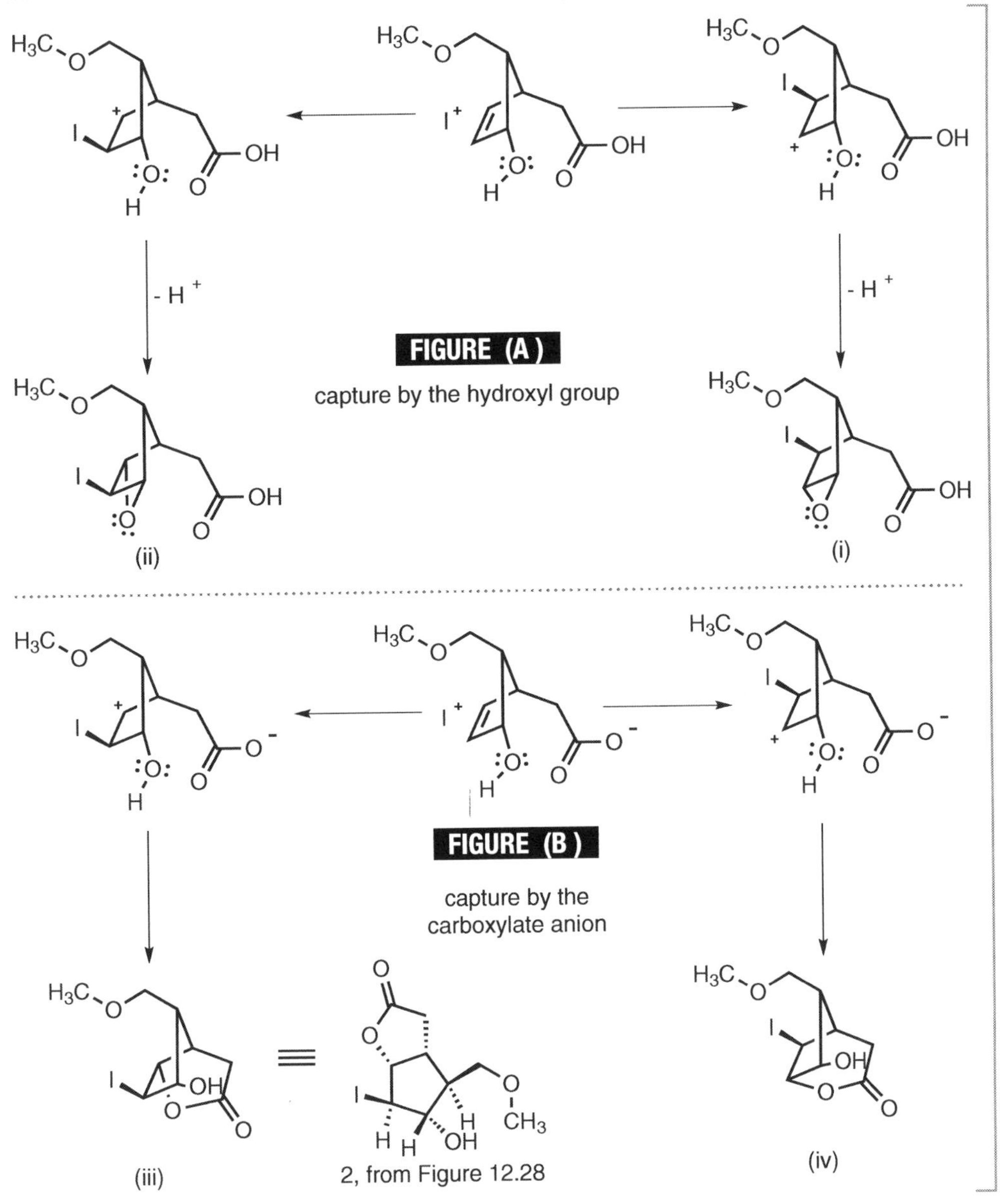

FIGURE (A)

capture by the hydroxyl group

FIGURE (B)

capture by the carboxylate anion

2, from Figure 12.28

◄ **FIGURE 12.29 (A & B)**

A) and (B) – The addition of a positively charge iodine atom adds to a double bond setting up a series of competitive remarkable intramolecular reactions of which one is predominately chosen, leading to the critical intermediate noted in figure 12.28.

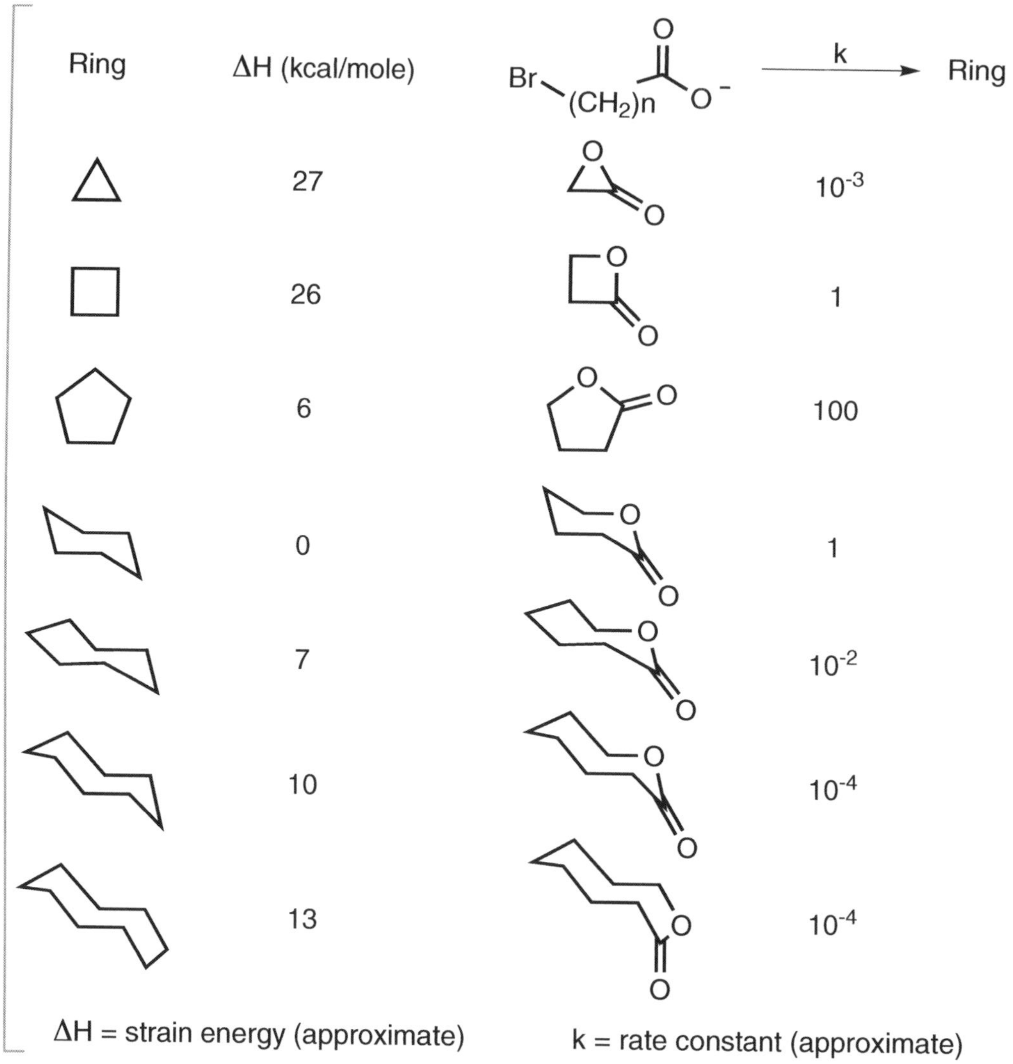

FIGURE 12.30 ▶

Thermodynamic and kinetic data for ring closing reactions of model compounds reveals the driving forces at work for closing rings and allows understanding of fhe successful ring closing shown in figure 12.29 (B).

ΔH = strain energy (approximate)   k = rate constant (approximate)

was certainly correct in predicting that the treatment of **1** would yield the desired product, the appended five member lactone (**2**, Figure 12.28, **iii** in Figure 12.29 (B)).

There's an interesting relationship between the thermodynamic and kinetic data sets in Figure 12.30. The strain energies in Figure 12.30 are thermodynamic parameters determined by calorimetric measurements, related to the kinds of measurements made to judge the stability of double bonds and benzene using heats of hydrogenation (section 6.7). Saturated carbocyclic rings that contain only CH_2 groups can be combusted with oxygen to yield CO_2 and H_2O and the heat released carefully measured and compared to the heat released by long chain hydrocarbons (without rings). The heat released per CH_2 group is then compared. The greater the heat released, the higher is the strain energy.

One can conclude from the ΔH data in Figure 12.30 that ring strain is severe for three and four membered rings, drops significantly for five and six membered rings, and rises again for larger rings until tapering off for very large rings. Now let's compare the strain energies with the kinetic data in Figure 12.30 for formation of lactones of various ring sizes.

The very slow rate of formation of the three and four membered ring lactones compared to the formation of five and six membered rings is certainly reasonable considering the relative severe strain energies of these rings. However, the data in Figure 12.30 show that any correlation between ring strain and rate of ring closure falls off when considering the larger rings, which is reinforced by the comparison between formation of the five and six membered rings. If ring strain were the entire story, then the six membered ring should be greatly preferred over the five membered ring, that is, **iv** should have been formed in preference to **iii** (Figure 12.29) based on strain energy alone. There must be another factor at work here.

In discussing the comparison between polymerization of ethylene and steam cracking of naphtha fractions of petroleum we discovered the important role played by entropy. This thermodynamic parameter comes to the fore again in understanding why ring strain alone does not control the rate of formation of rings. Considering entropy change, smaller rings are formed with a more favorable entropy change than larger rings. The shorter chain precursors to the small rings may be seen as less disorderly so that formation of the ring does not cause a large decrease in disorder, and therefore a large decrease in entropy. Larger rings must be formed from longer open chains so that the larger the ring formed, the larger the decrease in entropy, an important unfavorable factor for formation of the ring (as occurs in vulcanization, section 11.6).

Although formation of the six membered ring is attended by less strain than the formation of the five membered ring (Figure 12.30), the entropic cost for forming the five membered ring is less than the entropic cost for forming the six membered ring. In other words the magnitude of ΔS is less negative for forming **iii** than **iv** (Figure 12.29) to such an extent as to "overpower" the ring strain, which would have favored **iv**. In a more quantitative way, the entropy term, $-T(-\Delta S)$, for formation of the five membered ring **iii** is less positive. Since $\Delta G = \Delta H - T\Delta S$, it follows that the free energy change, ΔG, will be more negative and therefore more favorable to the ring closing. Although this argument is thermodynamic, these factors find parallel expression in the activation parameters (section 6.11) controlling the ring closing rate. The details of absolute reaction rate theory, concerned with enthalpic and entropic activation parameters, will be a subject to be covered in your studies of physical chemistry.

At any rate, Corey got it right. He now has a clear unimpeded view to the prostaglandin structures and especially using the synthetic power of phosphorus-based reagents, which we'll look into in the next section.

Although Corey did not make prominent use of boron-based chemistry in the road to the prostaglandins, let's take the opportunity to also look at this area. The creators of these fields involving phosphorus and boron stood on the stage at Stockholm together.

PROBLEM 12.53
Can you suggest a reason why I^+ favors approach to the double bond from the side opposite the OH and CO_2H groups as shown in Figure 12.29 (A) and (B)?

PROBLEM 12.54
Use curved arrows and show all electrons involved in the reactions leading to **i, ii, iii** and **iv** in Figure 12.29.

PROBLEM 12.55
Redraw the starting structure in Figure 12.29 with the five member ring in the plane of the page and with each pendant group shown up or down using the bold and slashed nomenclature commonly used in the text. In addition show all atoms including hydrogen atoms in the structure. Now the cis or trans relationships of the pendant groups can be clearly seen. Using this representation, show the reactions leading to **i, ii, iii** and **iv**.

PROBLEM 12.56
Outline the components (angle, torsion, steric) of the strain energies shown in Figure 12.30.

PROBLEM 12.57
Make models or three dimensional drawings of the cyclic hydrocarbons in Figure 12.30 showing all atoms to evaluate your answer to Problem 12.56.

12.14

Boron and Phosphorus: Useful Elements in Synthetic Chemistry

■ Georg Wittig

GEORG WITTIG, A GERMAN CHEMIST, who died in 1987 at the age of 90, shared a Nobel Prize in 1979 with **Herbert C. Brown**, an American chemist, who died in 2004 at the age of 88. Wittig was born in Berlin and Germany was his permanent home where he eventually came to be professor at the University of Heidelberg. Brown had a more complex geographic background. He was born in London, England to parents who were refugees from the Ukraine in 1908. This was a time when the Jewish populations in that part of Europe were under attack in the recurring pogroms, driving many to emigrate. His father's original name was Charles Brovarnik, which had been anglicized to Brown, the family name in their subsequent migration to Chicago in the United States, where Brown grew up and went to school and eventually settled as a professor at Purdue University. Wittig and Brown shared the Nobel Prize for demonstrating that two elements, not normally associated with in vitro organic chemistry, could in fact be of great use to this science, phosphorus and boron respectively. Although it was a reaction based on phosphorus, named appropriately the Wittig reaction, which proved critically important to Corey's path to the prostaglandins, this seems an opportune time to inform you of how helpful boron can be to synthetic work, especially because both reactions involve double bonds.

In the first volume of "Classics in Total Synthesis," by Nicolaou and Sorenson, published in 1996, a comprehensive accounting of synthetic work in organic chemistry from the last half of the 20th century, in which Corey wrote the introduction, one finds a certain kind of interesting measure of the importance and age of these classes of reactions: hydroboration—twenty one notations to its use in the index; the Wittig reaction—twenty six notations in the index, which may be compared to the Diels-Alder reaction with thirty three notations.

12.15

The Wittig Reaction

■ Herbert C. Brown

WITTIG'S REACTION ALLOWED ORGANIC CHEMISTS TO synthesize double bonds with a control previously not possible as seen in Figure 12.23. The Wittig reaction (one example shown in **Figure 12.31**) produced double bonds from aldehydes and ketones so that the double bond was reliably placed precisely where the carbon-oxygen double bond had been. The first step (reaction **1**, Figure 12.31) is displacement of a leaving group, Br⁻ in this situation, via an S_N2 reaction (section 10.6) to produce the positively charged phosphorus intermediate. A strong base, for example, n-butyl lithium in dimethyl sulfoxide solvent (DMSO), reacts to abstract a proton form the carbon atom adjacent to the charged phosphorus atom as shown in Figure 12.31 to produce what is called an ylid. A carbanion is produced adjacent to the phosphorus cation. Reaction of this ylid (reaction **2**) with a ketone or aldehyde produces the alkene specifically replacing the double bond of the carbonyl group (reaction **3** Figure 12.31).

We'll discuss the mechanism of reaction **3** shortly. But for now lets compare a sequence of reactions that could have been used to gain the alkene structure produced in reaction **3** before the Wittig reaction became available, namely the Grignard reaction followed by acid catalyzed loss of water. The Grignard reaction (section 12.6) produces the alcohol (Figure 12.31).

There are several ways (Figure 12.23) in which an alcohol can be transformed to an alkene. The OH group could be transformed to the bromide by reaction with phosphorus tribromide, PBr_3, for one example, and then elimination of HBr with base as in Figure 12.23. Or H_2O could be eliminated directly from the hydroxyl compound using acid catalysis. In either situation it would be difficult to avoid producing the mixture of constitutional isomers and stereoisomers portrayed in the the Grignard section of Figure 12.31.

Here we see the power of the Wittig reaction. Only one alkene constitutional isomer is produced in step **3**. But we also observe a limitation of the Wittig reaction.

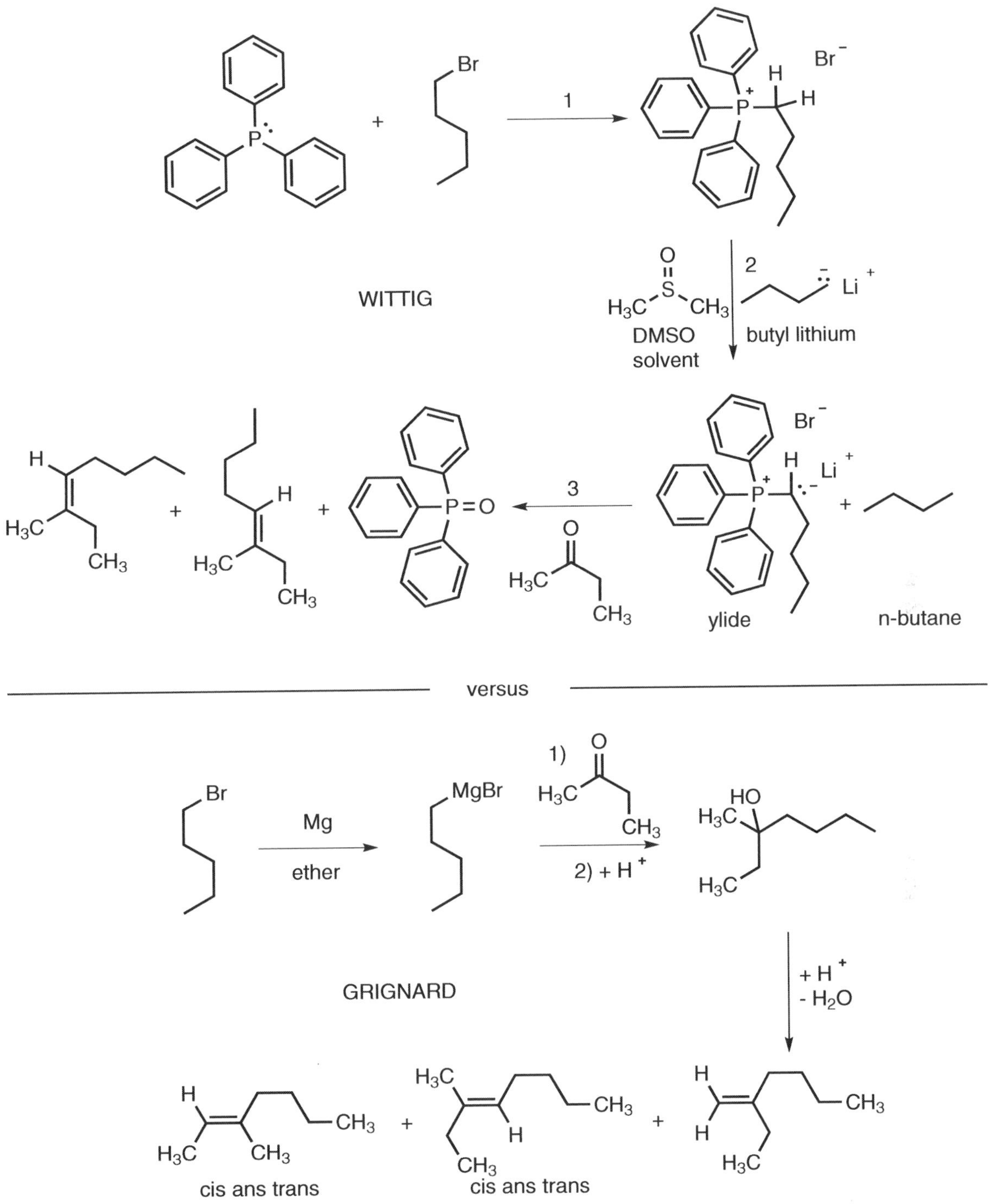

WITTIG

versus

GRIGNARD

cis ans trans   cis ans trans

Both stereoisomers of this constitutional isomer, cis (Z) and trans (E), are produced. The Wittig reaction in its general use as shown in Figure 12.31 is regiospecific (section 12.5) but not stereospecific.

You could imagine that organic chemists would try to find ways to control the stereochemical outcome of such a useful reaction and there has been success in this endeavor. A great deal of effort has led to some insight, which has been summarized by Carey and Sundberg in Part B of their valuable and comprehensive book, "Advanced Organic Chemistry," noted earlier in section 12.9. They point out that the stereochemical result of the Wittig reaction depends strongly on the nature of the ylid, its stability as measured by the strength of the base necessary to form the ylid, the reaction conditions, the solvent and temperature and salt concentration and other factors – a complex situation. But, the Wittig reaction does have the potential to be

▲ **FIGURE 12.31**

The Wittig and Grignard reactions are compared for formation of a double bond in a system where regioselectivity and stereochemistry are variable.

stereospecific, that is, to control the stereochemistry of the double bond as well as the placement of the double bond, that is, to be both regiospecific and stereospecific as Corey needed..

Figure 12.32 shows two examples of the Wittig reaction and how variation in the structure of one of the reactants largely controls the stereochemical outcome. Stereochemical control is an important matter for Corey's synthesis of the prostaglandins. From **5** in Figure 12.28, Corey had still to introduce two double bonds to gain the prostaglandins, one trans (E) and the other cis (Z) and for both synthetic steps, Corey used the Wittig reaction or a variation of this reaction, as we'll see shortly.

FIGURE 12.32 ▶

The stereochemical outcome of the Wittig reaction depends greatly on the stability of a negatively charged Intermediate.

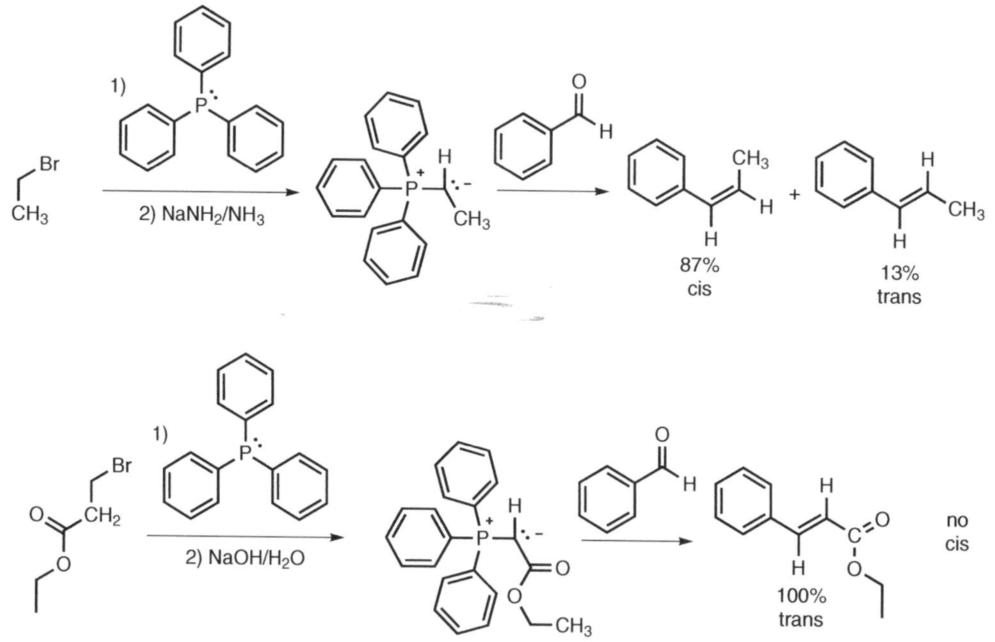

The results of the two variations on the Wittig reaction in Figure 12.32 represent what has been discovered about this reaction, that, all other things being equal, the stronger the base necessary to form the ylid the greater the amount of cis stereoisomer formed. The far weaker base necessary in the second example in Figure 12.32 arises from the adjacent ester carbonyl group in the bromide reactant. Both the positive phosphorus atom (the phosphonium ion) formed in step **1** and the adjacent carbonyl work together to enhance the acidity of the CH_2 group so that aqueous sodium hydroxide in step 2 is an adequate base. In the first example in Figure 12.32 an extraordinarily strong base, sodium hydride in liquid ammonia, is necessary (step 2) to form the ylid. Both ylides then react with the identical electrophile, benzaldehyde. What's going on?

The answer is certainly connected to a question you might have already had. What happened to the phosphorus moiety in the reactions shown in Figures 12.31 and 12.32? The answer to the latter question is shown in **Figure 12.33**, which outlines the approximate mechanism of the Wittig reaction applied to case 1 in Figure 12.32.

Nucleophilic attack at the carbonyl group by the carbanion of the ylid forms what are considered to be betaine intermediates (step **1**, Figure 12.33), which are free to rotate about the newly formed carbon-carbon bond. There is experimental evidence for the four member ring oxaphosphetane intermediate, which forms by closing of the phosphorus to oxygen bond (step **2**) and which then blocks the conformational motion allowed by the betaine. The reaction is driven by the strength of the phosphorus to oxygen double bond in the triphenyl phosphine oxide formed as a byproduct of the reaction. Such bonds between phosphorus and oxygen are exceptionally strong and are certainly the driving force for the formation of the alkene. The oxaphosphetane

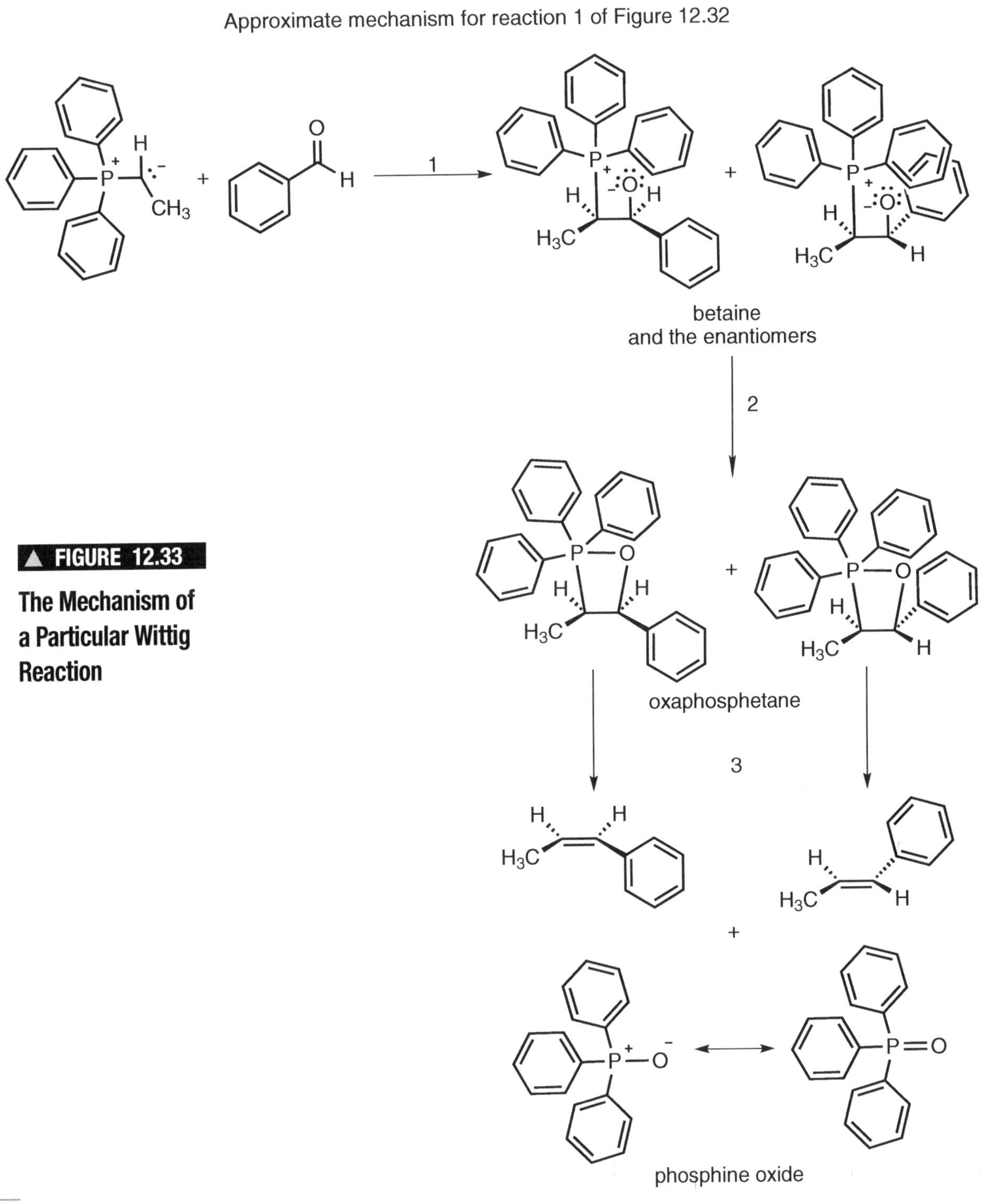

Approximate mechanism for reaction 1 of Figure 12.32

betaine
and the enantiomers

▲ FIGURE 12.33

**The Mechanism of
a Particular Wittig
Reaction**

oxaphosphetane

phosphine oxide

intermediate (Figure 12.33) then reveals the source of the stereochemical mixtures often observed in the Wittig reaction because the groups that end up on the double bond reside cis or trans on the four member intermediate. It is the stereochemistry of this intermediate, as seen in Figure 12.33, which is affected by the nature of the ylid and many other factors we can not go into here, which can control the stereochemistry of the resultant alkene. Corey was well aware of these factors and planned to make use of them to control the stereochemistry of the alkenes formed.

Before we see how Corey used the Wittig reaction for his synthetic route to the prostaglandins, lets look first at the work of the other chemist who participated in that Nobel Prize winning ceremony.

PROBLEM 12.58

One limitation of the Wittig reaction is that primary alkyl halides are necessary in the formation of the initial phosphonium salts, as for all the examples in Figures 12.31 and 12.32. What does this fact reveal about the mechanism of this step,which belongs to the category of nucleophilic substitution?

PROBLEM 12.59

How could the Wittig reaction be used in the synthesis of isobutylene ($CH_2=C(CH_3)_2$)?

PROBLEM 12.60

In step **3** in Figure 12.33 does the connection between the stereochemistry of the oxaphosphetane and the derived alkene require a concerted mechanism? Show the movement of the electrons for this step using the curved arrow nomenclature.

PROBLEM 12.61

Is the stereochemistry of the product alkene determined by the oxaphosphetane or the betaine intermediate? Can conformational motion about the carbon-carbon bond in the betaine intermediate change the stereochemical outcome?

PROBLEM 12.62

If the betaine intermediate controls the stereochemistry of the product alkenes, then what controls the stereochemistry of the betaine intermediates? Try to make a three dimensional drawing of the approach of the ylid to the carbonyl group for the reactants in Figure 12.33 that demonstrates how the betaine intermediate is formed.

PROBLEM 12.63

The reagent phosphorus tribromide, PBr_3, was noted in this section, whose reaction with an alcohol could convert an OH group to a bromide. Propose a mechanism for conversion of n-butanol ($CH_3CH_2CH_2CH_2OH$) to n-butyl bromide in which the first step for this reaction is conversion of the -OH group to $-O-PBr_2$ + Br⁻. Our study of the Wittig reaction has demonstrated the strength of the phosphorus to oxygen bond. How does this information help in creating a mechanism for substitution of Br for OH? How might your mechanism be altered if the reacting alcohol were t-butanol ($(CH_3)_3C-OH$)?

12.16

Hydroboration and Oxymercuration

THE TRANSFORMATION FROM an alkene to an alcohol involves addition of the elements of water to a double bond. This could be accomplished by simply adding water and an acid catalyst to the alkene but is rarely carried out in this manner because of the likely intermediacy of a carbocation and therefore the possibility of rearrangement (section 4.4). An example of this outcome is shown in reaction **1 Figure 12.34**.

A more controllable method in laboratory work uses chemistry based on mercury, oxymercuration, which is shown in reaction **2** Figure 12.34. In oxymercuration taking place in water solvent, divalent mercury adds to the least substituted terminus of the double bond with water minus a proton adding to the most substituted end of the double bond. In a separate step, a reducing agent, sodium borohydride, $NaBH_4$, is added (section 10.3). The mercury carbon bond is reduced to add a hydrogen atom to the carbon atom and releasing metallic mercury. I well remember the startling experience in running this reaction and seeing a small ball of shiny mercury suddenly appear at the bottom of the flask

If one uses acid catalyzed addition of water to an unsymmetrical double bond, as shown in Figure 12.34 or uses oxymercuration, the fundamental addition of the OH group to the most stable carbocation site is the overwhelming result (**1** and **2** in Figure 12.34). In other words, the addition follows Markovnikov's rule, which we've seen multiple times in the text and was introduced in section 4.10. The hydroxyl group is added to the carbon atom of the double bond, which can best support a positive charge.

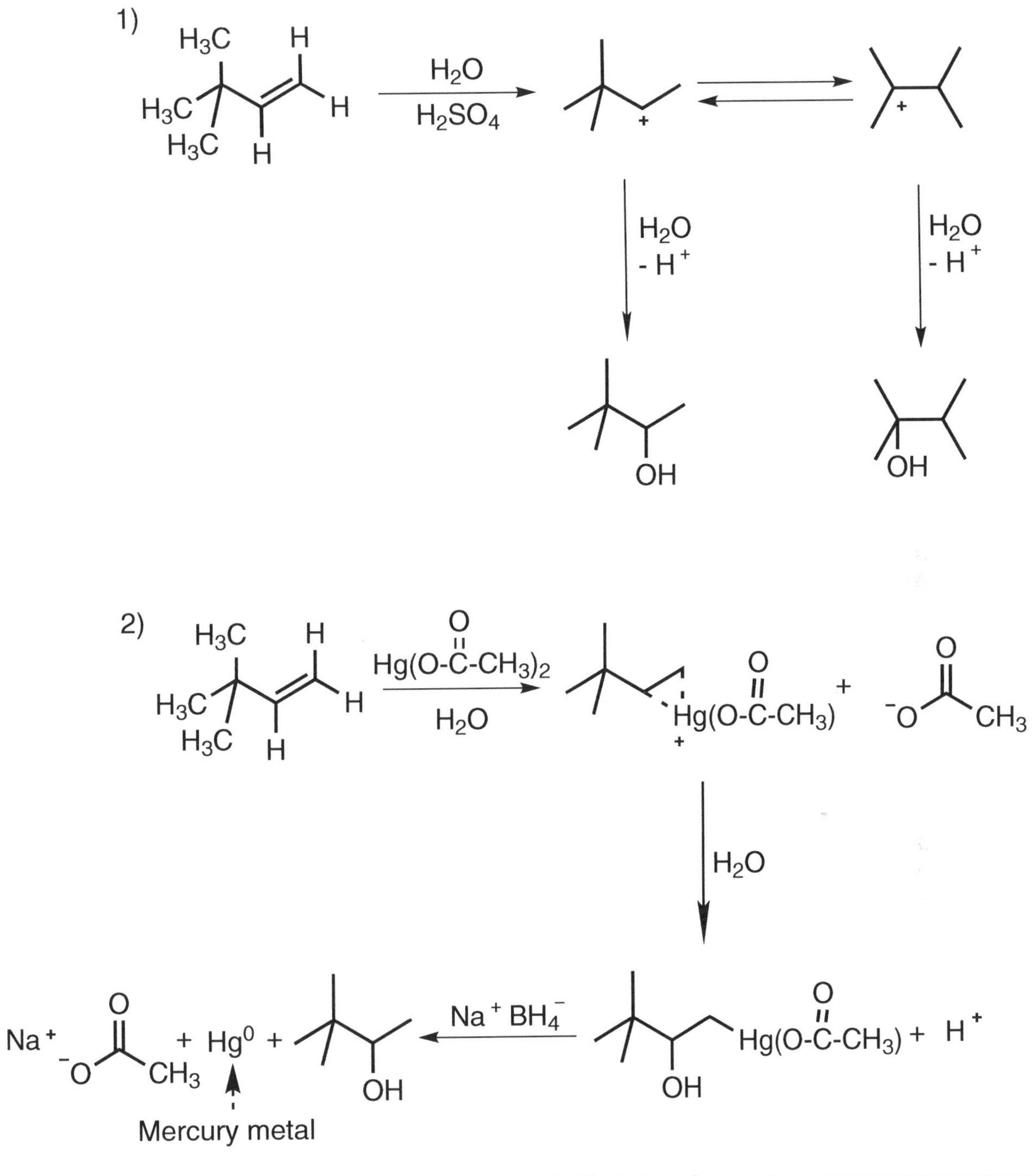

The use of boron chemistry allows addition of the elements of water to a double bond in a chemical reaction termed hydroboration. The OH group is added reliably to the carbon atom at a terminus of the double bond, which can *least* support a positive charge, which is least substituted, what is called anti-Markovnikov addition. This reaction and the mechanism are shown in **Figure 12.35**, as they are applied to the same alkene used for the compared processes in Figure 12.34.

Although boron hydride exists as a dimer, B_2H_6, in the gas phase, dissolving the gas in tetrahydrofuran (THF) dissociates the two boron trihydride (BH_3) moieties to form complexes with the THF (**1**, Figure 12.35). The driving force for dimer formation is certainly associated with the empty p orbital on trivalent boron, an orbital, which although involved in the complex with THF nevertheless maintains an electrophilic

▲ **FIGURE 12.34**

Addition of Water to Double Bonds: the Mechanism of Oxymercuration

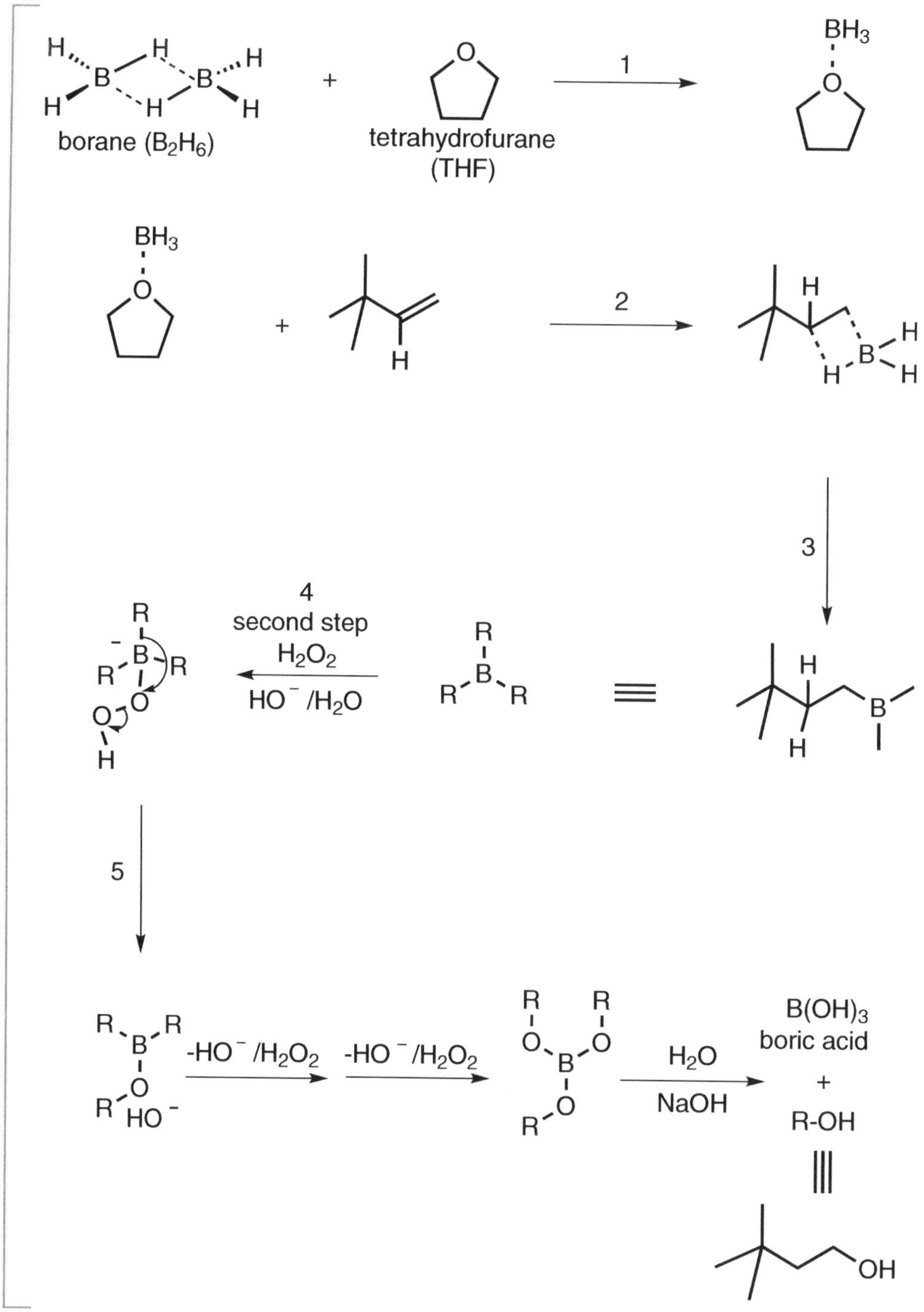

▲ **FIGURE 12.35**

Addition of Water to Double Bonds: the Mechanism of Hydroboration

character that is adequate to disrupt the π electrons of a double bond (step **2**).

As shown in step **3** (Figure 12.35) the interaction between BH$_3$ and the π bond breaks down to add a hydrogen atom to the most substituted end of the double bond with the boron bonded to the least substituted end. There are multiple reasons given for this mode of addition, one of them being that the more crowded boron atom is placed at the least substituted site for steric reasons.

The BH$_3$ group may add to three molecules of alkene to form what is shown as R$_3$B, or depending on the nature of the alkene and the proportions of reactant, may only add to one alkene to form, RBH$_2$ or R$_2$BH. The multiplicity of addition, however, has no effect on the details of the mechanistic picture portrayed in Figure 12.35.

In the next step, the experimenter then adds hydrogen peroxide and aqueous base. Hydrogen peroxide, H$_2$O$_2$, reacts with the HO$^-$ to form H-O-O$^-$, which acts as a nucleophile, driven to react at the empty p orbital on boron, to form the intermediate produced by step **4**. The step that quickly follows is driven by two factors: the weak –O–O– peroxide bond; and the strong bond that forms between boron and oxygen. Step **5** simultaneously breaks the weak bond and forms the strong bond leading to the rearranged structure. The aqueous base then hydrolyzes the boron-oxygen bond formed, to create boric acid and the product alcohol with addition of water to the double bond occurring as shown (Figure 12.35) in an anti-Markovnikov manner.

In synthetic work, hydroxyl is a commonly encountered functional group in natural products, as seen in the structures of both cholesterol and prostaglandin. OH often also plays a critical role because of its ease of transformation into the carbonyl functional group, which then opens opportunities for forming carbon-carbon bonds via the various carbonyl based reactions we have learned about largely in this chapter and in their biological roles in Chapter 7 and 8.

In laboratory work, the ability to control the site of the OH group is essential with hydroboration offering what had not been possible by any other method (Figure 12.34).

One interesting fact connecting the two men who stood together on the stage at Stockholm in 1979 is that the driving force for the Wittig reaction and for Brown's hydroboration is the large bond strength between phosphorus and oxygen and between boron and oxygen. We've seen how this driving force works in Figure 12.33 for the Wittig reaction. Figure 12.35 shows the mechanism for hydroboration illustrating this bond strength factor for the boron based reaction, a driving force that applies equally for many other boron based synthetically valuable reactions created by H. C. Brown, which we won't be showing here but can be found in more advanced treatises on organic chemistry.

PROBLEM 12.64
Addition of the elements of water to form the alcohol from 5-methyl-2-hexene, $CH_3-CH=CH-CH_2-CH(CH_3)_2$, using any method yields a mixture of constitutional isomers. Explain this result.

PROBLEM 12.65
Changing the structure of the alkene used in Problem 12.64 to 3-methyl-2-hexene, $CH_3-CH=C(CH_3)-CH_2-CH_2-CH_3$, causes hydroboration to yield a single isomer, and oxymercuration a single different isomer but with acid catalyzed addition of water yielding a mixture of isomers. Show the isomers that are formed and offer an explanation for the results.

PROBLEM 12.66
Using the Wittig reactions and any five-carbon or less saturated or unsaturated hydrocarbon, suggest a synthetic approach for the syntheses of both alkenes in the previous two problems.

12.17

The Importance of the Wittig Reaction to Corey's Synthesis of the Prostaglandins

IN FIGURE 12.24 IN THE PROSTAGLANDIN STRUCTURE shown, there are two hydrocarbon chains pendant to the five membered ring, one with a cis and the other with a trans double bond. Phosphorus-based Wittig reactions were used by Corey to introduce both of these alkene structural elements. Let's follow Corey's chemical path to see how this was done. For this purpose, let's pick up from structure **5** in Figures 12.24 and 12.28.

Corey's intent, shown in Figure 12.28, was to convert the CH_2OCH_3 in **2** to the aldehyde group, CH=O, in **5** as the site for the attack of the ylid in the Wittig reaction that would lead to the alky chain with the trans double bond in the final structure (Figure 12.24).

Although we are not looking at the intervening steps in detail it is clear that removing the methyl ether group from **2** would yield a structure (not shown in Figure 12.28)

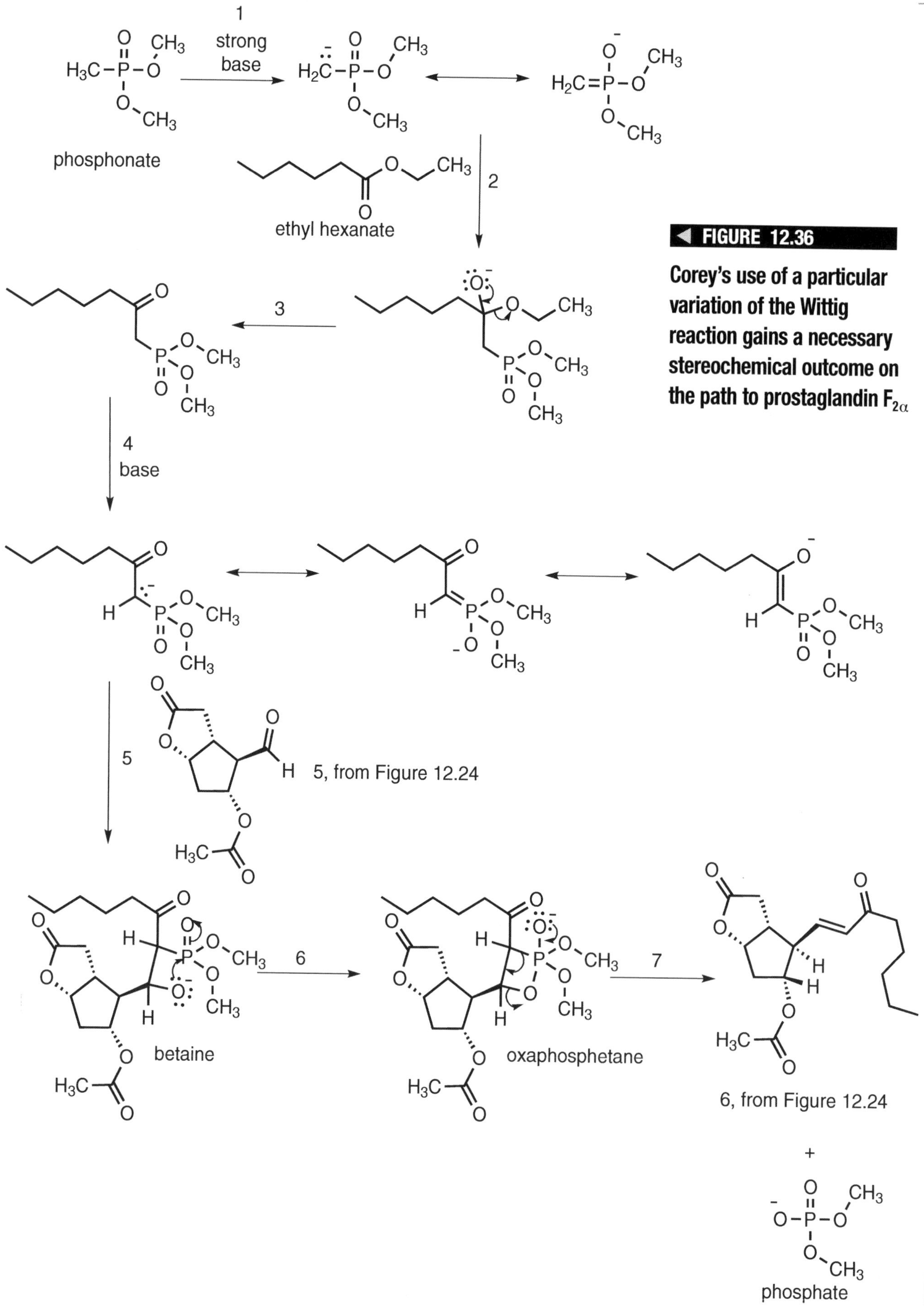

◄ FIGURE 12.36

Corey's use of a particular variation of the Wittig reaction gains a necessary stereochemical outcome on the path to prostaglandin F$_{2\alpha}$

with two hydroxyl groups, primary, RCH_2OH and secondary, R_2CHOH. The necessary oxidation of the primary hydroxyl group to the aldehyde would then simultaneously oxidize the secondary hydroxyl group to the unwanted ketone. For this reason the synthetic steps were organized so that the secondary hydroxyl group in 2 (Figure 12.28) was converted to the acetate ester to protect it while the necessary chemistry was carried out on the primary hydroxyl group. In the course of this chemistry a reagent, we'll not discuss, was used to replace the carbon iodine bond in **2** (Figure 12.28) with a C-H bond. In this manner 5 (Figure 12.24) was produced from 2 (Figure 12.28).

The Wittig intermediate used in Figure 12.36 (reaction **1**) is a variation of the ylid intermediate we've seen before. Here a phosphonate is used, which is converted to a phosphate (Figure 12.36), in contrast to the original formulation in which a phosphine is converted to a phosphine oxide (Figure 12.33). This exceptionally useful variation of the Wittig reaction (Figure 12.36) was created by W. S. Wadsworth, Jr, and W. D. Emmons at the Rohm and Haas Company in Philadelphia and published in the Journal of the American Chemical Society in 1961.

In those times, much basic research that was not of specific immediate use to commercial purposes was carried out in industrial laboratories, certainly stimulated by the discoveries in polymer science arising from such basic work, as discussed in Chapter 9 (section 9.1). However, in an article I read, published in 1983, Wadsworth, who at that time had become a professor at South Dakota State University, pointed out the possible practical implication of their work. He and Emmons were looking for a molecule containing both a double bond and a phosphorus based functional group with the possibility of forming fire resistant polymers.

The resonance-stabilized carbanion phosphonate (reaction **1**) is a powerful nucleophile and the reaction with ethyl hexanoate is no surprise, yielding first the expected tetracoordinate intermediate (reaction **2**), which then ejects the ethoxide anion (reaction **3**). The product of reaction 3 (**Figure 12.36**) plays the equivalent role as the intermediate produced in reaction **1**, shown in the outline of the Wittig reaction (Figure 12.31).

However, the compared intermediates differ greatly in their acidity. The loss of the proton (reaction **4** Figure 12.36) requires a far weaker base than that required for the equivalent Wittig intermediate produced by reaction **1** in Figure 12.31. The electron delocalization of the negatively charged phosphonate produced by reaction **4**, represented by the three resonance structures exhibited, is the source of both the ease of proton abstraction and, most importantly to Corey, the control of the stereochemistry of the alkene produced (reactions **5**, **6** and **7** Figure 12.36).

Let's return for a moment to Figure 12.32, where we observed that the percentage of trans over cis alkene produced by the Wittig reaction correlates with the resonance stabilization of the intermediate ylid. The equivalent intermediate in this variation of the Wittig reaction produced by reaction **4** (Figure 12.36) is greatly stabilized in this manner by the carbanion site flanked by both a C=O and a P=O group. As planned by Corey, this factor produces not only the desired trans double bond (reaction **7**) (structure **6** in Figure 8.24) but, as well, a carbonyl group placed so that the necessary OH group at this site can be easily gained by a simple reduction by a variant of sodium borohydride (sections 10.3 and 12.16).

Beautiful—you get two for the price of one, one carbonyl group yields: control of the stereochemistry of the double bond via stabilization of the ylid; the functional group necessary for a step toward the synthetic goal.

We're not far now from Prostaglandin $F_{2\alpha}$. For this result we need another Wittig reaction, which, however, has to be preceded by a few simple steps, which yield still another example of the importance and wide variability of protecting groups in synthetic work.

PROBLEM 12.67

In Figure 12.36, the betaine product of step **5** and the oxaphosphetane of step **6** are not drawn in the figiure to show the stereochemistry leading to the formation of the trans double bond. Try to improve on these structural drawings.

PROBLEM 12.68
Trace the electron distribution changes in the three resonance structures in Figure 12.36.

PROBLEM 12.69
In the discussion in this section, protecting groups played an essential role in the precursor molecules. What protective groups were used?

12.18 Protection groups are necessary.

THE PATH FROM **6**, where we are now, to **7** (Figure 12.24), involves some transformations of hydroxyl functional groups, with which we are already familiar in other chemical processes. First, Corey took care of the carbonyl group left over from the modified Wittig reaction (Figure 12.36). Reduction with zinc borohydride ($ZnBH_4$) formed the hydroxyl group at this site. I don't know why sodium borohydride was not used. At any rate it was not possible at that time to invent a way to form only the desired (S) configuration of **1** (**Figure 12.37**). A mixture of epimers (diastereomers) was formed.

But this problem could be easily solved by separation of the epimers, **1**(R) and **1**(S), followed by oxidation of the unwanted epimer **1**(R) back to the ketone, which could then be reduced to the mixture of epimers. Repeating this procedure several times was sufficient to gain an acceptable yield of **1**(S). Little was wasted.

The absence of stereochemical control in forming the epimers of **1** (Figure 12.37) is the exception that proves the rule. To help to appreciate the synthesis designed by Corey, take notice of the stereochemical control. In the final prostaglandin structure there are five chiral stereocenters and two double bonds with specific cis and trans configurations. All these stereochemical choices were perfectly controlled with the sole exception of the stereocenter produced in the formation of **1** (Figure 12.37).

Esters can be hydrolyzed to their precursor alcohols and carboxylic acids (termed saponification for fatty esters (sections 7.3 and 7.4) and treatment with potassium carbonate in water, a weak base, accomplishes this hydrolysis only for the acetate ester and not for the lactone, producing structure **2** in Figure 12.37. A stronger base under different conditions, or maybe the same base under the same conditions but for a longer time or at higher temperature, would have also hydrolyzed the lactone, which had to be avoided, for a reason we'll see shortly.

In the synthesis of a complex molecule the situation often arises when a necessary reactant intended for a specific purpose might have the capacity to react elsewhere in a molecule than at the site that is intended. We've just seen this situation in the hydrolysis of the acetate ester while avoiding the hydrolysis of the lactone (structure **1**, Figure 12.37).

In many other situations, where it is possible, which was not the situation for **1**, synthetic chemists use protecting groups to take care of this possibility. The unintended target of the reactant must be converted to a derivative, a protecting group, which blocks the unintended reaction.

However, such a protecting group must be formed and then removed in a manner that does not interfere with the intended synthetic path. We saw the use of a protecting group in section 12.5 when Woodward was faced with a carbanion that would have interfered with his intended synthetic path and another use of a protecting group for OH groups that, in themselves, were protecting a double bond, in section 12.7 (Figure 12.18).

Corey was faced with the problem of protecting the two hydroxyl groups in structure **2** in Figure 12.37. The Wittig reagent he intended to use has a carbanion site, which is not only nucleophilic but, also, strongly basic. The basic strength would have caused abstraction of a proton from a hydroxyl group therefore making the Wittig reaction in the presence of an OH group impossible.

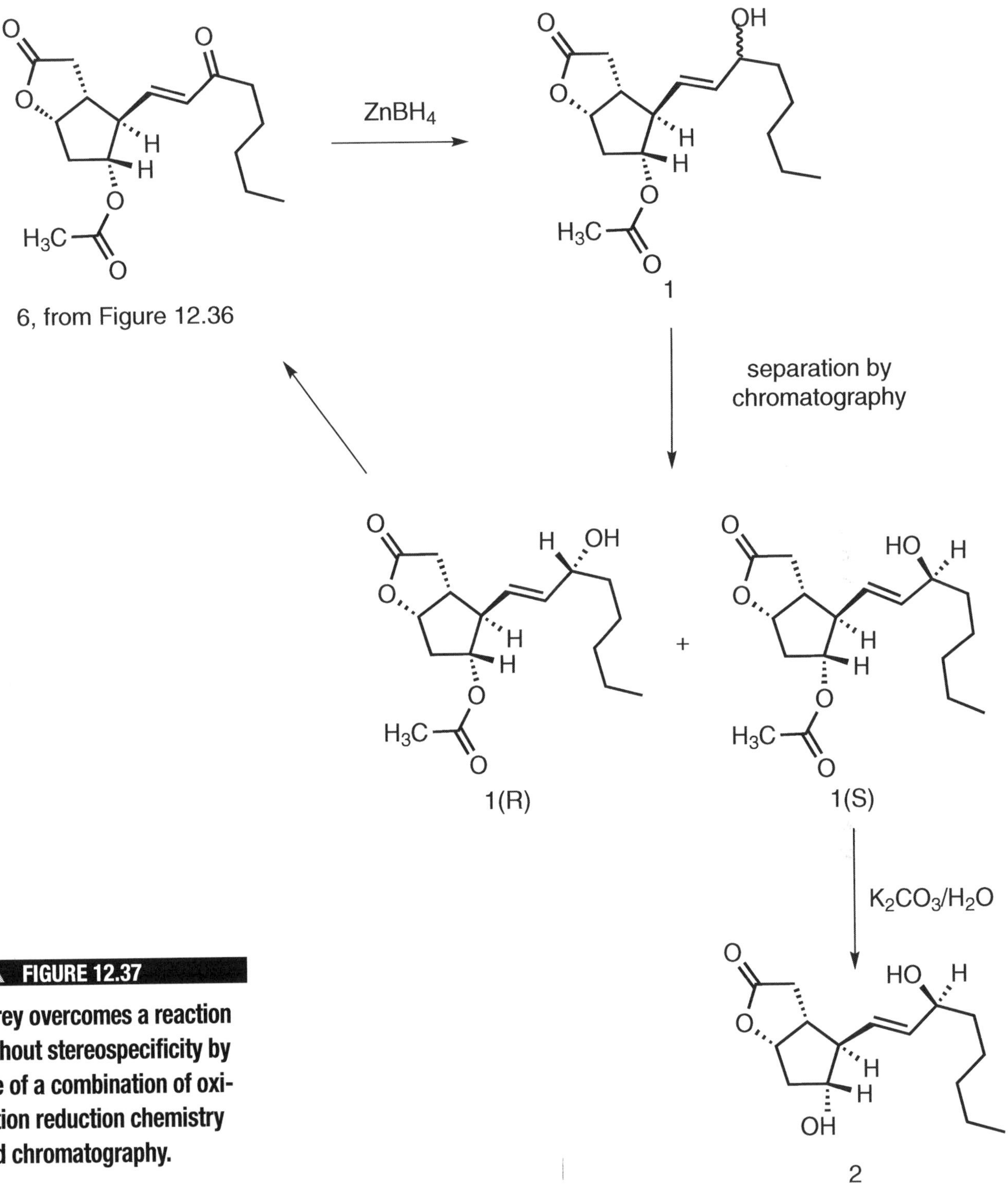

6, from Figure 12.36

1

separation by
chromatography

1(R)

+

1(S)

▲ **FIGURE 12.37**

Corey overcomes a reaction without stereospecificity by use of a combination of oxidation reduction chemistry and chromatography.

K_2CO_3/H_2O

2

Why not have simply left the acetate group unhydrolyzed, after all it is protecting one of the OH groups. And then he could have made an acetate ester of the other OH group, in **1**(S). The problem with having acetate esters is that the chemistry for their removal as the structure comes closer to the final prostaglandin is a bit too severe for the exceptional sensitivity of the prostaglandin structure. Corey needed a protecting group that could be put on and taken off under very mild acid conditions and would be stable in the presence of base. The protection of these two OH groups was best done with a classic reagent for this purpose, dihydropyran, as shown in Figure 12.38.

The structures protecting the two OH groups of the prostaglandin intermediate shown in structure **1** in Figure 12.38 can be recognized as acetals. We've looked at acetals and ketal

An acetal protecting group is used in the synthesis of prostaglandin F$_{2\alpha}$ by use of dihydropyran.

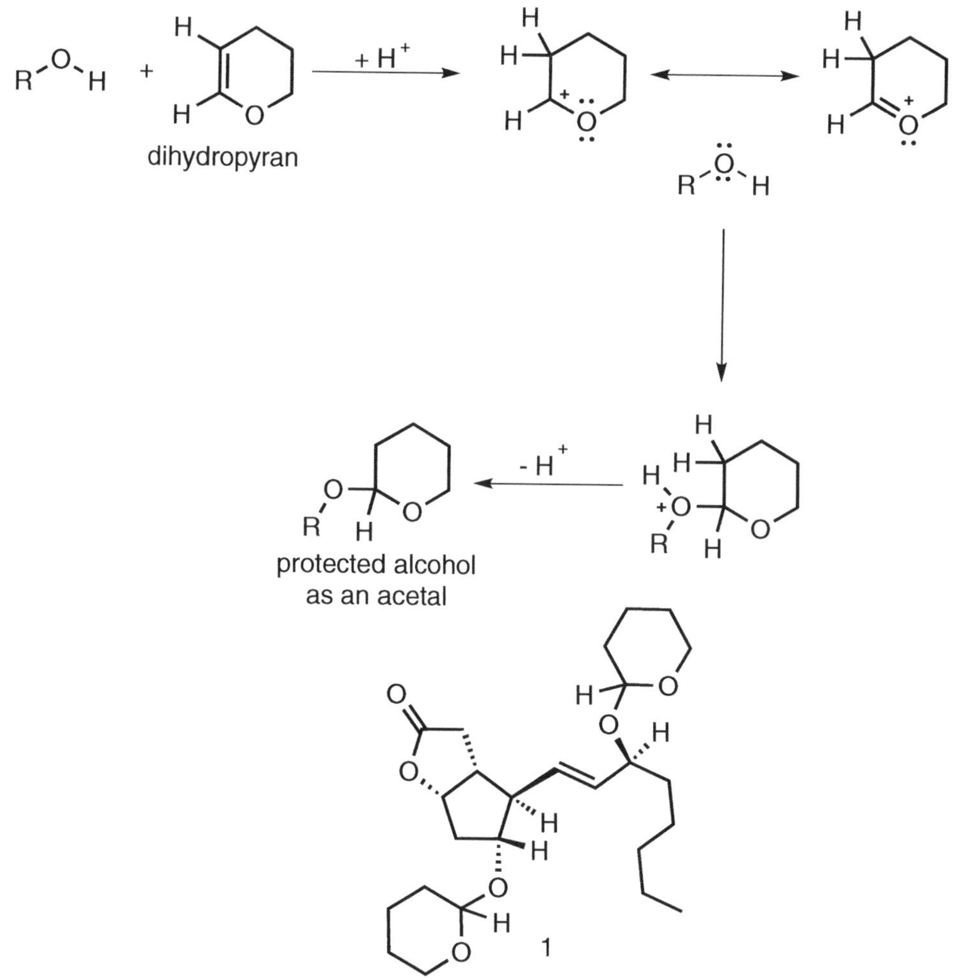

dihydropyran

protected alcohol as an acetal

1

before with regard to Woodward's use of protecting groups (Figure 12.17) and long before that in the structures of cellulose and starch, where, although not noted at that point, the units are linked by acetal groups (Figure 1.1). In fact, acetal and hemiacetal groups are commonly found functional groups in all sugar chemistry and are found repeatedly all through Chapters 1 and 2. In sucrose (Figure 3.9) we came across an acetal and a ketal functional group in one molecule, although our focus was elsewhere at that point.

An interesting point about the OH protecting group derived from dihydropyran is that although it is certainly an example of an acetal, there was no aldehyde involved in its formation. In a manner of speaking the acetal has been formed by a trick with no carbonyl involved. The mechanism is shown in Figure 12.38 where the power of the dihydropyran structure to capture a nucleophilic hydroxyl group is shown to derive from addition of a proton to form a resonance stabilized carbocation.

PROBLEM 12.70

The Wittig reaction can not be conducted in the presence of active hydrogen, such as in the presence of water. What other active hydrogen sources can you list and what other reactions, such as the Grignard reaction for one example, would also be sensitive to the presence of active hydrogen sources?

PROBLEM 12.71

The following molecule, 5-hydroxypentanal (O=CH-CH$_2$CH$_2$CH$_2$CH$_2$OH), in the presence of an alcohol, ROH, would form the same kind of protecting group as formed in Figure 12.38 although water would be eliminated. Outline a mechanism for this transformation.

PROBLEM 12.72

The acid catalyzed formation of the protecting group in Figure 12.28 does not eliminate water and therefore no Dean-Stark trap (section 12.7) is necessary. However, if so then how can one avoid the reverse of the steps shown in this figure?

PROBLEM 12.73

The protecting group shown in Figure 12.38 when used for an alcohol like 1-butanol offers no stereochemical complexity, which is not the situation for 2-butanol or for the prostaglandin intermediate shown in the figure. Why?

STRUCTURE **1** IN FIGURE 12.38 (structure **7** in Figure 12.24) differs from the prostaglandin structure (Figure 12.24) in several ways, but most notably by the absence of the second unsaturated chain that is pendant to the five membered ring. Corey used a Wittig reaction to add this chain but, as we've seen for this reaction, a carbonyl group must be present to react with the nucleophilic Wittig reagent (Figures 12.31–12.33).

In 1963, six years before Corey published the synthesis we are studying here, two chemists, J. Schmidlin and A. Wettstein, working at CIBA, the prominent Swiss pharmaceutical company, published a paper in the distinguished journal Helvetica Chimica Acta with a chemical reaction that likely influenced Corey's plans. The new reaction demonstrated that the lactone that was formed in the step using KI_3 (Figures 12.28 and 12.29) could now serve the purpose Corey had been planning. The reaction Corey found so useful was developed as part of CIBA's research in the

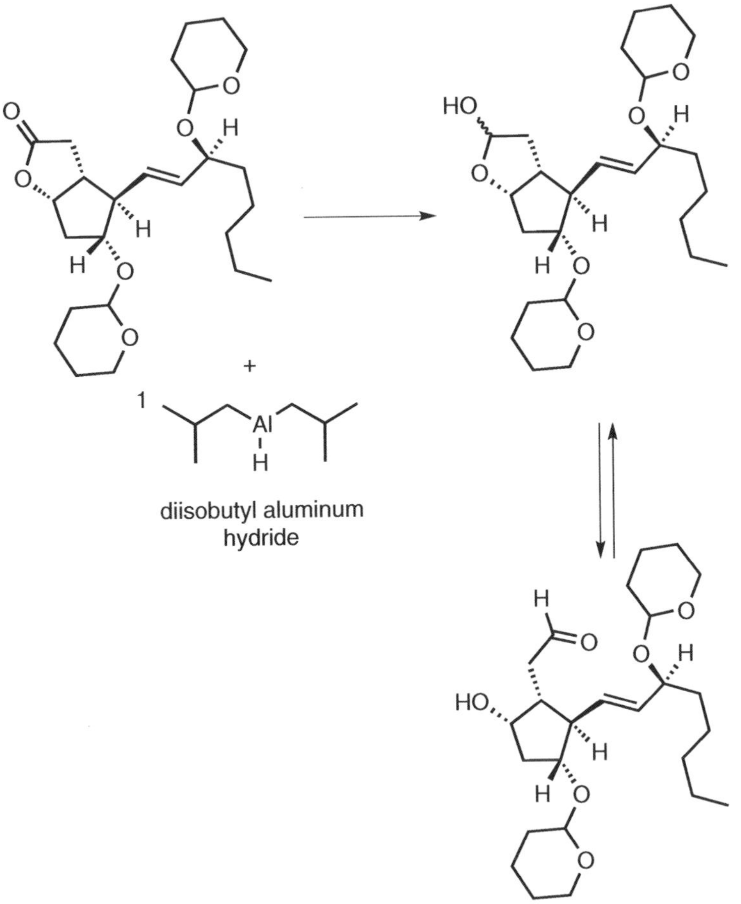

12.19

The End Game–the Wittig Reaction one more Time and a Protecting Group in Disguise

diisobutyl aluminum hydride

◀ **FIGURE 12.39**

A sterically hindered reducing agent reveals the hidden nature of a protecting group we were unaware of, producing an aldehyde that can be used for another Wittig reaction.

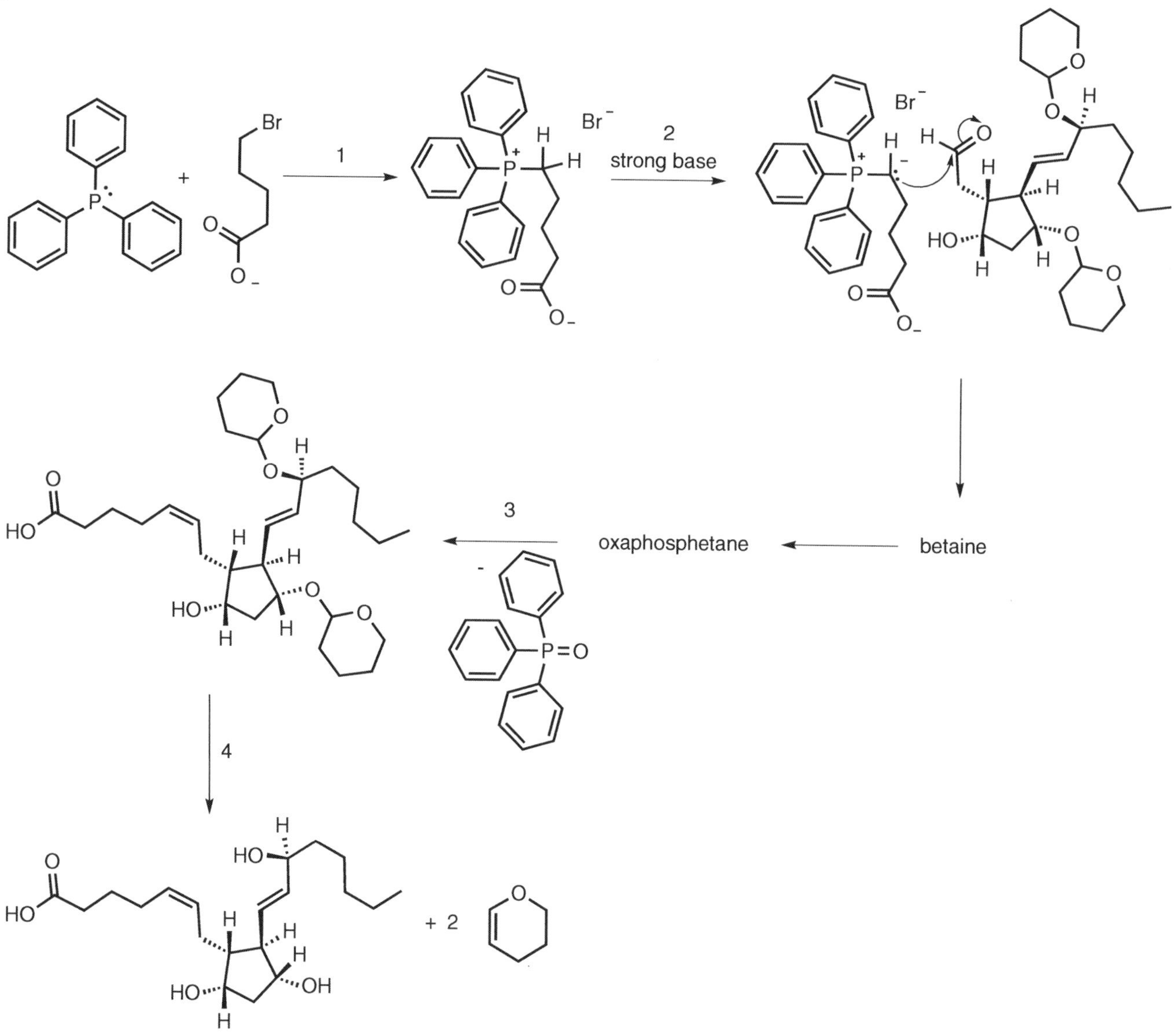

FIGURE 12.40 ▶

The Wittig reaction with the aldehyde group revealed in figure 12.39 is chosen to yield the desired stereochemistry by choosing the stabilization characteristics of the negatively charged intermediate, reaching therefore the final goal, prostaglandin F$_{2\alpha}$

steroid area in the 1950s and 60s for which much commercial value was expected. This effort was in competition with the development of what turned out to be the birth control pill pioneered by Carl Djerassi, whom we met in section 9.3 and who worked with the California pharmaceutical company, Syntex. Wettstein, at this time, was the director of research in this area for CIBA. This is another of the many many examples demonstrating that the results of research are often of unpredictable benefit.

The CIBA chemists' method was able to convert a lactone to a hemiacetal by reducing the carbonyl group with diisobutyl aluminum hydride, as shown for the transformation in Figure 12.39. It makes sense that the two isobutyl groups in their reagent were too sterically hindered to reduce the lactone further, as would have occurred with lithium aluminum hydride.

All hemiacetals are in equilibrium with the aldehydes and alcohols from which they are formed, as shown in **Figure 12.39**. We first encountered this equilibrium process in the mutarotation of glucose (Figure 3.11) but until this chapter we had not looked closely at the mechanism or used the name hemiacetal (section 12.7, Figure 12.17).

The lactone produced in the KI$_3$ step in Figure 12.28, we now realize was a protecting group (section 12.18) for an aldehyde group, but in disguise, with its true identity revealed by the reaction published by the CIBA group. This revealed aldehyde group (Figure 12.39) is now the partner for the Wittig reaction (Figure 12.40).

Figure 12.32 exhibited the alkene stereochemistry produced by two exemplary Wittig reactions in which the ylid intermediate varied in stability. The results followed the general rule that resonance stabilization of the ylid intermediate favored trans

prostaglandin E₂   prostaglandin E₁

stereochemistry of the alkene produced. In the use of the analog of the Wittig reaction used in Figure 12.36, the negatively charged intermediate was resonance stabilized as a consequence of both the adjacent C=O and P=O functions and the general rule worked beautifully to yield overwhelmingly the trans double bond.

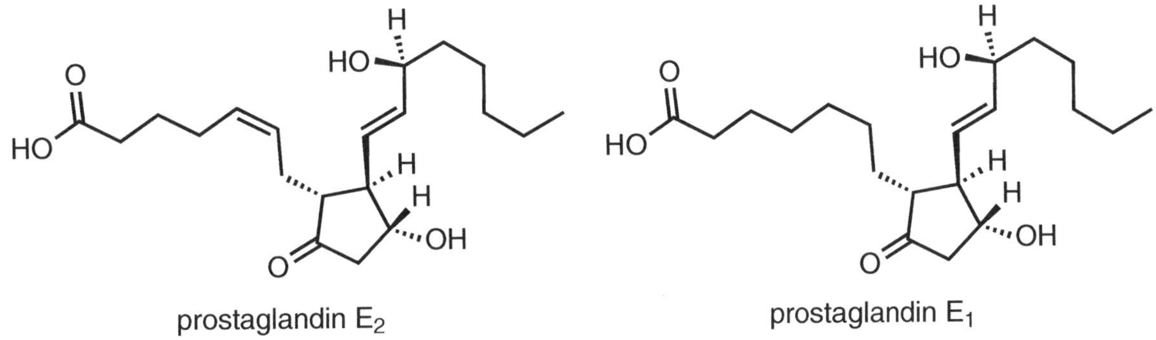

As shown in **Figure 12.40**, Corey produced (reaction **2**) an ylid that was not resonance stabilized, in addition to using a solvent, dimethyl sulfoxide ($(CH_3)S=O(CH_3)$), which was known to influence the Wittig reaction to form the cis double bond. The cis double bond was, in fact, produced (reaction **3**) almost exclusively. Both the ylid stability and the solvent were directed to gaining the required cis stereochemistry of the alkene bond.

At this point there must have been a thrill in the laboratory with the realization by the hard working group responsible for traveling this beautifully designed synthetic path, full of challenges overcome, that a minor step remained – removal of the two pyran protecting groups – to attain their goal (Figure 12.40). Mild dilute aqueous acid did the trick and Prostaglandin F$_{2\alpha}$ (Figures 12.24 and 12.40) was in hand, identical to the natural material.

To add to the success, Corey also applied the same method, with minor changes at the end of the synthesis, to yield two other important prostaglandins, the structures shown in **Figure 12.41**.

▲ **FIGURE 12.41**

The clever nature of the approach to prostaglandin F$_{2\alpha}$ allows easy synthetic access to both prostaglandin E$_1$ and E$_2$

PROBLEM 12.74
What would have been the result of using lithium aluminum hydride rather than diisobutyl aluminum hydride for the reduction of the lactone in Figure 12.39?

PROBLEM 12.75
Why might it have been necessary to waste an equivalent of the strong base used to synthesize the ylid in Figure 12.40?

PROBLEM 12.76
The solvent used was dimethyl sulfoxide for the Wittig reaction in Figure 12.40 and the base was derived from this solvent, the dimsyl anion, $(CH_3)S=O(CH_2:)$. Describe both dimethyl sulfoxide and the derived conjugate base using hybridization of orbitals and by incorporating all electrons and determining the presence of formal charges in these structures.

PROBLEM 12.77
In the classical Wittig reaction one end product is always triphenyl phosphine oxide as shown for one example in Figure 12.40. The structure of this molecule can be expressed in different ways with and without a phosphorus to oxygen double bond. Analyze these two resonance structures using hybridization of orbitals and considerations of formal charge.

PROBLEM 12.78

Suggest a way to alter the steps in Figure 12.40 to attain Prostaglandin E$_2$.

PROBLEM 12.79

Cis double bonds are generally more unstable than trans double bonds as we've come across a few times in the book (section 6.7, problem 6.18 and answer). What are the examples that demonstrate this relative stability of double bonds? Taking account of the fact that energies of activation are generally smaller for similar reactions according to the stability of the starting molecule, offer a synthetic approach altering Figure 12.41 to produce prostaglandin E$_1$ from prostaglandin E$_2$.

PROBLEM 12.80

Draw three dimensional structures for both the betaine and oxaphosphetane intermediates that would be necessary to yield the cis alkene product formed in Figure 12.40 (reaction **3**).

12.20

Retrosynthesis

E. J. COREY PRODUCED many beautifully designed synthetic paths to biologically important molecules in his continuing long career and the pharmaceutical industry has been significantly affected by his work. In addition, Corey has formalized for the world of synthetic chemistry the retrosynthetic method that guided his work. The fundamental idea was not to look at the starting material and imagine how to gain the final structure, but rather to work backwards in determining which reactions would lead to simpler and eventually easily available starting materials. Corey spoke of this method in his Nobel Prize winning lecture and chemists think consciously in this way for virtually all synthetic work. The retrosynthetic method has been formalized as the essential tool of synthetic

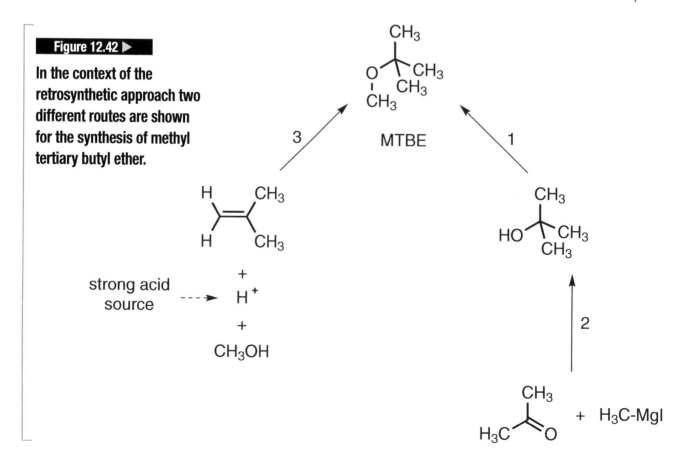

Figure 12.42 ▶

In the context of the retrosynthetic approach two different routes are shown for the synthesis of methyl tertiary butyl ether.

organic chemistry. Let's try it out on molecules, whose complexities would likely not be beyond the repertoire of reactions available at this stage of your studies and also will act to introduce you to new reactions.

Figure 12.42 shows the structure of a gasoline additive, methyl tertiary butyl ether (MTBE), which has played an important role in acting as an octane number enhancer when lead additives were removed from gasoline (section 4.8). This figure also outlines two retrosynthetic paths to MTBE. The two paths differ greatly but both demonstrate that the prerequisite for using the retrosynthetic method effectively is knowledge of chemical reactions and the mechanism of chemical reactions. Nothing can replace the hard work and experience associated with gaining this knowledge.

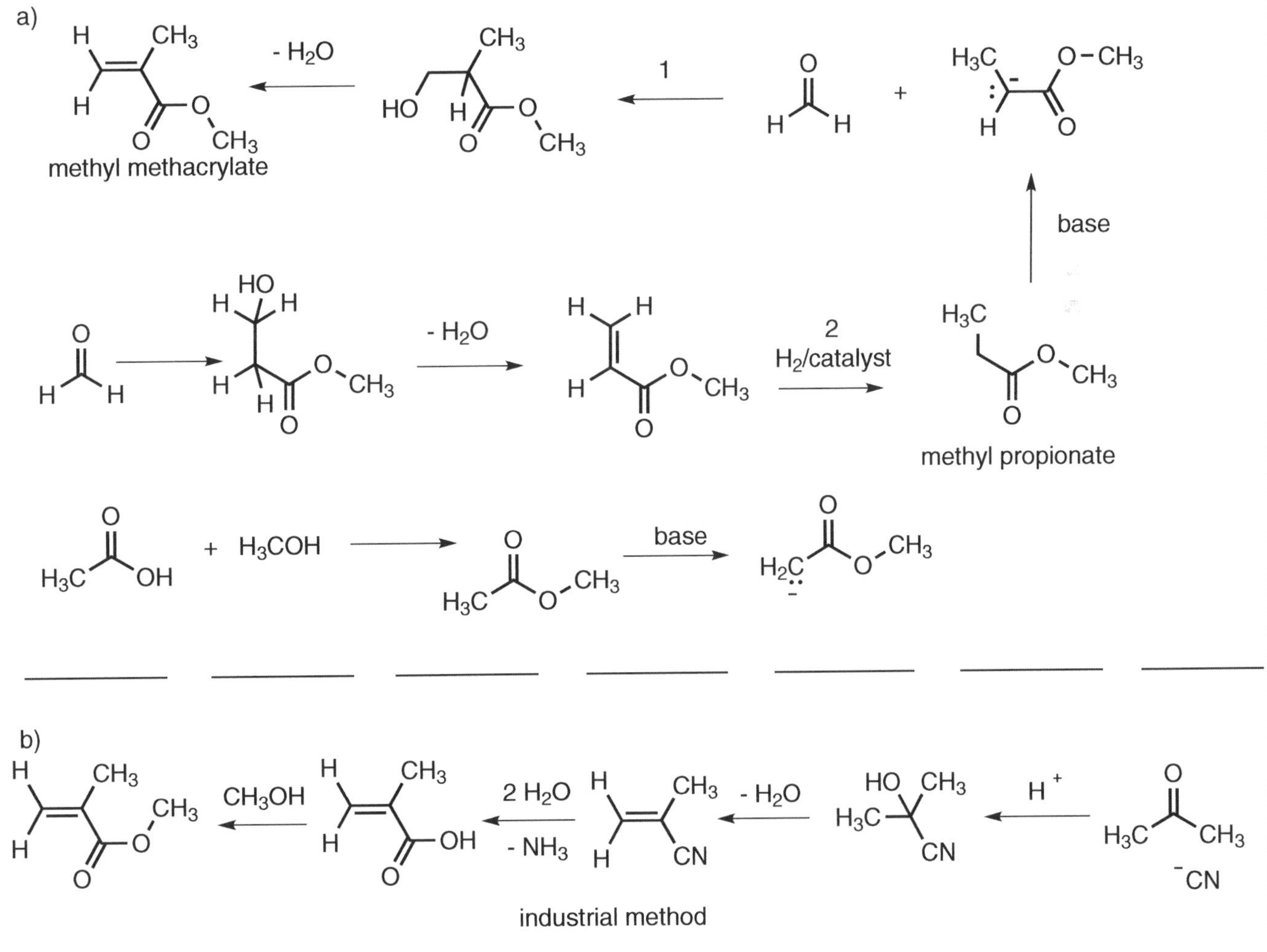

FIGURE 12.43

Two retrosynthetic routes to the precursor to Plexiglas, methyl methacrylate, are compared.

The first retrosynthetic path in Figure 12.42 imagines the direct precursor to MTBE to be the alcohol shown, tertiary butyl alcohol,, which then could gain the desired ether product by reaction of this alcohol with methyl bound to a leaving group, such as CH_3Br, via an S_N2 reaction (**1**).

Here we might mention **Alexander Williamson** who was born to a wealthy family in England and was brought up in Germany and France. Williamson was greatly respected among nineteenth century chemists, with his importance, by a certain kind of measure, judged by incorporation of his photograph and eight pages about him in Partington's classic book (section 4.9). Nevertheless, as Partington points out, Williamson's effective work as a chemist spanned only five years, between 1850 and 1855 as professor at University College London, although he held that position until 1887. During the active time in his professorial role, he created a method for synthesizing ethers, which carries his name, the Williamson ether synthesis and which is the reaction called to mind in our retrosynthetic path to MTBE (reaction **1**).

Continuing to work backwards, we ask how to gain the structure of the alcohol. A reasonable source would be the Grignard reaction between methyl iodide and acetone (reaction **2**, Figure 12.42). Here we have come to the end of our retrosynthetic path considering that both these reactants are readily available. The retrosynthetic path is certainly reasonable and would work admirably to gain the MTBE structure.

However, this is not the path that industry uses for this molecule, which is a single step reaction (**3** Figure 12.42) in which isobutylene, available in large amounts from catalytic cracking of petroleum (section 4.1 and following sections), when catalyzed by acid, reacts with methanol, to produced MTBE. In comparing the two retrosynthetic paths to this molecule one has to evaluate the conditions available. In a laboratory, handling the gaseous isobutene might be troublesome and if efficiency and economic considerations are not paramount, the Grignard route might be best.

Figure 12.43 shows the structure of methyl methacrylate, the monomer from which Plexiglas is made. This is a very important molecule to the chemical industry synthesized in the billions of pounds each year around the world. You might find it interesting to see the various methods used and considered by industry to gain this structure. Even an economic advantage of fractions of a penny per pound makes a big difference when such large volumes are involved.

This quest for the most economical synthesis of methyl methacrylate is the focus of a chapter in the Green and Wittcoff text (section 6.12). But here, let's try our hand using the retrosynthetic path. In any attempt at this task, there is no absolutely correct answer, only that the answer involves reasonable chemistry. All retrosynthetic paths are necessarily limited by the repertoire of chemical reactions and the conditions of their use known to the creator of the path.

The retrosynthetic path in Figure 12.43 (a) imagines gaining the final structure by elimination of water from the product of an Aldol reaction between the methyl ester of propionic acid and formaldehyde (reaction **1**). Although methyl propionate is certainly easily available, we can imagine it produced, as shown by an identical sequence to reaction **1**, involving elimination of water from the Aldol reaction between methyl acetate and formaldehyde followed by hydrogenation of the resulting double bond (reaction **2**). And if we wish, we could take another step backwards in producing methyl acetate from acetic acid and methanol (reaction **3**).

An important proposal made by the major chemical company BASF resembles a retrosynthetic path involving propanal (H_3C-CH_2-CH=O) and formaldehyde, which hardly needs comment as it so closely resembles the route we've proposed (**1-3**), so we won't show it. BASF is attracted to this synthetic approach because they have developed a catalyst allowing addition of hydrogen and carbon monoxide to ethylene – hydroformylation – which economically produces propanal.

The classic and still widely used industrial route to methyl methacrylate derives from something we learned in section 12.5 in Woodward's synthesis of cholesterol, which is the easy conversion of nitriles, R-C≡N, to carboxylic acids, R-CO_2H. We had learned earlier that nitrile anion, the conjugate base of HCN, is an excellent nucleophile (section 10.4 and following sections). These two facts stand behind the industrial route to methyl methacrylate (in broad outline) shown in Figure 12.43 (b).

Nucleophilic attack of CN⁻ at the carbonyl group of acetone produces a hydroxyl group. Loss of water produces the necessary double bond while the CN group produces the necessary carboxyl group, which is then converted to the methyl ester with methanol. In the actual industrial method, the details of these conversions differ but the general path is as shown in Figure 12.43 (b).

Figure 12.44 shows the structure of an industrially important plasticizer, 2-ethylhexanol, which is widely used to improve the commercial properties of poly(vinyl chloride), PVC. You can struggle quite a bit with a retrosynthetic path to 2-ethylhexanol before improving upon the path taken by industry, a simple Aldol condensation between two identical molecules as shown in Figure 12.44. What could

▪ **Alexander William Williamson**

be simpler than an aldol condensation between two identical aldehydes, butanal?

Let's try the well known molecule 1,4-butanediol shown in **Figure 12.45** and create a retrosynthetic path for its synthesis. We imagine the direct precursor of our target to be the unsaturated diol, which is hydrogenated in step **1**. The dialdehyde precursor to the unsaturated diol could be the unsaturated dialdehyde, which is easily reduced, for example, by sodium borohydride (section 10.3) (step **2**). Taking another backward step from our target, 1,4-butanediol, we come to the alcohol, which loses water in step **3**, a molecule that is the product of an Aldol condensation between two readily available chemicals, glyoxal and acetaldehyde, step **4** (**Figure 12.45**).

However, if we step back and give the problem just a bit more thought we could imagine a path so simple it is not worth making a figure for it. Succinic acid (Figure 8.20), a commonly available dicarboxylic acid (the "such" in "oh my such good apple pie") which we first encountered in our study of the citric acid cycle (section 8.6), could be the direct precursor of 1,4-butandiol simply by reduction of both carboxylic acid groups with lithium aluminum hydride (section 10.3).

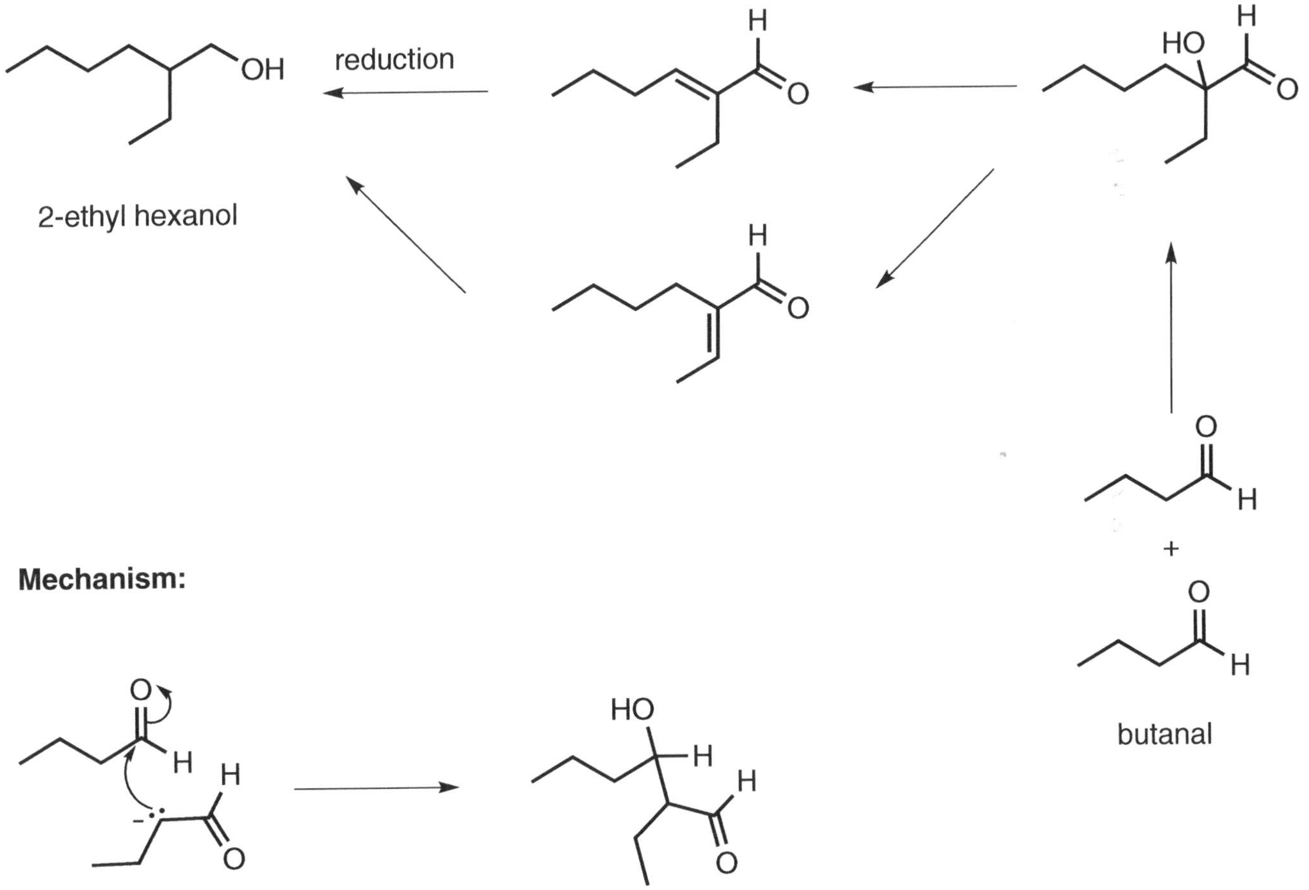

2-ethyl hexanol

butanal

Mechanism:

The importance of 1,4-butanediol to the chemical industry as a precursor to the engineering plastic poly(butylene terephthalate), focused much attention on the best route to this diol. For such a large scale chemical, industrial chemists and chemical engineers like to avoid the kind of multiple step processes shown in Figure 12.45 and also the use of industrially difficult to use reactants like lithium aluminum hydride. Instead industry has turned its attention to a molecule generally avoided for its explosive properties, acetylene.

Acetylene is an ideal precursor to 1,4-butanediol, so much so that its use for this industrial process is perhaps its last use for any heavily produced chemical. The process developed is shown in **Figure 12.46**. Before we discuss the acetylene route to 1,4-butanediol, however, let's take a moment to look into the unusual nature of the two C-H bonds in acetylene or for that matter the single C-H bond in terminal alkynes,

▲ **FIGURE 12.44**

Synthetic paths to the important industrial plasticizer, 2-ethyl hexanol, are compared.

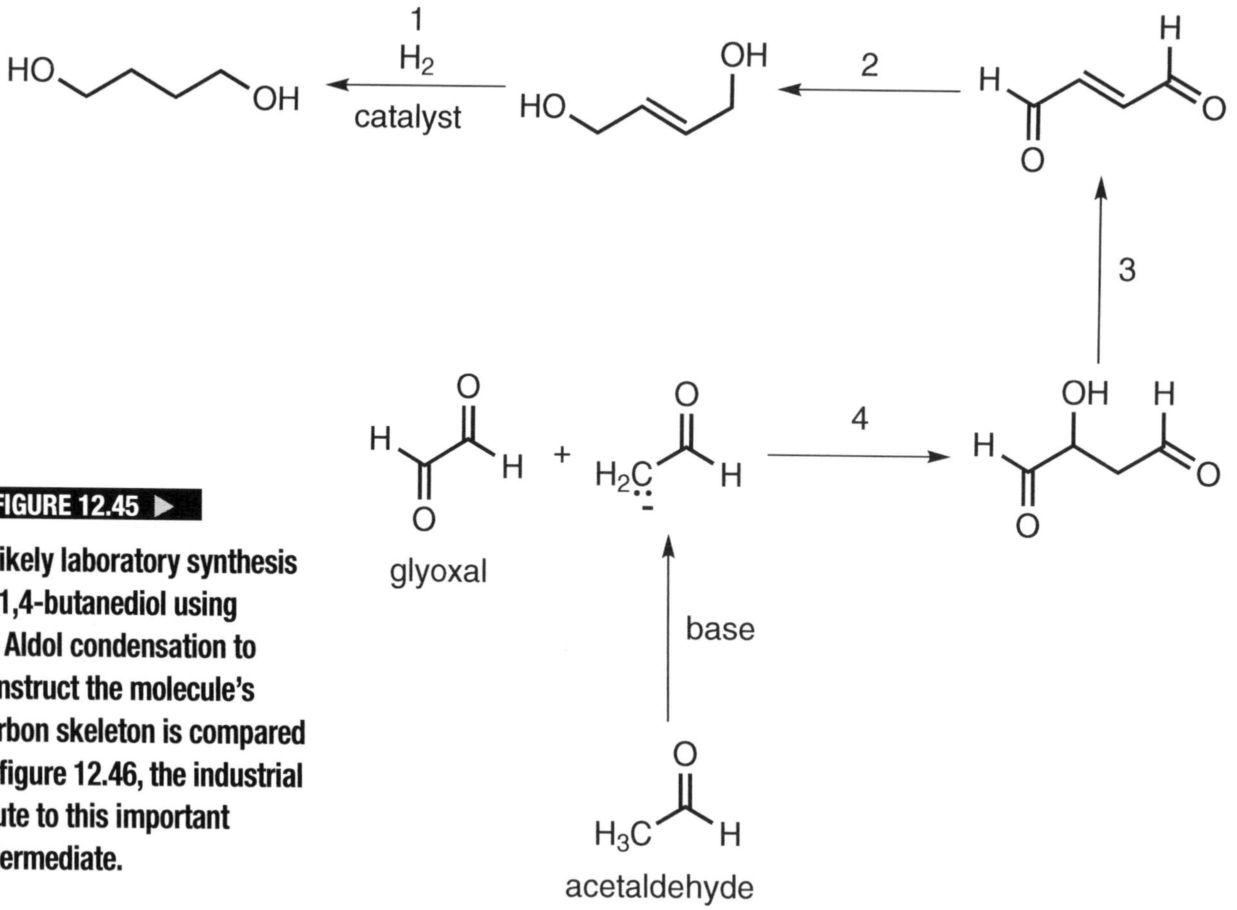

FIGURE 12.45 ▶

A likely laboratory synthesis of 1,4-butanediol using an Aldol condensation to construct the molecule's carbon skeleton is compared to figure 12.46, the industrial route to this important Intermediate.

R-C≡C-H. The table below, comparing acetylene, ethylene and methane, informs us (remembering that each pK_a unit is a factor of ten in the equilibrium constant (section 4.9)) of the relatively high acidity of this C-H bond. It is generally taken that the lowering of the pK_a, as shown, correlates with the increased s-character at the triple bonded carbon atom. Why do you think that is so?

Hydrocarbon	Approximate pK_a	% S-Character in C
H-C C-H	25	50
$H_2C=CH_2$	44	33
H_4C	48	25

We've come across the mention of triple bonds several times, both carbon-carbon (sections 7.1 and 9.4) and carbon-nitrogen in the nitrile group introduced in Chapter 10, and discovered the reactivity of the two π bonds allowing reactions that disrupt this linkage. However, the sp hybridization of the carbon atom involved in a triple bond yields another reactive property of the triply bonded carbon atom bearing a hydrogen atom, a reactive property which does not involve breaking of the π bond but rather easing the loss of a proton: R-C≡C-H forms R-C≡C:⁻ (an acetylide anion) exceptionally easily compared to C-H bonds from sp² and sp³ hybridized carbon (table above).

The sp hybridization of the triply bonded carbon atom means that the electron density at this carbon atom resides in a hybrid orbital with 50% s character. The s orbital is closest to the positive nucleus giving coulombic stabilization to a negative

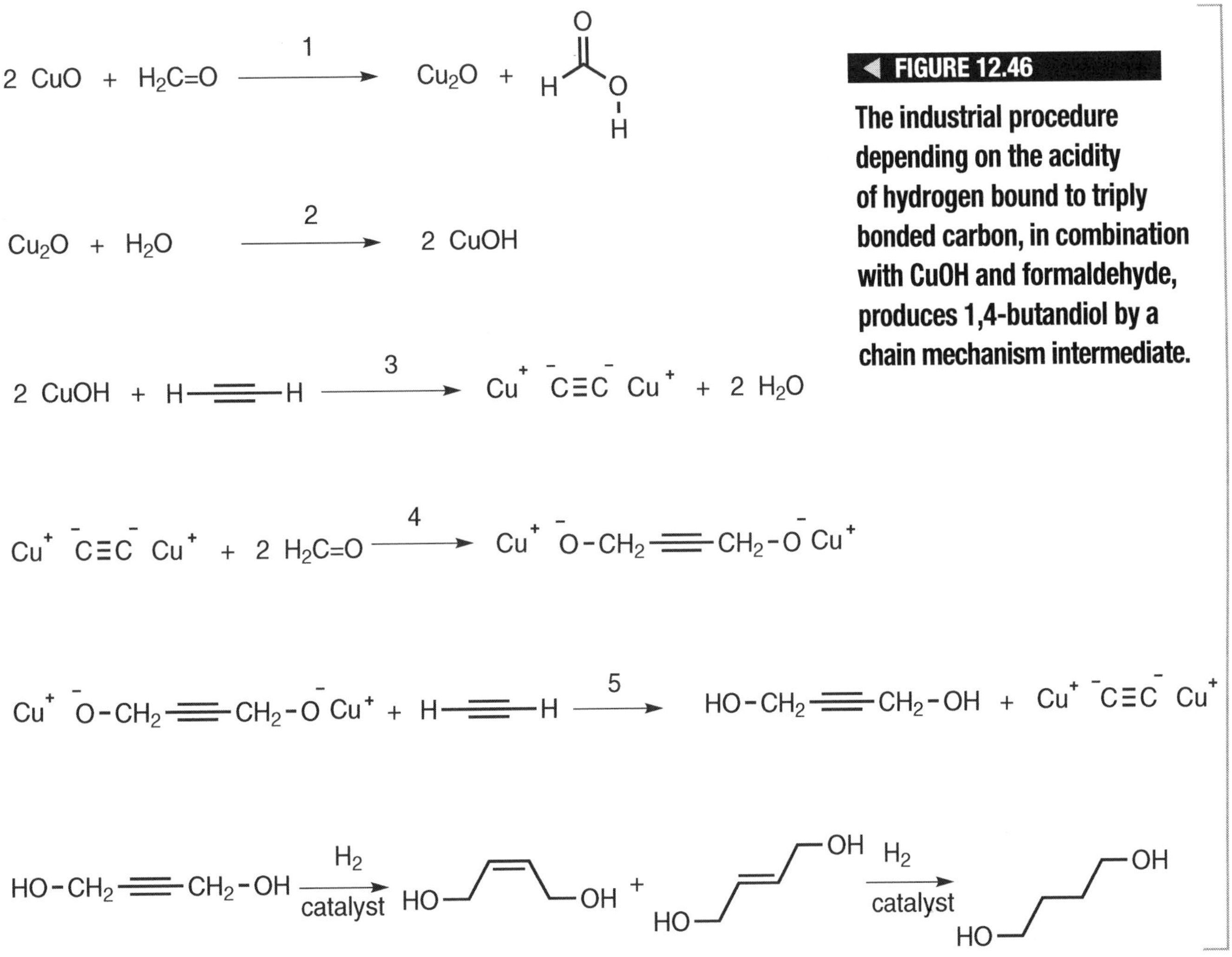

$$2\ CuO\ +\ H_2C{=}O\ \xrightarrow{\ 1\ }\ Cu_2O\ +\ HCOOH$$

$$Cu_2O\ +\ H_2O\ \xrightarrow{\ 2\ }\ 2\ CuOH$$

$$2\ CuOH\ +\ H{-}{\equiv}{-}H\ \xrightarrow{\ 3\ }\ Cu^+\ {}^-C{\equiv}C^-\ Cu^+\ +\ 2\ H_2O$$

$$Cu^+\ {}^-C{\equiv}C^-\ Cu^+\ +\ 2\ H_2C{=}O\ \xrightarrow{\ 4\ }\ Cu^+\ {}^-O{-}CH_2{\equiv}CH_2{-}O^-\ Cu^+$$

$$Cu^+\ {}^-O{-}CH_2{\equiv}CH_2{-}O^-\ Cu^+\ +\ H{-}{\equiv}{-}H\ \xrightarrow{\ 5\ }\ HO{-}CH_2{\equiv}CH_2{-}OH\ +\ Cu^+\ {}^-C{\equiv}C^-\ Cu^+$$

$$HO{-}CH_2{\equiv}CH_2{-}OH\ \xrightarrow[\text{catalyst}]{H_2}\ \text{HO}\diagup\diagdown\text{OH}\ +\ \text{HO}\diagup\diagdown\text{OH}\ \xrightarrow[\text{catalyst}]{H_2}\ \text{HO}\diagup\diagdown\text{OH}$$

◄ **FIGURE 12.46**

The industrial procedure depending on the acidity of hydrogen bound to triply bonded carbon, in combination with CuOH and formaldehyde, produces 1,4-butandiol by a chain mechanism intermediate.

charge on the sp hybridized carbon atom. We see this in the table and as well in the acidity of H-C≡N, where the hybridization factor combines with the electronegativity of nitrogen to give rise to a relatively high acidity to HCN (pK$_a$ close to 9).

The industrial process shown in Figure 12.46 takes advantage of the acidity of acetylene to produce the kind of mechanism industry loves best, a chain mechanism, which we've seen several times throughout the text (sections 4.10, 9.3, 9.4, 11.7).

There are two initiating steps in Figure 12.46 (**1** and **2**) to produce CuOH, a base powerful enough to abstract a proton from acetylene to produce the copper salt of the acetylide dianion in reaction **3**. The acetylide anion is nucleophilic and attacks the carbonyl group of formaldehyde in reaction **4** to produce the copper salt of the dialkoxide, a base in itself, which reacts with acetylene to regenerate the product of step **3** and so carries on the chain. As long as no termination step occurs the chain mechanism will continue until one of the reactants is consumed, a beautifully designed process.

The other product of the chain maintaining step **5** is a diol, which, as shown, on hydrogenation can be easily converted to the desired product, 1,4-butanediol (Figure 12.46).

This property of acetylene and terminal alkynes acting as Brønsted-Lowry acids to form acetylide anions, which then act as nucleophiles to make carbon-carbon bonds is hardly restricted to industrial processes or to attack at carbonyl carbon. Acetylide anions are well known nucleophiles via S$_N$2 reactions (section 10.6), but which are restricted to attacking primary carbon atoms bearing appropriate leaving groups. An excellent demonstration of this property can be found in another of E. J. Corey's synthetic triumphs, the path to *Cecropia* Juvenile Hormone. **Figure 12.47** shows the details of one of the steps in Corey's synthesis which uses this property of sp hybridized carbon.

Step **1** in Figure 12.47 shows the protection of the hydroxyl group using the same protecting group we encountered in Corey's synthesis of the prostaglandins (Figure 12.38). You can imagine for yourself the single step that produced the precursor alkyne alcohol that needed protection via step **1**.

In step **2**, the hydroxyl group on another unsaturated alcohol is reacted with a reagent, para-toluene sulfonyl chloride, not to protect this OH group but rather to convert the OH group to a leaving group, a tosylate, to set up the reaction with the acetylide anion, which was formed in step **3**. We've seen related leaving groups in our study of the S_N1 and S_N2 reactions but with Br instead of CH_3 in the para position (Figure 10.21, **1** and **2**).

All the synthetic efforts in steps **1-3** are preparation for the carbon-carbon bond forming step **4**, in which the tosylate is displaced by the acetylide anion in an S_N2 reaction. This is a classic use of the acetylide anion to form carbon-carbon bonds, a reaction that is generally restricted to CH_2 groups bearing appropriate leaving groups such as typically, tosylate or halide.

The product of this nucleophilic displacement is then treated with mild acid

Cecropia Juvenile Hormone

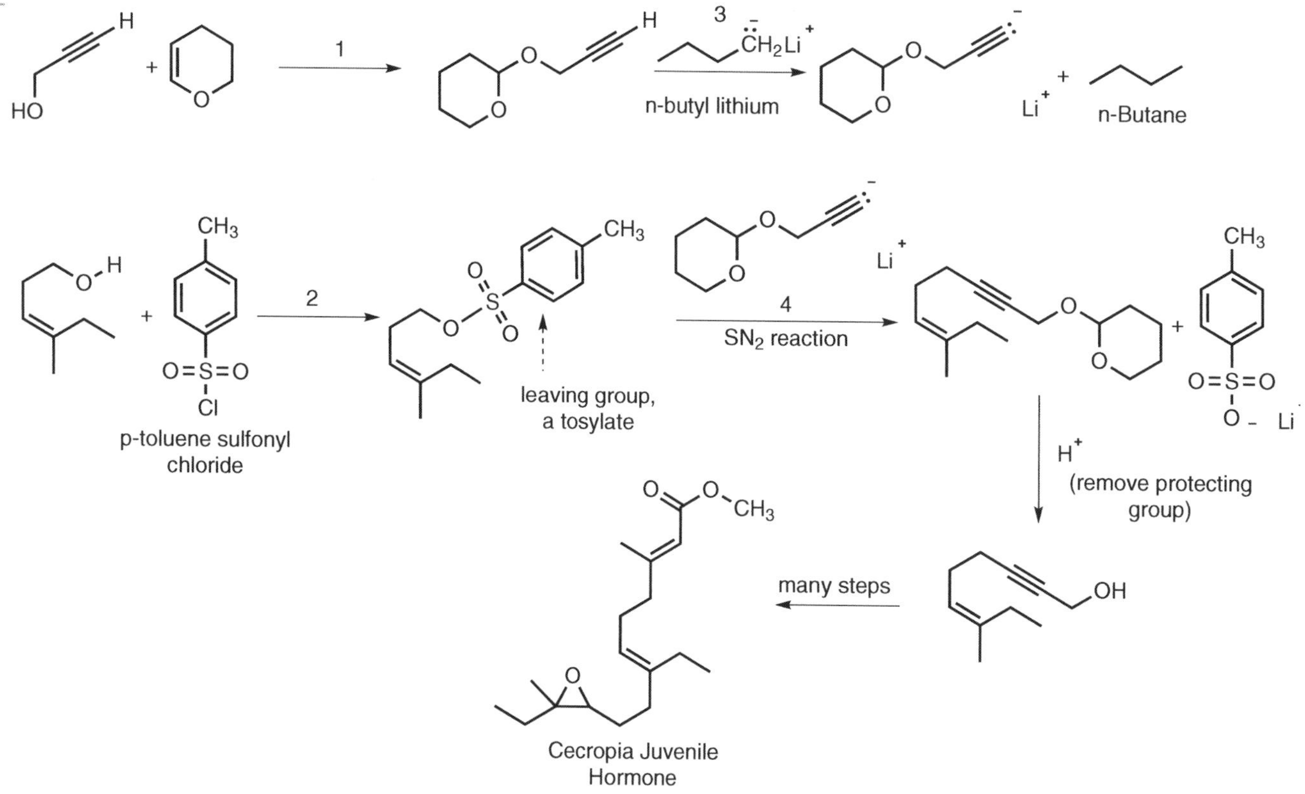

▲ FIGURE 12.47

Corey used the acidity of hydrogen bound to triply bonded carbon in combination with S$_N$2 chemistry to develop a synthesis of Cecropia Juvenile Hormone.

conditions to remove the protecting group, which is followed by many steps not described here to produce *Cecropia* Juvenile Hormone. If you want to follow up on how this was done and how the steps preceding those shown here were carried out there is no better source I am aware of than Fleming'book (section 12.1).

....

FOR THE PROBLEMS TO FOLLOW:

Consider that you have available all chemicals, gases, liquids and solids that could be used in a laboratory (in vitro methods), including, however, only organic molecules with six or fewer carbon atoms. For example, your laboratory has reducing and oxidizing agents, acids and bases of various strengths, phosphorus based reagents such as phosphorus tribromide and triphenyl phosphorus, boron based reagents such as hydroborane, aluminum trichloride, magnesium based chemicals and more.

PROBLEMS–Synthesis of:

12.81 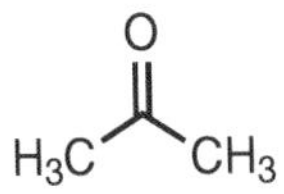 from

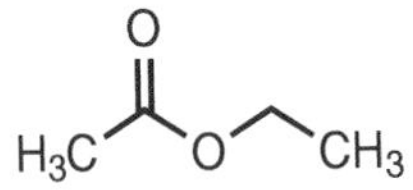

12.82 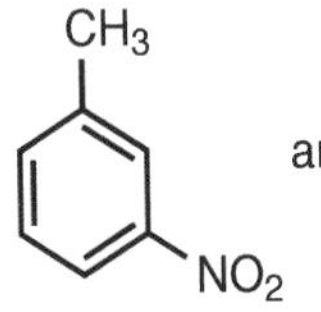 and a mixture of 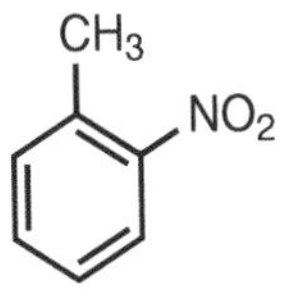 and

12.83 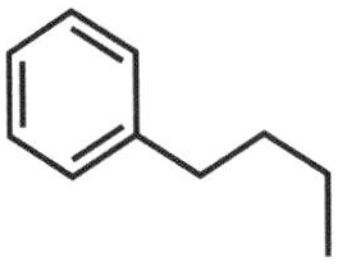 and also

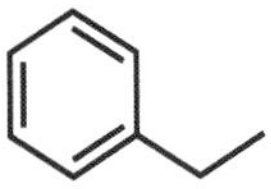

12.84

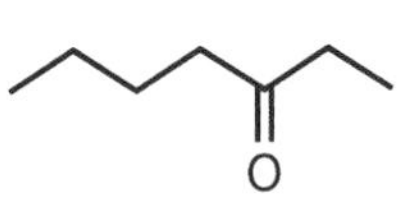

12.85

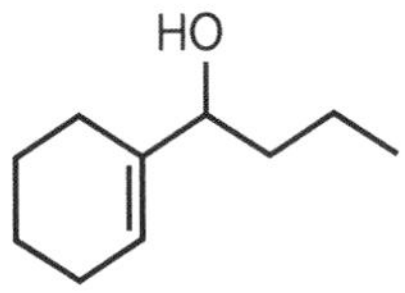

12.86 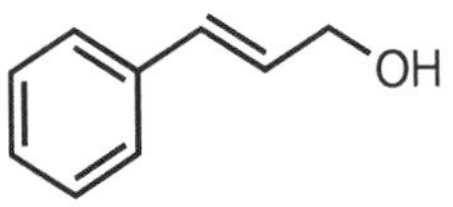 mixture of stereomers

12.87

12.88

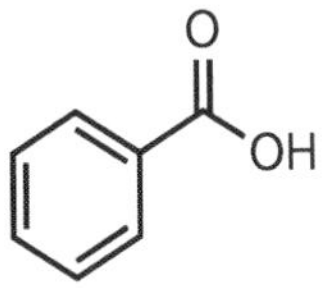

12.89 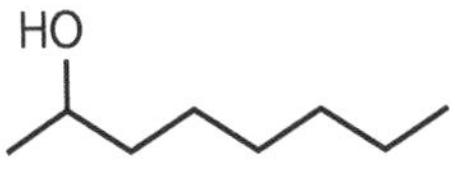 racemic

12.90

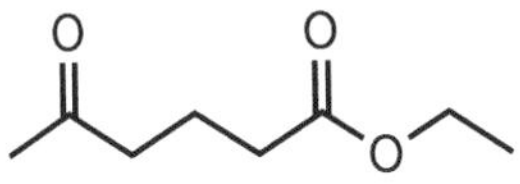

12.91

12.92

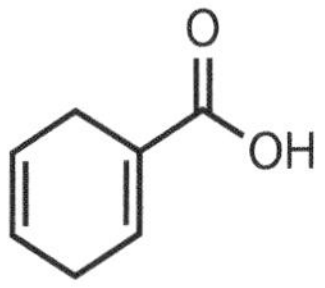

12.93

12.94 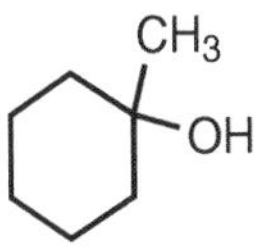 and also

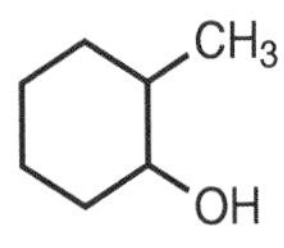

12.95

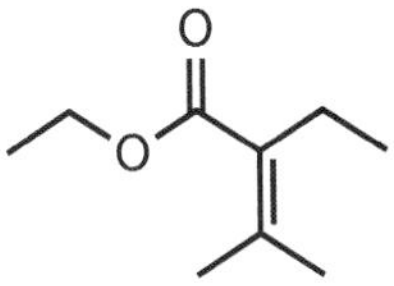

12.96

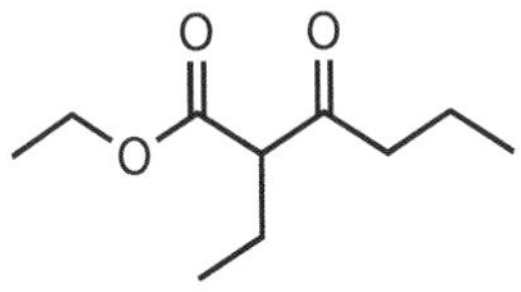

I N A PAPER PUBLISHED IN THE SCIENTIFIC journal, Tetrahedron, in 1962, William Doering whom we have met before regarding his discovery of Hückel's book (section 6.8), and his postdoctoral student at Yale, W. R. Roth, wrote about the quandary associated with "no-mechanism" reactions: "*No-mechanism is the designation given, half in jest, half in desperation, to "thermo-reorganization" reactions like the Diels-Alder and the Claisen and* **Cope** *rearrangements in which modern, mechanistic scrutiny discloses insensitivity to catalysis, little response to changes in medium and no involvement of common intermediates, such as carbanions, free radicals, carbonium ions and carbenes.*"

Doering and Roth succinctly summarized a major problem in the understanding of a large number of important chemical reactions little knowing that around the time of their writing, a theory was being developed to forever rid organic chemistry of the no-mechanism designation for these and many related reactions. This was the

12.21

The Mechanism of No-Mechanism Reactions

work of Woodward and a junior fellow at Harvard, **Roald Hoffmann**. Their work in turn was related to that of a Japanese scientist from Kyoto University, **Kenichi Fukui** who with his colleagues had published a paper in 1952 entitled: "A Molecular Orbital Theory of Aromatic Hydrocarbons." The effort of these scientists demonstrated that considerations of the energies and symmetry properties of molecular orbitals could yield understanding of the chemical reactions of organic molecules, results that led to the Nobel Prize for Hoffmann and Fukui in 1981. Woodward who certainly would have shared in the prize died in 1979.

Applying molecular orbital ideas to the Diels-Alder reaction illustrates the essential features of this approach, which takes several forms including "correlation diagrams," "aromatic and anti-aromatic transition states" and "frontier orbitals." The latter is the approach we will focus on here. An excellent treatment of the field with a discussion of all approaches to the use of molecular orbital theory applied to concerted percyclic reactions can be found in Chapter 11 of Part A of the book by Carey and Sundberg we have earlier refered to (section 12.15).

Hückel's insight to focus only on the p-electrons of aromatic molecules allowed understanding of the special stability of these cyclic molecules (section 6.8). Application of Hückel's approach in focusing on the p-electrons of the double bonds in linear chains or rings that are not aromatic, such as seen in the reactants undergoing the Diels-Alder reaction (Figure 12.2), and other reactions to be discussed below, leads to

FIGURE 12..48 ▶

The molecular orbital symmetries of an extended π-array can be described by the symmetries of the wave functions of a particle in a box.

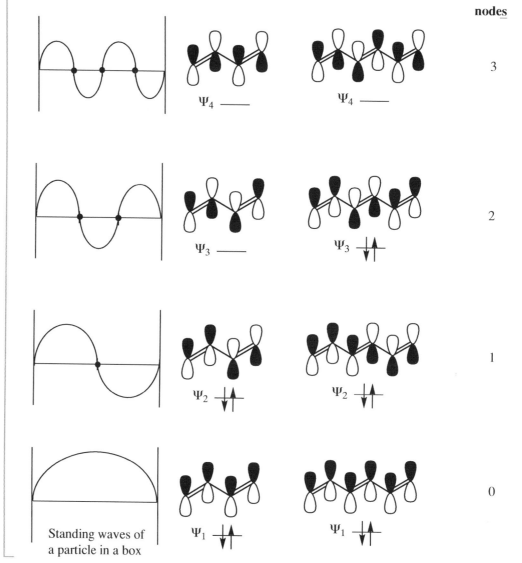

nodes

Ψ_4 —— Ψ_4 —— 3

Ψ_3 —— Ψ_3 ⥮ 2

Ψ_2 ⥮ Ψ_2 ⥮ 1

Standing waves of a particle in a box Ψ_1 ⥮ Ψ_1 ⥮ 0

understanding of the symmetry characteristic of the molecular orbitals occupied or occupiable by these p-electrons. These orbital characteristics then determine if these reactions are possible and their detailed mechanisms.

In 1970, Woodward and Hoffmann published "The Conservation of Orbital Symmetry," a detailed discussion of their insights into what had been called "no-mechanism reactions." (The essential ideas were first published in the Journal of the American Chemical Society in the mid-1960s). Within the first pages of the book the following sentence can be found, "The envelopes of polyene orbitals coincide with the curve of the wave function of a particle in a one-dimensional box." We've seen the simplest example of a polyene in sections 10.4 and 10.5 in the discussion of industry's use of 1,3-butadiene. Higher polyenes would then follow the pattern of this butadiene as in 1,3,5-hexatriene and so on. The relationship between the allowed standing waves of a particle in a box and the molecular orbitals that can be occupied by the electrons of these two polyenes, ψ_1 through ψ_4 are found in **Figure 12.48**. As for atomic orbitals discussed in sections 1.4 and 1.5, each molecular orbital can hold two electrons of opposite spin, as shown in Figure 12.48.

The p orbitals shown in the discussion of hybridization in Chapter 1 (sections 1.4 and 1.5) and which are used throughout the book, and shown to form the double bonds in the two structures in Figure 12.48, contain what is called a node, which is in the plane of the sigma bonds that form the two structures. The nodal plane is devoid of electron density and separates the two lobes of the p orbital, which differ in their symmetries. This difference is often designated by a plus and a minus sign for each lobe of each p orbital or in Figure 12.48 by being shaded and unshaded, which works equally well.

Observing the p orbitals in the π system, which is perpendicular to the sigma framework of both 1,3-butadiene and 1,3,5- hexatriene, one discovers shaded and unshaded p orbital lobes on adjacent carbon atoms. The differing symmetries of these p orbital lobes translates to the absence of electron density between them and therefore to a node which is now perpendicular to the sigma (σ) framework of the molecule.

The number of nodes perpendicular to the σ framework for each molecular orbital, ψ_1 through ψ_4, correspond to the number of nodes in the standing waves for the particle in a box, with the energy of the molecular orbitals increasing in going from ψ_1 to ψ_4. As for atomic orbitals, we fill each molecular orbital in turn, with increasing energy, with two electrons. The highest occupied molecular orbital, HOMO, therefore for 1,3-butadiene with four p electrons is ψ_2 with the lowest unoccupied molecular orbital, LUMO, ψ_3. The two more electrons in the p system of 1,3,5-hexatriene mean that HOMO is ψ_3 while LUMO is ψ_4. The symmetries of these molecular orbitals are determined by the number of nodes as shown in Figure 12.48. The larger are the number of nodes the higher the energy of the molecular orbital. The highest occupied and lowest unoccupied molecular orbitals, HOMO and LUMO are what are called the frontier orbitals, those that Fukui had pointed attention to in his 1952 publication, and it is these orbitals that we will focus on for our analysis of the Diels-Alder (Figure 12.2) and other reactions discussed in this section.

Experience with the Diels-Alder reaction informs us that the isolated double bond is most reactive when conjugated with electron withdrawing groups as seen in the first example and other examples in Figure 12.2. In one way of thinking we therefore imagine the HOMO of the diene pouring electrons into the LUMO of the ene, although as we'll see whichever way the electrons are imagined the symmetry characteristics of the molecular orbitals produce the same results. **Figure 12.49** shows the molecular orbital symmetry properties of HOMO and LUMO of ethylene and 1,3-butadiene and the interactions of these molecular orbitals in the Diels-Alder reaction. .

The examples of the Diels-Alder reaction in Figure 12.2 demonstrate that the stereochemical relationships of the pendant groups on the double bonds of the ene and the diene (cis or trans (Z or E)) is maintained in the product of the reactions. The stereochemical result requires that the ene, that is ethylene in the example in Figure

■ **Roald Hoffmann**

■ **Arthur C Cope**

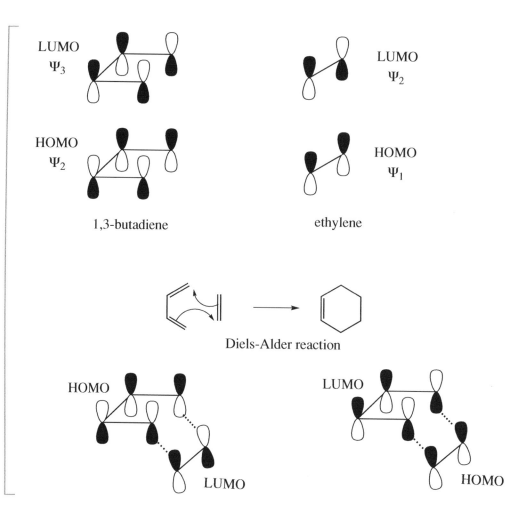

The highest and lowest occupied orbital symmetries of the reactants in the Diels-Alder reaction explain the characteristics of this reaction class.

LUMO Ψ_3

HOMO Ψ_2

1,3-butadiene

LUMO Ψ_2

HOMO Ψ_1

ethylene

Diels-Alder reaction

HOMO

LUMO

LUMO

HOMO

12.49, must approach the diene, that is 1,3-butadiene, so that the terminal carbon atoms of ethylene bond to the same face of the butadiene. This is precisely the result required by the overlap of orbitals of the same symmetry as shown in Figure 12.49, which is independent, as shown, of using the HOMO or LUMO of either reactant. The only requirement is that the frontier orbitals of the reactants are HOMO for one and LUMO for the other. All Diels-Alder reactions can be understood by the orbital symmetry characteristics shown in Figure 12.49. A no-mechanism reaction is turned into a reaction that can be understood by consideration of orbital symmetries.

FIGURE 12.50 ▶

The highest and lowest occupied orbital symmetries of ethylene allow understanding of the otherwise difficult to explain failure of two ethylene molecules to thermally form cyclobutane.

HOMO

node

LUMO

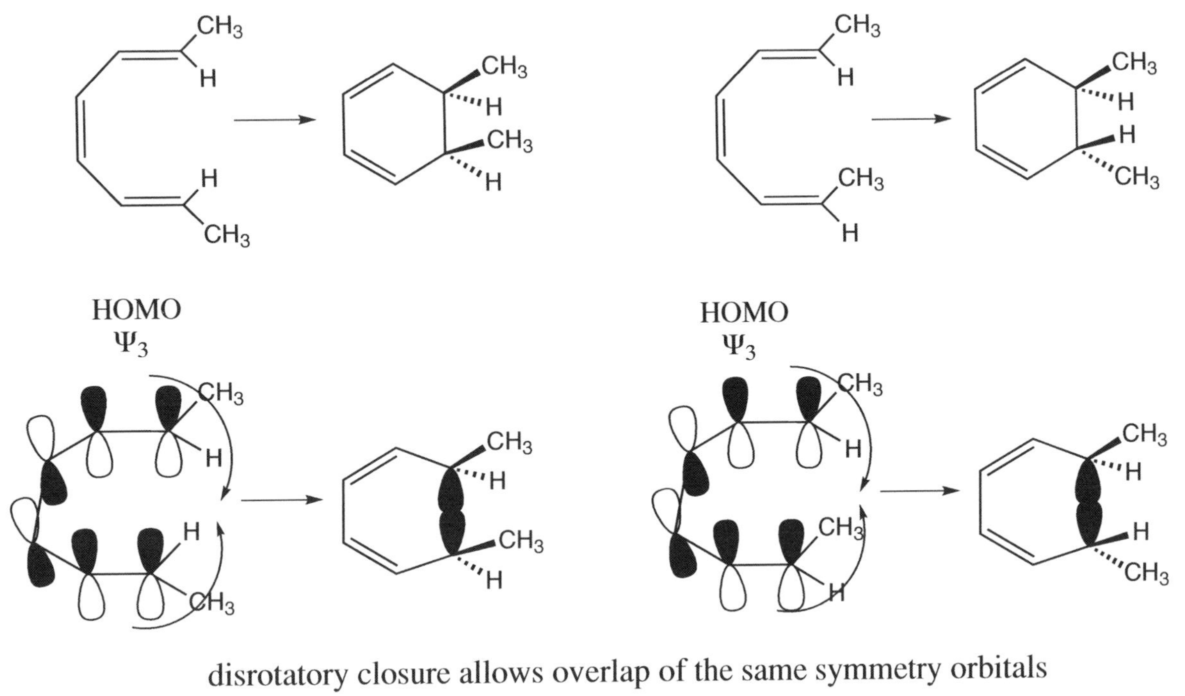

disrotatory closure allows overlap of the same symmetry orbitals

Orbital symmetries also yield understanding of why two molecules of an alkene such as ethylene refuse to undergo a pericyclic reaction to produce the four membered ring cyclobutane although simply moving electrons, as shown in Figure 12.50, would appear to allow this reaction path. Analysis of the symmetries of the HOMO and LUMO orbitals of ethylene (**Figure 12.50**) yield understanding. The simultaneous overlap of the orbitals would require a shaded orbital to overlap with an unshaded orbital producing a node. No bond could form. In molecular orbital terms this is an anti-bonding situation.

Another class of pericyclic reactions is termed electrocyclic reactions such as the cyclization of 1,3,5-hexatriene to cyclohexadiene as shown in **Figure 12.51** for substituted 1,3,5-hexatrienes. The substitution reveals the specific stereochemical consequences of the reaction, stereochemical consequences that could not be understood until orbital symmetries and frontier orbitals were considered.

In Figure 12.51 we see that each diastereomeric triene shown yields specifically only one cyclohexadiene diastereomer via formation of the sigma bond by rotation of the terminal atoms of the triene toward each other. This is termed disrotatory motion since the pendant groups are moving in opposite directions. The reason for this motion can be seen by inspection of the symmetry of the orbitals in the HOMO, which must form the new sigma bond. Many reactions of this kind are known in which the symmetry of the HOMO orbital controls the necessary motion. In 1,3-butadiene and its ring closed form, cyclobutene, the motion that transposes the ring opened and ring closed form is conrotatory. Here the HOMO, ψ_2, of 1,3-butadiene (Figure 12.48) has the p-orbital lobes of opposite symmetry facing in the same direction requiring an opposite motion to that seen in 1,3,5-hexatriene.

Another class of pericyclic reactions, sigmatropic rearrangements that could not be understood, becomes comprehensible by considering the symmetries of frontier orbitals. Let's look at two examples among many in this class of reaction, including an important reaction critical to the biochemistry of vitamin D which we won't cover here but you are encouraged to look up. Rearrangements that appear possible in both propylene and 1,3-pentadiene are shown in **Figure 12.52**. In both processes a hydrogen atom is transferred from the terminal methyl group with shifting of the positions of the double bonds.

Consider that the rearrangements shown in Figure 12.52 involve a hydrogen

▲ **FIGURE 12.51**

The geometries of the ring closing and ring opening reactions of extended π systems, which occur and do not occur, can be understood by the symmetries of the highest and lowest occupied molecular orbitals.

atom in reaction (1) and a propylene radical (three electrons), and in reaction (2) a hydrogen atom and a 1,3-pentadiene radical (five electrons). The HOMO for the propylene radical therefore must be ψ_2 while the HOMO for the 1,3-pentadiene radical must be ψ_3. As seen in Figure 12.48 ψ_2 has one node and ψ_3, two nodes with these nodal planes shown in Figure 12.52. Whatever may be the symmetries of the carbon framework, the hydrogen atom to be transferred, with only the s-orbital occupied, must be of one-symmetry. As shown therefore in Figure 12.52 a smooth (concerted) transfer of the hydrogen atom in the three carbon system (reaction (1)) is not possible while that in the five carbon system (reaction (2)) is consistent with the orbital symmetry. Only the latter rearrangement is experimentally observed.

The discussion in this section is only the top of the iceberg of this beautiful correspondence between an aspect of quantum mechanics and a class of organic chemical reactions, which, as noted above, is well covered in Chapter 11 of Part A of the book by Carey and Sundberg (sections 12.9 and 12.15).

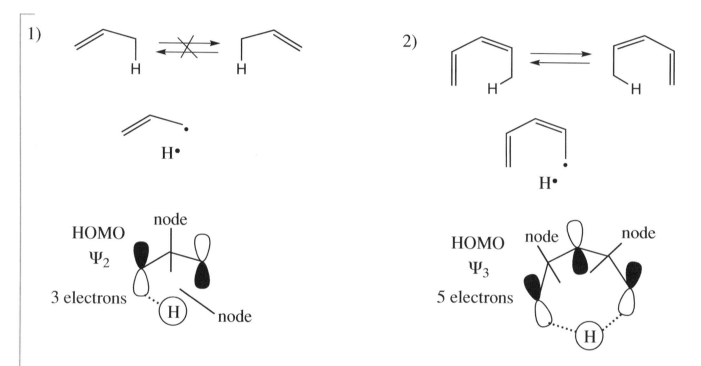

▲ FIGURE 12.52

Why some molecules allow hydrogen atoms to be intramolecularly transferred from sp³ carbon atoms to the terminus of double bonds, and only with specific geometries, can be understood by consideration of highest and lowest occupied molecular orbitals.

PROBLEM 12.97
Considering that a photochemical process can promote an electron from the HOMO to the LUMO, analyze the possibility that light could produce cyclobutane from two ethylene molecules.

PROBLEM 12.98
If the sigmatropic rearrangements presented in Figure 12.52 were instead presented as the transfer of a carbon atom with a positive charge instead of the hydrogen atom how would your analysis of the process be changed?

PROBLEM 12.99
Look up the chemical changes associated with the in vivo production of vitamin D and its precursors and identify all pericyclic reactions and determine their orbital symmetry characteristics.

PROBLEM 12.100
Analyze the sigmatropic rearrangement of a hydrogen atom from carbon-7 to carbon-1 of 1,3,5-hepatriene (CH₂=CH-CH=CH-CH=CH-CH₃). Would it be allowed under any conditions?

PROBLEM 12.101

Consider the electrocyclic reaction of 1,3,5,7-octatetraene to an eight member-ring with three double bonds. Would it occur via a conrotatory or disrotatory motion and justify your answer by considering frontier orbitals.

CHAPTER TWELVE SUMMARY of the Essential Material

THE FOCUS OF THIS CHAPTER IS MULTI-STEP SYNTHESIS or, in other words, the methods organic chemists use to make molecules that are not easily connected to the starting materials, which are generally commercially available. Here we look at the work of two of the masters of this field, R. B. Woodward and E. J. Corey and the approaches they took in the syntheses of two molecules that offered considerable challenge to the ability of organic chemists to synthesize complex molecules. Their success in these endeavors convinced the field of the power of the science of organic chemistry to accomplish multistep synthesis essentially without boundaries.

In Part I of this chapter we follow Woodward in his synthesis of cholesterol, a feat that opened the door in this field and we come to see that his success was based on the science reaching a high level of understanding of structure and mechanism. Every step Woodward took was based on considerations of the theory that supported the science at that time. He began with a reaction long interesting to him, the Diels-Alder reaction, which forms ring compounds from acyclic precursors and then how understanding of the nature of carbanions and equilibration processes and the stereochemistry of six-member rings allowed attaining fused rings that could later on his path lead to the construction of two of the rings of cholesterol. Taking the synthesis further along required adding a third fused ring, which was accomplished using three reactions familiar to us from our study of in vivo syntheses, the Michael, the Claisen, and the aldol condensations.

In adding the fourth fused ring Woodward was faced with the problem that the reaction he wanted to use could take place to yield multiple products instead of the single molecule he sought. This arose from various sources within the starting molecule of enolate formation, which Woodward dealt with by using the reactivity of the most favored enolate in a reaction with an amine to form an eneamine, which then acted as a protecting group. In this manner we learn some things about the reactions of amines and carbonyl chemistry and the nature of eneamines and Schiff bases. Woodward now could use the enolate he wanted in a Michael reaction with acrylonitrile, a conjugated alkene we can add to the Michael reactants we've seen before. The nitrile group was then hydrolyzed to the carboxylic acid, a reaction we've seen before in our study of polymers. The carboxylic acid is then activated by formation of a mixed anhydride, a method of increasing the reactivity of carboxylic acids we've also seen before, and this anhydride then reacts with the OH group produced by the enol formation of a ketone carbonyl in the third fused ring.

The synthetic route is now set up to form the fourth fused ring using one of the most famous reactions in organic chemistry, the Grignard reaction, in which carbon bound to magnesium is nucleophilic in a manner that causes reaction with various classes of carbonyl functional groups. We take the opportunity to learn something of this reaction and see the scope of its use. Using this Grignard chemistry Woodward was able to close the fourth ring by giving rise to a dicarbonyl compound that could undergo an Aldol condensation. Woodward is now set up with nearly the carbon skeleton of cholesterol with however, one of the four rings as a six rather than a five membered ring. To accomplish the necessary transformation of this ring Woodward had to remove a protecting group based on a vicinal diol forming a ketal with acetone. Here we are reminded of the chemistry of hemiacetals and hemiketals and full acetals and ketals, which brings us back to our studies of sugar chemistry. When the ketal is removed, Woodward is able to oxidize the resulting vicinal diol to a dialdehyde breaking open the six membered ring. Another Aldol condensation then mostly yields the five member ring he wanted with one aldehyde group placed to continue the synthesis.

At this point in the synthesis Woodward no longer wanted to work with racemic materials and we discover the elegant way he used a natural product to distinguish the enantiomers based on the principle we first came across in Chapter 1. Again our attention is turned to the Grignard reaction as the alkyl pendant group is placed on the five membered ring that had been formed by the Aldol condensation. However, the product of the Grignard reaction produces an OH group that has to be removed. This result leads us to have to look into a class of reactions not previously focused on, elimination, and we spend a bit of time looking into these double bond-forming reactions and then see how the absence of specificity to form constitutionally isomerically pure alkenes is of no problem to Woodward because of his planned approach. Simple hydrolysis of an acetate ester then produces cholesterol, which is found by mixed melting point to be identical to natural material.

We are now prepared to study the work of the other master in this field of multi-step synthesis, E. J. Corey. The target now is another biologically important class of molecules, the prostaglandins. Again, as in Woodward's beginning toward cholesterol, the Diels-Alder reaction proves critically important. But in this use of the reaction the diene structure is such as to produce fused rings that are bicyclic in a different way. We look into this structure and discover an understanding based on the conformational properties of cyclohexane. Again, the nitrile group has a role in being part of the structure of one of the Diels-Alder reactants but here this group is not hydrolyzed but, rather, its leaving group properties are utilized. The resulting ketone allows an introduction to another reaction, the Baeyer-Villager reaction, which enables conversion of ketones to lactones. The resulting bicyclic structure including the lactone sets up a structural situation in which formation of an iodonium ion allows a ring closing to take place that, remarkably forms a key prostaglandin intermediate in which several stereochemical centers simultaneously take the proper configurations. This chemistry causes us to further our understanding of how ring closing reactions take place in organic chemistry and we see how thermodynamic and kinetic parameters play a role and, specifically, ring strain and entropy.

Corey is then faced with the pendant alkyl groups in the prostaglandin structure, each of which contains a double bond but one cis and the other trans and, specifically, placed in the alkyl chains. We see why Corey turns to a reaction based on phosphorus invented by Wittig to solve the problem and understand how this approach accomplished the task at hand. We look into the Wittig reaction and one of its variants and discover that the presence or absence of resonance stabilization of the ylid intermediate in this reaction is the source of the stereochemical control and how a cyclic intermediate, the oxaphosphetane, determines the stereochemical outcome for the double bond. We also see how the strength of the phosphorus oxygen bond drives the reaction forward and why regiochemical control is the rule.

This point seemed than an excellent time to look at another important class of synthetically useful reactions in which a strong bond to oxygen is the driving force, but this time to boron. And our taking this diversion to boron-based chemistry is encouraged by the fact that the Nobel committee saw the parallel to the phosphorus-based chemistry in awarding the prize to both originators of the chemistries and that both reaction classes involve double bonds. In hydroboration however, the double bond is not made but rather water can be added to the double bond in a controlled manner and with the OH group added to the least substituted carbon – anti-Markovnikov. Hydroboration is compared to oxymercuration and we study the mechanism of both and discover the source of their different results.

All this brings us to discover that Corey had been using a protecting group in disguise, one that was not easily recognized but turned out, by using a special reducing agent, to allow a lactone to be turned into a hemiacetal. The equilibration of this hemiacetal exposes, then, the aldehyde group to undergo another Wittig reaction to form the second alkyl chain with its controlled regiospecificity and stereospecificity. And, in the course of this chemistry, we see still another protecting group that works with hydroxyl groups, a necessity in the synthetic path to avoid the presence of active hydrogen. Finally we appreciate the brilliance of Corey's synthetic work in seeing how his path brings him to all three prostaglandin structures by simple maneuvers toward the end of the synthesis.

We've seen that both Woodward and Corey made critical use of a reaction that had early on fascinated Woodward, the Diels-Alder reaction. In this reaction, one of the most well known and used in organic chemistry, we've seen how rings can be formed by the interaction between a diene and an alkene and the specific requirements for this reaction to be successful, and the precise stereochemical consequences of the reaction. We end the formal part of the explanation of the synthetic work of these two masters with a section referring to no-mechanism reactions, which before the mid-1960s referred to the Diels-Alder reaction among others coming under the general heading of concerted pericyclic reactions. Here we discover that the approach Hückel took in understanding the special characteristics of aromatic molecules (Chapter 6) by focusing only on the orbital characteristics of the p-electrons can be applied to understanding a broad class of reactions allowing organic chemists to discard the term "no-mechanism reactions." The key to understanding arises from the combination of the symmetry of orbitals and the importance of frontier orbitals to chemical reactions.

The two part chapter discusses Corey's formal proposal of the retrosynthetic idea in which synthetic plans are made not by considering the starting material but rather the end product and then working backward. We apply this approach to several industrial products and see how the methods used fit. Here we see the value of the relatively low pK_a of hydrogen atoms bound to triply bound carbon atoms and how the carbanion produced by loss of this proton can be used synthetically in important ways both in industry and in Corey's synthesis of Cecropia juvenile hormone. We also come to understand the hybridization source of this surprising acidity.

And finally you were invited to try your hand at some synthetic tasks in the problems that follow the chapter.

THE END

INDEX

-A-

absolute configuration
—D,L assignment for glucose
  Fischer guess 91
—(R) (S) assignment 35

acetal 393f
—and ketal, formation 394
—linkage cellulose and starch 393

acetaldehyde 215
—eneamine formation 385
—Schiff base from 384

acetic acid
—compare to ethanol,
  resonance and pK$_a$ 141f
—glucose 22
—pK$_a$ 121

acetone 215
—enol/keto forms 251
—ketal formation 393f

acetyl acetone
—enol/keto forms 251

acetyl coenzyme A
—product of fat catabolism 240, 244
—role in citric acid cycle 265
—source of isopentenyl
  diphosphate 255ff
—structure 134, 254

acetylene 11, 431ff
—acidity compared and source of 432
—Bronsted-Lowry acid 433
—explosive 284

acetylide anion 433

acids and bases, Bronsted-Lowry 120ff

acid strength, defined 121f

acrylonitrile
—Michael reaction 386

acyl
—anhydride, source of ester, amide and
  carboxylic acid 352
—chloride, compare carboxylic
  and sulfonyl 348ff
—chloride, source of ester, amide
  and carboxylic acid 351
—leaving group 350

acylium ion
—electrophilic aromatic substitution 197

adamantane 108f
—from tetrahydrocyclopentadiene 109

Adams, Roger 293

addition
—1,2 to butadiene source of

neoprene 313f
—1,4 butadiene-source
  of nylon 6,6 314
—of chlorine to 1,3-butadiene 313
—to alkenes 190
—compared to substitution,
  elimination 400
—of water to double bonds, acid
  catalyzed 416

adenine 186

adenosine
—monophosphate 270
—relation to leaving groups 269
—thioester formation 223f
—triphosphate ATP 187

S-adenosyl methionine
—formation via S$_N$2 mechanism 262

adipic acid
—amide derivative 308
—polyester formation 294
—synthesis from benzene 306

adiponitrile
—route to hexamethylene diamine 308

adrenaline 325

aldehyde, properties, hybridization 95ff
—ketone – glucose to fructose 247
—source of reactivity 96ff

Alder, Kurt 373

Aldol condensation 256f
—elimination of water 402
—fourth fused ring cholesterol 392
—third fused ring cholesterol 378

alkene
—addition of HCl 190
—addition of water, oxymercuration 417
—cis and trans 334
—constitutional isomers cholestero
—formation from acetate 399
—formation, lack of control 401f
—Markovnikov addition by
  oxymercuration 417
—stability in 1,2 versus 1,4 addition 316
—synthesis 398

allene 126

allyl radical, 289

aluminum trichloride
—electrophilic aromatic substitution **191**

ammonia
—ammonium bonding 13
—carboxylic acid to amide 308

ammonium cyanate
—conversion to urea 23

Andrussow, Leonid 313

amylose 7ff

anhydride
—mixed 388

antiaromatic 180ff

aromatic rings
—coenzymes 187
—FAD 228
—nucleotides 186

Arrhenius, Svante 194

Arrhenius equation 194
—pre-exponential factor 195

arrows, double headed, resonance 142

aspartate, in vivo hydrolysis 218f

atomic theory, multiple proportions 167f

ATP
—consumed by human activity 272
—formation of S-adenosyl methionine 323
—hydrolysis 269ff
—isopentenyl diphosphate synthesis 258f

azeotrope
—benzene and water 394

-B-

backbiting, polyethylene branches 282

backside approach
—nucleophilic substitution 326f

Baeyer, strain theory 72ff

Baeyer-Villiger reaction 406f

Barton, Derek 75f

Barton reaction 282

bent bonds, three membered rings 159

benzaldehyde
—Wittig reaction 414

benzene
—cumene, source of 189
—evolution of structure 170
—equivalent historical idea
  to resonance 173
—history of discovery 164 ff
—Hückel molecular orbitals 182
—Hückel theory 179ff
—hydrogenation compared to alkenes 176
—multialkylation 189ff
—nomenclature 172f
—nylon 6,6, source of carbon atoms in 300
—objection to ring structure 173ff
—proof of hexagonal structure by x-ray 175
—rearrangement- acid chlorides 197
—resonance control of multialkylation 198f
—resonance structures 199
—source of nylon monomers 306
—structure 168ff
—structure of derivatives 171
—toluene 200
—trichloromethyl benzene 200

Berson, Jerome 184

Berthelot, Pierre Marcelin 293
—hydrogenation heat of benzene 177

betaine and oxaphosphetane
—control double bond in
  Wittig reaction 420
—Wittig intermediate 415

[2.2.1] bicycloheptane 405

Bivoet, Johannes Martin 35

Biot, Jean-Baptiste 28ff

bisphenol A 188

Blatt, Albert H. 181

Bloch, Felix 52

Bloch, Konrad 156

block copolymer 360ff

blood-brain barrier 18

bond dissociation energies, effect on
—carbon-hydrogen 346f
—chemical reactions 345f

bond rotation 40

bond strength
—boron oxygen 419
—drives Wittig and hydroboration
  reactions 419

borane, 418

Borodin, Alexander Porfiryevich 257

Breslow, Ronald 180

Bridgeman, Percy W. 276

Bronstead, Johannes Nicolaus 120

Bronsted-Lowry acid 120ff

Branch, G.E.K. 128

Brown, Herbert C. 412

Bruhl, Julius Wilhelm 250

1,3-butadiene
—addition of chlorine 316f
—1,2 versus 1.4 addition 313ff

butanal 430

n-butane 434
—conformation relationship to
  cyclohexane 79

butanediol 431ff, 433

butenes, steam cracking 285

2-butene, 2-methyl 341

1-butene, 2,2-dimethyl
—hydroboration 418

Butlerow(v), Alexander Michailovich 34

n-butyl lithium 434

-C-

Cahn, Ingold, Prelog nomenclature 35

Cahn, R.S. 34

camphene 108

camphor, structure 135

carbanion, stabilization 226f
—role in carbon-carbon bond breaking 259

carbocation 106ff
—addition to double bonds 123ff
—empty p orbital 110
—formed via leaving groups 138
—hybridization 111
—intermediate in vulcanization 342f
—intermediates 108ff
—in living processes 132ff
—NMR study in superacids 112
—resonance stabilized 324
—1,2 shift 111
—1,2 shift in the synthesis of lanosterol 162
—stability in 1,2 versus 1.4 addition 316f
—synthesis of lanosterol 160f
—stability measured by ionization 116, 117

carbohydrates 7

carbon
—electrons in isolated atom,
  atomic number 9ff
—tetravalent 9ff

carbon dioxide,
—elimination in citric acid cycle 265
—in vivo elimination isopentenyl
  diphosphate synthesis 258f

carbon-hydrogen ratios by structure 165f

carbon monoxide
—source of phosgene 353

carbonyl
—differences between kind of 213ff
—nucleophilic substitution vs. addition 215

carboxylate anion
—anhydride leaving group 352

carboxylic acid chloride
—compare to sulfonyl chloride 350

Carothers, Wallace
—prerequisites for polymerization 293

Carey and Sundberg 402, 413

catabolism
—fat and sugar compare mechanism 260
—source of acetyl coenzyme A 254

catalytic cracking 104ff

Cecropia Juvenile hormone 434

cell membranes, 208

cellulose 6ff, 11
—source of difference to starch 101

cephalosporin C 372

chain mechanism
—formation of polyethylene 281
—free radical 346
—initiation, propagation 123f
—LDPE with SO_2 and Cl_2 346
—polymerization 289
—steam cracking 285
—vulcanization 343

Chevreul, Michel Eugene 132

chiral 25ff
—CIP nomenclature 35ff
—nomenclature 34ff
—separation by chromatography 27
—structural prerequisite 33ff

chloride
—acyl leaving group 353

chlorine
—bond dissociation energy 346f
—silver salts, action of light 210

chloromethyl methyl ether 404

meta-chloroperbenzoic acid 406f

chlorophyll-a 372

chloroprene, source of neoprene 314

cholesterol 133, 372
—overall synthetic path 376
—synthesis 372ff

cis double bonds
—compare rubber and fatty acids

citric acid 263
—Aldol condensation 265
—analogous mechanism to fat and
  sugar catabolism 266
—chirality 267
—Cycle 262
—enantiotopic groups 268
—organic chemistry of 264

Claisen condensation
—isopentenyl diphosphate synthesis 255

cobalt, oxidized and reduced states 306f

condensation polymerization 298f

conrotatory motion 439

constitutional isomers
—cholesterol synthesis 396

Cope, Arthur C. 437

cracking, catalytic 104ff
—1,2 shifts 115
—thermal 104ff

citronellol, citronella
—rose oil 136

Claisen, Rainer Ludwig 238

Claisen condensation 238f
—distillation head 239

coal tar 71

Conant, James B. 181

Condamine, de la Charles Marie 339

conformation
−anti, gauch, eclipsed 40ff
−isomers 38ff
−versus configuration 83

Conforth, John W. 232

conjugate bases, leaving groups 139

conjugate addition 231f

constitutional isomers
−addition to 1,3-butadiene 316f

Corey, Elias J. 403

Couper, Archibald Scott 33

covalent bond 9ff

Crafts, James 189

crosslinking
−Hypalon 354

crude oil 21

crystallization
−production of nylon 6,6 302
−sugars compared 359

crystals, left and right handed 28ff

cumene, source of industrial plastics 188

cupric hydroxide 433

curved arrows 97ff

cyclobutane
−disallowed synthesis from ethylene 438

cyclooctatetraene
−dianion 181
−tub shape 180

cyclohexane
−axial, equatorial, energy difference78ff
−chair form 73ff
--conformation, Newman projection 77ff
−reduction to benzene 306
−ring flip 77ff

cyclohexanol
−from cyclohexane 306

cyclohexanone
−from cyclohexane 306

cyclopentadiene 404

cyclopropane, strain 159

cysteine, fat catabolism mechanism 240
−nucleophile 237

cytosine 186

-D-

Dalton, John 167

Davy, Humphrey 352

Davy, John 352

Dean-Stark trap 394

dialdehyde
−vicinal diol cleavage 395

diaminoethane (ethylene diamine) 354

1,6-diamonohexane 300

diamond 73

diastereomers 19ff
−conformational 80
−separated by chromatography 423
−synthetic path to cholesterol 387, 398

dicarboxylic acids 262

1,3-dichloro-1-butene, chloroprene 314

Diels-Alder reaction 373
−enhanced at low temperature 405
−examples, 375
−orbital symmetry control 438
−synthesis of prostaglandin 405ff

Diels, Otto 373

Diesel fuel 21, 120

diethyl ether 389

digitalis purpurea 397

digitonin 397

diglyme
−solvent for LiAlH$_4$ 311

dihydropyran
−protecting group 424

dihydroxy phosphate 247

diisobutyl aluminum hydride 425

dimethylallyl diphosphate 144f, 320

2,2 dimethylbutane, mass spectrum 44

dimethyl sulfoxide 427

dipropyl ether, infrared spectrum 50

disrotatory motion
--orbital symmetry control 439

Djerassi, Carl 282, 426

Doering, William 184, 435

Dole, Malcolm 48

Domagyk, Gerhard 354

double bond
−cis and trans, Z and E 333f
−restricted rotation 151, 334

DuPont 293f
−Hypalon 345
−Spandex 363
−synthetic fiber commerce 296

dyes 71

-E-

Edgar, Graham 119

Einstein, Albert
−quote 276

elastomer
−crosslinked 336
−no double bonds 356ff
−stretched and coiled 336
−understanding properties 336ff

electrocyclic reactions
−orbital symmetry control 439

electron diffraction 74

electronegativity 9
−source of acyl reactivity 351
−source of NMR chemical shift 57

electrophilic aromatic substitution 189ff
−acylium ion 197
−carbonyl substitution 237
−nitrobenzene 201
−rate determining step 196
−reaction coordinate diagram 195, 197ff

elimination reactions 400
−exmples 401
−mechanisms: E-1; E-2; E-1cb 402
−stereochemistry 402

enantiomeric excess and ratio 27

enantiomers19ff
−odor differences 20, 153

enantiotopic
−assignment of configuration 147
−groups in citric acid cycle 268
−groups in enzyme catalysis 147ff

endothermic process
−steam cracking 287

eneamine 384ff
−protecting group 386

enediol
−glucose to fructose 247
−product of glucose catabolism 245

energy of activation193ff
−higher energy of activation 289f

engine, internal combustion 104ff

enol/keto
−enol forms a lactone 388
−equilibrium 250
−tautomers, separated 251
−tautomerization, catalyzed by base 252
−various structures 251

enolate
−cholesterol synthesis 377, 383

enthalpy
−polymerization versus steam cracking 287
−role in ring closing 411

entropy
—of mixing 361
—polymerization versus steam cracking 287
—role in ring closing 411

enzyme, lipase folded state 217
—geometric precision 218ff
—isomerizing 145
—lanosterol cynthesis 157

epimer 376f

epimerization 377

epinephrine 325

epoxy resin 188

equilibrium
—ester compared to amide 297

ester, hydrolysis
—transesterification 213ff

estradiol 132

ethylene
—diamine 365
—ethylene propylene rubber 356f
—polymer from 276

ethyl hexanoate 420

2-ethylhexanol 430f

ethylidenenorbornene
—crosslinking EPDM rubber 358

ethyl vinyl ketone 380

ethanol
—biofuel 21
—hybridization 123
—pK$_a$ 121

ethyl acetoacetate 238
—enol/keto 251

ethylene oxide, strain 159

Eucalyptus forest, terpene source 135

eukaryotic cells 261

evolution, Krebs' view 263f

exothermic
—polymerization 287
—reaction, relate to bond dissociation
    energy 346f

-F-

Faraday, Michael 164

Farnesol, structure 135

Fankuchen, Isidor 180

fatty acids 206
—catabolism 225ff
—edible 209
—in vivo double bond introduction 226f
—labelling 22
—saturated and unsaturated 206ff

Fieser, Louis 75

Fenn, John 46

fermentation 21

femtochemistry 327

Fischer, Emil 31, 89ff
—crystallization of sugars 359
—projections 90

Flavin adenine dinucleotide (FAD) 227f

formal charge 12ff
—zeolite 105ff

Fleming, Ian
—"Selected Organic Syntheses "370f

formaldehyde 433

formula, history of development 167f

formaldehyde 11
—bonding 13
—probe for chirality 25f

foxglove 397

free radicals
—addition to double bonds 281
—compared to carbocations 280
—crosslinking by 357
—formation 279
—hydrogen abstraction 282
—structure and hybridization 279

free rotation 39ff

Friedel, Charles 183

Friedel-Crafts reaction 196

fructose
—phosphorylated 246
—symmetrical catabolic fragmentation 246

Fukui, Kenichi 436

fumaric acid 263
—conversion to maleic acid 334

functional groups, definition 87
—isocyanate 364f

fused rings 405

-G-

galactosemia
—cause of 100ff
—galactose 70ff

gasoline 21, 104ff

Gay-Lussac, Joseph Louis 22

Gazzetta Chimica Italiana
—founding 384

geraniol, structure 136

Gila monster, exendin-4 297

Gillispie, Ronald 11

glass
—common for polymers 359ff
—transition temperature 361

glyoxal 431

glucose 6ff, 11, 70ff
—catabolism 244ff
—conversion to fructose 245
—diastereomers, axial/equatorial 93
—electrospray mass spectrum 47
—enantiomers and diastereomers 89ff
—mutarotation 88f
—α,β ring flip 81
—ring closing 98
—stereochemistry 17ff
—structural determination 85ff
—unsymmetrical catabolism 245

glutamic acid 263

glutaric acid 262

glycation, source of galactosemia 100f

glyceraldehyde, probe for chirality 27
—3-phosphate 247

glycerol 208ff
—structure 133

Goodyear, Charles 340

Grignard reagent
—cholesterol synthesis 398
—experimental procedure 389f
—various reactions 391

Grignard, Victor 388f

guanine 186

gutta percha 333ff
—trans double bonds 336f

-H-

Haber, Fritz 308f

halonium ion 315f, 341
—intermediate resonance structures 316

Hancock, Thomas 340

handedness 25ff

Hassel, Odd 74

Havea Brasiliensis 332

Hayward, Nathaniel 340

Hayyan, Jabin ibn 212

hemiacetal 393f

hemiketal 393f

Hermes, Matthew 300

hevea rubber, cis double bonds 333ff

hexahydrophthalic acid
—cis and trans 335

hexamethylene diamine 300
—adipic acid source 308
—from 1,3-butadiene 317f
—industrial route 312ff

1,6-hexanediol polyester formation 294

n-hexane, mass spectrum 44

hexanol
—proton NMR 61
—infrared spectrum 50

high energy bonds
—relationship to leaving groups 139

Hill, Julian, with Carothers 295f, 303

histidine 36
—fat catabolism 237
—triglyceride hydrolysis 218f

Hoffmann, Roald 373, 435

Houdry, Eugene 104

Hückel, Erich 179, 373

Hughes, Edward 320

Hughes and Ingold, 1935 paper 324

hybridization of orbitals 10ff
—acetylene 432

hydride
—from laboratory reducing agents 309f
—from NADH 310
—transfer to carbocation vulcanization 342f

hydroboration 412ff, 416
—mechanism 418

hydrocarbons
—branched structure and
  octane number 114
—close packing 337

hydrochloric acid, pK_a 121

hydrocyanic acid
—ammonia oxidation 313

hydrogen
—evolution from Grignard 389

hydrogen abstraction
—abort polymerization of propylene 289f

hydrogen bonds
—compared to covalent bonds 303
—nylon and silk 303f

hydrogen peroxide
—hydroboration reaction 418
—proteins and nylons 298f

hydrolysis, triglyceride 211ff

hydroxyacetaldehyde, probe for
chirality 26

hydrogenation, free energy change 177
—heats of…176f

Hypalon, synthetic elastomer 345

-I-

ICI 276f

infrared
—frequencies of structural groups 51f
—spectrum 49ff

Ingold, Christopher 12, 34, 319ff

intermediates, carbocation 110

isocyanate
—reactive intermediates from 364ff

isoborneol 108

isobutane 124

isobutene 124

isocitrate 265

isomerization 23ff

isomers, constitutional 22ff

isopentenyl diphosphate
—loss of carbon dioxide 258
—source of carbon atoms 255
—structure 134, 320

isopentenyl pyrophosphate
—isomerase 144

isopropyl benzene 187

isotopic tagging 156

-K-

Katz, T. J. 181

Kauzmann, Walter 361
—paradox for glass formation 361
—theory of protein folding 361

Kekulé, Friedrich August 33
—benzene structure 168ff

ketal
—protecting group 393f

α-ketoglutarate 265

Kettering, Charles Franklin 118

kinetic control
—ring closing 410
—versus thermodynamic 315ff

Knoop, Franz 225

Kolbe, Adolph 39

Kraton 362

Krebs, Hans Adolf 261

-L-

labelling
—citric acid cycle 268

lactic acid 36, 263

lactone
—cholesterol synthesis 388
—from a ketone 407
—protecting group 426
—reduced to hemiacetal 425

lactose 70f

lauric acid 206

LDPE
—low density polyethylene 278

lead118f

leaving groups
—conjugate bases 139
—formation of carbocations 138
—Lowry-Brønsted acidity 214
—phosphate correlated with loss of
  carbon dioxide 258f
—resonance stabilized, ATP 270
—transesterification 214ff

Le Bel, Joseph Achille 34

Le Blanc, Nicolas 212

Le Chatelier, Henri Louis 318

Lewis acid/base reactions 129
—acids and bases 127

Lewis, Gilbert Newton 15, 127

light, plane polarized 29

limonene, structure 136

Lipmann, Fritz 269

lithium aluminum hydride 309f, 378

living polymerization 361

Lonsdale, Kathleen 175

Lord Raleigh 49

Lowry, Thomas Martin 120

Lynen, Feodor 156

lysine
—Schiff base from 385

-M-

macromolecules 6

Macintosh, Charles 340

malic acid 36, 263

malonic acid 262

Marggraf, Andreas 84

Mark, Herman 181

Markovnikov, Vladimir Vassilyevich 125

Markovnikov's rule 123
—addition of HCl to alkenes 190
—anti via hydroboration 417
—controls addition to double bonds 315
—in enzyme catalyzed reaction 146
—oxymercuration 417

Marvel, Carl Shipp 293

Maxwell-Boltsmann distribution 193f

mass spectrometry
−electrospray 45
−electron impact 44ff
−matrix assisted laser desorption 49

Mechanism
−addition of Br_2 to double bond 341
−addition of chlorine to
  double bonds 315f, 341
−addition of sulfur to double bonds 341
−addition of water to conjugated
  double bond in vivo 231f
−Aldol route to isopentenyl diphosphate 256f
−Aldol, cholesterol synthesis 381
−Aldol, 2-ethylhexanol 431
−Aldol, intramolecular 396
−attempted propylene polymerization 289f
−Baeyer-Villiger reaction 407
−carbocation addition to double bond 124
−chain 123ff
−Claisen reaction, cholesterol 379
−dimethyl allyl diphosphate 145
−electrophilic addition to alkenes 190
−electrophilic aromatic substitution 191
−elimination reactions to alkenes 402
−enzyme retro-Claisen reaction 240
−formation of amide bond in nylon 302
−formation of LDPE 279ff
−geranyl diphosphate to limonene 152
−geranyl and farnesyl diphosphate 150
−glucose ring closing 99
−Grignard reaction, lactone 392
−hydroboration 418
−hydrolysis of ATP 271
−Hypalon crosslinking 354
−isocyanate to carbamate and a urea 365
−isopentenyl diphosphate in vivo 255
−ketal formation 394
−keto-enol tautomerism 252
−lanosterol synthesis 160ff
−Michael reaction, cholesterol 380
−NADH reduction of thioester 310
−norepinephrine to adrenoline 325
−oxymercuration 417
−polycarbonate formation 353
−polyester formation and hydrolysis 295
−reaction coordinate diagram for
  S-adenosyl methionine formation
  via S_N2 reaction 323
−reaction coordinate diagram S_N1 320
−reduction of nitrile by $LiAlH_4$ 311
−resonance factors 1,2 versus 1,4
  addition 316f
−retro-Claisen, cholesterol 381
−ring closing, prostaglandin 409
−saponification in vitro 211
−β-scission 286
−S_N1 formation of geranyl diphosphate 320
−S_N2 for S-adenosylmethionine 321ff
−steam cracking 285

−S_N2 opening of tetrahydrofuran 364
−sulfonyl chloride substituted LDPE 346
−thermodynamic versus kinetic control 316f
−thioester formation in fat catabolism 221ff
−triglyceride enzyme hydrolysis 216ff
−vulcanization 342f
−Wittig reaction 415
−Wittig reaction, prostaglandin 420f, 426

McMurry and Begley 321

Meerwein, Hans 108

melting point 24
−fatty acids 206
−mixed, test for purity 402

menthol, structure 136

mercury
−product of oxymercuration 417

mesomerism 319

methane 11, 13

methyl acetate 238

methyl amine, Schiff base 384

methylenediphenyl diisocyanate 364f

methyl methacrylate 429

methyl phenyl amine 386

methyl propionate 429

methyl tertiary butyl ether 428

Meyer, Kurt Heinrich 250f

Michael, Arthur 232

Michael reaction
−cholesterol synthesis 378

Midgley, Thomas 118

Milliken, Robert S. 183

mirror images 17ff

Mislow, Kurt 148, 211

mitochondria 261

mixed oil 306

Mohr, Ernst 73

molecular distillation, polyester 294

molecular orbitals
−aromaticity 182ff
−conjugated double bonds 436
−frontier 436

molecular stick models 8

Monsanto 288

Montecatini 291

Moore, James 211

Morawetz, Herbert 276, 298

motion, molecular 38ff

mutarotation 88f

myristic acid 206

-N-

NADH, FADH$_2$
−produced in citric acid cycle 261ff

NADH
−synthesis of isopentenyl diphosphate 258

NAD$^+$/NADH oxidation-reduction 234

naphtha, light 21
−fraction of petroleum 287

Natta, Giulio 290

neoprene 314

neral, structure 136

Newman, Melvin S. 41

Newman projections 40ff

Nicolaou and Sorenson
−"Classics in Total Synthesis" 412

Nicotinamide adenine dinucleotide 187
−fatty acid catabolism (NAD$^+$) 228, 233ff

NMR, nuclear magnetic resonance 52
−chemical shift 56ff
−2,2-dimethylbutane, proton and carbon 54
−electronegativity 56ff
−n-hexane, proton and carbon 53
−2-methylpentane, proton and carbon 55
−Pascal's triangle spin spin coupling 59ff

nitric acid
−mixed oil to adipic acid 306f

nitrile
−conversion to carboxylic acid 387
−laboratory reduction 309f,
−reduction to amine 387
−stabilization of carbanion 386f

nitrogen
−fixation 308

nodes, orbital 435ff

no-mechanism reactions 373, 435

nomenclature, hydrocarbons 113

Noller, Carl 98

nonbonding electron 14

norepinephrine 325

nucleophilic 188
−attack at carbonyl and sulfonyl 349ff
−attack reviewed 349ff
−intermolecular attack 101
−intramolecular attack 97ff
−substitution 12
−substitution at acyl carbon, nylon 301
−substitution, rate expressions
  for S_N1 and S_N2 321f
−substitution S_N1 and S_N2 319ff
−substitution stereochemistry 328ff
−substitution, steric, electronic and
  solvent effects for S_N1 and S_N2 322

Nyholm, Ronald 11

nylon
—comparison to silk 303f
—various kinds compared 300
—versus neoprene industrial route 317f

-O-

octane number
—1,2 shift source of 114ff
—various hydrocarbon 119

3-octene, 3-methyl 413

octet rule 12ff

Ogston, Alexander 232, 267

Olah, George 12, 112

oleic acid, structure 133, 206

oligomers
—from propylene 288f

optical activity, 28ff

optical rotation 24
—glucose 88

orbital
—HOMO and LUMO 438ff
—nodes 436
—symmetry control over electrocyclic reactions 439
—symmetry control over sigmatropic rearrangements 439f

orthophosphate 270

ortho, meta, para, 200ff

Ourisson, Guy 138, 154

oxalic acid 262

oxaloacetic acid, citric acid cycle 264

oxalosuccinic acid, citric acid cycle 264

oxaphosphetane
—Wittig intermediate 415

oxidation, hydroxyl to ketone (NAD⁺) 235

oxidation-reduction FAD 227

oxidized and reduced
—alcohol, aldehyde, carboxylic acid 86
—comparison between inorganic ions and biological systems 307

oxidosqualene 158
—lanosterol cyclase 157

oxygen, bonding 14

oxymercuration 416

ozone
—polymer degradation 345

-P-

Palaquium 332

para-aminobenzene sulfonate 355

palmitic acid 206

pancreatic lipase 216ff

particle in a box 436

Partington
—discussion of Scheele 313

Pelletier, Pierre-Joseph 179

Parr Shaker 178

Pasteur, Louis 28ff
—quote 276

Patterson, Clair C. 119

Pauling, Linus 9, 11, 39
—resonance 140ff

Pelletierne
—source of cyclooctatetraene 180

peptide nucleic acid
—electrospray mass spectrum 48

peracid
—oxidation 407

pericyclic reactions 435ff

periodic acid 395

Perkin Jr., William Henry 72

Peroxide
—source of free radicals 357
—weak bond 279f

petroleum 21, 104ff

Pfaundler, Leopold 193

phenol, enol/keto 251

phosgene, discovery 352f

phosphate, acyl leaving group 350

phosphonate
—resonance stabilized, Wittig 420f

phosphorus-oxygen bond strength 419

phosphorylation
—isopentenyl diphosphate synthesis 258

phthalic acid 335

pimelic acid 262

plane polarized light 29ff

Plexiglas 429

poly(1,3-butadiene) 361f

polycarbonate
—mechanism of formation 353

polyester
—fiber discovery 295
—hydrolysis problem 293ff

polyether
—component of Spandex 364

polyethylene 276
—branched 277
—low density 277
—spaghetti like entangled 282

polymers 6
—addition compared to condensation 298
—blocked crystallization 359
—controversy about structure 298
—degradation by ozone 345

polystyrene 361f

polyvinyl chloride 430

Pond, Caroline, "Fats of Life" 206

polycarbonate 188

potassium triiodide
—add to double bonds 407

Prelog, Vladimir 34

Priestley, Joseph 332f

Prontosil, discovery 355

prokaryotes 261

propanal 430

propylene 189
—addition of chlorine 315
—attempt to polymerize 288ff
—EPDM rubber 358
—from steam cracking 285f
—polymer tacticity 291

prostaglandin
—Cory synthesis 403ff
—E-1, E-2 427
—F$_{2\alpha}$ last synthetic steps 426
—overall synthetic path 404
—synthesis, protecting groups 423ff
—synthesis, use of Wittig 420ff

protecting groups 422ff
—dihydropyran 424
—ketal 394
—use of eneamine 386
—Woodward strategy 383

Protein
—tryp-cage 297 (see enzyme)

Prussian blue
—source of HCN 312
—value in painting 312

Pub, head for the 402

Purcell, Edward M. 52

Purity Hall, Carothers' laboratory 299

pyridoxal phosphate 187
—Schiff base from 384

pyrophosphate 270

pyruvic acid 254

-Q-

quantum level 11ff

quinine 372

quinone 376

-R-

racemic 32

Rabi, Isidor Issac 52

rate constant 193ff
—ring closing 410

rate determining step 196

reactions, coupled to ATP 269ff

reaction coordinate diagram 195
—attempt to polymerize propylene 290
—nucleophilic substitution, S_N2 323

reaction rate theory 193

rearrangement, carbon skeleton 107ff

refractive index 250

reserpine 372

resolution
—cholesterol synthesis 397

resonance
—effect on attempted propylene
polymerization 288f
—mesomerism 319

resonance versus equilibrium 252f

retro-Claisen condensation 238

retro-synthesis 428ff

ring
—closing, limonene 152f
—closing favors five and six-membered
rings 410
—closing, prostaglandin synthesis 409
—closing, sulfur addition to
double bonds 341
—Hückel molecular orbitals 182
—thermodynamic and kinetic control 410
—compounds 71ff
—strain 72
—unsaturated, aromatic, anti-aromatic 183

Ritthaler, Alexander 153

Robinson, Sir Robert 157, 319
—ring annelation 379

rotational motion 40ff

Roth, W. R. 435

rubber, structure 135
—cis double bonds 336
—crosslinking 339ff
—EP rubber 356ff
—history 339f

—natural 336
—stiffness as a function of temperature 338f
—stretching, work 338f
—temperature effect 336f
—thermodynamic considerations 338f
—vulcanization 341ff

Ruzicka, Lavoslav Stjepan 136

-S-

Sabatier, Paul 178

Sachse, Hermann 73

saponification, in vitro 211

sawhorse projections 40ff

Scheele, Carl Wilhelm 210
—discovery of HCN 312

Schiff, Hugo 382

Schiff base 384

Schoenheimer, Rudolph 156
—scission 285

segments
—block copolymer 363f

separation
—diasteromers by chromatography 423
—1,2 versus 1,4 addition products
to 1,3-butadiene 313ff
—water and ethanol 21

serine 36
—triglyceride hydrolysis 218f

Shakespeare
—quote applied to industrial process 318

sigmatropic rearrangements 439f

soap 211f

sodium borohydride 309f

sodium wire, dry ether 389

Spandex 363
—formation from components 366
—source of flexibility 364

specific rotation 30ff

squalene, relationship to lanosterol 154

starch 6ff, 11

steam cracking 284ff
—source of ethylene 278

stearic acid 206

stereochemistry
—difference between Havea rubber and
gutta percha 333
—S_N1 and S_N2 328ff

stereospecificity, enzyme 232

steric effects 39ff

stereoisomers 19ff

Stork, Gilbert 385

strain, torsional, steric 41
—ring closing 410

Streitwieser, Andrew 329
—and Heathcock 402

strychnine 372

Stryer, Lubert 272

styrene-butadiene block copolymer 361

succinic acid 262

sucrose 85
—source of glucose and fructose 246
—structure 246

sugar, history 84f

sulfa drugs 355

sulfur
—reactivity 341
—ring 341

sulfur dioxide 346f

sun 9

super acid 12

Szwarc, Michael 361

-T-

Tarbell, Dean Stanley 328

tacticity, polypropylene 291

Tanaka, Koichi 46

tartaric acid 30ff
—electrospray mass spectrum 47
—infrared spectra of stereoisomers 51
—meso, d,l 36ff

tautomerism, enol/keto, enediol 249

temperature
—effect on reaction rate 193ff
—steam cracking versus polymerization 287

Terebinthos, source of terpenes 135

terpenes 134ff
—formation coupled to ATP 272

terpene rule 137

testosterone 132

tetracoordinate carbon proposal 33ff

tetracoordinate intermediate
—ester hydrolysis, transesterification 214ff
—fat catabolism 240

tetrahedral carbon
—mirror image isomerism 34
—tetrahedron 11ff

tetrahydrofuran, 418
−mechanism of polymer from 364
−solvent for lithium aluminum hydride 311

thermodynamic
−control of ring closing 410
−polymerization versus steam cracking 287
−versus kinetic control 315ff

thermoplastic elastomer 362

thiamine diphosphate 187

thiol, enzyme fat catabolism 237

thioester formation, ATP 223f
−in vivo 221ff

Thomsen, Julius 177

Thomson, Thomas 167

Thunberg, Torsten 262

tire failure 336

thymine 186

p-toluene sulfonyl chloride 434

torsional motions 41

toxicity
−blocks industrial path 317f

trans double bond
−gutta percha compared to HDPE 337

transition state
−theory 196
−six-membered ring 282
−S$_N$2 326

2,2,4-trimethylpentane
−industrial synthesis 124

triphenyl phosphine, Wittig reaction 414

turpentine 29

-U-

uracil 186

urea, from ammonium cyanate 23

Urey, Harold C. 155

-V-

Vagner, Egor Egorevich 107

van der Waals, Johannes Diderik 80

van der Waal radii 80

van Rague Schleyer, Paul 108

van't Hoff, Henricus Jocobus 34

vicinal diol, cleavage 395

vinyl acetylene
−source of chloroprene 314

vital force 23

vitamin A, structure 135

vitamin B-1 266
−B-12 372

van Baeyer, Johann Friedrich Adolph 71f, 406

VSEPR theory 11f

vulcanization
−crosslinking mechanism 342

-W-

Wadsworth and Emmons
−variation of Wittig reaction 421

Walden, Paul 328

Warburg, Otto Heinrich 263

water
−addition to conjugated double bond 231f
−enzyme role in triglyceride hydrolysis 219

−ester formation 297
−exclusion from Grignard reagent 389f
−hydrogen bonding 303
−role in keto/enol tautomerization 252

wave functions 436ff

Westheimer, Frank H. 371

Wheland, George W. 196

Wickham, Henry 332

Williamson, Alexander W. 430

Williamson ether synthesis 429f

Wislicenus, Johannes 39

Wittcoff, Harold 199

Wittig, Georg 412

Wittig reaction 412ff
−and Grignard reactions compared 413
−base strength effect on
  stereochemistry 414f, 426
−mechanism 415
−phosphonate intermediate 420

Wöhler, Friedrich 23
−Schiff's teacher 383

Woodward, Robert Burns 41, 370ff
−and Hoffmann, "The Conservation
  of Orbital Symmetry" 437

Wurtz, Charles Adolph 33

-Z-

Zaytsev, Alexander 107

zeolites 104ff
−silica and alumina 105ff

Zewail, Ahmed H. 327

Ziegler, Karl 288, 356
−catalyst 290, 356

Zyklon B 312

23235495R00255

Made in the USA
Lexington, KY
02 June 2013